SIGNAL AND IMAGE PROCESSING IN NAVIGATIONAL SYSTEMS

THE ELECTRICAL ENGINEERING AND APPLIED SIGNAL PROCESSING SERIES

Edited by Alexander Poularikas

The Advanced Signal Processing Handbook: Theory and Implementation for Radar, Sonar, and Medical Imaging Real-Time Systems
Stergios Stergiopoulos

The Transform and Data Compression Handbook
K.R. Rao and P.C. Yip

Handbook of Multisensor Data Fusion
David Hall and James Llinas

Handbook of Neural Network Signal Processing
Yu Hen Hu and Jenq-Neng Hwang

Handbook of Antennas in Wireless Communications
Lal Chand Godara

Noise Reduction in Speech Applications
Gillian M. Davis

Signal Processing Noise
Vyacheslav P. Tuzlukov

Digital Signal Processing with Examples in MATLAB®
Samuel Stearns

Applications in Time-Frequency Signal Processing
Antonia Papandreou-Suppappola

The Digital Color Imaging Handbook
Gaurav Sharma

Pattern Recognition in Speech and Language Processing
Wu Chou and Biing-Hwang Juang

Propagation Handbook for Wireless Communication System Design
Robert K. Crane

Nonlinear Signal and Image Processing: Theory, Methods, and Applications
Kenneth E. Barner and Gonzalo R. Arce

Smart Antennas
Lal Chand Godara

Mobile Internet: Enabling Technologies and Services
Apostolis K. Salkintzis and Alexander Poularikas

Soft Computing with MATLAB®
Ali Zilouchian

Wireless Internet: Technologies and Applications
Apostolis K. Salkintzis and Alexander Poularikas

Signal and Image Processing in Navigational Systems
Vyacheslav P. Tuzlukov

SIGNAL AND IMAGE PROCESSING IN NAVIGATIONAL SYSTEMS

Vyacheslav P. Tuzlukov

CRC PRESS

Boca Raton London New York Washington, D.C.

Library of Congress Cataloging-in-Publication Data

Tuzlukov, V. P. (Vyacheslav Petrovich)
Signal and image processing in navigational systems / Vyacheslav Tuzlukov.
p. cm. -- (The electrical engineering and applied signal processing series)
Includes bibliographical references and index.
ISBN 0-8493-1598-0 (alk. paper)
1. Electronics in navigation 2. Signal processing. 3. Image processing. I. Title. II. Series.

VK560.T88 2004
623.89′3′0285—dc22 2004049669

Direct all inquiries to CRC Press, 2000 N.W. Corporate Blvd., Boca Raton, Florida 33431.

International Standard Book Number 0-8493-1598-0
Library of Congress Card Number 2004049669
Printed in the United States of America 1 2 3 4 5 6 7 8 9 0
Printed on acid-free paper

Preface

Noise immunity is the main problem in navigational systems. At the present time, there are many books and journal articles devoted to signal and image processing in noise in navigational systems, but many important problems remain to be solved. New approaches and study of complex problems allow us not only to summarize investigations but also to derive a better quality of signal and image processing in noise in navigational systems.

In the functioning of many navigational systems, reflections from the Earth's surface and the rough or rippled sea, hydrometeors (storm clouds, rain, shower, snow, etc.), the ionosphere, clouds of artificial scatterers, etc., play a large role. We can observe these reflections in the detection and tracking of low-flying, surface- or sea-surface-moving targets against the background of highly camouflaged reflections from the underlying surface. In these cases, reflections play the role of passive interference, and there is the need to construct specific methods and techniques for increasing the noise immunity of navigational systems.

There are many navigational systems in which reflections from the Earth's or sea's surface and hydrometeors are the information signal and not an interference. Examples are autonomous navigational systems for aircraft with the Doppler analyzer of velocity and drift angle, analyzer of vertical velocity of take-off and landing, height-finding radar; navigational systems with ground surveillance radar, scatter meters in which reflections from the Earth's surface and rough and rippled sea are used to obtain the detailed information about surface structure and state; storm-warning radar; and weather radar.

This book is devoted to the study of fluctuations of parameters of the target return signals and signal and image processing problems in navigational systems constructed on the basis of the generalized approach to signal and image processing in noise based on a seemingly abstract idea: the introduction of an additional noise source that does not carry any information about the signal with the purpose of improving the qualitative performances of complex navigational systems. Theoretical and experimental study carried out by the author leads to the conclusion that the proposed generalized approach to signal and image processing in noise in navigational systems allows us to formulate a decision-making rule based on the determination of the *jointly sufficient statistics of the* likelihood function (or functional) *mean and variance*. The use of classical and modern signal and image processing approaches in navigational systems allows us to define only the *sufficient statistic of the* likelihood function (or functional) *mean*.

The presence of additional information about the statistical characteristics of the likelihood function (or functional) leads to better qualitative performances of signal and image processing in navigational systems compared to the optimal signal and image processing algorithms of classical and modern theories. The generalized approach to signal and image processing in navigational systems allows us to extend the well-known boundaries of the potential noise immunity set up by classical and modern signal and image processing theories. The use of complex navigational systems based on the generalized approach allows us to obtain better detection performances and definition of object coordinates with high accuracy, in particular, in comparison with navigational systems constructed on the basis of optimal and asymptotic optimal signal and image processing algorithms of classical and modern theories.

To understand better the fundamental statements and concepts of the generalized approach, the reader is invited to consult my earlier books: *Signal Processing in Noise: A New Methodology* (IEC, Minsk, 1998), *Signal Detection Theory* (Springer-Verlag, New York, 2001), and *Signal Processing Noise* (CRC Press, Boca Raton, FL, 2002).

I would like to thank my many colleagues in the field of signal and image processing for very useful discussions about the main results, in particular, Professors V. Ignatov, A. Kolyada, I. Malevich, G. Manshin, B. Levin, D. Johnson, B. Bogner, Yu Sedyshev, J. Schroeder, Yu Shinakov, A. Kara, X.R. Lee, Yong Deak Kim, Won-Sik Yoon, V. Kuzkin, A. Dubey, and O. Drummond.

A special word of thanks to Ajou University, Suwon, South Korea, for allowing me to complete this project.

A lot of credit also needs to go to Nora Konopka, Jessica Vakili, Gail Renard, and the staff at CRC Press for their encouragement and support of this project.

Last, but definitely not least, I would like to thank my family, my lovely wife and sons and my dear mother, for putting up with me during the completion of the manuscript; without their support it would not have been possible!

I also wish to express my lifelong, heartfelt gratitude to Peter Tuzlukov, my father and teacher, who introduced me to science.

Vyacheslav Tuzlukov

The Author

Vyacheslav Tuzlukov, Ph.D., is Invited Full Professor at Ajou University, Suwon, South Korea, and Chief Research Fellow at the United Institute on Informatics Problems of the National Academy of Sciences, Belarus. He is also a Full Professor in the Electrical and Computer Engineering Department of the Belarussian State University in Minsk. During 2000 to 2002, Dr. Tuzlukov was a Visiting Full Professor at the University of Aizu, Aizu-Wakamatsu, Japan. He is actively engaged in research on radar, communications, and signal and image processing, and has more than 25 years' experience in these areas. Dr. Tuzlukov is the author of more than 120 journal articles and conference papers and five books — one in Russian and four in English — on signal processing.

Introduction

The main functioning principle of any navigational system is based on the comparison of the moving image of the Earth's surface or the totality of landmarks with the reference image or model image. The moving and model images are formed using various natural and manmade physical fields. As an illustration of these fields, we can use optical, radar, radio heat, electromagnetic, gravitational, and other fields. For example, using an airborne radar, we can obtain the moving radar images of the Earth's surface that are compared with the predetermined images corresponding to the required airborne flight track. The measure of deviation of the flight track from the predetermined or required airborne flight track is characterized by mutual noncoincidence between the moving image and the model image relative to each other. Coincidence between the moving and model images is used to restore an airborne flight track to the true flight track. This working principle of navigational systems is called the searching-free principle.

Another example allows us to consider a navigational system based on the searching principle. Let us assume there are data of model images corresponding to all possible airborne flight tracks. Each model image corresponds to the definite coordinate system. Maximum coincidence between the moving image and a model image allows us to define the true airborne flight track. Comparison between moving and model images is made by the functional that is at its maximum in the coincidence between the moving and model images. The mutual correlation function can be considered as a functional with some limitations. In the coincidence between the moving and model images, the correlation function must be maximum and its derivative must be minimum.

Due to the stochastic character of elementary scatterers, the amplitude and phase of the target return signal at the receiver or detector input in navigational systems are random variables. For many reasons, such as scatterer moving under the stimulus of the wind, the radar moving, radar antenna scanning, etc., the target return signal at the receiver or detector input is a stochastic process with fluctuating parameters. Therefore, the target return signal at the receiver or detector input in navigational systems can be defined by the probability distribution density, correlation function, and power spectral density that depend on some parameters of radar and navigational system devices and peculiarities of scatterers: the shape of the directional diagram, orientation of the directional diagram with respect to the velocity vector of the radar moving, localization of scatterers in space, the shape of the searching signal, laws of the radar moving and radar antenna scanning, and the nature and character of scatterer moving. The fluctuations of parameters of the target return signal at the receiver or detector input are caused, as a rule, by simultaneous stimulus of noise and interference sources. For example, in the use of navigational systems based on aircraft radar, we should take into consideration the radar moving, radar antenna scanning, and direction of the wind. In some cases, we have to add instability in the frequency of the transmitting radar antenna and (or) rotation of the polarization plane of the radar antenna caused by radar antenna scanning and rotation in navigational systems to the noise and interference sources mentioned previously. Furthermore, the various power spectral densities of fluctuations of the target return signal at the receiver or detector input can be formed under the stimulus of the same noise and interference sources in accordance with the specific input stochastic process because the radar moving, radar antenna scanning and rotation, and scatterer moving are different in different cases in practice. Because of this, the correlation functions and power spectral densities of fluctuations of the target return signal at the receiver or detector input in navigational systems are specific characteristics of the input stochastic process and must be defined for specific conditions in practice.

At the same time, signal processing in navigational systems depends greatly on the correlation features of the target return signal at the receiver or detector input. In particular, the definition of the shape of the power spectral density of passive interferences has a fundamental significance in solving the problem of optimal signal processing and in defining the effectiveness of signal processing in navigational systems. In the case of autonomous navigational systems in which the target return signal from the underlying surface of the Earth or sea possesses information regarding measured parameters, such as velocity, distance, and direction, it is necessary to take into consideration the probability distribution laws, the effective bandwidth, and shape of the power spectral density of the target return signal at the receiver or detector input because all of these allow us to choose the true signal processing technique and algorithms and to ensure a high accuracy of definition in the measured parameters of the target return signal.

To construct navigational systems with high noise immunity, we have to define with high accuracy the power spectral density of the target return signal at the receiver or detector input because the effectiveness of the navigational system functioning depends on the knowledge of, for example, the rate of decrease, the length of remainders, deviation from the axis of symmetry of the power spectral density, etc. The function between the power spectral density and various factors can be complex and not always clear. Therefore, in theoretical investigations, the power spectral density with the Gaussian, resonant, and square waveform shape are used for simplicity and convenience of analysis. The real power spectral densities of the target return signal at the receiver or detector input in navigational systems, taking into consideration specific conditions of their forming in practice, would be defined in a rigorous form. This book summarizes investigations carried out by the author over the last 20 years.

The book consists of two parts. The first part discusses fluctuations of the target return signal parameters in navigational systems based on the generalized approach to signal and image processing in noise. Discussed results presenting the majority of cases in practice take into consideration almost all possible sources of fluctuations of the target return signal parameters. The second part is concerned with navigational systems based on the generalized approach to space–time signal and image processing. Detailed attention is paid to the employment of optical detectors, optical direction finders, and optical coordinate analyzers constructed on the basis of the generalized approach to signal and image processing in noise.

The book comprises 14 chapters. Chapter 1 discusses the problems of definition of the probability distribution density of the target return signal amplitude and phase. Two-dimensional probability distribution density is defined. Based on the two-dimensional probability distribution density, we are able to obtain the following particular cases: the probability distribution density of the amplitude and the probability distribution density of the phase. The parameters of the probability distribution density are defined as a function of the distribution law of amplitudes and phases of elementary signals. Particular cases, namely, the uniform, "triangular," and Gaussian probability distribution densities of the phase are considered.

Chapter 2 deals with the study of the correlation function of the target return signal. Physical sources of fluctuations of parameters are investigated. The space–time correlation function and power spectral density are defined. The correlation function with the searching signal of arbitrary shape, for example, the narrow-band searching signal and pulsed searching signal, is discussed. The correlation function in scanning the three-dimensional (space) target is defined in the cases of the pulsed searching signal and the simple harmonic searching signal. The correlation function in angle scanning the two-dimensional (surface) target is studied in the cases of the pulsed searching signal and the simple harmonic searching signal. The correlation function of the target return signal is defined under vertical scanning of the two-dimensional (surface) target.

Chapter 3 is concerned with the definition of fluctuations of the target return signal parameters in scanning the three-dimensional (space) target by the moving radar. The cases of slow and rapid fluctuations are considered. Additionally, Doppler fluctuations of the target return signal parameters in navigational systems with the high-deflected antenna are investigated in the cases of arbitrary directional diagrams, Gaussian directional diagrams, and **sinc**-directional diagrams. Doppler fluctuations with the arbitrarily deflected radar antenna are discussed, and the total power spectral density of the target return signal in the case of the pulsed searching signal is defined.

Chapter 4 focuses on the definition of fluctuations of the target return signal parameters in scanning the two-dimensional (surface) target by the moving radar. The continuous nonmodulated searching and pulsed searching signals in the stationary radar are considered as initial premises for the following cases: arbitrary vertical-coverage directional diagrams — Gaussian pulsed and square waveform searching signals, and Gaussian vertical-coverage directional diagrams — square waveform searching signal. In the case of the pulsed searching signal in the moving radar, the angle correlation function is defined under the following conditions: the arbitrary vertical-coverage directional diagram — Gaussian pulsed searching signal, and the Gaussian vertical-coverage directional diagram — Gaussian pulsed searching signal. The azimuth correlation function in the case of the pulsed searching signal in the moving radar is defined at the high- and low-deflected radar antenna. The total correlation function and power spectral density of fluctuations of the target return signal parameters at the receiver or detector input are defined for the following cases: Gaussian directional diagrams and the Gaussian pulsed searching signal, the square waveform searching signal, and the pulsed searching signal with short duration. The minimum radar range is defined under the conditions described earlier. The vertical scanning of the two-dimensional (surface) target is investigated. Examples of determination of the power spectral density of the target return signal are presented.

Chapter 5 explores problems of definition of fluctuations of the target return signal parameters caused by radar antenna scanning. The correlation functions in space and surface scanning are defined. The general definition of the power spectral density is discussed. The line radar antenna scanning is investigated for the following cases: one-line ***T***-scanning, multiple-line ***T***-scanning, line segment scanning, and line ***T***-scanning for various directional diagrams in the transmitter–receiver block of the navigational system. Conical antenna scanning is studied in the cases of three-dimensional (space) and two-dimensional (surface) targets. Moreover, conical radar antenna scanning is considered in the case of circular polarization.

Chapter 6 is devoted to the definition of fluctuations of the target return signal parameters caused simultaneously by the moving radar and radar antenna scanning. The correlation functions of the target return signal in space and surface scanning are defined and the problems associated with the moving radar with the line and conical radar antenna scanning are discussed. The problems of space and surface scanning with the Gaussian directional diagram are considered. The minimum radar range of navigational systems for the case of the Gaussian directional diagram is investigated. The **sinc**^{2}-directional diagram is studied and the instantaneous and average power spectral densities of the target return signals are discussed. Theoretical study is strengthened by computer modeling and experimental results.

Chapter 7 deals with the fluctuations of the target return signal parameters caused by moving reflectors of the radar antenna under the stimulus of the wind. The following cases are discussed: the deterministic motion of radar antenna reflectors under the stimulus of the layered wind, the stochastic motion and rotation of radar antenna reflectors, and the simultaneous deterministic and stochastic rotation of radar antenna reflectors.

Chapter 8 is concerned with the study of the fluctuations of the target return signal parameters in navigational systems in scanning the two-dimensional (surface) target by the continuous frequency-modulated searching signal. Searching signals with linear frequency and the searching nonsymmetric saw-tooth frequency-modulated signals are discussed. The problems of scanning in definite angle and vertical scanning and moving are investigated. Additionally, the searching symmetric saw-tooth frequency-modulated signals and the searching harmonic frequency-modulated signals are studied. Phase characteristics of the target return signals in harmonic frequency modulation are discussed.

Chapter 9 focuses on the study of fluctuations of the target return signal parameters in scanning the three-dimensional (space) target by the continuous searching signal with varying frequency. Nontransformed and transformed searching signals are considered. As a particular case, the nonperiodic and periodic frequency-modulated searching signals are investigated. The average power spectral density of the target return signal in the periodic frequency-modulated searching signal is defined.

Chapter 10 discusses the problems of the target return signal parameter fluctuations caused by the change in the frequency from searching pulse to searching pulse. In the case of scanning the three-dimensional (space) target, the nonperiodic change in the frequency of the searching signals, the interperiodic fluctuations of parameters of the target return signals, the average power spectral density of the target return signal, and the periodic frequency modulation of searching signals are investigated. Moreover, the problems of scanning the two-dimensional (surface) target are discussed. The classification of stochastic target return signals is discussed.

Chapter 11 focuses on the main theoretical principles of the generalized approach to signal processing in the presence of additive Gaussian noise. The basic concepts of the signal detection problem are discussed. The criticism of classical and modern signal processing theories from the viewpoint of defining the jointly sufficient mean and variance statistics of the likelihood function (or functional) is explored and modifications and initial premises of the generalized approach to signal processing in noise are considered. The likelihood function (or functional) possessing the jointly sufficient mean and variance statistics in the generalized approach to signal processing in noise is investigated. The engineering interpretation of the generalized approach to signal processing in noise is discussed and the model of the generalized detector in the cases of both slow and rapid fluctuating noise is studied.

Chapter 12 is devoted to the main principles of the use of the generalized approach to the space–time signal and image processing. The basic concepts and foundations are considered. The problems of pattern recognition are discussed and the singularities of the generation of optical signals and radar images of the Earth's surface are investigated.

Chapter 13 focuses on the use of the generalized approach to the space–time signal and image processing in specific navigational systems.

The generalized space–time signal and image processing algorithms are considered and compared with the classical correlation space–time signal and image processing algorithms. The difference generalized image processing algorithm is investigated and the generalized phase image processing algorithm is discussed. The invariant moments, amplitude ranking, gradient vector summing, structural methods, and bipartite functions are defined and investigated. The hierarchy generalized image processing algorithm is considered. The problems of the use of the more informative image area, coding of images, and superposition of point images are investigated in the use of the generalized image processing algorithm in specific navigational systems. The multichannel generalized image processing algorithm is discussed.

Chapter 14 explores the use of the generalized approach in image preprocessing. The problems of image distortions, geometric transformations, image intensity distribution, detection of boundary edges, and sampling of images are discussed.

The content of the book shows us that it is possible to raise higher the upper boundary of the potential noise immunity for complex and specific navigational systems in various areas of applications in the use of the generalized approach to signal and image processing in comparison with the noise immunity defined by classical and modern signal and image processing algorithms used in navigational systems.

Contents

Part I
Theory of Fluctuating Target Return Signals in Navigational Systems

Chapter 1 Probability Distribution Density of the Amplitude and Phase of the Target Return Signal 3

1.1 Two-Dimensional Probability Distribution Density of the Amplitude and Phase 3

1.2 Probability Distribution Density of the Amplitude 8

1.3 Probability Distribution Density of the Phase 11

1.4 Probability Distribution Density Parameters of the Target Return Signal as a Function of the Distribution Law of the Amplitude and Phase of Elementary Signals 15

1.4.1 Uniform Probability Distribution Density of Phases 22

1.4.2 "Triangular" Probability Distribution Density of Phases 24

1.4.3 Gaussian Probability Distribution Density of Phases 25

1.5 Conclusions 26

References 27

Chapter 2 Correlation Function of Target Return Signal Fluctuations 29

2.1 Target Return Signal Fluctuations 29

2.1.1 Physical Sources of Fluctuations 29

2.1.2 The Target Return Signal: A Poisson Stochastic Process........ 30

2.2 The Correlation Function and Power Spectral Density of the Target Return Signal 36

2.2.1 Space–Time Correlation Function 36

2.2.2 The Power Spectral Density of Nonstationary Target Return Signal Fluctuations 41

2.3 The Correlation Function with the Searching Signal of Arbitrary Shape 45

2.3.1 General Statements 45

2.3.2 The Correlation Function with the Narrow-Band Searching Signal 46

2.3.3 The Correlation Function with the Pulsed Searching Signal 48

2.3.4 The Average Correlation Function 49

2.4 The Correlation Function under Scanning of the Three-Dimensional (Space) Target ... 51
2.4.1 General Statements ... 51
2.4.2 The Correlation Function with the Pulsed Searching Signal ... 53
2.4.3 The Target Return Signal Power with the Pulsed Searching Signal ... 58
2.4.4 The Correlation Function and Power of the Target Return Signal with the Simple Harmonic Searching Signal ... 61
2.5 The Correlation Function in Angle Scanning of the Two-Dimensional (Surface) Target ... 63
2.5.1 General Statements ... 63
2.5.2 The Correlation Function with the Pulsed Searching Signal ... 65
2.5.3 The Target Return Signal Power with the Pulsed Searching Signal ... 68
2.5.4 The Correlation Function and Power of the Target Return Signal with the Simple Harmonic Searching Signal ... 74
2.6 The Correlation Function under Vertical Scanning of the Two-Dimensional (Surface) Target ... 75
2.7 Conclusions ... 82
References ... 84

Chapter 3 Fluctuations under Scanning of the Three-Dimensional (Space) Target with the Moving Radar ... 89
3.1 Slow and Rapid Fluctuations ... 89
3.1.1 General Statements ... 90
3.1.2 The Fluctuations in the Radar Range ... 91
3.1.2.1 The Square Waveform Target Return Signal without Frequency Modulation ... 95
3.1.2.2 The Gaussian Target Return Signal without Frequency Modulation ... 95
3.1.2.3 The Smoothed Target Return Signal without Frequency Modulation ... 96
3.1.2.4 The Square Waveform Target Return Signal with Linear-Frequency Modulation ... 98
3.1.2.5 The Gaussian Target Return Signal with Linear-Frequency Modulation ... 101
3.1.3 The Doppler Fluctuations ... 101
3.2 The Doppler Fluctuations of a High-Deflected Radar Antenna ... 103

3.2.1 The Power Spectral Density for an Arbitrary Directional Diagram ... 103
3.2.2 The Power Spectral Density for the Gaussian Directional Diagram ... 111
3.2.3 The Power Spectral Density for the Sinc-Directional Diagram ... 112
3.2.4 The Power Spectral Density for Other Forms of the Directional Diagram ... 116
3.3 The Doppler Fluctuations in the Arbitrarily Deflected Radar Antenna ... 119
3.3.1 General Statements ... 119
3.3.2 The Gaussian Directional Diagram ... 122
3.3.3 Determination of the Power Spectral Density ... 126
3.4 The Total Power Spectral Density with the Pulsed Searching Signal ... 128
3.4.1 General Statements ... 128
3.4.2 Interperiod Fluctuations in the Glancing Radar Range ... 129
3.4.3 Interperiod Fluctuations in the Fixed Radar Range ... 129
3.4.4 Irregularly Moving Radar ... 135
3.5 Conclusions ... 136
References ... 137

Chapter 4 Fluctuations under Scanning of the Two-Dimensional (Surface) Target by the Moving Radar ... 141
4.1 General Statements ... 141
4.2 The Continuous Searching Nonmodulated Signal ... 143
4.3 The Pulsed Searching Signal with Stationary Radar ... 148
4.3.1 General Statements ... 148
4.3.2 The Arbitrary Vertical-Coverage Directional Diagram: The Gaussian Pulsed Searching Signal ... 151
4.3.3 The Arbitrary Vertical-Coverage Directional Diagram: The Square Waveform Pulsed Searching Signal ... 152
4.3.4 The Gaussian Vertical-Coverage Directional Diagram: The Square Waveform Pulsed Searching Signal ... 154
4.4 The Pulsed Searching Signal with the Moving Radar: The Aspect Angle Correlation Function ... 160
4.4.1 General Statements ... 160
4.4.2 The Arbitrary Vertical-Coverage Directional Diagram: The Gaussian Pulsed Searching Signal ... 162
4.4.3 The Gaussian Vertical-Coverage Directional Diagram: The Gaussian Pulsed Searching Signal ... 165
4.4.4 The Wide-Band Vertical-Coverage Directional Diagram: The Square Waveform Pulsed Searching Signal ... 167

4.5 The Pulsed Searching Signal with the Moving Radar: The Azimuth Correlation Function 170
4.5.1 General Statements 170
4.5.2 The High-Deflected Radar Antenna 174
4.5.3 The Low-Deflected Radar Antenna 176
4.6 The Pulsed Searching Signal with the Moving Radar: The Total Correlation Function and Power Spectral Density of the Target Return Signal Fluctuations 178
4.6.1 General Statements 178
4.6.2 The Gaussian Directional Diagram: The Gaussian Pulsed Searching Signal 179
4.6.3 The Gaussian Directional Diagram: The Square Waveform Pulsed Searching Signal 184
4.6.4 The Pulsed Searching Signal with Low Pulse Period-to-Pulse Duration Ratio 188
4.7 Short-Range Area of the Radar Antenna 189
4.8 Vertical Scanning of the Two-Dimensional (Surface) Target 193
4.8.1 The Intraperiod Fluctuations in Stationary Radar 193
4.8.2 The Interperiod Fluctuations with the Vertically Moving Radar 198
4.8.3 The Interperiod Fluctuations with the Horizontally Moving Radar 202
4.9 Determination of the Power Spectral Density 208
4.10 Conclusions 212
References 214

Chapter 5 Fluctuations Caused by Radar Antenna Scanning 219
5.1 General Statements 219
5.1.1 The Correlation Function under Space Scanning 219
5.1.2 The Correlation Function under Surface Scanning 222
5.1.3 The General Power Spectral Density Formula 225
5.2 Line Scanning 226
5.2.1 One-Line Circular Scanning 226
5.2.2 Multiple-Line Circular Scanning 233
5.2.3 Line Segment Scanning 235
5.2.4 Line Circular Scanning with Various Directional Diagrams under Transmitting and Receiving Conditions 241
5.3 Conical Scanning 243
5.3.1 Three-Dimensional (Space) Target Tracking 243
5.3.2 Two-Dimensional (Surface) Target Tracking 248
5.4 Conical Scanning with Simultaneous Rotation of Polarization Plane 249

5.5 Conclusions 251
References 252

Chapter 6 Fluctuations Caused by the Moving Radar with Simultaneous Radar Antenna Scanning 255
6.1 General Statements 255
6.1.1 The Correlation Function in the Scanning of the Three-Dimensional (Space) Target 255
6.1.2 The Correlation Function in the Scanning of the Two-Dimensional (Surface) Target 258
6.2 The Moving Radar with Simultaneous Radar Antenna Line Scanning 259
6.2.1 Scanning of the Three-Dimensional (Space) Target: The Gaussian Directional Diagram 259
6.2.2 Scanning of the Two-Dimensional (Surface) Target: The Gaussian Directional Diagram 265
6.2.3 Short-Range Area: The Gaussian Directional Diagram 265
6.2.4 The Sinc^2-Directional Diagram 267
6.3 The Moving Radar with Simultaneous Radar Antenna Conical Scanning 270
6.3.1 The Instantaneous Power Spectral Density 270
6.3.2 The Averaged Power Spectral Density 273
6.4 Conclusions 277
References 278

Chapter 7 Fluctuations Caused by Scatterers Moving under the Stimulus of the Wind 281
7.1 Deterministic Displacements of Scatterers under the Stimulus of the Layered Wind 281
7.1.1 The Radar Antenna Is Deflected in the Horizontal Plane 285
7.1.2 The Radar Antenna Is Deflected in the Vertical Plane 286
7.1.3 The Radar Antenna Is Directed along the Line of the Moving Radar 287
7.1.4 The Stationary Radar 288
7.2 Scatterers Moving Chaotically (Displacement and Rotation) 289
7.2.1 Amplitudes of Elementary Signals Are Independent of the Displacements of Scatterers 291
7.2.2 The Velocity of Moving Scatterers Is Random but Constant 293
7.2.3 The Amplitude of the Target Return Signal Is Functionally Related to Radial Displacements of Scatterers 295
7.2.4 Chaotic Rotation of Scatterers 296

7.2.5 Simultaneous Chaotic Displacements and Rotations of Scatterers.....298
7.3 Simultaneous Deterministic and Chaotic Motion of Scatterers.....301
7.3.1 Deterministic and Chaotic Displacements of Scatterers.....301
7.3.2 Chaotic Rotation of Scatterers and Rotation of the Polarization Plane.....304
7.3.3 Chaotic Displacements of Scatterers and Rotation of the Polarization Plane.....306
7.4 Conclusions.....308
References.....310

Chapter 8 Fluctuations under Scanning of the Two-Dimensional (Surface) Target with the Continuous Frequency-Modulated Signal.....313
8.1 General Statements.....313
8.2 The Linear Frequency-Modulated Searching Signal.....317
8.3 The Asymmetric Saw-Tooth Frequency-Modulated Searching Signal.....324
8.3.1 Sloping Scanning.....329
8.3.2 Vertical Scanning and Motion.....340
8.3.3 Vertical Scanning: The Velocity Vector Is Outside the Directional Diagram.....345
8.4 The Symmetric Saw-Tooth Frequency-Modulated Searching Signal.....350
8.5 The Harmonic Frequency-Modulated Searching Signal.....358
8.6 Phase Characteristics of the Transformed Target Return Signal under Harmonic Frequency Modulation.....366
8.7 Conclusions.....370
References.....372

Chapter 9 Fluctuations under Scanning of the Three-Dimensional (Space) Target by the Continuous Signal with a Frequency that Varies with Time.....375
9.1 General Statements.....375
9.2 The Nontransformed Target Return Signal.....378
9.2.1 The Searching Signal with Varying Nonperiodic Frequency.....378
9.2.2 The Periodic Frequency-Modulated Searching Signal.....383
9.2.3 The Average Power Spectral Density with the Periodic Frequency-Modulated Searching Signal.....388
9.3 The Transformed Target Return Signal.....392
9.3.1 Nonperiodic and Periodic Frequency-Modulated Searching Signals.....392

9.3.2 The Average Power Spectral Density with the Periodic Frequency-Modulated Searching Signal 394
9.4 Conclusions 397
References 399

Chapter 10 Fluctuations Caused by Variations in Frequency from Pulse to Pulse 401
10.1 Three-Dimensional (Space) Target Scanning 401
10.1.1 Nonperiodic Variations in the Frequency of the Searching Signal 401
10.1.2 The Interperiod Fluctuations 405
10.1.3 The Average Power Spectral Density 407
10.1.4 Periodic Frequency Modulation 410
10.2 Two-Dimensional (Surface) Target Scanning 414
10.3 Conclusions 417
References 418

Part II
Generalized Approach to Space–Time Signal and Image Processing in Navigational Systems

Chapter 11 Foundations of the Generalized Approach to Signal Processing in Noise 421
11.1 Basic Concepts 421
11.2 Criticism 425
11.3 Initial Premises 427
11.4 Likelihood Ratio 428
11.5 The Engineering Interpretation 433
11.6 Generalized Detector 436
11.6.1 The Case of the Slow Fluctuations 443
11.6.2 The Case of the Rapid Fluctuations 446
11.7 Conclusions 449
References 451

Chapter 12 Theory of Space–Time Signal and Image Processing in Navigational Systems 457
12.1 Basic Concepts of Navigational System Functioning 457
12.2 Basics of the Generalized Approach to Signal and Image Processing in Time 463
12.2.1 The Signal with Random Initial Phase 466
12.2.2 The Signal with Stochastic Amplitude and Random Initial Phase 469

12.3 Basics of the Generalized Approach to Space–Time Signal and Image Processing 472
12.4 Space–Time Signal Processing and Pattern Recognition Based on the Generalized Approach to Signal Processing 492
12.5 Peculiarities of Optical Signal Formation 499
12.6 Peculiarities of the Formation of the Earth's Surface Radar Image 509
12.7 Foundations of Digital Image Processing 513
12.8 Conclusions 519
References 523

Chapter 13 Implementation Methods of the Generalized Approach to Space–Time Signal and Image Processing in Navigational Systems 527

13.1 Synthesis of Quasioptimal Space–Time Signal and Image Processing Algorithms Based on the Generalized Approach to Signal Processing 527
13.1.1 Criterial Correlation Functions 529
13.1.2 Difference Criterial Functions 531
13.1.3 Spectral Criterial Functions 532
13.1.4 Bipartite Criterial Functions 533
13.1.5 Rank Criterial Functions 536
13.2 The Quasioptimal Generalized Image Processing Algorithm 536
13.3 The Classical Generalized Image Processing Algorithm 555
13.4 The Difference Generalized Image Processing Algorithm 561
13.5 The Generalized Phase Image Processing Algorithm 562
13.6 The Generalized Image Processing Algorithm: Invariant Moments 565
13.7 The Generalized Image Processing Algorithm: Amplitude Ranking 567
13.8 The Generalized Image Processing Algorithm: Gradient Vector Sums 570
13.9 The Generalized Image Processing Algorithm: Bipartite Functions 573
13.10 The Hierarchical Generalized Image Processing Algorithm 574
13.11 The Generalized Image Processing Algorithm: The Use of the Most Informative Area 577
13.12 The Generalized Image Processing Algorithm: Coding of Images 578
13.13 The Multichannel Generalized Image Processing Algorithm 580
13.14 Conclusions 581
References 582

Chapter 14 Object Image Preprocessing **585**
14.1 Object Image Distortions 585
14.2 Geometrical Transformations 585
14.2.1 The Perspective Transformation 586
14.2.2 Polynomial Estimation 589
14.2.3 Transformations of Brightness Characteristics 592
14.3 Detection of Boundary Edges 595
14.4 Conclusions 598
References 599

Appendix I Classification of Stochastic Processes **601**
References 612

Appendix II The Power Spectral Density of the Target Return Signal with Arbitrary Velocity Vector Direction of the Moving Radar in Space and with the Presence of Roll and Pitch Angles **613**
References 620

Notation Index **621**

Index **631**

Part I

Theory of Fluctuating Target Return Signals in Navigational Systems

1

Probability Distribution Density of the Amplitude and Phase of the Target Return Signal

1.1 Two-Dimensional Probability Distribution Density of the Amplitude and Phase

Signals reflected by various natural objects, for example, the Earth or sea surface, hydrometeors (snow, rain, hail, and fog), manmade metallic reflectors, and so on, have some general statistical features, which allow us to consider target return signals from unified theoretical propositions. The target return signal of any kind can be thought of as a great number of signals reflected by independent elementary scatterers distributed in a stochastic manner on a surface or in space. A similar model of the target return signal is widely used in radar and communications and works very well.[1-10]

Statistical characteristics of the target return signal are defined by their probability distribution densities and functions of the amplitude and phase. Definition of the probability distribution densities of the amplitude and phase of the target return signal in navigational systems is reduced, in the general case, to defining the probability distribution densities of module and argument of the sum of coplanar vectors and is one of the main fundamental problems of statistical signal processing. Particular decisions of this problem were found by Rayleigh and Rice and the general decision is discussed in References 11 and 12.

Let us consider scattering of signals by the two-dimensional (surface) or three-dimensional (space) target (see Figure 1.1). In the general case, the transmitter and receiver can be considered as a diversity system and take an arbitrary position in space. However, in practice, particularly in radar navigational systems, the transmitter and receiver are matched at the same point of space — for example, an aircraft. Let us assume the transmitter generates a simple harmonic signal, and all elements of the considered system — the transmitter, receiver, and elementary scatterers — are stationary. Each elementary target return signal is an exact copy in shape of the

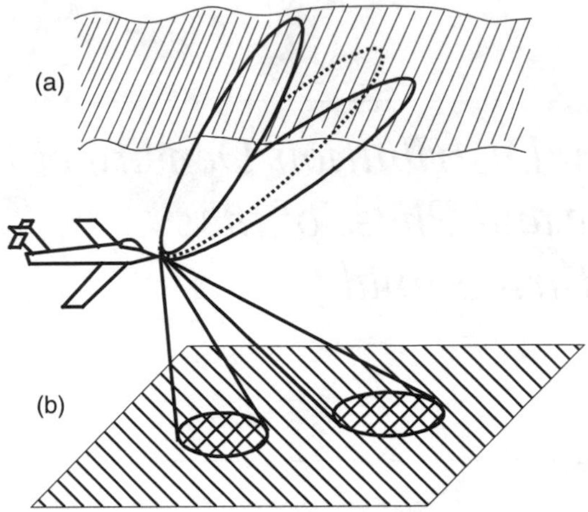

FIGURE 1.1
(a) The three-dimensional (space) target and (b) the two-dimensional (surface) target.

transmitted signal. However, the amplitude and phase of the target return signal are stochastic because there is dependence on some random factors: size, shape, orientation, nature of scatterers, position within the radar antenna directional diagram, distance, etc.[13-15]

The resulting signal W is the sum of a large number of sinusoidal oscillations w_i at the same frequency ω but with different amplitudes s_i and phases ϕ_i. For this reason, the resulting signal W is the sinusoid at the same frequency ω, the amplitude S and phase ϕ of which do not vary with time but depend on the amplitudes and phases of elementary signals:

$$W(t) = \sum_{i=-\infty}^{\infty} w_i = \sum_{i=-\infty}^{\infty} s_i(t)\cos(\omega t + \phi_i) = S(t)\cos(\omega t + \phi), \quad = X(t)\cos\omega t + Y(t)\sin\omega t \tag{1.1}$$

where

$$X(t) = \sum_{i=-\infty}^{\infty} s_i(t)\cos\phi_i = S(t)\cos\phi \tag{1.2}$$

and

$$Y(t) = \sum_{i=-\infty}^{\infty} s_i(t)\sin\phi_i = S(t)\sin\phi \tag{1.3}$$

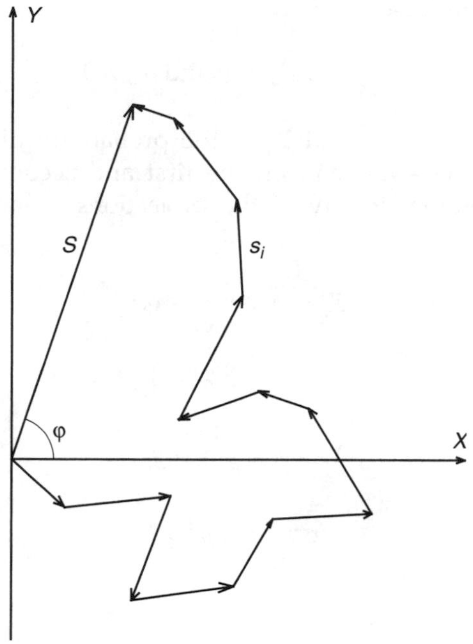

FIGURE 1.2
The resulting vector as a sum of stochastic vectors.

are the quadrature components of the received signal; $S^2(t) = X^2(t) + Y^2(t)$; and $\text{tg}\,\phi = \frac{Y(t)}{X(t)}$.

The amplitude w_i can be considered as a projection of the vector with length s_i and argument ϕ_i. The amplitude and phase of the resulting vector (see Figure 1.2) being functions of a large number of random variables (elementary vectors), are also random variables, defined by the corresponding probability distribution densities. The definition of these probability distribution densities is the main problem of the two-dimensional theory of Brownian motion. In addition, we can assume that the elementary stochastic vector defines the motion of some point on the plane. Then, the resulting vector corresponds to the position of this point after n stochastic displacements on the plane.[16,17]

For a large number of elementary vectors, the values of which have approximately the same order, projections of the resulting vector obey the Gaussian probability distribution law, parameters of which depend on the probability distribution densities of amplitudes and orienting angles of elementary vectors:[18]

$$f(X,Y) = \frac{1}{2\pi\sigma_X\sigma_Y} \cdot e^{-\frac{(X-X_0)^2}{2\sigma_X^2} - \frac{(Y-Y_0)^2}{2\sigma_Y^2}}, \tag{1.4}$$

where, in the general case,

$$\sigma_X \neq \sigma_Y,\ X_0 \neq 0, \text{ and } Y_0 \neq 0. \tag{1.5}$$

The parameters σ_X, σ_Y, X_0, and Y_0 of the probability distribution density given by Equation (1.4) depend on the first and second moments of the probability distribution density of the projections x and y of elementary vectors:

$$X_0 = n < x > + x_0; \tag{1.6}$$

$$\sigma_X^2 = n < x^2 >; \tag{1.7}$$

$$Y_0 = n < y > + y_0; \tag{1.8}$$

$$\sigma_Y^2 = n < y^2 >, \tag{1.9}$$

where

$$< x > = \int_{-\infty}^{\infty}\int_{-\infty}^{\infty} x f(x,y)\, dx\, dy; \tag{1.10}$$

$$< y > = \int_{-\infty}^{\infty}\int_{-\infty}^{\infty} y f(x,y)\, dx\, dy; \tag{1.11}$$

$$< x^2 > = \int_{-\infty}^{\infty}\int_{-\infty}^{\infty} x^2 f(x,y)\, dx\, dy; \tag{1.12}$$

$$< y^2 > = \int_{-\infty}^{\infty}\int_{-\infty}^{\infty} y^2 f(x,y)\, dx\, dy; \tag{1.13}$$

$<x>$ and $<y>$ are the mean of projections of elementary vectors, respectively; $<x^2>$ and $<y^2>$ are the mean square of projections of elementary vectors, respectively; $f(x, y)$ is the two-dimensional probability distribution density of the projections x and y; n is the number of stochastic elementary vectors;

x_0 and y_0 are the projections of the steady vector that can be added to the sum of stochastic elementary vectors in the general case.

Equation (1.4)–Equation (1.13) contain sufficient information to investigate the probability distribution densities of the amplitude and phase of the resulting vector (target return signal) under different assumptions regarding the character of the probability distribution density of elementary vectors. From Equation (1.4)–Equation (1.13) it is apparent that the problem considered here is divided into two subproblems. The first is to define the probability distribution density of the amplitude and phase of the vector, the projections X and Y of which are independent and obey the Gaussian probability distribution density with the means X_0 and Y_0 and the variances σ_X^2 and σ_Y^2, respectively. The second is to define a function between these parameters and the probability distribution density $f(x, y)$ of projections of elementary vectors given before. Let us consider the first problem. A general solution to this problem under the assumption that the projections of the resulting vector (target return signal) have the mean differed from zero and different variances and are correlated with each other is discussed in more detail in Feldman.[19] We discuss only the main results, assuming that correlation between the quadrature components $X(t)$ and $Y(t)$ is absent.

Going from the X and Y coordinates to the polarity coordinates S and ϕ and following the well-known rules of coordinate transformation, we can define the two-dimensional probability distribution density $f(S, \phi)$ of the amplitude and phase of the resulting vector using Equation (1.4). For this purpose, let us introduce the normalized amplitude $S = \frac{S}{\sqrt{\sigma_X \sigma_Y}}$ of the resulting vector instead of the amplitude S and the coefficient of asymmetry $\chi = \frac{\sigma_X}{\sigma_Y}$. Then

$$f(S,\phi) = \frac{S}{2\pi} \cdot e^{-0.25(\chi + \chi^{-1})\left[S^2 - 2SS_0\cos(\phi - \phi_0) + S_0^2\right]} \times e^{0.25(\chi + \chi^{-1})\left[S^2\cos 2\phi - 2SS_0\cos(\phi + \phi_0) + S_0^2\cos 2\phi_0\right]}, \tag{1.14}$$

where $S_0 = \frac{S_0}{\sqrt{\sigma_X \sigma_Y}} = \sqrt{\frac{X_0^2 + Y_0^2}{\sigma_X \sigma_Y}}$ and $\phi_0 = \text{arctg}\,\frac{Y_0}{X_0}$ are the normalized amplitude and phase of the average vector. The probability distribution density $f(S, \phi)$ in Equation (1.14) is defined by the following parameters: χ, S_0, and ϕ_0. Using Equation (1.14), we can define the one-dimensional probability distribution density of the amplitude and phase of the resulting vector (target return signal).

1.2 Probability Distribution Density of the Amplitude

Integrating Equation (1.14) with respect to the parameter ϕ within the limits of the interval $[0, 2\pi]$, we can define the probability distribution density of the amplitude of the resulting vector (target return signal). In the general case, when the parameters χ, S_0, and ϕ_0 are arbitrary, the determination of the integral is very difficult. However, using the following series expansion[20]

$$e^{x\cos\alpha} = I_0(x) + 2\sum_{k=1}^{\infty} I_k(x)\cos k\alpha\ , \tag{1.15}$$

where $I_k(x)$ is the modified Bessel function, the probability distribution density of the amplitude of the resulting vector (target return signal) can be defined in the following form:[11,12,18,19]

$$f(S) = S \cdot e^{-0.25(\chi+\chi^{-1})S^2 - 0.5(\chi\sin^2\phi_0 + \chi^{-1}\cos^2\phi_0)S_0^2} \times \sum_{k=0}^{\infty} \varepsilon_k I_k\left[0.25(\chi-\chi^{-1})S^2\right] I_{2k}(\gamma\, S_0 S)\cos 2k\phi'\ , \tag{1.16}$$

where

$$\varepsilon_k = \begin{cases} 1 & \text{at } k=0; \\ 2 & \text{at } k \geq 1; \end{cases} \tag{1.17}$$

$$\gamma = \sqrt{\chi^2\sin^2\phi_0 + \chi^{-2}\cos^2\phi_0}; \tag{1.18}$$

$$\text{tg}\,\phi' = \chi^2\,\text{tg}\,\phi_0. \tag{1.19}$$

Using Equation (1.16)–Equation (1.19), we can obtain the following particular cases. The first case is $\phi_0 = 0$ or the steady vector is parallel to the axis x. In this case, we can write:[11,12]

$$f(S) = S \cdot e^{-0.25(\chi+\chi^{-1})S^2 - 0.5\chi^{-1}S_0^2} \cdot \sum_{k=0}^{\infty} \varepsilon_k I_k\left[0.25(\chi-\chi^{-1})S^2\right] I_{2k}\left(\tfrac{S_0 S}{\chi}\right). \tag{1.20}$$

This is the Beckman probability distribution density. The second case is $\chi = 1$ or the variances are the same: $\sigma_X = \sigma_Y$. Then[21,22]

$$f(S) = S \cdot e^{-0.5(S_0^2+S^2)} \cdot I_0(S_0 S) \tag{1.21}$$

is the well-known Rice probability distribution law. The third case is $S_0 = 0$ or the steady vector is absent but $\chi \neq 1$. For this case, we obtain:[23,24]

$$f(S) = S \cdot e^{-0.25(\chi+\chi^{-1})S^2} \cdot I_0\left[0.25(\chi - \chi^{-1})S^2\right]. \tag{1.22}$$

This is the Hoyt probability distribution density. The fourth case is $S_0 = 0$ and $\chi = 1$. Then[25]

$$f(S) = S \cdot e^{-0.5S^2} \tag{1.23}$$

is the Rayleigh probability distribution law. Thus, it is necessary and sufficient to satisfy two conditions $\sigma_X = \sigma_Y$ and $S_0 = 0$ for the Rayleigh probability distribution density to be held true. Examples of the probability distribution densities of the amplitude shown in Figure 1.3–Figure 1.8 are obtained under the following conditions: $\chi \neq 1$ and $0 \leq \phi_0 \leq 0.5\pi$. If with the predetermined ϕ_0, it is necessary to define the probability distribution density of the amplitude for some values $\chi^* > 1$, then the corresponding probability distribution density coincides with that probability distribution density, the parameters of which are determined by $(\chi^*)^{-1}$ and $\phi_0^* = 0.5\pi - \phi_0$.

For any values of the parameter ϕ_0, an increase in the value of the parameter S_0 leads to a shift to the right of the probability distribution density, as well as in the case of the Rice probability distribution density $\chi = 1$. The shift and shape of the probability distribution density of the amplitude are significantly different at various values of the parameter ϕ_0. At $\chi^2 = 0.01$, we can clearly see this effect. In this case, the shape of the probability distribution density is not Gaussian in spite of an increase in the value of the parameter S_0. It is interesting to note that the well-known statement that follows, for example, from Equation (1.21) regarding the probability distribution density of the amplitude of the target return signal becoming Gaussian with an unlimited increase in the value of the steady vector S_0, is not completely general. This statement has one exception for the singular case when the variance σ_X or σ_Y is equal to zero and the steady vector of the amplitude is directed along the axis *x*.[19]

The problem when the probability distribution density of the amplitude tends to approach Gaussian is discussed in more detail in Feldman and Klimenkova.[26] As was proved therein, the probability distribution density of the amplitude can tend to approach Gaussian — we consider only the main part of the probability distribution density, not its residues and for the case of the Rice probability distribution density of the amplitude, if and only if the length of the steady vector of the amplitude tends to approach infinity and the variances σ_X and σ_Y are not the same except the singular case

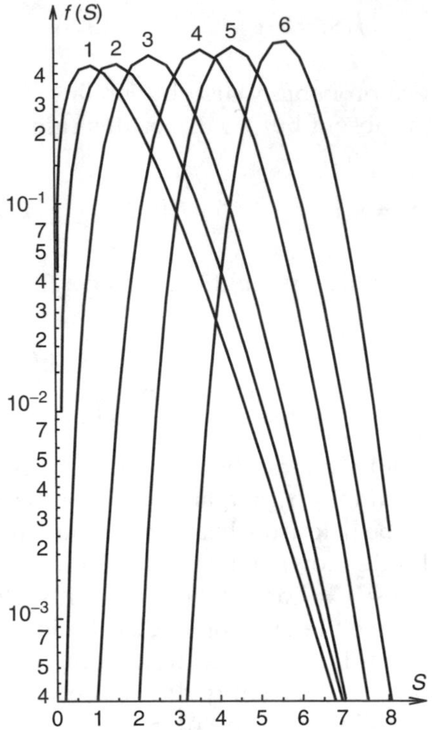

FIGURE 1.3
The probability distribution density of the amplitude of the target return signal at $S_0 = 0, \ldots, 5$, $\chi^2 = 0.1$, and $\phi_0 = 0°$: (1) $\rho_0 = 0$; (2) $\rho_0 = 1$; (3) $\rho_0 = 2$; (4) $\rho_0 = 3$; (5) $\rho_0 = 4$; (6) $\rho_0 = 5$.

mentioned in the preceding text. Parameters of the limiting Gaussian probability distribution density are defined by both the direction of the steady vector of the amplitude and a ratio between the variances of projections.

In the general case, the initial and central moments of the probability distribution density given by Equation (1.16) can only be defined numerically.[26] The first, second, third, and fourth normalized initial moments m_n of the probability distribution density of the amplitude as a function of the steady vector S_0 and coefficient of asymmetry χ are shown in Figure 1.9 and Figure 1.10. In the case of the nonnormalized probability distribution density, the initial moments must be multiplied by the value $(\sigma_X\sigma_Y)^{0.5n}$ and the steady vector S_0 must be replaced by the value $\frac{S_0}{\sqrt{\sigma_X\sigma_Y}}$.

The regions are shaded within the limits of which the values of the moments m_n vary as the parameter ϕ_0 changes from 0 to 0.5π. For the second initial moment ($n = 2$) of the probability distribution density of the amplitude, these regions are compressed to a line because dependence between the second initial moment m_2 and the parameter ϕ_0 is very weak. Under the condition $n > 2$, the variation of the parameter ϕ_0 within the limits of the

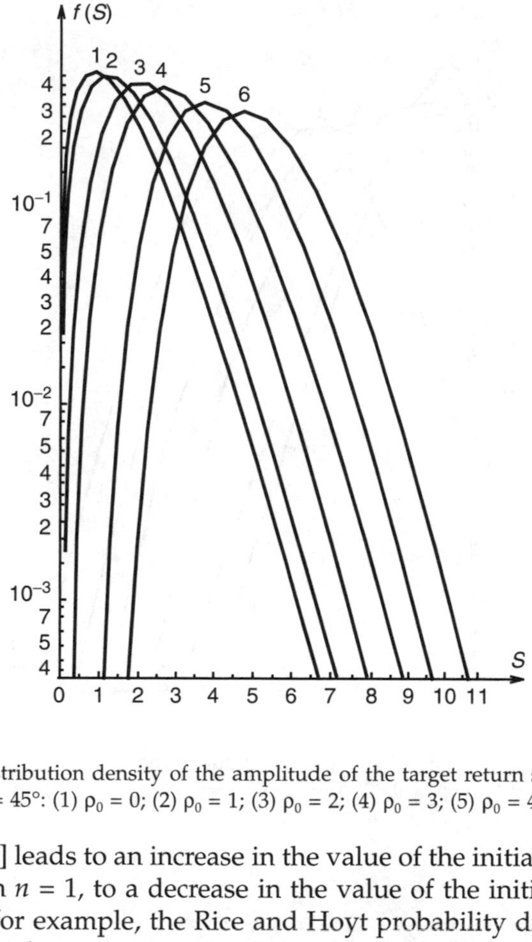

FIGURE 1.4
The probability distribution density of the amplitude of the target return signal at $S_0 = 0, \ldots, 5$, $\chi^2 = 0.1$, and $\phi_0 = 45°$: (1) $\rho_0 = 0$; (2) $\rho_0 = 1$; (3) $\rho_0 = 2$; (4) $\rho_0 = 3$; (5) $\rho_0 = 4$; (6) $\rho_0 = 5$.

interval $[0, 0.5\pi]$ leads to an increase in the value of the initial moments, and at the condition $n = 1$, to a decrease in the value of the initial moments. In some cases — for example, the Rice and Hoyt probability distribution densities — the initial moments can be defined by specific or elementary functions.[27,28] Numerical values of the initial moments at various magnitudes of the parameters χ, S_0, and ϕ_0 are discussed in Feldman and Klimenkova.[26]

1.3 Probability Distribution Density of the Phase

Integrating Equation (1.14) with respect to the normalized amplitude of the target return signal S within the limits of the interval $[0, \infty]$, we can determine the probability distribution density of the phase of the target return signal in a general form:[29]

$$f(\phi) = \varsigma\{1 + \sqrt{\pi}\,\eta \cdot e^{\eta^2} \cdot [1 + \Phi(x)]\}, \tag{1.24}$$

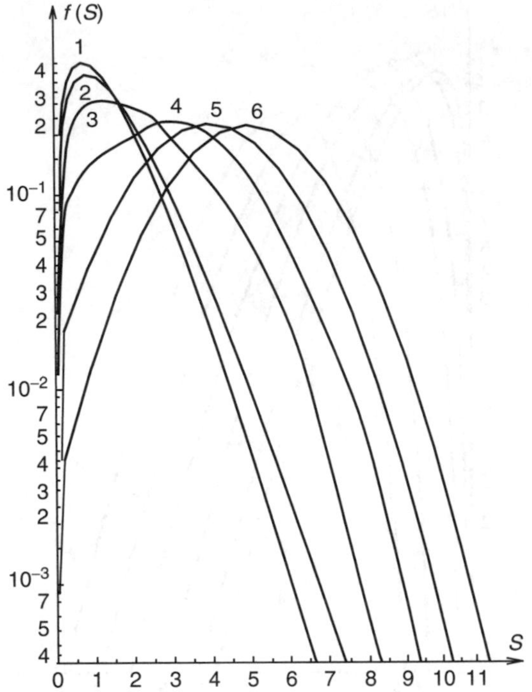

FIGURE 1.5
The probability distribution density of the amplitude of the target return signal at $S_0 = 0, \ldots, 5$, $\chi^2 = 0.1$, and $\phi_0 = 90°$: (1) $\rho_0 = 0$; (2) $\rho_0 = 1$; (3) $\rho_0 = 2$; (4) $\rho_0 = 3$; (5) $\rho_0 = 4$; (6) $\rho_0 = 5$.

where

$$\varsigma = \frac{e^{-0.5(\chi \sin^2 \phi_0 + \chi^{-1} \cos^2 \phi_0) S_0^2}}{2\pi(\chi \sin^2 \phi_0 + \chi^{-1} \cos^2 \phi_0)}; \tag{1.25}$$

$$\eta = \frac{(\chi \sin \phi_0 \sin \phi + \chi^{-1} \cos \phi_0 \cos \phi)\, S_0}{\sqrt{2(\chi \sin^2 \phi_0 + \chi^{-1} \cos^2 \phi_0)}}; \tag{1.26}$$

$$\Phi(x) = \frac{1}{\sqrt{2\pi}} \int_{-\infty}^{x} e^{-0.5t^2}\, dt \tag{1.27}$$

is the error integral. The probability distribution density of the phase of the target return signal is defined by the parameters χ, S_0, and ϕ_0 as well as the probability distribution density of amplitudes. In particular, if the following conditions $\chi = 1$ and $\sigma_X = \sigma_Y$ are satisfied (the case of the Rice probability

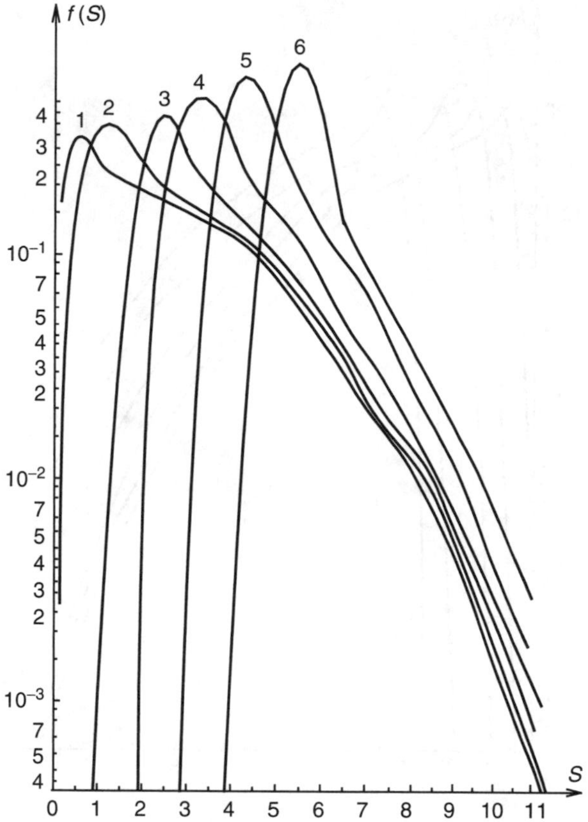

FIGURE 1.6
The probability distribution density of the amplitude of the target return signal at $S_0 = 0, \ldots, 5$, $\chi^2 = 0.01$, and $\phi_0 = 0°$: (1) $\rho_0 = 0$; (2) $\rho_0 = 1$; (3) $\rho_0 = 2$; (4) $\rho_0 = 3$; (5) $\rho_0 = 4$; (6) $\rho_0 = 5$.

distribution density of the amplitude of the target return signal), then reference to Equation (1.24) shows that

$$f(\phi) = \frac{1}{2\pi} \cdot e^{-0.5S_0^2}$$
$$\times \left\{1 + \sqrt{0.5\pi}\; S_0 \cos(\phi - \phi_0) \cdot e^{0.5S_0^2 \cos^2(\phi-\phi_0)} \cdot \left[1 + \Phi\left[2^{-0.5} S_0 \cos(\phi - \phi_0)\right]\right]\right\} \quad (1.28)$$

The probability distribution density of the phase of the target return signal given by Equation (1.28) is symmetric with respect to the parameter ϕ_0[30] (see Figure 1.11). For another particular case, when $S_0 = 0$ but $\chi \neq 1$ — the case of the Hoyt probability distribution density of the amplitude of the target return signal — we can write

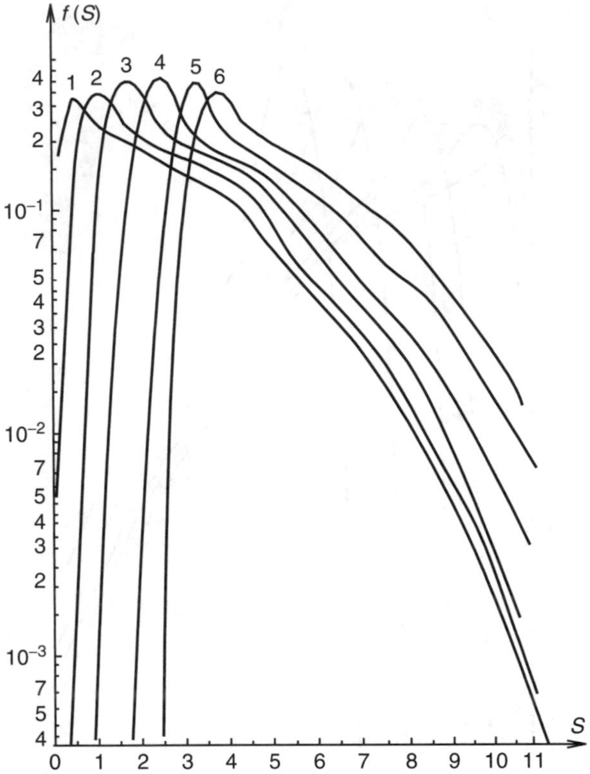

FIGURE 1.7
The probability distribution density of the amplitude of the target return signal at $S_0 = 0, \ldots, 5$, $\chi^2 - 0.01$, and $\phi_0 = 45°$: (1) $\rho_0 = 0$; (2) $\rho_0 = 1$; (3) $\rho_0 = 2$; (4) $\rho_0 = 3$; (5) $\rho_0 = 4$; (6) $\rho_0 = 5$.

$$f(\phi) = \frac{1}{2\pi(\chi \sin^2 \phi_0 - \chi^{-1} \cos^2 \phi_0)} \tag{1.29}$$

(see Figure 1.12). Finally, if the conditions $\chi = 1$ and $S_0 = 0$ are satisfied — the case of the Rayleigh probability distribution density of the amplitude of the target return signal — we can obtain the uniform probability distribution density of the phase within the limits of the interval [0, 2π], i.e., $f(\phi) = 0.5\pi$. The functions that are determined for the general case by the formula in Equation (1.24) at the conditions $S_0 = 1$ and χ = *variable* are shown in Figure 1.13, Figure 1.14, and Figure 1.15. The probability distribution density of the phase for the singular case at $\sigma_X = 0$ is defined in Feldman.[19]

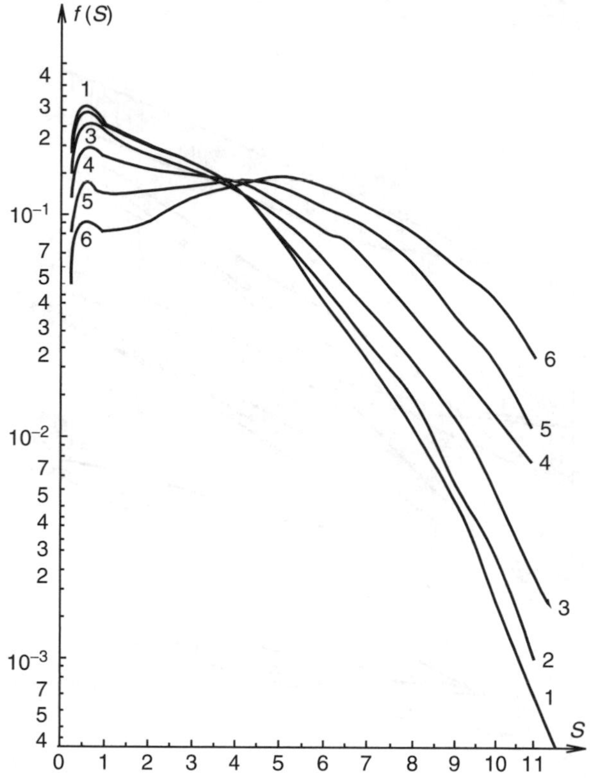

FIGURE 1.8
The probability distribution density of the amplitude of the target return signal at $S_0 = 0, \ldots, 5$, $\chi^2 = 0.01$, and $\phi_0 = 90°$: (1) $\rho_0 = 0$; (2) $\rho_0 = 1$; (3) $\rho_0 = 2$; (4) $\rho_0 = 3$; (5) $\rho_0 = 4$; (6) $\rho_0 = 5$.

1.4 Probability Distribution Density Parameters of the Target Return Signal as a Function of the Distribution Law of the Amplitude and Phase of Elementary Signals

The parameters χ, S_0, ϕ, and S_0, in their turn, are the functions of parameters of the probability distribution densities of elementary vectors and the initial vector. General formulae can be written based on Equation (1.6)–Equation (1.9) in the following form:

$$\chi = \frac{\sigma_X}{\sigma_Y} = \sqrt{\frac{< x^2 >}{< y^2 >}}; \tag{1.30}$$

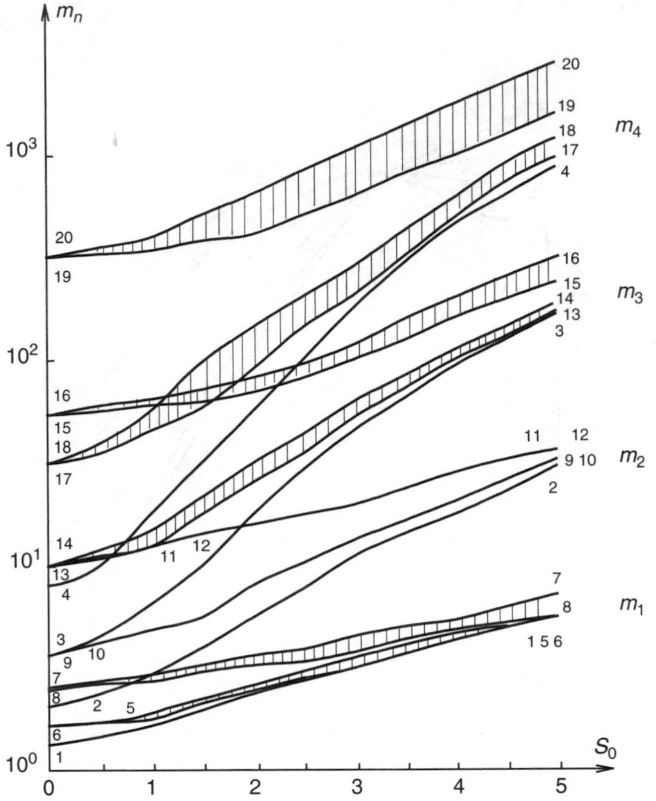

FIGURE 1.9
The initial moment of the probability distribution density of the amplitude of the target return signal as a function of the parameter S_0: (1,2,3, and 4) $\chi^2 = 1.0$; (5,9,13, and 17) $\chi^2 = 0.1$, $\phi_0 = 0°$; (6,10,14, and 18) $\chi^2 = 0.1$, $\varphi_0 = 90°$; (7,11,15, and 19) $\chi^2 = 0.01$, $\phi_0 = 0°$; (8,12,16, and 20) $\chi^2 = 0.01$, $\phi_0 = 90°$.

$$S_0 = \sqrt{\frac{X_0^2 + Y_0^2}{\sigma_X \sigma_Y}} = \sqrt{\frac{(n<x>+x_0)^2 + (n<y>+y_0)^2}{n\sqrt{<x^2>\cdot<y^2>}}}; \tag{1.31}$$

$$\phi_0 = \text{arctg}\,\frac{Y_0}{X_0} = \text{arctg}\,\frac{n<y>+y_0}{n<x>+x_0}; \tag{1.32}$$

$$\mathcal{S}_0 = S_0\sqrt{\sigma_X \sigma_Y} = \sqrt{X_0^2 + Y_0^2} = \sqrt{(n<x>+x_0)^2 + (n<y>+y_0)^2}. \tag{1.33}$$

The function between the parameters χ, S_0, ϕ, and $\mathcal{S}_0$ and the probability distribution densities of elementary vectors for some particular examples is of prime interest to us. In some cases, it is more convenient to use the

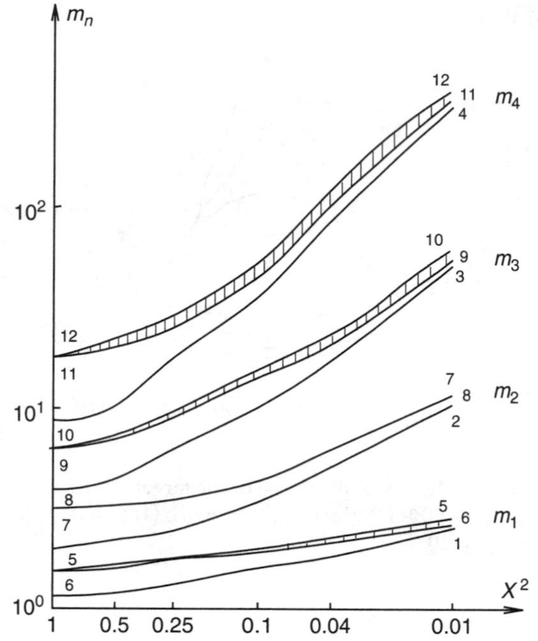

FIGURE 1.10
The initial moment of the probability distribution density of the amplitude of the target return signal as a function of the parameter χ^2: (1,2,3, and 4) $S_0 = 0$; (5,7,9, and 11) $S_0 = 1$, $\varphi_0 = 0°$; (6,8,10, and 12) $S_0 = 1$, $\phi_0 = 90°$.

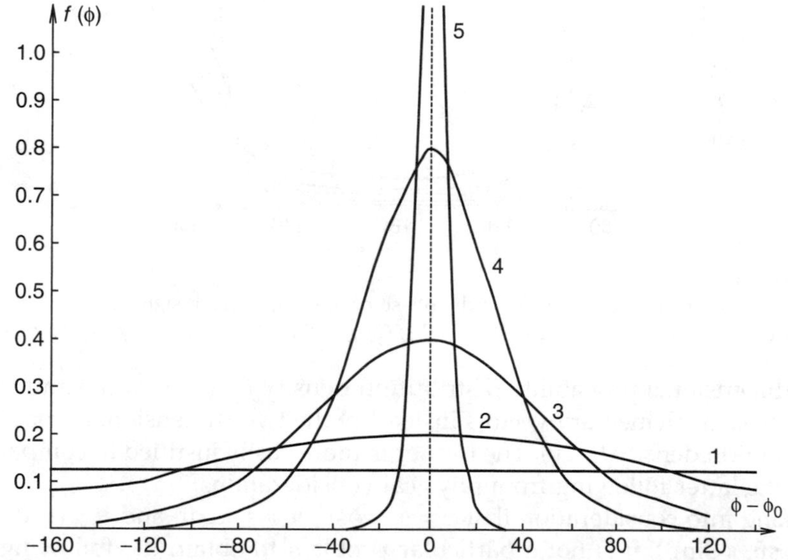

FIGURE 1.11
The probability distribution density of the phase of the target return signal at $\chi^2 = 1$ (the phase is shown in degrees): (1) $S_0 = 0$; (2) $S_0 = 0.32$; (3) $S_0 = 1$; (4) $S_0 = 2$; (5) $S_0 = 10$.

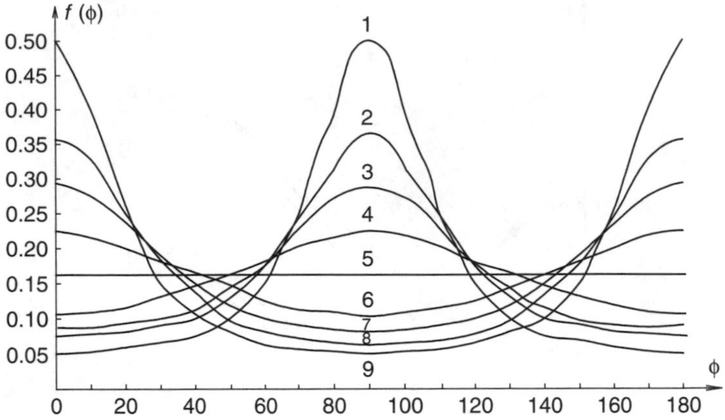

FIGURE 1.12
The probability distribution density of the phase of the target return signal at $S_0 = 0$ (the phase is shown in degrees): (1) $\chi^2 = 0.1$; (2) $\chi^2 = 0.2$; (3) $\chi^2 = 0.3$; (4) $\chi^2 = 0.5$; (5) $\chi^2 = 1$; (6) $\chi^2 = 2$; (7) $\chi^2 = 3.33$; (8) $\chi^2 = 5$; (9) $\chi^2 = 10$.

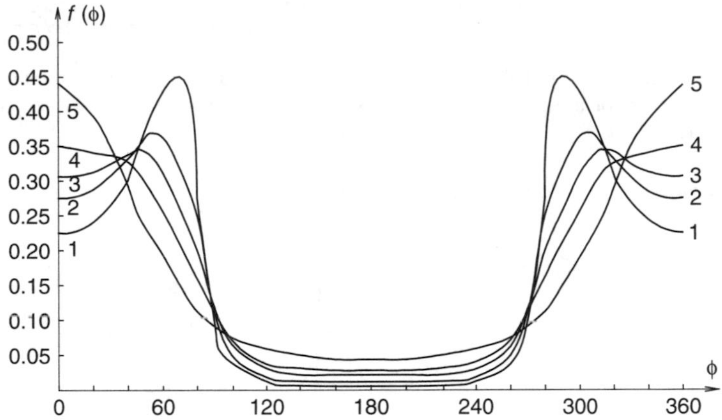

FIGURE 1.13
The probability distribution density of the phase of the target return signal at $S_0 = 1$ and $\phi_0 = 0°$ (the phase is shown in degrees): (1) $\chi^2 = 0.1$; (2) $\chi^2 = 0.2$; (3) $\chi^2 = 0.3$; (4) $\chi^2 = 0.5$; (5) $\chi^2 = 1$.

two-dimensional probability distribution density $f_{s,\vartheta}(s, \vartheta)$ of the amplitude and phase of elementary vectors instead of the two-dimensional probability distribution density $f(x, y)$. The former is more easily justified in comparison with the latter following from physical considerations.[31]

Taking into consideration that $x = s \cdot \cos\vartheta$, $y = s \cdot \sin\vartheta$ and $f_{s,\vartheta}(s, \vartheta) = s \cdot f(s \cos\vartheta, s \sin\vartheta)$, it is not a particular problem to obtain the following formulae for the values $\langle x \rangle$, $\langle y \rangle$, $\langle x^2 \rangle$, and $\langle y^2 \rangle$:

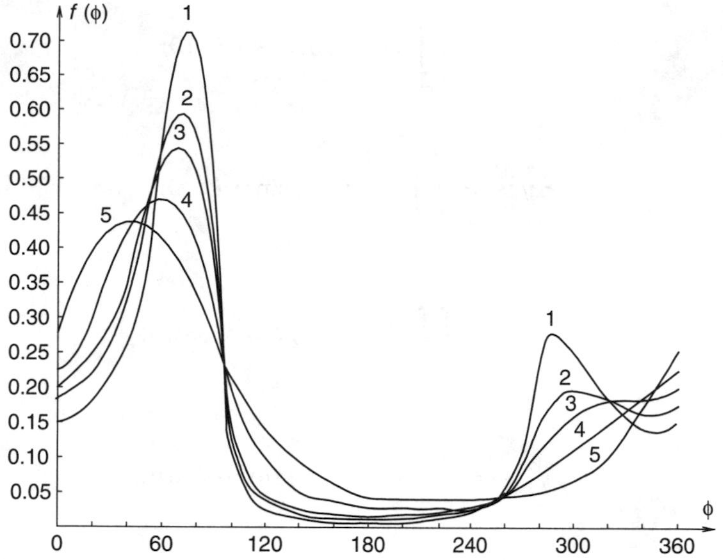

FIGURE 1.14
The probability distribution density of the phase of the target return signal at $S_0 = 1$ and $\phi_0 = 45°$ (the phase is shown in degrees): (1) $\chi^2 = 0.1$; (2) $\chi^2 = 0.2$; (3) $\chi^2 = 0.3$; (4) $\chi^2 = 0.5$; (5) $\chi^2 = 1$.

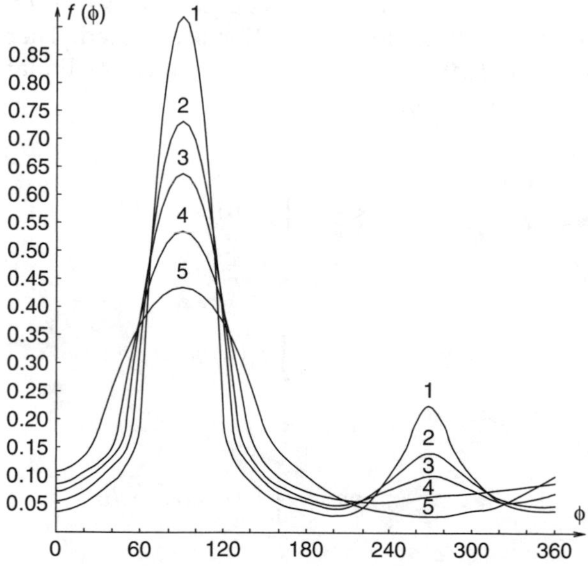

FIGURE 1.15
The probability distribution density of the phase of the target return signal at $S_0 = 1$ and $\phi_0 = 90°$ (the phase is shown in degrees): (1) $\chi^2 = 0.1$; (2) $\chi^2 = 0.2$; (3) $\chi^2 = 0.3$; (4) $\chi^2 = 0.5$; (5) $\chi^2 = 1$.

$$< x > = \int_0^{\infty}\int_{-\infty}^{\infty} s f_{s,\vartheta}(s,\vartheta)\cos\vartheta \, ds\, d\vartheta; \tag{1.34}$$

$$< y > = \int_0^{\infty}\int_{-\infty}^{\infty} s f_{s,\vartheta}(s,\vartheta)\sin\vartheta \, ds\, d\vartheta; \tag{1.35}$$

$$< x^2 > = \int_0^{\infty}\int_{-\infty}^{\infty} s^2 f_{s,\vartheta}(s,\vartheta)\cos^2\vartheta \, ds\, d\vartheta; \tag{1.36}$$

$$< y^2 > = \int_0^{\infty}\int_{-\infty}^{\infty} s^2 f_{s,\vartheta}(s,\vartheta)\sin^2\vartheta \, ds\, d\vartheta, \tag{1.37}$$

where we assume, for the sake of generality, that the parameter ϑ is within the limits of the interval $(-\infty, \infty)$.

Substituting Equation (1.34)–Equation (1.37) in Equation (1.30)–Equation (1.33), respectively, we can obtain the general formulae for the parameters χ, S_0, ϕ_0, and S_0. However, in the majority of cases, the amplitudes and phases of elementary vectors are mutually independent because they are defined by various parameters that are not functionally related. Therefore, we can write $f_{s,\vartheta}(s, \vartheta) = f_s(s) \cdot f_\vartheta(\vartheta)$. In this case, Equation (1.34)–Equation (1.37) can be rewritten in the following form:

$$< x > = < s > \int_{-\infty}^{\infty} f_\vartheta(\vartheta)\cos\vartheta \, d\vartheta ; \tag{1.38}$$

$$< y > = < s > \int_{-\infty}^{\infty} f_\vartheta(\vartheta)\sin\vartheta \, d\vartheta ; \tag{1.39}$$

$$< x^2 > = < s^2 > \int_{-\infty}^{\infty} f_\vartheta(\vartheta)\cos^2\vartheta \, d\vartheta ; \tag{1.40}$$

$$< y^2 > = < s^2 > \int_{-\infty}^{\infty} f_\vartheta(\vartheta)\sin^2\vartheta \, d\vartheta , \tag{1.41}$$

where $< s > = \int_0^{\infty} s f_s(s)\, ds$ and $< s^2 > = \int_0^{\infty} s^2 f_s(s)\, ds$ are the mean and variance of the amplitude of elementary vectors.

Let us assume that the steady vector (x_0, y_0) is equal to zero. This assumption narrows down insignificantly the generality of reasoning because the vector of the mean $X_0 = n <x>$ and $Y_0 = n <y>$ is not assumed to be equal to zero. However, the formulae will be more compact and obvious. The abandonment of this limitation leads us to very complex formulae. Substituting Equation (1.38)–Equation (1.41) in Equation (1.30)–Equation (1.33), we can write

$$\chi^2 = \frac{1 + \int_{-\infty}^{\infty} f_\vartheta(\vartheta) \cos 2\vartheta\, d\vartheta}{1 - \int_{-\infty}^{\infty} f_\vartheta(\vartheta) \sin 2\vartheta\, d\vartheta}; \tag{1.42}$$

$$S_0^2 = 2n \cdot \frac{< s >^2}{< s^2 >}$$

$$\times \frac{\int_{-\infty}^{\infty} f_\vartheta(\vartheta) \cos\vartheta\, d\vartheta \cdot \int_{-\infty}^{\infty} f_\vartheta(\vartheta) \cos\vartheta\, d\vartheta + \int_{-\infty}^{\infty} f_\vartheta(\vartheta) \sin\vartheta\, d\vartheta \cdot \int_{-\infty}^{\infty} f_\vartheta(\vartheta) \sin\vartheta\, d\vartheta}{\sqrt{1 - \int_{-\infty}^{\infty} f_\vartheta(\vartheta) \cos 2\vartheta\, d\vartheta \cdot \int_{-\infty}^{\infty} f_\vartheta(\vartheta) \cos 2\vartheta\, d\vartheta}};$$

$$(1.43)$$

$$tg\, \phi_0 = \frac{\int_{-\infty}^{\infty} f_\vartheta(\vartheta) \sin\vartheta\, d\vartheta}{\int_{-\infty}^{\infty} f_\vartheta(\vartheta) \cos\vartheta\, d\vartheta}; \tag{1.44}$$

$$S_0^2 = n^2 < s >^2$$

$$\times \left[\int_{-\infty}^{\infty} f_\vartheta(\vartheta) \cos\vartheta\, d\vartheta \cdot \int_{-\infty}^{\infty} f_\vartheta(\vartheta) \cos\vartheta\, d\vartheta + \int_{-\infty}^{\infty} f_\vartheta(\vartheta) \sin\vartheta\, d\vartheta \cdot \int_{-\infty}^{\infty} f_\vartheta(\vartheta) \sin\vartheta\, d\vartheta\right].$$

$$(1.45)$$

Reference to Equation (1.42) and Equation (1.44) shows that if the amplitudes and phases of elementary vectors are independent, then the parameters χ and φ_0 do not depend on the probability distribution density of the amplitude $f_s(s)$ of elementary signals and the number of vectors n. Naturally, the vector length of the mean S_0 depends on both the probability distribution density of the amplitude $f_s(s)$ of elementary signals and the number of vectors n. The vector length of the mean S_0 is proportional to the value $\sqrt{\frac{n\langle s\rangle^2}{\langle s^2\rangle}}$. In particular, the lengths of elementary vectors can be constant nonrandom variables s_0. Then $f_s(s) = \delta(s - s_0)$, where $\delta(x)$ is the delta function, and $\frac{\langle s\rangle^2}{\langle s^2\rangle} = 1$. If for the chosen coordinate system the probability distribution density of the phase is the even function, i.e., $f_\vartheta(-\vartheta) = f_\vartheta(\vartheta)$, then $y = 0$. In this case, as it follows from Equation (1.44), $\phi_0 = 0$ and Equation (1.43) and Equation (1.45) determining the parameters S_0 and S_0, respectively, are simplified. Let us consider the specific probability distribution density of elementary signals.

1.4.1 Uniform Probability Distribution Density of Phases

Let the probability distribution density of the phase be uniform within the limits of the interval $[\vartheta_1, \vartheta_2]$, i.e.,

$$f_\vartheta(\vartheta) = (\vartheta_2 - \vartheta_1)^{-1} \quad \text{at} \quad \vartheta \in [\vartheta_1, \vartheta_2] \tag{1.46}$$

and

$$f_\vartheta(\vartheta) = 0 \quad \text{at} \quad \vartheta \notin [\vartheta_1, \vartheta_2]. \tag{1.47}$$

In doing so,

$$\phi_0 = 0.5(\vartheta_1 + \vartheta_2); \tag{1.48}$$

$$\chi^2 = \frac{1 + \mathbf{sinc}\ \Delta\vartheta \cdot \cos 2\phi_0}{1 - \mathbf{sinc}\ \Delta\vartheta \cdot \cos 2\phi_0}; \tag{1.49}$$

$$S_0^2 = \frac{2n \cdot \frac{<s>^2}{<s^2>} \cdot \mathbf{sinc}^2(0.5\Delta\vartheta)}{\sqrt{1 - \left(\mathbf{sinc}\ \Delta\vartheta \cdot \cos 2\phi_0\right)^2}}; \tag{1.50}$$

$$S_0 = n <s> \cdot \ |\mathbf{sinc}\ (0.5\Delta\vartheta)|, \tag{1.51}$$

where

$$\Delta\vartheta = \vartheta_2 - \vartheta_1 \tag{1.52}$$

and **sinc** x is the sinc-function. Reference to Equation (1.48)–Equation (1.52) shows that if the following condition $\Delta\vartheta = 2k\pi$ at $k = 1, 2, \ldots$, is satisfied, then $\chi = 1$ and $S_0 = S_0 = 0$ and the amplitude of the resulting vector (target return signal) obeys the Rayleigh probability distribution density given by Equation (1.23). This is natural because the probability distribution density of the phase is uniform within the limits of the interval $[0, 2k\pi]$ or $[0, 2\pi]$.

If the following equality $\Delta\vartheta = (2k + 1)\pi$ is true, the conditions $\chi = 1$ and $S_0 = \frac{2n<s>^2}{<s^2>} \neq 0$ are satisfied, and we obtain the Rice probability distribution density of the amplitude with the parameter $S_0 = \frac{n<s>}{(k+0.5)\pi}$. In other words, the Rice probability distribution density of the amplitude has the same character and features as well as the probability distribution density of sum of the stochastic null vectors and the steady vector with the length S_0 that is directed at the angle $\phi_0 = \vartheta_1 + 0.5\pi$. With an increase in the value of k, the length of the steady vector S_0 is decreased and tends to approach zero as $k \to \infty$. This is natural because the value of k is higher, and the probability distribution density of the phase $f_\vartheta(\vartheta)$ is closer to the uniform probability distribution density within the limits of the interval $[0, 2\pi]$.

The Rice probability distribution density of the amplitude will occur for any interval $\Delta\vartheta$ if it is symmetric with regard to the value $\phi_0 = 0.25\pi$. Then, reference to Equation (1.49) shows that the coefficient of asymmetry of the probability distribution density is equal to unit ($\chi = 1$) and

$$S_0^2 = 2n \cdot \frac{<s>^2}{<s^2>} \cdot \mathbf{sinc}^2(0.5\Delta\vartheta). \tag{1.53}$$

When the phases of elementary signals are uniformly distributed symmetrically with regard to the values $\phi_0 = 0$ or $\phi_0 = 0.5\pi$, then we can write

$$\chi^2 = \frac{1 \pm \mathbf{sinc}\,\Delta\vartheta}{1 \mp \mathbf{sinc}\,\Delta\vartheta} \tag{1.54}$$

and

$$S_0^2 = \frac{2n \cdot \frac{<s>^2}{<s^2>} \cdot \mathbf{sinc}^2(0.5\Delta\vartheta)}{\sqrt{1 - \mathbf{sinc}^2\Delta\vartheta}} \tag{1.55}$$

The upper signs in Equation (1.55) are born a relation to the condition $\phi_0 = 0$ and the lower signs in Equation (1.55) are born a relation to the condition $\phi_0 = 0.5\pi$. Reference to Equation (1.54) and Equation (1.55) shows us that under the conditions $\phi_0 = 0$ and $\phi_0 = 0.5\pi$, the interval is narrowed down, i.e., $\Delta\vartheta \to 0$, then $\chi \to \infty$ or $\chi \to \infty$ 0, respectively, and $S_0 \to \infty$. In other words, the singular case $\sigma_X = 0$ or $\sigma_Y = 0$ occurs. Here, it should be borne in mind that the nonnormalized vector S_0 is independent of the initial phase ϕ_0.

1.4.2 "Triangular" Probability Distribution Density of Phases

Let us assume that the probability distribution density of phases of elementary signals takes the form of the isosceles triangle with the center at the point φ_0 and is determined in the following form:

$$f_{\vartheta}(\vartheta) = \begin{cases} \vartheta_m^{-1}\left(1 - \frac{|\vartheta - \vartheta_0|}{\vartheta_m}\right) & \text{at} \quad |\vartheta - \vartheta_0| \le \vartheta_m; \\ 0 & \text{at} \quad |\vartheta - \vartheta_0| > \vartheta_m, \end{cases} \tag{1.56}$$

where $2\vartheta_m$ is the interval, within the limits of which the phase is varied. Then

$$\phi_0 = \vartheta_0; \tag{1.57}$$

$$\chi^2 = \frac{1 + \mathbf{sinc}^2\vartheta_m \cdot \cos 2\phi_0}{1 - \mathbf{sinc}^2\vartheta_m \cdot \cos 2\phi_0}; \tag{1.58}$$

$$S_0^2 = 2n \cdot \frac{<s>^2}{<s^2>} \cdot \frac{\mathbf{sinc}^4 0.5\vartheta_m}{\sqrt{1 - \mathbf{sinc}^4\vartheta_m \cdot \cos 2\phi_0}}; \tag{1.59}$$

$$S_0 = n <s> \cdot \mathbf{sinc}^2 0.5\vartheta_m. \tag{1.60}$$

Dependences χ and S_0 on ϑ_m and ϕ_0 are similar to the corresponding dependences with the uniform probability distribution density of phases of elementary signals. Actually, under the condition $\vartheta_m = k\pi$, $k = 1, 2, \ldots$, the equality $\chi = 1$ is satisfied. In this case, the Rayleigh probability distribution density of the amplitude occurs if k is even and $S_0 = S_0 = 0$ or the Rice probability distribution density of the amplitude occurs if k is odd and

$$S_0 = 4n \cdot \frac{<s>}{(2k+1)^2\pi^2}. \tag{1.61}$$

Thus, formally the uniform probability distribution density of phases of the elementary signals is not a sufficient condition for the existence of the Rayleigh probability distribution density of the amplitude of the resulting signal. It is not difficult to show that for the considered case the "triangular" probability distribution density of phases within the limits of the interval $[-2k\pi, 2k\pi]$ is reduced to the uniform probability distribution density of phases within the limits of the interval $[0, 2k\pi]$ because an addition of the phase, which is multiple to 2π, is admissible. Therefore, the case considered under the condition $\vartheta_m = 2k\pi$ is reduced to the case discussed in Section 1.4.1. The Rice probability distribution density of the amplitude of the resulting vector will also occur for all values ϑ_m if the following conditions $\phi_0 = 0.25\pi$, $\chi = 1$, and $S_0 \neq 0$ are satisfied.

1.4.3 Gaussian Probability Distribution Density of Phases

Let us assume that the phases of elementary signals obey the Gaussian probability distribution density with the mean ϑ_0 and the variance σ_ϑ:

$$f_\vartheta(\vartheta) = \frac{1}{\sqrt{2\pi}\,\sigma_\vartheta} \cdot e^{-\frac{(\vartheta-\vartheta_0)^2}{2\sigma_\vartheta^2}} . \tag{1.62}$$

Then

$$\phi_0 = \vartheta_0 ; \tag{1.63}$$

$$\chi^2 = \frac{1 + e^{-2\sigma_\vartheta^2} \cdot \cos 2\phi_0}{1 - e^{-2\sigma_\vartheta^2} \cdot \cos 2\phi_0} ; \tag{1.64}$$

$$S_0^2 = 2n \cdot \frac{<s>^2}{<s^2>} \cdot \frac{e^{-\sigma_\vartheta^2}}{1 - e^{-4\sigma_\vartheta^2} \cdot \cos 2\phi_0} ; \tag{1.65}$$

$$S_0 = n <s> \cdot e^{-\sigma_\vartheta^2} . \tag{1.66}$$

Reference to Equation (1.63)–Equation (1.66) shows that if the conditions $\phi_0 = 0.25\pi$, $\chi = 1$, and $S_0 \neq 0$ are satisfied, we obtain the Rice probability distribution density of the amplitude of the resulting vector (target return signal). The Rayleigh probability distribution density of the amplitude of the resulting vector (target return signal) will not occur for all values of the parameters ϑ_0 and σ_ϑ if we do not take into consideration the case $\sigma_\vartheta \to \infty$, when the probability distribution density of phases changes to the uniform probability distribution density.

1.5 Conclusions

In this chapter, we considered briefly the probability distribution density of the amplitude and phase of the target return signal in navigational systems. We investigated the target return signal as the resulting vector of elementary signals. We defined the two-dimensional probability distribution density $f(S, \phi)$ of the amplitude and phase of the resulting vector under the assumption that the projections of the resulting vector are independent and obey the Gaussian probability distribution density. The two-dimensional probability distribution density $f(S, \phi)$ is defined by the normalized amplitude, coefficient of asymmetry, and the initial phase. Based on the two-dimensional probability distribution density of the amplitude and phase we are able to define the one-dimensional probability distribution density of the amplitude and phase of the resulting vector (target return signal).

The one-dimensional probability distribution density of the amplitude of the resulting vector was defined. Some particular cases were investigated: the Beckman, Rice, Hoyt, and Rayleigh probability distribution densities of the amplitude of the resulting vector (target return signal). The conditions under which the one-dimensional probability distribution density can tend to approach Gaussian are discussed. Parameters of the limiting Gaussian probability distribution density are defined by both the direction of the steady vector of the amplitude and the ratio between the variances of projections of the resulting vector. Under the condition $n > 2$, the variation of the initial phase within the limits of the interval $[0, 0.5\pi]$ leads to an increase in the value of the initial moments of the distribution law, and at the condition $n = 1$, to a decrease in the value of the initial moments of the probability distribution density. In some cases — for example, the Rice and Hoyt probability distribution densities of the amplitude — the initial moments can be defined by specific or elementary functions. The probability distribution density of the phase of the resulting vector is defined by the same parameters as well as the probability distribution density of the amplitude. The conditions under which the probability distribution density of the phase of the resulting vector is reduced to the uniform probability distribution density are defined.

The parameters χ, ϕ, S_0, and S_0, in their turn, are the functions of the parameters of the probability distribution densities of elementary vectors and the initial vector. Some specific probability distribution densities of the phase of elementary vectors — uniform, "triangular," and Gaussian — are investigated. In these cases, there the conditions under which the amplitude of the resulting vector obeys the Rayleigh or Rice probability distribution density were defined.

References

1. Scharf, L., *Statistical Signal Processing: Detection, Estimation, and Time Series Analysis,* Addison-Wesley, Reading, MA, 1991.
2. Poor, H., *An Introduction to Signal Detection and Estimation,* Springer-Verlag, New York, 1988.
3. Johnson, D. and Dudgeon, D., *Array Signal Processing: Concepts and Techniques,* Prentice-Hall, Englewood Cliffs, NJ, 1993.
4. Farina, A., *Antenna-Based Signal Processing Techniques for Radar Systems,* Artech House, Norwood, MA, 1992.
5. Kassam, S., *Signal Detection in Non-Gaussian Noise,* Springer-Verlag, New York, 1988.
6. Kay, S., *Fundamentals of Statistical Signal Processing: Estimation Theory,* Prentice Hall, Englewood Cliffs, NJ, 1993.
7. Skolnik, M., *Introduction to Radar Systems,* McGraw-Hill, New York, 1980.
8. Proakis, J., *Digital Communications,* 3rd ed., McGraw-Hill, New York, 1995.
9. Varshney, P., *Distributed Detection and Data Fusion,* Springer-Verlag, New York, 1997.
10. Helstrom, C., *Elements of Signal Detection and Estimation,* Prentice Hall, Englewood Cliffs, NJ, 1995.
11. Beckman, P., Statistical distribution of the amplitude and phase of a multiple scattered field, *J. Res. NSB,* Vol. 66D, No. 2, 1962, pp. 231–240.
12. Beckman, P. and Spizzichino, A., *The Scattering of Electromagnetic Waves from Rough Surfaces,* Pergamon Press, Oxford, 1963.
13. Capon, J. and Goodman, N., Probability distributions for estimators of the frequency wave-number spectrum, *Proceedings of the IEEE,* Vol. 58, 1970, pp. 1785–1786.
14. Kelly, E., Finite-sum expressions for signal detection probabilities, Technical Report 566, Lincoln Laboratory, MIT, Cambridge, MA, 1981.
15. Li, J., Liu, G., Jiang, N., and Stoica, P., Moving target feature extraction for airborne high-range resolution phased array radar, *IEEE Trans.,* Vol. SP-49, No. 2, 2001, pp. 277–289.
16. Field, E. and Lewinstein, M., Amplitude probability distribution model for VLF/ELF atmospheric noise, *IEEE Trans.,* Vol. COM-26, No. 1, 1978, pp. 83–87.
17. Wehner, D., *High Resolution Radar,* Artech House, Norwood, MA, 1987.
18. Tikhonov, V., *Statistical Radio Engineering,* Radio and Svyaz, Moscow, 1982 (in Russian).
19. Feldman, Yu, Probability distribution density of amplitude and phase of the vector, projections of which are normal and correlated with each other: a general case, *Problems in Radio Electronics,* Vol. OT, No. 1, 1964, pp. 78–98 (in Russian).
20. Silverman, B., *Density Estimation,* Chapman & Hall, London, 1996.
21. Kay, S., Nuttall, A., and Baggenstoss, P., Multidimensional probability density function approximations for detection, classification, and model order selection, *IEEE Trans.,* Vol. SP-49, No. 10, 2001, pp. 2240–2252.
22. Verdu, S., *Multiuser Detection,* Cambridge University Press, New York, 1998.
23. Mensa, D., *High Resolution Radar Cross-Section Imaging,* Artech House, Norwood, MA, 1991.

24. Porat, B., *Digital Processing of Random Signals,* 5th ed., Prentice Hall, Englewood Cliffs, NJ, 1994.
25. Schwartz, M., Benett, W., and Stein, S., *Communication Systems and Techniques,* IEEE Press, New York, 1996.
26. Feldman, Yu and Klimenkova, T., Probability distribution density of amplitude of stochastic vector, the projections of which are Gaussian: a general case, *Problems in Radio Engineering,* Vol. OT, No. 7, 1964, pp. 3–21 (in Russian).
27. Nobel, A., Morvai, G., and Kulkarni, S., Density estimation from an individual numerical sequence, *IEEE Trans.,* Vol. IT-44, No. 2, 1998, pp. 537–541.
28. Ward, K., Baker, F., and Watts, S., Maritime surveillance radar — part 1: radar scattering from the ocean surface, *Radar Signal Process, IEE Proceedings F,* Vol. 137, No. 2, 1990, pp. 51–62.
29. Davies, R., Brennan, L., and Reed, I., Angle estimation with adaptive array in external noise fields, *IEEE Trans.,* Vol. AES-12, No. 3, 1976, pp. 179–186.
30. Pozdnyak, S. and Melititzky, V., *Introduction to Statistical Theory of Polarization of Radio Waves,* Sov. Radio, Moscow, 1975 (in Russian).
31. Conte, E., Bisceglie, M., Galdi, C., and Ricci, G., A procedure for measuring the coherence length of sea texture, *IEEE Trans.,* Vol. IM-46, No. 8, 1997, pp. 836–841.

2

Correlation Function of Target Return Signal Fluctuations

2.1 Target Return Signal Fluctuations

2.1.1 Physical Sources of Fluctuations

The target return signal, being a stochastic process, is subject to random variations in parameters, which are called fluctuations.[1–3] As the target return signal is the sum of a large number of elementary signals (see Figure 2.1), sources of fluctuations can be considered as variations in the amplitude, phase, or frequency of elementary signals that lead to corresponding variations in these parameters in the resulting target return signal. For example, scatterers can move and rotate under the stimulus of the wind. Radial components of motion can give rise to phase changes in elementary signals. Tangential components of motion can give rise to amplitude changes in elementary signals if these changes are comparable with the width Δ_a of the radar antenna directional diagram. Rotational components, if scatterers do not have spherical symmetry, can give rise to both amplitude and phase changes in elementary signals.[4–6]

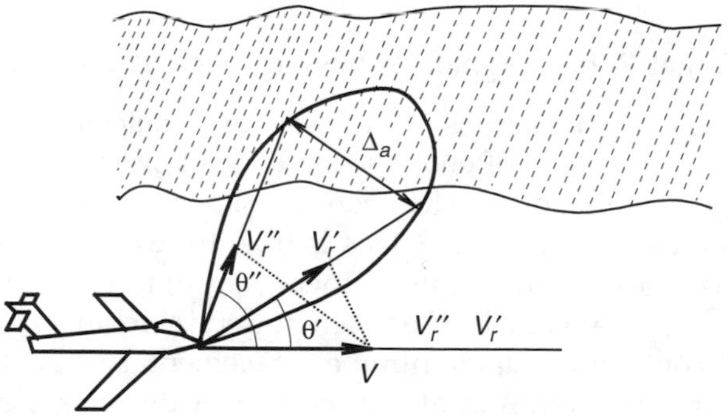

FIGURE 2.1
Doppler spectrum formation.

In the case of radar moving relative to the two-dimensional (surface) or three-dimensional (space) target too, phase changes in elementary signals occur. Let us suppose that the two-dimensional (surface) or three-dimensional (space) target has large angle dimensions, then various scatterers are observed at different angles within the directional diagram with respect to the direction of moving radar (see Figure 2.1). So, phase changes in elementary signals are the sources of fluctuations in the target return signal. Instead, we can also say that a moving radar can give rise to the Doppler shift in the frequency of elementary signals, because relative radial velocities of various scatterers differ within the searching area, due to differences in tracking angles (Figure 2.1).[1] The received target return signal is not a simple signal and contains an entire frequency spectrum corresponding to the energy spectrum of radial components of various elementary scatterer velocities. Beats between various frequencies of the energy spectrum manifest themselves as target return signal fluctuations, which are called *Doppler beats;* the phenomenon is called the *secondary Doppler effect*.[7]

If the transmitter, receiver, or detector in navigational systems and scatterers are stationary, then fluctuations can arise due to radar antenna scanning or rotation of the radar antenna polarization plane, because both these give rise to amplitude changes in elementary signals (scanning and polarization fluctuations).[8] Fluctuations of the received target return signal can be due to the nonstationary state of the searching frequency. Variations in frequency give rise to phase changes in elementary signals. Therefore, these variations are different for various scatterers and depend on the radar range. Unequal phase changes in elementary signals can give rise to fluctuations of the target return signal parameters (for example, frequency fluctuations). Peculiarities of the interaction between frequency fluctuations and target return signal Doppler fluctuations are discussed in more detail in References 9 to 11.

2.1.2 The Target Return Signal: A Poisson Stochastic Process

Target return signal fluctuations at the receiver or detector input in navigational systems can be very often considered as a Poisson stochastic process caused by superposition of nonstochastic (in shape) elementary signals arising at random instants of time.[1,12] This is true, for example, when the two-dimensional (surface) or three-dimensional (space) target is scanned by the pulsed searching radar signal. After each pulsed searching signal, the target return signal, containing a large number of elementary signals reflected by individual scatterers, comes in at the receiver or detector input in navigational systems. Thus, elementary signals are high-frequency pulses with the same shape and duration as the pulsed searching radar signal.

Incoming pulses possess stochastic amplitudes and, what is more important, the receiver or detector input in navigational systems arrives at random times that depend on the position of scatterers in space or on the surface.

Superposition of these deterministic (in shape) pulsed signals generates a Poisson stochastic process, which represents the target return signal fluctuations in the radar range. Thus, the response of the two-dimensional (surface) or three-dimensional (space) target to the pulsed searching radar signal is the interval of the stochastic process arising in the propagation of the target return signal.[13,14]

If the radar and scatterers are mutually stationary and the parameters of the radar equipment are stable, the target return signal caused by each pulsed searching signal is an exact copy of the previous target return signal, and the stochastic process becomes periodic (see the solid line in Figure 2.2a). In the case when the radar moves or radar antenna scans, the rigorous periodicity of the stochastic process is broken and the target return signal is shifted from period to period. The interperiod or slow fluctuations appear in contrast to the intraperiod or rapid fluctuations in the radar range (see the dotted line in Figure 2.2a).[15,16]

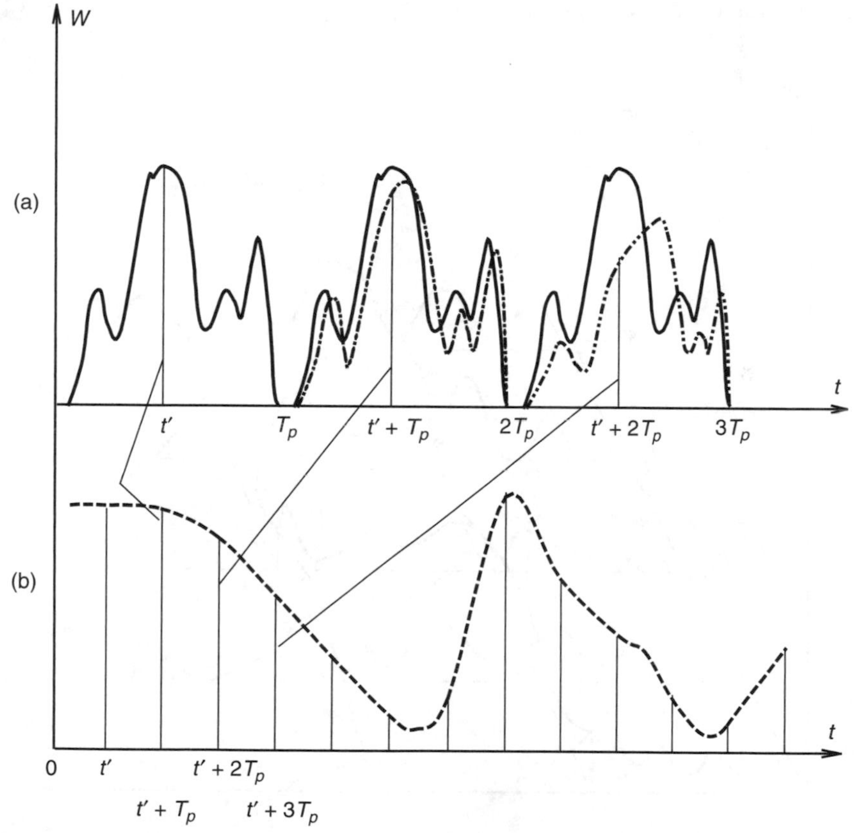

FIGURE 2.2
(a) The intraperiod fluctuations; (b) the slow fluctuations.

Let us consider the instantaneous values of the target return signal at neighboring periods at the instants of time t, which are fixed with respect to the origin of the period. Then, the instantaneous values of the target return signal will have the shape of slowly fluctuating pulsed signals (see Figure 2.2b), the envelope of which is the stochastic process caused by the changing state of the system *radar–scatterer*. The totality of these instants of time

$$\{t', t' + T_p, \ldots, t' + nT_p\} \equiv \mathbf{t'} \tag{2.1}$$

is called the *time section* according to Zukovsky et al.[17] General interperiod and intraperiod fluctuations are shown in Figure 2.3. The interperiod fluctuations can be caused by other sources in addition to the moving radar and antenna scanning, for example, by the nonstationary state of frequency of the signal transmitter (or signal generator) in navigational systems, rotation of the radar antenna polarization plane, displacements of scatterers under the stimulus of the wind, and so on.

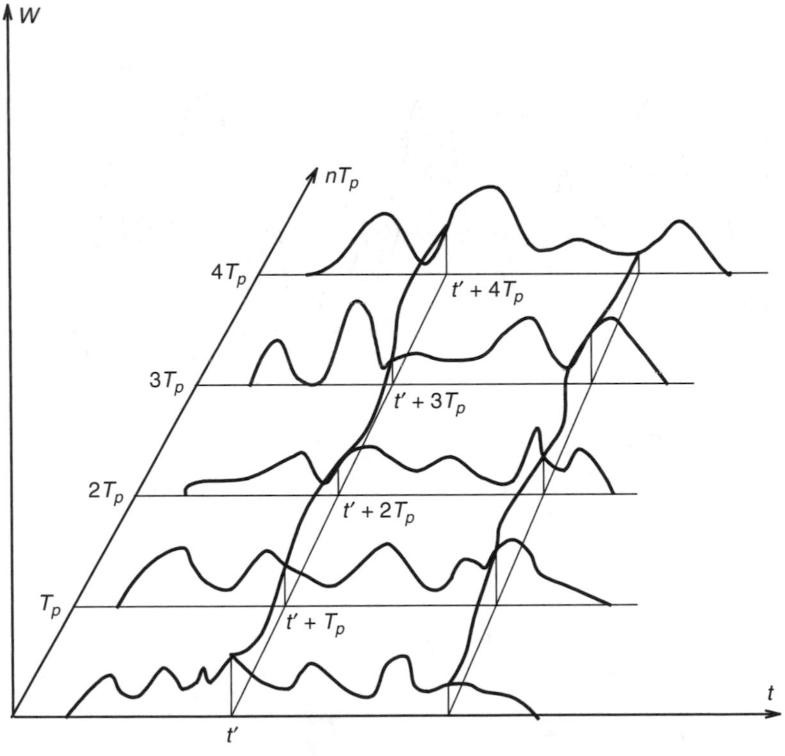

FIGURE 2.3
The intraperiod and interperiod fluctuations.

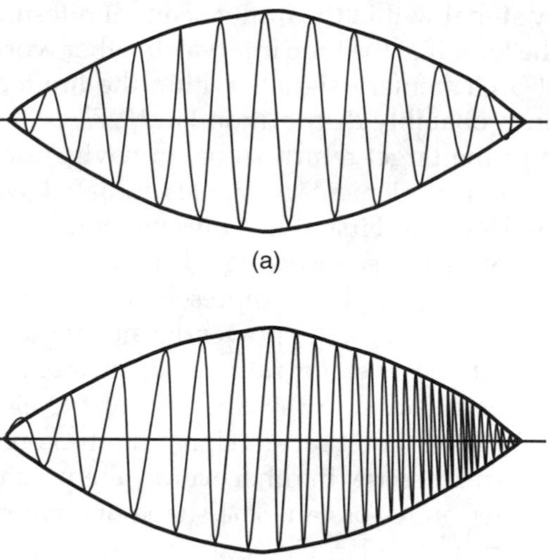

FIGURE 2.4
Elementary signals: (a) amplitude modulation; (b) amplitude–frequency modulation.

Target return signal fluctuations caused by the moving radar and antenna scanning with the searching simple signal can be called *slow fluctuations*. Unlike interperiod fluctuations, slow fluctuations also generate a Poisson stochastic process. The sources of fluctuations are the elementary signals reflected from elementary scatterers, which are modulated by the radar antenna directional diagram. Under scanning, this modulation is a pure amplitude modulation (see Figure 2.4a), whereas with moving radar, this modulation is an amplitude–frequency modulation (see Figure 2.4b). Therefore, frequency changes are caused by the Doppler shift, which is proportional to the radial component of scatterer relative velocity and varies during radar motion due to changes in the scanning angle. The target return signal is the sum of a large number of equivalent elementary signals that are nonstochastic (in shape) but arise at random instants of time.[18-20]

The Poisson stochastic process as a function of time can be determined by the sum of the deterministic functions $w(t, t_i) = s_i w(t - t_i)$ with random parameters s_i and t_i:

$$W(t) = \sum_{i=-\infty}^{\infty} w(t, t_i) = \sum_{i=-\infty}^{\infty} s_i w(t - t_i), \tag{2.2}$$

where s_i is the amplitude of the i-th elementary signal; t_i is the instant of time when the i-th elementary signal has arisen; t is the observed instant of time. The parameters s_i and t_i are statistically independent random variables. Let us assume that the time instants of individual elementary signal appearances are independent events and that the probability of appearance of a single elementary signal within the limits of an infinitesimal time interval is proportional to the length of the time interval. In other words, the probability of appearance of n elementary signals within the limited time interval Δt obeys the Poisson probability distribution law.[12,21,22]

Representation of the target return signal in navigational systems, in the form of superposition of elementary signals initiated with irregularity, is appropriate for two reasons. First, this representation allows us to define the statistics of the incoming target return signal at the receiver or detector input in navigational systems. Second, this representation also allows us to define the correlation function and spectral power density of the input target return signal fluctuations. If the average number of elementary signals arising within the time interval equal to the duration of the elementary signal is sufficiently high, then the target return signal can be considered as a Gaussian process, in a narrow sense. In other words, the probability distribution densities of any order are Gaussian. The shape of elementary signals does not play any role; in particular, elementary signals can be considered as pulses with a high radio-frequency carrier. This condition implies that a sufficiently large number of elementary targets must be within the scanning area, and the region filled by scatterers must be larger than the scanning area, so with scanning area displacement, some scatterers enter this area and others leave it, forming in this way a process of superposition of elementary signals.[23–25]

It is significant that in this case, displacements of individual scatterers are not assumed to be independent. Only the time instants of initiation of elementary signals can be considered to be independent. Displacements of scatterers can be correlated or may even be hardly dependent on each other, for example, in the case of moving radar or antenna scanning. Although the target return signal is considered Gaussian in a narrow sense, complete information about it is held in the correlation function or in the corresponding power spectral density of the target return signal fluctuations. The first part of this book is devoted to the investigation of the correlation function and power spectral density of target return signal fluctuations at the receiver or detector input in navigational systems.

From Equation (2.2), we are able to define the main features of the Poisson stochastic process characterizing the target return signal. The main relationships are given here without proofs and will be used at a later time. If elementary signals are written in the complex-valued form, the correlation function of target return signal fluctuations can be determined as follows:[12,21]

$$R(\tau) = n_1 \overline{s^2} \int_{-\infty}^{\infty} w(t) w^*(t+\tau)\, dt = R(0) \cdot R(\tau) \tag{2.3}$$

where n_1 is the number of elementary signals per time; * denotes a complex conjugate value;

$$p = R(0) = n_1 \overline{s^2} \int_{-\infty}^{\infty} |w(t)|^2 dt \tag{2.4}$$

is the power of the stochastic process;

$$R(\tau) = \frac{\int_{-\infty}^{\infty} w(t) w^*(t+\tau) dt}{\int_{-\infty}^{\infty} |w(t)|^2 dt} \tag{2.5}$$

is the normalized correlation function.

Using the Fourier transform in Equation (2.3) and the convolution theorem, we can define the power spectral density:[23,26]

$$S(\omega) = \frac{\frac{p}{\pi} \left| \int_{-\infty}^{\infty} w(t) \cdot e^{-j\omega t} dt \right|^2}{\int_{-\infty}^{\infty} |w(t)|^2 dt}. \tag{2.6}$$

The power spectral density obtained in Equation (2.6) coincides in shape with any individual elementary signal, clearly because all elementary signals are the same and differ only in amplitude and the time instant of initiation. Furthermore, their power spectral densities are identical and only the coefficient of proportionality and phase factor differ. The resulting power spectral density, equal to the average sum of a large number of identical elementary signals, coincides with the power spectral density of an individual elementary signal.[27]

From the previous considerations, it follows that with the target return signal represented as a Poisson stochastic process, the high-frequency pulsed signal (radio pulse) and the track of the directional diagram moving along the two-dimensional (surface) or three-dimensional (space) target with the velocity of radar antenna scanning or with the velocity of moving radar can play the role of an elementary signal. In the first case, if time is an independent variable of the stochastic process (time fluctuations), then in the second case, the angle and linear coordinates characterizing the position of the scanning area in the scattering environment are also independent variables of the stochastic process (space fluctuations). As all these cases can take place simultaneously, a multidimensional definition of elementary signals is necessary (instead of the one-dimensional definition) to generalize the concept of the pulsed stochastic process for the multidimensional space.[1,27]

2.2 The Correlation Function and Power Spectral Density of the Target Return Signal

2.2.1 Space–Time Correlation Function

Let an elementary signal reflected by an individual scatterer at the fixed instant of time t be in the following form:

$$w(\mathbf{x}, t) = w(x_1, x_2, \ldots, x_n, t), \tag{2.7}$$

where $\mathbf{x} \equiv \{x_1, \ldots, x_n\}$ is the set of variables describing the mutual position of the transmitter, receiver, or detector in navigational systems, and the scatterer. The function given by Equation (2.7) takes into consideration all essential factors: the shape and orientation of the transmitted and received radar antenna directional diagram, the position of the scatterer, the shape of the searching signal, and so on. The amplitude and phase of an elementary signal depend on these factors.[28,29]

Let us consider the n-dimensional space, where the variables $\mathbf{x}$ form any coordinate system. This space is then divided by coordinate surfaces into a large number of n-dimensional elementary volumes, the dimensions of which are infinitesimal, so that we can neglect the variation of the function $w(\mathbf{x}, t)$ within each volume; but there are a great number of scatterers in each volume. The target return signal of the i-th elementary volume can be written in the form: $w_i = m_i w(\mathbf{x}_i, t)$, where m_i is the number of scatterers inside the elementary volume and $\mathbf{x}_i = \{x_{1i}, x_{2i}, \ldots, x_{ni}\}$ are the relative coordinates of the elementary volume. Let us assume that the resulting target return signal determined by

$$W(\mathbf{x},t) = \sum_{i=0}^{\infty} m_i w(\mathbf{x}_i, t) \tag{2.8}$$

is a uniform field, which can be nonstationary.[30]

At the instant of time $t = t_1$, the target return signal can be written in the following form:

$$W_1(\mathbf{x},t) = \sum_{i=0}^{\infty} m_i w(\mathbf{x}_i, t_1)\,. \tag{2.9}$$

If the variables $\mathbf{x}_i$ change in value by some differential $\Delta\mathbf{x}$ as a consequence of the moving directional diagram, then the target return signal can be written in the following form:

$$W_2(\mathbf{x},t)=\sum_{i=0}^{\infty} m_i w(\mathbf{x}_i+\Delta\mathbf{x},t_2). \tag{2.10}$$

In the general case, the differentials $\Delta\mathbf{x}$ differ for various scatterers and are functions of the coordinate $\mathbf{x}_i$:

$$\Delta\mathbf{x} \equiv \{\Delta x_1(\mathbf{x}_i), \Delta x_2(\mathbf{x}_i), \ldots, \Delta x_n(\mathbf{x}_i)\}. \tag{2.11}$$

The correlation function of the space and time fluctuations has the following form:

$$R_{\Delta\mathbf{x}}(t_1,t_2)=\overline{\left[W_1(\mathbf{x},t)-\overline{W_1}(\mathbf{x},t)\right]\left[W_2^*(\mathbf{x},t)-\overline{W_2^*}(\mathbf{x},t)\right]}, \tag{2.12}$$

where

$$\overline{W_1}(\mathbf{x},t)=\sum_{i=0}^{\infty} \overline{m}_i w(\mathbf{x}_i,t_1); \tag{2.13}$$

$$\overline{W_2}(\mathbf{x},t)=\sum_{i=0}^{\infty} \overline{m}_i w(\mathbf{x}_i+\Delta\mathbf{x},t_2); \tag{2.14}$$

and the index $\Delta\mathbf{x}$ indicates that the correlation function given by Equation (2.12) is determined for the fluctuations that are functionally related to changes in the variable $\mathbf{x}$. Substituting Equation (2.13) and Equation (2.14) in Equation (2.12), we can write

$$\begin{aligned} R_{\Delta\mathbf{x}}(t_1,t_2)&=\overline{\left[\sum_{i=0}^{\infty}(m_i-\overline{m}_i)w(\mathbf{x}_i,t_1)\right]\left[\sum_{j=0}^{\infty}(m_j-\overline{m}_j)w^*(\mathbf{x}_j+\Delta\mathbf{x},t_2)\right]},\\ &=\sum_{i=0}^{\infty}\overline{(m_i-\overline{m}_i)^2}\,w(\mathbf{x}_i,t_1)w^*(\mathbf{x}_i+\Delta\mathbf{x},t_2) \end{aligned} \tag{2.15}$$

because $(m_i-\overline{m}_i)\cdot w(\mathbf{x}_i,t)$ and $(m_j-\overline{m}_j)\cdot w(\mathbf{x}_j,t)$ are independent random variables under the condition $i \neq j$. The total sum is equal to zero under the condition $i \neq j$. Here the value $\overline{(m_i-\overline{m}_i)^2}$ is the variance of the number of scatterers at the i-th elementary volume.

Let us assume that the number of scatterers in any fixed elementary volume obeys the Poisson distribution law, i.e., the probability that any scatterer can appear in a given elementary volume depends on the dimensions of the volume and is independent of the position of the elementary volume in space. Then we can write $\overline{(m_i - \overline{m}_i)^2} = \overline{m}_i = m_0 \times \delta\mathbf{x}$, where $m_0 = const$– is the mean (with respect to an ensemble) of the number of scatterers per unit volume; and $\delta\mathbf{x}$ is the elementary volume.[31] Going to the limit as $\delta\mathbf{x} \to 0$ and replacing the sum with the integral in Equation (2.15), we can write

$$R_{\Delta\mathbf{x}}(t_1, t_2) = m_0 \int_{-\infty}^{\infty} w(\mathbf{x}, t_1) \cdot w^*(\mathbf{x} + \Delta\mathbf{x}, t_2)\, d\mathbf{X}\,, \tag{2.16}$$

where $d\mathbf{X} = |J|\, dx_1 \ldots dx_n$; J is the Jacobian corresponding to the chosen coordinate system, and integration is carried out over the whole n-dimensional domain, where the integrand differs from zero. In the general case, if the stochastic process is nonstationary, the result of the integration depends on the variables t_1 and t_2. If the stochastic process is stationary, the result of the integration is defined by the difference $\tau = t_2 - t_1$.

In the general case, the differentials $\Delta\mathbf{x}$, as was noted, are not the same for all scatterers. The differentials $\Delta\mathbf{x}$ can be fixed-point or stochastic functions of multidimensional space coordinates. When the differentials $\Delta\mathbf{x}$ are random, as in fluctuations caused by the stimulus of the wind, Equation (2.16) must be additionally averaged in accordance with the multidimensional probability distribution density of the differentials $\Delta\mathbf{x}$:

$$\overline{R}_{\Delta\mathbf{x}}(t_1, t_2) = \int R_{\Delta\mathbf{x}}(t_1, t_2) f(\Delta\mathbf{x})\, d(\Delta\mathbf{x})\,, \tag{2.17}$$

where $f(\Delta\mathbf{x}) = f(\Delta x_1, \ldots, \Delta x_n)$ is the multidimensional probability distribution density of the differentials $\Delta\mathbf{x}$ and $d(\Delta\mathbf{x}) = d(\Delta x_1) \ldots d(\Delta x_n)$ Equation (2.16) and Equation (2.17) are space–time correlation functions, because they represent correlation characteristics of the received target return signals with respect to both the time and space coordinates defining the geometry of the radar–scatterer system.

The correlation function given by Equation (2.16) is functionally related with the space–time (frequency) power spectral density of the field using the multidimensional Fourier transform with respect to the coordinates $\mathbf{x}$, t_1, and t_2.[30] Using space coordinates, this Fourier transform gives us the n-dimensional space power spectral density. Using time coordinates, we obtain the generalized two-dimensional power spectral density that takes into consideration the correlation relationships between the spectral power densities at frequencies ω_1 and ω_2. Henceforth, the space variables $\Delta\mathbf{x}$ defining the dynamics of the radar–scatterer system will be represented by functions of time. Because of this, the space–time fluctuations can be represented by time fluctuations only, to define which we use the time correlation function and the frequency–time power spectral density.

The nonstationary correlation function given by Equation (2.16) can be written in the following form:

$$R_{\Delta x}(t_1, t_2) = \sigma(t_1) \cdot \sigma(t_2) \cdot R_{\Delta x}(t_1, t_2), \tag{2.18}$$

where

$$\sigma^2(t_i) = m_0 \int \left|w(\mathbf{x}, t_i)\right|^2 d\mathbf{X}, \qquad i = 1, 2 \tag{2.19}$$

is the variance (power) of the stochastic process at the instants of time t_1 and t_2;

$$R_{\Delta x}(t_1, t_2) = \frac{R_{\Delta x}(t_1, t_2)}{\sigma(t_1) \cdot \sigma(t_2)} = \frac{\int w(\mathbf{x}, t_1) \cdot w^*(\mathbf{x} + \Delta\mathbf{x}, t_2)\, d\mathbf{X}}{\sqrt{\int \left|w(\mathbf{x}, t_1)\right|^2 d\mathbf{X} \cdot \int \left|w(\mathbf{x}, t_2)\right|^2 d\mathbf{X}}} \tag{2.20}$$

is the nonstationary normalized correlation function. Equation (2.18)–Equation (2.20) are extensions of Equation (2.3)–Equation (2.5).

In the analysis of nonstationary stochastic processes, we use the instantaneous correlation function in parallel with the correlation function given by Equation (2.16):

$$R_{\Delta x}(t, \tau) = m_0 \int w(\mathbf{x}, t - 0.5\tau) \cdot w^*(\mathbf{x} + \Delta\mathbf{x}, t + 0.5\tau) d\mathbf{X} \tag{2.21}$$

results from Equation (2.16) by using the following transformations:

$$t_1 = t - 0.5\tau; \tag{2.22}$$

$$t_2 = t + 0.5\tau; \tag{2.23}$$

$$\tau = t_2 - t_1; \tag{2.24}$$

$$t = 0.5(t_1 + t_2). \tag{2.25}$$

Equation (2.16) can be written in the more symmetric form:

$$R_{\Delta x}(t, \tau) = m_0 \int w(\mathbf{x}' - 0.5\Delta\mathbf{x},\ t - 0.5\tau) \cdot w^*(\mathbf{x}' + 0.5\Delta\mathbf{x}, t + 0.5\tau) d\mathbf{X}, \tag{2.26}$$

where $\mathbf{x}' = \mathbf{x} + 0.5\Delta\mathbf{x}$. Due to uniformity of the considered stochastic field, Equation (2.21) and Equation (2.26) are equivalent; so, the symbol "′" of the variable $\mathbf{x}$ in Equation (2.26) can be omitted. The functions given by Equation (2.16), Equation (2.21), and Equation (2.26) are equivalent too, to fit the stationary stochastic process.

The transformation of Equation (2.16) in Equation (2.21) or Equation (2.26), i.e., the transformation of coordinates determined by Equation (2.22)–Equation (2.25) from the plane (t_1, t_2) into the plane (t, τ) rotated at 45°, allows us to separate, wherever possible, the stationary and nonstationary components of the stochastic process. When the target return signal is stationary, the correlation function $R(t, \tau)$ is independent of time t and becomes the ordinary stationary correlation function $R(\tau)$.

In spite of the fact that the correlation function determined by Equation (2.21) contains complete information regarding both the power (or variance) of the target return signal and characteristics of the power spectral density as a function of time, it is convenient to separate the correlation function into components having the following form:

$$R_{\Delta\mathbf{x}}(t,\tau) = R_{\Delta\mathbf{x}}(t,0)\big|_{\Delta\mathbf{x}=0} \cdot R_{\Delta\mathbf{x}}(t,\tau), \tag{2.27}$$

where

$$R_{\Delta\mathbf{x}}(t,0)\big|_{\Delta\mathbf{x}=0} = p(t) = \sigma^2(t) = m_0 \int |w(\mathbf{x},t)|^2 d\mathbf{X} \tag{2.28}$$

and

$$R_{\Delta\mathbf{x}}(t,\tau) = \frac{R_{\Delta\mathbf{x}}(t,\tau)}{\sigma^2(t)} = \frac{\int w(\mathbf{x},t-0.5\tau)\cdot w^*(\mathbf{x}+\Delta\mathbf{x},t+0.5\tau)\,d\mathbf{X}}{\int |w(\mathbf{x},t)|^2 d\mathbf{X}}. \tag{2.29}$$

The normalized correlation function given by Equation (2.29), as well as the correlation function in Equation (2.26) can be written in a more symmetric form.

The peculiarity of the nonstationary target return signal when the variables t and τ of the correlation function are separated, i.e., when the correlation function is separated according to[12]

$$R(t, \tau) = R_1(\tau) \cdot R_2(t), \tag{2.30}$$

is of prime interest to us. The power (or variance) of the target return signal depends on time: $R(t, 0) = R_1(0) \times R_2(t)$. The normalized correlation function and characteristics of the power spectral density are independent of time:

$$R(t,\tau)=\frac{R_1(\tau)}{R_1(0)}. \tag{2.31}$$

In cases where there is no need to get information regarding the power (or variance) of the target return signal, we can limit the study only to the normalized correlation function.[32,33]

Henceforth, in the majority of cases, the normalized correlation function will be written in a simpler form. For example, instead of Equation (2.29) we can write

$$R_{\Delta x}(t,\tau)=N\int w(\mathbf{x},t-0.5\tau)\cdot w^*(\mathbf{x}+\Delta\mathbf{x},t+0.5\tau)\,d\mathbf{X}\,, \tag{2.32}$$

where

$$N=\frac{1}{\int\left|w(\mathbf{x},t)\right|^2\,d\mathbf{X}} \tag{2.33}$$

is unity divided by the normalized correlation function of the target return signal fluctuations under the condition $\Delta\mathbf{x} = \tau = 0$. Specific formulae for N are different and depend on the form of the normalized correlation function.

2.2.2 The Power Spectral Density of Nonstationary Target Return Signal Fluctuations

Because in the general case, the nonstationary correlation function of target return signal fluctuations is a function of two variables t_1 and t_2 or t and τ, the power spectral density of target return signal fluctuations, defined by the Fourier transform with respect to the variable τ, is a function depending both on frequency and on time. This statement does not agree with the usual concept of the power spectral density as a sum of harmonic elementary signals independent of time. Because of this, the more general definition of the power spectral density, which clearly allows us to determine it at the output of a linear system, uses the following form

$$S^{out}(\omega) = |Q(j\omega)|^2\cdot S^{in}(\omega) \tag{2.34}$$

or an analogous form, where $Q(j\omega)$ is the frequency response of the linear system, and is introduced to fit nonstationary stochastic processes.[34-36]

The concept of the two-dimensional (generalized) power spectral density $S(\omega_1, \omega_2)$,[6,8] which is functionally related to the nonstationary correlation function $R(t_1, t_2)$ by the following Fourier transforms

$$S(\omega_1,\omega_2)=\int_{-\infty}^{\infty}\int_{-\infty}^{\infty} R(t_1,t_2)\cdot e^{j(\omega_1 t_1-\omega_2 t_2)}dt_1\,dt_2\,, \tag{2.35}$$

where

$$R(t_1,t_2)=\frac{1}{4\pi^2}\int_{-\infty}^{\infty}\int_{-\infty}^{\infty} S(\omega_1,\omega_2)\cdot e^{-j(\omega_1 t_1-\omega_2 t_2)}d\omega_1\,d\omega_2\,, \tag{2.36}$$

is universally adopted.

We can show that the generalized power spectral density is the mean of the product of the power spectral densities that are shifted in frequency, $S(\omega_1)$ and $S(\omega_2)$:

$$S(\omega_1,\omega_2)=\overline{S(\omega_1)\cdot S(\omega_2)}, \tag{2.37}$$

where $S(\omega)$ is the Fourier transform of a sample of the stochastic process $x(t)$, the nonstationary correlation function of which has the following form:

$$R(t_1,t_2)=\overline{x(t_1)\cdot x(t_2)}. \tag{2.38}$$

Comparing Equation (2.37) and Equation (2.38), one can see that the nonstationary power spectral density of the stochastic process $x(t)$ is the nonstationary correlation function of the stochastic process $S(\omega)$ in frequency space. In the general case, this correlation function differs from zero. This circumstance indicates that there is a correlation function between the power spectral densities at the frequencies ω_1 and ω_2. As is well known, any frequencies satisfying the condition $\omega_1 \neq \omega_2$ are not correlated to the power spectral density of the stationary stochastic process.

The statement that a stationary stochastic process becomes nonstationary when there is an interspectral correlation relationship, is clear on the basis of physical reasoning. For example, if the stationary stochastic process is modulated in amplitude by the frequency Ω, then this process becomes nonstationary and all pairs of the spectral components not tuned on the frequency Ω exchange all their side-lobe components containing information about the amplitude and phase. Due to this fact, there is a correlation between these side-lobe spectral components. This correlation is strong, and the coefficient of modulation is high. Actually, the analogous effect can occur under frequency and phase modulation. If the modulation law is more complex, in particular, nonperiodic, then there is an exchange of spectral side-lobe components between all components of the initial power spectral density and between all correlated frequencies that the power spectral density can give rise to. So, the power spectral density will be fuzzy on the plane

(ω_1, ω_2). In the case of the stationary stochastic process, we can write $S(\omega_1, \omega_2) = S(\omega_1) \times \delta(\omega_1 - \omega_2)$. In this case, the domain of the power spectral density definition contracts to the line $\omega_1 = \omega_2$. Furthermore, it is obvious that the power spectral densities are not correlated under the condition $\omega_1 \neq \omega_2$.

The concept of the instantaneous (time-frequency) power spectral density of the nonstationary stochastic process,[12] which is the Fourier transform of the instantaneous correlation function $R(t, \tau)$ with respect to the variable τ

$$S(t,\omega) = \int_{-\infty}^{\infty} R(t,\tau) \cdot e^{-j\omega\tau} d\tau \,, \tag{2.39}$$

is the second universally accepted concept, which is very convenient for theoretical investigations. The inverse Fourier transform in the form

$$R(t,\tau) = \frac{1}{2\pi} \int_{-\infty}^{\infty} S(t,\omega) \cdot e^{j\omega\tau} d\omega \tag{2.40}$$

with the condition $\tau = 0$ allows us to write

$$R(t,0) = \overline{x^2(t)} = \frac{1}{2\pi} \int_{-\infty}^{\infty} S(t,\omega) d\omega \,. \tag{2.41}$$

In other words, the instantaneous power spectral density defines the probability distribution law of power (or variance) of the stochastic process $x(t)$ in the coordinate system (t, ω). The integral over all frequencies gives the mean $\overline{x^2(t)}$. With some values of ω and t, the power spectral density $S(t, \omega)$ can be negative.[12] The instantaneous power spectral density $S(t, \omega)$, depending on the parameter t, yields very important information about the character of the stochastic process as a function of time. It is precisely this power spectral density, as a rule, that will subsequently be determined in the study of nonstationary stochastic processes.

The deterministic dependence of the instantaneous power spectral density as a function of time can be shown in the frequency region if we are able to define the Fourier transform for the power spectral density $S(t, \omega)$ with respect to the parameter t:

$$S(\omega,\Omega) = \int_{-\infty}^{\infty} S(t,\omega) \cdot e^{-j\Omega t} dt \,. \tag{2.42}$$

One can see that the formula in Equation (2.42) is the double Fourier transform of the correlation function $R(t, \tau)$ of fluctuations of the target return signal

$$S(\omega,\Omega)=\int_{-\infty}^{\infty}\int_{-\infty}^{\infty}R(t,\tau)\cdot e^{-j(\omega\tau+\Omega t)}d\tau\, dt\,. \tag{2.43}$$

From previous considerations, the two-dimensional power spectral density determined by Equation (2.43) is equivalent to the power spectral density given by Equation (2.35) with coordinates ω and Ω related functionally to the coordinates ω_1 and ω_2 by relationships that are analogous to the ones given in Equation (2.22)–Equation (2.25):

$$\Omega = \omega_1 - \omega_2; \tag{2.44}$$

$$\omega = 0.5(\omega_1 + \omega_2); \tag{2.45}$$

$$\omega_1 = \omega - 0.5\Omega; \tag{2.46}$$

$$\omega_2 = \omega + 0.5\Omega. \tag{2.47}$$

The relationships determined by Equation (2.44)–Equation (2.47) are a rotation of the coordinate axes at an angle equal to 45°. Substituting Equation (2.44)–Equation (2.47) in Equation (2.43), we can obtain Equation (2.35). In an analogous way [see Equation (2.22)–Equation (2.25)], the correlation functions $R(t, \tau)$ and $R(t_1, t_2)$ [see Equation (2.36)] are functionally related.

In the case of the stationary target return signal, the two-dimensional power spectral density $S(\omega, \Omega)$ is equal to zero within the plane (ω, Ω), except for the line $\Omega = 0$. This is the one-dimensional power spectral density. The appearance of a nonstationary state leads to the spreading of the power spectral density with respect to the coordinate Ω, giving rise to the two-dimensional power spectral density. If the correlation function is separable [see Equation (2.30)], then we can write

$$S(\omega,\Omega)=\int_{-\infty}^{\infty}\int_{-\infty}^{\infty}R_1(\tau)R_2(t)\cdot e^{-j(\omega\tau+\Omega t)}d\tau\, dt=S_1(\omega)\cdot S_2(\Omega)\,, \tag{2.48}$$

where

$$S_1(\omega)=\int_{-\infty}^{\infty}R_1(\tau)\cdot e^{-j\omega\tau}dt \tag{2.49}$$

and

$$S_2(\Omega) = \int_{-\infty}^{\infty} R_2(t) \cdot e^{-j\Omega\tau} dt \,. \tag{2.50}$$

Obviously, the instantaneous power spectral density $S(\omega, t)$ given by Equation (2.39) can be written in the form: $S(t, \omega) = S_1(\omega) \cdot R_2(t)$. In the case of the stationary target return signal under the condition $R_2(t) = const$, we can write $S_2(\Omega) = \delta(\Omega)$, and if $R(t, \tau) = const$, we obtain $S(\omega, \Omega) = S_1(\omega) \cdot \delta(\Omega)$. This fact proves that in the case of the stationary target return signal, the power spectral density $S(\omega, \Omega)$ is defined only along the axis $\Omega = 0$ and has the shape of the power spectral density $S_1(\omega)$. Peculiarities of the correlation functions and power spectral densities of nonstationary target return signal fluctuations are discussed in more detail in References 2, 21, and 37.

2.3 The Correlation Function with the Searching Signal of Arbitrary Shape

2.3.1 General Statements

Let us consider for simplicity that a single-position aircraft radar navigational system is generating the searching signal $U(t, \omega)$.[38,39] Then, the target return signal from an individual scatterer takes the following form:

$$w(\mathbf{x}, t) = S(\rho, x_2, \ldots, x_n) \cdot U\left(t - \tfrac{2\rho}{c}, \omega\right) = S(\rho, \mathbf{x}) \cdot U\left(t - \tfrac{2\rho}{c}, \omega\right), \tag{2.51}$$

where

$$\mathbf{x} \equiv \{x_2, \ldots, x_n\}; \tag{2.52}$$

$S(\rho, \mathbf{x})$ is the amplitude of the received target return signal; ρ is the radar range (in the case of navigational systems, ρ is the distance between the radar antenna and scatterer); and c is the velocity of light. The amplitude of the received target return signal depends on the distance ρ, mutual positions of the radar, scatterer, shape and orientation of the radar antenna directional diagram, position of the radar antenna polarization plane, effective scattering area of the scatterer, position of the scatterer in space, etc. This dependence is represented a function of the arguments ρ and $\mathbf{x}$. Here and subsequently, the coordinate of the distance $x_1 \equiv \rho$ is extracted from the general totality, because the coordinate x_1 has the special property of being a component of two factors simultaneously [see Equation (2.52)].

With the fixed parameter t, the target return signal at the radar receiver or detector input in navigational systems is the sum of elementary signals received from all scatterers at random points of the n-dimensional space within the scanned area. This target return signal depends on the time parameter t, because the function $U\left(t - \frac{2\rho}{c}, \omega\right)$ [see Equation (2.51)] slides along the distance axis $x_1 \equiv \rho$ with a velocity equal to half the velocity of light as a result of changing the parameter t. Because the amplitude $S(\rho, x_2, \ldots, x_n)$ of the target return signal depends also on the radar range ρ (or the distance between the radar antenna and scatterer), the functional dependence on the time parameter t leads to the nonstationary state of the stochastic process, in the general case.

Substituting Equation (2.51) in Equation (2.26), we can obtain the general formula for the space–time correlation function in the symmetrical form:

$$\begin{aligned} R_{\Delta\rho,\Delta\mathbf{x}}(t,\tau,\Delta\omega) = m_0 \int\int & S(\rho - 0.5\Delta\rho, \mathbf{x} - 0.5\Delta\mathbf{x}) \cdot S(\rho + 0.5\Delta\rho, \mathbf{x} + 0.5\Delta x) \\ & \times U\left(t - \tfrac{2\rho}{c} - 0.5\tau + \tfrac{\Delta\rho}{c}, \omega - 0.5\Delta\omega\right) \\ & \times U^*\left(t - \tfrac{2\rho}{c} + 0.5\tau - \tfrac{\Delta\rho}{c}, \omega + 0.5\Delta\omega\right) d\rho \, d\mathbf{X} \end{aligned}, \tag{2.53}$$

where $\Delta\mathbf{x} = \{\Delta x_2, \ldots, \Delta x_n\}$ and $d\mathbf{X} = |J|\, dx_2 \ldots dx_n$.

2.3.2 The Correlation Function with the Narrow-Band Searching Signal

As a rule, the searching signals generated by radar in navigational systems have a moderately narrow power spectral density with respect to the carrier frequency. So, these searching signals can be written in the following form:[40,41]

$$U(t) = S(t) \cdot e^{-j\int_0^t \omega(t)dt} = S(t) \cdot e^{-j[\omega_0 t + \Psi(t)]}, \tag{2.54}$$

where $\omega(t) = \omega_0 + \Omega(t)$; $\Omega(t) = \frac{d\Psi(t)}{dt}$ is the instantaneous frequency; and $\Psi(t)$ is the phase modulation law, which is not taken into consideration by the term $\omega_0 t$. Substituting Equation (2.54) in Equation (2.53), we can write

$$R_{\Delta\rho,\Delta\mathbf{x}}(t,\tau,\Delta\omega) = R^{en}_{\Delta\rho,\Delta\mathbf{x}}(t,\tau,\Delta\omega) \cdot e^{j\omega_0\tau}, \tag{2.55}$$

where

$$
\begin{aligned}
R^{en}_{\Delta\rho,\Delta\mathbf{x}}(t,\tau,\Delta\omega) = m_0 \iint & S(\rho - 0.5\Delta\rho, \mathbf{x} - 0.5\Delta\mathbf{x}) \cdot S(\rho + 0.5\Delta\rho, \mathbf{x} + 0.5\Delta\mathbf{x}) \\
& \times S\left(t - \tfrac{2\rho}{c} - 0.5\tau + \tfrac{\Delta\rho}{c}\right) \cdot S\left(t - \tfrac{2\rho}{c} + 0.5\tau - \tfrac{\Delta\rho}{c}\right) \cdot \\
& \times e^{j\left[\Psi\left(t - \frac{2\rho}{c} + 0.5\tau - \frac{\Delta\rho}{c}\right) - \Psi\left(t - \frac{2\rho}{c}\, 0.5\tau + \frac{\Delta\rho}{c}\right)\right]} \cdot e^{-j\left[2\omega \cdot \frac{\Delta\rho}{c} \Delta\omega\left(t - \frac{2\rho}{c}\right)\right]} dp\, d\mathbf{X}
\end{aligned} \tag{2.56}
$$

The formula in Equation (2.56) defines the envelope of the high-frequency correlation function of space and time fluctuations of the nonstationary target return signal. The correlation function $R^{en}_{\Delta\rho,\Delta\mathbf{x}}(t, \tau, \Delta\omega)$ in Equation (2.56), with respect to the delay (radar range or distance) and the Doppler frequency (radial velocity),[42] can be considered as an extension of the correlation function used in the theory of communications to define a signal passing through a channel with two-dimensional scattering, and the amplitude $S(\rho,\mathbf{x})$ is the multidimensional function of scattering with respect to the coordinates ρ and $\mathbf{x}$. In this case, the amplitude $S(\rho,\mathbf{x})$ is reduced to the two-dimensional function of scattering with respect to the delay and Doppler frequency.

The high-frequency correlation function $R_{\Delta\rho,\Delta\mathbf{x}}(t, \tau, \Delta\omega)$ given by Equation (2.55), as well as by Equation (2.54), consists of two factors: the rapidly varying function $e^{j\omega_0\tau}$ and the slowly varying function $R^{en}_{\Delta\rho,\Delta\mathbf{x}}(t, \tau, \Delta\omega)$ [see Equation (2.56)] that is the complex envelope of the correlation function or the low-frequency correlation function of the target return signal fluctuations. It is precisely this envelope $R^{en}_{\Delta\rho,\Delta\mathbf{x}}(t, \tau, \Delta\omega)$ of the high-frequency correlation function that is of prime interest to us, because it characterizes features of the fluctuations. The cofactor

$$
e^{-2j\omega\cdot\frac{\Delta\rho}{c}} = e^{-4j\pi\cdot\frac{\Delta\rho}{\lambda}} \tag{2.57}
$$

has appeared in Equation (2.56). This cofactor, determined by Equation (2.57), plays a very important role.

If the radar generates simple searching signals, for example, $S(t) \equiv 1$, $\Psi(t) \equiv 0$, $\omega = \omega_0$, and $\Delta\omega = 0$, then on the basis of Equation (2.56) we can write

$$
\begin{aligned}
R^{en}_{\Delta\rho,\Delta\mathbf{x}}(t,\tau) = m_0 \iint & S(\rho - 0.5\Delta\rho, \mathbf{x} - 0.5\Delta\mathbf{x}) \\
& \times S(\rho + 0.5\Delta\rho, \mathbf{x} + 0.5\Delta\mathbf{x}) \cdot e^{-2j\omega_0\frac{\Delta\rho}{c}}\, d\rho\, d\mathbf{X}
\end{aligned} \tag{2.58}
$$

The correlation function given by Equation (2.58) is the correlation function of the slow with the continuous searching signal. In other words, this is the correlation function of the fluctuations caused by the moving radar, antenna scanning, and displacements of scatterers — i.e., the space fluctuations.

2.3.3 The Correlation Function with the Pulsed Searching Signal

Let us assume that the radar generates a sequence of identical, coherent pulsed searching signals with the duration $\tau_p << T_p$, where T_p is the period of recurrence:

$$U(t,\omega)=\sum_{k=0}^{\infty} P\left(t-kT_p\right)\cdot e^{-j\omega_0\left(t-kT_p\right)}, \tag{2.59}$$

where

$$P(t) = \Pi(t)\cdot e^{-j\Psi(t)}; \tag{2.60}$$

$\Pi(t)$ is the envelope of the high-frequency pulse (video-signal); $\Psi(t)$ is the phase modulation law within the limits of the pulsed searching signal duration, which is not taken into consideration by the term ω_0. For example, $\Psi(t)$ is caused by frequency changes within the pulsed searching signal duration.[43,44]

To determine the correlation function of target return signal fluctuations, it is necessary to substitute Equation (2.59) in Equation (2.53). At first, we must define the product of two pulse sequences shifted with respect to each other:

$$\begin{aligned} U\left(t,\tau,\Delta\omega\right)\cdot U^*\left(t,\tau,\Delta\omega\right)&=U\left(t-\tfrac{2\rho}{c}-0.5\tau_1,\omega_0-0.5\Delta\omega\right)\\ &\times U^*\left(t-\tfrac{2\rho}{c}+0.5\tau_1,\omega_0+0.5\Delta\omega\right)\end{aligned}, \tag{2.61}$$

where $\tau_1 = \tau - \frac{2\Delta\rho}{c}$. Thus, $|\tau| = |t_2 - t_1| \in [0,\infty)$. Taking into consideration that $\tau_p << T_p$ and denoting the index of summing the signal $U(t, \tau, \Delta\omega)$ by k and the index of summation of the signal $U^*(t, \tau, \Delta\omega)$ by m and assuming $n = m - k$, we can write

$$\begin{aligned} U\left(t,\tau,\Delta\omega\right)\cdot U^*\left(t,\tau,\Delta\omega\right)&=e^{-j\left[2\omega_0\cdot\frac{\Delta\rho}{c}+\Delta\omega\left(t-\frac{2\rho}{c}\right)\right]}\\ &\times\sum_{m=0}^{\infty}\sum_{n=0}^{\infty}P\left(t-\tfrac{2\rho}{c}-0.5\tau_1+nT_p-mT_p\right)\\ &\times P^*\left(t-\tfrac{2\rho}{c}+0.5\tau_1-nT_p-mT_p\right)\cdot e^{i\omega_0\left(\tau-nT_p\right)}\end{aligned} \tag{2.62}$$

In the case of coherent pulsed searching signals, we can assume that $e^{-j\omega_0 nT_p} = 1$. For symmetry, we can write the expression for the double summation in Equation (2.62) in the following form:

$$\sum_{m=0}^{\infty}\sum_{n=0}^{\infty} \boldsymbol{P}\left[t - \tfrac{2\rho}{c} - 0.5(\tau_1 - nT_p) - mT_p\right] \cdot \boldsymbol{P}^*\left[t - \tfrac{2\rho}{c} + 0.5(\tau_1 - nT_p) - mT_p\right] \tag{2.63}$$

Here $|\tau_1 - nT_p| \leq T_p$. The function given by Equation (2.62) is periodic with respect to two variables: $\tau_1 - nT_p$, where nT_p is the relative shift between two pulsed searching signal sequences, and $t - mT_p$. Summing with respect to the index, m can be omitted. Then, substituting Equation (2.62) in Equation (2.53), we can write

$$R_{\Delta\rho,\Delta\mathbf{x}}(t,\tau,\Delta\omega) = R^{en}_{\Delta\rho,\Delta\mathbf{x}}(t,\tau,\Delta\omega) \cdot e^{j\omega_0\tau} \tag{2.64}$$

where

$$\begin{aligned} R^{en}_{\Delta\rho,\Delta\mathbf{x}}(t,\tau,\Delta\omega) = m_0 \iint & \boldsymbol{P}\left[t - \tfrac{2\rho}{c} - 0.5(\tau - nT_p - \Delta\rho)\right] \\ & \times \boldsymbol{P}^*\left[t - \tfrac{2\rho}{c} + 0.5(\tau - nT_p - \Delta\rho)\right] \\ & \times S(\rho - 0.5\Delta\rho, \mathbf{x} - 0.5\Delta\mathbf{x}) \\ & \times S(\rho + 0.5\Delta\rho, \mathbf{x} + 0.5\Delta\mathbf{x}) \cdot e^{-j\left[2\omega_0 \cdot \frac{\Delta\rho}{c} + \Delta\omega\left(t - \frac{2\rho}{c}\right)\right]} d\rho\, d\mathbf{X} \end{aligned} \tag{2.65}$$

is the envelope of the high-frequency correlation function $R_{\Delta\rho,\Delta\mathbf{x}}(t, \tau, \Delta\omega)$.

The high-frequency correlation function $R_{\Delta\rho,\Delta\mathbf{x}}(t, \tau, \Delta\omega)$ given by Equation (2.64) defines both fluctuations in the radar range (the intraperiod fluctuations) and fluctuations from period to period (the interperiod fluctuations). In the case of the nonstationary target return signal, $R_{\Delta\rho,\Delta\mathbf{x}}(t, \tau, \Delta\omega)$ depends on time. So long as the differentials $\Delta\rho$ and $\Delta\mathbf{x}$ are thought of as independent, i.e., independent of the parameters τ and t, the envelope $R^{en}_{\Delta\rho,\Delta\mathbf{x}}(t, \tau, \Delta\omega)$ is a periodic function of the parameters τ and t. If we assume that $\boldsymbol{P}(t) \equiv 1$ and $\Delta\omega = 0$, then the envelope $R^{en}_{\Delta\rho,\Delta\mathbf{x}}(t, \tau, \Delta\omega)$ transforms to the envelope of the correlation function determined by Equation (2.58). Thus, the envelope $R^{en}_{\Delta\rho,\Delta\mathbf{x}}(t, \tau, \Delta\omega)$ can be used with both pulsed searching signals and with continuous searching signals in navigational systems.[45,46]

2.3.4 The Average Correlation Function

In the case of the periodic searching signal, the correlation function of target return signal fluctuations too, as was noted previously, is a periodic function with respect to time. To determine the average (with respect to time) correlation function, we must average the time correlation function within the limits of the period of the pulsed searching signal:

$$\overline{R(t,\tau)} = \frac{1}{T_p} \int_{-0.5T_p}^{0.5T_p} R(t,\tau)\,dt\,. \tag{2.66}$$

To define the average correlation function given by Equation (2.66), we sometimes use the following technique. Represent the searching signal given by Equation (2.59) in the Fourier series form:

$$U(t) = \sum_{k=0}^{\infty} N_k \cdot e^{-j(\omega_0 + k\Omega_p)t}\,, \tag{2.67}$$

where

$$N_k = \frac{1}{T_p} \int_{-0.5T_p}^{0.5T_p} U(t) \cdot e^{j(\omega_0 + k\Omega_p)t}\,dt \tag{2.68}$$

and Ω_p is the frequency of the pulsed searching signal.

Substituting Equation (2.67) in Equation (2.53) under the condition $\Delta\omega = 0$, we can write

$$R_{\Delta\rho,\Delta\mathbf{x}}(t,\tau) = m_0 \sum_{k=0}^{\infty}\sum_{n=0}^{\infty} N_k N_n^* \cdot e^{0.5j[\omega_0 + (n+k)\Omega_p]\tau} \cdot \iint e^{-j\left\{(k-n)\Omega_p\left(t - \frac{2\rho}{c}\right) + [\omega_0 + 0.5(n+k)\Omega_p]\cdot\frac{2\Delta\rho}{c}\right\}}$$
$$\times S(\rho - 0.5\Delta\rho, \mathbf{x} - 0.5\Delta\mathbf{x}) \cdot S(\rho + 0.5\Delta\rho, \mathbf{x} + 0.5\Delta\mathbf{x})\,d\rho\,d\mathbf{X}\,. \tag{2.69}$$

The correlation function determined by Equation (2.69) is equivalent to the correlation function given by Equation (2.56), but the variables t and τ are separable. This fact allows us to average the correlation function given by Equation (2.66) in a simpler form. Integrating with respect to the variable t and taking into consideration the fact that the terms in Equation (2.68) differ from zero only under the condition $k = n$, we can write

$$\overline{R_{\Delta\rho,\Delta\mathbf{x}}(t,\tau)} = \sum_{n=0}^{\infty} |N_n|^2 \cdot R_{con}(\tau, \omega_0 + n\Omega_p), \tag{2.70}$$

where $R_{con}(\tau, \omega_0 + n\Omega_p)$ is the correlation function in the case of the continuous nonmodulated searching signal given by Equation (2.58), in which the frequency ω_0 is replaced with $\omega_0 + n\Omega_p$. Reference to Equation (2.70) shows we can conclude that the average power spectral density with respect to time of the target return signal fluctuations for any kind of modulation law is defined by the sum of the power spectral densities $S_{con}(\omega)$ at frequencies $\omega_0 + n\Omega_p$:

$$\overline{S(\omega,t)} = \sum_{n=0}^{\infty} |N_n|^2 \cdot S_{con}(\omega - \omega_0 - n\Omega_p)\,. \tag{2.71}$$

2.4 The Correlation Function under Scanning of the Three-Dimensional (Space) Target

2.4.1 General Statements

Using Equation (2.64), we can find the correlation function of the target return signal fluctuations in the cases of continuous and pulsed searching signals. The amplitude of the target return signal from an arbitrary scatterer can be determined using the main radar equation

$$S(\rho) = \frac{\sqrt{P_s} G_0 \lambda\, g(\varphi,\psi) q(\xi,\zeta)}{8\sqrt{\pi^3}\rho^2}, \tag{2.72}$$

where P_s is the power of the searching signal; λ is the wavelength; φ and ψ are the angles defining the position of the scatterer relative to the radar antenna directional diagram axis in the main planes (see Figure 2.5); G_0 is the radar antenna amplifier coefficient; $g(\varphi, \psi)$ is the normalized two-dimensional directional diagram under the condition $g(0,0) = 1$; ξ and ζ are the angles defining the position of the scatterer in space relative to the radar antenna polarization plane and direction of beam, as shown in Figure 2.6, where it is assumed that scatterers possess axis symmetry and that two angles exactly define the orientation of scatterers; $q(\xi, \zeta)$ is the function representing the dependence of the target return signal amplitude on the orientation of the scatterer in space and $q^2(\xi, \zeta)$ is the effective scattering area with fixed values of the angles ξ and ζ; and ρ is the radar range.

The volume element using the coordinates ρ, φ, and ψ is equal to $\rho^2\, d\rho\, d\varphi\, d\psi$ and the corresponding Jacobian is equal to ρ^2. The angles ξ and ζ belong to the spherical coordinate system. Under integration over all orientations of scatterers, the surface element can be written in the form: $\sin\zeta\, d\xi\, d\zeta$. If the directional diagrams are different under conditions of transmission and reception, we can write

$$g(\varphi, \psi) = g_t(\varphi, \psi) \cdot g_r(\varphi, \psi), \tag{2.73}$$

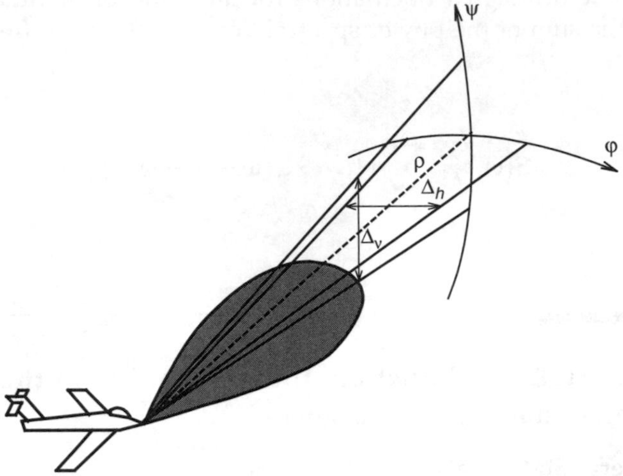

FIGURE 2.5
The coordinate system using the variables φ and ψ.

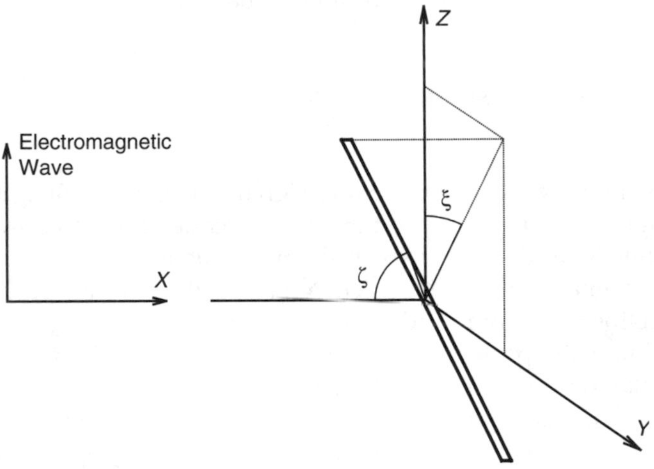

FIGURE 2.6
The coordinate system using the variables ξ and ζ.

where $g_t(\varphi, \psi)$ and $g_r(\varphi, \psi)$ are the directional diagrams under conditions of transmission and reception by voltage for the general coordinate system φ and ψ, respectively.

Thus, the amplitude of the target return signal is the function of five variables: ρ, φ, ψ, ξ, and ζ. Substituting Equation (2.72) in Equation (2.64) and using the condition $\Delta\omega = 0$ (which is equivalent to the fact that a rotation of scatterers $\Delta\xi$ and $\Delta\zeta$ is independent of their positions within the directional diagram, in other words, independent of the coordinates ρ, φ, and ψ, and that the angle differentials $\Delta\varphi$ and $\Delta\psi$ are independent of the orientations of scatterers, in other words, independent of the coordinates ξ and ζ, and using the asymmetric form, we obtain[47,48]

$$R^{en}_{\Delta\rho,\Delta\varphi,\Delta\psi,\Delta\xi,\Delta\zeta}(t,t+\tau) = p(t)\cdot R^{en}_{\Delta\rho,\Delta\varphi,\Delta\psi,\Delta\xi,\Delta\zeta}(t,t+\tau)\,, \tag{2.74}$$

where

$$R^{en}_{\Delta\rho,\Delta\varphi,\Delta\psi,\Delta\xi,\Delta\zeta}\left(t,t+\tau\right) = R_{\Delta\rho,\Delta\varphi,\Delta\psi}\left(t,t+\tau\right)\cdot R_{\Delta\xi,\Delta\zeta}\left(\Delta\xi,\Delta\zeta\right) \tag{2.75}$$

is the total normalized correlation function of the fluctuations;

$$R_{\Delta\rho,\Delta\varphi,\Delta\psi}\left(t,t+\tau\right) = N\cdot\sum_{n=0}^{\infty}\iiint P\left(t-\tfrac{2\rho}{c}\right)\cdot P^*\left(t-\tfrac{2\rho}{c}+\tau-nT_p-\tfrac{2\Delta\rho}{c}\right)$$
$$\times\frac{g\left(\varphi,\psi\right)\cdot g\left(\varphi+\Delta\varphi,\psi+\Delta\psi\right)}{\left(\rho+\Delta\rho\right)^2}\cdot e^{-j\omega_0\cdot\frac{2\Delta\rho}{c}}\,d\rho\,d\varphi\,d\psi \tag{2.76}$$

and

$$R_{\Delta\xi,\Delta\zeta}\left(\Delta\xi,\Delta\zeta\right) = N\iint q\left(\xi,\zeta\right)\cdot q\left(\xi+\Delta\xi,\zeta+\Delta\zeta\right)\sin\zeta\,d\xi\,d\zeta \tag{2.77}$$

are the normalized correlation functions: $R_{\Delta\rho,\Delta\varphi,\Delta\omega}(t, t+\tau)$ is the normalized correlation function of the fluctuations caused by the moving radar, displacements of scatterers, and radar antenna scanning; $R_{\Delta\xi,\Delta\zeta}(\Delta\xi, \Delta\zeta)$ is the normalized correlation function of the fluctuations caused by rotation of scatterers and radar antenna scanning; N is the normalized coefficient [see Equation (2.32)];

$$p\left(t\right) = \frac{m_0 P_S G_0^2\lambda^2}{64\,\pi^3}\cdot\int\frac{\Pi^2\left(t-\frac{2\rho}{c}\right)}{\rho^2}d\rho\iint g^2\left(\varphi,\psi\right)d\varphi\,d\psi\iint q^2\left(\xi,\zeta\right)\sin\zeta\,d\xi\,d\zeta \tag{2.78}$$

is the received target return signal power at the instant of time t.

2.4.2 The Correlation Function with the Pulsed Searching Signal

The total normalized correlation function of the fluctuations given by Equation (2.75) is the product of two normalized correlation functions. The first function defines the slow fluctuations caused by the moving radar, displacements of scatterers ($\Delta\rho$), and antenna scanning ($\Delta\varphi$ and $\Delta\psi$) and the rapid fluctuations caused by propagation of the searching signal in space (τ). The second function defines the slow fluctuations caused by rotation of scatterers and the radar antenna polarization plane ($\Delta\varphi$ and $\Delta\zeta$). Reference to Equation (2.76) shows that with moving radar ($\Delta\rho \neq 0$), there are three sources giving rise to target return signal fluctuations: phase changes in elementary signals –$\Delta\rho$ in the exponent of the exponential function; amplitude changes in elementary signals caused by the duration and shape of the searching signal envelope –$\Delta\rho$ is the argument of the function $\boldsymbol{P}(t)$; and amplitude changes in elementary signals caused by searching signal attenuation as a function of distance between the radar and scatterer –$\Delta\rho$ in the sum $\rho + \Delta\rho$. Fluctuations caused by phase changes in elementary signals are the most rapid fluctuations. Amplitude changes in elementary signals caused by the searching signal envelope give rise to the slow fluctuations. Attenuation of the searching signal caused by increasing the distance between the radar and the scatterer gives rise to slower fluctuations.[49,50] Because of this, we can neglect the differential $\Delta\rho$ in the sum $\rho + \Delta\rho$. This statement will be rigorously proved in Section 2.4.4.

To define the position of scatterers with respect to the radar and to the direction of moving radar, we introduce the spherical coordinate system that is functionally related to the radar: the distance ρ, the azimuth β, and the aspect angle γ (see Figure 2.7). The direction of moving radar in the spherical coordinate system is characterized by the angles β and γ, and the position of the directional diagram axis is characterized by the angles β_0 and γ_0. We must orient the coordinate system (φ, ψ) functionally related to the directional diagram axis in such a manner that the plane $\psi(\varphi = 0)$ can be matched with the plane γ. Then we can write

$$\varphi = (\beta - \beta_0)\cos\gamma \quad \text{and} \quad \psi = \gamma - \gamma_0. \tag{2.79}$$

When scatterers are stationary, the differentials $\Delta\varphi$ and $\Delta\psi$ of the angle coordinates φ and ψ are caused, in the general case, by two sources: antenna scanning ($\Delta\varphi_{sc}$ and $\Delta\psi_{sc}$) and moving radar ($\Delta\varphi_{rm}$ and $\Delta\psi_{rm}$). The differentials $\Delta\varphi_{sc}$ and $\Delta\psi_{sc}$ are defined by the shifts $\Delta\beta_0$ and $\Delta\gamma_0$ of the directional diagram axis during antenna scanning. Reference to Equation (2.79) shows that

$$\Delta\varphi_{sc} = \varphi(\beta_0) - \varphi(\beta_0 + \Delta\beta_0) = \Delta\beta_0\cos\gamma \tag{2.80}$$

and

$$\Delta\psi_{sc} = \psi(\gamma_0) - \psi(\gamma_0 + \Delta\gamma_0) = \Delta\gamma_0. \tag{2.81}$$

The differentials $\Delta\varphi_{rm}$ and $\Delta\psi_{rm}$ are caused by changes in the azimuth of scatterers when the radar moves over a distance $\Delta\ell$ (see Figure 2.8):

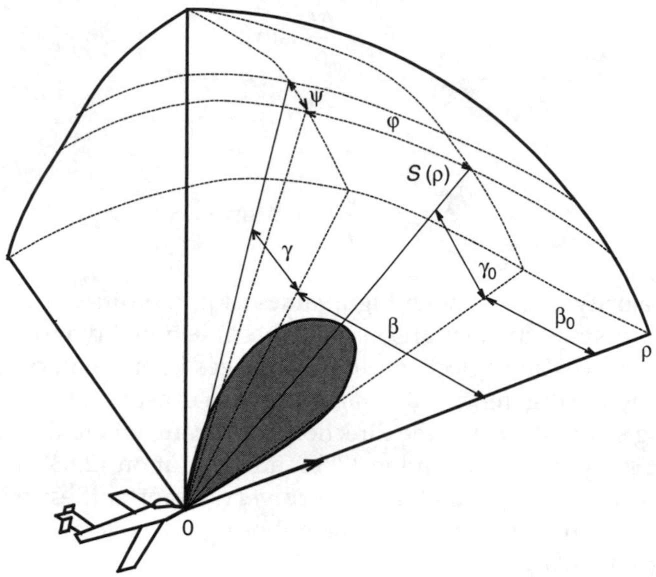

FIGURE 2.7
The coordinate system using the variables ρ, β, and γ for the three-dimensional (space) target.

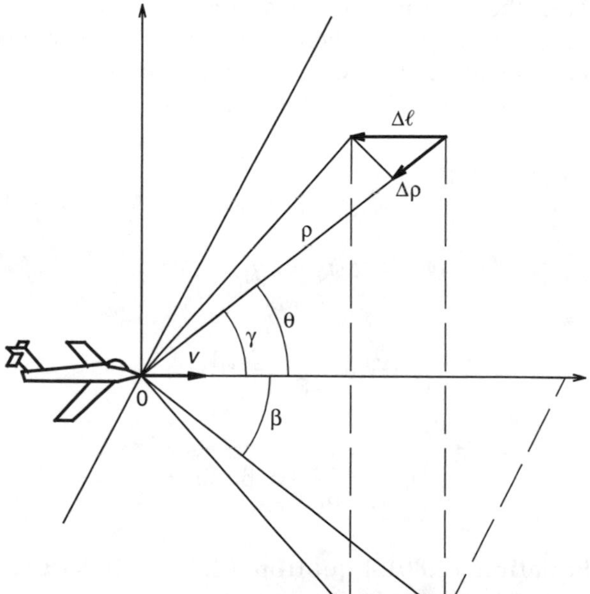

FIGURE 2.8
Radial displacement of scatterers with moving radar.

$$\Delta\varphi_{rm} = \frac{\Delta\ell}{\rho}\cdot\sin\beta \tag{2.82}$$

and

$$\Delta\psi_{rm} = \frac{\Delta\ell}{\rho}\cdot\cos\beta\,\sin\gamma\,. \tag{2.83}$$

In the majority of cases, with high values of ρ, the differentials $\Delta\varphi_{rm}$ and $\Delta\psi_{rm}$ are very small in comparison with the directional diagram width.[51,52] Consequently, we can neglect amplitude changes in elementary signals that are caused by the differentials $\Delta\varphi_{rm}$ and $\Delta\psi_{rm}$. An exception to this rule is the neighboring effective area of the directional diagram, where the differentials $\Delta\varphi_{rm}$ and $\Delta\psi_{rm}$ given by Equation (2.82) and Equation (2.83), respectively, must be taken into consideration. This is very important, especially in the use of laser antenna, for which the neighboring effective area of the directional diagram is large.

As can be seen from Equation (2.80)–Equation (2.83), the differentials $\Delta\varphi$ and $\Delta\psi$ differ for various scatterers because they depend on the coordinates ρ, β, and γ. Let us assume that the range of changes for the coordinates ρ, β, and γ within the resolution area forming the resulting target return signal is not so large. Then the coordinates ρ, β, and γ can be changed to the coordinates ρ_*, β_0, and γ_0, respectively, where ρ_* is the distance between the center of the pulse volume and radar. Also, we can assume that the differentials $\Delta\varphi$ and $\Delta\psi$ are the same for all scatterers. In other words, we can write the following formulae

$$\Delta\varphi_{sc} = \Delta\beta_0\cos\gamma_0\,; \tag{2.84}$$

$$\Delta\psi_{sc} = \Delta\gamma_0\,; \tag{2.85}$$

$$\Delta\varphi_{rm} = \frac{\Delta\ell}{\rho_*}\cdot\sin\beta_0\,; \tag{2.86}$$

$$\Delta\psi_{rm} = \frac{\Delta\ell}{\rho_*}\cdot\cos\beta_0\,\sin\gamma_0 \tag{2.87}$$

instead of Equation (2.80)–Equation (2.83). Thus, using Equation (2.84)–Equation (2.87), we can write

$$\Delta\varphi = \Delta\varphi_{sc} + \Delta\varphi_{rm} = \Delta\beta_0\cos\gamma_0 + \frac{\Delta\ell}{\rho_*}\cdot\sin\beta_0 \tag{2.88}$$

and

$$\Delta\psi = \Delta\psi_{sc} + \Delta\psi_{rm} = \Delta\gamma_0 + \frac{\Delta\ell}{\rho_*} \cdot \cos\beta_0 \sin\gamma_0 \,. \tag{2.89}$$

This representation of the differentials $\Delta\varphi$ and $\Delta\psi$ makes determination more simple, and allows us to define the normalized correlation function $R_{\Delta\rho,\Delta\varphi,\Delta\psi}(t, t+\tau)$ given by Equation (2.76) as a function of the shifts $\Delta\beta_0$ and $\Delta\gamma_0$ of the directional diagram axis and the displacement $\Delta\ell$ of the radar.

When the radar moves over a distance $\Delta\ell$ the radial displacements of scatterers can be determined as follows (see Figure 2.8):

$$\Delta\rho = \Delta\ell\cos\theta - \frac{\Delta\ell^2}{2\rho} \cdot \sin^2\theta_0 \,, \tag{2.90}$$

where θ is the angle between the velocity vector and the direction of scatterers, which is functionally related to the angles β and γ by

$$\cos\theta = \cos\beta \cdot \cos\gamma. \tag{2.91}$$

The second term in Equation (2.90) is infinitesimal and we can neglect it.

Using Equation (2.79) and Equation (2.91), we can write

$$\Delta\rho = \Delta\ell \cdot \cos\left(\beta_0 + \tfrac{\varphi}{\cos\gamma}\right) \cdot \cos\left(\gamma_0 + \psi\right) = \Delta\rho_0 + \Delta\rho_{\varphi,\psi} \,, \tag{2.92}$$

where $\Delta\rho_0 = \Delta\ell \cos\beta_0 \cos\gamma_0$ is the radial shift along the directional diagram axis; $\Delta\rho_{\varphi,\psi} = \Delta\ell \times f(\varphi, \psi)$ is the deviation of radial shifts for various scatterers within the directional diagram. Thus, the displacement $\Delta\rho$ is a function of the angles φ and ψ. In other words, the displacement $\Delta\rho$ depends on the position of scatterers within the resolution area. This function should be taken into account in the exponent $\frac{j\omega_0\Delta\rho}{c}$ of the exponential function in Equation (2.76), which defines phase changes in elementary signals with displacements of scatterers because variations in the value of $\Delta\rho$ can be compared with the wavelength λ. However, we can neglect the dependence between the parameters $\Delta\rho$, φ, and ψ in the argument of the pulsed function $\boldsymbol{P}(t)$ because variations in the value of $\Delta\rho$ are small in comparison with the length of the function $\boldsymbol{P}(t)$ and we can assume that $\Delta\rho = \Delta\rho_0$.

This assumption allows us to divide the triple integral in Equation (2.76) between the double integral with limits of integration over the variables φ and ψ and the simple integral with limits of integration over the variable ρ. As usual, the interval of changes in distance within the duration of the pulsed searching signal is not so large. Hence, we can replace thc variable ρ in the denominator of Equation (2.76) with the mean of the variable ρ_* and factor the variable ρ_* outside the integral sign. We introduce a variable $z = t - \frac{2\rho}{c}$. Then

$$R_{\Delta\rho,\Delta\varphi,\Delta\psi}(t, t+\tau) = R_g(\Delta\ell, \Delta\beta_0, \Delta\gamma_0) \cdot R_p(\tau, \Delta\ell), \tag{2.93}$$

where

$$R_g\left(\Delta\ell, \Delta\beta_0, \Delta\gamma_0\right) = N\iint g\left(\varphi,\psi\right)\cdot g\left(\varphi+\Delta\varphi,\psi+\Delta\psi\right)\cdot e^{-\frac{4j\pi\Delta\rho(\varphi,\psi)}{\lambda}}\,d\varphi\,d\psi \tag{2.94}$$

and

$$R_p\left(\tau,\Delta\ell\right) = N\sum_{n=0}^{\infty}\int \boldsymbol{P}\left(z\right)\cdot \boldsymbol{P}^*\left(z+\tau-nT_p-\tfrac{2\Delta\rho_0}{c}\right)dz\,. \tag{2.95}$$

Thus, the normalized correlation function of the target return signal fluctuations in navigational systems caused by the moving radar and antenna scanning is defined by the product of two normalized correlation functions with the pulsed searching signal. The first normalized correlation function defines the slow fluctuations caused by the different Doppler shifts in the frequency of elementary signals within the resolution area of pulse volume with moving radar, amplitude changes in elementary signals due to antenna scanning, and variation of the aspect angle during radar motion; these are the interperiod fluctuations. The second normalized correlation function is a periodic function of the variable τ and defines the rapid fluctuations caused by propagation of the pulsed searching signal in space with respect to simultaneous radar motion; these are the intraperiod fluctuations. For the considered approximation, the second normalized correlation function is independent of time or the radar range.

2.4.3 The Target Return Signal Power with the Pulsed Searching Signal

The power $p(t)$ of the target return signal is determined by Equation (2.78), for which

$$S_t = \iint q^2(\xi,\zeta)\cdot \sin\zeta\,d\xi\,d\zeta \tag{2.96}$$

is the effective scattering area of a scatterer, which is averaged over all possible positions of the scatterer in space. Thus, $\boldsymbol{S}^\circ = m_0\boldsymbol{S}_t$ is the specific effective scattering area. In other words, this is the effective scattering area of the pulse unit volume filled by a scatterer. In the determination of the double integral, we can assume that

$$g(\varphi, \psi) = g_h(\varphi) \cdot g_v(\psi), \tag{2.97}$$

where $g_h(\varphi)$ and $g_v(\psi)$ are the normalized radar antenna directional diagrams by power for two mutually perpendicular (main) planes, the conditionally horizontal plane and the vertical plane. Then

$$\iint g^2(\varphi,\psi)d\varphi\, d\psi = \int g_h^2(\varphi)d\varphi \cdot \int g_v^2(\psi) = \Delta_h^{(2)} \cdot \Delta_v^{(2)}, \tag{2.98}$$

where $\Delta_h^{(2)}$ and $\Delta_v^{(2)}$ are the effective widths of the squares of the directional diagram by power for the horizontal and vertical planes.

Usually, the radar antenna is characterized by the effective widths Δ_h and Δ_v of the first order (not of the second order) of the directional diagram by power: $\Delta = \int g(x)dx$.[53–55] The widths Δ_h and Δ_v of the first and second orders are functionally related:

$$\Delta_h^{(2)} = k_a^h \Delta_h \quad \text{and} \quad \Delta_v^{(2)} = k_a^v \Delta_v, \tag{2.99}$$

where k_a^h and k_a^v are coefficients of the shape of the directional diagram, respectively, which differ from unity. Henceforth, we will consider two kinds of directional diagram: the Gaussian and sinc^2 models.

The two-dimensional Gaussian directional diagram model takes the following form:

$$g(\varphi,\psi) = e^{-\pi\left(\frac{\varphi^2}{\Delta_h^{(2)}} + \frac{\psi^2}{\Delta_v^{(2)}}\right)} = g_h(\varphi) \cdot g_v(\psi). \tag{2.100}$$

It defines exactly the shape of the major lobe, and is also very convenient mathematically. Unlike the actual directional diagrams, the two-dimensional Gaussian directional diagram model does not have side-lobes. If the two-dimensional Gaussian directional diagram model has axis symmetry $\Delta_h = \Delta_v = \Delta_a$, then

$$g(\varphi,\psi) = e^{-\pi \cdot \frac{\varphi^2 + \psi^2}{\Delta_a^{(2)}}} = e^{-\pi \cdot \frac{\theta^2}{\Delta_a^{(2)}}} = g(\theta), \tag{2.101}$$

where $\theta^2 = \varphi^2 + \psi^2$. For the two-dimensional Gaussian directional diagram model, the following equality

$$k_a^h = k_a^v = k_a = 2^{-0.5} \tag{2.102}$$

is true. The two-dimensional sinc^2 directional diagram model takes the following form:

$$g(\varphi,\psi)=g_h(\varphi)\cdot g_v(\psi)=\text{sinc}^2\frac{\pi\varphi}{\Delta_h}\cdot\text{sinc}^2\frac{\pi\psi}{\Delta_v} \tag{2.103}$$

and corresponds approximately to uniform radiation of a rectangle aperture and has very large side-lobes. This model of the directional diagram does not possess axis symmetry even if the condition $\Delta_h = \Delta_v$ is true. In the case of the two-dimensional sinc^2 directional diagram model, the equality $k_a^h = k_a^v = k_a = \frac{2}{3}$ is true. The two-dimensional Gaussian and sinc^2 directional diagram models given by Equation (2.100) and Equation (2.103), respectively, can be considered, in principle, as two extreme cases limiting a set of directional diagram models with various levels of side radiation. The coefficient k_a is approximately the same for various directional diagram models in spite of differences in shapes. In other words, in the determination of target return signal power, the shape of the directional diagram model plays only a minor role. In the case of the rectangle directional diagram model, we can write $k_a = 1$. But the rectangle directional diagram model is not realizable in practice and, for this reason, is not studied further.

When the pulsed searching signal has a short duration, the third integral can be written in the following form:

$$\int\frac{\Pi^2\left(t-\frac{2\rho}{c}\right)}{\rho^2}d\rho\approx\frac{c}{2\rho_*^2}\cdot\int\Pi^2(t)\,dt=\tau_p^{(2)}\cdot\frac{c}{2\rho_*^2}=k_p\tau_p\cdot\frac{c}{2\rho_*^2}, \tag{2.104}$$

where $\tau_p^{(2)} = k_p\tau_p$ is the effective duration of the square of the envelope of the pulsed searching signal; k_p is the coefficient of the shape of the pulsed searching signal. In the case of the square waveform pulsed searching signal, the equality $k_p = 1$ is true. In the case of the Gaussian pulsed searching signal, we can write

$$\Pi(t)=e^{-\pi\cdot\frac{t^2}{\tau_p^{(2)}}}, \tag{2.105}$$

where τ_p is the effective duration of the pulsed searching signal and $k_p = 2^{-0.5}$.

Substituting Equation (2.96), Equation (2.98), Equation (2.99), and Equation (2.104) in Equation (2.78), we can write

$$p(t)=\frac{P_S G_0^2\lambda^2\mathbf{S}^\circ\Delta_h\Delta_v c\tau_p k_a^h k_a^v k_p}{128\pi^3\rho_*^2}, \tag{2.106}$$

where $\rho_* = 0.5ct$. Comparing Equation (2.106) with the radar equation for the point target with effective scattering area equal to $\mathbf{S}_t$

$$p(t) = \frac{P_S G_0^2 \lambda^2 \mathbf{S}_t}{64\pi^3 \rho^4}, \tag{2.107}$$

one can see that the total target return signal power from a three-dimensional target can be defined using the radar equation for a single target, if the effective scattering area has the form:[56,57] $S_t^{space} = S^\circ \cdot Q$, where

$$Q = 0.5 \cdot c\tau_p^{(2)} \Delta_h^{(2)} \Delta_v^{(2)} \rho_*^2 = 0.5 \cdot k_p k_a^h k_a^v c\tau_p \Delta_h \Delta_v \rho_*^2 \tag{2.108}$$

is the pulse or resolution volume in space. Thus, under scanning of the three-dimensional (space) target, the target return signal power at the receiver or detector input in navigational systems is inversely proportional to ρ^2, but not to ρ^4, as in the case of the point target. This can be explained by the fact that with an increase in distance between the radar and the three-dimensional (space) target, the pulse volume and effective scattering area increase proportionally to ρ^2.

2.4.4 The Correlation Function and Power of the Target Return Signal with the Simple Harmonic Searching Signal

In the considered case, Equation (2.65) is true if the condition

$$P(t) \equiv 1 \tag{2.109}$$

is satisfied. Formally, in the following formulae, the distance ρ in integrands is not equal to $\rho_* = const$ because the differentials $\Delta\varphi$ and $\Delta\psi$ in Equation (2.88) and Equation (2.89) depend on the distance ρ, which is a parameter of the functions given by Equation (2.76) and Equation (2.78). For this reason, it is necessary to integrate over the entire area occupied by scatterers but, as before, we can neglect the differential $\Delta\rho$ in the sum $\rho + \Delta\rho$ or, with rigorous analysis, we can assume that $\Delta\rho = \Delta\rho_0 = \Delta\ell \cos\beta_0 \cos\gamma_0$.

On the basis of the normalized correlation function of the target return signal fluctuations given by Equation (2.75), we can, as before, isolate the normalized correlation function $R_{\Delta\xi,\Delta\zeta}(\Delta\xi, \Delta\zeta)$ determined by Equation (2.77) in the form of an individual cofactor. The remaining continuous normalized correlation function $R^{con}_{\Delta\rho,\Delta\varphi,\Delta\psi}(t, t+\tau)$ takes the following form:

$$R^{con}_{\Delta\rho,\Delta\varphi,\Delta\psi}(t, t+\tau) = N \iiint \frac{g(\varphi,\psi) \cdot g(\varphi + \Delta\varphi, \psi + \Delta\psi)}{(\rho + \Delta\rho_0)^2} \cdot e^{-2j\omega_0 \cdot \frac{\Delta\rho}{c}} d\rho\, d\varphi\, d\psi\,. \tag{2.110}$$

In the long range of the radar antenna directional diagram, which is of prime interest to us, as a rule, the differentials $\Delta\varphi$ and $\Delta\psi$ are independent of the distance ρ.[58] Then,

$$R^{con}_{\Delta\rho,\Delta\varphi,\Delta\psi}(t, t+\tau) = R_g(\Delta\ell, \Delta\beta_0, \Delta\gamma_0) \cdot R_\rho(\Delta\ell), \tag{2.111}$$

where $R_g(\Delta\ell, \Delta\beta_0, \Delta\gamma_0)$ is determined by Equation (2.94) under the condition $\Delta\varphi_{rm} = \Delta\psi_{rm} = 0$ and

$$R_\rho(\Delta\ell) = \frac{\int \frac{d\rho}{(\rho + \Delta\rho_0)^2}}{\int \frac{d\rho}{\rho^2}}. \tag{2.112}$$

If the parameter ρ_1 is the minimum distance to receive the target return signal, then integrating the formula in Equation (2.112) within the limits of the interval $[\rho_1, \infty)$, we can write

$$R_\rho(\Delta\ell) = \frac{1}{1 + \frac{|\Delta\rho_0|}{\rho_1}} = \frac{1}{1 + \frac{|\Delta\ell \cos\beta_0 \cos\gamma_0|}{\rho_1}}. \tag{2.113}$$

The correlation interval of the corresponding fluctuations has the following form

$$\Delta\ell_\rho = \int_{-\infty}^{\infty} R_\rho(\Delta\ell)\, d(\Delta\ell) \to \infty, \tag{2.114}$$

which indicates a very small bandwidth for the power spectral density of these fluctuations, allowing us to neglect them. Thus, in the long range of the directional diagram, the normalized correlation function of the fluctuations with a continuous nonmodulated searching signal has the same shape as the normalized correlation function of the fluctuations given by Equation (2.94), which are caused by the moving radar and antenna scanning with the pulsed searching signal.

The target return signal power from the three-dimensional (space) target with the continuous searching signal can be determined by assuming in Equation (2.78) that the condition given by Equation (2.109) is true within the limits of the interval $[\rho_1, \rho_2]$ occupied by scatterers. Instead of Equation (2.104) we can write

$$\int_{\rho_1}^{\rho_2} \frac{d\rho}{\rho^2} = \rho_1^{-1} - \rho_2^{-1}. \tag{2.115}$$

Under the condition $\rho_2 >> \rho_1$, we can neglect the second term in Equation (2.115). Then the target return signal power with the simple harmonic searching signal has the following form:

$$p(t) = \frac{P_S G_0^2 \lambda^2 S^\circ \Delta_h \Delta_v k_a^h k_a^v}{64\pi^3 \rho_1} . \tag{2.116}$$

Reference to Equation (2.116) shows that the target return signal power with the simple harmonic searching signal decreases proportionally to the first order of the distance ρ_1 between the radar and front border of the three-dimensional (space) target. The region around the point $\rho_1 = 0$ is not considered. As $\rho_1 \to 0$, the target return signal power tends to approach ∞, i.e., $p(t) \to \infty$. This is called the "effect of dazzley." Comparing Equation (2.107) and Equation (2.116), one can see that the effective scattering area of the three-dimensional (space) target with the continuous pulsed searching signal has the following form:

$$S_t^{space} = S^\circ \Delta_h^{(2)} \Delta_v^{(2)} \rho_1^3 = S^\circ \Delta_h \Delta_v k_a^h k_a^v \rho_1^3 . \tag{2.117}$$

2.5 The Correlation Function in Angle Scanning of the Two-Dimensional (Surface) Target

2.5.1 General Statements

In scanning of the two-dimensional (surface) target, the radar range is not an independent coordinate.[59,60] The parameter ρ is a function of the altitude h and aspect angle γ (see Figure 2.9) and is determined as follows:

$$\rho = \frac{h}{\sin \gamma} = \frac{h}{\sin(\gamma_0 + \psi)} , \tag{2.118}$$

where γ_0 is the aspect angle of the radar antenna directional diagram axis. The element of the searched two-dimensional (surface) target can be determined using the coordinates φ and ψ as follows:

$$d\mathbf{X} = |J|\, d\varphi\, d\psi = \frac{h^2}{\sin^3(\gamma_0 + \psi)} d\varphi\, d\psi . \tag{2.119}$$

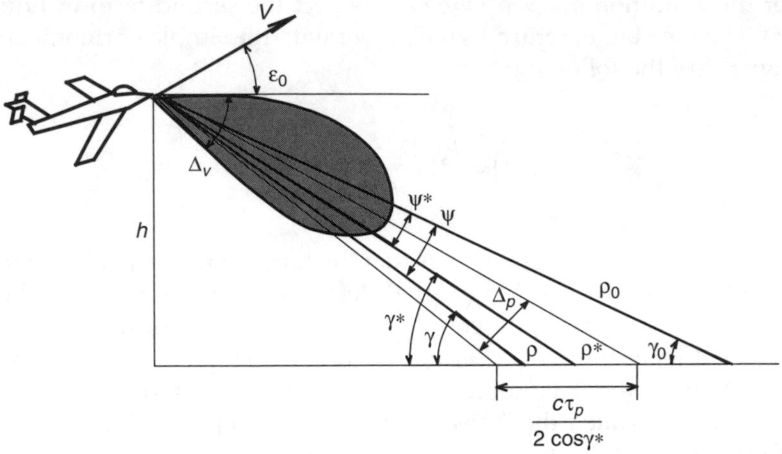

FIGURE 2.9
The coordinate system under scanning of the three-dimensional (space) target.

Furthermore, it is necessary to take into consideration the dependence of the effective scattering area $S°$ of the two-dimensional (surface) target as a function of the aspect angle γ and azimuth β. Hence, we can write

$$m_0 q^2(\xi, \eta) = S°(\beta, \gamma) = S°(\varphi + \beta_0, \psi + \gamma_0) \tag{2.120}$$

instead of the function $q^2(\xi, \eta)$. The function $S°(\beta, \gamma)$ can be considered as the back-scattering diagram of the two-dimensional (surface) target. These representations are equivalent.

As a rule, the surface of the two-dimensional target is assumed to be weakly nonisotropic. The reflectance of the surface of the two-dimensional target is a very slow function of the azimuth.[61,62] Because of this, we can neglect this function within the directional diagram. Then we can replace the variable $\beta = \beta_0 + \varphi$ with the variable β_0 in Equation (2.120). The majority of real surfaces of two-dimensional targets satisfy this condition. The target return signal amplitude under this condition takes the following form:

$$S(\rho) = \frac{\sqrt{P_S} G_0 \lambda \, g(\varphi, \psi) \sqrt{S°(\beta_0, \psi + \gamma_0)} \sin^2(\psi + \gamma_0)}{8\sqrt{\pi^3} h^2} \tag{2.121}$$

instead of Equation (2.72).

Using Equation (2.118)–Equation (2.121), assuming $\Delta\omega = 0$, and neglecting the displacement $\Delta\rho$ in the function $\sin(\psi + \gamma_0)$ [which is equivalent to neglecting the component $\Delta\rho$ in the sum $\rho + \Delta\rho$ in Equation (2.76)], on the basis of Equation (2.65) we can write

$$R^{en}_{\Delta\rho,\Delta\varphi,\Delta\psi}(t,\tau) = p_0 \sum_{n=0}^{\infty} \iint \boldsymbol{P}\left[t - \tfrac{2\rho(\psi)}{c} - 0.5(\tau - nT_p) + \tfrac{\Delta\rho}{c}\right]$$
$$\times \boldsymbol{P}^*\left[t - \tfrac{2\rho(\psi)}{c} + 0.5(\tau - nT_p) - \tfrac{\Delta\rho}{c}\right]$$
$$\times g(\varphi - 0.5\Delta\varphi, \psi - 0.5\Delta\psi)\cdot g(\varphi + 0.5\Delta\varphi, \psi + 0.5\Delta\psi)$$
$$\times e^{-2j\omega_0 \cdot \frac{\Delta\rho}{c}} S^{\circ}(\beta_0, \psi + \gamma_0)\sin(\psi + \gamma_0)\, d\varphi\, d\psi \tag{2.122}$$

where

$$p_0 = \frac{P_S G_0^2 \lambda^2}{64\pi^3 h^2}\,. \tag{2.123}$$

Assuming that $\Delta\rho = \Delta\varphi = \Delta\psi = \tau = 0$ in Equation (2.122) and omitting the summation sign, we can determine the target return signal power from the two-dimensional (surface) target at the instant of time t in the following form:

$$p(t) = p_0 \iint \Pi^2\left[t - \tfrac{2\rho(\psi)}{c}\right]\cdot g^2(\varphi, \psi)\cdot S^{\circ}(\beta_0, \psi + \gamma_0)\sin(\psi + \gamma_0)\, d\varphi\, d\psi\,. \tag{2.124}$$

2.5.2 The Correlation Function with the Pulsed Searching Signal

Suppose the angle between the direction of moving radar and the horizon is equal to ε_0 (see Figure 2.9). One can see that the differentials of coordinates can be determined as follows:

$$\Delta\varphi = \Delta\varphi_{sc} + \Delta\varphi_{rm} = \Delta\beta_0 \cos\gamma_* + \frac{\Delta\ell}{\rho_*}\cdot\cos\varepsilon_0 \sin\beta_0\,; \tag{2.125}$$

$$\Delta\psi = \Delta\psi_{sc} + \Delta\psi_{rm} = \Delta\gamma_0 + \frac{\Delta\ell}{\rho_*}(\cos\varepsilon_0 \cos\beta_0 \sin\gamma_* + \sin\varepsilon_0 \cos\gamma_*)\,; \tag{2.126}$$

$$\Delta\rho = \Delta\ell\left[\cos\varepsilon_0 \cos\left(\beta_0 + \tfrac{\varphi}{\cos\gamma_*}\right)\cos(\psi + \gamma_0) - \sin\varepsilon_0 \sin(\psi + \gamma_0)\right], \tag{2.127}$$

where $\gamma_* = \gamma_0 + \psi_*$ is the aspect angle of the middle of the two-dimensional (surface) target resolution element; and $\varepsilon_0 > 0$ if the altitude is increased with moving radar. Here, we can assume that the differentials $\Delta\varphi$ and $\Delta\psi$ are approximately the same for all scatterers within the resolution element, as in Section 2.4.2.

On the basis of the preceding observations, and taking Equation (2.95) into consideration, the differential $\Delta\rho$ (displacement), which is an argument of the function $\boldsymbol{P}(t)$, can be thought of as identical for all scatterers and has the following form:

$$\Delta\rho_* = \Delta\ell(\cos\varepsilon_0 \cos\beta_0 \cos\gamma_* - \sin\varepsilon_0 \sin\gamma_*). \tag{2.128}$$

We cannot neglect the dependence of the differential $\Delta\rho$ on the coordinates φ and ψ of scatterers in the exponential function $e^{-\frac{2j\omega_0\Delta\rho}{c}}$ in Equation (2.127) because differences in phase changes of various scatterers (differences in the Doppler frequency) are the main source of fluctuations with moving radar, as a rule.

Equation (2.122) can be simplified if we assume that the approximate equalities $S^\circ(\beta_0, \psi + \gamma_0) \approx S^\circ(\beta_0, \gamma_*)$ and $\sin(\psi + \gamma_*) \approx \sin\gamma_*$ are true within the two-dimensional (surface) target resolution element.

Unlike the case of the three-dimensional (space) target, the normalized correlation function of target return signal fluctuations given by Equation (2.122) cannot be expressed as the product of the normalized correlation functions of the intraperiod and interperiod fluctuations. However, using the fact that the interval of variations of the angle ψ within the duration of the pulsed searching signal is infinitesimal, we can determine this normalized correlation function as the product of two normalized correlation functions with alternative physical meanings.

Under scanning of the two-dimensional (surface) target by short-duration pulsed searching signals, so that the angle dimension of the resolution element on the plane γ is satisfied by the condition $\Delta_p << \Delta_v$ (see Figure 2.9), variables of the function $g(\varphi, \psi)$ can always be separated. So, we can write

$$g(\varphi, \psi) = g(\varphi, \psi + \kappa) \cong g_h(\varphi, \psi_*) \cdot g_v(\psi_* + \kappa) = g_h(\varphi, \psi_*) \cdot g_v(\psi), \tag{2.129}$$

where $\kappa = \psi - \psi_*$ is the angle on the plane ψ, which is determined within the resolution element, and the origin of the angle κ is the mean of the angle ψ^*. Thus, the rigorous condition given by Equation (2.127) need not be satisfied. So, we can write

$$\Delta\rho(\varphi, \psi) \cong \Delta\rho_\beta(\varphi) + \Delta\rho_\gamma(\psi), \tag{2.130}$$

where

$$\Delta\rho_\beta(\varphi) = \Delta\ell\left[\cos\varepsilon_0 \cos\left(\beta_0 + \tfrac{\varphi}{\cos\gamma_*}\right)\cos\gamma_* - \sin\varepsilon_0 \sin\gamma_*\right], \tag{2.131}$$

and

$$\Delta\rho_\gamma(\psi) = -\Delta\ell(\psi - \psi_*)\,(\cos\varepsilon_0 \cos\beta_0 \sin\gamma_* + \sin\varepsilon_0 \cos\gamma_*). \qquad (2.132)$$

Using Equation (2.129) and Equation (2.130), on the basis of Equation (2.122) we can write

$$\begin{aligned} R^{en}_{\Delta\rho,\Delta\varphi,\Delta\psi}(t,\tau) &= R^{en}(\Delta\ell,\Delta\beta_0,\Delta\gamma_0,\tau,t) = p(t)\cdot R^{en}(\Delta\ell,\Delta\beta_0,\Delta\gamma_0,\tau,t) = \\ &= p(t)\cdot R_\beta(\Delta\ell,\Delta\beta_0)\cdot R_\gamma(\Delta\ell,\Delta\gamma_0,\tau,t) \end{aligned}, \qquad (2.133)$$

where

$$R_\beta(\Delta\ell,\Delta\beta_0) = \mathrm{N}\int g_h(\varphi - 0.5\Delta\varphi,\psi_*)\cdot g_h(\varphi + 0.5\Delta\varphi,\psi_*)\cdot e^{-2j\omega_0\cdot\frac{\Delta\rho_\beta(\varphi)}{c}}\,d\varphi\,; \qquad (2.134)$$

$$\begin{aligned} R_\gamma(\Delta\ell,\Delta\gamma_0,\tau,t) = \mathrm{N}\sum_{n=0}^{\infty}\int & \boldsymbol{P}\left[t - \tfrac{2\rho(\varphi)}{c} - 0.5(\tau - nT_p) + \tfrac{\Delta\rho^*}{c}\right] \\ & \times \boldsymbol{P}^*\left[t - \tfrac{2\rho(\varphi)}{c} + 0.5(\tau - nT_p) - \tfrac{\Delta\rho^*}{c}\right] \\ & \times g_v(\psi - 0.5\Delta\psi)\cdot g'_v(\psi + 0.5\Delta\psi)\cdot e^{-2j\omega_0\cdot\frac{\Delta\rho(\psi)}{c}}\,d\psi \end{aligned} \qquad (2.135)$$

$$p(t) = p_0 S^\circ(\beta_0,\gamma_*)\sin\gamma_*\int(\varphi,\psi_*)\,d\varphi\cdot\int \Pi^2\left[t - \tfrac{2\rho(\psi)}{c}\right]\cdot g_v^2(\psi)\,d\psi\,, \qquad (2.136)$$

and the distance ρ is functionally related to the angle ψ by Equation (2.118).

Here, the space normalized correlation function $R_\beta(\Delta\ell, \Delta\beta_0)$ defines the slow fluctuations caused by the moving radar with varying elementary signal phase changes $\Delta\rho_\beta(\varphi)$ in the azimuth plane and by the rotation $\Delta\beta_0$ of the radar antenna directional diagram axis. The space–time normalized correlation function $R_\gamma(\Delta\ell, \Delta\gamma_0, t, \tau)$ defines the slow fluctuations caused by the moving radar with varying elementary signal phase changes $\Delta\rho_\gamma(\psi)$ in the plane with the aspect angle γ, plane and rotation $\Delta\gamma_0$ of the radar antenna axis, and the rapid fluctuations caused by propagation of the pulsed searching signal along the two-dimensional (surface) target. We will call the correlation function $R_\beta(\Delta\ell, \Delta\beta_0)$ as the azimuth-normalized correlation function of the target return signal fluctuations and the correlation function $R_\gamma(\Delta\ell, \Delta\gamma_0, t, \tau)$ as the aspect-angle-normalized correlation function of the target return signal fluctuations.

2.5.3 The Target Return Signal Power with the Pulsed Searching Signal

The target return signal power as a function of time is given by Equation (2.136). Taking into consideration that

$$\int g_h^2(\varphi, \psi_*)\, d\varphi = \Delta_h^{(2)}(\varphi_*) = \Delta_{h*}^{(2)} \tag{2.137}$$

and using the generalized theorem about the mean, we can write

$$p = \frac{P_S G_0^2 \lambda^2 \Delta_{h*}^{(2)} \Delta_p g_v^2(\psi_*) S^\circ(\beta_0, \gamma_*) \sin\gamma_*}{64\,\pi^3 h^2} = \frac{P_S G_0^2 \lambda^2 k_a^h \Delta_{h*} k_p c\tau_p g_v^2(\gamma_* - \gamma_0) S^\circ(\beta_0, \gamma_*)}{128\,\pi^3 \rho_*^3 \cos\gamma_*}, \tag{2.138}$$

where

$$\rho_* = \frac{h}{\sin\gamma_*} = \frac{h}{\sin(\gamma_0 + \psi_*)} = 0.5\,ct; \tag{2.139}$$

$$\Delta_p = \int \Pi^2\left[t - \tfrac{2\rho(\psi)}{c}\right] d\psi = \frac{k_p c\tau_p \sin^2\gamma_*}{2h\cos\gamma_*} = \frac{k_p c\tau_p \operatorname{tg}\gamma_*}{2\rho_*} \tag{2.140}$$

is the interval of variations of the angle ψ or γ within the effective duration of the pulsed searching signal squared envelope

$$\tau_p^{(2)} = k_p \tau_p; \tag{2.141}$$

τ_p is the effective duration of the pulsed searching signal and k_p is the coefficient of the pulsed searching signal shape. Under the condition $\gamma_* \approx 90^\circ$ the equality in Equation (2.141) is not true. In the determination of Δ_p, we used the following approximation

$$t - \frac{2\rho}{c} = \frac{2(\rho_* - \rho)}{c} \approx \frac{2\rho(\psi - \psi_*)}{c} \cdot \operatorname{ctg}\gamma_*. \tag{2.142}$$

Comparing Equation (2.107) and Equation (2.138), we can define the effective scattering area of the two-dimensional (surface) target as follows:

$$S_t^{surface} = g_v^2(\gamma_* - \gamma_0) \cdot S^\circ(\beta_0, \gamma_*) \cdot S_{area}, \tag{2.143}$$

where

$$S_{area} = \frac{k_a^h \Delta_{h_*} \rho_* k_p c \tau_p}{2\cos\gamma_*} \tag{2.144}$$

is the geometrical area of the two-dimensional (surface) target resolution element.

Thus, under scanning of the two-dimensional (surface) target by short-duration pulsed searching signals, i.e., when the condition $\Delta_p << \Delta_v^{(2)}$ is satisfied, the target return signal power at the instant of time t, for the distance ρ (the radar range)

$$\rho_* = 0.5\, ct, \tag{2.145}$$

is proportional to the amplification factor of the vertical-coverage radar antenna directional diagram under the condition $\psi_* = \gamma_* - \gamma_0$, and with the fixed angles γ_* and γ_0 is inversely proportional to ρ_*^3. This can be explained by the fact that the area of a two-dimensional (surface) target resolution element is proportional to ρ (the radar range). Dependences of the received target return signal power on the radar range are shown in Figure 2.10 when the directional diagram is considered Gaussian and $S° = const$. The obtained results are true for a wide range of angles, including angles that are very close to 90°.

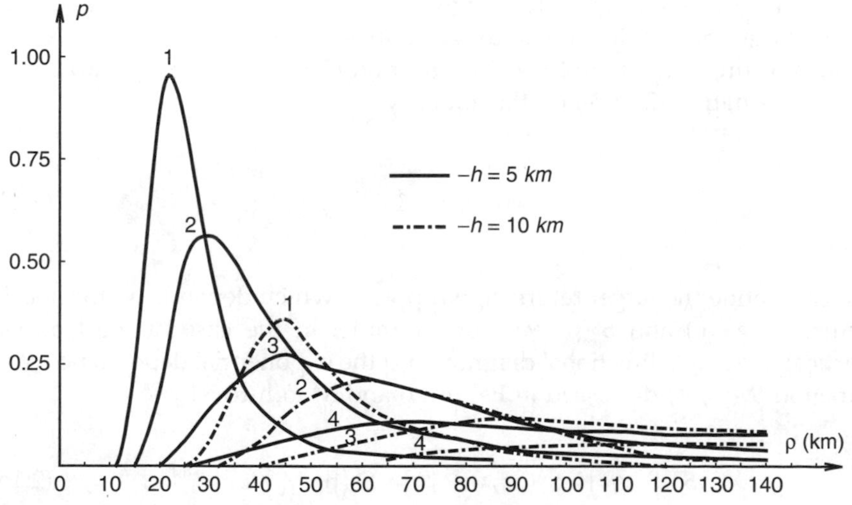

FIGURE 2.10
The target return signal power as a function of radar range: (1) $\gamma_0 = 12°$; (2) $\gamma_0 = 8°$; (3) $\gamma_0 = 4°$; (4) $\gamma_0 = 0°$.

During the process of propagation of the pulsed searching signal, the power $p(t)$ of the target return signal is a very complex function of time $t = \frac{2\rho_*}{c}$. This is one source of the nonstationary state of the target return signal at the receiver or detector input in navigational systems. It is not difficult to define exactly the target return signal power on the basis of Equation (2.138). For this purpose, it is necessary to introduce the $(\gamma_*) = \arcsin \frac{2h}{ct}$. But in many cases, the simple approximation is sufficient; when the vertical-coverage directional diagram is not large, i.e., when the condition $\sin \Delta_v \approx \Delta_v$ is satisfied, and when the angle γ_0 is not small, i.e., the condition $\Delta_v \operatorname{ctg} \gamma_0 << 1$ is true, the linear function

$$\psi_* = \gamma_* - \gamma_0 = \frac{T_d - t}{T_d \operatorname{ctg} \gamma_0} \tag{2.146}$$

or

$$t = T_d(1 - \psi_* \operatorname{ctg} \gamma_0) \tag{2.147}$$

is true, where

$$T_d = \frac{2\rho_0}{c} = \frac{2h}{c \sin \gamma_0} \tag{2.148}$$

is the delay of the target return signal from the surface of the two-dimensional target on the directional diagram axis.[63,64]

Substituting Equation (2.147) in Equation (2.138) and taking into consideration Equation (2.145) and the equality

$$\cos \gamma_* = \sqrt{1 - \frac{h^2}{\rho_*^2}}, \tag{2.149}$$

we can define the target return signal power, which depends on the specific functions $g_h(\psi_*)$ and $S^\circ(\beta_0, \gamma_*)$. For example, in the case of the Gaussian vertical-coverage directional diagram and the exponential dependence of the function $S^\circ(\beta_0, \gamma_*)$ discussed in Feldman and Mandurovsky[27]

$$S^\circ(\beta_0, \gamma_*) = S^\circ(\beta_0, \psi + \gamma_0) = S^\circ(\beta_0, \gamma_0) \cdot e^{k_1(\beta_0, \gamma_0)\psi}, \tag{2.150}$$

where k_1 is a coefficient defined by the landscape and the angles β_0 and γ_0, the target return signal power has the following form:

$$p(t) = p(T_d)\cdot\frac{T_d^3}{t^3}\cdot\frac{e^{-\pi\cdot\frac{(t-T_d)^2}{T_\tau^2} - k_1\Delta_h\cdot\frac{t-T_d}{\sqrt{2}T_\tau}}}{\sqrt{1-\frac{T_d^2}{t^2}\cdot\sin^2\gamma_0}}, \qquad (2.151)$$

where

$$T_\tau = \frac{\Delta_h}{\sqrt{2}}\cdot T_d \operatorname{ctg}\gamma_0 \qquad (2.152)$$

is the duration of the target return signal, which is defined as the difference of signal delays from the points of the vertical-coverage directional diagram corresponding to the angles $\gamma_0 \pm 0.5\Delta_h^{(2)}$, and

$$p(T_d) = \frac{P_S G_0^2 \lambda^2 \Delta_h c\tau_p k_a^h k_p S^\circ(\beta_0, \gamma_0)}{128\,\pi^3 \rho_0^3 \cos\gamma_0}. \qquad (2.153)$$

The target return signal power given by Equation (2.151) has the shape of the vertical-coverage directional diagram, but is deformed under the influence of the functions $S^\circ(\psi)$ and ρ^{-3}, and its maximum is shifted. When Equation (2.146) and Equation (2.147) are true, the value of $\frac{T_d}{t}$ is not so high, and we can write

$$p(t) \approx p_1(T_d)\cdot e^{-\pi\cdot\frac{(t-T_d')^2}{T_\tau^2}}, \qquad (2.154)$$

where

$$p_1(T_d) = p(T_d)\cdot e^{\frac{Q^2\Delta_h^{(2)}}{8\pi}}; \qquad (2.155)$$

$$T_d' = T_d\left(1 - \frac{Q\Delta_h^{(2)}}{4\pi}\cdot\operatorname{ctg}\gamma_0\right); \qquad (2.156)$$

$$Q = k_1 + 3\operatorname{ctg}\gamma_0 + \operatorname{tg}\gamma_0. \qquad (2.157)$$

The relative shift in maximum of the target return signal power is determined as follows:

$$\delta_p = \frac{T_d - T_d'}{T_d} = \frac{\Delta_v^{(2)}}{4\pi}\cdot(k_1 + 3\operatorname{ctg}\gamma_0 + \operatorname{tg}\gamma_0)\cdot\operatorname{ctg}\gamma_0. \qquad (2.158)$$

TABLE 2.1

Various Kinds of Surfaces of the Two-Dimensional Target

Surface	k_1	Note
Forest	0.62	—
Plow and dry snow	3.3	$\lambda = 3$ cm
Dry sand	5	$\gamma_0 = 50 \ldots 80°$
Ice	6.5	
Sea under conditions of wind velocity:		
48 km/h	6.7	—
18 … 28 km/h	10.7	—
0 … 11 km/h	16	—

The shift δ_p of the target return signal power can be defined on the basis of Equation (2.138) and Equation (2.139) under the condition $p'(\gamma_*) = 0$, when the maximum of the function $p(\gamma_*)$ is known. The first term in Equation (2.158) is defined by the function $S°(\gamma)$. The second and third terms in Equation (2.158) are defined by the radar range. When the surface of the two-dimensional target is smooth (see Table 2.1) and the condition $k_1 > 3 \text{ ctg } \gamma_0 + \text{tg } \gamma_0$ is satisfied, the main reason for the shift in the maximum target return signal power is the function $S°(\gamma)$. When the surface of the two-dimensional target is very rough, the main reason for the shift in maximum power is a function of the radar range. For example, at $\gamma_0 = 45°$, $\Delta_h = 120°$, and $k_1 = 16$, which corresponds to the quiescent surface of the sea, we obtain $\delta_p = (5 + 1 + 0.3)\%$ and at $k_1 = 0.6$, which corresponds to the forest, we obtain $\delta_p = (0.2 + 1 + 0.3)\%$. When the directional diagram is Gaussian and the function $S°(\gamma)$ is given by Equation (2.150), the target return signal power can be defined on the basis of Equation (2.124) without assuming that the pulsed searching signal has a short duration in comparison with T_r.

Consider the condition $\tau_p >> T_r$. In this case, the target return signal has a shape that is very close to a square waveform. The duration of the target return signal is equal to τ_p. The durations of the leading and trailing edges of the target return signal are very close to T_r. The maximum target return signal power, when the condition $\sin\gamma_0 \approx \sin(\psi + \gamma_0)$ is satisfied, is determined as follows:

$$P_{\max} = p\left(T_d''\right) = \frac{P_S G_0^2 \lambda^2 \Delta_h \Delta_v S°\left(\beta_0, \gamma_0\right) \sin\gamma_0}{128\,\pi^3 h^2} \cdot e^{\frac{k_1^2 \Delta_v^{(2)}}{8\pi}}, \tag{2.159}$$

where

$$T_d'' = T_d\left(1 - \tfrac{k_1 \Delta_v^{(2)}}{4\pi} \cdot \text{ctg } \gamma_0\right). \tag{2.160}$$

Under the conditions $h = variable$ and $\gamma_0 = const$, the maximum target return signal power is inversely proportional to h^2 or ρ^2. When the conditions $h = const$ and $\gamma_0 = variable$ are satisfied, the maximum target return signal power is inversely proportional to ρ. The error in the definition of the target return signal power caused by the functional dependence in Equation (2.151) is usually not high for high values of k_1. For example, at $k_1 = 16$ and $\Delta_v = 6°$, this error is approximately equal to 10%.

It is not difficult to define the target return signal power for arbitrary values τ_p and T_r if the conditions in Equation (2.142) and Equation (2.150) are satisfied and the vertical-coverage directional diagram and the pulsed searching signal are Gaussian. Neglecting Equation (2.150) for simplicity, we can write

$$p = \frac{P_S G_0^2 \lambda^2 \Delta_h g_v^2(\psi_*) S^\circ(\beta_0, \gamma_0) \sin\gamma_0}{128\,\pi^3 h^2} \cdot I\,, \tag{2.161}$$

where

$$I = \frac{\Delta_v \Delta_p}{\sqrt{\Delta_v^{(2)} + \Delta_p^{(2)}}} \cdot e^{-2\pi \cdot \frac{\psi_*}{\Delta_v^{(2)} + \Delta_p^{(2)}}}\,. \tag{2.162}$$

Then, in the case of the short-duration pulsed searching signal, i.e., when the condition $\Delta_p << \Delta_v$ is satisfied, we can write

$$I = \Delta_p \cdot e^{-2\pi \cdot \frac{\psi_*^2}{\Delta_v^{(2)}}} \tag{2.163}$$

and Equation (2.161) is transformed to Equation (2.138). In the case of the long-duration pulsed searching signal, i.e., when the condition $\Delta_p >> \Delta_v$ is satisfied, we can assume that $\gamma_* = \gamma_0$ or $\psi_* = 0$ and $I = \Delta_v$. In this case, Equation (2.161) is transformed to Equation (2.159), taking into consideration the condition $k_1 = 0$.

Thus, in the case of the short-duration pulsed searching signal, the target return signal power is inversely proportional to ρ_*^3 where ρ_* is the distance between the radar and the center of a two-dimensional (surface) target resolution element [see Equation (2.138)]. In the case of the long-duration pulsed searching signal, when the signal completely covers the scanned surface of the two-dimensional target, the target return signal power is inversely proportional to ρ_0 where ρ_0 is the distance between the radar and center of the scanned surface, since [see Equation (2.159)]

$$\sin\gamma_0 = \frac{h}{\rho_0}. \tag{2.164}$$

2.5.4 The Correlation Function and Power of the Target Return Signal with the Simple Harmonic Searching Signal

Assuming that the condition given by Equation (2.109) is satisfied in Equation (2.122) and omitting the summation sign, we can write

$$R^{en}_{\Delta\rho,\Delta\varphi,\Delta\psi}(t,\tau) = R^{en}(\Delta\ell, \Delta\beta_0, \Delta\gamma_0) = p_0 \iint g(\varphi - 0.5\Delta\varphi, \psi - 0.5\Delta\psi) \times g(\varphi + 0.5\Delta\varphi, \psi + 0.5\Delta\psi) \times S^\circ(\beta_0, \psi + \gamma_0)\sin(\psi + \gamma_0)\cdot e^{-2j\omega_0 \cdot \frac{\Delta\rho(\varphi,\psi)}{c}}\, d\varphi\, d\psi \tag{2.165}$$

where the differentials are given by Equation (2.125)–Equation (2.127), respectively, and it is necessary to replace γ_* with γ_0 and ρ_* with ρ_0, respectively, using Equation (2.164). Equation (2.165), as well as Equation (2.133), can be represented as the product of the azimuth and aspect angle normalized correlation functions of target return signal fluctuations if the condition in Equation (2.97) is satisfied. If the condition $\Delta\rho = \Delta\varphi = \Delta\psi = 0$ is satisfied in Equation (2.165), we can define the target return signal power in the following form:

$$p(t) = p_0 \iint g^2(\varphi,\psi)\cdot S^\circ(\beta_0, \psi + \gamma_0)\sin(\psi + \gamma_0)\, d\varphi\, d\psi\,. \tag{2.166}$$

If the radar antenna directional diagram width is not so large and the variables in the function $g(\varphi, \psi)$ are separated, then when the conditions $S^\circ(\beta_0, \psi + \gamma_0) \approx S^\circ(\beta_0, \gamma_*)$ and $\sin(\psi + \gamma) \approx \sin\gamma_*$ are satisfied, we can write

$$p = \frac{P_S G_0^2 \lambda^2 \Delta_h^{(2)} \Delta_v^{(2)} S^\circ(\beta_0, \gamma_0)\sin\gamma_0}{64\,\pi^3 h^2}. \tag{2.167}$$

Comparing Equation (2.167) and Equation (2.107), we can define the effective scattering area of the two-dimensional (surface) target in the following form:

$$S_t^{surface} = S^\circ(\beta_0, \gamma_0)\cdot S^{surface}, \tag{2.168}$$

where

$$S^{surface} = \frac{\Delta_h^{(2)} \Delta_v^{(2)} \rho_0^2}{\sin\gamma_0} \tag{2.169}$$

is the surface covered by the directional diagram. Equation (2.167) is equivalent to Equation (2.159). Equation (2.167) is true within the wide range of variation of the angle γ_0 except for very small values of the angle γ_0, when the top edge of the directional diagram breaks away from the surface of the two-dimensional target. Equation (2.167) can also be used for high values of the directional diagram width, but the directional diagram should be uniformly contrasting.

2.6 The Correlation Function under Vertical Scanning of the Two-Dimensional (Surface) Target

Under scanning of the two-dimensional (surface) target at angles that are very close to 90° it is convenient to introduce a new coordinate system, in which the position of some scatterer *D* on the surface is given by the angles α and θ (see Figure 2.11). Assuming that the deviation θ_0 of the radar antenna directional diagram axis from the vertical line and the directional diagram width are infinitesimal, we can use the approximate equality: $\sin\theta \approx \theta$.[65,66] The origin of the coordinate system *OXYZ* is matched with a view of the phase center of the radar antenna in navigational systems so that $z = h$, and the axes *OX* and *OY* are directed in parallel to orthogonal straight lines that are formed at the intersection of the surface by the main planes φ and ψ of the two-dimensional directional diagram. Under these conditions, we can write

$$\varphi = \theta \cos \alpha - \varphi_0 \quad \text{and} \quad \psi = \theta \sin \alpha - \varphi_0; \tag{2.170}$$

$$\rho = \frac{h}{\cos\theta} \approx h\left(1 + 0.5\theta^2\right), \tag{2.171}$$

where

$$\varphi_0 = \theta_0 \cos \alpha_0; \ \psi_0 = \theta_0 \sin \alpha_0 \quad \text{and} \quad d\varphi d\psi = \sin\theta \, d\theta \, d\alpha. \tag{2.172}$$

Let us assume that the radar moves uniformly and linearly with velocity *V* the direction of which is given by the azimuth angle β_0 in the horizontal plane with respect to the axis *OX* and the trajectory angle ε_0 in the vertical plane.[67] Then

$$\Delta\rho(\alpha, \theta) = -V_r(\alpha, \theta)\,\tau = -V \cdot [\cos \varepsilon_0 \cos(\alpha - \beta_0) \sin \theta + \sin \varepsilon_0 \cos \theta]\tau. \tag{2.173}$$

The product of the pulsed integrands in Equation (2.122) for the given instant of time *t* defines the interval of integration with respect to the variable

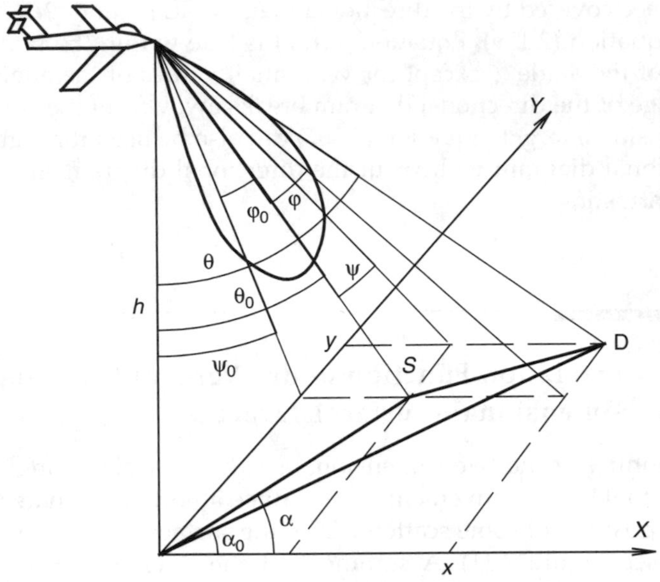

FIGURE 2.11
The coordinate system using the variables α and θ.

θ. With square waveform pulsed searching signals of duration τ_p as follows from the condition of overlapping

$$-0.5\left(\tau_p-\tau\right)\le t-\frac{2\rho}{c}\le 0.5\left(\tau_p-\tau\right), \tag{2.174}$$

the limits of integration can be determined as follows

$$\theta_{1,2}=\sqrt{\theta_*^2\mp\left(\tau_p-\mu\tau+nT_p\right)\cdot\frac{c}{2h}}=\sqrt{\frac{2\left(t-T_d\right)\mp\left(\tau_p-\mu\tau+nT_p\right)}{T_d}}, \tag{2.175}$$

where

$$\theta_*^2\approx\frac{2\left(\rho_*-h\right)}{h}\approx\frac{2\left(t-T_d\right)}{T_d} \tag{2.176}$$

and

$$\rho_*=\frac{h}{\cos\theta_*}\approx h\left(1+0.5\theta_*^2\right) \tag{2.177}$$

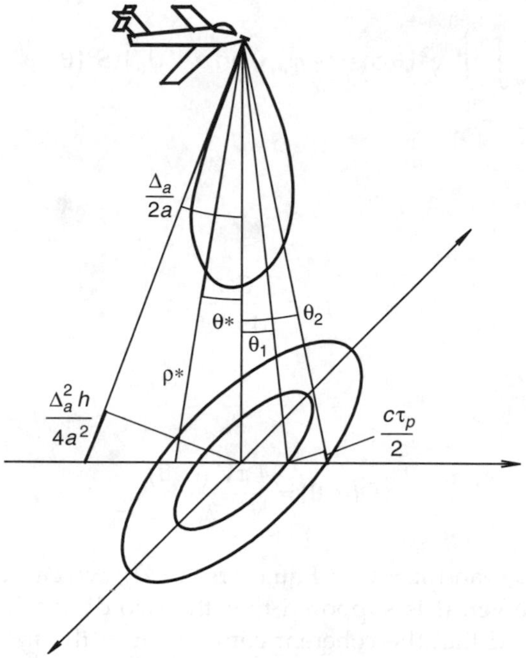

FIGURE 2.12
Space relationships under vertical scanning of surface.

are the angle and distance defining the position of the considered resolution element at the instant of time $t = \frac{2\rho_*}{c}$ (see Figure 2.12); $T_d = \frac{2h}{c}$ is the delay of the target return signal;

$$\mu = 1 + \frac{2V}{c} \cdot \cos\theta_* . \tag{2.178}$$

Clearly, the following condition

$$t - T_d \geq 0.5(\tau_p - \mu\tau + nT_p) \tag{2.179}$$

should be satisfied. For the instants of time that are within the interval

$$[T_d - 0.5(\tau_p - \mu\tau + nT_p),\ T_d + 0.5(\tau_p - \mu\tau + nT_p)], \tag{2.180}$$

the condition $\theta_1 = 0$ is true. If the condition

$$t < T_d - 0.5(\tau_p - \mu\tau + nT_p) \tag{2.181}$$

is satisfied, we can write $\theta_1 = \theta_2 = 0$.

Finally, on the basis of Equation (2.122) we can write

$$R^{en}(t,\tau)=p_0\sum_{n=0}^{\infty}\int_0^{2\pi}\int_{\theta_1(n)}^{\theta_2(n)} g^2(\theta\cos\alpha-\varphi_0,\theta\sin\alpha-\psi_0)\cdot S^{\circ}(\theta)\cdot e^{j\Omega(\alpha,\theta)\tau}\sin\theta\, d\theta\, d\alpha\,, \tag{2.182}$$

where

$$p_0=\frac{P_S G_0^2\lambda^2}{64\pi^3h^2} \tag{2.183}$$

and

$$\Omega(\alpha,\theta)=\frac{4\pi V_r(\alpha,\theta)}{\lambda}. \tag{2.184}$$

Assuming $\tau = 0$ and $n = 0$ in Equation (2.182), we can define the target return signal power. It is supposed that the two-dimensional target has a rough surface and that the coherent component of the target return signal is absent.[68] So,

$$p(t)=p_0\int_0^{2\pi}\int_{\theta_1}^{\theta_2} g^2(\theta\cos\alpha-\varphi_0,\theta\sin\alpha-\psi_0)\cdot S^{\circ}(\theta)\sin\theta\, d\theta\, d\alpha\,, \tag{2.185}$$

where

$$\theta_{1,2}=\sqrt{\theta_*^2\mp\frac{c\tau_p}{2h}}=\sqrt{\frac{2(t-T_d)\mp\tau_p}{T_d}}\,,\qquad t>T_d+0.5\tau_p\,. \tag{2.186}$$

Let us assume that the directional diagram has symmetric axes and obeys the Gaussian law. Then, the effective scattering area as a function of the angle θ is Gaussian too,[27] and

$$S^{\circ}(\theta)=S_N^{\circ}\cdot e^{-k_2\theta^2}\,, \tag{2.187}$$

where S_N° is the effective scattering area under vertical scanning. The function in Equation (2.188), when the angle θ is not so high in value, is equivalent to the function determined by:[69]

$$S^{\circ}(\theta)=S_N^{\circ}\cdot e^{-k_2 \mathrm{tg}^2\theta}\,. \tag{2.188}$$

The parameter k_2 characterizes the roughness of the surface of the two-dimensional target and increases as the roughness of the surface decreases. Then we can write[27]

$$p(t) = p_0 S_N^{\circ} \int_0^{2\pi} \int_{\theta_1}^{\theta_2} e^{-2\pi \cdot \frac{a^2\theta^2 + \theta_0^2 - 2\theta\theta_0 \cos\alpha}{\Delta_a^{(2)}}} \theta \, d\theta \, da \, , \qquad (2.189)$$

where

$$a^2 = 1 + \frac{k_2 \Delta_a^{(2)}}{2\pi} \qquad (2.190)$$

and Δ_a is the directional diagram width in navigational systems.

Integrating with respect to the variable a and introducing a new variable

$$\theta = \frac{\Delta_a}{2\sqrt{\pi\alpha}} \cdot x \qquad \text{or} \qquad x = \frac{2\sqrt{\pi a}}{\Delta_a} \cdot \theta \, , \qquad (2.191)$$

we can determine the target return signal power in the following form:

$$p(t) = \frac{P_S G_0^2 \lambda^2 \Delta_a^{(2)} S_N^{\circ}}{128 \pi^3 a^2 h^2} \cdot F(t) \, , \qquad (2.192)$$

where

$$F(t) = \int_{x_1(t)}^{x_2(t)} x \cdot e^{-0.5\left(x^2 + b^2\right)} I_0(bx) dx \, ; \qquad (2.193)$$

$$x_{1,2}(t) = 2\sqrt{\pi a} \cdot \frac{\theta_{1,2}}{\Delta_a} = \sqrt{2\pi \cdot \frac{t - T_d \mp 0.5\tau_p}{T_p^{con}}} \, , \qquad t > T_d + 0.5\tau_p \, ; \qquad (2.194)$$

$$b = 2\sqrt{\pi} \cdot \frac{\theta_0}{a\Delta_a} \, . \qquad (2.195)$$

One important point to remember is that

$$\text{if} \quad T_d - 0.5\tau_p < t < T_d + 0.5\tau_p, \quad \text{then} \quad x_1 = 0 \qquad (2.196)$$

and

$$\text{if} \quad t < T_d - 0.5\tau_p, \quad \text{then} \quad x_1 = x_2 = 0 \quad \text{and} \quad F(t) = 0. \tag{2.197}$$

Taking into consideration Equation (2.196) and Equation (2.197), we can write

$$T_p^{con} = \frac{\Delta_a^{(2)}}{4a^2} \cdot T_d = \frac{h\Delta_a^{(2)}}{2a^2 c}, \tag{2.198}$$

where T_p^{con} is the conditional duration of the target return signal, equal to the time during which the pulse leading edge is propagated along a surface from the first point of tangency $t = T_d - 0.5\tau_p$ to the circle observed under the angle $\frac{\Delta_a}{a}$ (see Figure 2.13 and Figure 2.14); $\frac{\Delta_a}{a}$ is the width of the equivalent directional diagram that is more narrow because of the function $S^\circ(\theta)$ given by Equation (2.187); a is the coefficient of narrowing given by Equation (2.190). Changes in the conditional duration T_p^{con} of the target return signal given by Equation (2.198) can occur in the case of two-dimensional targets with smooth surfaces. For example, at $\Delta_a = 12°$ and $k_2 = 200$ — the case of weak sea waves — T_p^{con} is decreased 2.5 times.[27,69]

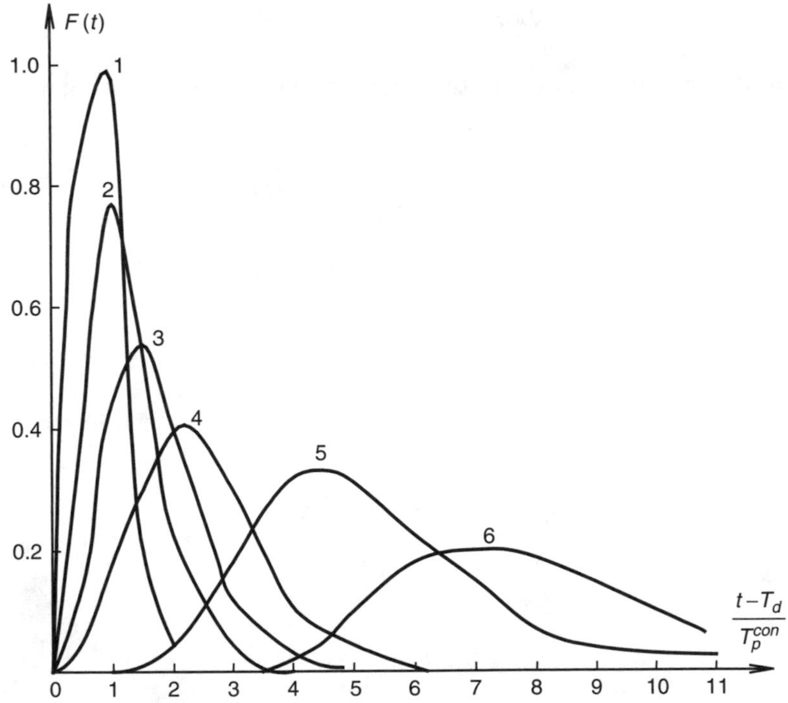

FIGURE 2.13
The function $F(t)$ vs. the ratio $\frac{\theta_0}{\Delta_a}$ at $\frac{\tau_p}{T_p^{con}} = 2$: (1) $\frac{\theta_0}{\Delta_a} = 0$; (2) $\frac{\theta_0}{\Delta_a} = 0.5$; (3) $\frac{\theta_0}{\Delta_a} = 0.75$; (4) $\frac{\theta_0}{\Delta_a} = 1$; (5) $\frac{\theta_0}{\Delta_a} = 1.5$; (6) $\frac{\theta_0}{\Delta_a} = 2$.

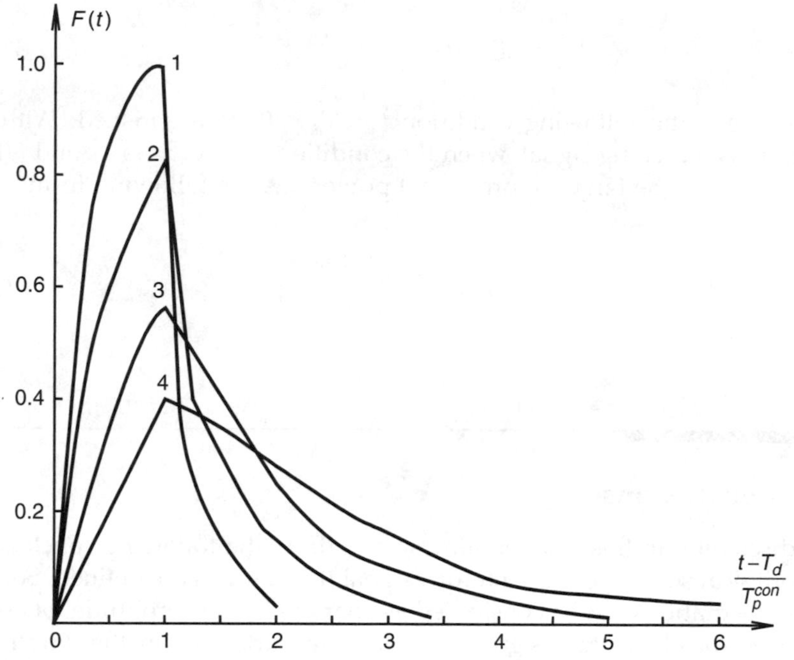

FIGURE 2.14
The function $F(t)$ vs. the ratio $\frac{\tau_p}{T_p^{con}}$ at $\theta_0 = 0$: (1) $\frac{\tau_p}{T_p^{con}} = 2$; (2) $\frac{\tau_p}{T_p^{con}} = 1$; (3) $\frac{\tau_p}{T_p^{con}} = 0.5$; (4) $\frac{\tau_p}{T_p^{con}} = 0.33$.

The integral in Equation (2.193) can be determined using the incomplete Toronto function:[27]

$$F(t) = \boldsymbol{T}_{\frac{x_2}{\sqrt{2}}}\left(1, 0, \frac{b}{\sqrt{2}}\right) - \boldsymbol{T}_{\frac{x_1}{\sqrt{2}}}\left(1, 0, \frac{b}{\sqrt{2}}\right). \tag{2.199}$$

In the particular case where $\theta_0 = 0$,[27] we can write

$$1 - e^{-\frac{x_2^2}{\sqrt{2}}} = 1 - e^{-\pi \cdot \frac{t - T_d + 0.5\tau_p}{T_p^{com}}} \tag{2.200}$$

if the condition in Equation (2.196) is satisfied and

$$F(t) = e^{-0.5x_1^2} - e^{-0.5x_2^2} = e^{-\pi \cdot \frac{t - T_d - 0.5\tau_p}{T_p^{com}}} - e^{-\pi \cdot \frac{t - T_d + 0.5\tau_p}{T_p^{com}}} \tag{2.201}$$

if the condition $t > T_d + 0.5\tau_p$ is true. The function $F(t)$ the shape of the target return signal from the two-dimensional (surface) target during the time interval $t - T_d$. The function $F(t)$ is shown in Figure 2.13 and Figure 2.14 for some values of $\frac{\theta_0}{\Delta_a}$ and $\frac{\tau_p}{T_p^{con}}$. In particular, for $\theta_0 = 0$, the maximum of the function $F(t)$ is determined by

$$F_{\max}(t) = 1 - e^{-\pi \cdot \frac{\tau_p}{T_p^{com}}} \tag{2.202}$$

and exists if the following condition $t = T_d + 0.5\tau_p$ is satisfied. With the continuous searching signal, when the conditions $x_1 = 0$, $x_2 \to \infty$, and $F(t) \to 1$ are satisfied, the target return signal power has the following form:

$$p(t) = \frac{P_S G_0^2 \lambda^2 \Delta_a^{(2)} S_N^o}{128\, \pi^3 a^2 h^2} \ . \tag{2.203}$$

2.7 Conclusions

The discussion in this chapter allows us to draw the following conclusions. The physical sources of target return signal fluctuations are defined. Sources of the fluctuations can be considered as changes in the amplitude, phase, or frequency of elementary signals that lead to variations in the amplitude, phase, or frequency of the resulting target return signal. The amplitude, phase, or frequency of elementary signals can be changed as a consequence of: displacements and rotation of scatterers under the stimulus of the wind; high angle dimensions of the two-dimensional (surface) or three-dimensional (space) target; Doppler frequencies and the secondary Doppler effect; antenna scanning or rotation of the scanning polarization plane of radar antenna; and the nonstationary state of the searching signal frequency. The following are forms of target return signal fluctuations: fluctuations in the radar range; interperiod fluctuations; intraperiod fluctuations; slow fluctuations; rapid fluctuations; time fluctuations; and space fluctuations. The target return signal can be characterized by the normalized correlation function of time and space fluctuations. The two-dimensional spectral power density can define the nonstationary target return signal. The one-dimensional spectral power density defines the stationary target return signal. In the case of the narrow-band searching signal, the correlation function of the target return signal fluctuations is the correlation function of the slow fluctuations with the continuous searching signal. In other words, this is the correlation function of the fluctuations caused by the moving radar, antenna scanning, and displacements of scatterers — i.e., the space fluctuations. With the pulsed searching signal, the correlation function of target return signal fluctuations is defined both by the fluctuations in the radar range — the intraperiod fluctuations — and by the interperiod fluctuations.

The generalized normalized correlation function of target return signal fluctuations under scanning of the three-dimensional (space) target is defined by the product of two normalized correlation functions. The first normalized correlation function defines the slow fluctuations caused by the moving radar, displacements of scatterers, and antenna scanning and the rapid fluctuations caused by propagation of the searching signal in space. The second normalized correlation function defines the slow target return signal fluctuations caused by rotation of scatterers and the radar antenna polarization plane. In the case of the pulsed searching signal, the normalized correlation function of the fluctuations that are caused by the moving radar and antenna scanning, too, is defined by the product of two normalized correlation functions. The first normalized correlation function defines the slow fluctuations caused by the different Doppler shifts in the frequency of elementary signals within the pulsed searching signal resolution area with moving radar, and amplitude changes in elementary signals caused by antenna scanning, and variations of the aspect angle during radar motion — the interperiod fluctuations. The second normalized correlation function is a periodic function with respect to τ and defines the rapid fluctuations caused by propagation of the pulsed searching signal in space with respect to simultaneous radar motion — i.e., the intraperiod fluctuations.

Under scanning of the three-dimensional (space) target as well as a point target by the pulsed searching signal, the target return signal power is inversely proportional to ρ^2, where ρ is the radar range, but not inversely proportional to ρ^4. This can be explained by the fact that with an increase in the radar range, the area of the pulsed searching signal and the effective scattering area increase proportionally to ρ^2. Under angle scanning of the two-dimensional (surface) target by the pulsed searching signal, the normalized correlation function of the fluctuations is defined by the product of the azimuth-normalized correlation function $R_\beta(\Delta\ell, \Delta\beta_0)$ and the aspect-angle-normalized correlation function $R_\gamma(\Delta\ell, \Delta\gamma_0, t, \tau)$. The azimuth-normalized correlation function defines the slow fluctuations caused by the moving radar with varying phase changes in elementary signals in the azimuth plane and by rotation of the radar antenna axis. The aspect-angle-normalized (or space–time) correlation function defines the slow fluctuations caused by the moving radar with varying phase changes in elementary signals in the aspect angle plane, and by rotation of the radar antenna axis, and defines the rapid fluctuations caused by propagation of the pulsed searching signal along the two-dimensional (surface) target. Under vertical scanning of the two-dimensional (surface) target by the pulsed searching signal, the normalized correlation function of the fluctuations and power of the target return signal are defined on the basis of the Toronto function. The main assertions that are true under angle scanning of the two-dimensional (surface) target are also true under vertical scanning.

References

1. Feldman, Yu, Gidaspov, Yu, and Gomzin, V., *Moving Target Tracking*, Soviet Radio, Moscow, 1978 (in Russian).
2. Bacut, P. et al., *Problems of Statistical Radar*, Soviet Radio, Moscow, 1963 (in Russian).
3. Dulevich, V. et al., *Theoretical Foundations of Radar*, Soviet Radio, Moscow, 1978 (in Russian).
4. Ferrara, E. and Parks, T., Direction finding with an array of antennas having diverse polarizations, *IEEE Trans.*, Vol. AP-31, No. 3, 1983, pp. 231–236.
5. Roy, R. and Kailath, T., ESPRIT — Estimation of signal parameters via rotational invariance techniques, *IEEE Trans.*, Vol. ASSP-37, No. 7, 1989, pp. 984–995.
6. Wong, K. and Zoltowski, M., High accuracy 2D angle estimation with extended aperture vector sensor arrays, in *Proceedings of the ICASSP*, Vol. 5, May 1996, pp. 2789–2792.
7. Kolchinsky, V., Mandurovsky, I., and Konstantinovsky, M., *Doppler Devices and Navigational Systems*, Soviet Radio, Moscow, 1975 (in Russian).
8. Li, J., Direction and polarization estimation using arrays with small loops and short dipoles, *IEEE Trans.*, Vol. AP-41, No. 3, 1993, pp. 379–387.
9. Winitzky, A., *Basis of Radar under Continuous Generation of Radio Waves*, Soviet Radio, Moscow, 1961 (in Russian).
10. Farina, A., Gini, F., Greco, M., and Lee, P., Improvement factor for real-sea clutter Doppler frequency spectra, in *Proc. Inst. Elect. Eng. F*, Vol. 123, October, 1996, pp. 341–344.
11. Cirban, H. and Tsatsanis, M., Maximum likelihood blind channel estimation in the presence of Doppler shifts, *IEEE Trans.*, Vol. SP-47, No. 5, 1999, pp. 1559–1569.
12. Rytov, S., *Introduction to Statistical Radio Physics. Part I: Stochastic Processes*, Nauka, Moscow, 1976 (in Russian).
13. Blackman, S., *Multiple-Target Tracking with Radar Applications*, Artech House, Norwood, MA, 1986.
14. Gardner, W., *Introduction to Random Processes with Application to Signals and Systems*, 2nd ed., McGraw-Hill, New York, 1989.
15. Richaczek, A., *Principles of High-Resolution Radar*, Peninsula, San Francisco, CA, 1985.
16. Abarband, H., *Analysis of Observed Chaotic Data*, Springer-Verlag, New York, 1996.
17. Zukovsky, A., Onoprienko, E., and Chizov, V., *Theoretical Foundations of Radio Altimetry*, Soviet Radio, Moscow, 1979 (in Russian).
18. Verdu, S., *Multiuser Detection*, Cambridge University Press, Cambridge, U.K., 1988.
19. Rappaport, T., *Wireless Communications: Principles and Practice*, Prentice Hall, Englewood Cliffs, NJ, 1996.
20. Papoulis, L., *Signal Analysis*, McGraw-Hill, New York, 1977.
21. Tikhonov, V., *Statistical Radio Engineering*, Radio and Svyaz, Moscow, 1982 (in Russian).
22. Yaglom, A., *Correlation Theory of Stationary Stochastic Functions*, Hidrometeoizdat, Saint Petersburg, 1981 (in Russian).

23. Kay, S., *Fundamentals of Statistical Signal Processing: Detection Theory,* Vol. 2, Prentice Hall, Englewood Cliffs, NJ, 1998.
24. Macchi, O., *Adaptive Processing,* John Wiley & Sons, New York, 1995.
25. Scharf, L., *Statistical Signal Processing: Detection, Estimation and Time Series Analysis,* Addison-Wesley, Reading, MA, 1991.
26. Stoica, P. and Moses, R., *Introduction to Spectral Analysis,* Prentice Hall, Upper Saddle River, NJ, 1997.
27. Feldman, Yu and Mandurovsky, I., *Theory of Fluctuations of Radar Signals,* Radio and Svyaz, Moscow, 1988 (in Russian).
28. Widrow, B. and Stearns, S., *Adaptive Signal Processing,* Prentice Hall, Englewood Cliffs, NJ, 1985.
29. Poor, H., *An Introduction to Signal Detection and Estimation,* 2nd ed., Springer-Verlag, New York, 1994.
30. Rytov, S., Kravtzov, Yu, and Tatarsky, V., *Introduction to Statistical Radio Physics. Part II: Stochastic Fields,* Nauka, Moscow, 1978 (in Russian).
31. Cohen, L., *Time-Frequency Analysis,* Prentice Hall, Englewood Cliffs, NJ, 1995.
32. Shuster, H., *Deterministic Chaos,* VCH, New York, 1989.
33. Hannan, E. and Deistler, M., *The Statistical Theory of Linear Systems,* John Wiley & Sons, New York, 1988.
34. Jain, A., *Fundamentals of Digital Image Processing,* Prentice Hall, Englewood Cliffs, NJ, 1989.
35. Haykin, S., *Adaptive Filter Theory,* 3rd ed., Prentice Hall, Upper Saddle River, NJ, 1996.
36. Kay, S., *Modern Spectral Estimation: Theory and Application,* Prentice Hall, Englewood Cliffs, NJ, 1988.
37. Paholkov, G., Cashinov, V., and Ponomarenko, B., *Variational Technique for Synthesis of Signals and Filters,* Radio and Svyaz, Moscow, 1981 (in Russian).
38. Kassam, S., *Signal Detection in Non-Gaussian Noise,* Springer-Verlag, New York, 1988.
39. Gerlach, K. and Steiner, M., Adaptive detection of range distributed targets, *IEEE Trans.,* Vol. SP-47, No. 7, 1999, pp. 1844–1851.
40. Oppenheim, A. and Willsky, A., *Signals and Systems,* Prentice Hall, Englewood Cliffs, NJ, 1983.
41. Proakis, J., *Digital Communications,* 3rd ed., McGraw-Hill, New York, 1995.
42. Van Trees, H., *Detection, Estimation and Modulation Theory. Part III: Radar-Sonar Signal Processing and Gaussian Signals in Noise,* John Wiley & Sons, New York, 1972.
43. Porat, B., *Digital Processing of Random Signals: Theory and Methods,* Prentice-Hall, Englewood Cliffs, NJ, 1994.
44. Davies, K., *Ionospheric Radio,* Peter Peregrinns, London, 1990.
45. Friedlander, B. and Francos, A., Estimation of amplitude and phase parameters of multicomponent signals, *IEEE Trans.,* Vol. SP-43, No. 4, pp. 917–927.
46. Jeruchim, M., Balaban, P., and Shanmugan, K., *Simulation of Communication Systems,* Plenum, New York, 1992.
47. McNamara, L., *The Ionosphere: Communications, Surveillance, and Direction Finding,* Krieger, Malabar, FL, 1991.
48. Gingras, D., Gerstoft, P., and Gerr, N., Electromagnetic matched-field processing: basic concepts and tropospheric simulations, *IEEE Trans.,* Vol. AP-45, No. 10, 1997, pp. 1536–1545.

49. Micka, O. and Weiss, A., Estimating frequencies of exponentials in noise using joint diagonalization, *IEEE Trans.*, Vol. SP-47, No. 2, 1999, pp. 341–348.
50. Francos, A. and Porat, M., Analysis and synthesis of multicomponent signals using positive time–frequency distributions, *IEEE Trans.*, Vol. SP-47, No. 2, 1999, pp. 493–504.
51. Conte, E., Di Bisceglie, M., Longo, M., and Lops, M., Canonical detection in spherically invariant noise, *IEEE Trans.*, Vol. COM-43, No. 2, 1995, pp. 347–353.
52. Therrien, C., *Discrete Random Signals and Statistical Signal Processing*, Prentice Hall, Englewood Cliffs, NJ, 1992.
53. Kapoor, S., Marchok, D., and Huang, Y.-F., Adaptive interference suppression in multiuser wireless OFDM systems using antenna arrays, *IEEE Trans.*, Vol. SP-47, No. 12, 1999, pp. 3381–3391.
54. Trump, T. and Ottersten, B., Estimation of nominal direction of arrival and angular spread using an array of sensors, *Signal Process.*, Vol. 50, No. 1–2, 1996, pp. 57–69.
55. Anderson, C., Green, S., and Kingsley, S., HF skywave radar: estimating aircraft heights using super-resolution in range, in *Proc. Inst. Elect. Eng. Radar Sonar Navigat.*, Vol. 143, August 1996, pp. 281–285.
56. Nehorai, A., Ho, K.-C., and Tan, B., Minimum-noise-variance beam former with an electromagnetic vector sensor, *IEEE Trans.*, Vol. SP-47, No. 3, 1999, pp. 601–618.
57. Anderson, T., *An Introduction to Multivariate Statistical Analysis*, 2nd ed., John Wiley & Sons, New York, 1984.
58. Leung, H. and Lo, T., Chaotic radar signal processing over the sea, *IEEE J. Oceanic Eng.*, Vol. 18, 1993, pp. 287–295.
59. Godard, D., Self-recovering equalization and carrier tracking in two-dimensional data communication systems, *IEEE Trans.*, Vol. COM-32, No. 6, 1975, pp. 679–682.
60. Agafe, C. and Iltis, R., Statistics of the RSS estimation algorithm for gaussian measurement noise, *IEEE Trans.*, Vol. SP-47, No. 1, 1999, pp. 22–32.
61. Wong, K. and Zoltowski, M., Univector-sensor ESPRIT for multi-source azimuth-elevation angle-estimation, in *Antennas and Propagation Society International Symposium*, Vol. 2, AP-S. Digest, 1996, pp. 1368–1371.
62. Papazoglou, M. and Krolik, J., Matched-field estimation of aircraft altitude from multiple over-the-horizon radar revisits, *IEEE Trans.*, Vol. SP-47, No. 4, 1999, pp. 966–975.
63. Barbarossa, S. and Scaglione, A., Adaptive time-varying cancellation of wideband interferences in spread-spectrum communications based on time-frequency distributions, *IEEE Trans.*, Vol. SP-47, No. 4, pp. 957–965.
64. Frenkel, L. and Feder, M., Recursive expectation maximization (EM) algorithms for time-varying parameters with applications to multiple target tracking, *IEEE Trans.*, Vol. SP-47, No. 2, 1999, pp. 306–320.
65. Tufts, D. and Kumaresan, R., Estimation of frequencies of multiple sinusoids: making linear prediction perform like maximum likelihood, in *Proceedings of the IEEE*, Vol. 70, September 1982, pp. 975–989.
66. Nehorai, A. and Paldi, E., Vector-sensor array processing for electromagnetic source localization, *IEEE Trans.*, Vol. SP-42, No. 2, 1994, pp. 376–398.
67. Krolik, J. and Anderson, R., Maximum likelihood coordinate registration for over-the-horizon radar, *IEEE Trans.*, Vol. SP-45, No. 4, 1997, pp. 945–959.

68. Johnson, N., Kotz, S., and Balakrishnan, N., *Continuous Univariate Distributions*, Vol. 2, John Wiley & Sons, New York, 1995.
69. Zubkovich, S., *Statistical Characteristics of Radio Signals Reflected by the Earth Surface*, Soviet Radio, Moscow, 1968 (in Russian).

3

Fluctuations under Scanning of the Three-Dimensional (Space) Target with the Moving Radar

3.1 Slow and Rapid Fluctuations

In the study of target return signal fluctuations caused by the moving radar, we suppose:

- The radar antenna is stationary,

$$\Delta\varphi_{sc} = \Delta\psi_{sc} = 0 . \tag{3.1}$$

- Fluctuations caused by the wind are absent,

$$\Delta\xi = \Delta\zeta = 0 . \tag{3.2}$$

- Scatterers are in the long range of the radar antenna directional diagram,

$$\Delta\varphi_{rm} = \Delta\psi_{rm} = 0 . \tag{3.3}$$

Let us assume that the radar moves uniformly and linearly with velocity V:

$$\Delta\ell = -V\cdot\tau \qquad \text{and} \qquad \Delta\rho = -V_r\cdot\tau , \tag{3.4}$$

where V_r is the radial component of scatterer velocity relative to the radar. When the radar is brought closer to scatterers, the condition $V > 0$ is true. Changeover from the space displacement $\Delta\rho$ to the shift in time τ implies a changeover from the space fluctuations to the time fluctuations and, consequently, a changeover from the space–time correlation function to the time correlation function.[1–3]

3.1.1 General Statements

Using Equation (2.18), Equation (2.74), and Equation (2.92)–Equation (2.95) under the previously mentioned conditions, we can write the total correlation function of target return signal fluctuations in the following form:

$$R(t,\tau) = p(t)\cdot R(\tau) = p(t)\cdot R_p(\tau)\cdot R_g(\tau)\cdot e^{j\omega_0\tau}, \tag{3.5}$$

where

$$R_p(\tau) = \sum_{n=0}^{\infty} R_p^{en}(\tau - nT_p'); \tag{3.6}$$

$$R_p^{en}(\tau) = \frac{\int \mathbf{P}(z)\cdot \mathbf{P}^*(z+\mu\tau)\,dz}{\int \Pi^2(z)\,dz}; \tag{3.7}$$

$$R_g(\tau) = \frac{\iint g^2(\varphi,\psi)\cdot e^{j\Omega(\varphi,\psi)\,\tau}\,d\varphi\,d\psi}{\iint g^2(\varphi,\psi)\,d\varphi\,d\psi}; \tag{3.8}$$

$$\Omega(\varphi,\psi) = \frac{2\,V_r\omega_0}{c} = \frac{4\,\pi\,V_r}{\lambda} = \Omega_{\max}\cos\left(\beta_0 + \tfrac{\varphi}{\cos\gamma_0}\right)\,\cos(\psi+\gamma_0) \tag{3.9}$$

is the Doppler frequency for a scatterer with the coordinates β and γ;

$$\Omega_{\max} = \frac{2\,V\omega_0}{c} = \frac{4\,\pi\,V}{\lambda} \tag{3.10}$$

is the maximum Doppler frequency;

$$\mu = 1 + \frac{2\,V_{r_0}}{c} = 1 + \frac{2\,V}{c}\cdot\cos\beta_0\cos\gamma_0 \quad \text{and} \quad T_p' = \frac{T_p}{\mu}. \tag{3.11}$$

Thus, in scanning of the three-dimensional (space) target with pulsed searching signals, the total normalized correlation function is defined as the product of two normalized correlation functions

$$R(\tau) = R_p(\tau)\cdot R_g(\tau)\cdot e^{j\omega_0\tau}. \tag{3.12}$$

The first normalized correlation function takes into consideration the target return signal fluctuations caused by propagation of periodic pulsed searching signals inside the target, i.e., the fluctuations in the radar range or the rapid target return signal fluctuations. The second normalized correlation function takes into consideration the target return signal fluctuations caused by differences in the radial velocities of scatterers moving relative to the radar antenna, i.e., the Doppler or the slow target return signal fluctuations.[4–6] The target return signal defined by the correlation function given in Equation (3.5) is a nonstationary stochastic process, but it is a separable process; the normalized correlation function of target return signal fluctuations determined by Equation (3.12) is independent of the parameter t (time). All information about the nonstationary state is included in the function $p(t)$ given by Equation (2.78), which was studied in more detail in Section 2.4.3. Because of this, our consideration is limited only to the study of the normalized correlation function given by Equation (3.12).

With the continuous nonmodulated pulsed searching signal, the condition $R_p(\tau) \equiv 1$ is true and the total normalized correlation function of the fluctuations coincides with the normalized correlation function of the slow fluctuations, as the rapid fluctuations are absent. If the radar is stationary, the condition $R_g(\tau) \equiv 1$ is true and there are only the rapid (intraperiod) fluctuations under the condition $\mu = 1$. The slow interperiod fluctuations are absent.[7–9] Thus, the normalized correlation functions given by Equation (3.6) and Equation (3.8) have a well-founded physical meaning. Let us consider each normalized correlation function individually and, after that, investigate the total normalized correlation function given by Equation (3.12).

3.1.2 The Fluctuations in the Radar Range

The normalized correlation function $R_p(\tau)$ given by Equation (3.6) is a periodic function with respect to the variable τ, the period T_p consisting of narrow waves, the width of which is defined by the duration of the function $\Pi(t)$, i.e., by the pulsed signal duration τ_p (see Figure 3.1a). The meaning of this dependence is as follows. Under the condition $\tau < \tau_p$, the normalized correlation function $R_p(\tau)$ differs from zero because the pulsed signals shifted by τ are partially overlapped. As τ is increased, the pulsed signal overlapping is decreased and a correlation dies out gradually. If the condition $\tau > \tau_p$ is true, the pulsed signal overlapping is absent and the normalized correlation function $R_p(\tau)$ is equal to zero, i.e., $R_p(\tau) = 0$. In other words, the correlation interval of the target return signal fluctuations in the radar range becomes very close to the pulsed signal duration. With an increase in the value of τ, so long as the value of τ becomes close or equal to the period T_p, a correlation between the pulsed signals appears again, as the pulsed signal of the next period is an exact copy of the pulsed signal in the previous period.[10,11]

For an unchanged "radar–scatterer" system, in which the radar, radar antenna, and scatterers are stationary, the pulsed searching signals are exact

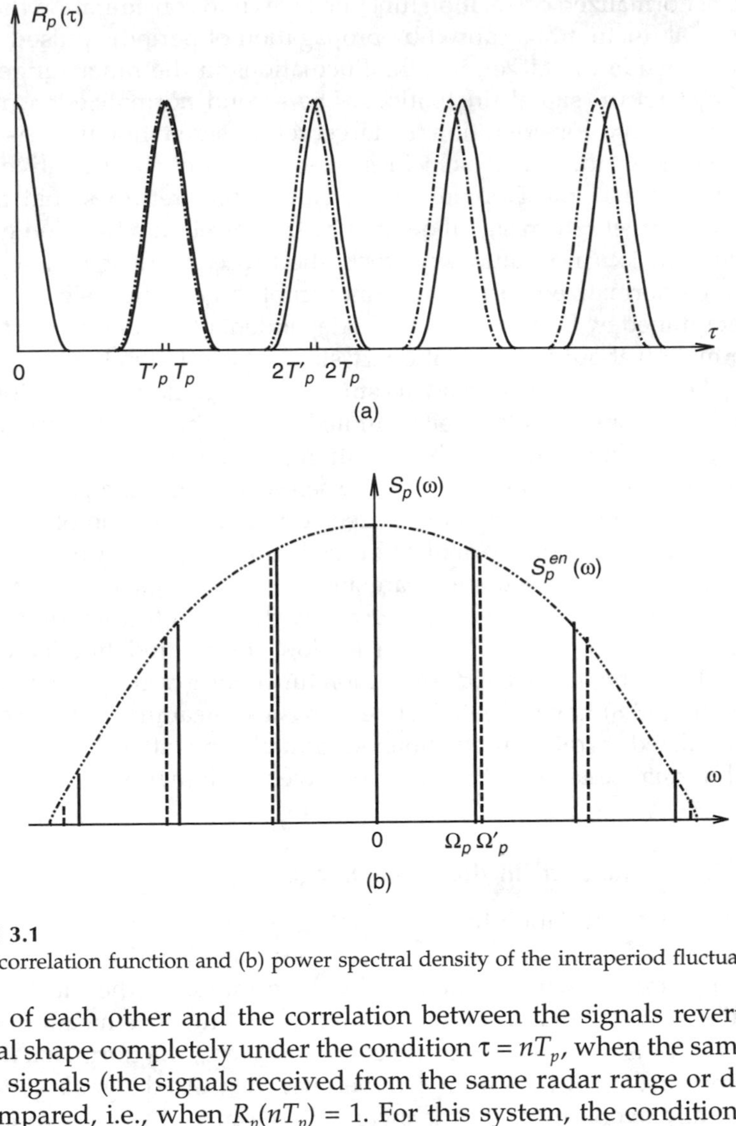

FIGURE 3.1
(a) The correlation function and (b) power spectral density of the intraperiod fluctuations.

copies of each other and the correlation between the signals reverts to its original shape completely under the condition $\tau = nT_p$, when the same target return signals (the signals received from the same radar range or distance) are compared, i.e., when $R_p(nT_p) = 1$. For this system, the condition $R_g(\tau) = R_{\Delta\xi,\Delta\zeta}(\tau) \equiv 1$ is true and correlation properties are completely defined by the normalized correlation function $R_p(\tau)$ under the condition $\mu = 1$ (see the solid line in Figure 3.1a).

When the radar moves, i.e., when $\mu \neq 1$, the correlation between the pulsed searching signals becomes maximum under the condition

$$\tau = nT'_p \qquad \text{or} \qquad R_p(nT'_p) = 1 \tag{3.13}$$

when the target return signals received from the same pulse volume removed during the time interval nT'_p with the distance determined by

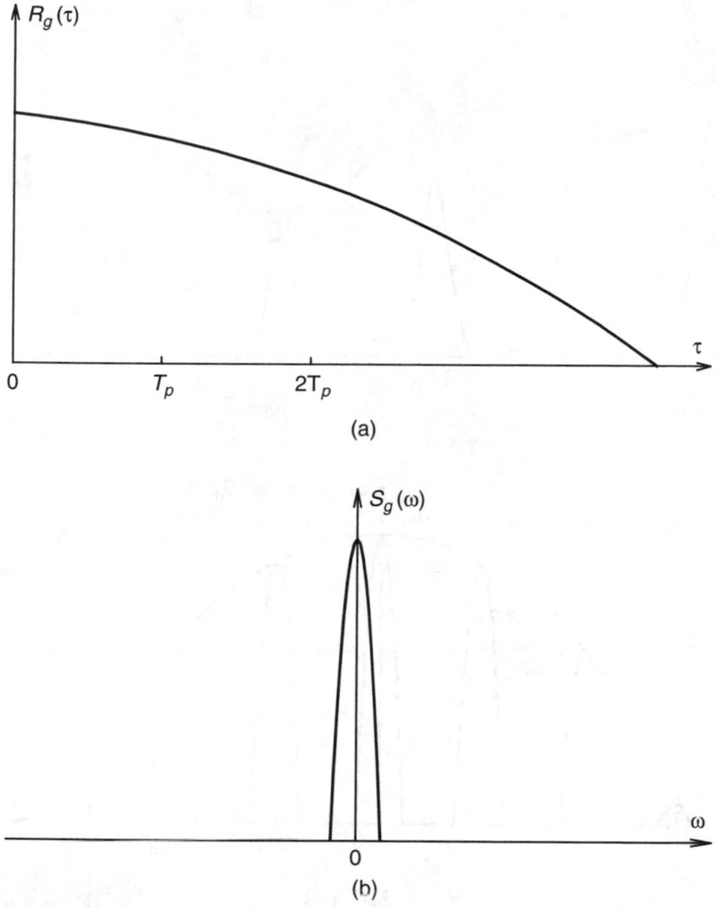

FIGURE 3.2
(a) The correlation function and (b) power spectral density of the interperiod fluctuations.

$$\Delta\rho_0 = nT_p'\, V \cos\beta_0 \cos\gamma_0 \tag{3.14}$$

are compared (see the dotted line in Figure 3.1a). In other words, the radar motion, which is defined by replacing the argument τ with the argument $\mu\tau$ in the normalized correlation function $R_p(\tau)$ given by Equation (3.6), implies compression under the condition $\mu > 1$ or expansion under the condition $\mu < 1$ of the time scale μ times. Therefore, the period T_p and the width of the waves are decreased (or are increased) μ times. This is a natural manifestation of the Doppler effect, which is accompanied by changes in pulsed searching signal parameters such as duration, frequency of signal iteration, and changes in the carrier frequency of the signal under radar antenna scanning.[12–14]

Naturally, when the radar moves, the correlation resumes its original shape incompletely under the condition given by Equation (3.13). Correlation

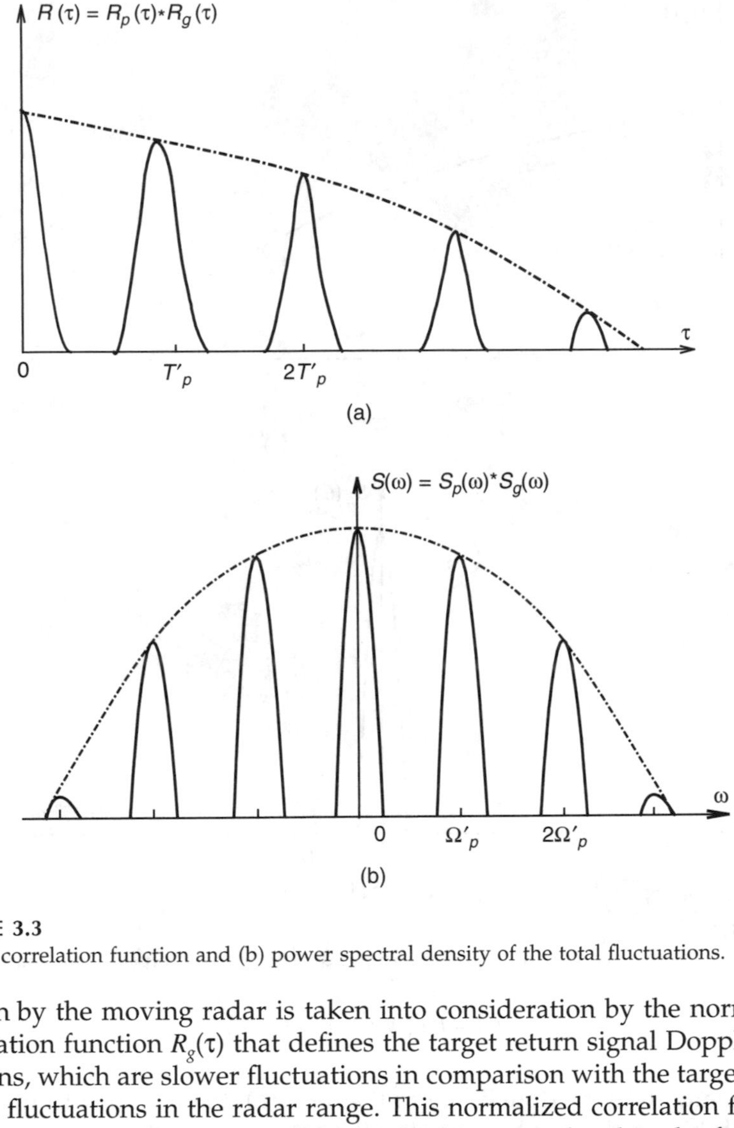

FIGURE 3.3
(a) The correlation function and (b) power spectral density of the total fluctuations.

broken by the moving radar is taken into consideration by the normalized correlation function $R_g(\tau)$ that defines the target return signal Doppler fluctuations, which are slower fluctuations in comparison with the target return signal fluctuations in the radar range. This normalized correlation function of the Doppler fluctuations will be studied in more detail in the following sections.

The power spectral density of intraperiod target return signal fluctuations, which correspond to the periodic normalized correlation function $R_p(\tau)$, has the following form:

$$S_p(\omega) = S_p^{en}(\omega) \cdot \sum_{n=0}^{\infty} \delta(\omega - n\Omega'_p) = \sum_{n=0}^{\infty} S_p^{en}(n\Omega'_p) \cdot \delta(\omega - n\Omega'_p)\,, \qquad (3.15)$$

where

$$\Omega'_p = \frac{2\pi}{T'_p} = \mu\,\Omega_p\,. \tag{3.16}$$

The envelope of the power spectral density of intraperiod fluctuations has the following form [see Equation (2.6)]:

$$S_p^{en}(\omega) = \frac{p}{\pi}\int R_p^{en}(\mu\tau)\cdot e^{-j\omega\tau}d\tau = \frac{p}{\pi}\cdot\frac{\left|\int P(t)\cdot e^{-\frac{j\omega t}{\mu}}dt\right|^2}{\int \Pi^2(t)\,dt}\,. \tag{3.17}$$

The power spectral density given by Equation (3.15) is a regulated function. This power spectral density is a sequence of discrete harmonics separated from each other by the frequency Ω'_p. The envelope of the power spectral density given by Equation (3.17) is the energy spectrum of the pulsed searching signal expanded or compressed μ times (see the dotted line in Figure 3.1b). The shape of the normalized correlation function and the effective power spectral density bandwidth of intraperiod fluctuations are completely defined by the shape and direction of the pulsed searching signal.

In the following examples, we consider only the envelope of the normalized correlation function $R_p^{en}(\tau)$ determined by Equation (3.7) and the envelope of the power spectral density $S_p^{en}(\omega)$ given by Equation (3.17). The normalized correlation function $R_p(\tau)$ and the regulated power spectral density $S_p(\omega)$ can be defined without any difficulty using Equation (3.6) and Equation (3.15), respectively.[15,16]

3.1.2.1 The Square Waveform Target Return Signal without Frequency Modulation

In this case, we can write

$$R_p^{en}(\tau) = 1 - \frac{|\mu\tau|}{\tau_p} = 1 - \frac{|\tau|}{\tau'_p}\,, \qquad |\tau| \le \tau'_p\,; \tag{3.18}$$

$$S_p^{en}(\omega) \approx \mathbf{sinc}^2(0.5\omega\,\tau'_p)\,, \tag{3.19}$$

where $\tau'_p = \frac{\tau_p}{\mu}$.

3.1.2.2 The Gaussian Target Return Signal without Frequency Modulation

In this case, we can write

$$R_p^{en}(\tau) \cong e^{-0.5\pi\cdot\left(\frac{\mu\tau^2}{\tau_p}\right)} = e^{-\pi\cdot\frac{\tau^2}{\tau_c^2}}\,; \tag{3.20}$$

$$S_p^{en}(\omega) \cong e^{-\pi \cdot \frac{\omega^2}{\Delta\Omega_p^2}}, \tag{3.21}$$

where

$$\tau_c = \sqrt{2}\,\tau'_p \quad \text{and} \quad \Delta\Omega_p = 2\,\pi\,\Delta F_p = \frac{2\,\pi}{\tau_c} = \frac{\sqrt{2}\,\pi}{\tau'_p}. \tag{3.22}$$

Here ΔF_p is the effective power spectral density bandwidth of the fluctuations.

3.1.2.3 The Smoothed Target Return Signal without Frequency Modulation

In this case, the shape of the target return signal can be thought as intermediate between the square waveform and the Gaussian shape:

$$\Pi(t) = \frac{\Phi\left(\frac{t+0.5\tau_0}{0.5\tau_{fr}}\right) - \Phi\left(\frac{t-0.5\tau_0}{0.5\tau_{fr}}\right)}{2\Phi\left(\frac{\tau_0}{\tau_{fr}}\right)}, \tag{3.23}$$

where $\Phi(x)$ is the error integral given by Equation (1.27); τ_{fr} is the duration of the pulsed searching signal front between levels 0.08 and 0.92, respectively; τ_0 is the pulsed searching signal duration at the level 0.5 under the condition $\tau_{fr} \le \tau_0$. Equation (3.23) covers, in particular, the cases of the square waveform pulsed searching signal under the condition $\frac{\tau_0}{\tau_{fr}} \to \infty$ and the Gaussian pulsed searching signal under the condition $\frac{\tau_0}{\tau_{fr}} \to 0$. The effective pulsed searching signal duration given by

$$\tau_{ef} = \frac{\tau_0}{\Phi\left(\frac{\tau_0}{\tau_{fr}}\right)} \tag{3.24}$$

differs from τ_0 by not more than 6%. For the limiting case of the Gaussian pulsed signal, we have $\tau_{ef} = 0.6\tau_0$. The normalized correlation function of the target return signal fluctuations in the case of the pulsed searching signal with the function $\Pi(t)$ given by Equation (3.23) is very cumbersome and is omitted here. The envelope of the power spectral density has the following form:

$$S_p^{en}(\omega) \cong \mathbf{sinc}^2(0.5\omega\,\tau'_0) \cdot e^{-\left(\frac{\omega\,\tau'_{fr}}{2\sqrt{2}}\right)^2}. \tag{3.25}$$

The envelope of the power spectral density given by Equation (3.25) is the product of the envelope of the power spectral density with the square waveform pulsed searching signal having a duration τ_0' and the envelope of the power spectral density with the Gaussian pulsed searching signal having a duration $0.5\sqrt{\pi}\,\tau_{f_r}'$, where

$$\tau_0' = \frac{\tau_0}{\mu} \quad \text{and} \quad \tau_{f_r}' = \frac{\tau_{f_r}}{\mu}. \tag{3.26}$$

The effective bandwidth of the power spectral density envelope takes the following form

$$\Delta F_p = \Phi(x) - \frac{1-e^{-x^2}}{\sqrt{\pi}\,x} \cdot \frac{1}{\tau_0'}, \tag{3.27}$$

where

$$x = \sqrt{2} \cdot \frac{\tau_0'}{\tau_{f_r}'} \tag{3.28}$$

(see Figure 3.4). For two limiting cases, as $\tau_{f_r} \to 0$ and $\tau_0 \to 0$, Equation (3.25) coincides with Equation (3.19) and Equation (3.21).

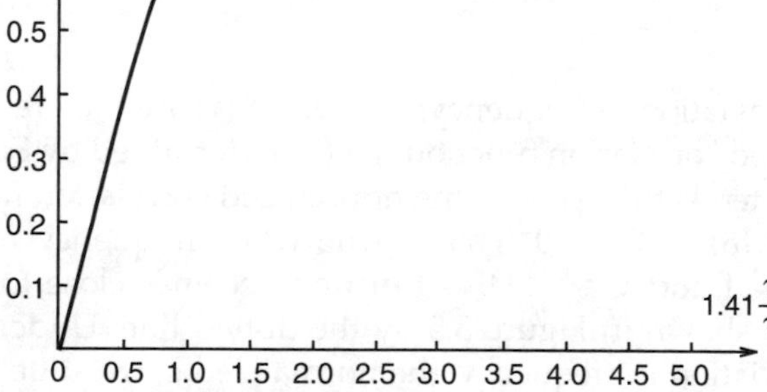

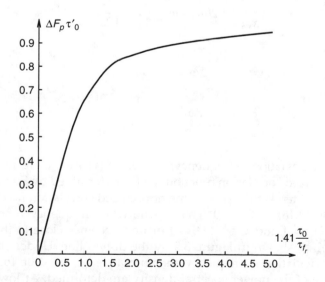

FIGURE 3.4
The normalized effective bandwidth of envelope of the power spectral density of the fluctuations in the radar range with the smoothed pulsed searching signal.

3.1.2.4 The Square Waveform Target Return Signal with Linear-Frequency Modulation

In this case, we can write:

$$\omega(t) = \omega_0 + \frac{\Delta\omega_M}{\tau_p} \cdot t \; ; \tag{3.29}$$

$$\Psi(t) = \int_0^t (\omega - \omega_0)\, dt = \frac{\Delta\omega_M}{2\,\tau_p} \cdot t^2 \, , \tag{3.30}$$

where $\Delta\omega_M$ is the deviation in frequency during the time interval T_p. Therefore,

$$R_p^{en}(\tau) = \frac{2}{\Delta\omega'_M \tau} \cdot \sin\left[0.5\Delta\omega'_M \tau\left(1 - \tfrac{|\tau|}{\tau'_p}\right)\right] ; \tag{3.31}$$

$$S_p^{en}(\omega) \cong \frac{p}{\Delta\omega'_M} \cdot \left\{\left[\mathbf{C}(x_2) - \mathbf{C}(x_1)\right]^2 + \left[\mathbf{S}(x_2) - \mathbf{S}(x_1)\right]^2\right\} , \tag{3.32}$$

where

$$x_{1,2} = \left[\frac{2(\omega_0 - \omega)}{\Delta\omega'_M} \mp 1\right] \cdot \sqrt{0.25\pi\, D} \; ; \tag{3.33}$$

$$\Delta\omega'_M = \mu\, \Delta\omega_M \, ; \tag{3.34}$$

$$D = \frac{\Delta\omega'_M \tau'_p}{2\,\pi} = \Delta f_M \cdot \tau_p \tag{3.35}$$

is the relative deviation in frequency; $\mathbf{C}(x)$ and $\mathbf{S}(x)$ are the Fresnel integrals.

The normalized correlation function $R_p^{en}(\tau)$ determined by Equation (3.31) (see Figure 3.5) tends to approach the normalized correlation function given by Equation (3.18) as D → 0. This is true when frequency modulation is absent. At $D >> 1$ and $\tau << \tau_p$, this function becomes close to the function $\mathbf{sinc}^2(0.5\Delta\omega'_M \tau)$, shown in Figure 3.5 by the dotted line. Under changing of the relative deviation in frequency, the normalized correlation function and the envelope of the power spectral density are deformed. At low values of D, the envelope of the power spectral density is approximately close to one for the case of the square waveform pulsed searching signal given by Equation (3.19), with the effective bandwidth equal to $(\tau'_p)^{-1}$. With high frequency deviation, the envelope of the power spectral density tends to approach the

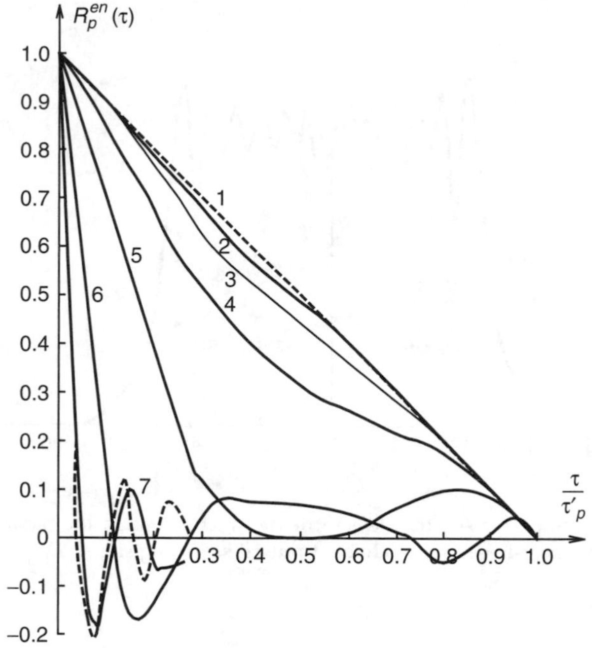

FIGURE 3.5
The normalized correlation function of the fluctuations in the radar range with the square waveform linear-frequency modulated pulsed searching signal, $n = 0$: (1) $D = 0$; (2) $D = 0.5$; (3) $D = 1$; (4) $D = 2$; (5) $D = 4$; (6) $D = 10$; (7) $D = 20$.

square waveform, one with the bandwidth equal to deviation in frequency $\Delta f'_M$ during the pulsed searching signal duration (see Figure 3.6).

In the general case, the correlation interval of the target return signal, which is inversely proportional to the effective bandwidth of the power spectral density envelope, has the following form:

$$\frac{\tau_c}{\tau_p} = \frac{2}{D} \cdot \left[\mathbf{C}^2(0.5\sqrt{\pi D}) + \mathbf{S}^2(0.5\sqrt{\pi D})\right] \tag{3.36}$$

and is shown in Figure 3.7. We see that

$$\tau_c \to \tau'_p \quad \text{as} \quad D \to 0 \quad \text{or} \quad \Delta f_M \to 0\,. \tag{3.37}$$

This statement can be proved using the L'Hospital rule twice in Equation (3.36).[17,18] As $D \to \infty$ or as $\Delta f_M \to \infty$, the correlation interval tends to approach $\tau_c \to (\Delta f'_M)^{-1}$ and is defined by the deviation in frequency during the pulsed searching signal duration. Because the correlation interval of the target return signal is equal to the signal duration when frequency modulation is absent, we can say that the linear-frequency modulation from the standpoint of the

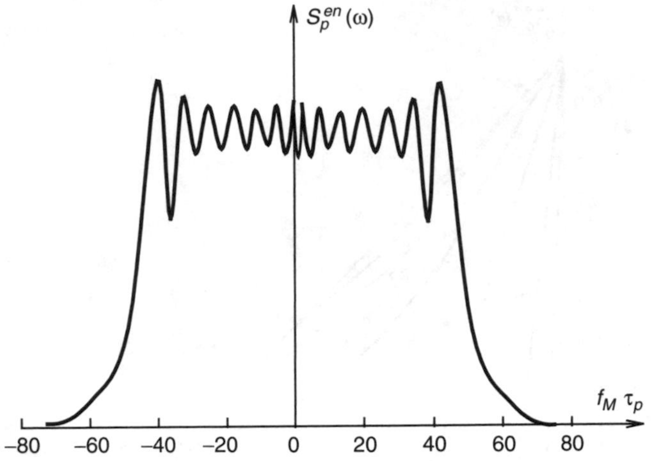

FIGURE 3.6
The envelope of the power spectral density of the fluctuations in the radar range with the square waveform linear-frequency modulated pulsed searching signal, $D = 100$.

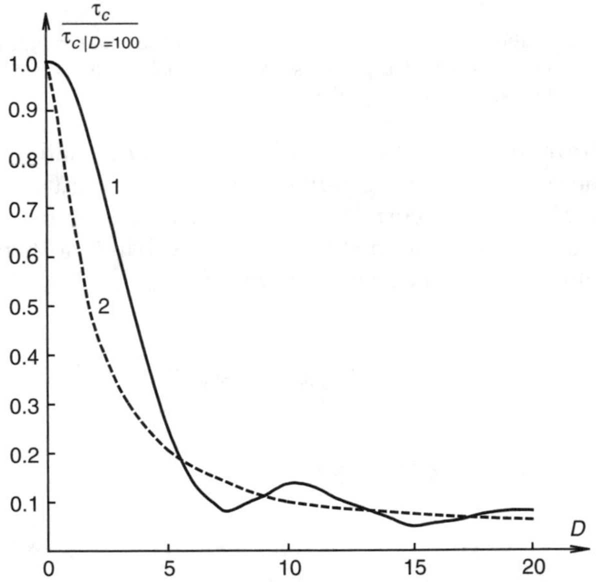

FIGURE 3.7
The normalized correlation interval of the fluctuations in the radar range as a function of the parameter $D = \Delta f_M \cdot \tau_p$: (1) the linear-frequency modulated pulsed searching signal; (2) the Gaussian pulsed searching signal.

correlation interval of the target return signal is equivalent to the truncation of the pulsed searching signal D times, i.e., prior to the duration $(\Delta f'_M)^{-1}$. This is to fit the well-known phenomenon of improvement in the radar range resolution with high-frequency modulation.

3.1.2.5 The Gaussian Target Return Signal with Linear-Frequency Modulation

In this case, the normalized correlation function and the envelope of the power spectral density of target return signal fluctuations are determined by Equation (3.20) and Equation (3.21), in which

$$\tau_c = \frac{\sqrt{2}\,\tau'_p}{\sqrt{1+\Delta f_M^2 \tau_p^2}}\,, \tag{3.38}$$

where

$$\Delta f_M \tau_p = \Delta f'_M \tau'_p\,. \tag{3.39}$$

If

$$\Delta f_M \tau_p << 1\,, \qquad \text{then} \qquad \tau_c = \sqrt{2}\,\tau'_p \tag{3.40}$$

and if

$$\Delta f_M \tau_p >> 1\,, \qquad \text{then} \qquad \tau_c = (\Delta f'_M)^{-1}\,. \tag{3.41}$$

The function

$$\frac{\tau_c}{\sqrt{2}\,\tau'_p} = \frac{1}{\sqrt{1+D^2}} \tag{3.42}$$

given by Equation (3.38)–Equation (3.41) is shown in Figure 3.7 by the dotted line.

3.1.3 The Doppler Fluctuations

The target return signal Doppler fluctuations are defined by the normalized correlation function $R_g(\tau)$ given by Equation (3.8). This function $R_g(\tau)$ is slow in comparison with the normalized correlation function $R_p^{en}(\tau)$ of the fluctuations given by Equation (3.7).[19,20] The function $R_g(\tau)$ is the envelope of the normalized correlation function $R(\tau)$ given by Equation (3.12) and characterizes the interperiod fluctuations (see Figure 3.2a). When the target return signal at the receiver or detector input in navigational systems is a continuous, nonmodulated stochastic process, if the fluctuations in the radar range are absent, the normalized correlation function $R_g(\tau)$ defines the fluctuations completely.[21,22]

Let us define the power spectral density of the Doppler fluctuations corresponding to the normalized correlation function $R_g(\tau)$. Multiplying the normalized correlation function $R_g(\tau)$ by the factor $e^{j\omega_0\tau}$ and using the Fourier transform, we can write

$$S_g(\omega) \cong \iint g^2(\varphi,\psi)\, d\varphi\, d\psi \int\limits_{-\infty}^{\infty} e^{j[\omega_0+\Omega(\varphi,\psi)-\omega]\,\tau} d\tau \cong \iint g^2(\varphi,\psi) \times \delta\big[\omega_0 + \Omega(\varphi,\psi) - \omega\big]\, d\varphi\, d\psi \tag{3.43}$$

Using the filtering properties of the delta function, the double integral in Equation (3.43) can be reduced to a simple integral. For this purpose, the condition

$$\Omega(\varphi,\psi) = \Omega_{\max} \cos\big(\beta_0 + \tfrac{\varphi}{\cos\gamma_0}\big)\, \cos(\psi + \gamma_0) = \omega - \omega_0 \tag{3.44}$$

must be satisfied [see Equation (3.9)]. In the general case, determination of the integral in Equation (3.43) could require the use of numerical techniques. In the majority of important cases in practice, we can determine the integral in Equation (3.43) without using numerical techniques, which is very important.[23,24]

Equation (3.44) is equivalent to the formula

$$\cos\theta = (\omega - \omega_0) \cdot (\Omega_m)^{-1} = \nu\,, \tag{3.45}$$

where θ is the angle between the vector of a velocity of a moving aircraft (radar), for example, and the direction of radar antenna scanning [see Equation (2.91) and Equation (2.92)]. This means that the delta function in Equation (3.43) differs from zero when the angles φ and ψ satisfy the following condition: $\theta = \arccos \nu = const$. In other words, the power spectral density at the relative Doppler frequency is formed by summing the powers of the target return signals from those scatterers that are placed on the surface of the cone defined near the velocity vector of a moving aircraft (radar) with apex angle 2θ. This geometric representation of the formation of the Doppler fluctuations with moving radar can be used for determination of the power spectral density of Doppler fluctuations.[25,26]

Assuming that $\theta \in [0, \pi]$ in Equation (3.45), we can define that the spectrum given by Equation (3.44) is always within the limits of the interval

$$\omega_0 - \Omega_{\max} \leq \omega \leq \omega_0 + \Omega_{\max}. \tag{3.46}$$

3.2 The Doppler Fluctuations of a High-Deflected Radar Antenna

3.2.1 The Power Spectral Density for an Arbitrary Directional Diagram

Equation (3.43) can be fundamentally simplified and easily studied if we assume that the radar antenna directional diagram axis is deflected from the direction of moving radar so that at least one of the angles β_0 or γ_0 is greater than the directional diagram width in the corresponding plane. In this case, reasoning that the width is small, we can use the Taylor-series expansion[27,28] for the function $\Omega(\varphi, \psi)$ in Equation (3.9), limiting terms to the first order:

$$\Omega'(\varphi, \psi) \cong \Omega_0 - \varphi\,\Omega_h - \psi\,\Omega_v, \tag{3.47}$$

where

$$\Omega_0 = \Omega_{\max} \cos\beta_0 \cos\gamma_0; \tag{3.48}$$

$$\Omega_{\max} = \frac{2\,\omega_0 V}{c} = \frac{4\,\pi\,V}{\lambda}; \tag{3.49}$$

$$\Omega_h = \Omega_{\max} \sin\beta_0; \tag{3.50}$$

$$\Omega_v = \Omega_{\max} \cos\beta_0 \sin\gamma_0. \tag{3.51}$$

Here, Ω_0 is the Doppler frequency corresponding to the center of the pulse volume — a direction along the directional diagram axis.

At first, we assume that the variables φ and ψ in the function $g(\varphi, \psi)$ are separable. Substituting Equation (3.47) in Equation (3.8), we can write

$$R_g(\tau) = R_g^h(\tau) \cdot R_g^v(\tau) \cdot e^{j(\omega_0 + \Omega_0)\tau}, \tag{3.52}$$

where

$$R_g^h(\tau) = \frac{\int g_h^2(\varphi) \cdot e^{-j\varphi\,\Omega_h \tau} d\varphi}{\int g_h^2(\varphi)\, d\varphi} \tag{3.53}$$

and

$$R_g^v(\tau) = \frac{\int g_v^2(\psi) \cdot e^{-j\psi\,\Omega_v \tau} d\psi}{\int g_v^2(\psi)\, d\psi} \tag{3.54}$$

are the normalized correlation functions with the corresponding power spectral densities

$$S_g^h(\omega) \cong g_h^2(-\tfrac{\omega}{\Omega_h}) \quad \text{and} \quad S_g^v(\omega) \cong g_v^2(-\tfrac{\omega}{\Omega_v})\,. \tag{3.55}$$

This is obvious, as Equation (3.53) and Equation (3.54) are the Fourier transforms of the power spectral densities given by Equation (3.55). The power spectral densities of target return signal Doppler fluctuations given by Equation (3.55) coincide in shape with the square of the directional diagram in the horizontal and vertical planes. The total power spectral density corresponding to the normalized correlation function given by Equation (3.52) is defined by the convolution of the power spectral densities determined by Equation (3.55) and a shift of convolution by $\omega_0 + \Omega_0$:

$$S_g(\omega) = S_g^h(\omega) * S_g^v(\omega) * \delta(\omega - \omega_0 - \Omega_0) \cong \int g_h^2(-\tfrac{x}{\Omega_h}) \cdot\, g_v^2\left(\tfrac{\omega_0 + \Omega_0 - \omega + x}{\Omega_v}\right) dx \tag{3.56}$$

If both angles β_0 and γ_0 are very small, the formulae obtained based on Equation (3.47) are not true. If $\beta_0 = \gamma_0 = 0$, the equality $R_g^h(\tau) = R_g^v(\tau) \equiv 1$ follows from these formulae. This means that the Doppler fluctuations are absent. However, this statement, rigorously speaking, is not true because under the conditions $\beta_0 \to 0$ and $\gamma_0 \to 0$, the effective power spectral density bandwidth of the Doppler fluctuations is decreased straight away but does not equal zero. This will be proved in Section 3.3. If one of the angles β_0 and γ_0 is equal to zero, for example, if $\gamma_0 = 0$, we can write that $R_g^v(\tau) \equiv 1$ and $S_g^v(\omega) = \delta(\omega)$. Thus, the power spectral density $S_g(\omega)$ given by Equation (3.56) coincides with the power spectral density $S_g^h(\omega)$ shifted by

$$\omega_0 + \Omega_0 = \omega_0 + \Omega_{max} \cos\beta_0\,. \tag{3.57}$$

So, we can write

$$S_g(\omega) = S_g^h(\omega - \omega_0 - \Omega_0) \approx g_h^2\left(\tfrac{\omega_0 + \Omega_0 - \omega}{\Omega_h}\right) . \quad (3.58)$$

In an analogous way, under the condition $\beta_0 = 0$, we can write $R_g^h(\tau) \equiv 1$, $S_g^h(\omega) = \delta(\omega)$, and

$$S_g(\omega) = S_g^v(\omega - \omega_0 - \Omega_0) \approx g_v^2\left(\tfrac{\omega_0 + \Omega_0 - \omega}{\Omega_v}\right) , \quad (3.59)$$

where

$$\Omega_0 = \Omega_{\max} \cos \gamma_0 \quad (3.60)$$

and

$$\Omega_v = \Omega_{\max} \sin \gamma_0 . \quad (3.61)$$

In spite of the fact that Equation (3.58) and Equation (3.59) are approximate, as rigorously speaking, $S_g^v(\omega) \neq \delta(\omega)$ at $\gamma_0 = 0$ and $S_g^h(\omega) \neq \delta(\omega)$ at $\beta_0 = 0$, they can give us sufficient accuracy in the majority of cases. An exception to this rule is the case where $\Delta_h << \Delta_v$ at $\gamma_0 = 0$ and $\Delta_v << \Delta_h$ at $\beta_0 = 0$.

Let us continue to consider the general case when the variables φ and ψ in the directional diagram $g(\varphi, \psi)$ need not be separable. Substituting Equation (3.47) in Equation (3.43), we can write

$$S_g(\omega) \approx \int g^2(\varphi, \psi) \cdot \delta(\omega_0 + \Omega_0 - \varphi\, \Omega_h - \psi\, \Omega_v - \omega)\, d\varphi\, d\psi . \quad (3.62)$$

Using the filtering property of the delta function,[29,30] we can carry out integration with respect to only one variable, for example, φ:

$$S_g(\omega) \approx \int g^2\left(\tfrac{\omega_0 + \Omega_0 - \omega - \psi\, \Omega_v}{\Omega_h}, \psi\right) d\psi . \quad (3.63)$$

If the variables φ and ψ are separable, Equation (3.56) follows from Equation (3.63).

Let us introduce a new system of coordinates φ' and ψ' rotated by some angle κ relative to the coordinate system φ and ψ:

$$\varphi = \varphi' \cos \kappa - \psi' \sin \kappa ; \quad (3.64)$$

$$\psi = \varphi' \sin \kappa + \psi' \cos \kappa . \quad (3.65)$$

Choose the angle κ so that the conditions

$$\cos\kappa = \frac{\Omega_h}{G} \quad \text{and} \quad \sin\kappa = \frac{\Omega_v}{G} \tag{3.66}$$

are satisfied, where

$$G = \sqrt{\Omega_h^2 + \Omega_v^2} = \Omega_{\max}\sqrt{\sin^2\beta_0 + \cos^2\beta_0 \sin^2\gamma_0}\,. \tag{3.67}$$

Then, instead of Equation (3.62), we can write

$$S_g(\omega) \approx \iint \bar{g}^2(\varphi', \psi') \cdot \delta(\omega_0 + \Omega_0 - \omega - G\varphi')\, d\varphi'\, d\psi', \tag{3.68}$$

where

$$\bar{g}(\varphi', \psi') = g(\varphi, \psi) \tag{3.69}$$

is the directional diagram in the coordinate system φ' and ψ', or

$$S_g(\omega) \approx \int \bar{g}^2\left(\tfrac{\omega_0 + \Omega_0 - \omega}{G\psi'}\right) d\psi'. \tag{3.70}$$

The condition in Equation (3.66) means that we must choose the axes φ' and ψ' in such a manner that the plane $\varphi'(\psi' = 0)$ coincides with the plane crossing the velocity vector of moving aircraft (radar) and the directional diagram axis. It is possible to show that

$$\sqrt{\sin^2\beta_0 + \cos^2\beta_0 \sin^2\gamma_0} = \sin\theta_0 \quad \text{and} \quad \cos\beta_0 \cos\gamma_0 = \cos\theta_0, \tag{3.71}$$

where θ_0 is the angle between the directional diagram axis and the direction of moving radar. Because of this, we use the following forms in Equation (3.48), Equation (3.49), and Equation (3.70):

$$G = \Omega_{\max}\sin\theta_0 \quad \text{and} \quad \Omega_0 = \Omega_{\max}\cos\theta_0\,. \tag{3.72}$$

To make clear the physical meaning of Equation (3.70), we assume

$$\bar{g}(\varphi', \psi') = \overline{g_h}(\varphi') \cdot \overline{g_v}(\psi'), \tag{3.73}$$

which is analogous to Equation (2.97). Then Equation (3.70) can be represented in the simpler form:[31]

$$S_g(\omega) \approx \overline{g}_h^2\left(\frac{\omega_0+\Omega_0-\omega}{\Omega_{\max}\sin\theta_0}\right). \tag{3.74}$$

Thus, if the radar antenna axis is deflected from the direction of moving radar and the condition given by Equation (3.73) is satisfied, then the power spectral density of the Doppler fluctuations coincides in shape with the square of the directional diagram in the plane crossing the direction of radar motion and the directional diagram axis. The frequency corresponding to the maximum power spectral density has the following form:

$$\omega = \omega_0 + \Omega_0 = \omega_0 + \Omega_{\max}\cos\theta_0 = \omega_0 + \frac{4\pi V}{\lambda}\cdot\cos\beta_0\cos\gamma_0. \tag{3.75}$$

The power spectral density bandwidth given by Equation (3.74) is defined by the width $\tilde{\Delta}_a^{(2)}$ of the square of the directional diagram by power in the plane of deflection:

$$\Delta F = \frac{2V}{\lambda}\cdot\tilde{\Delta}_a^{(2)}\sin\theta_0. \tag{3.76}$$

Thus, ΔF can be considered as the effective power spectral density bandwidth of the Doppler fluctuations and $\tilde{\Delta}_a^{(2)}$ can be considered as the directional diagram width at some arbitrary level. If $\tilde{\Delta}_a^{(2)}$ denotes the effective width, then we can write $\tilde{\Delta}_a^{(2)} = k_a\tilde{\Delta}_a$, where k_a is a coefficient defining the shape of the directional diagram [see Equation (2.99)].

As the directional diagram width has the form $\Delta_a = \frac{\varepsilon\lambda}{d_a}$, where ε is a coefficient that is approximately equal to the unit and d_a is the radar antenna diameter, reference to Equation (3.76) allows us to arrive at a very important conclusion: With the deflected directional diagram, the effective power spectral density bandwidth of the Doppler fluctuations is independent of the wavelength λ. This can be explained as follows. With an increase (decrease) in the wavelength λ, the width Δ_a of the directional diagram increases (decreases) and the Doppler frequency $\frac{2V}{\lambda}$ decreases (increases) simultaneously. The functions of the effective bandwidth $\Delta\Omega = 2\pi\Delta F$ of the power spectral density of the Doppler fluctuations and frequency Ω_0 versus the angle θ_0 are shown in Figure 3.8 by the solid and dotted lines, respectively.

Reference to Equation (3.76) shows that the space correlation interval has the following form:

$$\Delta\ell_c = V\cdot\tau_c = \frac{V}{\Delta F} = \frac{d_a}{2k_a\varepsilon\sin\theta_0} \approx \frac{d_a}{\sin\theta_0}. \tag{3.77}$$

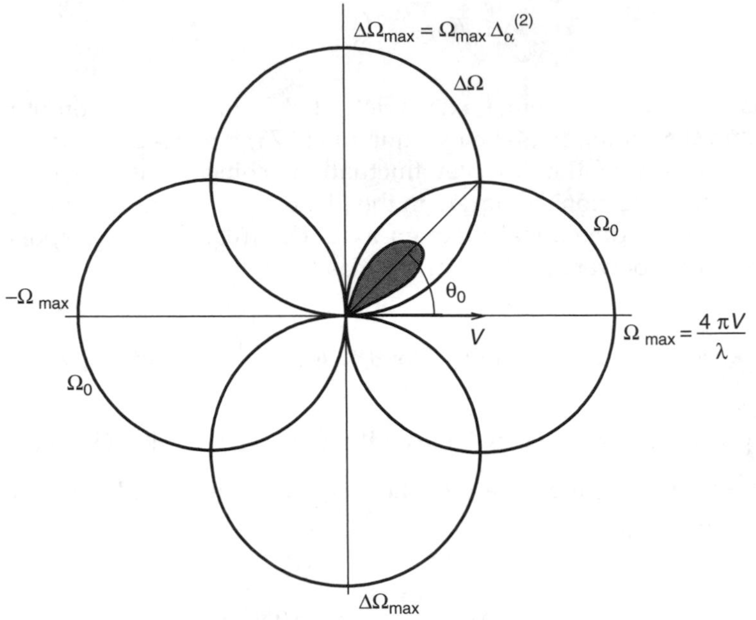

FIGURE 3.8
The bandwidth $\Delta\Omega$ and average frequency Ω_0 of the power spectral density of the Doppler fluctuations as a function of the radar antenna deflection angle θ_0.

Hence, it follows that if $\theta = 90°$, then $\Delta\ell_c \approx d_a$, i.e., with the deflection of a radar antenna by d_a, the target return signals become uncorrelated.[32] With a decrease in the angle θ_0, the correlation interval increases (see Figure 3.9) and tends to approach ∞ as $\theta_0 \to 0$, but the last statement is not true because under the condition $\theta_0 = 0$, Equation (3.76), rigorously speaking, is not true (see Section 3.3).

If the condition given by Equation (3.73) is not satisfied, then the power spectral density of the Doppler fluctuations is determined by Equation (3.70). The meaning of the formula in Equation (3.70) is the same as the meaning of the formula in Equation (3.74). A difference is that according to the definition of the power spectral density at some frequency ω, it is necessary to carry out integration with respect to the variable ψ' in Equation (3.70). The result of this integration differs at various frequencies because the condition given by Equation (3.73) is not satisfied.

For this case, rigorously speaking, we cannot state that the shape of the power spectral density coincides with the shape of the radar antenna directional diagram in the plane of radar antenna deflection, i.e., in the plane $\psi' = 0$, because for various values of the variable ψ', the directional diagrams are different. However, integration with respect to the variable ψ' means averaging with respect to the variable ψ' and, in practice, the right side of Equation (3.70) is the average directional diagram in the plane of radar antenna deflection. In particular, if the directional diagram is symmetrical

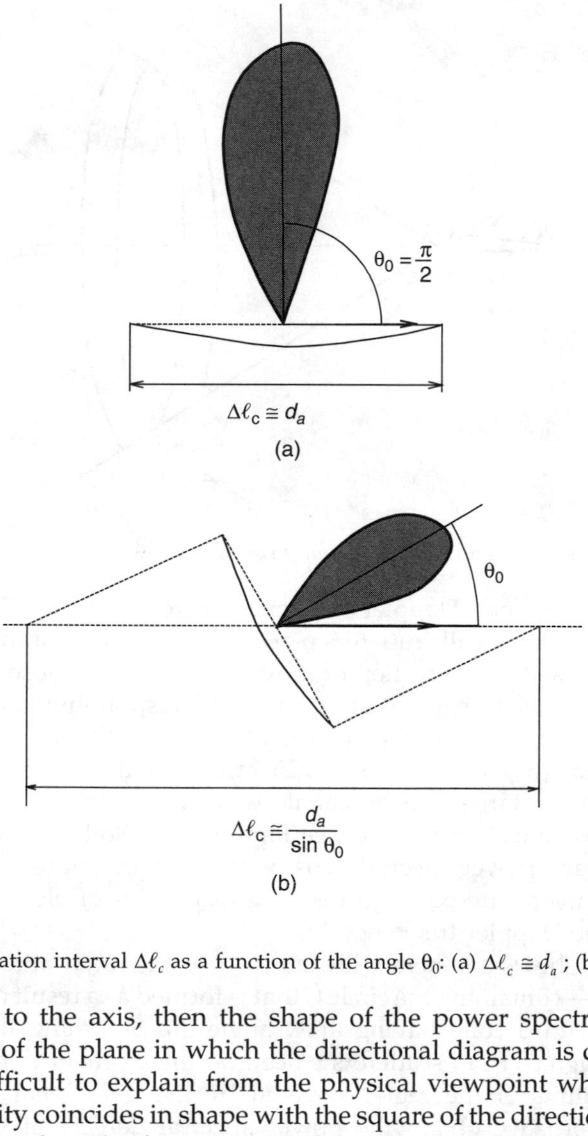

FIGURE 3.9
The space correlation interval $\Delta\ell_c$ as a function of the angle θ_0: (a) $\Delta\ell_c \cong d_a$; (b) $\Delta\ell_c \cong \frac{d_a}{\sin\theta_0}$.

with respect to the axis, then the shape of the power spectral density is independent of the plane in which the directional diagram is deflected.[33,34]

It is not difficult to explain from the physical viewpoint why the power spectral density coincides in shape with the square of the directional diagram with moving radar. The frequency ω of the target return signal from some elementary scatterers is different from the frequency ω_0 by the Doppler frequency

$$\Omega = \omega - \omega_0 = \Omega_{\max}\cos\theta = \Omega_{\max}\cos\beta\cos\gamma\,, \tag{3.78}$$

where ω_0 is the carrier frequency of the pulsed searching signal and θ is the angle between the direction of moving radar and that toward the given elementary scatterers.

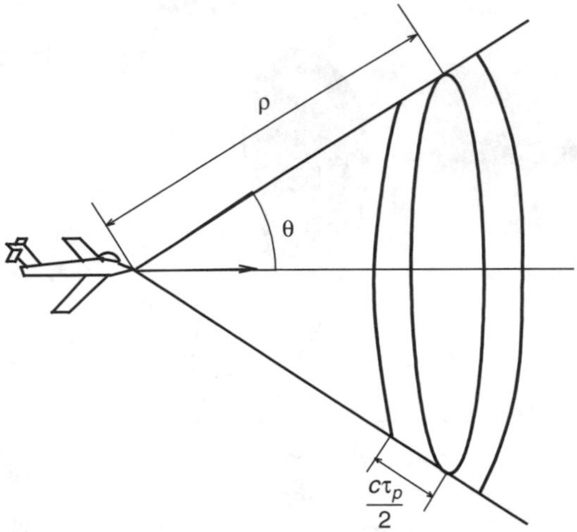

FIGURE 3.10
Formation of the signal with Doppler frequency $\Omega = \Omega_{\max} \cos\theta$.

Obviously, the same Doppler frequency corresponds to all scatterers in which $\theta = const$, i.e., to all scatterers placed on the surface of the cone whose axis coincides with the direction of moving radar and whose apex angle is 2θ. Scatterers placed on the surface of the corresponding cone, which are also in the pulse volume (i.e., are placed at a distance that is within the limits of the interval $[\rho - 0.25\ c\tau_p,\ \rho + 0.25\ c\tau_p]$ from the radar), take part in generation of the target return signals with the given Doppler frequency received at some instant of time (see Figure 3.10). Both the average power and the average power spectral density of the fluctuations with the given Doppler frequency are equal to the sum of powers of elementary signals with the same Doppler frequency.[35,36]

Figure 3.11 shows a sphere with radius ρ — the radar antenna is placed at the center — containing the circle C that is formed as a result of intersecting the sphere with a cone having apex angle 2θ. Summing the powers of elementary signals from scatterers placed on the segment C' of the circle C within the limits of the region covered by the directional diagram (the hatched region), we obtain the power spectral density at the frequency $\omega_0 + \Omega$, where Ω is determined by Equation (3.78).

Evidently, if a deflection of the directional diagram axis *OA* from the direction of moving radar *OD* is sufficiently large, so that the segment C' does not differ greatly from the straight line within the limits of the directional diagram (which is assumed to be narrow), then the power spectral density depends on the squared amplification coefficient along the direction *OB* as a function of the parameter. The total power spectral density coincides in shape with the square of the directional diagram by power within the cross section $B'B''$. If the radar antenna directional diagrams are the same at the planes that are perpendicular to the cross section $B'B''$ — i.e., the condition given by Equation

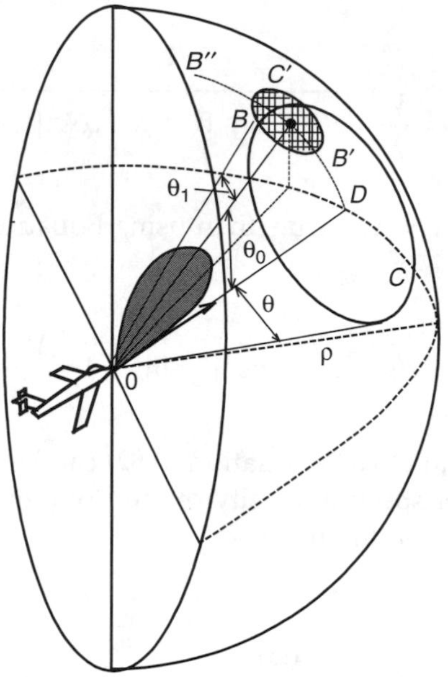

FIGURE 3.11
Formation of the power spectral density of the Doppler fluctuations.

(3.73) is true — then the results of integration over various circles at different values of the angle θ or frequency Ω are the same and do not impact the shape of the power spectral density. When these directional diagrams are not the same, then the power spectral density is different from the square of the radar antenna directional diagram at the cross section $B'B''$. The power spectral density will coincide with the square of the directional diagram averaged over all cross sections that are parallel to the cross section $B'B''$.

In practice, the real directional diagram consists of the main lobe and side lobes.[37,38] The side lobes of the radar antenna directional diagram form remainders of the power spectral density. These remainders must be taken into consideration when solving many problems that arise in practice.

3.2.2 The Power Spectral Density for the Gaussian Directional Diagram

When the condition given by Equation (3.47) is satisfied and the radar antenna directional diagram is Gaussian, reference to Equation (3.52) and Equation (3.56) shows that

$$R_g(\tau) = e^{-\pi \cdot \frac{\tau^2}{\tau_c^2}} \cdot e^{j(\omega_0 + \Omega_0)\tau}; \tag{3.79}$$

$$S_g(\omega) \cong e^{-\pi \cdot \frac{(\omega - \omega_0 - \Omega_0)^2}{4\pi^2 \Delta F^2}}, \tag{3.80}$$

where

$$\Delta F = \frac{1}{\tau_c} = \frac{V}{\lambda} \cdot \sqrt{2(\Delta_h^{(2)} \sin^2 \beta_0 + \Delta_v^{(2)} \cos^2 \beta_0 \sin^2 \gamma_0)} \ . \tag{3.81}$$

If the condition $\Delta_h = \Delta_v = \Delta_a$ is true, then using Equation (3.71) and Equation (3.81), we can write

$$\Delta F = \frac{V\Delta_a}{\lambda} \cdot \sqrt{2(\sin^2 \beta_0 + \cos^2 \beta_0 \sin^2 \gamma_0)} = \frac{V\Delta_a}{\lambda} \cdot \sqrt{2} \sin \theta_0 \ . \tag{3.82}$$

Formulae in Equation (3.80)–Equation (3.82) can be obtained also by convolution of the power spectral density of the Doppler fluctuations given by Equation (3.55) if the condition

$$S_g^{h,v}(\omega) \cong e^{-\pi \cdot \frac{\omega^2}{4\pi^2 \Delta F_{h,v}^2}} \tag{3.83}$$

is true, where

$$\Delta F_h = \frac{V\Delta_h}{\lambda} \cdot \sqrt{2} \sin \beta_0 \quad \text{and} \quad \Delta F_v = \frac{V\Delta_v}{\lambda} \cdot \sqrt{2} \cos \beta_0 \sin \gamma_0 \ . \tag{3.84}$$

Comparing Equation (3.81) and Equation (3.84), one can see that

$$\Delta F = \sqrt{\Delta F_h^2 + \Delta F_v^2} \ . \tag{3.85}$$

The following example can help us to estimate the effective power spectral density bandwidth: at $V = 300 \ \frac{\text{m}}{\text{sec}}$, $\lambda = 3$ cm, $\Delta_h = \Delta_v = 2°$, and $\beta_0 = \gamma_0 = 30°$, we obtain $\Delta F = 330$ Hz.

3.2.3 The Power Spectral Density for the Sinc-Directional Diagram

In particular cases that arise in practice, when the radar antenna directional diagram is deflected so that the velocity vector of moving radar is in the plane of one of the main cross sections of the directional diagram, and the condition given by Equation (3.73) is satisfied, there is no need to define the convolution of Equation (3.56), and the power spectral density of the Doppler fluctuations can be easily determined using Equation (2.103), Equation (3.53), Equation (3.54), Equation (3.58), and Equation (3.59). For example, under the conditions $\beta_0 \neq 0$ and $\gamma_0 = 0$ for the case of the **sinc**-diagram, we can write

$$S_g(\omega) = S_g^h(\omega) \approx \mathbf{sinc}^4\left(\tfrac{\omega-\omega_0-\Omega_0}{3\,\Delta F_h}\right), \tag{3.86}$$

where $\Omega_0 = \Omega_{max} \cos \beta_0$ and

$$\Delta F_h = \frac{4\,V\Delta_h}{3\,\lambda} \cdot \sin\beta_0 \tag{3.87}$$

is the effective power spectral density bandwidth. Comparing Equation (3.84) and Equation (3.87), one can see that with the same effective bandwidth of the directional diagram, the effective bandwidth of the Gaussian power spectral density is 6% more than the effective power spectral density bandwidth given by Equation (3.86).

For the case considered, the normalized correlation function has the following form:

$$R_g(\tau) = \begin{cases} 1-(1.5)^3\bar{\tau}^2+(1.5)^4\bar{\tau}^3 & \text{at} \quad \bar{\tau} \le 0.67; \\ 2[1-0.75\,\bar{\tau}]^3 & \text{at} \quad 0.67 < \bar{\tau} \le 1.33; \\ 0 & \text{at} \quad \bar{\tau} > 1.33, \end{cases} \tag{3.88}$$

where

$$\bar{\tau} = \frac{|\tau|}{\tau_c^h}; \tag{3.89}$$

$$\tau_c^h = (\Delta F_h)^{-1} \tag{3.90}$$

is the correlation interval.

The power spectral density given by Equation (3.86) is shown in Figure 3.12 (the solid line). The normalized correlation function $R_h(\tau)$ is shown in Figure 3.13 (the solid line). The normalized correlation function $R_h(\tau)$ is defined within the limits of the intervals $\bar{\tau} < 0.67$ and $\bar{\tau} > 0.67$ by various functions that have the same derivative at the point $\bar{\tau} = 0.67$. In the case of the Gaussian directional diagram, the power spectral density (see the dotted line in Figure 3.12) lacks side lobes and is very close to the power spectral density $S_g(\omega)$ given by Equation (3.86), within the main lobe. The normalized correlation functions are very close to each other. However, there is an essential difference between them.[39,40]

The normalized correlation function $R_h(\tau)$ given by Equation (3.88) is different from zero within the limits of the finite interval and the normalized correlation function $R_g(\tau)$ given by Equation (3.79) is different from zero within the limits of the infinite interval. This fact gives rise to a difference

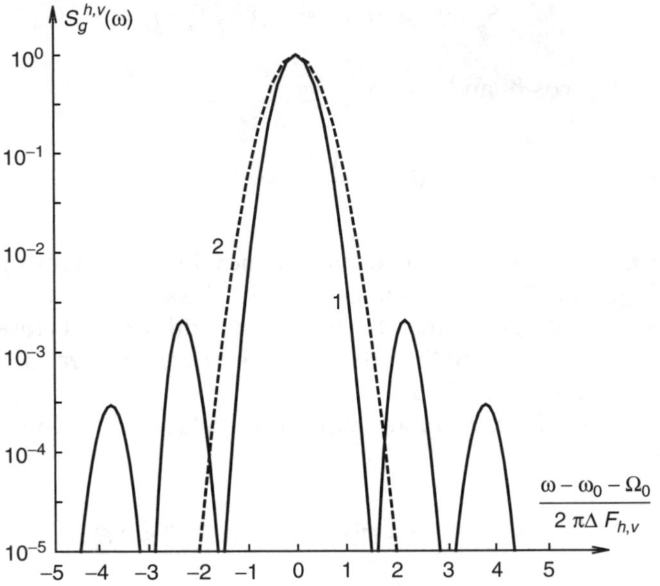

FIGURE 3.12
The power spectral density of the Doppler fluctuations at $\beta_0 = 0$ or $\gamma_0 = 0$: (1) the **sinc**-diagram; (2) the Gaussian directional diagram.

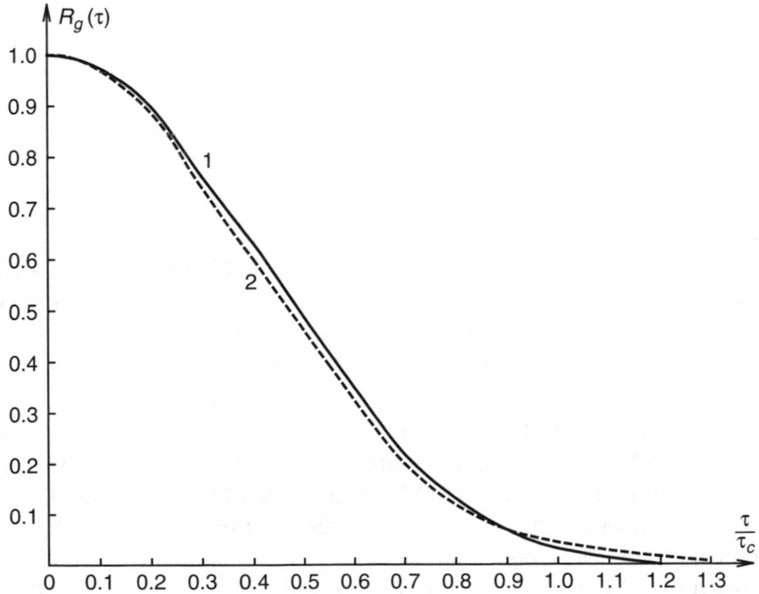

FIGURE 3.13
The normalized correlation function at $\beta_0 = 0$ or $\gamma_0 = 0$: (1) the **sinc**-diagram; (2) the Gaussian directional diagram.

in the power spectral density shown in Figure 3.12 — the power spectral density given by Equation (3.86) has side lobes.

In the general case, i.e., if the conditions $\beta_0 \neq 0$ and $\gamma_0 = 0$ are satisfied, the total normalized correlation function $R_g(\tau)$ given by Equation (3.52) is defined by the product of two normalized correlation functions $R_g^h(\tau)$ and $R_g^v(\tau)$ determined by Equation (3.88). In this case, a definition of the power spectral density in the form of the Fourier transform of the total normalized correlation function $R_g(\tau)$ [see Equation (3.52)] or convolution [see Equation (3.56)] of the power spectral densities [see Equation (3.86)] under the condition $\Delta F_h \neq \Delta F_v$ is very cumbersome.[41]

In accordance with Equation (2.103) and Equation (3.55), we can write

$$S_g^h(\omega) \cong \mathbf{sinc}^4\left(\tfrac{\omega}{3\,\Delta F_h}\right) \quad \text{and} \quad S_g^v(\omega) \cong \mathbf{sinc}^4\left(\tfrac{\omega}{3\,\Delta F_v}\right), \tag{3.91}$$

where

$$\Delta F_h = \frac{4\,V\Delta_h}{3\,\lambda} \cdot \sin\beta_0\,; \tag{3.92}$$

$$\Delta F_v = \frac{4\,V\Delta_v}{3\,\lambda} \cdot \cos\beta_0 \sin\gamma_0\,. \tag{3.93}$$

Under the condition $\Delta F_h = \Delta F_v = \Delta F$, which occurs if the condition

$$\mathrm{tg}\,\beta_0 = \frac{\Delta_v}{\Delta_h} \cdot \sin\gamma_0 \tag{3.94}$$

is satisfied, the convolution [see Equation (3.56)] of the power spectral densities given by Equation (3.91) gives us the following result:

$$S_g(\omega) \cong \frac{1}{\nu^4}\left[3 + \tfrac{90}{\nu^2} - \left(\tfrac{12}{\nu} + \tfrac{120}{\nu^3}\right)\sin\nu - \left(1 - \tfrac{60}{\nu^2}\right)\cos\nu - \tfrac{15}{\nu^3}\sin 2\nu\right], \tag{3.95}$$

where

$$\nu = \frac{2(\omega - \omega_0 - \Omega_0)}{3\,\Delta F}\,. \tag{3.96}$$

The power spectral density $S_g(\omega)$ given by Equation (3.95) is shown in Figure 3.14 by the solid line. If the condition $\Delta F_h \neq \Delta F_v$ is true, but the difference between ΔF_h and ΔF_v is not so high, we can use Equation (3.95) under the condition

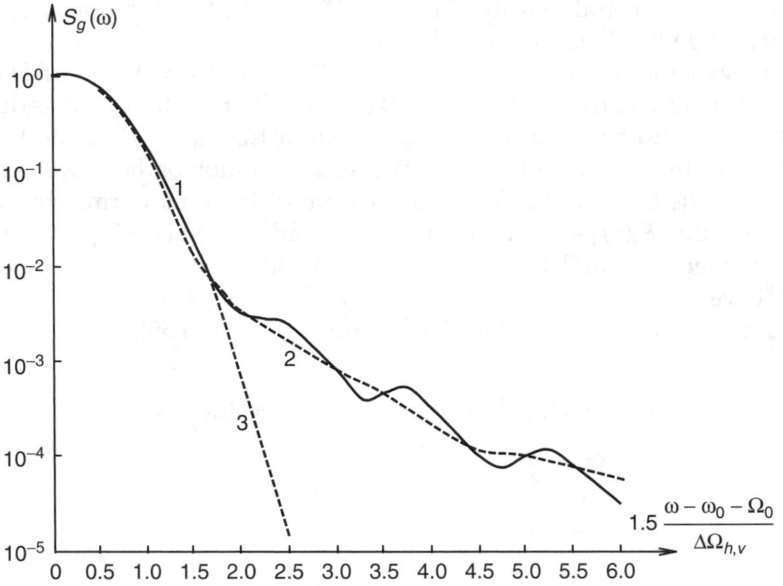

FIGURE 3.14
The exact and approximate power spectral densities of the Doppler fluctuations for **sinc**-diagram at $\beta_0 = 0$ or $\gamma_0 = 0$.

$$\Delta F = \sqrt{0.5(\Delta F_h^2 + \Delta F_v^2)}\,. \tag{3.97}$$

When the difference between ΔF_h and ΔF_v is very high, we can use approximate techniques to define a convolution of the power spectral densities in addition to computer modeling and computer calculation of integrals.[42]

3.2.4 The Power Spectral Density for Other Forms of the Directional Diagram

The approximate technique is based on representation of the squares of the radar antenna directional diagram $g_h^2(\varphi)$ and $g_v^2(\psi)$ and the power spectral densities $S_g^h(\omega)$ and $S_g^v(\omega)$ in the form of the sum of the main lobe (containing the main part of energy and obeying the Gaussian law) and of the remainders, taking into consideration the average side lobes:[43]

$$S_g^h(\omega) = S_h^{en}(\omega) + s_h(\omega) \quad \text{and} \quad S_g^v(\omega) = S_v^{en}(\omega) + s_v(\omega)\,. \tag{3.98}$$

Because the energy contained in the power spectral densities $s_h(\omega)$ and $s_v(\omega)$ is low, and the length of the power spectral densities $s_h(\omega)$ and $s_v(\omega)$ is large in comparison with that of the main lobe, the convolution has the following form:

$$S_g^h(\omega) * S_g^v(\omega) \cong S_h^{en}(\omega) * S_v^{en}(\omega) + S_h^{en}(\omega) * s_v(\omega) + S_v^{en}(\omega) * s_h(\omega)$$
$$\approx \int S_h^{en}(x) \cdot S_v^{en}(\omega - x)\, dx + s_v(\omega) \int S_h^{en}(x)\, dx + s_h(\omega) \int S_v^{en}(x)\, dx \; . \tag{3.99}$$

As the power spectral densities $S_h^{en}(\omega)$ and $S_v^{en}(\omega)$ are Gaussian, we can use the computer to calculate the formula in Equation (3.99) without any difficulty.

In many practical cases, the squared directional diagram with average near side lobes can be written in the following form[44]

$$g_{h,v}^2(\varphi) = e^{-2\pi\left(\frac{\varphi}{\varepsilon \Delta_{h,v}^{en}}\right)^2} + \frac{3}{8a^2} \cdot \left(\frac{\Delta_{h,v}^{en}}{\varphi}\right)^n \cdot \mathcal{L}_\varphi(b\varepsilon\, \Delta_{h,v}^{en}) \, , \tag{3.100}$$

where a, n, and b are coefficients characterizing energy or power of side-lobes and the rate of decrease of the side-lobes and depending on the distribution of current along the radar antenna diameter; $\Delta_{h,v}^{en} = \frac{\lambda}{d_{h,v}}$; $\varepsilon\, \Delta_{h,v}^{en} = \Delta_{h,v}$ is the effective width of the directional diagram by power; $d_{h,v}$ is the radar antenna diameter in the horizontal and vertical planes, respectively; ε is a coefficient ensuring equality between the width of the real squared directional diagram and the approximated Gaussian function at the level 0.5;

$$\mathcal{L}_\varphi(b\varepsilon\, \Delta_{h,v}^{en}) = \begin{cases} 0 & \text{at} \quad |\varphi| \; < \; b\varepsilon\, \Delta_{h,v}^{en} \; ; \\ 1 & \text{at} \quad |\varphi| \; > \; b\varepsilon\, \Delta_{h,v}^{en} \; . \end{cases} \tag{3.101}$$

Usually, in the determination of the power spectral density of the Doppler fluctuations, the shape of side lobes of the directional diagram need not be taken into consideration as they are highly smoothed in the formation of the convolution of the power spectral densities. In Table 3.1 the reader can find the exact formulae for the three forms of the directional diagrams used in practice and the corresponding power spectral densities obtained on the basis of Equation (3.55). In addition, approximate values of the parameters ε, a, n, and b are presented in Table 3.1.

Substituting Equation (3.100) in Equation (3.55), we can define the power spectral densities

$$S_g^{h,v}(\omega) \cong e^{-\pi \cdot \frac{\omega^2}{\Delta\Omega_{h,v}^2}} + G\left(\frac{\sqrt{2}\, \Delta\Omega_{h,v}}{\omega}\right)^n \cdot \mathcal{L}_\omega\left(\sqrt{2}\, b\Delta\Omega_{h,v}\right) , \tag{3.102}$$

TABLE 3.1

Radar Antenna Directional Diagrams and the Corresponding Spectral Power Densities

Distribution Law	$g_{h,v}(\varphi)$	$S_{h,v}(\omega)$	ε	a	n	b	G
Uniform	$\mathbf{sinc}^2\left(\frac{\pi\varphi}{\Delta_{h,v}^{en}}\right)$	$\mathbf{sinc}^4\left(\frac{\pi\omega}{\Delta\Omega}\right)$	0.96	10	4	1.1	$4\cdot10^{-3}$
$\cos\left(\frac{\pi\Delta\ell_c}{d_a}\right)$	$\left[\frac{\cos\left(\frac{\pi\varphi}{\Delta_{h,v}^{en}}\right)}{1-\left(\frac{2\varphi}{\Delta_{h,v}^{en}}\right)^2}\right]^2$	$\left[\frac{\cos\left(\frac{\pi\omega}{\Delta\Omega}\right)}{1-\left(\frac{2\omega}{\Delta\Omega}\right)^2}\right]^4$	1.28	16	8	1.4	$2\cdot10^{-4}$
$\cos^2\left(\frac{\pi\Delta\ell_c}{d_a}\right)$	$\left[\frac{\mathbf{sinc}\left(\frac{\pi\varphi}{\Delta_{h,v}^{en}}\right)}{1-\left(\frac{\varphi}{\Delta_{h,v}^{en}}\right)^2}\right]^2$	$\left[\frac{\mathbf{sinc}\left(\frac{\pi\omega}{\Delta\Omega}\right)}{1-\left(\frac{\omega}{\Delta\Omega}\right)^2}\right]^4$	1.55	10	12	1.5	$2\cdot10^{-5}$

where

$$\Delta\Omega_{h,v} = 2\,\pi\,\Delta F_{h,v} \tag{3.103}$$

and $\Delta F_{h,v}$ is determined by Equation (3.84), in which

$$\Delta_{h,v} = \frac{\varepsilon\,\lambda}{d_{h,v}} \quad \text{and} \quad G = \frac{3}{8\,a^2\varepsilon^n}\,, \tag{3.104}$$

(see Table 3.1). Using the formula in Equation (3.99), we can write

$$S_g(\omega) \cong \frac{e^{-\pi\cdot\frac{\omega^2}{\Delta\Omega^2}}}{\Omega} + 2^{2n}G\left[\frac{\Delta\Omega_h^{n-1}}{\omega^n}\cdot\mathcal{L}_\omega\left(\sqrt{2}\,b\Delta\Omega_h\right) + \frac{\Delta\Omega_v^{n-1}}{\omega^n}\cdot\mathcal{L}_\omega\left(\sqrt{2}\,b\Delta\Omega_v\right)\right], \tag{3.105}$$

where

$$\Delta\Omega = \sqrt{\Delta\Omega_h^2 + \Delta\Omega_v^2} = \frac{2\sqrt{2}\,\pi\,\varepsilon\,V}{\lambda}\cdot\sqrt{\Delta_h^{(2)}\sin^2\beta_0 + \Delta_v^{(2)}\cos^2\beta_0\sin^2\gamma_0}\,. \tag{3.106}$$

The accuracy of this technique can be estimated in the following manner. If the condition $\Delta\Omega_h = \Delta\Omega_v$ is true in Equation (3.105), we can apply this approximate procedure for the **sinc**-diagram (see the first row in Table 3.1). Curve 2 shown in Figure 3.14 is determined by Equation (3.105) (see the dotted line in Figure 3.14). Curve 2 is very close to the exact determination of the power spectral density (see curve 1 in Figure 3.14) given by Equation (3.95). The case of the Gaussian power spectral density is shown in Figure 3.14 too (curve 3). Comparative analysis made on the basis of Figure 3.14

shows that taking into consideration the real side lobes of the directional diagram is very important during estimation of the power spectral density in the peripheral region. In particular, we cannot approximate the radar antenna directional diagram by the rectangle function, as the rectangle function generates a square waveform power spectral density and cuts off all remainders of the power spectral density of the Doppler fluctuations.

3.3 The Doppler Fluctuations in the Arbitrarily Deflected Radar Antenna

3.3.1 General Statements

In this case, it is convenient to go from the coordinates β and γ to the polar coordinate α and θ (see Figure 3.15), in which the position of the radar antenna directional diagram axis is defined by the angles α_0 and θ_0. If the angle θ_0 and the directional diagram width are not so high in value, then $\sin\theta \approx \theta$, and we can write

$$\varphi = \beta - \beta_0 = \theta\cos\alpha - \beta_0\,; \tag{3.107}$$

$$\psi = \gamma - \gamma_0 = \theta\sin\alpha - \gamma_0\,; \tag{3.108}$$

$$d\varphi\, d\psi = \theta\, d\alpha\, d\theta\,, \tag{3.109}$$

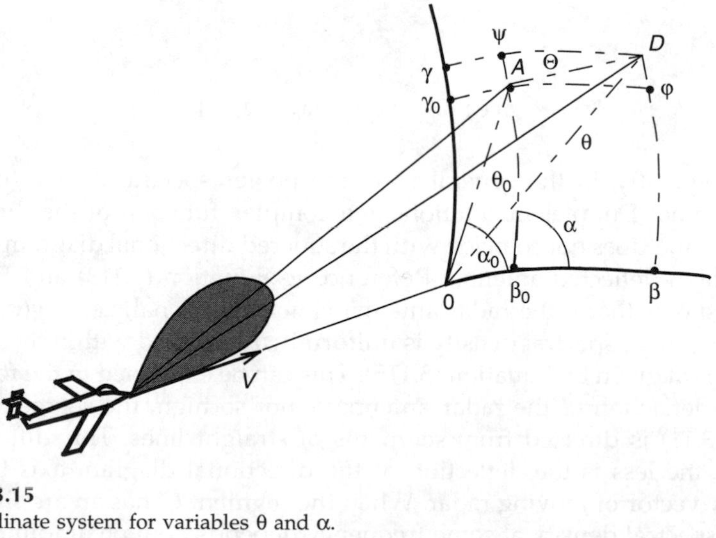

FIGURE 3.15
The coordinate system for variables θ and α.

where

$$\beta_0 = \theta_0 \cos\alpha_0 \qquad \text{and} \qquad \gamma_0 = \theta_0 \sin\alpha_0 . \tag{3.110}$$

Using Equation (3.78), we can write

$$\Omega(\varphi, \psi) = \Omega_{max} \cos\theta , \tag{3.111}$$

so based on Equation (3.8) we have

$$R_g(\tau) = N \int_0^{2\pi} \int_0^{\pi} g^2(\theta\cos\alpha - \beta_0,\ \theta\sin\alpha - \gamma_0) \cdot e^{j\Omega_{max}\tau\cos\theta}\theta \, d\theta \, d\alpha . \tag{3.112}$$

Multiplying Equation (3.112) by the factor $e^{j\omega_0\tau}$ and using the Fourier transform, we can write

$$S_g(\omega) \cong \int_0^{2\pi} g^2(\theta\cos\alpha - \beta_0,\ \theta\sin\alpha - \gamma_0) \, d\alpha , \tag{3.113}$$

where

$$\theta = \arccos \frac{\omega - \omega_0}{\Omega_{max}} \approx \sqrt{\frac{2(\omega_0 + \Omega_0 - \omega)}{\Omega_{max}}} \tag{3.114}$$

and

$$\omega \in [\, \omega_0 - \Omega_{max},\ \omega_0 + \Omega_{max}] . \tag{3.115}$$

Consequently, in the general case, the power spectral density of target return signal Doppler fluctuations is a complex function of the directional diagram and does not coincide with the squared directional diagram because of the high-deflected antenna. Reference to Equation (3.113) and Equation (3.114) shows that if the radar antenna is not directional, i.e., if $g(\varphi, \psi) \equiv 1$, then the power spectral density is uniformly distributed within the limits of the interval given by Equation (3.115). This can be explained in the following way. If deflection of the radar antenna is not so high, the segment C' (see Figure 3.11) is directed from segments of straight lines. This difference is greater, the less is the deflection of the directional diagram axis from the velocity vector of moving radar. When the segment C' has an arc shape, the power spectral density at some frequency depends both on amplification of the radar antenna along the direction OB and on amplification of the radar antenna in other directions. The shape of the power spectral density is changed due to this fact.

Consider the case when the directional diagram is symmetrical with respect to its own axis. In this case, the variables φ and ψ of the function $g(\varphi, \psi)$ are not separable, as a rule. An exception to this rule is the Gaussian directional diagram. So, in this case, we can write

$$g(\varphi, \psi) = g\left(\sqrt{\varphi^2 + \psi^2}\right) = g(\Theta), \tag{3.116}$$

where Θ is the angle between the directional diagram axis and the direction to scatterer (see Figure 3.15). The angle Θ is related to the angles θ and θ_0 by the following relationships:

$$\cos\Theta = \cos\theta_0 \cos\theta + \sin\theta_0 \sin\theta \cos(\alpha - \alpha_0). \tag{3.117}$$

When the directional diagram width is small, the following relationship:

$$\Theta^2 \cong 2\,[1 - \cos\theta_0 \cos\theta - \sin\theta_0 \sin\theta \cos(\alpha - \alpha_0)] \tag{3.118}$$

is true. Moreover, if the angle θ_0 is not large, we can write

$$\Theta^2 \cong \theta_0^2 + \theta^2 - 2\,\theta_0\theta\cos(\alpha - \alpha_0). \tag{3.119}$$

Substituting Equation (3.116)–Equation (3.119) in Equation (3.112) and Equation (3.113), we obtain various formulae for the correlation function and power spectral density, for example

$$S_g(\omega) \cong \int_0^{2\pi} g^2[\Theta(\alpha)]\, d\alpha \cong \int_0^{2\pi} g^2\left(\sqrt{\theta_0^2 + \theta^2 - 2\,\theta_0\theta\cos(\alpha - \alpha_0)}\right) d\alpha\,. \tag{3.120}$$

Under the conditions

$$\beta_0 = 0 \quad \text{and} \quad \gamma_0 = 0 \qquad \text{or} \qquad \theta_0 = 0 \quad \text{and} \quad \Theta = 0, \tag{3.121}$$

we can write

$$S_g(\omega) \cong g^2(\theta) = g^2\left(\arccos \tfrac{\omega - \omega_0}{\Omega_{\max}}\right) \cong g^2\left(\sqrt{2 \cdot \tfrac{\omega_0 + \Omega_{\max} - \omega}{\Omega}}\right). \tag{3.122}$$

The power spectral density is maximal under the condition $\omega = \omega_0 + \Omega_{\max}$ and is not symmetric. The effective bandwidth $\Delta F_{0.5}$ of this power spectral density at the level 0.5 is related to the width of the square of the directional diagram $\Delta_{0.5}^{(2)}$ at that level by the following relationship

$$\Delta F_{0.5} = \frac{V}{4\lambda} \cdot (\Delta_{0.5}^{(2)})^2 . \tag{3.123}$$

Because the width $\Delta_{0.5}^{(2)}$ is proportional to the wavelength λ, the effective power spectral density bandwidth in Equation (3.123) is also proportional to the wavelength λ, unlike in Equation (3.76).

Under the condition $\theta_0 >> \Delta_a$, we can write $\cos(\alpha - \alpha_0) \approx 1$ in Equation (3.120). Then

$$S_g(\omega) \cong g^2(\theta - \theta_0) = g^2\left(\frac{\omega_0 + \Omega_{max}\cos\theta_0 - \omega}{\Omega_{max}\sin\theta_0}\right) , \tag{3.124}$$

i.e., with the high-deflected radar antenna, the power spectral density coincides in shape with the squared radar antenna directional diagram, which is the expected result [compare with Equation (3.74)]. For example, in the case of the circular radar antenna with a uniform distribution of electromagnetic field, we can write

$$S_g(\omega) \cong \left[\frac{2J_1(\pi\nu)}{\pi\nu}\right]^4 , \tag{3.125}$$

where

$$\nu = \frac{\omega_0 + \Omega_{max}\cos\theta_0 - \omega}{\Omega_{max}\sin\theta_0} . \tag{3.126}$$

3.3.2 The Gaussian Directional Diagram

Let the radar antenna directional diagram be Gaussian. Then, from Equation (3.113) it follows that

$$S_g(\omega) \cong \int_0^{2\pi} e^{p(\Omega)\cos 2\alpha + q(\Omega)\cos(\alpha - \alpha') - r(\Omega)} d\alpha , \tag{3.127}$$

where

$$p(\Omega) = \frac{(\chi - \chi^{-1})\,\Omega}{2\,\Delta\Omega_0} ; \tag{3.128}$$

$$q(\Omega) = \sqrt{\frac{2\,\Omega_{\max}\Omega\,(\gamma_0^2\chi^2 + \beta_0^2\chi^{-2})}{\Delta\Omega_0^2}}\;; \tag{3.129}$$

$$r(\Omega) = \frac{(\chi + \chi^{-1})\,\Omega}{2\,\Delta\Omega_0} + \frac{(\gamma_0^2\chi + \beta_0^2\chi^{-1})\,\Omega_{\max}}{2\,\Delta\Omega_0}\;; \tag{3.130}$$

$$\operatorname{tg}\alpha' = \frac{\gamma_0}{\beta_0}\cdot\chi^2\;; \tag{3.131}$$

$$\chi = \frac{\Delta_h}{\Delta_v}\;; \tag{3.132}$$

$$\Omega = \omega_0 + \Omega_{\max} - \omega\;; \tag{3.133}$$

$$\Delta\Omega_0 = \frac{\Omega_{\max}\Delta_h\Delta_v}{4\,\pi} = \frac{V\Delta_h\Delta_v}{\lambda}\,. \tag{3.134}$$

To determine the power spectral density of the Doppler fluctuations, we can use the series expansion given by Equation (1.15). Then we can write

$$S_g(\omega) \cong \Big[I_0(p)\cdot I_0(q) + 2\sum_{m=1}^{\infty} I_m(p)\cdot I_m(q)\cos 2m\alpha'\Big]\cdot e^{-r}\,. \tag{3.135}$$

Equation (3.135) is similar to Equation (1.16) if the following conditions are applied to Equation (1.16)

$$S = \sqrt{\tfrac{2\Omega}{\Delta\Omega_0}}\;;\quad S_0 = \sqrt{\tfrac{\Omega_{\max}}{\Delta\Omega_0}}\cdot\theta_0\;;\quad \text{and}\quad \varphi_0 = \alpha'\,. \tag{3.136}$$

Because of this, we can use results discussed in Section 1.2 to determine the power spectral density given by Equation (3.127), as

$$S_g(\omega) \cong \frac{f\left(\sqrt{\frac{2\Omega}{\Delta\Omega_0}}\right)}{\sqrt{\frac{2\Omega}{\Delta\Omega_0}}}\,. \tag{3.137}$$

Using Figure 1.3–Figure 1.8 and Equation (3.137), we can easily construct some curves of the power spectral density for various conditions. In some cases, Equation (3.135) can be fundamentally simplified. If the directional diagram is symmetric with respect to its own axis, i.e., if $\Delta_h = \Delta_v = \Delta_a$, then $p(\Omega) = 0$ and[31]

$$S_g(\omega) \cong I_0\left(2\,\delta_0\sqrt{\tfrac{\Omega}{\Delta\Omega_0}}\right)\cdot e^{-\frac{\Omega}{\Delta\Omega_0}-\delta_0^2}, \tag{3.138}$$

where

$$\delta_0 = \frac{\sqrt{2\,\pi}\,\theta_0}{\Delta_a}. \tag{3.139}$$

Equation (3.138) is similar to Equation (1.21). It is not difficult to prove that under the condition $\theta_0 \leq 0.4\,\Delta_a$, the maximum power spectral density is defined at the frequency $\omega = \omega_0 + \Omega_{max}$, and their effective bandwidth has the following form:

$$\Delta\Omega = \frac{V\Delta_a^{(2)}}{\lambda}\cdot e^{\frac{2\,\pi\,\theta_0^2}{\Delta_a^{(2)}}}. \tag{3.140}$$

Under the condition $\theta_0 \geq 0.5\,\Delta_a$, we cannot use Equation (3.140), but we use Equation (3.76) and the error is less than 10%.

When $\theta_0 = 0$, Equation (3.138) can be rewritten in the following form:

$$S_g(\omega) \cong e^{-\frac{\Omega}{\Delta\Omega_0'}}. \tag{3.141}$$

The effective power spectral density bandwidth determined by Equation (3.141) has the following form:

$$\Delta\Omega = \Delta\Omega_0' = \frac{V\Delta_a^{(2)}}{\lambda}. \tag{3.142}$$

Usually, the effective bandwidth $\Delta\Omega_0'$ is very small but not equal to zero, which follows from Equation (3.82). For example, at $V = 300$ m/sec, $\Delta_a = 2°$, and $\lambda = 3$ cm, we obtain the effective bandwidth $\Delta F = 2$ Hz. Comparing Equation (3.82) and Equation (3.141), one can see that under the condition $\theta_0 = 90°$, the effective power spectral density bandwidth is $\frac{2\sqrt{2\,\pi}}{\Delta_a}$ times more than that at $\theta_0 = 0$. In other words, the difference is about some hundred times.

The normalized power spectral density given by Equation (3.138) is shown in Figure 3.16 at various deflections of the directional diagram axis: the angle θ_0 is varied from 0 to Δ_a. Under the condition θ_0 to Δ_a, the power spectral density given by Equation (3.80) is shown by the dotted line in Figure 3.16. This power spectral density is correct for high values of the angle θ_0 and coincides very well with the middle part of the exactly determined power spectral density. However, there is an essential difference in the remainders. If $\Delta_h \neq \Delta_v$, but $\theta_0 = 0$, then $q(\Omega) = 0$ and we can write

$$S_g(\omega) \cong I_0\left[\frac{(\chi-\chi^{-1})\,\Omega}{2\Delta\Omega_0}\right]\cdot e^{-\frac{(\chi+\chi^{-1})\,\Omega}{2\Delta\Omega_0}}. \tag{3.143}$$

Under the condition $\Delta_h \approx \Delta_v$, the power spectral density given by Equation (3.143) is not essentially different from that determined by Equation (3.141). The power spectral density given by Equation (3.143) is shown in Figure 3.17.

The rate of decrease of the power spectral density $S_g(\omega)$ depends mainly on the greatest values of Δ_h and Δ_v if the difference between Δ_h and Δ_v is high. The effective power spectral density bandwidth given by Equation (3.143) then takes the following form:

$$\Delta\Omega = \frac{V\Delta_h\Delta_v}{\lambda}. \tag{3.144}$$

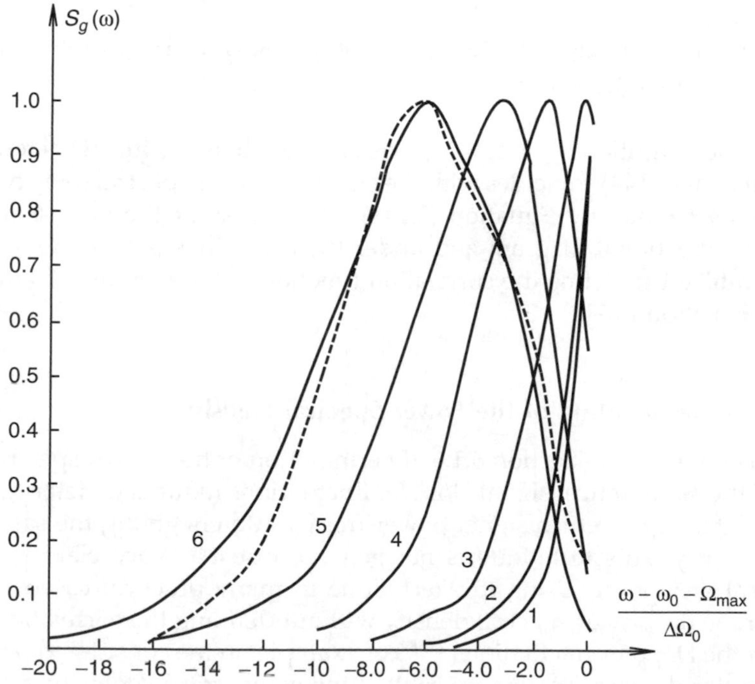

FIGURE 3.16
The power spectral density of the Doppler fluctuations. The Gaussian directional diagram is deflected, $\Delta_h = \Delta_v = \Delta_a$: (1) $\frac{\theta_0}{\Delta_a} = 0$; (2) $\frac{\theta_0}{\Delta_a} = 0.2$; (3) $\frac{\theta_0}{\Delta_a} = 0.4$; (4) $\frac{\theta_0}{\Delta_a} = 0.6$; (5) $\frac{\theta_0}{\Delta_a} = 0.8$; (6) $\frac{\theta_0}{\Delta_a} = 1$.

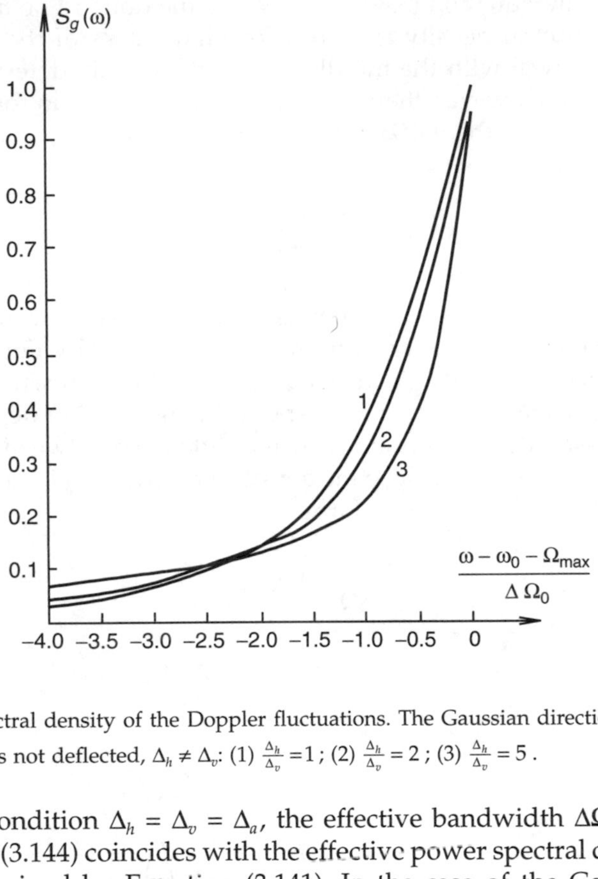

FIGURE 3.17
The power spectral density of the Doppler fluctuations. The Gaussian directional diagram of radar antenna is not deflected, $\Delta_h \neq \Delta_v$: (1) $\frac{\Delta_h}{\Delta_v} = 1$; (2) $\frac{\Delta_h}{\Delta_v} = 2$; (3) $\frac{\Delta_h}{\Delta_v} = 5$.

Under the condition $\Delta_h = \Delta_v = \Delta_a$, the effective bandwidth $\Delta\Omega$ determined by Equation (3.144) coincides with the effective power spectral density bandwidth determined by Equation (3.141). In the case of the Gaussian radar antenna directional diagram and under the conditions $\beta_0 = 0$ and $\gamma_0 = 0$, it is not difficult to define the correlation function of the Doppler fluctuations using Equation (3.112).

3.3.3 Determination of the Power Spectral Density

As was discussed in Section 3.1.3, determination of the power spectral density of the target return signal Doppler fluctuations reduces to determination of the total target return signal power from scatterers giving the same shift in frequency. This technique is not general, but it is very clear from the physical viewpoint. Using this technique in many practical cases, we can determine the power spectral density without defining the correlation function of the Doppler fluctuations.[25,45] For example, in the considered problem, an annular domain (see Figure 3.10) is the geometrical center of scatterers with the same Doppler shift in frequency. The target return signal power from the scatterer scanned under the angle Θ (see Figure 3.15) is proportional to the squared radar antenna directional diagram $g^2(\Theta)$. The total target return signal power from the annular domain (θ, $\theta + d\theta$; ρ, $\rho + 0.5\, c\tau_p$) has the following form:

$$dp = m_0 \int p_1 g^2(\Theta)\, dQ\,, \tag{3.145}$$

where m_0 is the average number of scatterers per volume unit; p_1 is the target return signal power from the individual scatterer with effective scattering area S_1 when the scatterer is placed on the directional diagram axis [see Equation (2.107)]; Q is the integrated domain [see Equation (2.108)];

$$dQ = \rho^2 \sin\theta\, d\rho\, d\theta\, d\alpha \tag{3.146}$$

is the volume element. Under the conditions $\rho = const$ and $\theta = const$, which are satisfied within the considered volume, we can write

$$dp = m_0 p_1 (0.5 c\tau_p)\rho^2 \sin\theta\, d\theta \int_0^{2\pi} g^2[\Theta(\alpha)]\, d\alpha\,. \tag{3.147}$$

Equation (3.147) together with Equation (3.78) is the parametric form of the power spectral density; the angle θ is the parameter. Reference to Equation (3.78) shows that

$$d\Omega = -\ \Omega_{max} \sin\theta\, d\theta\,. \tag{3.148}$$

Going from the power dp to the power spectral density

$$S(\Omega) = \frac{dp\,[\Theta(\Omega)]}{d\Omega}\,, \tag{3.149}$$

we obtain the well-known formula [see Equation (3.120)]. This technique can be successfully used in the determination of the power spectral density of the Doppler fluctuations under scanning of the two-dimensional (surface) target (see Section 4.8) and in the study of some forms of chaotic motion of scatterers (see Section 7.2).

3.4 The Total Power Spectral Density with the Pulsed Searching Signal

3.4.1 General Statements

Reference to Equation (3.12) shows that if the three-dimensional (space) target is scanned by the pulsed searching signal of moving radar, the total normalized correlation function $R(\tau)$ of target return signal fluctuations is defined by the product of the periodic normalized correlation function $R_p(\tau)$ of the fluctuations in the radar range, or in the distance between the radar and scatterer, and the nonperiodic normalized correlation function $R_g(\tau)$ of the Doppler fluctuations (see Figure 3.1–Figure 3.3). Naturally, in this case, the total normalized correlation function $R(\tau)$ is not a periodic function because with an increase in the value of $\tau = nT_p$, the waves of the normalized correlation function $R(\tau)$ decrease in value due to the Doppler fluctuations. This destruction of correlation is strong; the value of n is high (see Figure 3.3a).

The total power spectral density $S(\omega)$ is defined by convolution of the linear power spectral density $S_p(\omega)$ given by Equation (3.15) and the continuous power spectral density $S_g(\omega)$ given by Equation (3.43) (see Figure 3.1–Figure 3.3):

$$S(\omega) \cong \int S_p(x) * S_g(\omega - x)\, dx = \sum_{n=0}^{\infty} \int S_p^{en}(x) \cdot \delta(x - n\Omega_p') \cdot S_g(\omega - x)\, dx$$

$$= \sum_{n=0}^{\infty} S_p^{en}(n\Omega_p') \cdot S_g(\omega - n\Omega_p') \approx S_p^{en}(\omega) \sum_{n=0}^{\infty} S_g(\omega - n\Omega_p') \,. \tag{3.150}$$

We have to use the result of convolution at the frequency $\omega_0 + \Omega_0$ [see Equation (3.150)] if this fact has not been taken into consideration in the power spectral densities $S_p(\omega)$ or $S_g(\omega)$. As a rule, the power spectral density $S_g(\omega - n\Omega_p')$ is very narrow in comparison with the power spectral density $S_p^{en}(\omega)$. The total power spectral density $S(\omega)$ can be approximately considered as the product of the wedge-like Doppler power spectral densities $\sum_{n=0}^{\infty} S_g(\omega - n\Omega_p')$ and the envelope $S_p^{en}(\omega)$ of the power spectral density $S_p(\omega)$ of the fluctuations in the radar range (see Figure 3.3b). These power spectral densities have been determined in Section 3.1–Section 3.3 for various cases. Definition of the total power spectral density based on Equation (3.150) is not difficult. For example, with the square waveform pulsed searching signal and the Gaussian radar antenna directional diagram, the deflection of which is very high, reference to Equation (3.19) and Equation (3.80) shows that the total power spectral density of the target return signal fluctuations can be written in the following form:

$$S(\omega) \cong \sum_{n=0}^{\infty} \mathbf{sinc}^2\left(n\Omega'_p, 0.5\tau_p\right) \cdot e^{-\frac{\pi}{\Delta\Omega^2} \cdot (\omega - \omega_0 - \Omega_0 - n\Omega'_p)^2}, \tag{3.151}$$

where $\Delta\Omega$ is determined by Equation (3.81).

3.4.2 Interperiod Fluctuations in the Glancing Radar Range

Consider the total normalized correlation function $R(\tau)$ given by Equation (3.12). Reasoning that $\tau = nT'_p$, we can obtain the normalized correlation function of the interperiod target return signal fluctuations in the pure form based on the total normalized correlation function $R(\tau)$. Under the condition $\tau = nT'_p$, we can write $R_p(\tau) = 1$ and $R(\tau) = R_g(\tau)$. The condition $\tau = nT'_p$ means that the correlation function is determined for widely spaced instants of time on time periods that are compressed n times or expanded $\mu = 1 + \frac{2V_r}{c}$ times of periods.[46] This correlation function characterizes the Doppler fluctuations caused by scanning the same pulse volume with moving radar. In other words, this correlation function takes into consideration changes in distance between the radar and scatterers during the time nT'_p (see Figure 3.18). The power spectral density of these target return signal fluctuations coincides with the power spectral density $S_g(\omega)$ shifted in frequency by $\omega_0 + \Omega_0$. The power spectral density $S_g(\omega)$ is investigated in Section 3.2 and Section 3.3. In other words, we can state that the power spectral density of these fluctuations coincides with the main wave of the total power spectral density $S(\omega)$.

3.4.3 Interperiod Fluctuations in the Fixed Radar Range

It is worthwhile to consider the interperiod fluctuations not only in the glancing radar range, but also in the fixed radar range, i.e., under the condition $\tau = nT_p$ as well.[47,48] This means that the correlation function of the fluctuations is determined for widely spaced instants of time on n undistorted periods, i.e., for the instants of time fixed with respect to the instant of time of generation of the pulsed searching signal. In this case, the correlation function characterizes the fluctuations arising by scanning the pulse volume, which is located at a fixed distance from the moving radar. In other words, we can state that the pulse volume moves with the radar (or aircraft).

Unlike the previous case in Section 3.4.2, the fluctuations in the fixed radar range are caused by two reasons. The fluctuations caused by the moving pulse volume (the distance between the moving radar and the moving pulse volume

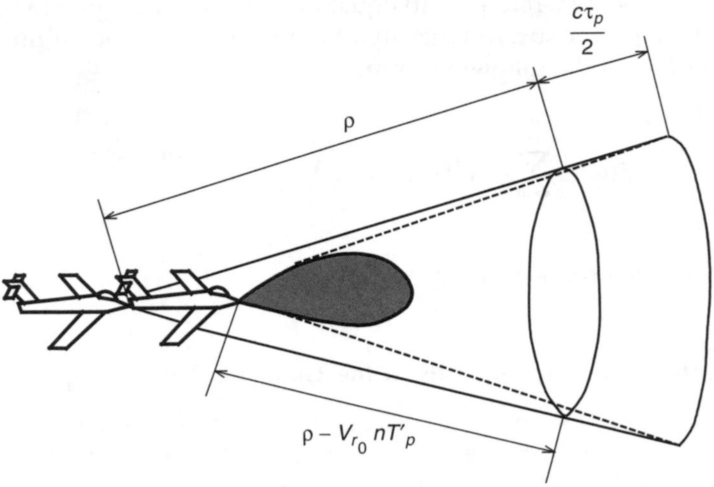

FIGURE 3.18
Fluctuations forming in the glancing radar range.

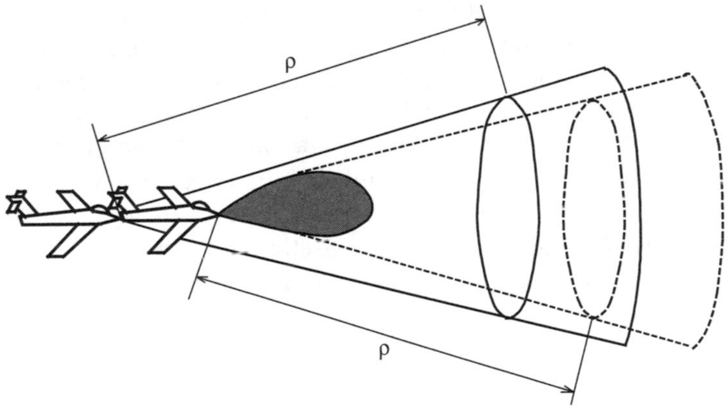

FIGURE 3.19
Fluctuations forming in the fixed radar range.

is fixed) arise in addition to the Doppler fluctuations. These fluctuations are, as a rule, identical to the fluctuations caused by exchange of scatterers in the case of square waveform pulsed searching signals. In the case of the pulsed searching signal with an arbitrary shape, the boundaries of the pulse volume are not clear. This can be explained by amplitude changes in the elementary signals due to their modulation by the envelope of the pulsed searching signal amplitude with moving radar.

It is necessary to consider the interperiod fluctuations both in the fixed radar range and in the glancing radar range because in interperiod signal processing by navigational systems, the target return signal at the receiver or detector input can be shifted in time both by the value $\tau = nT_p$ (radar motion is not taken into consideration) and by the value $\tau = nT_p'$ (radar motion is taken into consideration). Moreover, in some cases arising in practice, for example, when tracking a point target against a background of the three-dimensional (space) or the two-dimensional (surface) target, signal processing must be carried out at intervals coinciding with the value of T_p'', not of T_p or T_p', where the time interval T_p'' takes into consideration the moving tracking target. We do not consider this case. It is a continuation of the next case.

Replace the radial velocity V_{r0} with the approach velocity $V_{r0} + V_{r_t}$ of the radar and target. Assume that $\tau = nT_p$. Reference to Equation (3.5)–Equation (3.8) shows that

$$R^\circ(\tau) = R_p^{en}(\bar{\mu}\tau) \cdot R_g(\tau)\,, \tag{3.152}$$

where

$$\bar{\mu} = \mu - 1 = \frac{2\,V_{r_0}}{c} = \frac{2\,V}{c} \cdot \cos\beta_0 \cos\gamma_0 \quad << \quad 1\,; \tag{3.153}$$

$$R_p^{en}(\bar{\mu}\tau) = \frac{\int \boldsymbol{P}(z) \cdot \boldsymbol{P}^*\left(z + \frac{2\,V_{r0}}{c}\right) dz}{\int \boldsymbol{P}^2(z)\, dz}\,. \tag{3.154}$$

Thus, the normalized correlation function of the interperiod fluctuations in the fixed radar range is defined by the product of the normalized correlation function $R_g(\tau)$ of the interperiod fluctuations in the glancing radar range and the normalized correlation function $R_p^{en}(\bar{\mu}\tau)$ of the interperiod fluctuations with the highly expanded scale — $\bar{\mu}\tau$ instead of $\mu\tau$. The normalized correlation function $R_p^{en}(\bar{\mu}\tau)$ defines the fluctuations caused by exchange of scatterers.

The correlation interval of the target return signal fluctuations caused by exchange of scatterers is equal to the time required for the pulse volume — the resolution element — moving with velocity V_{r0} to be renewed, and the power spectral density of the fluctuations caused by exchange of scatterers depends on the pulsed searching signal shape and is determined by Equation (3.17), in which μ is replaced with $\bar{\mu}$:

$$S_p(\omega) \cong \left| \int P(t) \cdot e^{-\frac{j\omega t}{\bar{\mu}}}\, dt \right|^2\,. \tag{3.155}$$

For example, with the square waveform pulsed searching signal, the normalized correlation function given by Equation (3.154) takes the following form

$$R_p^{en}(\bar{\mu}\tau) = 1 - \frac{|\bar{\mu}\tau|}{\tau_p} = 1 - \frac{|\tau|}{\tau_c} \quad \text{at} \quad |\tau| \leq \tau_c, \tag{3.156}$$

where

$$\tau_c = \frac{\tau_p}{\bar{\mu}} = \frac{c\tau}{2\,V\cos\beta_0\cos\gamma_0} \quad >> \quad \tau_p \tag{3.157}$$

is the time required for the radar moving with velocity V to traverse a distance $\frac{c\tau_p}{2\cos\beta_0\cos\gamma_0}$.

The corresponding power spectral density of the fluctuations coincides with the power spectral density of the square waveform pulsed searching signal with length τ_c given by Equation (3.157):

$$S_p(\omega) \cong \mathbf{sinc}^2(0.5\omega\tau_c). \tag{3.158}$$

The effective bandwidth $\Delta F = \frac{1}{\tau_c}$ of the power spectral density of the target return signal fluctuations is independent both of the wavelength and the radar antenna directional diagram width and is defined by the pulsed searching signal duration, velocity of moving radar, and directional diagram orientation. When the directional diagram is not deflected, the effective bandwidth ΔF of the power spectral density is maximal, but the maximum is very steep.

Consider this example: at V = 300 m/sec, τ_p = 0.5 μsec, β_0 = 0°, and γ_0 = 0,we obtain the effective bandwidth ΔF = 4 Hz; if β_0 = 45° and γ_0 = 45° we obtain the effective bandwidth ΔF = 2 Hz. Usually, the fluctuations caused by exchange of scatterers are very slow. In the majority of cases, we can neglect these fluctuations in comparison with the Doppler fluctuations, which have a power spectral density bandwidth that is 10 or 10^2 times more [see the example in Section 3.2.2, Equation (3.85)]. However, there are exceptions to this rule; for example, if the pulsed searching signal duration is very low in value, or in the case of the frequency-modulated pulsed signal, when the power spectral density $S_p(\omega)$ in the radar range is expanded, or if the directional diagram is not deflected and the power spectral density $S_g(\omega)$ is narrowed down [see the example in Section 3.3.2, Equation (3.141)].

In the case of the square waveform linear-frequency modulated pulsed signals with the deviation $\Delta\omega_M$, we can write based on Equation (3.31)

$$R_p^{en}(\bar{\mu}\tau) = \frac{\sin\left[0.5\Delta\omega_M \bar{\mu}\tau\left(1 - \frac{|\bar{\mu}\tau|}{\tau_p}\right)\right]}{0.5\Delta\omega_M \bar{\mu}\tau}. \tag{3.159}$$

Reference to Equation (3.160) shows that as $\Delta\omega_M \to 0$, we obtain the normalized correlation function given by Equation (3.156) and, at $\Delta\omega_M >> \frac{2\pi}{\tau_p}$, the normalized correlation function is determined by the function $\mathbf{sinc}(0.5\Delta\omega_M \bar{\mu}\tau)$, with the correlation interval given by

$$\tau_c = \frac{1}{\bar{\mu}\,\Delta f_M} = \frac{c}{2\,V_{r_0}\Delta f_M} = \frac{c\tau_p}{2\,V_{r_0} D}, \tag{3.160}$$

where $D = \Delta f_M \tau_p$. Therefore, the correlation interval given by Equation (3.160) is D times less than the correlation interval given by Equation (3.157). Consequently, the effective bandwidth is D times greater than the effective power spectral density bandwidth given by Equation (3.158). The shape of the power spectral density tends to approach the square waveform shape. For an arbitrary value of D, the power spectral density coincides with the envelope of the power spectral density given by Equation (3.32) and the correlation interval is determined by Equation (3.36) if τ_p is replaced with $\frac{\tau_p}{\bar{\mu}}$. Under these conditions, we can use Figure 3.5–Figure 3.7.

In the cases of the Gaussian pulsed searching signal without linear-frequency modulation and with linear-frequency modulation, both the normalized correlation function and the power spectral density of the target return signal fluctuations are defined by the Gaussian law. The effective bandwidth and correlation interval of the power spectral density have the following form:

$$\Delta F = \frac{1}{\tau_c} = \frac{\sqrt{2\,(1+D^2)}\;\,V\cos\beta_0\cos\gamma_0}{c\tau_p}. \tag{3.161}$$

When $D = 0$, the effective bandwidth ΔF is $\sqrt{2}$ times less than the effective power spectral density bandwidth given by Equation (3.158). If $D >> 1$, the effective power spectral density bandwidth increases D times.

If the pulsed searching signal and directional diagram are Gaussian, then in the case of high-deflected radar antenna, where the power spectral density of the Doppler fluctuations is determined by Equation (3.80), the total power spectral density of the slow fluctuations given by Equation (3.152) is Gaussian too, with the effective bandwidth given by

$$\Delta F = \sqrt{\Delta F_1^2 + \Delta F_2^2}, \tag{3.162}$$

where the effective bandwidth ΔF_1 is determined by Equation (3.81) and the effective bandwidth ΔF_2 is given by Equation (3.161). Equation (3.162) allows us to estimate an extension of the power spectral density of the fluctuations caused by exchange of scatterers without any difficulty.

When the radar antenna is not deflected, the power spectral density of the Doppler fluctuations has the form of the exponential function given by Equation (3.141) and the convolution of this power spectral density and the Gaussian power spectral density of the fluctuations caused by exchange of scatterers has the following form

$$S(\omega) \cong \left\{1 + \Phi\left[\sqrt{\pi}\left(\frac{\Omega}{\Delta\Omega} - \frac{\Delta\Omega_2}{2\pi\Delta\Omega_0'}\right)\right]\right\} \cdot e^{-\frac{\Omega}{\Delta\Omega_0'}}, \tag{3.163}$$

where

$$\Omega = \omega_0 + \Omega_{\max} - \omega, \tag{3.164}$$

$\Delta\Omega_2$ is determined by Equation (3.161), and $\Delta\Omega_0'$ is given by Equation (3.141). The power spectral density given by Equation (3.163) is shown in Figure 3.20 under the condition $\Delta\Omega_2 = \Delta\Omega_0'$.

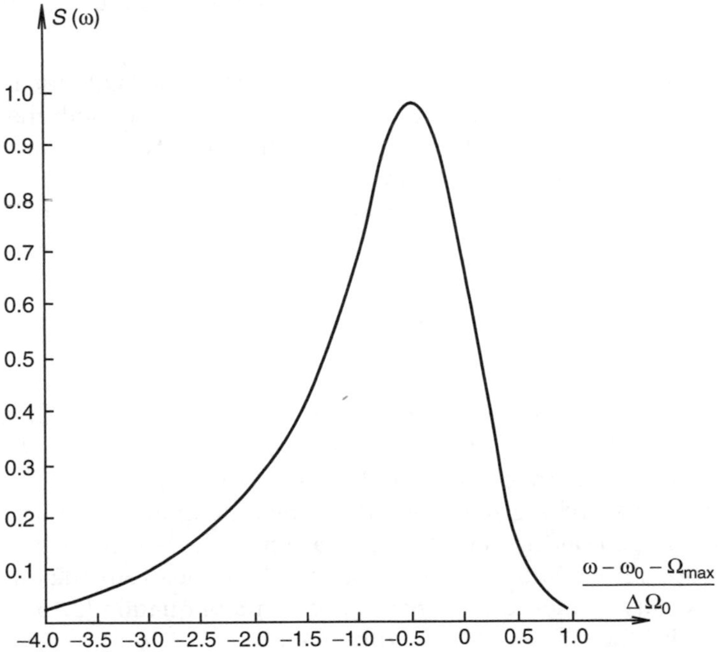

FIGURE 3.20
The total power spectral density. The Gaussian directional diagram, $\theta_0 = 0$.

3.4.4 Irregularly Moving Radar

Consider the case when the radar moves varying both in the value and direction of velocity. Because the Doppler effect is directly initiated by the moving radar, changes in the velocity of moving radar result in changes in all consequences of the effect. With the simple harmonic searching signal, the function $V(t)$ shows that the average Doppler frequency given by Equation (3.75) and the effective power spectral density bandwidth of the Doppler fluctuations given by Equation (3.76) depend on time. The process becomes nonstationary. With the pulsed searching signal, in addition to this, the values of the pulse duration $\tau'_p = \frac{\tau_p}{\mu}$ and the pulse period $T'_p = \frac{T_p}{\mu}$, where μ is determined by Equation (3.11), become functions of time. The first phenomenon is infinitesimal, as a rule, but the second phenomenon appears in the investigation of the interperiod fluctuations with fixed radar range as a function of time for the effective power spectral density bandwidth of the target return signal fluctuations caused by exchange of scatterers [see Equation (3.157)].

Changes in the velocity along a direction of moving radar lead us to amplitude modulation of elementary signals, which is similar to amplitude modulation caused by radar antenna scanning. Suppose the radar moves in a horizontal plane with constant velocity V along an arc with radius ρ. As a result, the angle shifts of scatterers with respect to the horizontal-coverage directional diagram $\Delta\varphi = \frac{V\tau}{\rho}$, where $\frac{V}{\rho}$ is the angular velocity of the directional diagram, arise. In this case, the formulae in Equation (3.5) and Equation (3.12) are true, but instead of Equation (3.8) we have to use the general formula in Equation (2.94). The general formula in Equation (2.94) takes into consideration the angle shifts of scatterers. The study procedure is the same as in the case of simultaneous radar movement and radar antenna scanning.

However, there is a distinction in principle. Because the position of the directional diagram with respect to the velocity vector is not variable for the considered case, the effective power spectral density bandwidth of the Doppler fluctuations is independent of the parameter t (time), and if the condition $\rho = const$ is satisfied, the process is stationary. With the deflected Gaussian directional diagram, the power spectral density of the target return signal fluctuations is Gaussian too and the effective power spectral density bandwidth has the following form

$$\Delta F = \sqrt{\Delta F_D^2 + \Delta F_\rho^2}\,, \tag{3.165}$$

where ΔF_D is the effective power spectral density bandwidth of the Doppler fluctuations given by Equation (3.82);

$$\Delta F_\rho = \frac{V \cos \gamma_0}{\sqrt{2}\,\rho\,\Delta_h} \tag{3.166}$$

is the effective power spectral density bandwidth of the target return signal fluctuations caused by curvature of the trajectory of aircraft, for example. As a rule, the value of ΔF_ρ is very low.

3.5 Conclusions

We summarize briefly the main results discussed in this chapter. In the study of the target return signal fluctuations caused by the moving radar, the total correlation function is a nonstationary separable process, i.e., the normalized correlation function is independent of the time parameter t. All information regarding a nonstationary state is included in the power of the target return signal given by Equation (2.78). With the continuous nonmodulated pulsed searching signal, the normalized correlation function $R_p(\tau) \equiv 1$ and the total normalized correlation function of target return signal fluctuations coincides with the normalized correlation function of the slow fluctuations. The rapid fluctuations are absent. If the radar is stationary, the condition $R_g(\tau) \equiv 1$ is true and there are only the rapid (intraperiod) fluctuations under the condition $\mu = 1$. The slow fluctuations are absent.

Radar motion defined by changeover from the argument τ to the argument $\mu\tau$ in the normalized correlation function $R_p(\tau)$ of the fluctuations given by Equation (3.6) implies a compression ($\mu < 1$) or expansion ($\mu > 1$) of the time scale μ times. This is a natural manifestation of the Doppler effect, which is accompanied by changes in the pulsed searching signal duration and iteration frequency, in addition to changes in the carrier frequency under radar antenna scanning. The shape of the normalized correlation function and the effective power spectral density bandwidth of intraperiod fluctuations are completely defined by the shape and duration of the pulsed searching signal.

The Doppler fluctuations caused by the moving radar are defined by the envelope of the normalized correlation function of the fluctuations in the radar range given by Equation (3.12). The normalized correlation function $R_g(\tau)$ characterizes the interperiod fluctuations. The power spectral density at the relative Doppler frequency is formed by summing the powers of the target return signals from those scatterers that are placed on the surface of the cone defined near the velocity vector of moving radar and with apex angle 2θ.

When the radar antenna axis is highly deflected from the direction of moving radar and the variables φ' and ψ' are separable in the radar antenna directional diagram, then the power spectral density of the Doppler fluctuations coincides in shape with the square of the directional diagram in the plane crossing a direction of moving radar and the directional diagram axis. The effective power spectral density bandwidth is defined by the squared directional diagram width by power in the plane of radar antenna deflection. If the radar antenna axis is deflected in an arbitrary way from the direction of moving radar, the power spectral density is a complex function of the directional diagram and does not coincide with the squared directional diagram.

If the three-dimensional (space) target is scanned by pulsed searching signals of moving radar, the total normalized correlation function $R(\tau)$ is defined by the product of the periodic normalized correlation function $R_p(\tau)$ of the fluctuations in the radar range and the nonperiodic normalized correlation function $R_g(\tau)$ of the Doppler fluctuations. In this case, the total normalized correlation function of the target return signal fluctuations is not a periodic function. The total power spectral density is defined by convolution of the linear power spectral density of the fluctuations in the radar range and the continuous power spectral density of the Doppler fluctuations. The normalized correlation function of the interperiod fluctuations in the fixed radar range is defined by the product of the normalized correlation function $R_g(\tau)$ of the interperiod fluctuations in the glancing radar range and the normalized correlation function $R_p^{en}(\bar{\mu}\tau)$ of the interperiod fluctuations.

References

1. Ward, K., Baker, C., and Watts, S., Maritime surveillance radar. Part 1: radar scattering from the ocean surface, in *Proc. Inst. Elect. Eng. F,* Vol. 137, No. 2, 1990, pp. 51–62.
2. Hall, H., A new model for impulsive phenomena: application to atmospheric-noise communication channel, *Technical Report 3412-8,* Stanford University, Stanford, CA, 1966.
3. Di Bisceglie, M. and Galdi, C., Random walk based characterization of radar backscatterer from the sea surface, in *Radar, Sonar, and Navigation, IEEE Proceedings,* Vol. 145, No. 4, 1993, pp. 216–225.
4. Levanon, N., *Radar Principles,* Wiley, New York, 1988.
5. Pentini, A., Farina, A., and Zirilli, F., Radar detection of targets located in a coherent K-distributed clutter background, in *Proceedings IEEE,* Vol. 139, June 1992, pp. 341–358.
6. Rihaczek, A., *Principles of High-Resolution Radar,* Peninsala, San Jose, CA, 1985.
7. Doisy, Y., Derauz, L., Beerens, P., and Been, R., Target Doppler estimation using wideband frequency modulated signals, *IEEE Trans.,* Vol. SP-48, No. 5, 2000, pp. 1213–1224.
8. Kramer, S., Doppler and acceleration tolerances of high-gain, wideband linear FM correlation sonars, in *Proceedings of the IEEE,* Vol. 55, No. 5, 1967, pp. 627–636.
9. Farina, A., *Antenna-Based Signal Processing Techniques for Radar Systems,* Artech House, Norwood, MA, 1992.
10. Baculev, P. and Slepin, V., *Methods and Apparatus of Selection for Moving Targets,* Soviet Radio, Moscow, 1986 (in Russian).
11. Parsons, J., *The Mobile Radio Propagation Channel,* John Wiley & Sons, New York, 1996.

12. Ali, I., Al-Dhair, N., and Hershey, J., Doppler characterization for LEO satellites, *IEEE Trans.*, Vol. COM-46, No. 3, 1998, pp. 309–313.
13. Ward, J., Space-time adaptive processing for airborne radar, Lincoln Labs Technical Report 1015, MIT, Cambridge, MA, 1994.
14. Collins, T. and Atteins, P., Doppler-sensitive active sonar pulse designs for reverberation processing, in *Radar, Sonar, and Navigation, IEEE Proceedings*, Vol. 145, No. 12, 1998, pp. 1215–1225.
15. Bretthorst, G., *Bayesian Spectrum Analysis and Parameter Estimation*, Springer-Verlag, New York, 1988.
16. Costas, J., A study of a class of detection waveforms having nearly ideal range-Doppler ambiguity properties, in *Proceedings of the IEEE*, Vol. 72, No. 6, 1984, pp. 996–1009.
17. Muirhead, R., *Aspects of Multivariate Statistical Theory*, John Wiley & Sons, New York, 1982.
18. Haykin, S., *Non-Linear Methods of Spectral Analysis*, Springer-Verlag, New York, 1979.
19. Kay, S., *Modern Spectral Estimation: Theory and Application*, Prentice Hall, Englewood Cliffs, NJ, 1988.
20. Rozenbach, K. and Ziegenbein, J., About the effective Doppler sensitivity of certain non-linear chirp signals (NLFM), in *Proceedings of the Low Frequency Active Sonar Conference*, La Spezia, Italy, May 24–28, 1993, pp. 571–579.
21. Shanmugan, K. and Breipohl, A., *Random Signals: Detection, Estimation, and Data Analysis*, John Wiley & Sons, New York, 1988.
22. Pillai, S., *Array Signal Processing*, Springer-Verlag, New York, 1998.
23. Pahlavan, K. and Levesque, A., *Wireless Information Networks*, John Wiley & Sons, New York, 1995.
24. Haykin, S., *Adaptive Filter Theory*, 3rd ed., Prentice Hall, Englewood Cliffs, NJ, 1996.
25. Feldman, Yu, Determination of spectrum of target return signals, *Problems in Radio Electronics*, Vol. OT, No 6, 1959, pp. 22–38 (in Russian).
26. Stoica, P. and Moses, R., *Introduction to Spectral Analysis*, Prentice-Hall, Englewood Cliffs, NJ, 1997.
27. Proakis, J. and Manolakis, D., *Digital Signal Processing Principles, Algorithms, and Applications*, Prentice Hall, Englewood Cliffs, NJ, 1995.
28. Brillinger, D., *Time Series: Data Analysis and Theory*, Holden-Day, San Francisco, CA, 1981.
29. Johnson, N. and Kotz, S., *Distributions in Statistics: Continuous Univariate Distributions*, Vol. 2, John Wiley & Sons, New York, 1970.
30. Papoulis, A., *Probability, Random Variables, and Stochastic Processes*, McGraw-Hill, New York, 1984.
31. Feldman, Yu, Gidaspov, Yu, and Gomzin, V., *Moving Target Tracking*, Soviet Radio, Moscow, 1978 (in Russian).
32. Armond, N., Correlation function of waves scattered by rough surfaces, *Radio Eng. Electron. Phys.*, Vol. 30, No. 7, 1985, pp. 1307–1311 (in Russian).
33. Watts, Ed., Radar clutter and multipath propagation, in *Proceedings of IEEF*, Vol. 138, April 1994, pp. 187–199.
34. Ward, K., Compound representation of high resolution sea clutter, *Electron. Lett.*, Vol. 17, No. 16, 1981, pp. 561–563.
35. Poor, V., *An Introduction to Signal Detection and Estimation*, Springer-Verlag, New York, 1988.

36. Carter, G., *Coherence and Time Delay Estimation,* IEEE Press, New York, 1993.
37. Tsao, J. and Steinberg, D., Reduction of side-lobe and speckle artifacts in microwave imaging: the CLEAN technique, *IEEE Trans.,* Vol. AP-36, No. 2, 1988, pp. 543–556.
38. Kalson, S., An adaptive array detector with mismatched signal detection, *IEEE Trans.,* Vol. AES-28, No. 1, 1992, pp. 195–207.
39. Rappaport, T., *Wireless Communications Principles and Practice,* Prentice Hall, Upper Saddle River, NJ, 1996.
40. Jourdain, G. and Henrioux, J., Use of large bandwidth-duration binary phase shift keying signals in target delay Doppler measurements, *J. Acoust. Soc. Amer.,* Vol. 90, No. 1, 1991, pp. 299–309.
41. Stark, H. and Woods, J., *Probability, Random Processes, and Estimation Theory for Engineers,* Prentice Hall, Englewood Cliffs, NJ, 1986.
42. Porat, B., *Digital Processing of Random Signals,* 5th ed., Prentice Hall, Englewood Cliffs, NJ, 1994.
43. Gotwols, B., Chapman, R., and Sterner II, R., Ocean radar backscatterer statistics and the generalized log normal distribution, in *Proceedings of PIERS94,* The Netherlands, July 11–15, 1994, pp. 1028–1031.
44. Feldman, Yu, Nonlinear transformations of Doppler spectra, *Problems in Radio Electronics,* Vol. OT, No. 5, 1980, pp. 3–14 (in Russian).
45. Borkus, M., Energy spectrum of target return signal from atmosphere aerosol scatterers, *Problems in Radio Electronics,* Vol. OT, No. 1, 1977, pp. 43–50 (in Russian).
46. Kroszczinski, J., Pulse compression by means of linear-period modulation, in *Proceedings of the IEEE,* Vol. 57, No. 7, 1969, pp. 1260–1266.
47. Tseng, C. and Giffiths, L., A unified approach to the design of linear constraints in minimum variance adaptive beamformers, *IEEE Trans.,* Vol. AP-40, No. 6, 1992, pp. 1533–1542.
48. Wax, M. and Anu, Y., Performance analysis of the minimum variance beamformer, *IEEE Trans.,* Vol. SP-44, No. 4, 1996, pp. 928–937.

4

Fluctuations under Scanning of the Two-Dimensional (Surface) Target by the Moving Radar

4.1 General Statements

As explained in Chapter 3, we assume that the radar antenna is stationary, $\Delta\varphi_{sc} = \Delta\psi_{sc} = 0$, and we consider the long-range area of the radar antenna directional diagram $\Delta\varphi_{rm} = \Delta\psi_{rm} = 0$. We assume that the radar moves rectilinearly and uniformly with the velocity V, i.e., $\Delta\ell = -V \cdot \tau$. Based on Equations (2.122), (2.127), (2.128), and (2.165), we can write

$$R^{en}(t,\tau) = p_0 \sum_{n=0}^{\infty} \iint \boldsymbol{P}\left[t - \tfrac{2\rho(\psi)}{c} - 0.5(\mu\tau - nT_p)\right] \cdot \boldsymbol{P}^*\left[t - \tfrac{2\rho(\psi)}{c} + 0.5(\mu\tau - nT_p)\right]$$

$$\times g^2(\varphi,\psi) \cdot S^\circ(\beta_0, \psi + \gamma_0)\sin(\psi + \gamma_0) \cdot e^{j\Omega(\varphi,\psi)\tau} d\varphi\, d\psi, \tag{4.1}$$

where

$$\Omega(\varphi,\psi) = \Omega_{\max}\left[\cos\varepsilon_0 \cos\left(\beta_0 + \tfrac{\varphi}{\cos\gamma}\right)\cos(\psi + \gamma_0) - \sin\varepsilon_0 \sin(\psi + \gamma_0)\right]; \tag{4.2}$$

$$\mu = 1 + 2V_r c^{-1} = 1 + 2Vc^{-1}(\cos\varepsilon_0 \cos\beta_0 \cos\bar{\gamma} - \sin\varepsilon_0 \sin\bar{\gamma}); \tag{4.3}$$

$\bar{\gamma} = \gamma_0$ with the continuous searching signal; $\bar{\gamma} = \gamma_*$ with the pulsed searching signal; the power p_0 of the target return signal is determined by Equation (2.122); and the maximum Doppler frequency $\Omega_{\max}$ is given by Equation (3.10).

Using Equation (4.1), we can investigate the case of both the searching simple harmonic signal and the pulsed searching signal.[1,2] At the pulsed searching signal and conditions $S^\circ(\beta_0, \psi + \gamma_0) \approx S^\circ(\beta_0, \gamma_*)$ and $\sin(\psi + \gamma_0) \approx \sin\gamma_*$, based on Equation (4.1), we can write

$$R^{en}(t,\tau) = p(t)\cdot R^{en}(t,\tau)\,, \tag{4.4}$$

where

$$\begin{aligned} R^{en}(t,\tau) = N\sum_{n=0}^{\infty}\iint & \boldsymbol{P}\left[t - \tfrac{2\rho(\psi)}{c} - 0.5(\mu\tau - nT_p)\right] \\ & \times \boldsymbol{P}^*\left[t - \tfrac{2\rho(\psi)}{c} + 0.5(\mu\tau - nT_p)\right]g^2(\varphi,\psi)\cdot e^{j\Omega(\varphi,\psi)\,\tau}d\varphi\, d\psi \end{aligned} \tag{4.5}$$

The power $p(t)$ of the target return signal was studied in more detail in Section 2.5.3. We assum that, with the pulsed searching signal, the conditions given by Equations (2.129) and (2.130) are satisfied and we can write

$$R^{en}(t,\tau) = R_\beta(t,\tau)\cdot R_\gamma(t,\tau) \tag{4.6}$$

instead of Equation (4.5), where

$$R_\beta(t,\tau) = N\int g_h^2(\varphi,\psi_*)\cdot e^{j\Omega_\beta(\varphi)\,\tau}d\varphi\,; \tag{4.7}$$

$$\begin{aligned} R_\gamma(t,\tau) = N\sum_{n=0}^{\infty}\int & \boldsymbol{P}\left[t - \tfrac{2\rho(\psi)}{c} - 0.5(\mu\tau - nT_p)\right]\cdot \boldsymbol{P}^*\left[t - \tfrac{2\rho(\psi)}{c} + 0.5(\mu\tau - nT_p)\right] \\ & \times g_v^2(\psi)\cdot e^{-j(\psi-\psi_*)\Omega_\gamma\tau}d\psi \end{aligned} \tag{4.8}$$

$$\Omega_\beta(\varphi) = \Omega_{\max}\left[\cos\varepsilon_0\cos\left(\beta_0 + \tfrac{\varphi}{\cos\gamma_*}\right)\cos\gamma_* - \sin\varepsilon_0\sin\gamma_*\right]; \tag{4.9}$$

$$\Omega_\gamma = \Omega_{\max}(\cos\varepsilon_0\cos\beta_0\sin\gamma_* - \sin\varepsilon_0\cos\gamma_*)\,; \tag{4.10}$$

$$\Omega(\varphi,\psi) \approx \Omega_\beta(\varphi) - (\psi - \psi_*)\Omega_\gamma\,. \tag{4.11}$$

Compare with Equation (4.2).

The azimuth-normalized correlation function $R_\beta(t, \tau)$ given by Equation (4.7) takes into consideration the slow target return signal fluctuations caused by differences in the Doppler frequencies in the azimuth plane. The aspect-angle normalized correlation function $R_\gamma(t, \tau)$ given by Equation (4.8) takes into consideration the slow target return signal fluctuations, which in turn are caused by differences in the Doppler frequencies in the aspect-angle plane, and the rapid target return signal fluctuations, which are caused by the propagation of the pulsed searching signal along the scanned surface of the two-dimensional target.[3–5]

4.2 The Continuous Searching Nonmodulated Signal

Assuming that the condition given by Equation (2.109) is satisfied in Equation (4.1), and omitting the summation sign, we can write the correlation function in the following form[6,7]

$$R^{en}(t,\tau) = p(t)\cdot R_g^{en}(\tau)\,, \tag{4.12}$$

where the power $p(t)$ of the target return signal is determined by Equation (2.166) and

$$R_g^{en}(\tau) = N\iint g^2(\varphi,\psi)\cdot S^\circ(\beta_0,\psi+\gamma_0)\sin(\psi+\gamma_0)\cdot e^{j\Omega(\varphi,\psi)\,\tau}d\varphi\, d\psi\,. \tag{4.13}$$

Comparing Equation (4.13) with Equation (3.8), one can see that Equation (4.13) follows from Equation (3.8) if we replace the function $g^2(\varphi,\psi)$ in Equation (3.8) with the function $\tilde{g}^2(\varphi,\psi)$ that can be determined in the following form

$$\tilde{g}^2(\varphi,\psi) = g^2(\varphi,\psi)\cdot S^\circ(\beta_0,\psi+\gamma_0)\sin(\psi+\gamma_0) \tag{4.14}$$

and assume that $\varepsilon_0 = 0$ in Equation (4.2). The function $\tilde{g}(\varphi,\psi)$ can be considered as the generalized radar antenna directional diagram.[8,9] For this reason, certain results obtained in Chapter 3 can be used in the investigations in this chapter, replacing the function $g(\varphi,\psi)$ with the generalized function $\tilde{g}(\varphi,\psi)$.

We consider that the vector of velocity of the moving radar is outside the limits of the directional diagram. Using the linear expansion for the frequency $\Omega(\varphi,\psi)$ given by Equation (4.2), as shown in Equation (3.47),

$$\Omega(\varphi,\psi) = \tilde{\Omega}_0 - \varphi\,\tilde{\Omega}_h - \psi\,\tilde{\Omega}_v\,, \tag{4.15}$$

where

$$\tilde{\Omega}_0 = \Omega_{\max}(\cos\varepsilon_0\cos\beta_0\cos\gamma_0 - \sin\varepsilon_0\sin\gamma_0) = \Omega_{\max}\cos\theta_0\,; \tag{4.16}$$

$$\tilde{\Omega}_h = \Omega_{\max} \cos\varepsilon_0 \sin\beta_0 \quad \text{and} \quad \tilde{\Omega}_v = \Omega_{\max}(\cos\varepsilon_0 \cos\beta_0 \sin\gamma_0 + \sin\varepsilon_0 \cos\gamma_0). \tag{4.17}$$

Under the condition $\varepsilon_0 = 0$, Equation (4.16) and Equation (4.17) coincide with Equations (3.48) to (3.51).

When the variables φ and ψ are separable in the function $g(\varphi, \psi)$, then Equation (4.13) can be written in the form of the product that is analogous to Equation (3.52). In this case, the normalized correlation function $R_g^v(\tau)$ of the target return signal Doppler fluctuations is different from that given by Equation (3.54), in which we use the function

$$\tilde{g}_v^2(\psi) = g_v^2(\psi) \cdot S^\circ(\beta_0, \psi + \gamma_0)\sin(\psi + \gamma_0) \tag{4.18}$$

instead of the function $g_v^2(\psi)$, and the parameters Ω_0, Ω_h, and Ω_v in Equations (3.48) through (3.51) are replaced with $\tilde{\Omega}_0, \tilde{\Omega}_h$, and $\tilde{\Omega}_v$ in Equation (4.16) and Equation (4.17), respectively. Consequently, the total power spectral density of target return signal fluctuations is formed by convolution between two power spectral densities

$$S_g^h(\omega) \cong g_h^2\left(-\tfrac{\omega}{\tilde{\Omega}_h}\right), \tag{4.19}$$

$$S_g^v(\omega) \cong \tilde{g}_v^2\left(-\tfrac{\omega}{\tilde{\Omega}_v}\right) = g_v^2\left(-\tfrac{\omega}{\tilde{\Omega}_v}\right) \cdot S^\circ\left(\beta_0, \gamma_0 - \tfrac{\omega}{\tilde{\Omega}_v}\right)\sin\left(\gamma_0 - \tfrac{\omega}{\tilde{\Omega}_v}\right) \tag{4.20}$$

and by using the result of the convolution at the frequency $\omega_0 + \tilde{\Omega}_0$.

Let the directional diagram be determined by the Gaussian distribution law, and the specific effective scattering area $S^\circ(\gamma)$ by the exponential law given by Equation (2.150).[10,11] Reference to Equation (4.19) and Equation (4.20) shows that

$$S_g^h(\omega) \cong e^{-\frac{\pi\omega^2}{\Delta\tilde{\Omega}_h^2}}; \tag{4.21}$$

$$S_g^v(\omega) \cong e^{-\frac{\pi\omega^2}{\Delta\tilde{\Omega}_v^2} - \frac{k_1\omega}{\tilde{\Omega}_v}} \cdot \sin\left(\gamma_0 - \tfrac{\omega}{\tilde{\Omega}_v}\right), \tag{4.22}$$

where

$$\Delta\tilde{\Omega}_{h,v} = \tilde{\Omega}_{h,v} \cdot \Delta_{h,v}^{(2)} = \frac{\tilde{\Omega}_{h,v} \cdot \Delta_{h,v}}{\sqrt{2}}. \tag{4.23}$$

The convolution between the power spectral densities $S_g^h(\omega)$ and $S_g^v(\omega)$ of the fluctuations is rigorously determined, but it looks very cumbersome. To obtain a simpler form of convolution between $S_g^h(\omega)$ and $S_g^v(\omega)$, we have to take into consideration the following circumstance. If the vertical-coverage directional diagram width is not so high in value and the angle γ_0 is not so low in value, the generalized directional diagram given by Equation (4.18) is approximately Gaussian:[12]

$$\tilde{g}_v(\psi) = g_0 \cdot e^{-\frac{\pi(\psi-\psi_0)^2}{\Delta_v^{(2)}}}. \tag{4.24}$$

The generalized vertical-coverage directional diagram has the same effective width Δ_v, but it acquires the shift ψ_0, which is low by value, and the coefficient of proportionality g_0:

$$\psi_0 = \frac{(k_1 + \text{ctg}\,\gamma_0)\Delta_v^{(2)}}{4\pi} \quad \text{and} \quad g_0^2 = S^\circ(\gamma_0)\sin\gamma_0 \cdot e^{\frac{2\pi\,\psi_0^2}{\Delta_v^{(2)}}}. \tag{4.25}$$

The shift ψ_0 consists of two terms. The first term is caused by the function $S^\circ(\gamma)$ given by Equation (2.150) — the specific effective scattering area. The second term is determined by the target return signal power as a function of the radar range. The power spectral density given by Equation (4.22) takes the following form in this case

$$S_g^v(\omega) \approx e^{-\pi \cdot \frac{(\omega+\delta\,\Omega)^2}{\Delta\tilde{\Omega}_v^2}} \tag{4.26}$$

where

$$\delta\,\Omega = \tilde{\Omega}_v \cdot \psi_0 = \tilde{\Omega}_v \cdot \frac{(k_1 + \text{ctg}\,\gamma_0)\Delta_v^{(2)}}{4\,\pi}. \tag{4.27}$$

In terms of the shift in the frequency of the value of $\omega_0 + \tilde{\Omega}_0$, the convolution between the power spectral densities $S_g^h(\omega)$ and $S_g^v(\omega)$, which are determined by Equation (4.21) and Equation (4.26), respectively, gives us the following result:[13]

$$S_g(\omega) = S_g^h(\omega) * S_g^v(\omega) * \delta(\omega - \omega_0 - \tilde{\Omega}_0) \approx e^{-\pi \cdot \frac{(\omega-\omega_0-\tilde{\Omega})^2}{\Delta\tilde{\Omega}^2}}, \tag{4.28}$$

where $\Delta\tilde{\Omega}^2 = \Delta\tilde{\Omega}_h^2 + \Delta\tilde{\Omega}_v^2$, and $\overline{\Omega} = \tilde{\Omega}_0 - \delta\Omega$ is the average frequency of the power spectral density $S_g(\omega)$ of the Doppler fluctuations given by Equation (4.28). The average frequency $\overline{\Omega}$ is different from the average Doppler frequency $\tilde{\Omega}_0$ given by Equation (4.16) for the value of $\delta\Omega$ (the error). The meaning and significance of the error $\delta\Omega$ can be defined under the condition $\varepsilon_0 = 0$.[14–16] Then

$$\delta\Omega = \Omega_0\psi_0 \text{tg}\,\gamma_0 = \Omega_0 \cdot \frac{(k_1 \text{tg}\,\gamma_0 + 1)\Delta_v^{(2)}}{4\,\pi}, \tag{4.29}$$

where $\Omega_0 = \Omega_{max} \cos\beta_0 \cos\gamma_0$. The relative value of the error $\delta\Omega$ given by Equation (4.29) has the following form:

$$\frac{\delta\Omega}{\Omega_0} = \delta_1 + \delta_2 = \frac{(k_1 \text{tg}\,\gamma_0 + 1)\Delta_v^{(2)}}{4\,\pi}. \tag{4.30}$$

The error δ_1 is caused by the specific effective scattering area $S°(\gamma)$. At k_1 = 3.3 and k_1 = 13, which corresponds to reflection from a plow in summer and rough sea of 1 (see Table 2.1), Δ_v = 6° and γ_0 = 60°, we obtain the error δ_1 equal to 0.5% and 2%, respectively. The error δ_2 is less, as a rule. At Δ_v = 6°, the error δ_2 is equal to 0.1%. The errors δ_1 and δ_2 depend on the characteristics and parameters of the radar antenna only in the vertical plane — parameters Δ_v and γ_0. Experimental results regarding the error δ_1 as a function of the rough sea are shown in Figure 4.1 at Δ_v = 6° and γ_0 = 65° when a rough sea is caused by the wind and there is a rippled sea. When there is a rippled sea, the error δ_1 is greater because the sea surface becomes smoother in spite of the waves being high, in comparison with the rough sea caused by the wind.[17,18] The error δ_1 given by the experiment for the Earth's surface is significantly less, as one would expect. At the same values of Δ_v and γ_0, the error δ_1 given by the experiment is equal to 0.35%, 0.45%, and 0.55% for the forest, field, and plow, respectively. These experimental values for the error δ_1 coincide very well with theoretical results.[19–22]

The realistic directional diagram is significantly different from its Gaussian directional diagram due to the presence of side lobes. Very often, a changeover from the directional to the Gaussian diagram does not ensure the required accuracy of determination and computer calculation. In this case, the power spectral density of the fluctuations must be defined using the exact approximation for the directional diagram $g(\varphi, \psi)$, taking into consideration the side lobes. For example, see Section 3.2, the numerical integration of the real two-dimensional directional diagram given by the experiment [see Equation (4.13)], or the technique of partial diagrams.[23,24]

Let us represent the square of the real two-dimensional directional diagram as a sum of partial diagrams of the main beam and side lobes. In doing so, each partial diagram is Gaussian:[25]

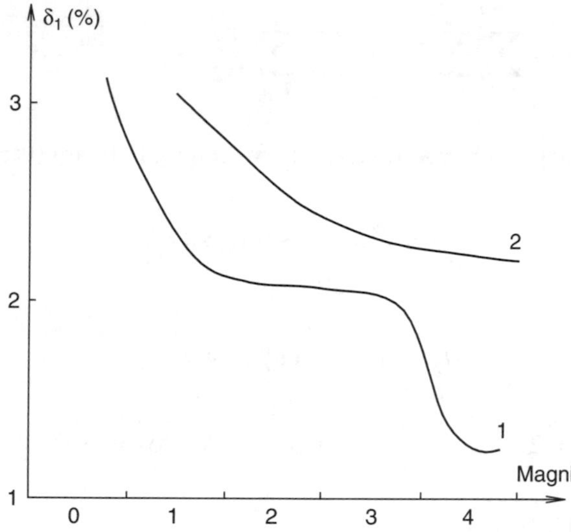

FIGURE 4.1
Shift in average Doppler frequency as a function of magnitude of the rough (the curve 1) and rippled (the curve 2) sea.

$$g^2(\varphi,\psi)=\sum_{i=0}^{n}\sum_{j=0}^{m} g_{ij}^2\cdot e^{-2\pi\cdot\left[\frac{(\varphi-\varphi_{ij})^2}{\Delta_{h_{ij}}^{(2)}}+\frac{(\psi-\psi_{ij})^2}{\Delta_{v_{ij}}^{(2)}}\right]}, \tag{4.31}$$

where g_{ij} is the relative power of the i-th, j-th side lobe; i is the number of side lobes of the partial diagram in the plane φ; j is the number of side lobes of the partial diagram in the plane ψ; φ_{ij} is the angle coordinate of the center of the side-lobe, relatively the center of the main beam in the plane φ; ψ_{ij} is the angle coordinate of the center of the side lobe, relatively the center of the main beam in the plane ψ; $\Delta_{h_{ij}}$ is the effective width of the i-th, j-th side lobe in the plane φ; $\Delta_{v_{ij}}$ is the effective width of the i-th, j-th side lobe in the plane ψ; $i = j = 0$ is the case of the main beam. All these parameters are determined using the experimental two-dimensional directional diagram. We can use the width at the level 0.5 from the maximum instead of the effective width of the main beam and side lobes. We can assume that the value of $S^\circ(\gamma)$ is constant within the limits of the side lobe and do not consider changes in the radar range.[26,27]

Each partial diagram can be considered independent of other partial diagrams. Consequently, the target return signals for partial diagrams are noncoherent. In this case, the total power spectral density of the fluctuations is equal to sum of the independent partial power spectral densities formed by individual side lobes. Using Equation (4.28) and taking into account the contribution in the energy of each side lobe, we can write the power spectral density in the following form:[28,29]

$$S(\omega)=\sum_{i=0}^{n}\sum_{j=0}^{m}S_{ij}(\omega)=\sum_{i=0}^{n}\sum_{j=0}^{m}\frac{p_{ij}}{\Delta\Omega_{ij}}\cdot e^{-\pi\cdot\frac{(\omega-\omega_0-\Omega_{ij})^2}{\Delta\Omega_{ij}^2}}, \tag{4.32}$$

where $S_{00}(\omega)$ is the power spectral density for the main beam (the main lobe);

$$p_{ij}=\frac{P\lambda^2 g_{ij}^2\Delta_{h_{ij}}\Delta_{v_{ij}}\sin\gamma_{ij}\ S^\circ(\gamma_{ij})}{128\ \pi^3 h^2}; \tag{4.33}$$

$$\Omega_{ij}=4\pi V\lambda^{-1}\cos\beta_{ij}\cos\gamma_{ij}; \tag{4.34}$$

$$\Delta\Omega_{ij}=2\sqrt{2}\ \pi\ V\lambda^{-1}\sqrt{(\Delta_{h_{ij}}\sin\beta_{ij})^2+(\Delta_{v_{ij}}\cos\beta_{ij}\sin\gamma_{ij})^2}; \tag{4.35}$$

$\beta_{ij}=\beta_0+\varphi_{ij}$; and $\gamma_{ij}=\gamma_0+\psi_{ij}$.

4.3 The Pulsed Searching Signal with Stationary Radar

4.3.1 General Statements

At $V = 0$, we obtain that $\Omega(\varphi,\psi)=0$ and $\mu = 1$. Introducing a new variable [see Equation (2.142)]

$$z=t-\tfrac{2\rho(\psi)}{c}\quad\text{and}\quad c_*dz=c\,d\psi \tag{4.36}$$

in Equation (4.5), we can write

$$R^{en}(t,\tau)=\sum_{n=0}^{\infty}R_0(t,\tau-nT_p), \tag{4.37}$$

where

$$R_0(t,\tau)=p_*\int \boldsymbol{P}(z-0.5\tau)\cdot\ \boldsymbol{P}^*(z+0.5\tau)\cdot g_v^2(\psi_*+c_*z)\,dz \tag{4.38}$$

is the correlation function at $n = 0$;

$$p_*=\frac{PG_0^2\lambda^2\Delta_h^{(2)}c_*S^\circ(\gamma_*)\sin\gamma_*}{64\ \pi^3 h^2}; \tag{4.39}$$

$c_* = 0.5c\rho^{-1}\text{tg}\,\gamma_*$; and $c_*z = \psi - \psi_*$. Here, a dependence on the time parameter t — a feature of the nonstationary state — is changed by dependence on the angles $\psi_*(t)$ and $\gamma_*(t) = \psi_*(t) + \gamma_0$. The angles $\psi_*(t)$ and $\gamma_*(t)$ are related to the time parameter t by the formula [see Equation (2.139)]:

$$t = \frac{2\,\rho_*}{c} = \frac{2\,h}{c\sin\gamma_*} = \frac{2\,h}{c\sin(\psi_* + \gamma_0)}\,. \tag{4.40}$$

The correlation function of the fluctuations given by Equation (4.37) is a periodic function of the parameter τ with the period T_p (see Figure 3.1a, the solid line). This correlation function defines the rapid fluctuations in the radar range. Because, in the general case, the variables $t(\psi_*)$ and τ are not separable, the correlation function is not separable either — the spectral characteristics depend on the time parameter t. The corresponding instantaneous power spectral density of the fluctuations is a regulated function.[30,31] The envelope of the power spectral density is the Fourier transform of the envelope of the correlation function given by Equation (4.38) and also depends on the time parameter t.

The instant of time t (or the angle ψ_*) defines the interval within the limits of which the variable $\psi_* + c_*z$ in the integrand function $g_v(\psi_* + c_*z)$ (see Figure 4.2) changes, and the form of the total part of the product of the functions $\mathbf{P}(t)$ and $\mathbf{P}^*(t)$ (see Figure 4.2, the hatched area). Thus, we can say that the shape of the waves of the correlation function given by Equation (4.38) and the envelope of the regulated power spectral density are defined by the instant of time t. Naturally, the target return signal power also depends on the instant of time t.

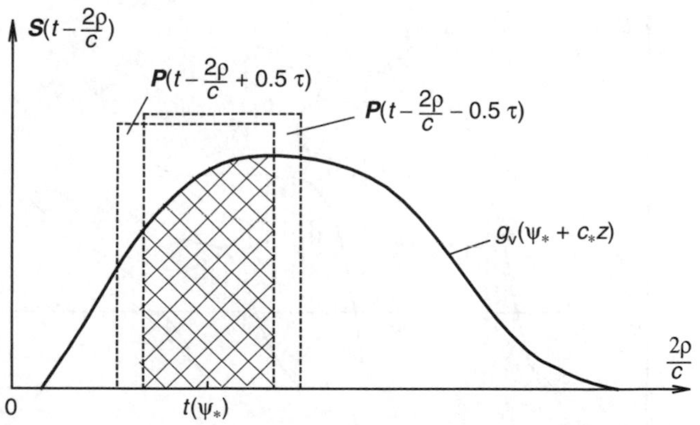

FIGURE 4.2
The instantaneous power spectral density of the target return signal fluctuations as a function of the delay t.

The instantaneous power spectral density of the fluctuations in the radar range is presented in Figure 4.3 as a function of time t in the case of the individual pulsed searching signal propagated along the surface of the two-dimensional target — the continuous power spectral density. The analogous power spectral density of the fluctuations in the case of the infinite periodic sequence of pulsed searching signals is shown in Figure 4.4 as a function of the time-cross-section $t(\rho_*)$ — the regulated power spectral density of the fluctuations [see Equation (2.1)].

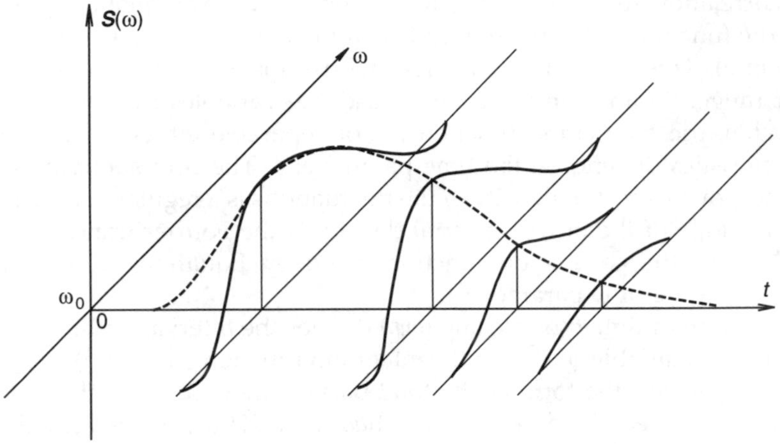

FIGURE 4.3
The instantaneous power spectral density of the target return signal fluctuations with the individual pulsed searching signal.

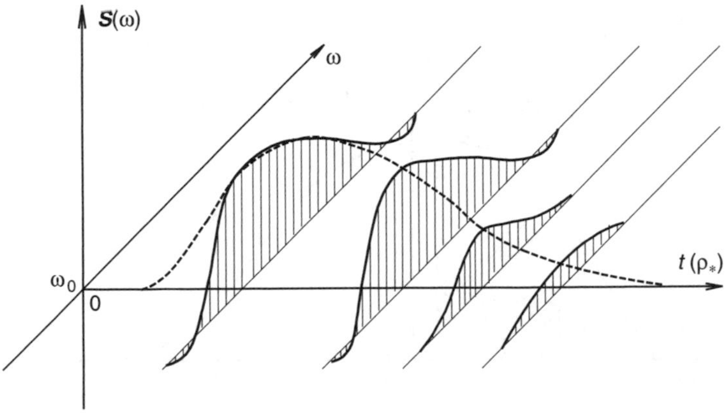

FIGURE 4.4
The instantaneous power spectral density of the target return signal fluctuations with a sequence of pulsed searching signals.

Reference to Equation (4.38) shows that if the vertical-coverage directional diagram is wide, i.e., the condition $\Delta_v >> \Delta_p$ (or $\tau_p << T_r$) is true, we can assume that $g_v(\psi_* + c^* z) \approx g_v(\psi_*)$. In this case, we can write

$$R_0(t,\tau) = p_* g_v^2(\psi_*) \int \mathbf{P}(z - 0.5\tau) \cdot \mathbf{P}^*(z + 0.5\tau)\, dz\,. \tag{4.41}$$

This correlation function of fluctuations in the radar range is equivalent to that given by Equation (3.3) for the three-dimensional (space) target under the condition $\mu = 1$, i.e., the radar is stationary. Consequently, all results obtained in Section 3.1 are true for the case considered here. However, it differs from the case discussed in Chapter 3 by the other dependence between the target return signal power and time t (or radar range) that was discussed in Section 2.4 and Section 2.5. In the discussed approximation procedure, the variables $t(\psi_*)$ and τ are separable, as shown in Equation (4.41). Consequently, the correlation function of the fluctuations is separable. Let us consider and discuss some examples in practice.

4.3.2 The Arbitrary Vertical-Coverage Directional Diagram: The Gaussian Pulsed Searching Signal

Reference to Equation (4.37) shows that if the pulsed searching signal is Gaussian [see Equation (2.105)], we can write

$$R^{en}(t,\tau) = p(t) \cdot \sum_{n=0}^{\infty} e^{-\pi \cdot \frac{(\tau - nT_p)^2}{2\tau_p^2}}\,, \tag{4.42}$$

where

$$p(t) = p_* \int_{-\infty}^{\infty} g_v^2(\psi_* + c_* z) \cdot e^{-\frac{2\pi z^2}{\tau_p^2}}\, dz\,. \tag{4.43}$$

Here the variables t and τ are separable, i.e., we can say that the correlation function of the target return signal fluctuations is a separable process for any shape and width of the vertical-coverage directional diagram. The power $p(t)$ of the target return signal as a function of the parameter t for the case of the Gaussian vertical-coverage directional diagram was discussed in Section 2.5. The normalized correlation function of the fluctuations following from Equation (4.42) coincides with that given by Equation (3.20) in the case of the three-dimensional (space) target when the radar is stationary, $\mu = 1$. This correlation function is represented as a comb of Gaussian waves with the effective width equal to $\sqrt{2}\tau_p$. The corresponding power spectral density is the regulated function, and the envelope of the power spectral density is Gaussian with the effective bandwidth equal to $(\sqrt{2}\,\tau_p)^{-1}$. The envelope of the power spectral density of the fluctuations is similar to that given by Equation (3.21) if we replace the parameter τ'_p with the parameter τ_p in Equation (3.21).

4.3.3 The Arbitrary Vertical-Coverage Directional Diagram: The Square Waveform Pulsed Searching Signal

The directional diagram is defined by continuous functions with continuous derivatives, and we can use the Taylor-series expansion[32–34]

$$g_v^2(\psi_* + c_* z) = \sum_{k=0}^{\infty} a_k z^k , \tag{4.44}$$

where $a_0 = g_v^2(\psi_*)$ and $a_k = \frac{c_*^k g_v^{2k}(\psi_*)}{k!}$. In the case of the square waveform pulsed searching signal, reference to Equation (4.38) and Equation (4.44) shows that all terms with the odd powers of the value z are equal to zero, and we can write

$$R_0(t,\tau) = p_* \int_{-0.5(\tau_p-\tau)}^{0.5(\tau_p-\tau)} g_v^2(\psi_* + c_* z)\,dz = p' \sum_{k=0}^{\infty} b_k \left(1 - \tfrac{|\tau|}{\tau_p}\right)^{2k+1} , \tag{4.45}$$

where $p' = p_* \tau_p g_v^2(\psi_*)$; $b_0 = 1$;

$$b_k = \frac{a_{2k}\tau_p^{2k}}{2^{2k}(2k+1)!\,g_v^2(\psi_*)} = \frac{\Delta_p^{(2k)}}{2^{2k}(2k+1)!} \cdot \frac{[g_v^2(\psi_*)]^{2k}}{g_v^2(\psi_*)} ; \tag{4.46}$$

$$\Delta_p = c_* \tau_p = \frac{c\tau_p}{2\,\rho_* \mathrm{ctg}\,\gamma_*} = \frac{c\tau_p \sin^2\gamma_*}{2\,h\cos\gamma_*} \tag{4.47}$$

is the angle dimension of the resolution element in the radar range [see Equation (2.140)].

Because the coefficients b_k depend on the time t through the parameters Δ_p and $g_v(\psi_*)$, the variables t and τ are not separable in the general case, and thus the correlation function of the target return signal fluctuations is not separable. As the delay or radar range ρ_* is increased, i.e., the parameter γ_* is decreased, the coefficients b_k are rapidly decreased, and the influence of the high-order terms so quickly weakened that the duration of the pulsed searching signal is short. In the pulsed searching signals with very short duration, or with low values of γ_*, we need consider only the first term. Then

$$R^{en}(t,\tau) \approx p'\left(1 - \tfrac{|\tau|}{T_p}\right). \tag{4.48}$$

For this approximation, the correlation function is separable. The instantaneous normalized correlation function and power spectral density are the same as in the case of scanning the three-dimensional (space) target with the stationary radar, $\mu = 1$ [see Equation (3.18) and Equation (3.19)].

In rigorous analysis, when it is necessary to consider high-order terms in Equation (4.45), the envelope of the power spectral density of the fluctuations can be written in the following form[35]

$$S_p^{en}(\omega,t) \cong \sum_{k=0}^{\infty} b_k(\psi_*)\cdot S_k(\omega)\,, \tag{4.49}$$

where

$$S_k(\omega) = (-1)^{k+1}\cdot \tfrac{2(2k+1)!}{(\omega\tau_p)^{2k+2}}\cdot\left[\cos\omega\tau_p - \sum_{m=0}^{k}(-1)^m\cdot\tfrac{(\omega\tau_p)^{2m}}{(2m)!}\right] \quad \text{and} \quad S_k(0) = \tfrac{1}{k+1}\,. \tag{4.50}$$

We can write the power spectral densities at $k = 0, 1, 2$ in the following form:

$$S_0(\omega) = \mathbf{sinc}^2(0.5\omega\tau_p) \quad \text{and} \quad S_0(0) = 1; \tag{4.51}$$

$$S_1(\omega) = 6(\omega\tau_p)^{-2}[1 - S_0(\omega)] \quad \text{and} \quad S_1(0) = 0.5\,; \tag{4.52}$$

$$S_2(\omega) = 10(\omega\tau_p)^{-2}[1 - 2S_1(\omega)] \quad \text{and} \quad S_2(0) \approx 0.33\,. \tag{4.53}$$

These power spectral densities are shown in Figure 4.5. The effective bandwidth of the k-th power spectral density increases with an increase in the number:

$$\Delta F_k = \int_{-\tau_p}^{\tau_p}\left(1 - \tfrac{|\tau|}{\tau_p}\right)^{2k+1} d\tau = \tfrac{k+1}{\tau_p}\,. \tag{4.54}$$

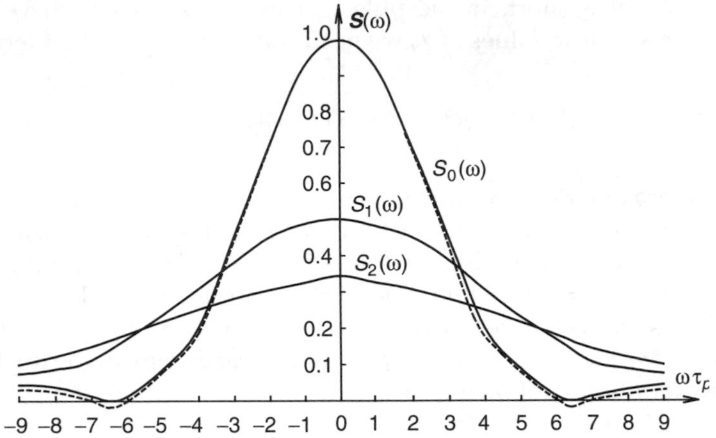

FIGURE 4.5
The components $S_0(\omega)$, $S_1(\omega)$, and $S_2(\omega)$ of the power spectral density of the target return signal fluctuations in the radar range (the solid line) and their sum (the dotted line) at $\psi_* = 0°$ and $\frac{\Delta_p^*}{\Delta_h} = 0.5$.

Because the contribution of these terms depends on the value of the current angle γ_*, the instantaneous power spectral density is a function of the delay of the target return signal or radar range ρ_*.

4.3.4 The Gaussian Vertical-Coverage Directional Diagram: The Square Waveform Pulsed Searching Signal

Reference to Equation (4.45) shows that

$$R_0(t,\tau) = \frac{\sqrt{\pi}}{2}\cdot p_*\left\{\Phi\left\{\frac{\sqrt{2\pi}}{\Delta_v}\left[\psi_* + \frac{\Delta_p}{2}\left(1-\frac{|\tau|}{\tau_p}\right)\right]\right\} - \Phi\left\{\frac{\sqrt{2\pi}}{\Delta_v}\left[\psi_* - \frac{\Delta_p}{2}\left(1-\frac{|\tau|}{\tau_p}\right)\right]\right\}\right\}, \tag{4.55}$$

where $\Phi(x)$ is the error integral given by Equation (1.27). Determination of the correlation function of the fluctuations given by Equation (4.55) is true for any relationship between Δ_v and Δ_p. Further, this determination does not have any limitations with regard to the width of the vertical-coverage directional diagram and the angle ψ_*. These values can be high.[36]

When the width of the vertical-coverage directional diagram is low in value, and the value of the angle γ_0 is not so low, we can use the linear relationship between the variables t and ψ_* [see Equation (2.142)], and the values of the angle ψ_*, Δ_v, and Δ_p in Equation (4.55) can be defined using the variables T_r, $t - T_d$, and τ_p:

$$\frac{\psi_k}{\Delta_v} \approx \frac{T_d - t}{\sqrt{2}\,T_r} \qquad \text{at} \qquad \psi_* \mathrm{ctg}\,\gamma_0 << 1\,; \tag{4.56}$$

$$\frac{\Delta_p}{\Delta_v} \approx \frac{\tau_p}{\sqrt{2}\,T_r} \qquad \text{at} \qquad \rho_* \mathrm{ctg}\,\gamma_* \approx \rho_0 \mathrm{ctg}\,\gamma_0\,, \tag{4.57}$$

where

$$T_r = \frac{T_d \Delta_v}{\sqrt{2}} \cdot \mathrm{ctg}\,\gamma_0 \tag{4.58}$$

is the effective duration of the target return signal [see Equation (2.151)] and

$$T_d = \frac{2\,\rho_0}{c} = \frac{2\,h}{c \sin\gamma_0} \tag{4.59}$$

is the delay of the target return signal from the scatterer placed on the axis of the directional diagram in the angle γ_0.

In this case, the correlation function of the fluctuations given by Equation (4.55) can be written in the explicit functional form of the parameter t:

$$R_0(t,\tau) \cong \tfrac{\sqrt{\pi}}{2} \cdot p_* \left\{\Phi\left\{\tfrac{\sqrt{\pi}}{T_r}\left[t - T_d + \tfrac{\tau_p}{2}\left(1 - \tfrac{|\tau|}{\tau_p}\right)\right]\right\} - \Phi\left\{\tfrac{\sqrt{\pi}}{T_r}\left[t - T_d - \tfrac{\tau_p}{2}\left(1 - \tfrac{|\tau|}{\tau_p}\right)\right]\right\}\right\}. \tag{4.60}$$

Equation (4.55) can be represented in the series expansion form [see Equation (4.45)], where

$$b_k = \frac{\left(\frac{\pi}{2}\right)^k \left(\frac{\Delta_p}{\Delta_v}\right)^{2k}}{(2k+1)!} \cdot \boldsymbol{H}_{2k}(x) \qquad \text{and} \qquad x = \sqrt{2\pi} \cdot \tfrac{\psi_*}{\Delta_v}\,; \tag{4.61}$$

$$\boldsymbol{H}_0(x) = 1\,; \quad \boldsymbol{H}_2(x) = -\,2 + 4x^2\,; \quad \text{and} \quad \boldsymbol{H}_4(x) = 12 - 48x^2 + 16x^4 \tag{4.62}$$

are the Hermite polynomials.

Because the coefficients b_k depend on the variable $\psi_*(t)$, the variables t and τ are not separable in the general case, and the shape of the envelope of the instantaneous power spectral density is changed during the propagation of the target return signal. One can see from Equation (4.61) that the ratio $\frac{\Delta_p}{\Delta_v}$ plays the main role in this formula. If the ratio $\frac{\Delta_p}{\Delta_v}$ is so low in value and for any values of the angle $\psi_*(t)$ we can neglect all forms of the series expansion except the first term, the correlation function of the fluctuations takes the form as in Equation (4.48) and is independent of the time variable t. In this approximation, the correlation function is separable. For example, at $\gamma_0 = 60°$, $\tau_p = 1$ μsec, $h = 3000$ m, $\Delta_v = 6°$, we obtain the ratio $\frac{\Delta_p}{\Delta_v} = 0.75$, i.e., it is not so low in value. In the condition $\frac{\Delta_p}{\Delta_v} > 1$, the series given by Equation (4.45) converges slowly.

At $\psi_* = 0$ or $t = T_d$, Equation (4.55) can be simplified and has the following form:

$$R_0(\tau, T_d) = \sqrt{\pi} \cdot \Phi\left[\sqrt{\tfrac{\pi}{2}} \cdot \tfrac{\Delta_p^\circ}{\Delta_v}\left(1 - \tfrac{|\tau|}{\tau_p}\right)\right], \tag{4.63}$$

where

$$\Delta_p^\circ = \frac{c\tau_p \sin^2\gamma_0}{2\,h\cos\gamma_0}\,. \tag{4.64}$$

The coefficients b_k given by Equation (4.61) take the following form:

$$b_k = \frac{\left(\frac{\pi}{2}\right)^k \left(\frac{\Delta_p^\circ}{\Delta_v}\right)^{2k}}{(2k+1)!} \cdot H_{2k}(0) = \frac{(-\pi)^k}{(2k)!!(2k+1)} \cdot \left(\frac{\Delta_p^\circ}{\Delta_v}\right)^{2k}. \tag{4.65}$$

The normalized correlation function of the fluctuations given by Equation (4.63) is shown in Figure 4.6 at various values of the ratio $\frac{\Delta_p^\circ}{\Delta_v}$. The periodic normalized correlation function given by Equation (4.37), with the waves given by Equation (4.63), is shown in Figure 4.7 at various values of the duration τ_p of the pulsed searching signal and at the constant value of Δ_v. Under the condition $\frac{\Delta_p^\circ}{\Delta_v} \ll 1$, the shape of waves of the periodic normalized correlation function is close to the triangular form [see Equation (4.48)]. As the duration τ_p of the pulsed searching signal is increased, the waves of the periodic normalized correlation function are naturally expanded. The width of the waves — the base of the triangular form — is always equal to $2\tau_p$ and the apex of waves is smoothed, but the duration and slope of the leading and trailing edges of the pulsed searching signal are the same.

In the condition $\tau_p \to T_p$, the waves of the periodic normalized correlation function of the fluctuations are merged in line with the constant level. This means that the intraperiod fluctuations are absent in passing from the pulsed searching signal to the continuous searching signal. The physical meaning of this characteristic of the correlation function is clear. When the pulsed

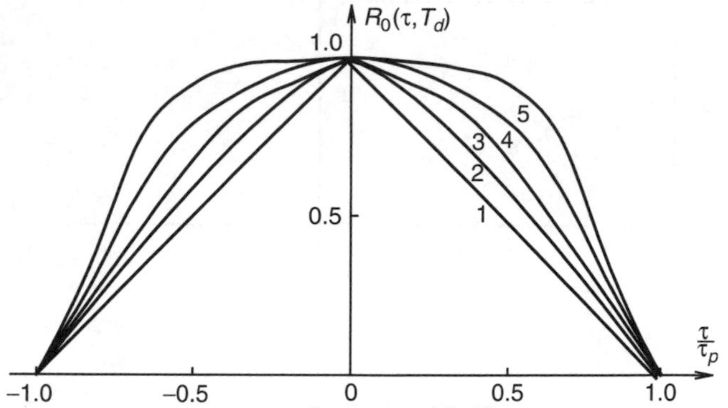

FIGURE 4.6
The normalized correlation function of the target return signal fluctuations in the radar range for individual waves at the stationary radar, $\psi_* = 0°$ and $\tau_p = const$: (1) $\frac{\Delta_p^\circ}{\Delta_h} = 0$; (2) $\frac{\Delta_p^\circ}{\Delta_h} = 0.5$; (3) $\frac{\Delta_p^\circ}{\Delta_h} = 1.0$; (4) $\frac{\Delta_p^\circ}{\Delta_h} = 1.5$; (5) $\frac{\Delta_p^\circ}{\Delta_h} = 2.0$.

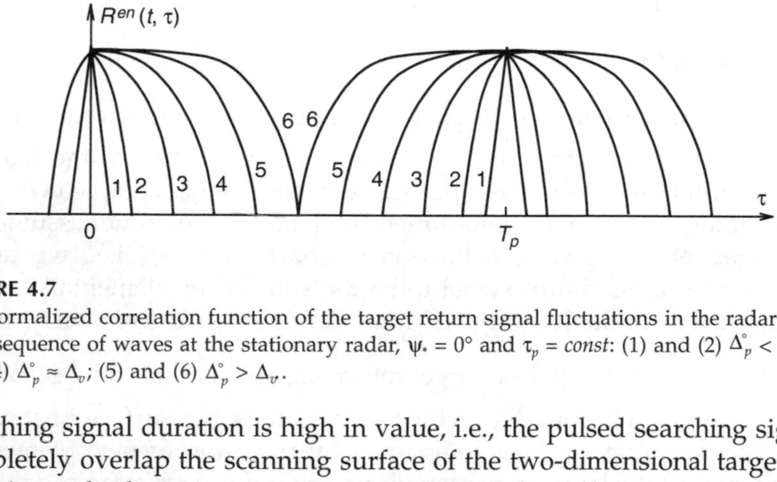

FIGURE 4.7
The normalized correlation function of the target return signal fluctuations in the radar range for a sequence of waves at the stationary radar, $\psi_* = 0°$ and $\tau_p = const$: (1) and (2) $\Delta_p^\circ < \Delta_v$; (3) and (4) $\Delta_p^\circ \approx \Delta_v$; (5) and (6) $\Delta_p^\circ > \Delta_v$.

searching signal duration is high in value, i.e., the pulsed searching signals completely overlap the scanning surface of the two-dimensional target, the resulting pulsed target return signal assumes a shape that is very similar to that of the pulsed searching signal (see Figure 4.8a). In this case, there are the fluctuations of the slope of the leading and trailing edges, which are formed in scanning the two-dimensional (surface) target (see Figure 4.8b). The top of the pulsed target return signal is flat and the fluctuations are absent. The power spectral density given by Equation (4.49) corresponds to the sum of the terms in Equation (4.50)–Equation (4.53) and is shown in Figure 4.5 by the dotted line in the conditions $\psi_* = 0°$ and $\frac{\Delta_p^\circ}{\Delta_v} = 0.5$.

However, in spite of the definition of the envelope of the correlation function given by Equation (4.55) being formally true for any values of Δ_p°

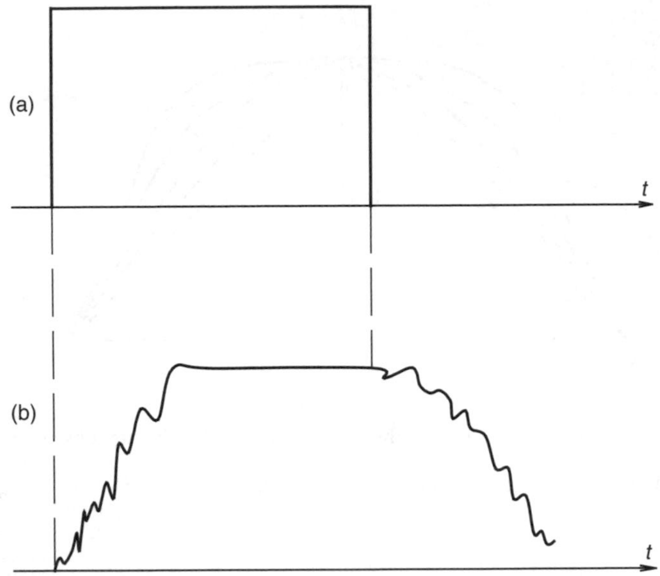

FIGURE 4.8
(a) Shape of the pulsed searching signal, and (b) shape of the target return signal at $\Delta_p^\circ >> \Delta_v$.

and Δ_v, we should be very careful in using Equation (4.55) at the condition $\frac{\Delta_p^\circ}{\Delta_v} > 1$, because the stochastic process degenerates when the duration of the pulsed searching signal is very high in value. Within the limits of the time interval in which the pulsed searching signal completely covers the scanning surface of the two-dimensional target, we cannot assume that the target return signal is a Poisson stochastic process, i.e., we cannot consider the target return signal to be the sum of the elementary signals arising at the random instants of time. As mentioned previously, in the condition $\frac{\Delta_p^\circ}{\Delta_v} >> 1$, the pulsed target return signal has a flat top because the propagation of the pulsed searching signal along the surface of the two-dimensional target is not accompanied by initiation of new elementary signals during the large time intervals. In other words, we can say that the stochastic process is converted from the nonsingular process, as in the case of the pulsed searching signal with the short duration, into the two-parametric singular process, for example, $A \cos(\omega t + \varphi)$, in which only the parameters A and φ are stochastic.[37] The correlation function of this process does not completely define the properties of the process because it has not ceased to exist as a Gaussian stochastic process.

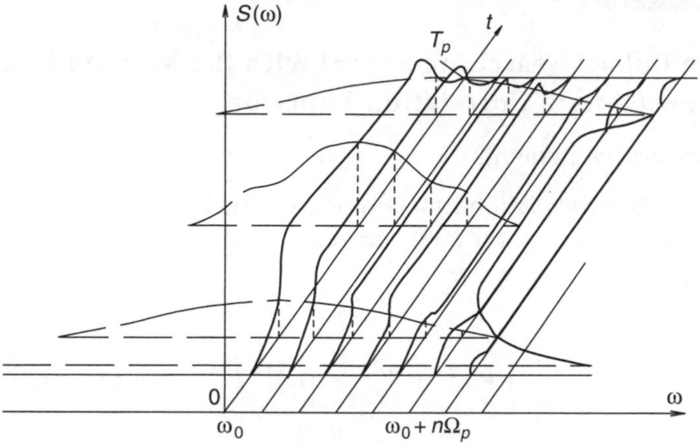

FIGURE 4.9
The instantaneous power spectral density of the target return signal fluctuations as a function of time under the condition $\Delta_p^\circ << \Delta_v$.

Under the condition $\frac{\Delta_p^\circ}{\Delta_v} >> 1$, in the act of the pulsed searching signal overlapping the scanned surface of the two-dimensional target, the fluctuations have a very short correlation interval at the start. Thereafter, it increases so that the fluctuations are absent at the top of the pulsed target return signal. As the pulsed searching signal is run down from the scanned surface of the two-dimensional target, the process goes into inverse sequence: in the beginning, the slow target return signal fluctuations arise, and thereafter, they become rapid (see Figure 4.8b).

The instantaneous power spectral density as a function of time t corresponding to the process described in the preceding text is shown in Figure 4.9 under the condition $\omega >> \omega_0$, where it is symmetric with respect to the frequency ω_0. In the beginning, the instantaneous power spectral density has an effective bandwidth high in value and low power. Thereafter, the effective bandwidth is decreased and the power is increased. At the end of the pulse, the spectral density is expanded again — i.e., the effective bandwidth is increased — and the power is decreased. Because the process is rigorous and periodic, i.e., the radar is stationary and the interperiod fluctuations are absent, we have the regulated power spectral density with the distance Ω_p between harmonics. The effective bandwidth in the stationary region is defined by the duration τ_p of the pulsed searching signal: $\Delta F \approx (\tau_p)^{-1}$.

4.4 The Pulsed Searching Signal with the Moving Radar: The Aspect Angle Correlation Function

4.4.1 General Statements

The aspect angle correlation function of the target return signal fluctuations is determined by Equation (4.8). Introducing a new variable z given by Equation (4.36), we can write

$$R_\gamma(t,\tau) = N\sum_{n=0}^{\infty} \int \boldsymbol{P}\,[z - 0.5(\mu\tau - nT_p)]\cdot \boldsymbol{P}^*[z + 0.5(\mu\tau - nT_p)] \\ \times g_v^2(\psi_* + c_* z)\cdot e^{-jc_*\Omega_\gamma z\tau}dz \quad , \tag{4.66}$$

where $c_* = \frac{c}{2\rho_*}\operatorname{tg}\gamma_*$; $c_* z = \psi - \psi_*$; and Ω_γ is determined by Equation (4.10). The aspect angle normalized correlation function given by Equation (4.66) has a very complex structure and will be investigated in more detail using specific examples. Here we study only the interperiod fluctuations. We can simplify some formulae in Equation (4.9) and Equation (4.10) that containing the parameters Ω_β and Ω_γ, assuming that the radar moves only in the horizontal plane, i.e., $\varepsilon_0 = 0$. We can always omit this limitation.

The normalized correlation function in the glancing radar range follows from Equation (4.66) at the condition $\tau = nT_p$:

$$R_\gamma(t,\tau) = N\int \Pi^2(z)\cdot g_v^2(\psi_* + c_* z)\cdot e^{-jc_*\Omega_\gamma z\tau}dz\,. \tag{4.67}$$

The corresponding power spectral density takes the following form:

$$S_\gamma[\omega, \psi_*(t)] \cong \Pi^2\left(-\frac{\omega}{c_*\Omega_\gamma}\right)\cdot g_v^2\left(\psi_* - \frac{\omega}{\Omega_\gamma}\right). \tag{4.68}$$

Reference to Equation (4.68) shows that the power spectral density depends essentially on the angle ψ_* because the parameters c_* and Ω_γ depend on it. The relative position of the target return signal and the vertical-coverage directional diagram also depends on the angle ψ_* because the normalized correlation function given by Equation (4.67) is not separable, i.e., the variables t and τ are not separable.

When the vertical-coverage directional diagram width — the beam width — is large in value, i.e., the condition $\Delta_v >> \Delta_p$ is satisfied, the shape of the power spectral density can be determined in the following form:[38]

$$S_\gamma(\omega, \psi_*) \cong \Pi^2\left(-\frac{\omega}{c_*\Omega_\gamma}\right). \tag{4.69}$$

The effective power spectral density bandwidth is determined as follows:

$$\Delta\Omega_\tau = k_p \tau_p c_* \cdot \Omega_\gamma = \Delta_p \cdot \Omega_\gamma = \frac{2\pi\, k_p c \tau_p V \sin^3\gamma_* \cos\beta_0}{\lambda\, h \cos\gamma_*}. \tag{4.70}$$

The effective bandwidth $\Delta\Omega_\tau$ depends essentially on the aspect angle γ_* (see Figure 4.10, the solid line; the dotted line will be discussed in Section 10.2). With high values of the aspect angle γ_*, the effective bandwidth $\Delta\Omega_\tau$ is very high. For example, at $V = 300\ \frac{\text{m}}{\text{sec}}$, $l = 3\text{cm}$, $\tau_p = 1\mu\text{sec}$, $\gamma_* = 45°$, $\beta_0 = 45°$, we obtain the effective bandwidth $\Delta F_\tau = \frac{\Delta\Omega_\tau}{2\pi} = 350$ Hz. Under the condition $\gamma_* \to 90°$, both Equation (4.70) and Equation (2.140) are not true.

If the condition $h = const$ is true, with an increase in the radar range ρ_*, the effective bandwidth $\Delta\Omega_\tau$ decreases sharply and tends to approach zero because the angle Δ_p given by Equation (2.140) is decreased. Deviation of the radar antenna from the direction of the moving radar ($\beta_0 \neq 0$) leads to decrease in the effective bandwidth $\Delta\Omega_\tau$ of the power spectral density. If the duration of the pulsed searching signal is high in value, i.e., the condition $\Delta_p >> \Delta_v$ is satisfied, then, for those time cross sections in which the pulsed searching signal completely overlaps the scanned surface of the two-dimensional target, the shape of the power spectral density is defined by the shape

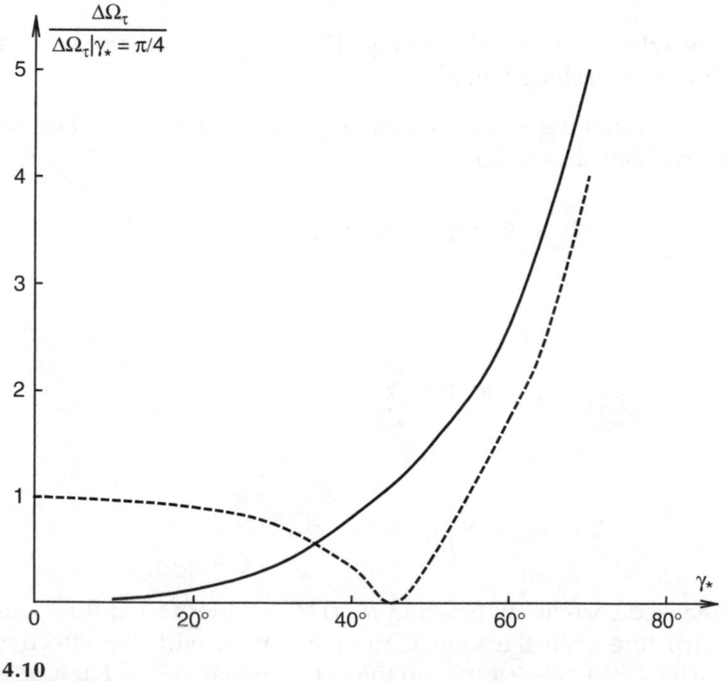

FIGURE 4.10
The bandwidth $\Delta\Omega_\tau$ of the power spectral density of the target return signal fluctuations as a function of the aspect angle γ_*: $\Delta\Omega_\tau$ — the solid line; $|\Delta\Omega_\tau - \Delta\Omega_\omega|$ — the dotted line.

of the square of the vertical-coverage directional diagram under the condition $\psi_* = 0$ or $\gamma_* = \gamma_0$:

$$S_\gamma(\omega, \psi_*) \cong g_v^2\left(-\frac{\omega}{\Omega_\gamma}\right). \tag{4.71}$$

Equation (4.71) is equivalent to Equation (3.55) and Equation (4.20) if the conditions $S^\circ(\beta_0, \psi + \gamma_0) \approx S^\circ(\beta_0, \gamma_*)$ and $\sin(\psi + \gamma_0) \approx \sin \gamma_*$ are satisfied in Equation (4.20) and $\varepsilon_0 = 0$.

The effective power spectral density bandwidth of the interperiod fluctuations given by Equation (4.71) has the following form:

$$\Delta\Omega_v = \Delta_v^{(2)} \cdot \Omega_\gamma = \frac{4\pi V}{\lambda} \cdot \Delta_v^{(2)} \cos\beta_0 \cos\gamma_0 \tag{4.72}$$

and coincides with Equation (3.84). The effective bandwidth given by Equation (4.72) is different from that given by Equation (4.70). We use the width $\Delta_v^{(2)}$ of the vertical-coverage directional diagram, i.e., the beam width, in Equation (4.72) instead of the angle Δ_p used in Equation (4.70). The normalized correlation function in the fixed radar range can be obtained based on Equation (4.66) under the condition $\tau = nT_p$.

4.4.2 The Arbitrary Vertical-Coverage Directional Diagram: The Gaussian Pulsed Searching Signal

If the pulsed searching signal is Gaussian [see Equation (2.105)], reference to Equation (4.66) shows that

$$R_\gamma(\tau, \psi_*) = R'_\gamma(\tau) \cdot R''_\gamma(\tau, \psi_*), \tag{4.73}$$

where

$$R'_\gamma(\tau) = \sum_{n=0}^{\infty} e^{-\pi \cdot \frac{(\mu\tau - nT_p)^2}{2\tau_p^2}}; \tag{4.74}$$

$$R''_\gamma(\tau, \psi_*) = N \int g_v^2(\psi_* + c_* z) \cdot e^{-\frac{2\pi z^2}{\tau_p^2} - jc_*\Omega_\gamma z\tau} dz. \tag{4.75}$$

The normalized correlation function $R'_\gamma(\tau)$ of the interperiod fluctuations has a comb structure with the same Gaussian waves, with the effective bandwidth equal to $\sqrt{2}\,\tau'_p = \sqrt{2}\,\tau_p\mu^{-1}$ and the period equal to $T'_p = T_p\mu^{-1}$. $R'_\gamma(\tau)$ coincides with the normalized correlation function $R_p(\tau)$ in the radar range scanning the three-dimensional (space) target [see Equation (3.6) and Equation (3.20); Figure 3.1a, the dotted line]. The corresponding power spectral density $S'_\gamma(\omega)$ is the regulated function given by Equation (3.15) and Equation (3.21) (see Figure 3.1b, the dotted line).

The normalized correlation function $R''_\gamma(\tau, \psi_*)$ defines the slow fluctuations caused by the difference in the Doppler frequencies in the aspect-angle plane. $R''_\gamma(\tau, \psi_*)$ characterizes the destruction of correlation from period to period because, under the condition $\tau = nT'_p$, the aspect-angle normalized correlation function $R_\gamma(\tau, \psi_*)$ is defined by the normalized correlation function $R''_\gamma(\tau, \psi_*)$ in the aspect angle plane: $R_\gamma(\tau, \psi_*) = R''_\gamma(\tau, \psi_*)$. In other words, $R''_\gamma(\tau, \psi_*)$ is the normalized correlation function in the glancing radar range. Thus, with the Gaussian pulsed searching signal, the aspect-angle normalized correlation function given by Equation (4.73) and the total normalized correlation function are defined by the product of the intraperiod and interperiod fluctuations, as in the case of scanning the three-dimensional (space) target. The power spectral density corresponding to the normalized correlation function $R''_\gamma(\tau,\psi_*)$, given by Equation (4.75), can be written in the following form:

$$S''_\gamma(\omega, \psi_*) \cong g_v^2\left(\psi_* - \tfrac{\omega}{\Omega_\gamma}\right) \cdot e^{-\pi \cdot \frac{\omega^2}{\Delta\Omega_\tau^2}}, \tag{4.76}$$

where the effective bandwidth $\Delta\Omega_\tau$ is determined by Equation (4.70) under the condition $k_p = (\sqrt{2})^{-1}$.

The power spectral density $S''_\gamma(\omega, \gamma_*)$ of interperiod fluctuations is the particular case of $S_\gamma[\omega, \psi_*(\tau)]$ in the glancing radar range given by Equation (4.68). When the vertical-coverage directional diagram width is large in value, i.e., the condition $\Delta_v >> \Delta_p$ is true, $S''_\gamma(\omega, \psi_*)$ takes the following form:

$$S''_\gamma(\omega, \psi_*) \cong e^{-\pi \cdot \frac{\omega^2}{\Delta\Omega_\tau^2}}. \tag{4.77}$$

This formula is a particular case of the power spectral density given by Equation (4.69). If the duration of the pulsed searching signal is high in value, i.e., the condition $\Delta_p >> \Delta_v$ is satisfied, and at the values of ψ_* in which the pulsed searching signal completely overlaps the scanned surface of the two-dimensional target, the power spectral density of the interperiod fluctuations is determined by Equation (4.71).

The power spectral density corresponding to the normalized correlation function given by Equation (4.73) is defined by the convolution between $S'_\gamma(\omega)$ and $S''_\gamma(\omega, \psi_*)$ (see Figure 3.3b),

$$S_\gamma(\omega, \psi_*) \cong \sum_{n=0}^{\infty} S''_\gamma(\omega - n\Omega'_p, \psi_*) \cdot e^{-\pi \cdot \left(\frac{n\Omega'_p}{\Delta\Omega'_p}\right)^2}, \tag{4.78}$$

where

$$\Delta\Omega'_p = 2\,\pi \cdot \Delta F'_p = \sqrt{2}\,\pi(\tau'_p)^{-1}. \tag{4.79}$$

Consider the interperiod fluctuations in the fixed radar range, assuming that the condition $\tau = nT_p$ in Equation (4.73)–Equation (4.75) is true. Then

$$R'_\gamma(\tau) = e^{-\frac{\pi}{2}\cdot\left(\frac{\bar{\mu}\,\tau}{\tau_p}\right)^2}, \tag{4.80}$$

where

$$\bar{\mu} = \mu - 1 = 2Vc^{-1}\cos\beta_0\cos\gamma_*. \tag{4.81}$$

The correlation interval of the normalized correlation function in the fixed radar range given by Equation (4.80) is very high in value:

$$\tau'_c = (\Delta F')^{-1} = \frac{c\tau_p}{\sqrt{2}\;V\cos\beta_0\cos\gamma_*}. \tag{4.82}$$

The correlation interval is defined by the time required for the radar moving with the velocity V to travel the distance equal to $\frac{c\tau_p}{\sqrt{2}\,\cos\beta_0\cos\gamma_*}$. Compared with Equation (3.157), that is equal to the effective bandwidth of the scanned element resolved in the radar range at the cross section, which is parallel to the direction of the moving radar. Within the limits of the time interval equal to the correlation length τ'_c, all scatterers filling the scanned element resolved in the radar range are exchanged.

The power spectral density in the fixed radar range corresponding to the normalized correlation function given by Equation (4.80) has the following form:

$$S'_\gamma(\omega) \cong e^{-\pi\cdot\left(\frac{\omega}{2\,\pi\,\Delta F'}\right)^2}. \tag{4.83}$$

The effective bandwidth $\Delta F' = (\tau'_c)^{-1}$ given by Equation (4.95) is very low in value, as a rule. The total power spectral density in the fixed radar range corresponding to the normalized correlation function given by Equation (4.73) is defined by the convolution between $S'_\gamma(\omega)$ and $S''_\gamma(\omega, \psi_*)$, given by Equation (4.83) and Equation (4.76), respectively.

4.4.3 The Gaussian Vertical-Coverage Directional Diagram: The Gaussian Pulsed Searching Signal

If the vertical-coverage directional diagram is Gaussian, Equation (4.73) is true when

$$R''_\gamma(\tau, \psi_*) = e^{-\pi(\Delta F''\tau)^2 + j\Omega''\tau}, \tag{4.84}$$

where

$$\Omega'' = \frac{\psi_* \Omega_\gamma}{1 + \frac{\Delta_v^{(2)}}{\Delta_p^{(2)}}}; \tag{4.85}$$

$$\Delta F'' = \frac{\Delta F_\tau \cdot \Delta F}{\sqrt{\Delta F_\tau^2 + \Delta F_v^2}} = \frac{\Delta F_\tau}{\sqrt{1 + \frac{\Delta_p^{(2)}}{\Delta_v^{(2)}}}}. \tag{4.86}$$

The power spectral density of the interperiod fluctuations, which can be obtained by the Fourier transform of the normalized correlation function given by Equation (4.84) or follows from Equation (4.76), takes the following form

$$S''_\gamma(\omega, \psi_*) \cong e^{-\pi \cdot \frac{(\omega - \Omega'')^2}{(\Delta\Omega'')^2}}, \tag{4.87}$$

where Ω'' and $\Delta\Omega''$ are determined by Equation (4.85) and Equation (4.86), respectively. Reference to Equation (4.86) shows that under the condition $\Delta_p << \Delta_v$, the effective bandwidth $\Delta F''$ is close to that given by Equation (4.70). If the condition $\Delta_p >> \Delta_v$ is true, $\Delta F''$ is close to that given by Equation (4.72). The effective bandwidth $\Delta F''$ is a complex function of the angle γ_* because both the effective bandwidth ΔF_τ and the angle dimension Δ_p of the resolved element depend on the angle γ_*.[39]

The ratio $\frac{\Delta_p}{\Delta_v}$ is also a function of the angle γ_*. With the same parameters of the radar antenna τ_p, λ, Δ_v and the fixed conditions of radar motion, for example, V and h, the condition $\Delta_p >> \Delta_v$ can be satisfied when the radar range is low in value — the aspect angle is high in value — and if the condition $\Delta_p << \Delta_v$ would be satisfied, the radar range is high in value. The average frequency Ω'' of the power spectral density $S''_\gamma(\omega, \psi_*)$ is proportional to the angle $\psi_* = \gamma_* - \gamma_0$ and depends on the parameters Ω_γ and Δ_p. In other words, the average frequency Ω'' is a function of the angle $\gamma_*(t)$. On the axis of the directional diagram, we have the equalities $\psi_* = 0°$ and $\Omega'' = 0$. Therefore, we can write

$$\begin{cases} \Omega'' = \dfrac{\psi_* \Omega_\gamma \Delta_p^{(2)}}{\Delta_v^{(2)}} \cong 0 & \text{at} \quad \Delta_p << \Delta_v\,; \\ \Omega'' \approx \psi_* \Omega_\gamma & \text{at} \quad \Delta_p >> \Delta_v\,. \end{cases} \tag{4.88}$$

The total aspect-angle normalized correlation function of the interperiod fluctuations is defined by the product of the normalized correlation functions given by Equation (4.73) and Equation (4.84) and can be written in the following form:

$$R''_\gamma(\tau, \psi_*) = e^{-\pi\,(\Delta F''\tau)^2 + j\Omega''\tau} \cdot \sum_{n=0}^{\infty} e^{-\pi \cdot \frac{(\mu\tau - nT_p)^2}{2\,\tau_p^2}}\,. \tag{4.89}$$

The power spectral density corresponding to the total normalized correlation function $R_\gamma(\tau, \psi_*)$ is determined in the following form according to Equation (4.78) and Equation (4.87)

$$S_\gamma(\omega) \cong \sum_{n=0}^{\infty} e^{-\pi \cdot \left(\frac{n\Omega'_p}{\Delta\Omega'_p}\right)^2} \cdot e^{-\pi \cdot \left(\frac{\omega - \Omega'' - n\Omega'_p}{\Delta\Omega''}\right)^2}, \tag{4.90}$$

where the effective bandwidth $\Delta\Omega'_p$ is determined by Equation (4.79), and the effective bandwidth $\Delta\Omega''$ can be written in the form $\Delta\Omega'' = 2\pi\Delta F''$. The power spectral density given by Equation (4.90) has a comb structure, in which the waves at the frequencies $\omega = n\Omega'_p + \Omega''$ have the effective bandwidth $\Delta\Omega''$. The envelope of the waves is Gaussian with the effective bandwidth $\Delta\Omega'_p$.

As we noted in Section 4.4.2, $R''_\gamma(\tau, \psi_*)$ and $S''_\gamma(\omega, \psi_*)$ are the normalized correlation function and the power spectral density in the glancing radar range. $R''_\gamma(\tau, \psi_*)$ and $S''_\gamma(\omega, \psi_*)$ are completely defined by Equation (4.84)–Equation (4.87). The normalized correlation function of the interperiod fluctuations in the fixed radar range is defined by the product of the normalized correlation functions $R'_\gamma(\tau)$ and $R''_\gamma(\tau, \psi_*)$ given by Equation (4.80) and Equation (4.84), respectively, and can be written in the following form:

$$R_\gamma(\tau, \psi_*) = e^{-\pi[(\Delta F''\tau)^2 + (\Delta F'\tau)^2] + j\Omega''\tau}\,. \tag{4.91}$$

The power spectral density is determined by Equation (4.87) with the effective bandwidth given by

$$\Delta F = \sqrt{(\Delta F'')^2 + (\Delta F')^2}\ . \tag{4.92}$$

4.4.4 The Wide-Band Vertical-Coverage Directional Diagram: The Square Waveform Pulsed Searching Signal

Let us assume that the condition $\Delta_p \ll \Delta_v$ is true and that we can neglect the dependence between the directional diagram $g_v(\psi_* + c_* z)$ and the parameter z within the limits of duration of the pulsed searching signal. Then, in the case of the square waveform pulsed searching signal with the duration τ_p, we can write

$$R_\gamma(\tau) = \sum_{n=0}^{\infty} \frac{\sin\left[\pi\,\Delta F_\tau \tau\left(1 - \frac{|\mu\tau - nT_p|}{\tau_p}\right)\right]}{\pi\,\Delta F_\tau \tau}\ . \tag{4.93}$$

Comparing Equation (3.31) and Equation (4.93), one can see that $R_\gamma(\tau)$ is the aspect-angle normalized correlation function of the target return signal fluctuations formed by covering the linearly frequency-modulated pulsed searching signals with the deviation $\Delta\overline{F}_\tau$ given by Equation (4.70) and duration $\tau'_p = \tau_p \mu^{-1}$. Unlike the normalized correlation function given by Equation (3.31), the one given by Equation (4.93), rigorously speaking, is not a periodic function despite this function possessing the properties of periodicity. The aspect-angle normalized correlation function given by Equation (4.93) is shown in Figure 4.11. This function consists of dissimilar narrow waves with the width equal to $2\tau'_p$ and the period T'_p. At $n = 0$, due to the condition $\Delta F_\tau \tau_p \ll 1$, as a rule, we can write

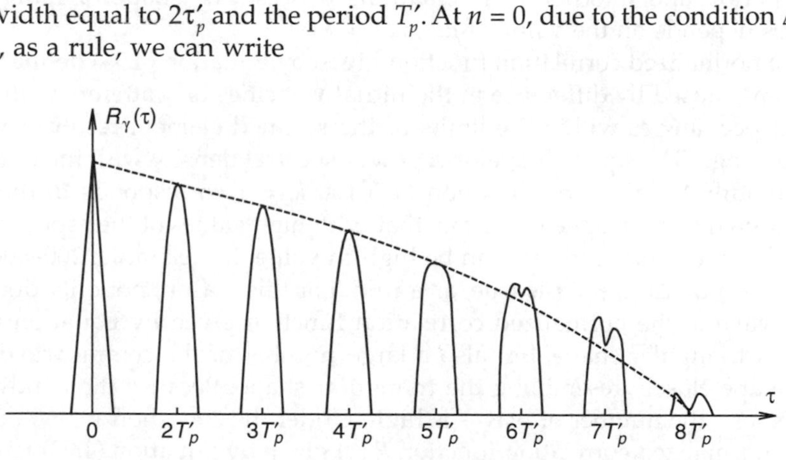

FIGURE 4.11
The aspect angle normalized correlation function of the target return signal fluctuations with the square waveform pulsed searching signal.

$$R_\gamma(\tau) = 1 - \frac{|\tau|}{\tau'_p}. \tag{4.94}$$

The function $R_\gamma(\tau)$ is the aspect-angle normalized correlation function within the limits of a single period of the square waveform pulsed searching signal. At $n \neq 0$, the shape of the waves is not triangular, and when the value of n is high, the difference between the shape of the waves and the triangular shape is greater. The duration of the waves is equal to $2\tau'_p$ independent of the value of n.

Under the condition $\Delta F_\tau \tau_p = \Delta F_\tau n\, T'_p = 0.5$, the waves have the shape of the cosine curve within the limits of the interval $[-0.5\pi, 0.5\pi]$. If $\Delta F_\tau n T'_p < 0$ is true, the waves have spiked tops. If $\Delta F_\tau n\, T'_p > 0.5$ is satisfied, the waves have crevasses. When $\Delta F_\tau n\, T'_p = 1$, the crevasse is maximal and reaches zero. When $\tau = n\, T'_p$, we can write

$$R_\gamma(\tau) = \mathbf{sinc}(\pi\, \Delta F_\tau \tau). \tag{4.95}$$

The function $R_\gamma(\tau)$ is the aspect-angle normalized correlation function of the interperiod fluctuations in the glancing radar range. The normalized correlation function given by Equation (4.95) is shown in Figure 4.11 by the dotted line. The aspect-angle normalized correlation function $R_\gamma(\tau)$ given by Equation (4.95) is not an envelope of the normalized correlation function $R_\gamma(\tau)$ given by Equation (4.93). Therefore, the normalized correlation function $R_\gamma(\tau)$ is not equal to the product of the normalized correlation functions of intraperiod and interperiod fluctuations because the shape of individual waves depends on the value of n.

The normalized correlation function given by Equation (4.95) defines fluctuations caused by difference in the radial velocities of scatterers with various aspect angles within the limits of the scanned element resolved in the radar range. The square waveform power spectral density with the effective bandwidth ΔF_τ given by Equation (4.70) at $k_p = 1$ corresponds to the case considered here. In spite of the fact that with high values of the aspect angle, the effective bandwidth ΔF_τ can be high in value, for example, 200–600 Hz, the condition $\Delta F_\tau \tau_p << 1$ is true, as a rule. For this reason, not only does the first wave of the normalized correlation function given by Equation (4.93) have a triangular shape, but also a large number of the central waves has the shape closely resembling the triangular shape. Because the bandwidth ΔF_τ is less, the number of waves is high. Under the condition $V \to 0$ or $\gamma_* \to 0$ the normalized correlation function $R_\gamma(\tau)$ given by Equation (4.93) tends to approach the following form

$$R_\gamma(\tau) = \sum_{n=0}^{\infty} \left(1 - \tfrac{|\mu\, \tau_p - nT_p\,|}{\tau_p}\right) \tag{4.96}$$

and the envelope of the power spectral density is determined by Equation (3.19).

Consider the normalized correlation function of the interperiod fluctuations in the fixed radar range.[40,41] If the condition $\tau = nT_p$ [see Equation (4.93)] is satisfied, we can write

$$R_\gamma(\tau) = \frac{\sin[\pi\, \Delta F_\tau \tau(1 - \Delta F'|\tau|)]}{\pi\, \Delta F_\tau \tau}, \tag{4.97}$$

where

$$\Delta F' = \frac{\bar{\mu}}{\tau_p} = \frac{2\,V}{c\tau_p} \cdot \cos\beta_0 \cos\gamma_* \,. \tag{4.98}$$

Compare this with Equation (4.82). The normalized correlation function $R_\gamma(\tau)$ given by Equation (4.97) is not periodic. Comparing Equation (3.31) and Equation (4.97) under the condition $\mu = 1$, one can see that $R_\gamma(\tau)$ is the aspect-angle normalized correlation function of the fluctuations formed by covering the square waveform linear-frequency-modulated pulsed signals with the duration $\tau'_c = \frac{1}{\Delta F'}$ and deviation ΔF_τ. These pulsed signals are formed by reflection from individual elementary scatterers moving relative to the moving radar and scanned with various values of the aspect angle within the limits of the element resolved in the radar range. Changes in the aspect angle during scatterer motion lead to changes in the Doppler frequency, i.e., to frequency modulation. The pulsed signal duration τ' is the time required so that the scatterer with the coordinates β_0 and γ_* and moving with the velocity V could cover the scanned element resolved in the radar range.

The character of the normalized correlation function $R_\gamma(\tau)$ of the interperiod fluctuations given by Equation (4.97) depends on the ratio

$$\frac{\Delta F_\tau}{\Delta F'} = \frac{c^2 \tau_p^2 \sin^3\gamma_*}{2\, h\lambda \cos^2\gamma_*} \,. \tag{4.99}$$

Under the condition $\Delta F' >> \Delta F_\tau$, based on Equation (4.97), we can write

$$R_\gamma(\tau) = 1 - \Delta F'|\tau| \,. \tag{4.100}$$

This takes into account the fluctuations caused by the exchange of scatterers within the limits of the scanned element resolved in the radar range with the moving radar (see Section 3.4). These fluctuations are caused by the interperiod fluctuations in the fixed radar range. The power spectral density

$$S_\gamma(\omega) \cong \mathbf{sinc}^2\left(\tfrac{\omega}{2\Delta F'}\right) \tag{4.101}$$

corresponds to $R_\gamma(\tau)$ given by Equation (4.100). The effective bandwidth of the power spectral density $S_\gamma(\omega)$ given by Equation (4.101) is equal to $\Delta F'$ [see Equation (4.98)]. Thus, we have the power spectral density of the square waveform target return signals with the duration $\tau'_c = \frac{1}{\Delta F'}$.

Under the condition $\Delta F_\tau >> \Delta F'$, we can obtain $R_\gamma(\tau)$ given by Equation (4.95) based on Equation (4.97). Under this condition, the normalized correlation function in the glancing radar range coincides with that in the fixed radar range because the value of the effective bandwidth $\Delta F'$ is infinitesimal. Comparing the effective bandwidths ΔF_τ and $\Delta F'$ given by Equation (4.70) and Equation (4.98), respectively, we can conclude that at high values of the aspect angle, the Doppler fluctuations play a main role, and with low values of the aspect angle, the fluctuations caused by the exchange of scatterers play a main role. For arbitrary values of the ratio $\frac{\Delta F_\tau}{\Delta F'}$, the power spectral density is given by Equation (3.32) if we replace the parameters ω'_M and τ'_p with the parameters $\Delta\Omega_\tau$ and $\frac{1}{\Delta F'}$, respectively. Under this condition, both the determination of τ_c in Equation (3.36) and Figure 3.5–Figure 3.7 are true.

4.5 The Pulsed Searching Signal with the Moving Radar: The Azimuth Correlation Function

4.5.1 General Statements

The azimuth-normalized correlation function is determined by Equation (4.7). For simplicity, we assume that the radar moves horizontally, i.e., $\varepsilon_0 = 0$ and the vertical-coverage directional diagram is independent of the angle ψ_*. In this case, the azimuth-normalized correlation function can be written in the following form:

$$R_\beta(\tau) = N \int g_h^2(\varphi) \cdot e^{j\tau\,\Omega_{\max}\cos\left(\beta_0 + \frac{\varphi}{\cos\gamma_*}\right)\cos\gamma_*}\, d\varphi\,. \tag{4.102}$$

The azimuth-normalized correlation function $R_\beta(\tau)$ defines the fluctuations caused by differences in the Doppler frequency of scatterers scanned with various values of the azimuth angle within the directional diagram in the horizontal plane, i.e., the horizontal-coverage directional diagram.[42] After multiplying by the factor $e^{j\omega_0\tau}$ and using the Fourier transform for $R_\beta(\tau)$, we can write the power spectral density of the Doppler fluctuations in the following form:

$$S_\beta(\omega) \cong \int g_h^2(\varphi)\cdot\delta\left[\omega-\omega_0-\Omega_{\max}\cos\left(\beta_0+\tfrac{\varphi}{\cos\gamma_*}\right)\cos\gamma_*\right]d\varphi\,. \quad (4.103)$$

Let us introduce a new variable

$$\Omega=\Omega_{\max}\cdot\cos\left(\beta_0+\tfrac{\varphi}{\cos\gamma_*}\right)\cos\gamma_*\,, \quad (4.104)$$

where Ω is the Doppler shift in frequency for the scatterer with the coordinates $\beta = \beta_0 + \frac{\varphi}{\cos\gamma_*}$ and $\gamma = \gamma_*$. Using the filtering property of the delta function,[43,44] we can write the power spectral density of Doppler fluctuations in the following form:

$$S_\beta(\omega) \cong \frac{g_h^2\left[\left(\arccos\frac{\omega-\omega_0}{\Omega_{\max}\cos\gamma_*}-\beta_0\right)\cos\gamma_*\right]}{\sqrt{1-\left(\frac{\omega-\omega_0}{\Omega_{\max}\cos\gamma_*}\right)^2}}+\frac{g_h^2\left[\left(-\arccos\frac{\omega-\omega_0}{\Omega_{\max}\cos\gamma_*}-\beta_0\right)\cos\gamma_*\right]}{\sqrt{1-\left(\frac{\omega-\omega_0}{\Omega_{\max}\cos\gamma_*}\right)^2}}\,, \quad (4.105)$$

where arccos (x) and β_0 have the same sign.

The power spectral density $S_\beta(\omega)$ given by Equation (4.105) — the power per unit bandwidth — is the ratio between the power dp of the target return signal from the scatterers, which are skewed with respect to the direction of the moving radar under the angle $\pm\beta$ with the same Doppler frequency,

$$dp \cong [g_h^2(\varphi)+g_h^2(-\varphi-2\beta_0\cos\gamma_*)]\,d\varphi\,. \quad (4.106)$$

In addition, the bandwidth $d\omega$ occupied by the target return signal is determined by

$$d\omega = d\left[\omega_0+\Omega_{\max}\cos\left(\beta_0+\tfrac{\varphi}{\cos\gamma_*}\right)\cos\gamma_*\right] = -\,\Omega_{\max}\sin\left(\beta_0+\tfrac{\varphi}{\cos\gamma_*}\right)d\varphi\,. \quad (4.107)$$

In this case, we can write $S_\beta(\omega)$ in the following form:

$$S_\beta(\omega) \cong \frac{g_h^2(\varphi)+g_h^2(-\varphi-2\beta_0\cos\gamma_*)}{\sin\left(\beta_0+\frac{\varphi}{\cos\gamma_*}\right)}\,, \quad (4.108)$$

where φ and Ω are related by Equation (4.104). Based on Equation (4.104), we can write

$$\varphi = \left(\arccos\tfrac{\omega-\omega_0}{\Omega_{\max}\cos\gamma_*}-\beta_0\right)\cos\gamma_*\,. \quad (4.109)$$

Substituting Equation (4.109) in Equation (4.108), we can obtain the power spectral density of the Doppler fluctuations given by Equation (4.105). This physical representation is the basis of the technique for determining the power spectral density of the Doppler fluctuations.

Reference to Equation (4.105) shows that we can obtain the following boundary values for frequency ω:

$$\omega_{min} = \omega_0 - \Omega_{max} \cos\gamma_* \leq \omega \leq \omega_0 + \Omega_{max} \cos\gamma_* = \omega_{max} \tag{4.110}$$

independently of the shape of the directional diagram. The power spectral density $S_\beta(\omega)$ at frequencies ω_{min} and ω_{max} tends to approach ∞ if a coefficient of amplification of the radar antenna in the corresponding directions does not equal zero. The frequency-modulated searching signal has the same power spectral density under the slow harmonic frequency modulation within the limits of the interval $[-\Omega_{max} \cos \gamma_*, \Omega_{max} \cos \gamma_*]$. This circumstance does not contradict the physical representations and can be explained by the fact that the bandwidth

$$|d\Omega| = \Omega_{max} \sin\beta \cos\gamma_* \; d\beta \tag{4.111}$$

of the target return signal within the limits of the azimuth angle $d\beta$ tends to approach zero as $\beta \to 0$, but the target return signal power

$$dp = 0.5c\tau_p \; d\beta \tag{4.112}$$

remains finite. The presence of two terms in the numerator in Equation (4.105) can be explained as follows. Scatterers disposed under the same angle at the left and right from the direction of the moving radar can generate the same Doppler frequency (see Figure 4.12). Consider, for example, the non-directed antenna in the horizontal plane radar antenna. Reference to Equation (4.105) shows that

$$S_\beta(\omega) \cong \frac{2}{\sqrt{1-\left(\frac{\omega-\omega_0}{\Omega_{max}\cos\gamma_*}\right)^2}}. \tag{4.113}$$

The power spectral density $S_\beta(\omega)$ given by Equation (4.113) is shown in Figure 4.13 by the dotted line. At the frequencies ω_{min} and ω_{max}, $S_\beta(\omega)$, as was mentioned previously, tends to approach ∞. The analogous power spectral density in scanning the three-dimensional (space) target is shown in Figure 4.13 by the horizontal dotted line. In the case of the three-dimensional (space) target, the maximum and minimum Doppler frequencies of $S_\beta(\omega)$ are determined by Equation (3.46) that is based on the use of Equation (4.110) under the condition $\gamma_* = 0$. $S_\beta(\omega)$ is finite at these frequencies. The corresponding

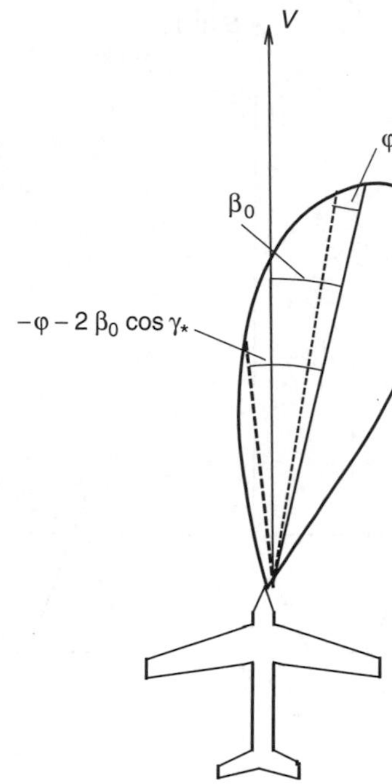

FIGURE 4.12
Formation of the power spectral density $S_\beta(\omega)$.

azimuth-normalized correlation function $R_\beta(\tau)$ is determined in the following form:

$$R_\beta(\tau) = J_0(\Omega_{max}\tau\cos\gamma_*)\cdot e^{j\omega_0\tau}, \tag{4.114}$$

where $J_0(x)$ is the Bessel function of the first order.

Reference to Equation (4.105) shows that if the horizontal-coverage directional diagram is narrow, then a small part of the power spectral density $S_\beta(\omega)$ of the Doppler fluctuations given by Equation (4.113) can be cut. The shape of the cut power spectral density depends both on the form of the horizontal-coverage directional diagram and on the position of the radar antenna axis with respect to the direction of the moving radar. Deformation of $S_\beta(\omega)$ for various positions of the pencil-beam radar antenna with respect to the direction of moving radar without consideration of the side lobes is shown in Figure 4.13.

If the horizontal-coverage directional diagram is narrow, the power spectral density $S_\beta(\omega)$ given by Equation (4.105) can be simplified in two important cases: (1) the radar antenna is high deflected from the direction of moving radar, and (2) it is low deflected. The simplest way to do this is to introduce an approximate factor in Equation (4.102):

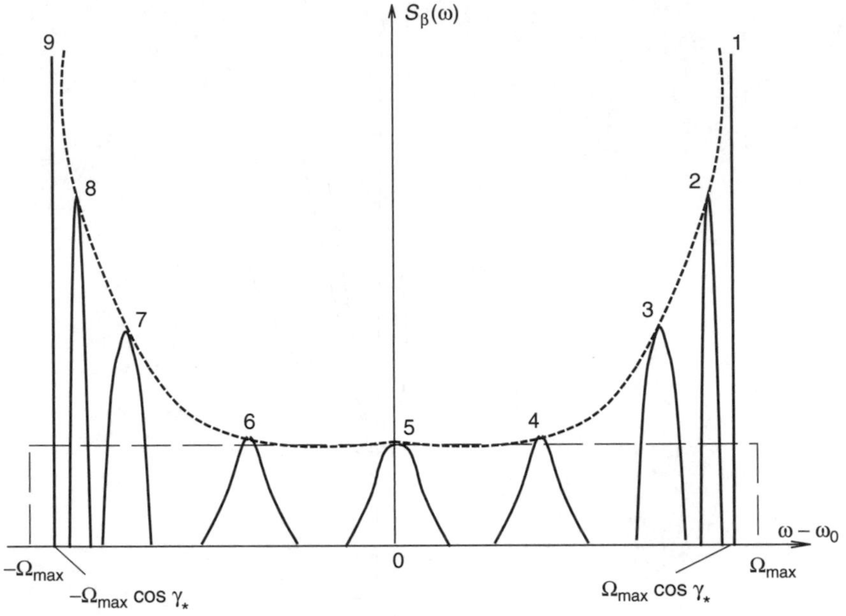

FIGURE 4.13
The power spectral density $S_\beta(\omega)$ with the nondirected (dotted line) and pencil-beam (solid line) radar antenna at various values of β_0: (1) $\beta_0 = 0°$; (2) $\beta_0 = 10°$; (3) $\beta_0 = 30°$; (4) $\beta_0 = 60°$; (5) $\beta_0 = 90°$; (6) $\beta_0 = 120°$; (7) $\beta_0 = 150°$; (8) $\beta_0 = 170°$; (9) $\beta_0 = 180°$.

$$\Omega(\varphi) \cong \Omega_* - \varphi\,\Omega_{max}\sin\beta_0 - \varphi^2\Omega_{max}\cos\beta_0, \tag{4.115}$$

where

$$\Omega_* = \Omega_{max}\cos\beta_0\cos\gamma_* \tag{4.116}$$

is the Doppler frequency corresponding to the middle of the scanned surface element of the two-dimensional target. Equation (4.116) is different from Equation (3.48) and Equation (3.49) by the exchange of the angle γ_0 for the angle γ_*.

4.5.2 The High-Deflected Radar Antenna

With the high-deflected radar antenna, if the value of $\sin\beta_0$ is not so low, we can neglect the third term in Equation (4.115). Then the azimuth-normalized correlation function $R_\beta(\tau)$ of the target return signal Doppler fluctuations can be written in the following form:

$$R_\beta(\tau) = N \cdot e^{j\Omega_* \tau} \int g_h^2(\varphi) \cdot e^{-j\varphi\,\Omega_{\max}\tau \sin\beta_0} d\varphi . \tag{4.117}$$

Multiplying Equation (4.117) with the factor $e^{j\omega_0 \tau}$, we can write[44]

$$S_\beta(\omega) \approx g_h^2\left(-\frac{\omega - \omega_0 - \Omega_*}{\Omega_{\max}\sin\beta_0}\right). \tag{4.118}$$

With the high-deflected radar antenna, the power spectral density $S_\beta(\omega)$ coincides approximately in shape with the square of the directional diagram by power — the receiving and transmission directional diagrams — at the cross section through the scanned surface element of the two-dimensional target resolved in the radar range.[45] The analogous conclusion was made by investigation of the three-dimensional (space) target, although the directional diagrams are different for these cases. This phenomenon can be easily explained based on the physical meaning.

Reference to Equation (4.118) shows that the effective bandwidth of the power spectral density $S_\beta(\omega)$ is related to the effective widths Δ_* and $\Delta_*^{(2)}$ of the directional diagram by power at the cross section ψ_* by the following relationships:

$$\Delta F_h = 2V\lambda\Delta_*^{(2)} \sin\beta_0 = 2V\lambda k_h \Delta_* \sin\beta_0 , \tag{4.119}$$

where k_h is the coefficient of the shape of the directional diagram (see Section 2.4). Equation (4.119) is similar to Equation (3.76) in the case of the three-dimensional (space) target.

Using Equation (4.115) and Equation (4.118) for the Gaussian and **sinc**-directional diagrams, we can obtain the same formulae as in Section 3.2: the power spectral densities of the Doppler fluctuations are determined by Equation (3.80) and Equation (3.86), respectively, and the normalized correlation functions are given by Equation (3.79) and Equation (3.88), respectively. The effective bandwidth of the power spectral density $S_\beta(\omega)$ and the correlation interval are determined by Equation (4.119), where $k_h = \frac{1}{\sqrt{2}}$ and $\frac{2}{3}$, respectively. Figure 3.12 and Figure 3.13 are true for the cases considered here. In particular, with the Gaussian directional diagram, we can write

$$S_\beta(\omega) \cong e^{-\pi \cdot \left(\frac{\omega - \omega_0 - \Omega_*}{2\pi\Delta F_h}\right)^2} ; \tag{4.120}$$

$$R_\beta(\tau) = e^{-\pi\Delta F_h^2 \tau^2 + j(\omega_0 + \Omega_*)\tau} . \tag{4.121}$$

4.5.3 The Low-Deflected Radar Antenna

Let us consider the opposite case, i.e., the radar antenna is low deflected from the direction of the moving radar. When the value of the angle β_0 is low and the beam width of the horizontal-coverage directional diagram is narrow, the power spectral density $S_\beta(\omega)$ of the Doppler fluctuations is close to the maximum Doppler frequency $\omega_0 + \Omega_{max} \cos \gamma_*$. We can assume that $\frac{\omega - \omega_0}{\Omega_{max} \cos \gamma_*} \cong 1$. Then we can write

$$\arccos \frac{\omega - \omega_0}{\Omega_{max} \cos \gamma_*} \cong \sqrt{2\left[1 - \frac{\omega - \omega_0}{\Omega_{max} \cos \gamma_*}\right]} = \sqrt{\frac{2\,\Omega}{\Omega_{max*}}}\,, \tag{4.122}$$

where

$$\Omega = \omega_0 + \Omega_{max} \cos \gamma_* - \omega \qquad \text{and} \qquad \Omega_{max*} = \Omega_{max} \cos \gamma_* \tag{4.123}$$

Reference to Equation (4.105) shows that at the low-deflected radar antenna, the power spectral density $S_\beta(\omega)$ can be written in the following form:

$$S_\beta(\omega) \cong \sqrt{\frac{\Omega_{max*}}{\Omega}} \cdot \left\{ g_h^2\left[\left(\sqrt{\frac{2\,\Omega}{\Omega_{max*}}} - \beta_0\right)\cos\gamma_*\right] + g_h^2\left[\left(-\sqrt{\frac{2\,\Omega}{\Omega_{max*}}} - \beta_0\right)\right]\cos\gamma_* \right\}, \tag{4.124}$$

where $\Omega \geq 0$, i.e., $\omega \leq \omega_0 + \Omega_{max} \cos \gamma_*$ If $\beta = 0°$ and the horizontal-coverage directional diagram is symmetric, we can write

$$S_\beta(\omega) \cong \sqrt{\frac{\Omega_{max*}}{\Omega}} \cdot g_h^2\left(\sqrt{\frac{2\,\Omega}{\Omega_{max*}}}\ \cos\gamma_*\right). \tag{4.125}$$

The shape of the power spectral density $S_\beta(\omega)$ at low values of the angle β_0 is greatly different from the shape of the square of the directional diagram. It tends to approach ∞ at the maximum frequency. $S_\beta(\omega)$ is asymmetric and is shown in Figure 4.14 with the Gaussian low-deflected directional diagram. In this case, we can write

$$S_\beta(\omega) \cong \sqrt{\frac{\Delta\Omega_0}{\Omega}} \cdot e^{-\frac{\Omega}{\Delta\Omega_0} - \eta^2} \operatorname{ch}\left(2\eta\sqrt{\frac{\Omega}{\Delta\Omega_0}}\right), \tag{4.126}$$

where

$$\Delta\Omega_0 = \frac{\Omega_{max*}\,\Delta_*^{(2)}}{4\,\pi \cos^2\gamma_*} = \frac{V\,\Delta_*^{(2)}}{\lambda} \cdot \cos\gamma_*\,; \tag{4.127}$$

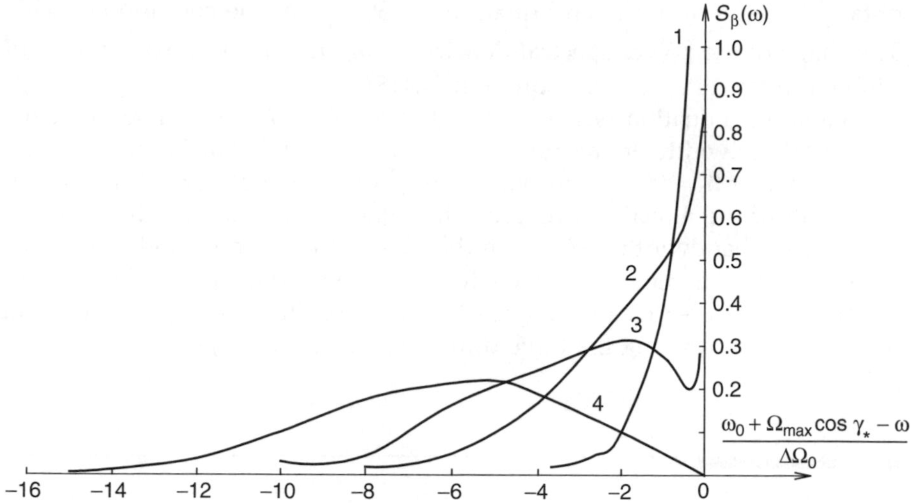

FIGURE 4.14

The power spectral density $S_\beta(\omega)$ with the low-deflected radar antenna: (1) $\frac{\beta_0}{\Delta_*}\cos\gamma_* = 0$; (2) $\frac{\beta_0}{\Delta_*}\cos\gamma_* = 0.5$; (3) $\frac{\beta_0}{\Delta_*}\cos\gamma_* = \frac{1}{\sqrt{2}}$; (4) $\frac{\beta_0}{\Delta_*}\cos\gamma_* = 1$.

$$\eta = \sqrt{2\pi}\cdot\frac{\beta_0}{\Delta_*}\cdot\cos\gamma_*. \tag{4.128}$$

At $\beta = 0°$, we can write

$$S_\beta(\omega) \cong \sqrt{\frac{\Delta\Omega_0}{\Omega}}\cdot e^{-\frac{\Omega}{\Delta\Omega_0}}. \tag{4.129}$$

As $S_\beta(\omega)$ tends to approach ∞ at the maximum frequency, it is impossible to define both the effective bandwidth and bandwidth of $S_\beta(\omega)$ at any level with respect to the maximum. It is necessary to define the bandwidth of $S_\beta(\omega)$ without normalization. Let us introduce, for example, the effective bandwidth ΔF_h, in which there is the same total power that is within the limits of the effective bandwidth of $S_\beta(\omega)$, for example, 80%, with Gaussian and **sinc**-directional diagrams. With this definition of the effective bandwidth, we can write

$$\Delta F_h = \frac{k_h^2 V\Delta_*^{(2)}}{4\lambda\cos\gamma_*}\cdot\left(1+\frac{2\beta_0\cos\gamma_*}{k_h\Delta_*}\right) \tag{4.130}$$

if $\beta_0 < \beta_0' = \frac{k_h\Delta_*}{2\cos\gamma_*}$, where k_h is the same as in Equation (4.119). Under the condition $\beta_0 \geq \beta_0'$, the definition of ΔF_h in Equation (4.119) is true and we obtain the same result as in Equation (4.130) under the condition $\beta_0 = \beta'_0$. The shape of the power spectral density $S_\beta(\omega)$, rigorously speaking, is very different from that given by Equation (4.118).

Comparing Equation (4.119) and Equation (4.130), one can see that the effective bandwidth of the power spectral density $S_\beta(\omega)$ of the Doppler fluctuations at $\beta_0 = 90°$ is approximately $\frac{10}{\Delta_*}$ times that at $\beta_0 = 0°$. The lower the width of the directional diagram in value, the greater the difference in the effective bandwidth of $S_\beta(\omega)$. In the case of the narrow-band directional diagram, this difference is nearly a few hundreds. The effective bandwidth is not more than 2–3 Hz or even 0.3–0.5 Hz in specific cases. The same result is observed in scanning the three-dimensional (space) target.

4.6 The Pulsed Searching Signal with the Moving Radar: The Total Correlation Function and Power Spectral Density of the Target Return Signal Fluctuations

4.6.1 General Statements

The total normalized correlation function $R(t, \tau)$ of the target return signal fluctuations with the pulsed searching signal is defined by the product of the aspect-angle normalized correlation function $R_\gamma(t, \tau)$ and the azimuth-normalized correlation function $R_\beta(\tau)$:

$$R(t,\tau) = R_\gamma(t,\tau)\cdot R_\beta(\tau)\,. \tag{4.131}$$

$R_\gamma(t, \tau)$, given by Equation (4.66), defines the rapid fluctuations in the radar range and the slow interperiod fluctuations caused by the difference in the radial components of the velocity of scatterers in the aspect-angle plane. The azimuth-normalized correlation function $R_\beta(\tau)$ given by Equation (4.102) defines only the interperiod fluctuations caused by the difference in the radial components of the velocity of scatterers in the azimuth plane. For this reason, the product of $R_\gamma(t, \tau)$ and $R_\beta(\tau)$ leads to changes only in the interperiod fluctuations and does not act on the fluctuations in the radar range.

The interperiod fluctuations in the glancing radar range, i.e., when the condition $\tau = nT'_p$ is true, are defined by the product of the aspect-angle normalized correlation function $R_\gamma(t, \tau)$ given by Equation (4.67) and the azimuth-normalized correlation function $R_\beta(\tau)$ given by Equation (4.102). The total power spectral density in the glancing radar range is defined by the convolution between the aspect-angle power spectral density $S_\gamma[\omega, \psi_*(t)]$ given by Equation (4.68) and the azimuth power spectral density $S_\beta(\omega)$ given by Equation (4.105). If the shape of the directional diagram in the vertical and horizontal planes is arbitrary, the shape of the pulsed searching signal is also arbitrary, and there is a linear approximation for the variable $\Omega(\varphi)$ in Equation (4.115), and the total power spectral density of the interperiod fluctuations is the convolution between $S_\gamma[\omega, \psi_*(t)]$ given by Equation (4.68) and $S_\beta(\omega)$ given by Equation (4.118):

$$S(\omega) \cong \int \Pi^2\Big(-\tfrac{x}{c_*\Omega_\gamma}\Big)\cdot g_h^2\Big(\psi_* - \tfrac{x}{\Omega_\gamma}\Big)\cdot g\Big(-\tfrac{\omega-\omega_0-\Omega_*-x}{\Omega_h}\Big)\,dx\,. \tag{4.132}$$

In particular, with the square waveform pulsed searching signal, we can write

$$S(\omega) \cong \int\limits_{-0.5\Delta\Omega_\tau}^{0.5\Delta\Omega_\tau} g_h^2\Big(\psi_* - \tfrac{x}{\Omega_\gamma}\Big)\cdot g\Big(-\tfrac{\omega-\omega_0-\Omega_*-x}{\Omega_h}\Big)\,dx\,. \tag{4.133}$$

The interperiod fluctuations in the fixed radar range are defined by the product of the aspect-angle normalized correlation function $R_\gamma(t, \tau)$ given by Equation (4.66) at the condition $\tau = nT_p$ and the azimuth-normalized correlation function $R_\beta(\tau)$ given by Equation (4.102). In the general case, we must apply various numerical techniques to define the power spectral densities $S_\beta(\omega)$ and $S_\gamma[\omega, \psi^*(t)]$.[46] Let us consider two particular cases in more detail at the Gaussian directional diagram: (1) the Gaussian pulsed searching signal and (2) the square waveform pulsed searching signal. During consideration of these particular cases, we do not use numerical techniques.

4.6.2 The Gaussian Directional Diagram: The Gaussian Pulsed Searching Signal

With the Gaussian directional diagram and when the pulsed searching signal is Gaussian, we can write the total normalized correlation function $R(t, \tau)$ of the target return signal fluctuations in the following form [see Equations (4.6), (4.73), (4.74), (4.84), and (4.121)]:

$$R[\tau,\psi(t)] = R'_\gamma(\tau)\cdot R''_\gamma(\tau,\psi_*)\cdot R_\beta(\tau)\cdot e^{j\omega_0\tau} = e^{-\pi\,\Delta F^2\tau^2 + j(\omega_0+\Omega_D)\,\tau}\cdot\sum_{n=0}^{\infty} e^{-\frac{\pi}{2}\cdot\left(\frac{\tau-nT_p}{\tau'_p}\right)^2}, \tag{4.134}$$

where

$$\Omega_D = \Omega_* + \Omega'' = \Omega_{\max}\cos\beta_0\cos\gamma_* + \psi_*\Omega_\gamma\left(1+\frac{\Delta_v^{(2)}}{\Delta_p^{(2)}}\right); \tag{4.135}$$

$$\Delta F^2 = \Delta F_h^2 + \frac{\Delta F_\tau^2}{1+\frac{\Delta_p^{(2)}}{\Delta_v^{(2)}}}. \tag{4.136}$$

The power spectral density of the fluctuations corresponding to the total normalized correlation function $R[\tau, \psi(t)]$, which is determined by Equation (4.134), is analogous to the spectral power density $S_\gamma(\omega, \psi_*)$ given by Equation (4.90):

$$S(\omega, \psi_*) \cong \sum_{n=0}^{\infty} e^{-\pi\cdot\left(\frac{n\Omega'_p}{\Delta\Omega'_p}\right)^2} \cdot e^{-\pi\cdot\left(\frac{\omega-\Omega_D-n\Omega'_p}{\Delta\Omega}\right)^2}, \tag{4.137}$$

where $\Delta\Omega'_p$ is given by Equation (4.79). As one can see from Equation (4.137), the power spectral density $S(\omega, \psi_*)$ of the fluctuations is different from the power spectral density $S_\gamma(\omega, \psi_*)$ given by Equation (4.90) by the width of the waves $\Delta\Omega = 2\pi\Delta F$, where the effective bandwidth ΔF is determined by Equation (4.136) instead of Equation (4.86), and by their position on the frequency axis; the shift with respect to the value $n\Omega'_p$ is given by $\Omega_D = \Omega_* + \Omega''$, where Ω_D is given by Equation (4.135), instead of the value Ω'' given by Equation (4.85). The envelope of comb waves is the same as for $S_\gamma(\omega, \psi_*)$ given by Equation (4.90).

The frequency Ω_D characterizes the Doppler shift in the frequency of the power spectral density of the slow fluctuations. The value of Ω_D depends on the ratio $\frac{\Delta_v}{\Delta_p}$. With the narrow-band pulsed searching signal, when the condition $\Delta_p << \Delta_v$ is true, we can neglect the second term in Equation (4.135), i.e., $\Omega_D = \Omega_*$. This means that the average frequency of the power spectral density $S(\omega, \psi_*)$ is defined by the direction to the center of the scanned element resolved in the radar range, i.e., $\cos\beta_0 \cos\gamma_* = \cos\theta_*$. In doing so, the vertical-coverage directional diagram does not act on the frequency Ω_D. In the case of the broadband pulsed searching signal, if the condition $\Delta_p >> \Delta_v$ is true, we can write

$$\Omega_D \to \Omega_0 = \Omega_{\max}\cos\beta_0\cos\gamma_0, \tag{4.138}$$

i.e., it is defined by the direction of the directional diagram axis: $\cos\beta_0 \cos\gamma_0 = \cos\theta_0$. In other cases, some intermediate value of the angle γ, which is within the limits of the interval $[\gamma_*, \gamma_0]$, is the definitive one.

Consider the interperiod fluctuations, the power spectral density of which is transformed as a result of the convolution. Under the condition $\tau = \frac{nT_p}{\mu}$, based on Equation (4.134), we can write the normalized correlation function of the interperiod fluctuations in the glancing radar range in the following form:

$$R(\tau, \psi_*) = e^{-\pi\,\Delta F^2 \tau^2 + j(\omega_0 + \Omega_D)\,\tau}\,. \tag{4.139}$$

The corresponding power spectral density takes the following form:

$$S(\omega, \psi_*) \cong \sum_{n=0}^{\infty} e^{-\pi \cdot \left(\frac{\omega - \Omega_D}{\Delta\Omega}\right)^2}\,. \tag{4.140}$$

The normalized correlation function $R(\tau, \psi_*)$ given by Equation (4.139) is the envelope of the comb waves of the normalized correlation function $R[\tau, \psi(t)]$ given by Equation (4.134). The power spectral density $S(\omega, \psi_*)$ given by Equation (4.140) is the central wave of the power spectral density $S(\omega, \psi_*)$ given by Equation (4.137).

The effective bandwidth of $S(\omega, \psi_*)$ given by Equation (4.140) depends on the ratio $\frac{\Delta_p}{\Delta_v}$. Under the condition $\Delta_p << \Delta_v$, we can write this in the following form:

$$\Delta F = \sqrt{\Delta F_h^2 + \Delta F_\tau^2} = \sqrt{2}\, V\lambda^{-1}\sqrt{\Delta_h^{(2)} \sin^2\beta_0 + \Delta_p^{(2)} \cos^2\beta_0 \sin^2\gamma_*}\,. \tag{4.141}$$

Under the condition $\Delta_p << \Delta_v$, the effective bandwidth can be written in the following form also:

$$\Delta F = \sqrt{\Delta F_h^2 + \Delta F_v^2} = \sqrt{2}\, V\lambda^{-1}\sqrt{\Delta_h^{(2)} \sin^2\beta_0 + \Delta_v^{(2)} \cos^2\beta_0 \sin^2\gamma_0}\,. \tag{4.142}$$

The effective bandwidth ΔF of $S(\omega, \psi_*)$ given by Equation (4.142) is the same as in the case of scanning the three-dimensional (space) target [see Equation (3.81)]. In particular, when the directional diagram is axial symmetric, i.e., when the condition $\Delta_h = \Delta_p = \Delta_a$ is satisfied, we can write this in the following form:

$$\Delta F = \sqrt{2}\, V\lambda^{-1} \cdot \Delta_a \sin\theta_0\,, \tag{4.143}$$

where θ_0 is the angle between the directional diagram axis and the vector of velocity of the moving radar. Equation (4.143) is similar to Equation (3.76).

Equation (4.141) and Equation (4.142) are equivalent. Reference to Equation (4.141) shows that the angle width Δ_p of the surface element of the two-dimensional target resolved in the radar range plays a role of the vertical-coverage directional diagram width Δ_v. The first term both in Equation

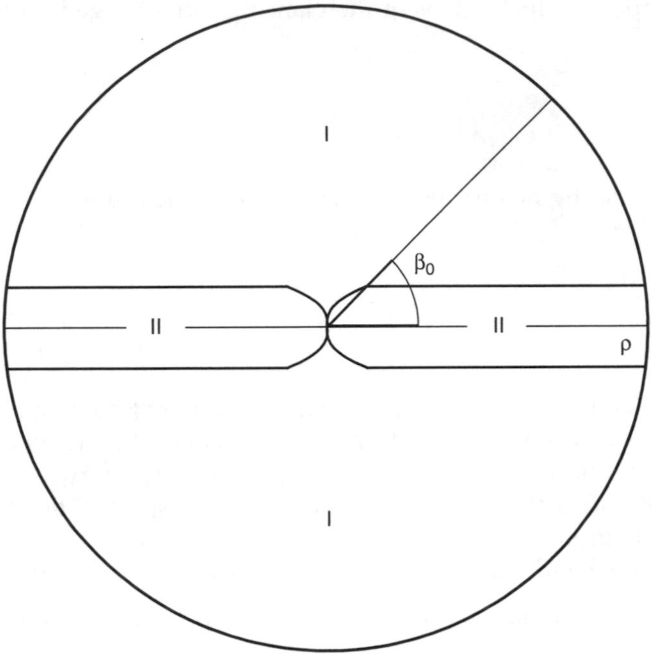

FIGURE 4.15
Regions of the interperiod target return signal fluctuations in the glancing radar range.

(4.141) and Equation (4.142) is defined by the difference in the Doppler frequencies — the radial components of velocity of the moving scatterers — in the horizontal plane within the horizontal-coverage directional diagram. The second term, both in Equation (4.141) and Equation (4.142), is defined by the difference in the Doppler frequencies in the vertical plane within the limits of the scanned surface element of the two-dimensional target resolved in the radar range [see Equation (4.141)] or of the vertical-coverage directional diagram [see Equation (4.142)]. The contribution of the first and second terms [see Equation (4.141) and Equation (4.142)] in the effective bandwidth ΔF of the power spectral density $S(\omega, \psi_*)$ of the interperiod fluctuations is shown in Figure 4.15, where bounds of "equal contributions" are on the plane (β_0, ρ). These bounds are determined using the following formula:

$$h^{-1}\rho = \operatorname{cosec}\gamma_* = k \operatorname{ctg}\beta_0 , \tag{4.144}$$

where $k = \frac{\Delta_p}{\Delta_h}$ for Equation (4.162) and $k = \frac{\Delta_v}{\Delta_h}$ for Equation (4.163). In the latter case, we must use the angle γ_0 instead of the angle γ_* in Equation (4.144). Near the boundaries, the target return signal fluctuations are caused by the difference in the Doppler frequencies both in the azimuth plane and in the aspect-angle plane. Under the fixed azimuth angle, with an increase in the radar range, the difference in the radial velocities in the azimuth plane is increased, and with a decrease in the radar range, the difference in the radial velocities in the aspect-angle plane is increased. The scanned surface of the two-dimensional target is divided on two regions: I and II. The main source of the fluctuations in region I is a difference in the radial velocities in the azimuth plane, whereas that in region II is a difference in the radial velocities in the aspect-angle plane. In the boundary region, it is necessary to take into consideration the two sources of fluctuations [see Equation (4.136) and Figure 4.15]. Under the condition $\tau = nT_p$, based on Equation (4.134), we can write the normalized correlation function of the interperiod fluctuations in the fixed radar range in the following form:

$$R(\tau,\psi_*) = e^{-\pi\,[\Delta F^2 + (\Delta F')^2]\tau^2 + j(\omega_0+\Omega_D)\,\tau}\,. \tag{4.145}$$

The normalized correlation function $R(\tau, \psi_*)$ in Equation (4.145) is the product of the aspect-angle normalized correlation function $R'_\gamma(\tau)$ given by Equation (4.80), the aspect-angle normalized correlation function $R''_\gamma(\tau, \psi_*)$ given by Equation (4.84), and the azimuth-normalized correlation function $R_\beta(\tau)$ given by Equation (4.121).

The power spectral density in the fixed radar range corresponding to $R(\tau, \psi_*)$ in Equation (4.145) is equivalent to the power spectral density $S(\omega, \psi_*)$ determined by Equation (4.140) if we change the effective bandwidth $\Delta\Omega$ for the effective bandwidth $\sqrt{\Delta\Omega^2 + (\Delta\Omega')^2}$, where $\Delta\Omega'$ is determined by Equation (4.82). The component $\Delta\Omega'$ takes into consideration the fluctuations caused by the exchange of scatterers and is very low in value, as a rule. For this reason, the value of $\Delta\Omega'$ is taken into account if the effective bandwidth is low in value, i.e., with low values of the angles β_0 and γ_* or γ_0. In this case, we have three regions (see Figure 4.16). Regions I and II are the same as in Figure 4.15. Region III defines the fluctuations caused by the exchange of scatterers in the scanned surface element of the two-dimensional target resolved in the radar range.

In more rigorous investigation, we should consider the following: as $\beta \to 0$, the effective bandwidth ΔF_h of the power spectral density $S(\omega, \psi_*)$ of the interperiod fluctuations in the fixed radar range, does not tend to approach zero and remains a finite value, but it is low [see Equation (4.130)]. Because of this, the fluctuations caused by the exchange of scatterers and those caused by the difference in the Doppler frequencies in the azimuth plane have approximately the same contribution. For example, at $\beta_0 = 0°$, $\Delta_h = 2°$, $\tau_p = 1$ μsec, $\lambda = 3$ cm, and if the angle γ_* is not so high in value, the effective bandwidth is defined by the form $\Delta F_h \cong \Delta F'$.

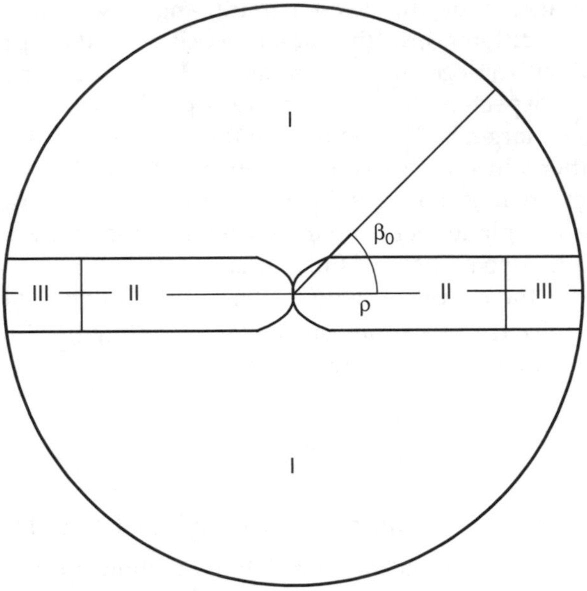

FIGURE 4.16
Regions of the interperiod target return signal fluctuations in the fixed radar range.

4.6.3 The Gaussian Directional Diagram: The Square Waveform Pulsed Searching Signal

Let us consider the total instantaneous normalized correlation function with the square waveform pulsed searching signal and the Gaussian directional diagram. Let us assume that the vertical-coverage directional diagram is wide, i.e., the condition $\Delta_p << \Delta_v$ is satisfied and the radar antenna is high deflected. Otherwise, if the condition $\Delta_p >> \Delta_v$ is satisfied, then, as was discussed in Section 4.3.4, the intraperiod fluctuations are absent in the central part of the pulsed target return signal. They are present only at the leading and trailing edges of the pulsed target return signal.[47] Then the total instantaneous normalized correlation function defines the intraperiod and interperiod target return signal fluctuations. Based on Equation (4.93) and Equation (4.121), we can write this in the following form:

$$R(\tau, \psi_*) = R_\beta(\tau) \cdot R_\gamma(\tau, \psi_*) = e^{-\pi \Delta F_h^2 \tau^2 + j(\omega_0 + \Omega_*)\tau} \cdot \sum_{n=0}^{\infty} \frac{\sin \pi \Delta F_\tau \tau \left(1 - \frac{|\tau - nT_p'|}{\tau_p'}\right)}{\pi \Delta F_\tau \tau}, \tag{4.146}$$

where the effective bandwidths ΔF_h and ΔF_τ are determined by Equation (4.70) and Equation (4.119), respectively. The total instantaneous normalized correlation function $R(\tau, \psi_*)$ of the fluctuations given by Equation (4.146) is shown in Figure 4.17 without the factor $e^{j(\omega_0 + \Omega_*)\tau}$.

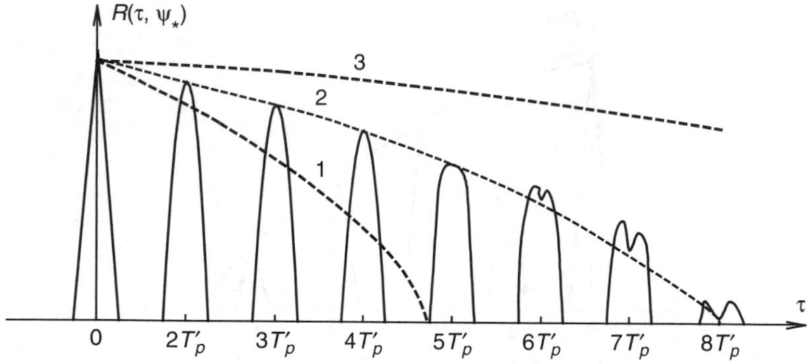

FIGURE 4.17
The total instantaneous correlation function of the intraperiod and interperiod target return signal fluctuations with the square waveform pulsed searching signal and Gaussian horizontal-coverage directional diagram of radar antenna — $R_\gamma(\tau, \psi_*)$ is shown by the solid line and $R_\beta(\tau)$ is shown by the dotted line: (1) $\Delta F_h >> \Delta F_\tau$; (2) $\Delta F_h \approx \Delta F_\tau$; (3) $\Delta F_h << \Delta F_\tau$.

To define the spectral power density of the fluctuations corresponding to the total instantaneous normalized correlation function given by Equation (4.146) is very difficult in the general case. However, under the condition $\Delta F_h >> \Delta F_\tau$, reference to Figure 4.17 (curve 1) shows that the condition

$$\Delta F_\tau \tau \leq \frac{\Delta F_\tau}{\Delta F_h} << 1 \tag{4.147}$$

is true within the limits of the correlation interval of the azimuth-normalized correlation function $R_\beta(\tau)$, i.e., if the condition $\tau \leq \frac{1}{\Delta F_h}$ is satisfied. If the condition given by Equation (4.147) is satisfied, the approximation made in Equation (4.96) is true for the aspect-angle normalized correlation function $R_\gamma(\tau)$. In terms of Figure 4.18, the total instantaneous normalized correlation function of the fluctuations can be written in the following form:

$$R(\tau, \psi_*) = e^{-\pi \Delta F_h^2 \tau^2 + j(\omega_0 + \Omega_*)\tau} \cdot \sum_{n=0}^{\infty} \left(1 - \frac{|\tau - nT_p'|}{\tau_p'}\right). \tag{4.148}$$

The condition given by Equation (4.147) means that

$$\frac{\Delta F_\tau}{\Delta \Gamma_h} = \frac{c\tau_p \sin^3 \gamma_*}{2h\Delta_* \mathrm{tg}\, \beta_0 \cos \gamma_*} << 1 \tag{4.149}$$

[see Equation (4.70) and Equation (4.119)]. Usually, this condition is satisfied when the directional diagram is high deflected from the direction of the

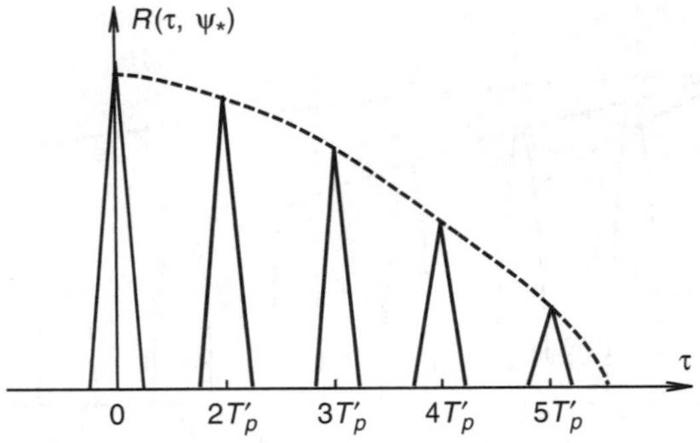

FIGURE 4.18
The total instantaneous correlation function of the intraperiod and interperiod target return signal fluctuations with the square waveform pulsed searching signal and Gaussian horizontal-coverage directional diagram of radar antenna — $R_\gamma(\tau, \psi_*)$ is shown by the solid line and $R_\beta(\tau)$ is shown by the dotted line: $\Delta F_h >> \Delta F_\tau$.

moving radar in the horizontal plane and the aspect angles are not so high in value within the vertical-coverage directional diagram.

The power spectral density of the fluctuations corresponding to the normalized correlation function $R(\tau, \psi_*)$ given by Equation (4.148) has the comb shape and consists of the partial Gaussian power spectral densities — the Gaussian waves — with the effective bandwidth ΔF_h and the envelope determined in the following form:

$$S(\omega) \cong \mathbf{sinc}^2[0.5(\omega - \omega_0 - \Omega_*)\,\tau_p] \sum_{n=-\infty}^{\infty} e^{-\pi \cdot \left(\frac{\omega - \omega_0 - \Omega_* - n\,\Omega'_p}{\Delta\Omega_h^2}\right)^2} . \qquad (4.150)$$

For the considered approximation, the average Doppler frequency Ω_* depends on the function $\psi_*(t)$ and is within the limits of the interval

$$\Omega_* \in [\Omega_{\max} \cos\beta_0 \cos(\gamma_0 - \Delta_v),\ \Omega_{\max} \cos\beta_0 \cos(\gamma_0 + \Delta_v)] . \qquad (4.151)$$

Let us consider the interperiod fluctuations in the glancing radar range. With the two-dimensional Gaussian directional diagram, and based on Equation (4.133), we can write the power spectral density in the following form:

$$S(\omega) \cong \int_{-0.5\Delta\Omega_\tau}^{0.5\Delta\Omega_\tau} e^{-2\pi \cdot \left[\left(\frac{\omega - \omega_0 - \Omega_* - x}{\Delta\Omega_h}\right)^2 + \left(\frac{\psi_*\Omega_\gamma - x}{\Delta\Omega_v}\right)^2\right]} dx . \qquad (4.152)$$

The integral in Equation (4.152) can be written using the error integrals. The particular cases can be obtained based on Equation (4.152).

If the duration of the pulsed searching signal is long, i.e., the condition $\Delta_p >> \Delta_v$ is true, the limits of the integral in Equation (4.152) can be considered as infinite. In terms of $\Omega_0 = \Omega_* + \psi_* \Omega_\gamma$, we can write

$$S(\omega) \cong e^{-2\pi \cdot \left(\frac{\omega - \omega_0 - \Omega_0}{\Delta\Omega_h^2 + \Delta\Omega_v^2}\right)}. \tag{4.153}$$

This power spectral density coincides with the power spectral density $S_g(\omega)$ of the slow fluctuations given by Equation (4.28) with the simple harmonic searching signal at the conditions $S^\circ(\gamma) \cong S^\circ(\gamma_0)$ and $\sin\gamma \cong \sin\gamma_0$. If the vertical-coverage directional diagram is wide, i.e., the condition $\Delta_p << \Delta_v$ is satisfied, we can neglect the second term in Equation (4.152). Then, the power spectral density of the fluctuations can be written in the following form:

$$S(\omega) \cong \Phi\left(\sqrt{2\pi} \cdot \frac{\omega - \omega_0 - \Omega_* + 0.5\Delta\Omega_\tau}{\Delta\Omega_h}\right) - \Phi\left(\sqrt{2\pi} \cdot \frac{\omega - \omega_0 - \Omega_* - 0.5\Delta\Omega_\tau}{\Delta\Omega_h}\right). \tag{4.154}$$

This power spectral density is the result of the convolution of the Gaussian power spectral density $S_\beta(\omega)$ given by Equation (4.120) and the square waveform power spectral density with the effective bandwidth $\Delta\Omega_\tau$ given by Equation (4.70).

In the arbitrary relationships between the values of $\Delta\Omega_\tau$ and $\Delta\Omega_h$, the power spectral density $S(\omega)$ given by Equation (4.154) has the shape of the "smoothed trapezium" placed symmetrically with respect to the frequency $\omega = \omega_0 + \Omega_*$. The basis of the "smoothed trapezium" is equal to $\Delta\Omega_\tau$. The slope of the leading and trailing edges of the "smoothed trapezium" is equal to $\Delta\Omega_h$. Under the condition $\Delta\Omega_\tau \geq 2\Delta\Omega_h$, the bandwidth of $S(\omega)$ given by Equation (4.154) at the level 0.5 — the level at which the bandwidth of $S(\omega)$ is equal to the effective bandwidth — is equal to $\Delta\Omega_\tau$. The width of the leading edge and trailing edge between the levels 0.8 and 0.92 is equal to $\sqrt{\frac{2}{\pi}}\,\Delta\Omega_\tau$. Under the condition $\Delta\Omega_\tau \leq \Delta\Omega_h$, the power spectral density of the fluctuations becomes Gaussian:

$$S(\omega) \cong e^{-2\pi \cdot \left(\frac{\omega - \omega_0 - \Omega_*}{\Delta\Omega_h}\right)^2}. \tag{4.155}$$

The effective bandwidth $\Delta\Omega_h$ and average frequency $\omega_0 + \Omega_*$ of the Gaussian power spectral density $S(\omega)$ given by Equation (4.155) are different from the effective bandwidth and average frequency of the power spectral density $S(\omega)$ given by Equation (4.153). The regions within the bounds of which one of the conditions — $\Delta\Omega_\tau \leq \Delta\Omega_h$ or $\Delta\Omega_\tau \geq \Delta\Omega_h$ — is satisfied are shown in Figure 4.15. These regions are the same as for the power spectral density $S(\omega, \psi_*)$ given by Equation (4.140). If the directional diagram is not deflected, i.e., the condition $\beta_0 = 0°$ is true, the total power spectral density of fluctuations is defined by the convolution of the square waveform power spectral density with the effective bandwidth $\Delta\Omega_\tau$ given by Equation (4.70) and the power spectral density $S_\beta(\omega)$ given by Equation (4.129).

4.6.4 The Pulsed Searching Signal with Low Pulse Period-to-Pulse Duration Ratio

The discussion up until now has been based upon the assumption that the pulse period-to-pulse duration ratio of the pulsed searching signal is large. Under this condition, not more than one scanned area of the Earth's surface or the surface of the two-dimensional target is simultaneously within the vertical-coverage directional diagram at any fixed instant of time. If the pulse period-to-pulse duration ratio is low in value and the duration of the pulsed searching signal is short, some simultaneously scanned areas of the Earth's surface or the surface of the two-dimensional target are within the vertical-coverage directional diagram at each instant of time (see Figure 4.19). Their number is determined by $f \cong \frac{2T_r}{T_p}$. Because the target return signals from these areas of the Earth's surface are independent, then the correlation function of the resulting target return signal fluctuations can be written in the following form:

$$R_\Sigma^{en}(t,\tau) = \sum_{i=1}^{f} R^{en}(t_i,\tau) = \sum_{i=1}^{f} R^{en}(t + iT_p,\tau)\ , \tag{4.156}$$

where $R^{en}(t_i, \tau)$ is the correlation function of the target return signal from the i-th scanned area of the Earth's surface or the surface of the two-dimensional

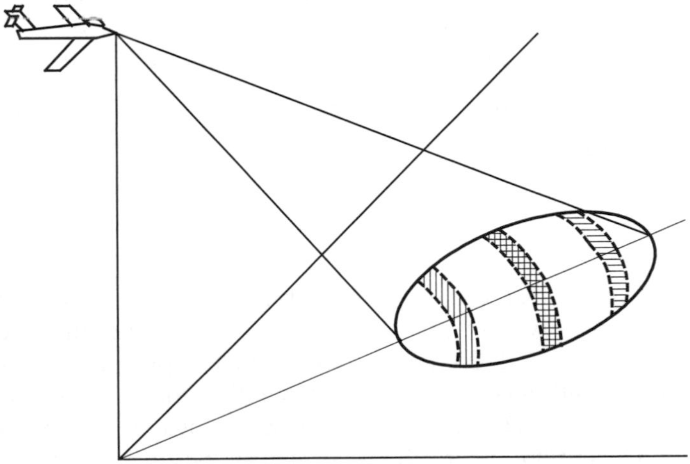

FIGURE 4.19
Simultaneously scanned areas by the pulsed searching signal with low pulse period-to-pulse duration ratio.

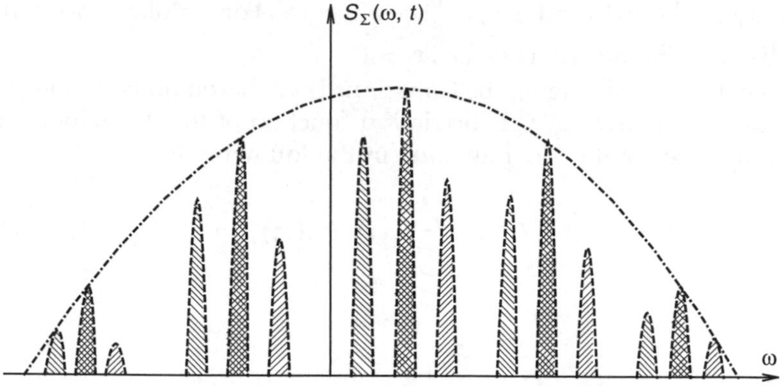

FIGURE 4.20
The power spectral density $S_\Sigma(\omega, t)$ of the resulting target return signal at simultaneous scanning of areas by the pulsed searching signal with the low pulse period-to-pulse duration ratio.

target [see Equation (2.133)]. The power spectral density with the comb structure and with different values of $p(t_i)$ and $\Omega_*(t_i)$, in the general case, with different widths of waves, corresponds to each term of the sum in Equation (4.156). The power spectral density of the resulting target return signal is shown in Figure 4.20.

4.7 Short-Range Area of the Radar Antenna

Let us consider the fluctuations in the short-range area of the radar antenna. In this case, it is necessary to account for the changes in values of the tracking angles $\Delta\varphi_{rm}$ and $\Delta\psi_{rm}$ of scatterers with the moving radar. We introduce a new variable z given by Equation (4.36). In this case, reference to Equation (2.122)–Equation (2.126) shows that the normalized correlation function of the fluctuations in the short-range area of the radar antenna can be written in the following form:

$$
\begin{aligned}
R_{sr}^{en}(t,\tau) = N\sum_{n=0}^{\infty}\iint & \boldsymbol{P}[z-0.5(\mu\tau-nT_p)]\cdot \boldsymbol{P}^*[z+0.5(\mu\tau-nT_p)] \\
& \times g\left(\varphi-0.5\Delta\varphi_{rm},\ \psi_* + c_* z - 0.5\Delta\psi_{rm}\right) \\
& \times g\left(\varphi+0.5\Delta\varphi_{rm},\ \psi_* + c_* z + 0.5\Delta\psi_{rm}\right) \\
& \times e^{-j\Omega(\varphi,\ \psi_* + c_* z)\,\tau} d\varphi\, dz\,,
\end{aligned}
\tag{4.157}
$$

where $\Delta\varphi_{rm} = \frac{V\tau}{\rho_*}\sin\beta_0$ and $\Delta\psi_{rm} = \frac{V\tau}{\rho_*}\cos\beta_0 \sin\gamma_*$. For simplicity, we consider the radar moving horizontally, i.e., $\varepsilon_0 = 0$.

Let the directional diagram be Gaussian. Then, based on Equation (4.157), we can write the normalized correlation function of the fluctuations in the short-range area of the radar antenna in the following form:

$$R_{sr}^{en}(t,\tau) = R_1(\tau)\cdot R_2(t,\tau), \tag{4.158}$$

where

$$R_1(\tau) = e^{-0.5\pi\tau^2\left(\frac{\Delta\varphi_{rm}^2}{\Delta_h^{(2)}} + \frac{\Delta\psi_{rm}^2}{\Delta_v^{(2)}}\right)} = e^{-\pi\Delta F_1^2\tau^2}; \tag{4.159}$$

$$\Delta F_1^2 = \frac{V^2}{2\rho_*}\cdot\left(\frac{\sin^2\beta_0}{\Delta_h^{(2)}} + \frac{\cos^2\beta_0 \sin^2\gamma_*}{\Delta_v^{(2)}}\right). \tag{4.160}$$

Here $R_1(\tau)$ is the normalized correlation function of the fluctuations caused by changes in the value of the aspect angle of scatterers. The function $R_2(t, \tau)$ is the normalized correlation function in the long-range area where the directional diagram $g(\varphi, \psi)$ is Gaussian. Naturally, $R_1(\tau)$ defines the interperiod fluctuations and not the intraperiod fluctuations.

The results discussed in Section 4.6, in which we consider the fluctuations in the long-range area when the directional diagram $g(\varphi, \psi)$ is Gaussian and the pulsed searching signal has the Gaussian and square waveform shapes, are true for the normalized correlation function $R_2(t, \tau)$. We can use the results discussed in Section 4.6.2 — the pulsed searching signal is Gaussian — for the normalized correlation function $R_1(\tau)$ given by Equation (4.159). In this case, we have to consider the following fact.

Taking into consideration changes in the values of the angles of tracking $\Delta\varphi_{rm}$ and $\Delta\psi_{rm}$, we should use the effective power spectral density bandwidth $\Delta\tilde{F}$ of the fluctuations instead of the effective bandwidth ΔF used in Section 4.6.2. $\Delta\tilde{F}$ can be determined in the following form:

$$\Delta\tilde{F} = \sqrt{\Delta F^2 + \Delta F_1^2}, \tag{4.161}$$

where the effective bandwidth ΔF_1 is given by Equation (4.160) and the effective bandwidth ΔF is given by Equations (4.136), (4.141), and (4.142). Under the condition $\Delta_h = \Delta_v = \Delta_a$, ΔF_1 can be written in the form $\Delta F_1 = \frac{V}{\sqrt{2}\,\rho_*\Delta_a} \times \sin\theta_*$, where θ_* is the angle between the vector of velocity of the moving radar and the direction (β_0, γ_*) in which the radar is moving.

Thus, for the case considered here, the normalized correlation function and power spectral density of the interperiod fluctuations are defined by the

Gaussian law, as in the case of the long-range area. The distinction is only in the case of the effective bandwidth ΔF_1, which is different from the effective bandwidth ΔF given by Equation (4.136) by the additional term [see Equation (4.160)] decreasing with an increase in the radar range ρ. This additional term takes into consideration changes in the amplitude of elementary signals when the scatterers are covered by the directional diagram, and is defined by the time of scanning the scatterers by the radar antenna beam — the time when the scatterers are within the limits of the radar antenna beam width. The value of ΔF_1, unlike ΔF, is inversely proportional to the directional diagram width or the radar antenna beam width.

Let us consider the particular cases. If the duration of the pulsed searching signal is short, i.e., the condition $\Delta_p << \Delta_v$ is satisfied, the effective bandwidth ΔF must be given by Equation (4.141), and the effective bandwidth $\Delta\tilde{F}$ can be determined in the following form:

$$\Delta\tilde{F} = V\cdot\sqrt{\left[\frac{2\Delta_h^{(2)}}{\lambda^2}+\frac{1}{2\rho_*^2\Delta_h^{(2)}}\right]\sin^2\beta_0+\left[\frac{2\Delta_p^{(2)}}{\lambda^2}+\frac{1}{2\rho_*^2\Delta_v^{(2)}}\right]\cos^2\beta_0\sin^2\gamma_*}\ . \tag{4.162}$$

Reference to Equation (4.162) shows that, if the radar range is determined by $\rho_* = \rho_*' = \frac{\lambda}{2\Delta_h^{(2)}}$, the first term in Equation (4.162) is doubled. If the radar range is given by $\rho_* = \rho_*'' = \frac{\lambda}{2\Delta_h\Delta_v}$, the second term in Equation (4.163) is doubled. The distance ρ_*' defines the conditional boundary of the short-range area of the horizontal-coverage directional diagram, as follows from the theory of antennas. The distance ρ_*'' can be determined in the following form:

$$\rho_*'' = \frac{\lambda}{2\Delta_v^{(2)}}\cdot\frac{\Delta_v}{\Delta_p}\ . \tag{4.163}$$

Reference to Equation (4.163) shows that the distance ρ_*'' is $\frac{\Delta_v}{\Delta_p} >> 1$ times more than the distance to the boundary of the short-range area of the horizontal-coverage directional diagram.

If the condition $\Delta_p >> \Delta_v$ is satisfied or the searching signal is a continuous process, and therefore the conditions $\gamma_* = \gamma_0$ and $\rho_* = \rho_0$ are true, the effective bandwidth must be given by Equation (4.142). Then the effective bandwidth $\Delta\tilde{F}$ can be determined in the following form:

$$\Delta\tilde{F} = V\cdot\sqrt{\left[\frac{2\Delta_h^{(2)}}{\lambda^2}+\frac{1}{2\rho_0^2\Delta_h^{(2)}}\right]\sin^2\beta_0+\left[\frac{2\Delta_v^{(2)}}{\lambda^2}+\frac{1}{2\rho_0^2\Delta_v^{(2)}}\right]\cos^2\beta_0\sin^2\gamma_*}\ . \tag{4.164}$$

The first term in Equation (4.164) is the same as in Equation (4.162). The second term in Equation (4.164) is doubled when the radar range ρ_0 is determined in the form $\rho_0 = \frac{\lambda^2}{\Delta_h^{(2)}}$, i.e., at the boundary of the short-range area of the vertical-coverage directional diagram. Under the condition $\Delta_h = \Delta_v = \Delta_a$, the effective bandwidth $\Delta\tilde{F}$ can be determined in the following form:

$$\Delta\tilde{F} = \frac{\sqrt{2}\, V\Delta_a}{\lambda} \cdot \sqrt{1 + \frac{\rho_{sr}^2}{\rho_0^2}} \cdot \sin\theta_0 = \sqrt{1 + \frac{\rho_{sr}^2}{\rho_0^2}} \cdot \Delta F\,, \tag{4.165}$$

where

$$\rho_{sr} = \frac{\lambda}{2\Delta_a^{(2)}} = \frac{d_a^2}{2\lambda} \tag{4.166}$$

is the conditional boundary of the short-range area, d_a is the diameter of radar antenna, and ΔF is the effective bandwidth given by Equation (4.143) in the long-range area. To assume the value of ρ_{sr}, consider the following instance: at $d_a = 3$ m, $\lambda = 3$ cm, we obtain the conditional boundary of the short-range area $\rho_{sr} = 150$ m, or at $d_a = 0.01$ m, $\lambda = 5 \cdot 10^{-7}$ m — in the case of the laser antenna — we obtain the conditional boundary of the short-range area $\rho_{sr} = 100$ m.

Note that within the limits of the short-range area we must be very careful in using the previously mentioned formulae because if the radar antenna is not focused, there is no time for its directional diagram to form. For this reason, rigorously speaking, the previously mentioned formulae are true only in the case of the focused radar antennas.

The power spectral density of the fluctuations in the short-range area with the continuous searching signal is determined[48] on the condition that the transmitting and receiving antennas are separated. The transmitting and receiving antennas are diverse during the distance ℓ_0 along the direction of the vector of velocity of the moving radar. The directional diagrams are the Gaussian axial symmetric directional diagrams with the same beam width. The axes of these directional diagrams are parallel, and being so, the target return signal is decreased due to the incomplete overlapping between the directional diagrams of the transmitting and receiving antennas, but the power spectral density of the fluctuations is Gaussian and has the effective bandwidth given in Equation (4.165), as when the transmitting and receiving antennas are matched. The shift in the average frequency is determined in the following form:

$$\Omega_0' = \Omega_0(1 - \delta_0)\,, \tag{4.167}$$

where

$$\Omega_0 = \frac{4\pi V}{\lambda}\cdot\cos\beta_0\cos\gamma_0 \qquad \text{and} \qquad \delta_0 = \frac{\ell_0^2}{2h^2}\cdot\sin^2\theta_0\sin^2\gamma_0 . \quad (4.168)$$

Thus, there is the additional shift δ_0 in frequency depending on the ratio $\frac{\ell_0}{h}$. For example, at $\gamma_0 = 60°$, $\theta_0 = 70°$, $\frac{\ell_0}{h} = 0.25$, we obtain the additional shift $\delta_0 = 0.5\%$. With an increase in the value of the ratio $\frac{\ell_0}{h}$, the additional shift δ_0 is sharply decreased. If the transmitting and receiving antennas are separated in the direction that is perpendicular to the vector of velocity of the moving radar, the additional shift δ_0 is absent. We have to note that even if the transmitting and receiving antennas are matched, the radar moving during the time of propagation of the pulsed searching signal leads to incomplete overlapping of tracks of the directional diagrams of the transmitting and receiving antennas on the Earth's surface or the surface of the two-dimensional target, making the target return signal more weak and giving rise to the shift in the average Doppler frequency. In the radar, these differences and values are infinitesimal, but in the sonar, these differences and values are considerable.[49]

4.8 Vertical Scanning of the Two-Dimensional (Surface) Target

4.8.1 The Intraperiod Fluctuations in Stationary Radar

In the vertical scanning of the two-dimensional (surface) target, the correlation function of the target return signal fluctuations is determined by Equation (2.182). Let us assume that the radar is stationary and consider the intraperiod target return signal fluctuations. The interperiod target return signal fluctuations are absent. Therefore, the exponential factor in Equation (2.182) transforms into 1 and the correlation function of the target return signal fluctuations becomes a periodic function with respect to the variable τ. For this reason, it is sufficient to consider a single term ($n = 0$):

$$R_0^{en}(t,\tau) = p_0\int_0^{2\pi}\int_{\theta_1}^{\theta_2} g^2(\theta\cos\alpha-\varphi_0,\ \theta\sin\alpha-\psi_0)\cdot S^{\circ}(\theta)\sin\theta\, d\theta\, d\alpha , \quad (4.169)$$

where the limits of integration with respect to the variable θ depend on the variables $t = T_d + \Delta t$ and τ:

$$\text{if} \quad \Delta t \geq 0.5\tau_p, \quad \text{then} \quad \theta \in [\theta_1, \theta_2] \quad \text{at} \quad 0 \leq 0.5|\tau| \leq 0.5\tau_p; \tag{4.170}$$

$$\text{if} \quad 0 \leq \Delta t \leq 0.5\tau_p, \quad \text{then} \quad \theta \in \begin{cases} [0, \theta_2] & \text{at} \quad 0 \leq 0.5|\tau| \leq \Delta t; \\ [\theta_1, \theta_2] & \text{at} \quad \Delta t \leq 0.5|\tau| \leq 0.5\tau_p; \end{cases} \tag{4.171}$$

$$\text{if} \quad -0.5\tau_p \leq \Delta t \leq 0, \quad \text{then} \quad \theta \in [0, \theta_2] \quad \text{at} \quad 0 \leq 0.5|\tau| \leq \Delta t + 0.5\tau_p; \tag{4.172}$$

$$\theta_{1,2} = \sqrt{\theta_*^2 \mp c \cdot \tfrac{\tau_p - |\tau|}{2h}} = \sqrt{2\,\Delta t \mp \tfrac{\tau_p - |\tau|}{T_d}}\,; \tag{4.173}$$

$$\Delta t = t - T_d \qquad \text{and} \qquad T_d = 2hc^{-1}. \tag{4.174}$$

Let us limit the case of the axial symmetric nondeflected Gaussian directional diagram. Let us suppose that the specific effective scattering area $S^\circ(\theta)$ is given by Equation (2.187). Then, in the region where the condition $\Delta t \geq 0.5\tau_p$ is satisfied [see Equation (4.170)], we can write

$$R_0^{en}(t,\tau) = 2\pi p_0 S_N^\circ \int_{\theta_1}^{\theta_2} e^{-2\pi \cdot \frac{\theta^2}{\Delta_a^{(2)}} - k_2^2\theta}\, \theta\, d\theta = \frac{p_0 \Delta_a^{(2)} S_N^\circ}{2\, a^2} \cdot e^{-\pi \cdot \frac{\Delta t}{T_r}} \cdot \left(e^{\frac{\pi}{2} \cdot \frac{\tau_p - |\tau|}{T_r}} - e^{-\frac{\pi}{2} \cdot \frac{\tau_p - |\tau|}{T_r}}\right), \tag{4.175}$$

where S_N° is the specific effective scattering area under vertical scanning;

$$T_r = \frac{\Delta_a^{(2)} h}{2\, c\, a^2} = \frac{\Delta_a^{(2)} T_d}{4\, a^2}\,; \tag{4.176}$$

$$a^2 = 1 + \frac{k_2 \Delta_a^{(2)}}{4\,\pi}\,; \tag{4.177}$$

and $|\tau| \leq \tau_p$.

When the duration of the pulsed searching signal is short, i.e., the condition $\tau_p << T_r$ is satisfied, reference to Equation (4.175) shows that

$$R_0^{en}(t,\tau) = p(t) \cdot \left(1 - \tfrac{|\tau|}{\tau_p}\right), \tag{4.178}$$

where

$$p(t) = \frac{p_0 \Delta_a^{(2)} S_N^{\circ}}{2\,a^2} \cdot \frac{\pi\,\tau_p}{T_r} \cdot e^{-\pi \cdot \frac{\Delta t}{T_r}}\,. \tag{4.179}$$

The normalized correlation functions of the fluctuations given by Equation (4.175) and Equation (4.178) are shown in Figure 4.21 for various values of the ratio $\frac{\tau_p}{T_r}$. The variables t and τ are separable for these normalized correlation functions, i.e., the normalized correlation function $R_0^{en}(t,\tau)$ is separable within the limits of the region given by Equation (4.170).

Because the normalized correlation function $R_0^{en}(t,\tau)$ is periodic, the power spectral density is linear. The envelope of the power spectral density of the fluctuations, with the accuracy of the power dependent on the time t, is defined by the Fourier transform with respect to Equation (4.175) for the arbitrary pulsed signal duration,

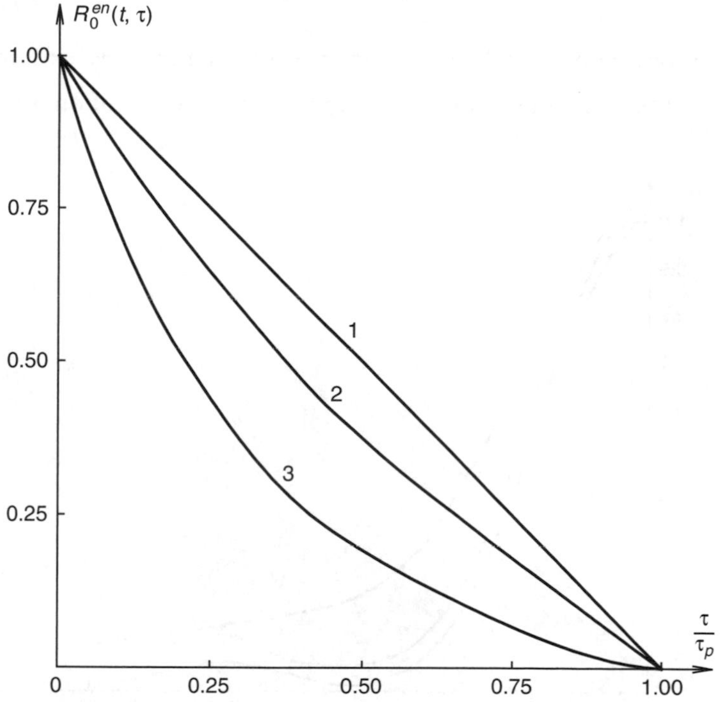

FIGURE 4.21

The normalized correlation function R_0^{en} (t, τ) of the intraperiod target return signal fluctuations at $n = 0$ and vertical scanning: (1) $\frac{\tau_p}{T_r} << 1$; (2) $\frac{\tau_p}{T_r} = 1.0$; (3) $\frac{\tau_p}{T_r} = 2.0$.

$$S^{en}(\omega) \cong \frac{\operatorname{ch} \frac{\pi \tau_p}{2 T_r} - \cos \omega \tau_p}{\left(\frac{\pi \tau_p}{2 T_r}\right)^2 + \omega^2 \tau_p^2}. \tag{4.180}$$

Under the condition $\frac{\tau_p}{T_r} \to 0$, the power spectral density $S^{en}(\omega)$ is defined by **sinc**^{2} $(0.5\omega\tau_p)$ and corresponds to the normalized correlation function given by Equation (4.178). The normalized power spectral density $S^{en}(\omega)$ given by Equation (4.180) is shown in Figure 4.22 for various values of the ratio $\frac{\tau_p}{T_r}$ and under the condition $\tau_p << T_r$. At high values of frequency, $S^{en}(\omega)$ is decreased in accordance with the law $\frac{1}{\omega^2}$.

Reference to Equation (4.175) shows that integrating with the variable τ, we can define the effective correlation interval as

$$\tau_c^{ef} = 4\, T_r \cdot \frac{\operatorname{ch} \frac{\pi \tau_p}{2T_r} - 1}{\pi \operatorname{sh} \frac{\pi \tau_p}{2T_r}}. \tag{4.181}$$

Under the condition $\frac{\tau_p}{T_r} \to 0$, we obtain $\tau_c^{ef} \to \tau_p$, which is what it must be. With an increase in the value of τ_p, the effective correlation interval is

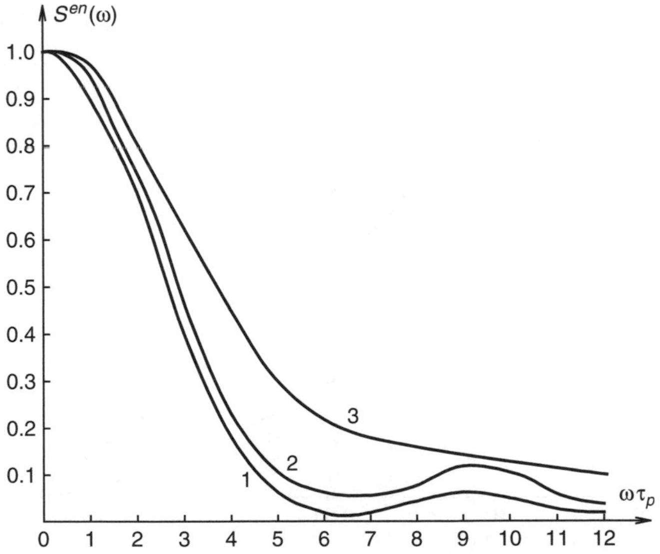

FIGURE 4.22
The power spectral density $S^{en}(\omega)$ of the intraperiod target return signal fluctuations under vertical scanning: (1) $\frac{\tau_p}{T_r}$ <<1; (2) $\frac{\tau_p}{T_r} = 1$; (3) $\frac{\tau_p}{T_r} = 2$.

increased and tends to approach the ratio $\frac{4T_r}{\pi}$. The zero correlation interval determined under the condition $R(\tau) = 0$ is always equal to $2\tau_p$, as follows from Equation (4.175).

Let us consider the intermediate region given by Equation (4.171) and Equation (4.172) when $\Delta t \in [-0.5\,\tau_p, 0.5\,\tau_p]$. For this region, the correlation function is defined by two functions. The first function is the normalized correlation function of the fluctuations given by Equation (4.175) that follows from Equation (4.169) under the conditions $\theta_1 > 0$ and $\theta_2 > 0$. The second function follows from Equation (4.175) under the condition $\theta_1 = 0$ and takes the following form:

$$R_0^{en}(t,\tau) = \frac{p_0 \Delta_a^{(2)} S_N^{\circ}}{2\,a^2} \cdot \left[1 - e^{-\pi \cdot \frac{2\Delta t + \tau_p - |\tau|}{2T_r}}\right]. \tag{4.182}$$

Within the limits of the region $0 \le \Delta t \le 0.5\tau_p$ given by Equation (4.171) under the condition $\Delta t \le 0.5\,|\tau| \le 0.5\tau_p$, the correlation function of the fluctuations is determined by Equation (4.175). Under the condition $0 \le 0.5\,|\tau| \le \Delta t$, it is given by Equation (4.182). Within the limits of the interval $-0.5\tau_p \le \Delta t \le 0$ [see Equation (4.172)], Equation (4.182) is true and the variable τ is within the limits of the interval $\tau \in [0, \Delta t + 0.5\tau_p]$.

Because the variables t and τ are not separable in the correlation function given by Equation (4.182), the correlation function is not separable and the instantaneous power spectral density depends essentially on the time t within the limits of the regions given by Equation (4.171) and Equation (4.172). The zero correlation intervals depend also on the time t. Under the condition $t \ge T_d$ or $\Delta t \ge 0$, the zero correlation interval is equal to τ_p, for the region given by Equation (4.170). If the condition $t \in [T_d - 0.5\,\tau_p, T_d]$ is satisfied, we can consider that

$$\tau_c = \tau_p + 2\,\Delta t \in [0,\ \tau_p], \tag{4.183}$$

i.e., the correlation interval is less than τ_p, and under the condition $\Delta t \to -0.5\tau_p$, the correlation interval tends to approach zero. This is clear because the instant of time $\Delta t = -0.5\tau_p$ corresponds to the leading edge of the pulsed searching signal reaching the scanned surface of the two-dimensional target. At the instants of time that are within the limits of the interval $[T_d - 0.5\tau_p, T_d]$, only those parts of the pulsed searching signal can reach the scanned surface of the two-dimensional target that are removed from the leading edge of the pulsed searching signal at the time equal to $\Delta t < 0.5\tau_p$. Evidently, with the shift in the time of two overlapping pulsed searching signals in the integrand in Equation (2.122) by the value $\pm\Delta t$, the product of these two pulsed searching signals becomes equal to zero and the correlation disappears. Thus, the effective instantaneous power spectral density bandwidth, which is inversely proportional to the correlation interval, tending to approach ∞ at the instant of time when the pulsed searching signal reaches the scanned surface of the two-dimensional target, is rapidly decreased as the pulsed searching signal is propagated and, at the instant of time $t = T_d$, approaches the value $\frac{1}{\tau_p}$, which is constant.

The correlation function of the fluctuations is determined by Equation (4.182) within the limits of the time interval $[T_d, T_d + 0.5\tau_p]$ and is not separable, as seen earlier. The instantaneous power spectral density of the fluctuations is deformed, although its effective bandwidth is approximately constant. The target return signal power increases and becomes maximal if the condition $\Delta t = 0.5\tau_p$ is satisfied (see Section 2.6).

4.8.2 The Interperiod Fluctuations with the Vertically Moving Radar

Let us consider the interperiod target return signal fluctuations in the glancing radar range when the vertically moving radar approaches the scanned surface of the two-dimensional target. We assume that the conditions $\tau = nT'_p$ and $\varepsilon_0 = -0.5\pi$ are true in Equation (2.182). In this case, the correlation function of the interperiod fluctuations can be written in the following form:

$$R_0^{en}(t,\tau) = p_0 \int_0^{2\pi} \int_{\theta_1}^{\theta_2} g^2(\theta\cos\alpha - \varphi_0, \ \theta\sin\alpha - \psi_0) \cdot S^{\circ}(\theta) \cdot e^{j\Omega_{\max}\cos\theta} \, d\theta \, d\alpha , \tag{4.184}$$

where

$$\theta_{1,2} = \sqrt{\theta_*^2 \mp \frac{c\tau_p}{2h}} = \sqrt{\frac{2\,(\rho_* - h \mp 0.25c\tau_p)}{h}} = \sqrt{\frac{2(\Delta t \mp 0.5\tau_p)}{T_d}} \ ; \tag{4.185}$$

$$\Delta t = t - T_d \geq 0.5\tau_p \, . \tag{4.186}$$

Reference to Equation (4.185) and Equation (4.186) shows that unlike in Equation (4.173), the limits of integration θ_1 and θ_2 are independent of the parameter τ. For the instants of time that are within the limits of the time interval $[-0.5\tau_p, 0.5\tau_p]$, we obtain the angle $\theta_1 = 90°$.

Multiplying Equation (4.184) with the factor $e^{j\omega_0\tau}$ and using the Fourier transform and the filtering property of the delta function, the power spectral density of the fluctuations can be written in the following form:

$$S(\omega,t) \cong f[\theta(\omega)] \cdot \Pi[\theta_1(\omega,t), \ \theta_2(\omega,t)] \, , \tag{4.187}$$

where

$$f[\theta(\omega)] = \int_0^{2\pi} g^2(\theta\cos\alpha - \varphi_0,\ \theta\sin\alpha - \psi_0)\cdot S^\circ(\theta)\,d\alpha\ ; \tag{4.188}$$

$\cos\theta = \frac{\omega-\omega_0}{\Omega_{max}}$, $\Pi = 1$ at $\theta \in [\theta_1, \theta_2]$, and $\Pi = 0$ at $\theta \notin [\theta_1, \theta_2]$. The boundary values of the angles θ_1 and θ_2 given by Equation (4.186) define the boundary frequencies of the instantaneous power spectral density at the instant of time $t = \frac{2\rho_*}{c}$ independently of the type of the function $f[\theta(\omega)]$.

Under the condition $\Delta t > 0.5\,\tau_p$ or $\rho_* > h + 0.25c\tau_p$, the boundary frequencies of the power spectral density have the following form:

$$\omega_{1,2} = \omega_0 + \Omega_{max}\cos\theta_{1,2} = \omega_0 + \frac{\Omega_{max}h}{\rho_{1,2}} = \omega_0 + \frac{\Omega_{max}h}{\rho_* \mp 0.25c\,\tau_p}\,. \tag{4.189}$$

If the condition $0.5c\,\tau_p \ll h$ is satisfied, then the effective bandwidth at the boundary frequencies can be written in the following form:

$$\Delta\Omega = \omega_1 - \omega_2 \cong \frac{\Omega_{max}c\tau_p h}{2\rho_*^2} = \frac{c\tau_p}{2h}\cdot\Omega_{max}\cos^2\theta_*\,. \tag{4.190}$$

If the condition $\theta_* < \Delta_a \ll 1$ is true, then, in practice, under the condition $\Delta t > 0.25\tau_p$, the effective power spectral density bandwidth is very slightly decreased because the average angle θ_* increases and tends to approach constant. The average frequency of the power spectral density given by

$$\overline{\omega} = \omega_0 + \Omega_{max}\cos\theta_* \tag{4.191}$$

is slowly decreased, also with an increase in the value of the angle θ_*.

To define the dependence of the parameters of the power spectral density as a function of the time, we can write

$$\cos\theta = h\rho_*^{-1} = T_d t^{-1} = T_d(T_d + \Delta t)^{-1} \cong 1 - \Delta t T_d^{-1}\,, \tag{4.192}$$

where

$$\Delta t = t - T_d = 0.5c(\rho_* - h) \geq 0.5\tau_p\,. \tag{4.193}$$

Substituting Equation (4.192) in Equation (4.190) and Equation (4.191), we can write

$$\Delta\Omega = 0.5c\tau_p h^{-1}\Omega_{\max}\left(1-\Delta\, tT_d^{-1}\right)^2 \quad \text{and} \quad \bar{\omega} = \omega_0 + \Omega_{\max}\left(1-\Delta\, tT_d^{-1}\right). \tag{4.194}$$

The angle $\theta_1 = 0°$ within the limits of the following intervals $\Delta\, t \in [-0.5\tau_p, 0.5\tau_p]$ or $\rho_* \in [h - 0.25c\tau_p, h + 0.25c\tau_p]$. Consequently, the following equality $\omega_1 = \omega_0 + \Omega_{\max}$ is true. In addition, we can assume $\rho_2 = h + 0.25c\tau_p + 0.5c\Delta t$, where $|\Delta\, t| = |t - T_d| \le 0.5\,\tau_p$. Then the effective bandwidth can be written in the following form:

$$\Delta\Omega = \Omega_{\max} - \Omega_{\max}\cos\theta_2 = \Omega_{\max}\left(1 - h\rho_2^{-1}\right) = 0.25c\tau_p h\Omega_{\max}\left(1 + 2\Delta t\,\tau_p^{-1}\right). \tag{4.195}$$

Under the condition $\Delta\, t < -0.5\tau_p$, i.e., $t < T_d - 0.5\tau_p$, the target return signal is as yet absent at the receiver or detector input in any navigational system.

The effective bandwidth is linearly increased from zero to the value of

$$\Delta\Omega_{\max} = 0.5\Omega_{\max}c\tau_p h^{-1} \tag{4.196}$$

within the limits of the time interval $\Delta\, t \in [-0.5\tau_p, 0.5\tau_p]$, and after that it is approximately constant until the target return signal disappears completely. This character of the power spectral density of the target return signal fluctuations is clear. At the first instants of time, when the pulsed searching signal could reach the scanned surface of the two-dimensional target, i.e., when an illuminated spot is very small in dimension and the difference $\rho_2 - h$ is low in value, changes in the dimensions of the scanned surface of the two-dimensional target or changes in the corresponding differences of phases of the target return signal, which are caused by the interperiod fluctuations, are very low in value. This fact can explain why the power spectral density is narrow-band. At the instant of time equal to $t = T_d - 0.5\tau_p$, when the pulsed searching signal could reach the scanned surface of the two-dimensional target, the illuminated spot is reduced to a point. The moving radar does not produce any fluctuations, i.e., the effective bandwidth is equal to zero at this instant of time.

Our interest is in comparing this conclusion with the conclusion made in Section 4.8.1. The effective power spectral density bandwidth of the interperiod fluctuations in the radar range was changed from infinity to minimal value, and after that had held the value constant within the limits of the same time interval. It follows from the results already discussed that if the duration of the pulsed searching signal is short and the function $f(\omega)$ is smoothed, the shape of the function $f(\omega)$ defines the target return signal power as a function of time t or the angle θ_* (see Section 2.6), but it does not define the shape of the power spectral density, which is close to that of a rectangle with a skewed top (see Figure 4.23). The shape of the rectangle top

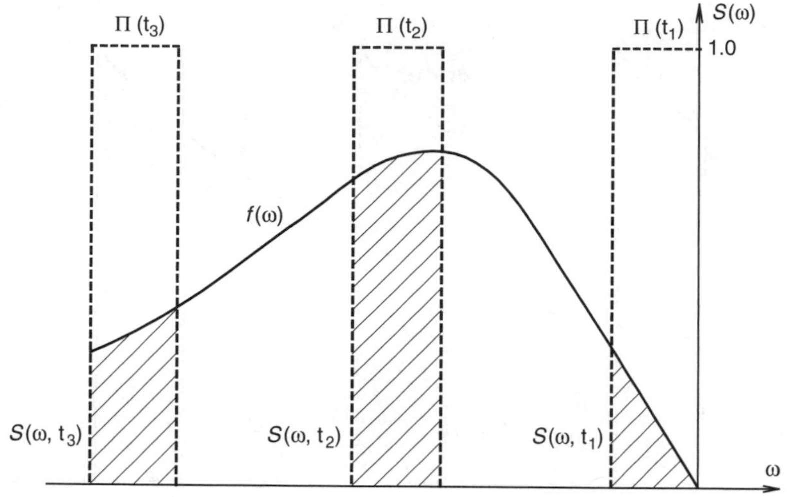

FIGURE 4.23
The shape of the power spectral density of the interperiod fluctuations as a function of appearance time of the target return signal under the conditions $\tau_p << T_r$ and $\theta_0 \neq 0°$.

is not important for analysis if it is not spiked. The main information regarding the power spectral density is contained in the values of the effective bandwidth and average frequency.

Because of this, the determination of the function $f(\omega)$ is of interest for us when the duration of the pulsed searching signal is large in value first, and especially when the pulsed searching signal is a continuous nonmodulated process. In this case, the power spectral density of the interperiod fluctuations is defined by the complete function $f[\theta(\omega)]$ given by Equation (4.188), as the condition given by Equation (2.109) is true. If the directional diagram is approximated by the Gaussian bottle and $S°(\theta) = const$ within the limits of the radar antenna beam width, then we can obtain the power spectral density $S_g(\omega)$ given by Equation (3.136) based on Equation (4.187) and Equation (4.188), in which Δ_h and Δ_v must be replaced with the notations Δ_φ and Δ_ψ, respectively, for obtaining the effective bandwidth of $S_g(\omega)$ in the angle planes φ and ψ. Finally, we can say that the main results discussed in Section 3.3 are true for the considered case here.

Under the condition $\theta_0 = 0°$, we can obtain a simple solution if the specific effective scattering area is determined in the form $S°(\theta) = e^{-k_2\theta^2}$. The power spectral density of the fluctuations with the continuous searching signal is given by Equation (3.144), in which the values Δ_φ and Δ_ψ must be replaced with the values $\frac{\Delta_\varphi}{a}$ and $\frac{\Delta_\psi}{a}$, respectively, where $a = 1 + \frac{k_2\Delta_a^{(2)}}{2\pi}$. If the condition $\Delta_\varphi = \Delta_\psi$ is satisfied, we can use the power spectral density $S_g(\omega)$ given by Equation (3.142).

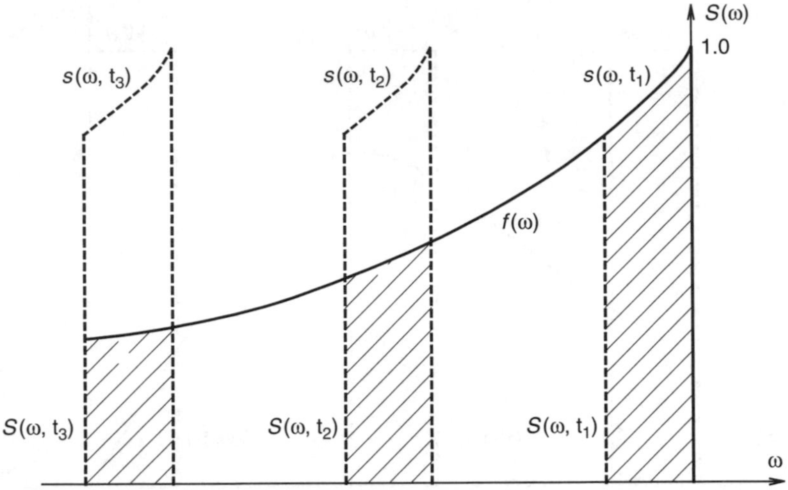

FIGURE 4.24
The shape of the power spectral density of the interperiod fluctuations as a function of appearance time of the target return signal under the conditions $\tau_p << T_r$ and $\theta_0 \neq 0°$.

The instantaneous power spectral densities $S(\omega, t)$ with various time cross sections t, if the following conditions $\theta_0 = 0°$, $\Delta_\varphi = \Delta_\psi$, and $\tau_p << T_r$ are true, are shown in Figure 4.24. Thus, according to Equation (3.142), we can write

$$f(\omega) = e^{-\frac{\omega_0 + \Omega_{max} - \omega}{\Delta\Omega}} \tag{4.197}$$

and only the power depends on the time-cross-section t, but not the shape of the power spectral density. The normalized power spectral density of the fluctuations can be written in the following form:

$$s(\omega, t) = \frac{S(\omega, t)}{S(\overline{\omega}, t)}, \tag{4.198}$$

where $\overline{\omega}$ is the average frequency and is shown in Figure 4.24 by the dotted line. The normalized power spectral density $s(\omega, t)$ of the fluctuations is independent of the time cross section t.

4.8.3 The Interperiod Fluctuations with the Horizontally Moving Radar

Let the radar motion be parallel to the underlying surface of the Earth, i.e., $\varepsilon_0 = 0$. The interperiod target return signal fluctuations in the glancing radar range ($\tau = nT_p'$) are defined by the correlation function determined in the following form:[50]

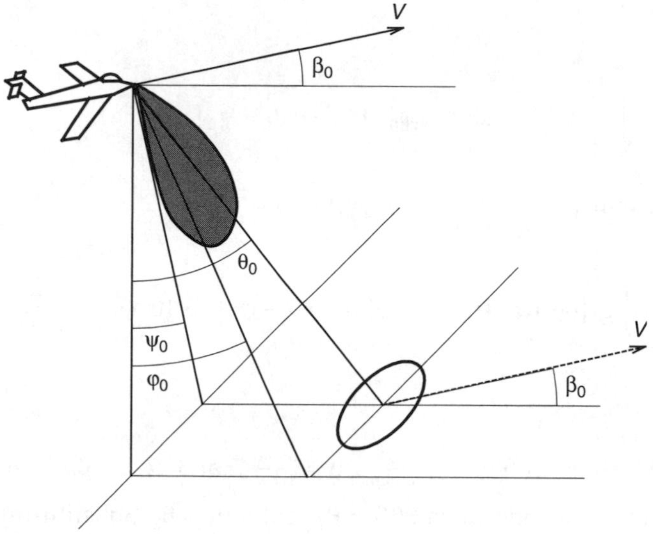

FIGURE 4.25
The position of the vector of velocity of the moving radar relative to the radar antenna directional diagram and two-dimensional (surface) target.

$$R^{en}(t,\tau) = p_0 \int_0^{2\pi} \int_{\theta_1}^{\theta_2} g^2(\theta\cos\alpha - \varphi_0,\ \theta\sin\alpha - \psi_0) \times S^\circ(\theta) \cdot e^{j\Omega_{\max}\tau\sin\theta\cos(\alpha-\beta_0)} \sin\theta\, d\theta\, d\alpha\,, \tag{4.199}$$

where θ_1 and θ_2 are given by Equation (4.184), β_0 is the angle characterizing the mutual position of the directional diagram and the vector of velocity of the moving radar (see Figure 4.25).

Multiplying Equation (4.199) with the factor $e^{j\omega_0\tau}$ and using the Fourier transform, the power spectral density $S(\omega, t)$ can be written in the following form:

$$S(\omega,t) \cong \int_0^{2\pi} \int_{\theta_1}^{\theta_2} g^2(\theta\cos\alpha - \varphi_0,\ \theta\sin\alpha - \psi_0) \times S^\circ(\theta)\sin\theta \cdot \delta[\omega_0 + \Omega_{\max}\sin\theta\cos(\alpha-\beta_0) - \omega]\, d\theta\, d\alpha \tag{4.200}$$

Assuming $\sin\theta \approx \theta$ and introducing a new variable

$$\theta = \frac{\omega - \omega_0 + x}{\Omega_{\max}\cos(\alpha - \beta_0)}, \tag{4.201}$$

we can write

$$S(\omega,t)\cong\int_{0}^{2\pi}\int_{\theta_1}^{\theta_2} g^2[\theta(x)\cos\alpha-\varphi_0,\ \theta(x)\sin\alpha-\psi_0]$$

$$\times S^\circ[\theta(x)]\cdot\frac{\omega-\omega_0+x}{\Omega_{\max}^2\cos^2(\alpha-\beta_0)}\cdot\delta(x)\,dx\,d\alpha= \qquad ,$$

$$=\int_{\alpha_1}^{\alpha_2} g^2[\theta(0)\cos\alpha-\varphi_0,\quad \theta(0)\sin\alpha-\psi_0]\cdot\ S^\circ[\theta(x)]\cdot\frac{\omega-\omega_0+x}{\Omega_{\max}^2\cos^2(\alpha-\beta_0)}\,d\alpha \tag{4.202}$$

where the values of $\theta(0)=\frac{\upsilon}{\cos(\alpha-\beta_0)}$, $\upsilon=\frac{\omega-\omega_0}{\Omega_{\max}}$, and $\alpha_{1,2}=\beta_0+\arccos\frac{\upsilon}{\theta_{1,2}}$ are defined under the conditions $\theta(0)=\theta_1$ and $\theta(0)=\theta_2$. Substituting the value of $\theta(0)$ in the integrand in Equation (4.202) and introducing a new variable $y=\text{tg}\,(\alpha-\beta_0)$, we can write

$$S(\omega,t)\cong\int_{y_1}^{y_2} g^2(\upsilon,y)\cdot\ S^\circ\left(\upsilon\sqrt{1+y^2}\right)dy\ , \tag{4.203}$$

where

$$g(\upsilon,y)=g[\upsilon(\cos\beta_0-y\sin\beta_0)-\varphi_0,\ \upsilon(\sin\beta_0+y\cos\beta_0)-\psi_0] \tag{4.204}$$

and $y_{1,2}=\sqrt{\theta_{1,2}^2\upsilon^{-2}-1}$.

If the duration of the pulsed searching signal is short, so that the length of the interval $[y_1, y_2]$ is low in value, using the theorem about the mean, we can write

$$S(\omega,t)\approx\Omega_{\max}^{-1}\upsilon\cdot g^2(\upsilon,y_*)\ \cdot S^\circ\left(\upsilon\sqrt{1+y_*^2}\right)\cdot(y_2-y_1)\ , \tag{4.205}$$

where $y_*=\sqrt{\theta_*^2\upsilon^{-2}-1}$ and

$$y_2-y_1=\upsilon^{-1}\left[\sqrt{\theta_*^2+0.5c\tau_p h^{-1}-\upsilon^2}-\sqrt{\theta_*^2-0.5c\tau_p h^{-1}-\upsilon^2}\right]. \tag{4.206}$$

We must keep in mind that $y_1=0$ if $\upsilon^2\in[\theta_*^2-\frac{c\tau_p}{2h},\ \theta_*^2+\frac{c\tau_p}{2h}]$. If the duration of the pulsed searching signal is very large in value, i.e., the condition $\tau_p >> T_r$ is satisfied or the pulsed searching signal is a continuous process, then integration with respect to the variable y is carried out within the limits of the interval $(-\infty, \infty)$.

Let us assume that the directional diagram and the specific effective scattering area $S^\circ(\theta)$ are defined by the Gaussian law. The power spectral density of the fluctuations given by Equation (4.202) can be defined with an arbitrary duration of the pulsed searching signal in the following form:

$$S(\omega,t) \cong \Omega_{\max}^{-1}\upsilon \int_{y_1}^{y_2} e^{-2\pi\cdot\left\{\frac{\left[\upsilon(\cos\beta_0 - y\sin\beta_0)-\varphi_0\right]^2}{\Delta_\varphi^2} + \frac{\left[\upsilon(\sin\beta_0 + y\cos\beta_0)-\psi_0\right]^2}{\Delta_\psi^2}\right\} - k_2\upsilon(1+y^2)} dy\,. \tag{4.207}$$

The integral in Equation (4.207) is defined using the error integral given in Equation (1.27). In the general case, detailed study and analysis of this are very cumbersome.[51] Because of this, we limit our consideration to particular cases. Let us suppose that the directional diagram axis is downward directed, i.e., the conditions $\theta_0 = 0°$, $\varphi_0 = 0°$, and $\psi_0 = 0°$ are satisfied, and $\beta_0 = 0°$. Then, we can write

$$S(\omega,t) \cong \left[\Phi\left(\sqrt{\frac{2\pi a_\psi^2(\theta_2^2-\upsilon^2)}{\Delta_\psi^{(2)}}}\right) - \Phi\left(\sqrt{\frac{2\pi a_\psi^2(\theta_1^2-\upsilon^2)}{\Delta_\psi^{(2)}}}\right)\right]\cdot e^{-2\pi\cdot\frac{a_\varphi^2}{\Delta_\varphi^{(2)}}\cdot\upsilon^2}, \tag{4.208}$$

where

$$a_\varphi^2 = 1 + 0.5\pi^{-1}k_2\Delta_\varphi^{(2)} \quad \text{and} \quad a_\psi^2 = 1 + 0.5\pi^{-1}k_2\Delta_\psi^{(2)}\,. \tag{4.209}$$

We need to compare this with Equation (4.175).

With the continuous searching signal and under the condition $\beta_0 = 0°$, we can write

$$S(\omega) \cong e^{-2\pi\cdot\frac{a_\varphi^2}{\Delta_\varphi^{(2)}}\cdot\frac{(\omega-\omega_0)^2}{\Omega_{\max}^2}}, \tag{4.210}$$

i.e., the power spectral density of the fluctuations coincides with the square of the directional diagram narrowed in a_φ times on the plane of the cross section φ, in which there is the vector of velocity of the moving radar. The effective bandwidth of the power spectral density $S(\omega)$ can be determined in the following form:

$$\Delta F = \frac{\sqrt{2}\,V}{\lambda}\cdot\frac{\Delta_\varphi}{a_\varphi} \tag{4.211}$$

and depends on the power of roughness of the scanned surface of the two-dimensional target using the coefficient k_2. With the scanned two-dimensional target having a very rough surface, i.e., under the condition $a_\varphi \approx 1$, the effective bandwidth of $S(\omega)$ is maximal. As the roughness of the scanned surface of the two-dimensional target is smoothed, the effective bandwidth of $S(\omega)$ is decreased.

In the short-range area, the effective bandwidth of the power spectral density $S(\omega)$ of the fluctuations can be determined in the following form:

$$\Delta F = V \cdot \sqrt{\frac{2\,\Delta_\varphi^{(2)}}{\lambda^2 a_\varphi^2} + \frac{a_\varphi^2}{2\,h^2 \Delta_\varphi^{(2)}}} \tag{4.212}$$

instead of Equation (4.211) (see Section 4.7). The first term in Equation (4.212) defines the fluctuations caused by the Doppler effect; the corresponding correlation interval is equal to the time during which the radar moves the distance equal to the diameter of the radar antenna. The second term in Equation (4.212) defines the fluctuations caused by the exchange of scatterers within the limits of the illuminated spot; the corresponding correlation interval is equal to the time during which the radar moves the distance equal to the dimensions of the illuminated spot. In the long-range area, we can neglect this effect.

If the duration of the pulsed searching signal is short and the condition $\beta_0 = 0°$ is satisfied, the power spectral density $S(\omega,t)$ can be written in the following form:

$$S(\omega,t) \cong \left(\sqrt{\theta_*^2 + 0.5c\tau_p h^{-1} - \upsilon^2} - \sqrt{\theta_*^2 - 0.5c\tau_p h^{-1} - \upsilon^2}\right) \cdot e^{-2\pi \cdot \left(\frac{a_\psi^2 \theta_*^2}{\Delta_\psi^{(2)}} + \frac{\upsilon^2 a_\varphi^2}{\Delta_\varphi^{(2)}} - \frac{\upsilon^2 a_\psi^2}{\Delta_\psi^{(2)}}\right)}. \tag{4.213}$$

Under the condition $\beta_0 = 90°$, we have the same results, but it is necessary to replace the parameters a_φ and Δ_φ with the parameters a_ψ and Δ_ψ, respectively, in Equation (4.208)–Equation (4.213). When the directional diagram is axial symmetric, the parameters a_φ, Δ_φ and a_ψ, Δ_ψ must be replaced with the parameters a_a and Δ_a, respectively.

Reference to Equation (4.213) shows that the power spectral density $S(\omega, t)$ of the fluctuations is symmetric relative to the frequency ω_0 and is within the limits of the frequency interval $[\omega_1, \omega_2]$, where

$$\omega_1 = \omega_0 - \Omega_{\max} \cdot \sqrt{\theta_*^2 + 0.5c\tau_p h^{-1}} \quad \text{and} \quad \omega_2 = \omega_0 + \Omega_{\max} \cdot \sqrt{\theta_*^2 + 0.5c\tau_p h^{-1}}\,. \tag{4.214}$$

For this reason, the effective bandwidth of $S(\omega, t)$ at zero frequencies under the condition $\Delta t \geq -0.5\tau_p$ is determined in the following form:

$$\Delta\Omega = 2\,\Omega_{\max}\cdot\sqrt{\theta_*^2 + 0.5c\tau_p h^{-1}} = 2\,\Omega_{\max}\cdot\sqrt{2(\Delta\,t + 0.5\tau_p)T_d^{-1}}\,. \quad (4.215)$$

Because the angle θ_* is increased as the pulsed searching signal is propagated, the effective bandwidth of $S(\omega, t)$ depends on the radar range or time cross section t. The average frequency of the effective bandwidth of $S(\omega, t)$ is independent of the parameter t and always coincides with the parameter ω_0. Reference to Equation (4.215) shows that under the condition $\Delta\,t = -0.5\tau_p$, the effective bandwidth $\Delta\Omega$ of $S(\omega, t)$ at zero frequencies is equal to zero: $\Delta\Omega = 0$. The reason for this is choosing the origin of the coordinate t, which is to be related to the middle of the pulsed searching signal — the instant of time when the leading edge of the pulsed searching signal reaches the scanned surface of the two-dimensional target: $t_0 = T_d - 0.5\tau_p$.

The normalized power spectral density $s(\omega, t)$ given by Equation (4.213) is shown in Figure 4.26 for the axial symmetry directional diagram $\Delta_h = \Delta_v$ with various values of the angle θ_* at the relative delay $2\cdot\frac{\Delta t}{T_r} = \theta_*^2\,\frac{T_r}{\tau_p}$. The power spectral density $s(\omega, t)$ is normalized, so that under the conditions $\omega = \omega_{\max}$ and $\omega = \omega_{\min}$, for all values θ_* the power spectral densities are equal to 1:

$$s(\omega,t) \cong \sqrt{\Delta\,t\tau_p^{-1} + 0.5 - 0.5\upsilon^2 T_r\tau_p^{-1}} - \sqrt{\Delta\,t\tau_p^{-1} - 0.5 - 0.5\upsilon^2 T_r\tau_p^{-1}}\,. \quad (4.216)$$

The power spectral density $S(\omega, t)$ given by the exact formula in Equation (4.208) is shown in Figure 4.26 by the dotted line. The main conclusions made in this section coincide with the conclusions presented in Jukovsky et al.[52]

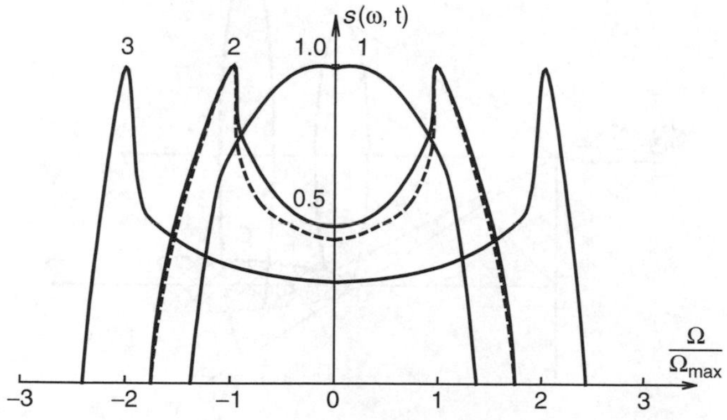

FIGURE 4.26
The power spectral density of the interperiod fluctuations as a function of appearance time of the target return signal under vertical scanning of the two-dimensional (surface) target and radar moving in the horizontal way: (1) $\frac{\Delta t}{\tau_p} = 0.5$; (2) $\frac{\Delta t}{\tau_p} = 1$; (3) $\frac{\Delta t}{\tau_p} = 2.5$.

4.9 Determination of the Power Spectral Density

To determine the power spectral density of the target return signal Doppler fluctuations we use the technique discussed in Section 3.3.3. Let us assume, for example, that the radar moves horizontally. The geometric locus with the same Doppler frequency is the line corresponding to the cross section of the Earth's surface or the scanned surface of the two-dimensional target by the lateral area of the cone, the axis of which coincides with the vector of velocity of the moving radar (see Figure 4.27). This line is often called the isodope and is the hyperbola determined in the form $h_2 = x^2\text{tg}^2\theta - \gamma^2$. The surface plane of the Earth or the two-dimensional target can be covered by the totality of these hyperbolae with the parameter θ (see Figure 4.28). The positions of these hyperbolae can be expressed by asymptotes in the form $y = \pm\,\text{tg}\theta$.[53,54] The power of the target return signal corresponding to the interval of the Doppler frequencies $[\Omega, \Omega + d\Omega]$ is equal to the power of the target return signal from the narrow zone of the Earth or the scanned two-dimensional target surface placed between the hyperbolae with the parameters θ and $\theta + d\theta$ (see Figure 4.29, the hatched area) within the limits of the area $\mathcal{F}$.

If the azimuth angle β_0 is sufficiently high by value (see Figure 4.29), the area of each narrow zone of the Earth or scanned two-dimensional target surface is the same for all hyperbolae. For this reason, the power of the target return signal from this area is defined only by the coefficient of amplification of the radar antenna — both the receiving and the transmitting radar antenna — for the given direction. The power spectral density of the fluctuations

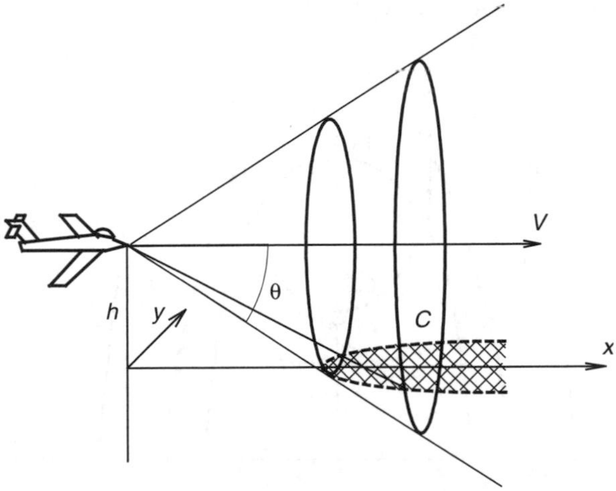

FIGURE 4.27
Formation of the isodope: the cross section of the cone of the constant Doppler frequencies by the Earth's surface.

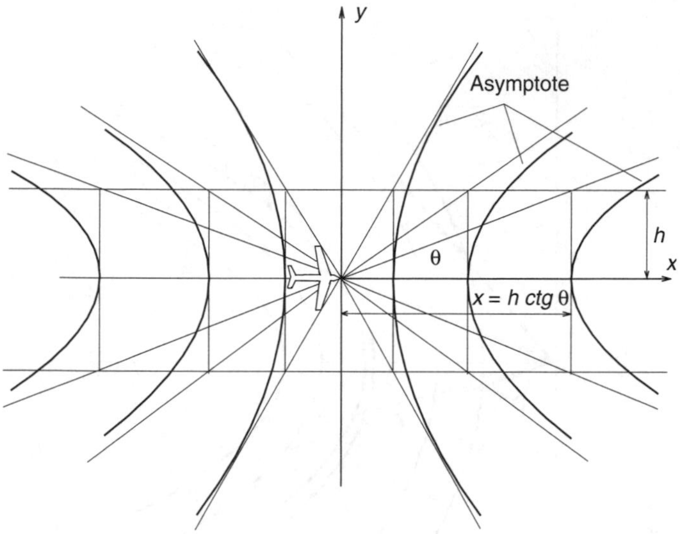

FIGURE 4.28
Formation of the isodope: the totality of isodopes and asymptotes of isodopes.

coincides in shape with the square of the horizontal-coverage directional diagram. The same result was discussed in Section 4.5.2 [see Equation (4.118)]. If the azimuth angle β_0 of the radar antenna is low in value and the aspect angle of the considered area is high in value, the hyperbolae have a large radius of curvature near the tops. Then the angle *AOB* (see Figure 4.30) contracting the segment of the hyperbola *AB* within the limits of the area $\mathcal{F}$ is more than the beam width of the horizontal-coverage directional diagram for the majority of hyperbolae. Because of this, the power of the target return signal corresponding to the given Doppler frequency depends on the coefficient of amplification of the radar antenna in the vertical plane, and not on the shape of the horizontal-coverage directional diagram, except in the case of small areas near the boundaries of the region $\mathcal{F}$. The shape of the power spectral density is defined, in the general case, by the shape of a part of the vertical-coverage directional diagram that covers the region $\mathcal{F}$. This phenomenon is defined by Equation (4.68). Thus, we can rigorously determine the power spectral density of the Doppler fluctuations based on the statements just discussed. Here we do not present these formulae because the obtained results coincide with the results we just defined.

As an exception, let us show the results for the exact determination of the power spectral density of the Doppler fluctuations under the condition $\beta_0 = 0°$ and various values of the aspect angle, as this case was investigated in a rather simple manner. Under the condition $\beta_0 = 0°$ and low values of the aspect angle, the angle *A′OB′* can be less than the beam width of the horizontal-coverage directional diagram (see Figure 4.30). Then the limits of integration with respect to the variable φ depend on the variable ψ near the boundaries of the region $\mathcal{F}$, and Equation (4.4), rigorously speaking, cannot

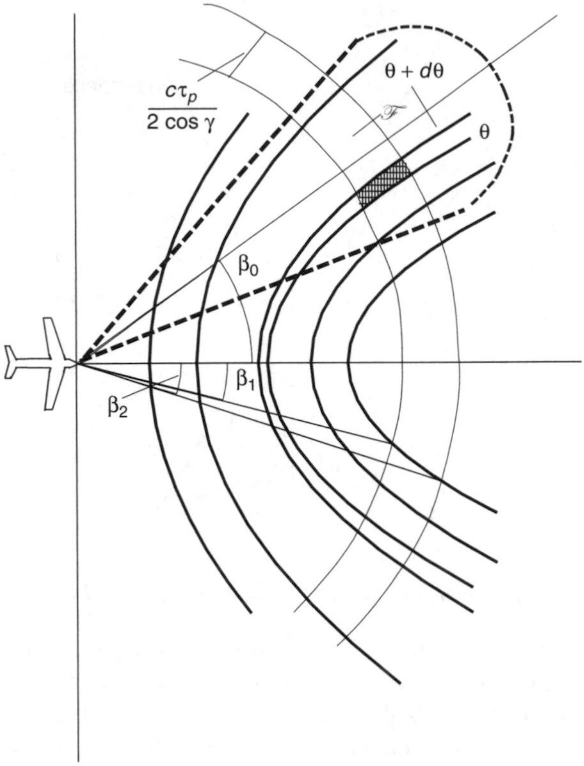

FIGURE 4.29
Space relations for determination of the power spectral density of the target return signal Doppler fluctuations with the high-deflected directional diagram of radar antenna.

be written in the form of the product, as in Equation (4.6). For this reason, the exact formula of the power spectral density is different from that given by Equation (4.68).

With the Gaussian vertical-coverage directional diagram, the power spectral density of the Doppler fluctuations takes the following form:

$$S(\Omega) = S_0\left[\Phi\left(\frac{2\sqrt{\pi}\,\Omega}{\Delta_h\Omega_{\max}}\cdot\sqrt{\frac{\Omega_2-\Omega}{\Omega_2}}\right)-\Phi\left(\frac{2\sqrt{\pi}\,\Omega}{\Delta_h\Omega_{\max}}\cdot\sqrt{\frac{\Omega_1-\Omega}{\Omega_1}}\right)\right], \qquad (4.217)$$

where

$$S_0 = \frac{PG_0^2\lambda^3\Delta_h S^\circ(\gamma_*)g_v^2(\gamma_*-\gamma_0)}{64\,\pi^3h^3V}; \qquad (4.218)$$

and

$$\Omega_{1,2} = \Omega_{\max}\cdot\cos\gamma_{1,2} = \Omega_{\max}\cdot\sqrt{1-\frac{h^2}{(\rho_*\mp 0.25c\tau_p)^2}} \qquad (4.219)$$

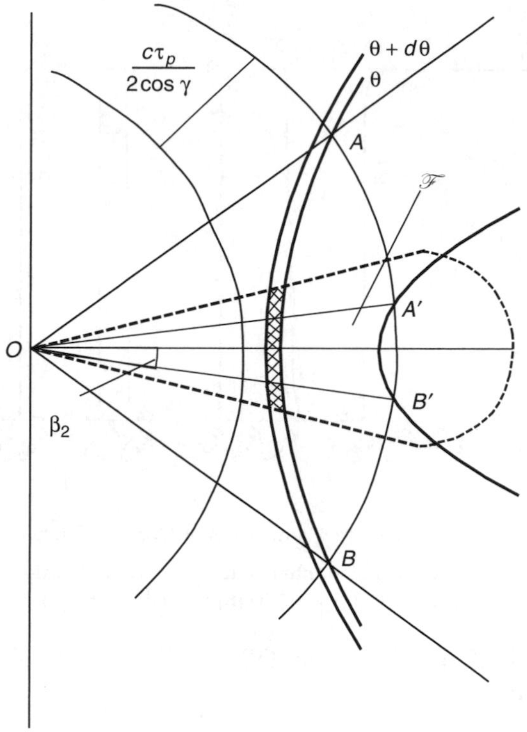

FIGURE 4.30
Space relations for determination of the power spectral density of the target return signal Doppler fluctuations with the nondeflected directional diagram of the radar antenna.

are the Doppler frequencies corresponding to the inside and outside boundaries of the region $\mathcal{F}$; Ω_2 is the maximum frequency: $\Omega_1 < \Omega_2$. Where $\Omega < \Omega_1$, we have to use Equation (4.217) to determine the power spectral density. Where $\Omega_1 < \Omega < \Omega_2$, we have to assume that the second term in Equation (4.217) is equal to zero, and where $\Omega > \Omega_2$, that both terms in Equation (4.217) are equal to zero.

As can be seen from Figure 4.31, the power spectral density of the Doppler fluctuations given by Equation (4.217) is significantly different from the square waveform power spectral density only at low values of the aspect angle. The power spectral density bandwidth at the level 0.5 — which in practice is the effective bandwidth — is determined by Equation (4.70). The frequency interval, within the limits of which the power spectral density is increased from 0 to 0.9 or decreased from 1 to 0.1, is equal to $\frac{0.1\,\Omega_{\max}\Delta_h^{(2)}}{\cos\gamma_*}$. In other words, we can say that this frequency interval decreases as the horizontal-coverage directional diagram is narrowed and the value of the aspect angle is lowered. This technique of power spectral density definition was used by Sokolov and Chadovich[55] to determine the power spectral density

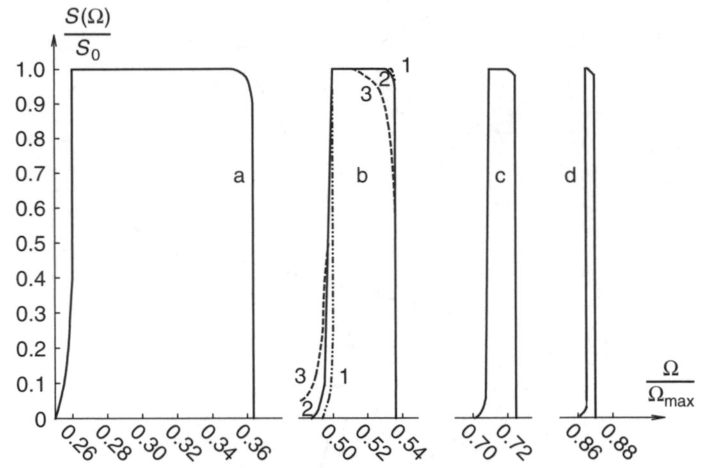

FIGURE 4.31
The power spectral density of the target return signal Doppler fluctuations with the nondeflected directional diagram of the radar antenna at $\frac{c\tau_p}{2h}$ = 0.3 and various values of Δ_h and γ_1: (1) $\Delta_h = 3°$; (2) $\Delta_h = 6°$; (3) $\Delta_h = 12°$; (a) $\gamma_1 = 75°$; (b) $\gamma_1 = 60°$; (c) $\gamma_1 = 45°$; (d) $\gamma_1 = 30°$.

with the radar moving nonhorizontally and vertically scanning the two-dimensional (surface) target.

4.10 Conclusions

Under scanning of the two-dimensional (surface) target by the moving radar, the total normalized correlation function $R^{en}(t, \tau)$ of the target return signal fluctuations can be expressed as the product of two normalized correlation functions: the azimuth-normalized correlation function $R_\beta(t, \tau)$ and the aspect-angle normalized correlation function $R_\gamma(t, \tau)$. $R_\beta(t, \tau)$ takes into consideration the slow target return signal fluctuations caused by the difference in the Doppler frequencies in the azimuth plane. $R_\gamma(t, \tau)$ takes into consideration the slow fluctuations caused by the difference in the Doppler frequencies in the aspect-angle plane and the rapid fluctuations, which in turn are caused by the propagation of the pulsed searching signal along the scanned surface of the two-dimensional target.

While scanning the two-dimensional (surface) target by the pulsed searching signal, when the radar is stationary, the correlation function $R^{en}[t(\psi_*), \tau]$ is a periodic function with respect to the parameter τ with the period T_p. $R^{en}[t(\psi_*), \tau]$ defines the rapid fluctuations in the radar range. Because, in the general case, the variables $t(\psi_*)$ and τ are not separable, the correlation function in the radar range is not separable, i.e., the spectral characteristics depend on the parameter τ. The corresponding instantaneous power spectral density is the regulated function. The envelope of the power spectral density is defined by the Fourier transform for the correlation function $R^{en}[t(\psi_*), \tau]$ and depends on the parameter t.

Under scanning of the two-dimensional (surface) target by the pulsed searching signal when the radar moves, the aspect-angle correlation function $R_\gamma(t, \tau)$ defines the interperiod fluctuations in the glancing radar range and is not separable. The power spectral density $S_\gamma[\omega, \psi_* (t)]$ of the interperiod fluctuations in the glancing radar range depends essentially on the value of the angle ψ_*. With the Gaussian pulsed searching signal, the aspect-angle normalized correlation function $R_\gamma(\tau, \psi_*)$ is separable and is defined by the product of the two normalized correlation functions $R'_\gamma(\tau)$ and $R''_\gamma(\tau, \psi_*)$. $R'_\gamma(\tau)$ defines the fluctuations in the radar range. $R''_\gamma(\tau, \psi_*)$ defines the slow fluctuations that are caused by the difference in the Doppler frequencies at the aspect-angle plane, and the interperiod fluctuations in the glancing radar range.

Under scanning of the two-dimensional (surface) target by the pulsed searching signal when the radar moves, the azimuth-normalized correlation function $R_\beta(\tau)$ defines the target return signal fluctuations caused by the difference in the Doppler frequencies of scatterers observed with various values of the azimuth angle within the horizontal-coverage directional diagram. The power spectral density $S_\beta(\omega)$ corresponding to $R_\beta(\tau)$ is defined for two cases: the high- and low-deflected radar antenna.

Under scanning of the two-dimensional (surface) target by the moving radar, the total normalized correlation function $R^{en}(t, \tau)$ of the fluctuations is defined by the product of the aspect-angle normalized correlation function $R_\gamma(t, \tau)$ and the azimuth-normalized correlation function $R_\beta(\tau)$. $R_\gamma(t, \tau)$ defines the rapid fluctuations in the radar range and the slow interperiod fluctuations caused by the difference in the radial components of the velocity of scatterers in the aspect-angle plane. $R_\beta(\tau)$ defines only the interperiod fluctuations caused by difference in the radial components of the velocity of scatterers in the azimuth plane. For this reason, the product of $R_\gamma(t, \tau)$ and $R_\beta(\tau)$ leads to changes only in the interperiod fluctuations and does not act on fluctuations in the radar range.

The interperiod fluctuations in the glancing radar range are defined by the product of the aspect-angle normalized correlation function $R_\gamma(t, \tau)$ and the azimuth-normalized correlation function $R_\beta(\tau)$. The total power spectral density in the glancing radar range is defined by the convolution between the aspect-angle power spectral density $S_\gamma[\omega, \psi_*(t)]$ and azimuth power spectral density $S_\beta(\omega)$. The total normalized correlation function in the fixed radar range is defined by the product of $R_\gamma(t, \tau)$ under the condition $\tau = nT_p$ and $R_\beta(\tau)$. In the general case, we have to use various numerical techniques to define the power spectral densities $S_\beta(\omega)$ and $S_\gamma(\omega)$.

In considering of the short-range area, the normalized correlation function and power spectral density are defined by the Gaussian law, as in the case of the long-range area. The difference is only in the effective bandwidths, between ΔF_1 and ΔF by the additional term decreasing with an increase in the radar range ρ. This additional term takes into consideration changes in the amplitudes of elementary signals when the scatterers are covered by the directional diagram and is defined by the time of scanning the scatterers by the radar antenna beam, i.e., the time when the scatterers are within the limits of the radar antenna beam width. The value of the effective bandwidth ΔF_1, unlike the value of the effective bandwidth ΔF, is inversely proportional to the directional diagram width or the radar antenna beam width.

Under vertical scanning of the two-dimensional (surface) target, three cases are possible: the radar is stationary; the radar moves only vertically; the radar moves only horizontally. If the radar is stationary, there are only the intraperiod fluctuations. In this case, the correlation function $R_0^{en}(t,\tau)$ is periodic with respect to the variable τ. The power spectral density is the regulated function. When the radar moves only vertically, there are the interperiod fluctuations in the glancing radar range. The effective power spectral density bandwidth is linearly increased from zero to $\Delta\Omega_{\max}$ within the limits of the time interval $[-0.5\tau_p, 0.5\tau_p]$, and after that it is approximately constant until the target return signal disappears completely. If the radar moves only horizontally, the interperiod fluctuations in the glancing range are defined by the correlation function $R^{en}(t, \tau)$ given by Equation (4.199) and the power spectral density $S(\omega, t)$ given by Equation (4.200).

References

1. Trump, T. and Ottersten, B., Estimation of nominal direction of arrival and angular spread using an array of sensors, *Signal Process.*, Vol. 50, No. 2, 1996, pp. 57–70.
2. Li, J., Liu, G., Jiang, N., and Stoica, P., Moving target feature extraction for airborne high-range resolution phased-array radar, *IEEE Trans.*, Vol. SP-49, No. 2, 2001, pp. 277–289.
3. Katkovnik, V., A new concept of beamforming for moving sources and impulse noise environment, *Signal Process.*, Vol. 80, No. 4, 2000, pp. 1863–1882.
4. Farina, A., *Antenna-Based Signal Processing Techniques for Radar Systems*, Artech House, Norwood, MA, 1992.
5. Ward, J., Space-time adaptive processing for airborne radar, Technical Report, 1015, Lincoln Laboratory, MIT, Cambridge, December 1994.
6. Abramovich, Yu, Spencer, N., and Gorokhov, A., Detection-estimation of more uncorrelated gaussian sources than sensors in non-uniform linear antenna arrays — part 1: Fully augmentable arrays, *IEEE Trans.*, Vol. SP-49, No. 5, 2001, pp. 959–971.
7. Gerlach, K. and Steiner, M., Adaptive detection of range distributed targets, *IEEE Trans.*, Vol. SP-47, No. 7, 1999, pp. 1844–1851.
8. Jacobs, S. and O'Sullivan, A., High-resolution radar models for joint tracking and recognition, in *Proceedings of the 1997 IEEE National Radar Conference*, Syracuse, NY, May 1997, pp. 99–104.
9. Scharf, L., *Statistical Signal Processing*, Addison-Wesley, Reading, MA, 1991.

10. Johnson, D. and Dudgeon, D., *Array Signal Processing: Concepts and Techniques*, Prentice Hall, Englewood Cliffs, NJ, 1993.
11. Moffet, A., Minimum-redundancy linear arrays, *IEEE Trans.*, Vol. AP-16, No. 2, 1968, pp. 172–175.
12. Besson, O., Vincent, F., Stoica, P., and Gershman, A., Approximate maximum likelihood estimators for array processing in multiplicative noise environments, *IEEE Trans.*, Vol. SP-48, No. 9, 2000, pp. 2506–2518.
13. Dedersen, K., Mogensen, P., and Flenry, B., Analysis of time, azimuth and Doppler dispersion in outdoor radio channels, in *Proceedings of the ACTS Mobile Communications Summit*, Aalbory, Denmark, Oct. 1997, pp. 308–313.
14. Besson, O. and Stoica, P., Decoupled estimation of DOA and angular spread for a spatially distributed source, *IEEE Trans.*, Vol. SP-48, No. 7, 2000, pp. 1872–1882.
15. Davies, R., Brennan, H., and Reed, I., Angle estimation with adaptive arrays in external noise fields, *IEEE Trans.*, Vol. AES-12, No. 3, 1976, pp. 179–186.
16. Reid, D., Zoubir, A., and Boashash, B., Aircraft flight parameter estimation based on passive acoustic techniques using the polynomial Wigner–Ville distribution, *J. Acoust. Soc. Amer.*, Vol. 102, No. 1, 1997, pp. 207–223.
17. Ward, K., Baker, F., and Watts, S., Maritime surveillance radar — Part 1: Radar scattering from the ocean surface, in *Radar and Signal Processing, IEE Proceedings F*, Vol. 137, No. 2, April 1990, pp. 330–340.
18. Conte, E., Di Bisceglie, M., Galdi, C., and Ricci, G., A procedure for measuring the coherence length of the sea texture, *IEEE Trans.*, Vol. IM-46, No. 4, 1997, pp. 836–841.
19. Gerlach, K. and Steiner, M., Fast converging adaptive detection of Doppler-shifted range distributed targets, *IEEE Trans.*, Vol. SP-48, No. 9, 2000, pp. 2686–2690.
20. Kullback, S., *Information Theory and Statistics*, Dover, Mineola, NJ, 1997.
21. Gardner, W., *Statistical Spectral Analysis: A Non-Probability Theory*, Prentice Hall, Englewood Cliffs, NJ, 1988.
22. Mendel, J., *Lessons in Estimation Theory for Signal Processing, Communications, and Control*, Prentice Hall, Englewood Cliffs, NJ, 1995.
23. Raich, R., Goldberg, J., and Messer, H., Bearing estimation for a distributed source via the conventional beamformer, in *Proceedings of the SSAP Workshop*, Portland, OR, September 1988, pp. 5–8.
24. Telatar, E., Capacity of multi-antenna Gaussian channels, *Europ. Trans. Telecommun.*, Vol. 10, No. 6, 1997, pp. 585–596.
25. Mandurovsky, I., Spectral performances of target return signals from the Earth surface under presence of side-lobes, *Problems in Radio Electronics*, Vol. OT, No. 9, 1978, pp. 3–12 (in Russian).
26. Dickey, F., Labitt, M., and Staudaher, F., Development of airborne moving target radar for long-range surveillance, *IEEE Trans.*, Vol. AES-27, No. 11, 1991, pp. 959–971.
27. Unghes, P., A high-resolution range radar detection strategy, *IEEE Trans.*, Vol. AES-19, No. 9, 1983, pp. 663–667.
28. Schleher, D., *MTI and Pulsed Doppler Radar*, Artech House, Norwood, MA, 1991.
29. Farina, A., Scannapieco, F., and Vinelli, F., Target detection and classification with very high range resolution radar, in *Proceedings of the International Conference on Radar*, Versailles, France, April 1989, pp. 20–25.

30. Jeruchim, M., Balaban, P., and Shanmugam, K., *Simulation of Communication Systems,* Plenum, New York, 1992.
31. Stuber, G., *Principles of Mobile Communication,* Kluwer, Norwell, MA, 1996.
32. Pahlavan, K. and Levesque, A., *Wireless Information Networks,* John Wiley & Sons, New York, 1995.
33. Robertson, P. and Kaisen, S., The effects of Doppler spreads in OFDM (A) mobile radio system, in *Proceedings of the Vehicular Technology Conference (VTC'99-Fall),* 1999, pp. 329–333.
34. Einarsson, G., *Principles of Lightwave Communications,* John Wiley & Sons, New York, 1996.
35. Stoica, P. and Moses, R., *Introduction to Spectral Analysis,* Prentice Hall, Upper Saddle River, NJ, 1997.
36. Griffiths, L. and Jim, C., An alternative approach to linearly constrained adaptive beamforming, *IEEE Trans.,* Vol. AP-30, No. 1, 1982, pp. 27–34.
37. Tzvetkov, A., *Non-Stationary Stochastic Processes,* Energy, Moscow, 1973 (in Russian).
38. Rappaport, T., *Wireless Communications Principle and Practice,* Prentice Hall, Englewood Cliffs, NJ, 1996.
39. O'Sullivan, J., De Vore, M., Kedia, V., and Miller, I., Automatic target recognition performance for SAR imagery using a conditionally Gaussian model, *IEEE Trans.,* Vol. AES-37, No. 1, 2001, pp. 91–108.
40. Hill, D. and Bodie, J., Carrier detection of PSK signals, *IEEE Trans.,* Vol. COM-49, No. 3, 2001, pp. 487–495.
41. Lance, E. and Kaleh, G., A diversity scheme for a phase-coherent frequency-hopping spread-spectrum system, *IEEE Trans.,* Vol. COM-45, No. 9, 1997, pp. 1123–1129.
42. Aschwartz, M., Bennet, W., and Stein, S., *Communication System and Techniques,* IEEE Press, New York, 1996.
43. Proakis, J., *Digital Communications,* 3rd ed., McGraw-Hill, New York, 1995.
44. Feldman, Yu, Fluctuations of the target return signal caused by moving radar, *Problems in Radio Electronics,* Vol. OT, No. 6, 1972, pp. 3–21 (in Russian).
45. Mensa, D., *High-Resolution Radar Cross Section Imaging,* Artech House, Norwood, MA, 1991.
46. Wicker, S., *Error Control Systems for Digital Communication and Storage,* Prentice Hall, Englewood Cliffs, NJ, 1995.
47. Stoica, P., Besson, O., and Gershman, A., Direction-of-arrival estimation of an amplitude-distorted wave-front, *IEEE Trans.,* Vol. SP-49, No. 2, 2001, pp. 269–276.
48. Kolchinsky, V., Mandurovsky, I., and Konstantinovsky, M., *Doppler Apparatus and Navigational Systems,* Soviet Radio, Moscow, 1975 (in Russian).
49. Simon, M., Hinedi, S., and Lindsey, W., *Digital Communication Technique,* Prentice Hall, Englewood Cliffs, NJ, 1995.
50. Wehner, D., *High Resolution Radar,* Artech House, Norwood, MA, 1987.
51. Bowman, A. and Azzalini, A., *Applied Smoothing Techniques for Data Analysis,* Oxford University Press, Oxford, U.K., 1997.
52. Jukovsky, A., Onoprienko, E., and Chijov, V., *Theoretical Foundations of Radar Altimetry,* Soviet Radio, Moscow, 1979 (in Russian).
53. Pillai, S., Bar-Ness, Y., and Haber, F., A new approach to array geometry for improved spatial spectrum estimation, in *Proceedings of the IEEE,* Vol. 73, Oct 1985, pp. 1522–1524.

54. Feldman, Yu, Determination of power spectral density of target return signals, *Problems in Radio Electronics*, Vol. OT, No. 6, 1959, pp. 22–38 (in Russian).
55. Sokolov, M. and Chadovich, I., Characteristics of Doppler spectrum under inclined flight of airborne, *News of the USSR University, Radio Electronics*, No. 12, 1975, pp. 61–66 (in Russian).

5

Fluctuations Caused by Radar Antenna Scanning

5.1 General Statements

If the radar is stationary, radar antenna scanning is the main source of slow target return signal fluctuations.[1,2] Let us investigate the correlation function and power spectral density of the target return signal fluctuations caused by the moving radar antenna in scanning the three-dimensional (space) and two-dimensional (surface) targets. Two of the most universally adopted forms of radar antenna scanning[3] are line scanning used, as a rule, in the detection of targets and conical scanning used in the tracking of moving targets. Conical radar antenna scanning with simultaneous rotation of radar antenna polarization plane will also be studied here.

5.1.1 The Correlation Function under Space Scanning

If under space scanning by the radar antenna, the position of the polarization plane is changed simultaneously with the moving radar antenna, it can be assumed that $\Delta\xi \neq 0$. The value of $\Delta\xi$ is the same for all scatterers. Using Equation (2.75)–Equation (2.77), Equation (2.84), Equation (2.85), and Equation (2.93)–Equation (2.95), and assuming that $\Delta\ell = \Delta\omega = \Delta\zeta = 0$, we can write the space–time normalized correlation function of the target return signal fluctuations in the following form:[4]

$$R^{en}_{\Delta\varphi,\Delta\psi,\Delta\xi}(t,t+\tau) = R_p(t,t+\tau)\cdot R_{sc}(\Delta\beta_0,\Delta\gamma_0)\cdot R_q(\Delta\xi)\,, \tag{5.1}$$

where

$$R_p(t,t+\tau) = N\cdot\sum_{n=0}^{\infty}\int \boldsymbol{P}\left[t-\tfrac{2\rho}{c}\right]\cdot\boldsymbol{P}^*\left[t-\tfrac{2\rho}{c}+\tau-nT_p\right]\rho^{-2}d\rho \tag{5.2}$$

is the normalized correlation function of the fluctuations in the radar range, which is caused by propagation of the pulsed searching signal;

$$R_{sc}(\Delta\beta_0, \Delta\gamma_0) = N \cdot \iint g(\varphi, \psi) \cdot g(\varphi + \Delta\beta_0 \cos\gamma_0,\ \psi + \Delta\gamma_0)\, d\varphi\, d\psi \tag{5.3}$$

is the normalized correlation function of the space fluctuations, which is caused by radar antenna scanning;

$$R_q(\Delta\xi) = N \cdot \iint q(\xi, \zeta) \cdot q(\xi + \Delta\xi, \zeta) \sin d\xi\, d\zeta \tag{5.4}$$

is the normalized correlation function of the space fluctuations, which is caused by the rotation of the radar antenna polarization plane.

With the pulsed searching signal, we can neglect variations in the variable ρ in the denominator of the integrand of the normalized correlation function $R_p(t, t + \tau)$ of the time fluctuations in the radar range, which is determined by Equation (5.2), and a new variable can be introduced:

$$z = t - 2\rho c^{-1} . \tag{5.5}$$

In this case, $R_p(t, t + \tau)$, which is given by Equation (5.2), differs from that given by Equation (3.7) only by the absence of compression in the time scale, i.e., $\mu = 1$, which happens with the moving radar. In terms of what was just discussed, all the main statements and conclusions made in Section 3.1 regarding the fluctuations in the radar range are true for the case considered here. With a simple searching harmonic signal, the normalized correlation function $R_p(t, t + \tau)$ becomes equal to 1.

A study of the normalized correlation functions $R_{sc}(\Delta\beta_0, \Delta\gamma_0)$ and $R_q(\Delta\xi)$ of the slow (space) fluctuations, which are determined by Equation (5.3) and Equation (5.4) and caused by radar antenna scanning and rotation of the polarization plane, respectively, is of prime interest. If the variables φ and ψ can be separated in the function $g(\varphi, \psi)$ — the radar antenna directional diagram — we can write

$$R_{sc}(\Delta\beta_0, \Delta\gamma_0) = R_{sc}^h(\Delta\beta_0) \cdot R_{sc}^v(\Delta\gamma_0) , \tag{5.6}$$

where

$$R_{sc}^h(\Delta\beta_0) = N \cdot \int g_h(\varphi) \cdot g_h(\varphi + \Delta\beta_0 \cos\gamma_0)\, d\varphi ; \tag{5.7}$$

$$R_{sc}^{v}(\Delta\gamma_0) = N \cdot \int g_v(\psi) \cdot g_v(\psi + \Delta\gamma_0)\, d\psi\,. \tag{5.8}$$

If we are able to write the directional diagram $g(\varphi, \psi)$, as the function $\bar{g}(\varphi', \psi')$ for a new coordinate system φ' and ψ', which is rotated with respect to the previous coordinate system φ and ψ by the angle

$$\alpha = \text{arctg}\, \frac{\Delta\gamma_0}{\Delta\beta_0 \cos\gamma_0}\,, \tag{5.9}$$

then, instead of Equation (5.3) we can write $R_{sc}(\Delta\beta_0, \Delta\gamma_0)$ in the following form:

$$R_{sc}(\Delta\beta_0, \Delta\gamma_0) = N \cdot \iint \bar{g}(\varphi', \psi') \cdot \bar{g}(\varphi' + \Delta\varphi', \psi')\, d\varphi' d\psi', \tag{5.10}$$

where

$$\Delta\varphi' = \sqrt{\Delta\beta_0^2 \cos^2\gamma_0 + \Delta\gamma_0^2}\,. \tag{5.11}$$

If the variables φ' and ψ' are separable in the function $\bar{g}(\varphi', \psi')$, i.e.,

$$\bar{g}(\varphi', \psi') = \bar{g}_1(\varphi') \cdot \bar{g}_2(\psi')\,, \tag{5.12}$$

$R_{sc}(\Delta\beta_0, \Delta\gamma_0)$ can be written in the following form:

$$R_{sc}(\Delta\beta_0, \Delta\gamma_0) = N \cdot \int \bar{g}_1(\varphi') \cdot \bar{g}_1(\varphi' + \Delta\varphi')\, d\varphi'\,, \tag{5.13}$$

where $\bar{g}_1(\varphi')$ is the directional diagram by power in the plane in which the directional diagram axis moves.

Equation (5.3) and Equation (5.6) allow us to determine the normalized correlation function $R_{sc}(\Delta\beta_0, \Delta\gamma_0)$ with an arbitrary orientation of the directional diagram. For instance, in the case of the Gaussian directional diagram, $R_{sc}(\Delta\beta_0, \Delta\gamma_0)$ takes the following form:

$$R_{sc}(\Delta\beta_0, \Delta\gamma_0) = e^{-0.5\pi\left[\frac{\Delta\beta_0^2 \cos^2\gamma_0}{\Delta_h^{(2)}} + \frac{\Delta\gamma_0^2}{\Delta_v^{(2)}}\right]}\,. \tag{5.14}$$

In the case of the directional diagram with the uniform distribution law, for instance, the **sinc**^{2}-directional diagram [see Equation (2.103)], $R_{sc}(\Delta\beta_0, \Delta\gamma_0)$ has the following form:

$$R_{sc}(\Delta\beta_0, \Delta\gamma_0) = \frac{36(1 - \textbf{sinc}\ X)(1 - \textbf{sinc}\ Y)}{X^2Y^2}\ , \tag{5.15}$$

where

$$X = 2\pi \cdot \frac{\Delta\beta_0}{\Delta_h}\cos\gamma_0 \qquad \text{and} \qquad Y = 2\pi \cdot \frac{\Delta\gamma_0}{\Delta_v}\ . \tag{5.16}$$

If radar antenna scanning is given as a function of time, we can transform $R_{sc}(\Delta\beta_0, \Delta\gamma_0)$ into the normalized correlation function of the time fluctuations without any problems. For this purpose, we need to define the angle shifts $\Delta\beta_0$ and $\Delta\gamma_0$ as a function of τ for the stationary stochastic target return signal, or as a function of t and τ for the nonstationary stochastic target return signal. With the parametric form of definition of scanning law $\beta_0 = f_h(t)$ and $\gamma_0 = f_v(t)$, we can write

$$\Delta\beta_0 = f_h(t + 0.5\ \tau) - f_h(t - 0.5\ \tau) \qquad \text{and} \qquad \Delta\gamma_0 = f_v(t + 0.5\ \tau) - f_v(t - 0.5\ \tau)\ . \tag{5.17}$$

In the case of the uniform radar antenna scanning, we can write

$$\beta_0 = \Omega_{sc}^h \cdot t\ ;\ \ \gamma_0 = \Omega_{sc}^v \cdot t\ ;\ \ \Delta\beta_0 = \Omega_{sc}^h \cdot \tau\ ; \qquad \text{and} \qquad \Delta\gamma_0 = \Omega_{sc}^v \cdot \tau\ , \tag{5.18}$$

where Ω_{sc}^h and Ω_{sc}^v are the angular velocities in the horizontal and vertical planes of radar antenna scanning, respectively. In this case, the slow fluctuations can be represented as a stationary stochastic process.

5.1.2 The Correlation Function under Surface Scanning

Consider the correlation function $R_{\Delta\rho,\Delta\varphi,\Delta\psi}^{en}(t,\tau)$ of the target return signal fluctuations, which is determined by Equation (2.122). Let us assume that

$$\Delta\rho = 0\ ; \qquad \Delta\varphi = \Delta\varphi_{sc} = \Delta\beta_0\cos(\psi + \gamma_0)\ ; \qquad \text{and} \qquad \Delta\psi = \Delta\psi_{sc} = \Delta\gamma_0\ . \tag{5.19}$$

$R_{\Delta\rho,\Delta\varphi,\Delta\psi}^{en}(t,\tau)$ can be written in the following symmetric form:

$$R_{\Delta\varphi,\Delta\psi}(t,\tau) = p(t) \cdot R_{\Delta\varphi,\Delta\psi}^{en}(t,\tau) \cdot e^{j\omega_0\tau}\ , \tag{5.20}$$

where the normalized correlation function $R^{en}_{\Delta\varphi,\Delta\psi}(t,\tau)$ is determined by

$$
\begin{aligned}
R^{en}_{\Delta\varphi,\Delta\psi}(t,\tau) = N \cdot \sum_{n=0}^{\infty} \iint & \boldsymbol{P}\left[t - \tfrac{2\rho(\psi)}{c} - 0.5(\tau - nT_p)\right] \\
& \times \boldsymbol{P}^*\left[t - \tfrac{2\rho(\psi)}{c} + 0.5(\tau - nT_p)\right] \\
& \times g[\varphi - 0.5\Delta\beta_0 \cos(\psi + \gamma_0),\ \ \psi - 0.5\Delta\gamma_0] \\
& \times g[\varphi + 0.5\Delta\beta_0 \cos(\psi + \gamma_0),\ \ \psi + 0.5\Delta\gamma_0] \\
& \times \boldsymbol{S}^\circ(\psi + \gamma_0)\sin(\psi + \gamma_0)\, d\varphi\, d\psi
\end{aligned} \tag{5.21}
$$

where the power $p(t)$ of the target return signal is given by Equation (2.136).

In the general case, $R_{\Delta\varphi,\Delta\psi}(t,\tau)$ given by Equation (5.20) cannot be expressed as the product of the correlation function of the fluctuations in the radar range and that of the fluctuations caused by radar antenna scanning, unlike the normalized correlation function $R^{en}_{\Delta\varphi,\Delta\psi,\Delta\xi}(t, t+\tau)$ given by Equation (5.1). This phenomenon can be explained by the fact that the radar range ρ becomes a function of the angle ψ, not an independent coordinate. However, with the short pulsed searching signals (the pulse duration is low in value), we can assume that

$$
\boldsymbol{S}^\circ(\psi + \gamma_0) \approx \boldsymbol{S}^\circ(\gamma_*) \qquad \text{and} \qquad \sin(\psi + \gamma_0) \approx \sin\gamma_* \, . \tag{5.22}
$$

Let us introduce a new variable given by Equation (5.5) that leads us to exchange the variable ψ for the variable $\psi_* + c_* z$. Here, we neglect variations in the directional diagram $g(\varphi, \psi)$ as a function of the parameter z within the limits of the pulsed searching signal period. Then, the normalized correlation function $R^{en}_{\Delta\varphi,\Delta\psi}(t,\tau)$ of the fluctuations can be written in the following form:

$$
R^{en}_{\Delta\varphi,\Delta\psi}(t,\tau) = R_p(\tau) \cdot R_{sc}(\Delta\beta_0, \Delta\gamma_0)\,, \tag{5.23}
$$

where

$$
R_p(\tau) = N \cdot \sum_{n=0}^{\infty} \int \boldsymbol{P}\,[z - 0.5(\tau - nT_p)] \cdot \boldsymbol{P}^*[z + 0.5(\tau - nT_p)]\, dz\,; \tag{5.24}
$$

$$R_{sc}(\Delta\beta_0, \Delta\gamma_0) = N \cdot \int_{\varphi_1}^{\varphi_2} g\left[\varphi - 0.5\Delta\beta_0 \cos\gamma_*, -0.5\Delta\gamma_0\right] \times g\left[\varphi + 0.5\Delta\beta_0 \cos\gamma_*, \psi_* + 0.5\Delta\gamma_0\right] d\varphi \quad (5.25)$$

Thus, in scanning the two-dimensional (surface) target by the pulsed searching signal, $R^{en}_{\Delta\varphi,\Delta\psi}(t,\tau)$ can be expressed as the product of the periodic normalized correlation function $R_p(\tau)$ of the fluctuations in the radar range (the comb function), which is given by Equation (5.24), and $R_{sc}(\Delta\beta_0, \Delta\gamma_0)$ of the slow fluctuations, which is given by Equation (5.25). These normalized correlation functions are caused by radar antenna scanning. The normalized correlation function $R_p(\tau)$ given by Equation (5.24) is an analog of $R_p(t, t + \tau)$ given by Equation (5.2). $R_{sc}(\Delta\beta_0, \Delta\gamma_0)$ given by Equation (5.25) depends on the angle ψ_* as a function of the parameter if the shape of the directional diagram in the plane of the angle ψ is different for various values of ψ_*.

If the variables in the function $g(\varphi, \psi)$ are separable, then the normalized correlation function of the slow fluctuations $R_{sc}(\Delta\beta_0, \Delta\gamma_0)$ caused by radar antenna scanning can be expressed as the product given by Equation (5.6), where

$$R^h_{sc}(\Delta\beta_0) = N \cdot \int_{\varphi_1}^{\varphi_2} g_h[\varphi - 0.5\Delta\beta_0 \cos\gamma_*] \cdot g_h[\varphi + 0.5\Delta\beta_0 \cos\gamma_*]\, d\varphi\ ; \quad (5.26)$$

$$R^v_{sc}(\Delta\gamma_0) = \frac{g_v(\psi_* - 0.5\Delta\gamma_0) \cdot g_v(\psi_* + 0.5\Delta\gamma_0)}{g_v^2(\psi_*)}\ . \quad (5.27)$$

The normalized correlation function $R^h_{sc}(\Delta\beta_0)$ given by Equation (5.26) is different from that determined by Equation (5.7) in the symmetric form of writing and the use of the angle γ_* instead of the angle γ_0. Comparison of the normalized correlation function $R^v_{sc}(\Delta\gamma_0)$ given by Equation (5.27) with the one determined by Equation (5.8) shows the essential distinction: an integration with respect to the variable ψ is not carried out in $R^v_{sc}(\Delta\gamma_0)$ given by Equation (5.27) because the interval of variation of the variable ψ is low in value. When the directional diagram is Gaussian, $R_{sc}(\Delta\beta_0, \Delta\gamma_0)$ of the slow fluctuations is determined by Equation (5.14) based on Equation (5.26) and Equation (5.27), in which the angle γ_* is used instead of the angle γ_0.

When the continuous searching signal, i.e., the condition given by Equation (2.109), is true, $R_{sc}(\Delta\beta_0, \Delta\gamma_0)$ follows from Equation (5.21) and can be written in the following form:

$$R_{sc}(\Delta\beta_0, \Delta\gamma_0) = N \cdot \int_{\varphi_1}^{\varphi_2}\int_{\psi_1}^{\psi_2} g\left[\varphi - 0.5\Delta\beta_0 \cos(\psi + \gamma_0), \psi - 0.5\Delta\gamma_0\right]$$
$$\times g\left[\varphi + 0.5\Delta\beta_0 \cos(\psi + \gamma_0), \psi + 0.5\Delta\gamma_0\right] \quad . \qquad (5.28)$$
$$\times S^\circ(\psi + \gamma_0)\sin(\psi + \gamma_0)\, d\varphi\, d\psi$$

5.1.3 The General Power Spectral Density Formula

Consider the normalized correlation function $R_{sc}(\Delta\beta_0, \Delta\gamma_0)$ of the slow target return signal fluctuations, which is determined by Equation (5.13). Let the radar antenna scanning be uniform in the plane of the angle φ' with the angular velocity Ω_{sc}. Then, we can write:[5]

$$\varphi' = \Omega_{sc} \cdot t \qquad \text{and} \qquad \Delta\varphi' = \Omega_{sc} \cdot \tau \, . \qquad (5.29)$$

In this case, the power spectral density $S(\omega)$ of the target return signal fluctuations can be written in the following form:

$$S(\omega) = \frac{p(t)}{\pi} \cdot \frac{\left| \int_{-\infty}^{\infty} \bar{g}_1(\Omega_{sc} t) \cdot e^{-j\omega t} dt \right|^2}{\int_{-\infty}^{\infty} \bar{g}_1^2(\Omega_{sc} t)\, dt} \, . \qquad (5.30)$$

The formula in Equation (5.30) has a simple physical meaning. This is the square of the power spectral density of the amplitude of the target return signal, which is the Fourier transform of the individual elementary signal covering all elementary scatterers by the directional diagram during radar antenna scanning. This formula corresponds to a general physical process of forming the power spectral density of the pulsed stochastic process (the target return signal) discussed in Section 2.1.

The directional diagram by power $g(\Omega_{sc} t)$, included in the formula in Equation (5.30), is the squared directional diagram by voltage or is the product of the directional diagrams by voltage for the transmitting and receiving conditions. For this reason, the amplitude power spectral density of the fluctuations caused by radar antenna scanning can be considered as the Fourier transform of the squared directional diagram by voltage which, in its turn, can be expressed as the convolution between functions that are the Fourier transform of the first order directional diagram by voltage. Let us denote this function by $E(y)$. The function $E(y)$ defines a distribution of the electromagnetic field over a radar antenna area. Because of this, the amplitude power spectral density of the fluctuations caused by radar antenna scanning can be determined in the following form:

$$s(\omega) = \int_{-\infty}^{\infty} E(y) \cdot E(\omega - y)\, dy\,, \tag{5.31}$$

where

$$y = 2\pi\lambda^{-1}\Omega_{sc}x \tag{5.32}$$

and x is the distance from the center of the radar antenna area. The power spectral density of the fluctuations is equal to the squared amplitude power spectral density $s(\omega)$ given by Equation (5.31).

5.2 Line Scanning

5.2.1 One-Line Circular Scanning

Let us assume that the radar antenna rotates uniformly around its vertical axis with the angular velocity Ω_{sc}^{h} and the constant aspect angle γ_0. In doing so,

$$\Delta\gamma_0 = 0 \qquad \text{and} \qquad \Delta\beta_0 = \Omega_{sc}^{h} \cdot \tau\,. \tag{5.33}$$

We assume that the variables φ and ψ in the function $g(\varphi, \psi)$ are separable. Then, the normalized correlation function $R_{sc}^{h}(\Delta\beta_0)$ of fluctuations caused by radar antenna scanning is defined by scanning the two-dimensional (surface) target in the same manner as in scanning the three-dimensional (space) target [see Equation (5.7) and Equation (5.26)].

In the case of the Gaussian directional diagram, it follows from Equation (5.14) that

$$R_{sc}(\tau) = e^{-\frac{\pi\tau^2}{\tau_c^2}}\,, \tag{5.34}$$

where

$$\tau_c = \frac{\sqrt{2}\,\Delta_h}{\Omega_{sc}^{h}\cos\gamma_0} \tag{5.35}$$

is the correlation interval;

$$\Omega_{sc}^{h} = \frac{2\,\pi}{T_{sc}} = 2\,\pi\,N_{sc}\;; \tag{5.36}$$

T_{sc} is the period of radar antenna scanning; and $N_{sc} = \frac{1}{T_{sc}}$ is the rotation per second. Using Equation (5.34), we can define the power spectral density in the following form:

$$S_{sc}^{en}(\omega) \cong e^{-\pi \cdot \frac{\omega^2}{(2\,\pi\,\Delta F_{sc})^2}}\,, \tag{5.37}$$

where

$$\Delta F_{sc} = \frac{1}{\tau_{sc}} = \frac{\Omega_{sc}^{h}\cos\gamma_0}{\sqrt{2}\,\Delta_h} \tag{5.38}$$

is the effective bandwidth of the power spectral density. For example, at $\Delta_h = 2°$, $\gamma_0 = 0°$, and $\Omega_{sc} = 1\,\frac{\text{radian}}{\text{sec}}$, we obtain that ΔF_{sc} = 20 Hz. Reference to Equation (5.35) and Equation (5.38) shows that when the horizontal-coverage directional diagram is wide, and its angular velocity is low in value, the power spectral density is narrow, i.e., the effective bandwidth is low in value.

In the case of the **sinc**-directional diagram, it follows from Equation (5.15) that[6]

$$R_{sc}^{h}(\tau) = 6\mathcal{F}^{-2}(1 - \mathbf{sinc}\,\mathcal{F})\,, \tag{5.39}$$

where

$$\mathcal{F} = \frac{3\,\pi\,\tau}{\tau_c} \qquad \text{and} \qquad \tau_c = \frac{3\Delta_h}{2\Omega_{sc}^{h}\cos\gamma_0}\,. \tag{5.40}$$

With the low values of τ, the normalized correlation function $R_{sc}^{h}(\tau)$ given by Equation (5.39) is very close to the Gaussian function [see Equation (5.34) and Figure 5.1]. Using the Fourier transform for $R_{sc}^{h}(\tau)$ given by Equation (5.39), the power spectral density $S_{sc}^{en}(\omega)$ can be determined in the following form:

$$S_{sc}^{en}(\omega) \cong \begin{cases} \left[1 - \frac{|\,\omega\,|}{3\,\pi\,\Delta F_{sc}}\right]^2 & \text{at} \quad |\,\omega\,| \le 3\,\pi\,\Delta F_{sc}\,; \\ 0 & \text{at} \quad |\,\omega\,| > 3\,\pi\,\Delta F_{sc}\,, \end{cases} \tag{5.41}$$

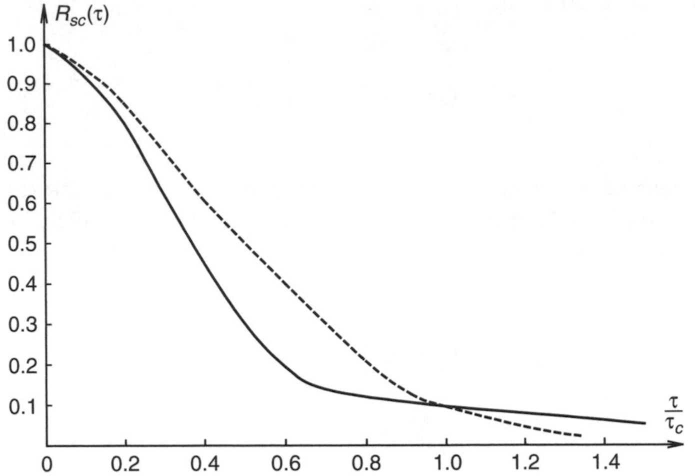

FIGURE 5.1
Normalized correlation functions with line circular scanning for the **sinc**- (solid line) and Gaussian (dotted line) directional diagrams of the radar antenna.

where

$$\Delta F_{sc} = \frac{1}{\tau_c} = \frac{2\,\Omega_{sc}^{h}\cos\gamma_0}{3\,\Delta_h} \tag{5.42}$$

is the effective bandwidth of $S_{sc}^{en}(\omega)$. $S_{sc}^{en}(\omega)$ given by Equation (5.41) is shown in Figure 5.2 by the solid line. It has clear bounds; the extreme frequencies are equal to

$$\omega_{ext} = \pm\, 3\,\pi\,\Delta F_{sc} = \pm\,\frac{2\,\pi\,\Omega_{sc}^{h}\cos\gamma_0}{\Delta_h}\;. \tag{5.43}$$

The formula in Equation (5.43) is to fit a physical representation with respect to the limitation of the power spectral density $S_{sc}^{en}(\omega)$ of fluctuations caused by radar antenna scanning.

Let us consider that with radar antenna scanning, $S_{sc}^{en}(\omega)$ is formed due to the Doppler shift in the frequency of the elementary emitters of the radar antenna, which move with different angular velocities.[7] The frequency ω_{ext} corresponds to emitters that are the most distant from the center of radar antenna area. Any directional diagram of a practicable radar antenna, which is a result of the Fourier transform of the electromagnetic field distribution law given within the limits of the radar antenna area, will have $S_{sc}^{en}(\omega)$ limited by frequencies

$$\omega_{ext} = 4\,\pi\,V_{max}\lambda^{-1}\,, \tag{5.44}$$

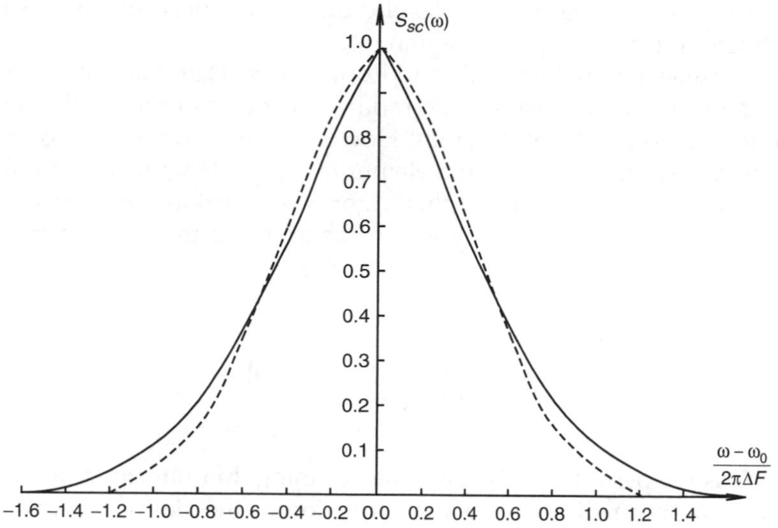

FIGURE 5.2
Power spectral densities with line circular scanning for the **sinc**- (solid line) and Gaussian (dotted line) directional diagrams of the radar antenna.

where V_{max}, the maximum velocity of moving radar antenna, ends with respect to the scanned target. This consequence follows immediately from Equation (5.31) and Equation (5.43).

With the Gaussian directional diagram, the power spectral density $S_{sc}^{en}(\omega)$ is unlimited. The Gaussian directional diagram is impracticable as it would be subject to the Gaussian distribution law of the electromagnetic field within the limits of the radar antenna area, which is impracticable due to the finite linear dimensions of the radar antenna. This effect appears more clearly if the directional diagram is approximated by a rectangle in the following form:

$$S_{sc}^{en}(\omega) \cong \mathbf{sinc}^2\left(\frac{\omega\,\Delta_h}{2\,\Omega_{sc}\cos\gamma_0}\right). \tag{5.45}$$

Thus, consideration of side lobes is very important for the definition of the power spectral density of fluctuations $S_{sc}^{en}(\omega)$ of the target return signal, both with moving radar (see Section 3.2.4) and under radar antenna scanning.[8,9] However, if in the first case we do not take into consideration the side lobes, the power spectral density $S_{sc}^{en}(\omega)$ deprives the remainders existing in practice. If we do the same in the second case, the false remainders of the power spectral density $S_{sc}^{en}(\omega)$ can appear. Under radar antenna scanning, the presence of side lobes in the directional diagram used in practice prompts us to smooth the stochastic process (the target return signal). Comparing Equation (5.35) and Equation (5.40), one can see that with the same effective width of the directional diagram, the correlation intervals and the bandwidths of the power spectral densities, respectively, differ from each other by 6% — as 1.5 and $\sqrt{2}$ — despite the power spectral densities being very different in the peripheral region.

The formulae in Equation (5.37) and Equation (5.41) are the power spectral densities of fluctuations caused by radar antenna scanning in the form of continuous power spectral densities. However, we do not take into account the periodic iteration of the same elementary signals with each revolution of the radar antenna. Because of this, rigorously speaking, we must use the periodic sequence of elementary signals as $w(t)$, not the elementary signal per one revolution of the radar antenna $g_h(\Omega^h_{sc}t)$:

$$w(t) = \sum_{n=0}^{\infty} g_h[\Omega^h_{sc}(t - nT_{sc})] , \tag{5.46}$$

where T_{sc} is the period of radar antenna scanning. In uniform radar antenna scanning, the normalized correlation function of the slow fluctuations can be given in the following form (see Figure 5.3a):

$$R^h_{sc}(\tau, nT_{sc}) = \sum_{n=0}^{\infty} R^h_{sc}(\tau - nT_{sc}) , \tag{5.47}$$

where $R^h_{sc}(\tau)$ is given by Equation (5.7) if $\Delta\beta_0 = \Omega^h_{sc}\tau$, and by Equation (5.34) and Equation (5.39), following from Equation (5.7).

As a result, the power spectral density of the slow fluctuations caused by radar antenna scanning is a regulated function, not continuous, with a distance between lines equal to $\Delta F_{sc} = \frac{1}{T_{sc}} = N_{sc}$ in hertz (see Figure 5.3b)

$$S_{sc}(\omega) = S^{en}_{sc}(\omega) \cdot \sum_{n=0}^{\infty} \delta(\omega - n\Omega^h_{sc}) = \sum_{n=0}^{\infty} S^h_{sc}(n\Omega^h_{sc}) \cdot \delta(\omega - n\Omega^h_{sc}) , \tag{5.48}$$

where $S^{en}_{sc}(\omega)$ is given by Equation (5.37) or Equation (5.41). For instance,

$$S_{sc}(\omega) \approx e^{-\pi \cdot \frac{\omega^2}{(2\pi \Delta F_{sc})^2}} \cdot \sum_{n=0}^{\infty} \delta(\omega - n\Omega^h_{sc}) . \tag{5.49}$$

With the continuous searching signal, the process of one-line scanning generates a stationary periodic stochastic process (the target return signal). The normalized correlation function $R^h_{sc}(\tau, nT_{sc})$ of the slow fluctuations caused by radar scanning, which is given by Equation (5.47), and the power spectral density $S^{en}_{sc}(\omega)$ given by Equation (5.48) result in an exhaustive representation regarding spectral properties of the target return signal.

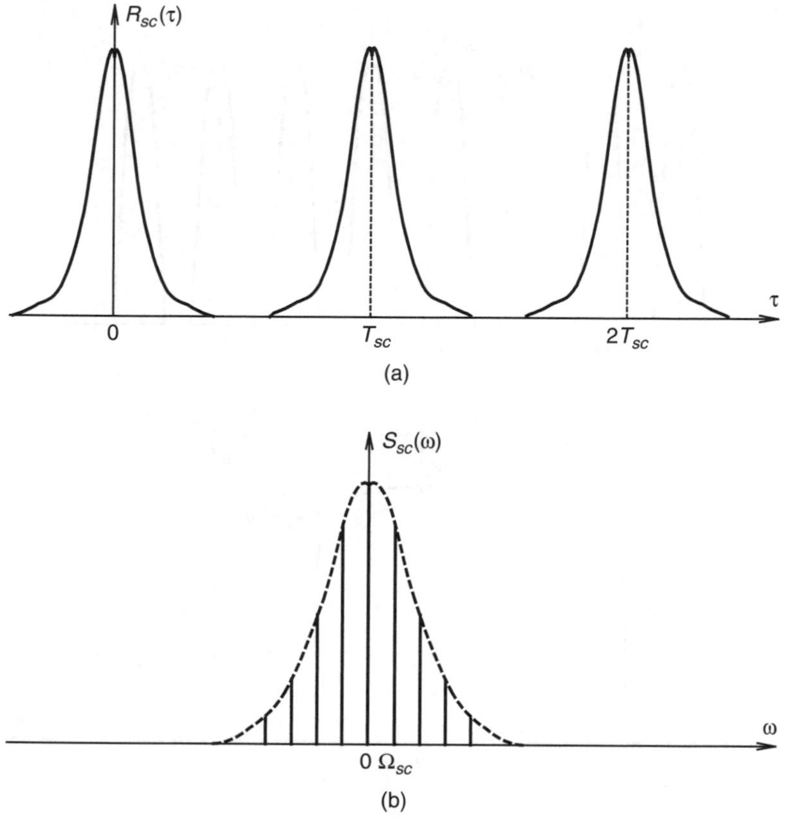

FIGURE 5.3
The normalized correlation function (a) and power spectral density (b) with the continuous searching signal — line scanning by the radar antenna.

With the pulsed searching signal, as follows from Equation (5.1), $R^{h}_{sc}(\tau, nT_{sc})$, which is determined by Equation (5.47), must be multiplied with $R_p(t, t + \tau)$ of fluctuations in the radar range, which is given by Equation (5.2). The normalized correlation function $R_p(t, t + \tau)$ and the corresponding power spectral density are shown in Figure 5.4. The normalized correlation function of fluctuations caused by radar antenna line scanning, which is obtained as the product of Equation (5.2) and Equation (5.47), and the corresponding power spectral density defined by the convolution of the power spectral densities given by Equation (3.15) and Equation (5.48), are shown in Figure 5.5 under the condition $m = \frac{T_{sc}}{T_p}$, where m is integer. Thus, the stochastic target return signal is rigorously periodic, but not stationary, because the normalized correlation function of the rapid fluctuations can be varied within the limits of the pulsed searching signal duration T_p, as discussed in Section 4.3, and the period of the normalized correlation function is equal to $T_{sc} = \tau$.

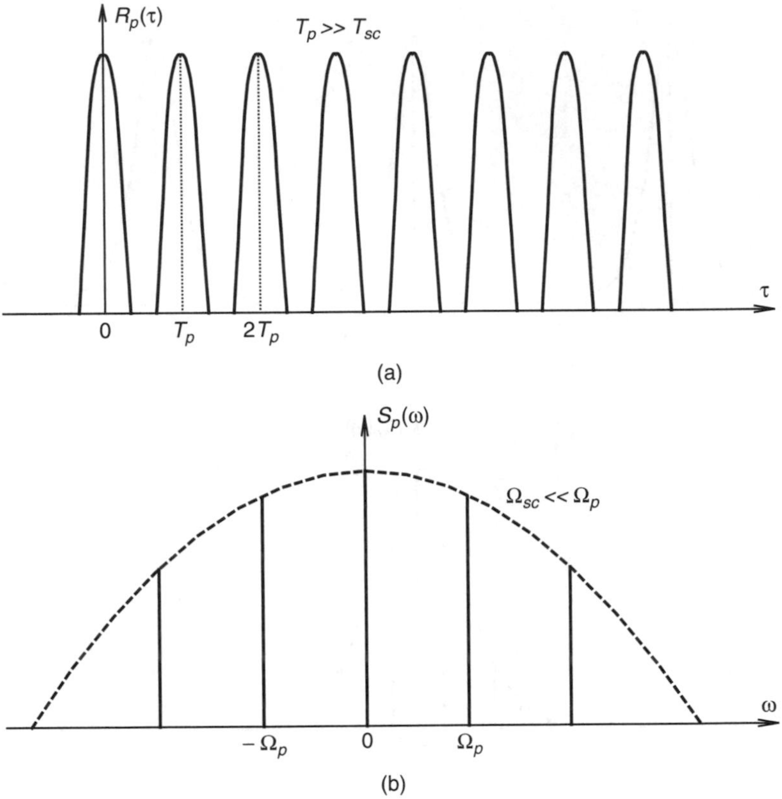

FIGURE 5.4
The normalized correlation function (a) and power spectral density (b) with the continuous searching signal — no line scanning by the radar antenna.

If the ratio $\frac{m_1}{m_2} = \frac{T_{sc}}{T_p}$ is not an integer and is a rational number, then the normalized correlation function of rapid fluctuations is not rigorously periodic, and its values will be modulated with the period equal to $m_2 = T_{sc}$. Thus, the power spectral density will have discrete components with frequencies multiple to $\frac{\Omega_{sc}}{m_2}$. The degree of modulation and values of these components are high, and the number of pulsed signals per beam is less. If the ratio $\frac{T_{sc}}{T_p}$ is a rational number, there are continuous remainders near the main harmonics of the power spectral density of the fluctuations.

As a rule, the process of radar antenna scanning is slow, so that the distance between spectral lines of the power spectral densities given by Equation (5.37) and Equation (5.41) is proportional from 10–800 Hz. Because of this, in many cases, we can neglect the regulated character of the power spectral density and assume that one is continuous, making simpler the

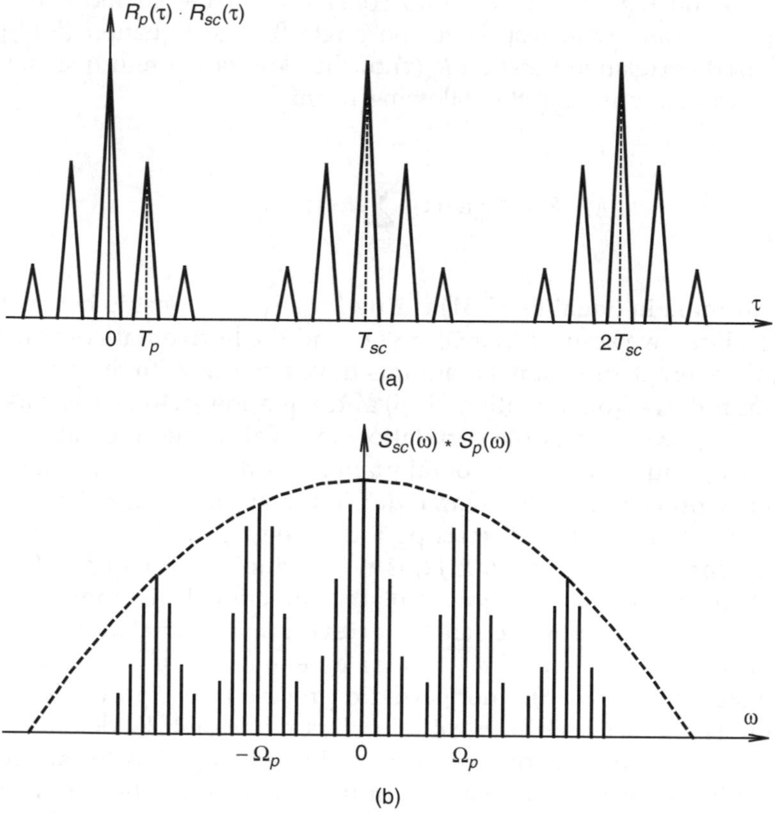

FIGURE 5.5
The normalized correlation function (a) and power spectral density (b) with the pulsed searching signal — line scanning by the radar antenna.

determination and computer calculation without high error. A different situation arises when the power spectral density is estimated with azimuth-tracking radar antenna scanning. These antennas function at about 100 r/sec.[10] In this case, we cannot neglect the regulated character of the power spectral density of the fluctuations (see Section 5.3).

5.2.2 Multiple-Line Circular Scanning

Let us assume that the radar antenna performs more complex (multiple-line or "spiral") scanning of space. For the case considered here, it can be written that

$$\beta_0 = \Omega_{sc}^h \cdot t\,; \quad \gamma_0 = \Omega_{sc}^v \cdot t\,; \quad \Delta\beta_0 = \Omega_{sc}^h \cdot \tau\,; \quad \text{and} \quad \Delta\gamma_0 = \Omega_{sc}^v \cdot \tau\,. \quad (5.50)$$

If the variables φ and ψ are separable in the function $g(\varphi, \psi)$, Equation (5.6)–Equation (5.8) are true. Substituting Equation (5.50) in Equation

(5.6)–Equation (5.8) and taking into consideration the character of radar antenna scanning with respect to the angle β_0 [see Equation (5.47)], the normalized correlation function $R_{sc}(\tau)$ of the slow target return signal fluctuations can be written in the following form:

$$R_{sc}(\tau) = R_{sc}^{v}(\tau) \cdot \sum_{n=0}^{\infty} R_{sc}^{h}(\tau - nT_{sc}^{h}) . \tag{5.51}$$

The formula in Equation (5.51) is true for any relationship between Ω_{sc}^{h} and Ω_{sc}^{v}. First, we assume that $\Omega_{sc}^{h} >> \Omega_{sc}^{v}$ and the horizontal-coverage and vertical-coverage directional diagrams have a width with the same order. Then, as follows from Equation (5.35) and Equation (5.40), we can assume that $\tau_c^v >> \tau_c^h$, where τ_c^h is the correlation interval of the fluctuations with the moving radar in the horizontal plane; τ_c^v is the correlation interval of the fluctuations with the moving radar in the vertical plane. The normalized correlation function $R_{sc}^{h}(\tau)$ is periodic (see Figure 5.7a) and the normalized correlation function $R_{sc}^{v}(\tau)$ is not periodic (see Figure 5.6a) with respect to the variable τ. Because of the directional diagram shift with respect to the aspect angle, an "interrevolution" correlation is broken from revolution to revolution of radar antenna scanning, and the target return signal becomes the nonperiodic process (see Figure 5.8a). If the value of Ω_{sc}^{v} is increased, the value of τ_c^v is decreased. Under the condition $\tau_c^v < \tau_c^h$, the "interrevolution" correlation is completely broken during one revolution of radar antenna scanning.[11] Therefore, the sum in Equation (5.51) has only a single term under the condition $n = 0$. The power spectral density of the fluctuations contains only a single domain near the zero frequency.

With the continuous searching signal, the previously discussed normalized correlation functions of the fluctuations and the corresponding power spectral densities (see Figure 5.6b–Figure 5.8b) give us full information about the properties of the frequency of the target return signal. With the pulsed searching signal, it is necessary to multiply the normalized correlation function $R_{sc}(\tau)$ given by Equation (5.51) with the normalized correlation function $R_p(t, t + \tau)$ of fluctuations in the radar range given by Equation (5.2) (see Figure 5.9a). In this case, the resulting power spectral density of the fluctuations is the convolution between the power spectral densities $S_{sc}(\omega)$ and $S_p(\omega)$ shown in Figure 5.9b. This is similar to the power spectral density shown in Figure 5.7b, but instead of discrete δ components with the frequency $\frac{1}{T_{sc}^{h}}$,this power spectral density is formed by narrow partial domains with the bandwidth equal approximately to $\frac{1}{\tau_c^v}$ (see Figure 5.10).

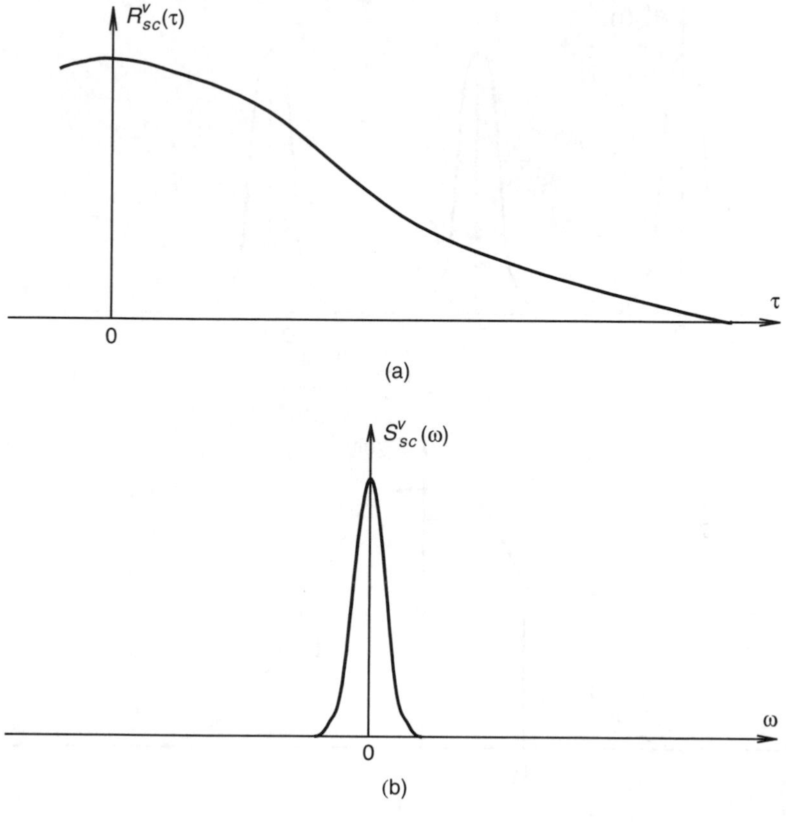

FIGURE 5.6
The normalized correlation function (a) and power spectral density (b) with the continuous searching signal — multiple-line ("spiral") scanning by the radar antenna in the aspect-angle plane.

5.2.3 Line Segment Scanning

Let us consider the case when the radar antenna, the directional diagram axis of which is under the constant angle γ_0, performs the line scanning in a segment limited by the angles $\pm\beta_m$. Let us assume that the radar antenna rotates according to the law

$$\beta_0 = \beta_m \sin \Omega_a t \, , \tag{5.52}$$

where $\Omega_a = \frac{2\pi}{T_a}$ and T_a is the period of radar antenna hunting. Using Equation (5.17), we can find that

$$\Delta\beta_0 = 2\beta_m \sin 0.5\Omega_a \tau \cos \Omega_a t \quad \text{and} \quad \Delta\gamma_0 = 0 \tag{5.53}$$

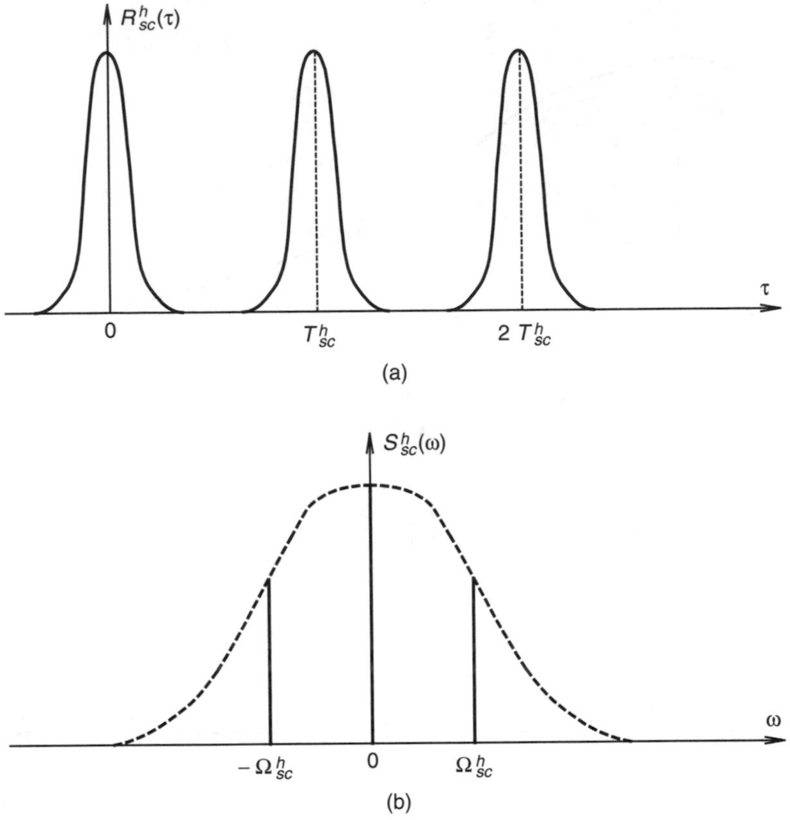

FIGURE 5.7
The normalized correlation function (a) and power spectral density (b) with the continuous searching signal — multiple-line ("spiral") scanning by the radar antenna in the azimuth plane.

or, using Equation (5.52) and excluding the time in an explicit form, we can write

$$\Delta\beta_0 = 2\sqrt{\beta_m^2 - \beta_0^2(t)}\;\sin 0.5\Omega_a\tau\,. \tag{5.54}$$

Reference to Equation (5.54) shows that the observed target return signal is a nonstationary periodic stochastic process because the normalized correlation function of the target return signal fluctuations used in Equation (5.53) will be a periodic function of time unlike the normalized correlation function with uniform circular radar antenna scanning, when the target return signal is also periodic but stationary because the normalized correlation function is independent of time. This is clearly under consideration, for instance, for the Gaussian directional diagram and substitution of Equation (5.53) and Equation (5.54) in Equation (5.14). Then, we can write the normalized correlation function $R_{sc}(t, \tau)$ of the fluctuations in the following form:

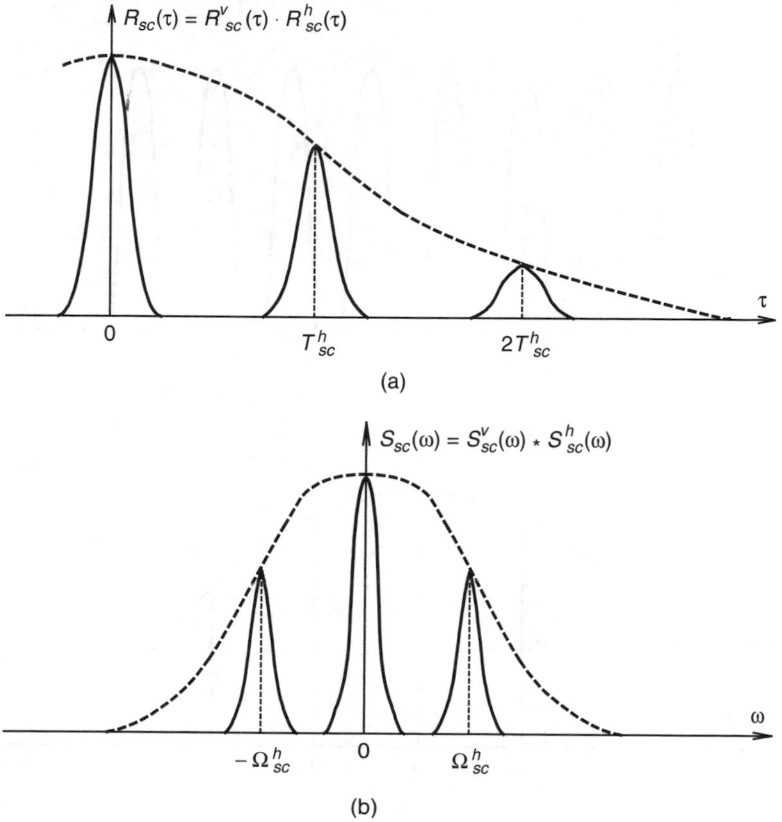

FIGURE 5.8
The normalized correlation function (a) and power spectral density (b) with the continuous searching signal — multiple-line ("spiral") scanning by the radar antenna in the aspect-angle and azimuth planes.

$$R_{sc}(t,\tau) = e^{-2\pi \frac{\beta_m^2}{\Delta_h^{(2)}} \cos^2\gamma_0 \sin^2 0.5\Omega_a\tau \cos^2\Omega_a t} = e^{-2\pi \frac{\beta_m^2 - \beta_0^2(t)}{\Delta_h^{(2)}} \cos^2\gamma_0 \sin^2 0.5\Omega_a\tau} . \tag{5.55}$$

$R_{sc}(t, \tau)$ of fluctuations caused by radar antenna scanning, which is given by Equation (5.55), is a periodic function with respect to the parameter τ with fixed values of t or $\beta_0(t)$. In this case, t or $\beta_0(t)$ is the parameter, and the function of t is also periodic. $R_{sc}(t, \tau)$, which is determined by Equation (5.55), is shown in Figure 5.11 with various values of the parameter L that can be given by the following form:

$$L = \frac{\beta_m^2 - \beta_0^2(t)}{\Delta_h^{(2)}} \cos^2\gamma_0 . \tag{5.56}$$

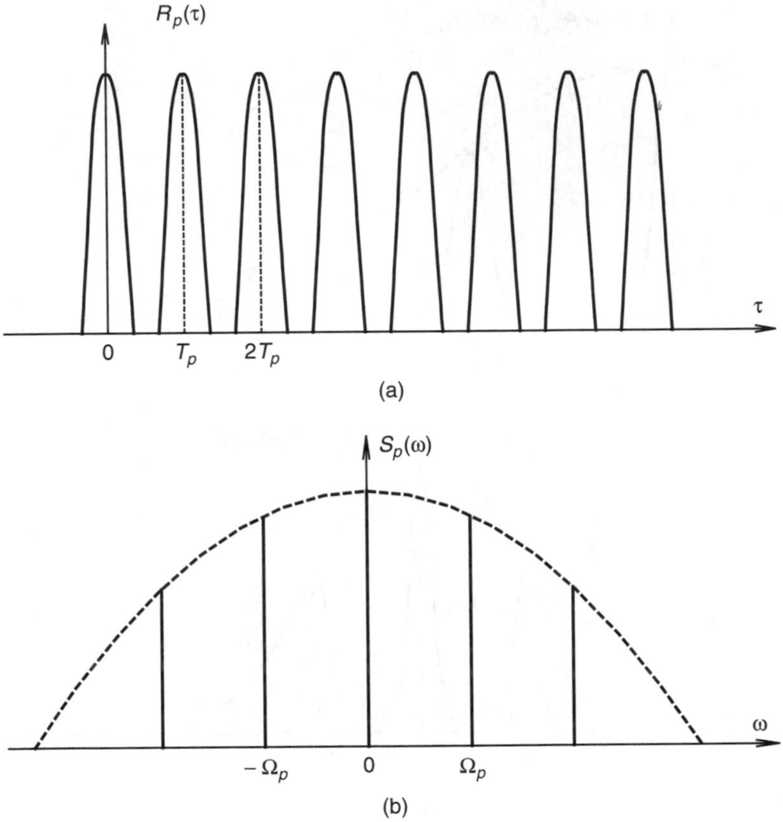

FIGURE 5.9
The normalized correlation function (a) and power spectral density (b) with the pulsed searching signal — no multiple-line ("spiral") scanning by the radar antenna.

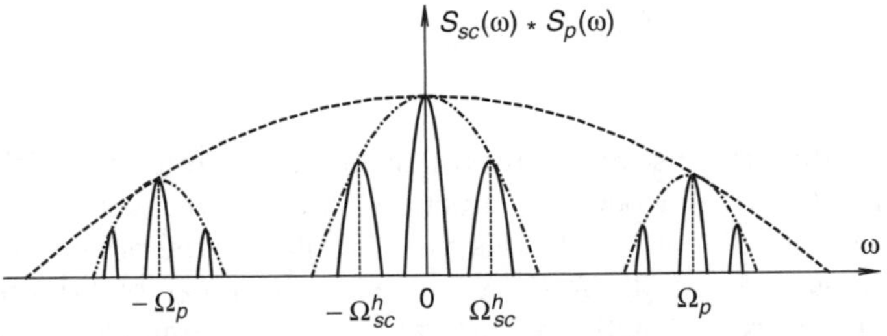

FIGURE 5.10
The power spectral density with the pulsed searching signal — multiple-line ("spiral") scanning.

The formula in Equation (5.55) is true for any relationships between the width of the scanned segment $2\beta_m$ and the horizontal-coverage directional diagram width, even if the condition $\beta_m < \Delta_h$ is satisfied. In particular, if the

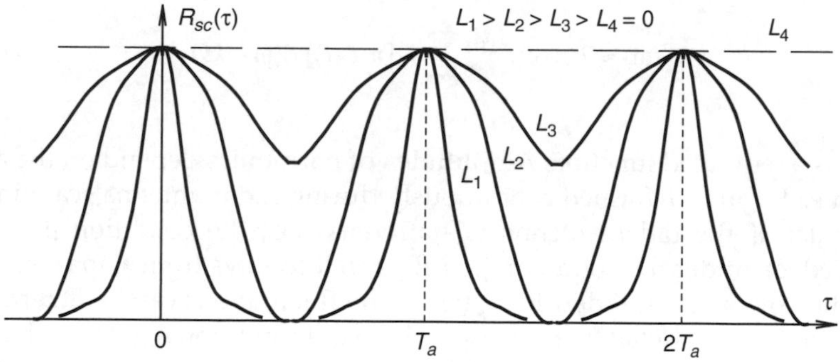

FIGURE 5.11
The correlation function — line segment scanning by the radar antenna.

radar antenna is stationary, i.e., the condition $\beta_m = 0$ is satisfied, the clear results follow from Equation (5.55): $R_{sc}(t, \tau) \equiv 1$. However, in the subsequent discussion, we assume that the condition $\beta_m >> \Delta_h$ is satisfied, as it is carried out in practice. Then, integrating Equation (5.55) with respect to the parameter τ within the limits of the scanning period T_a of the radar antenna, i.e., when the condition $T_a >> \tau_c$ is satisfied, we can determine the correlation interval of the observed target return signal in the following form:

$$\tau_c = T_a \cdot e^{-\pi L(t) I_0[\pi L(t)]} , \tag{5.57}$$

which is a function of the parameter $\beta_0(t)$. The bandwidth $\Delta F_{sc} = \frac{1}{\tau_c}$ of the instantaneous power spectral density of the fluctuations caused by radar antenna scanning is also a function of the moving position of radar antenna, i.e., $\beta_0(t)$.

Reference to Equation (5.57) shows that under the conditions $\beta_0 << \beta_m$, and $\beta_m >> \Delta_h$, the correlation interval can be determined in the following form:

$$\tau_c = \frac{T_a}{\pi\sqrt{2L(t)}} = \frac{1}{\Omega_a} \cdot \sqrt{\frac{2}{L(t)}} . \tag{5.58}$$

Defining the parameter β_0 as a function of t and using Equation (5.52), we can easily show that the correlation interval τ_c is equal to the time during which the directional diagram axis is moved in the angle equal to the effective width of the squared horizontal-coverage directional diagram. The physical meaning of this result is obvious.

Using the Fourier-series expansion for the periodic normalized correlation function $R_{sc}(t, \tau)$ of the fluctuations, which is determined by Equation (5.55), with respect to the modified Bessel function given by Equation (1.15), we can obtain the instantaneous power spectral density $S_{sc}(\omega)$ in the following form:

$$S_{sc}(\omega) \approx 2\,\pi \cdot e^{-\pi \cdot L(t)} \sum_{k=0}^{\infty} I_k[\pi\, L(t)]\; \delta(\omega - k\Omega_a)\;. \tag{5.59}$$

$S_{sc}(\omega)$ is a regulated function. Amplitudes of harmonics depend on the angle $\beta_0(t)$, i.e., $S_{sc}(\omega)$ is deformed continuously during radar antenna scanning. In particular, if the radar antenna is stationary, i.e., the condition $\beta_m = 0$ is satisfied, or under the condition $\beta_0 = \beta_m \neq 0$, it follows from Equation (5.59) that the power spectral density $S_{sc}(\omega)$ of the fluctuations can be determined in the form $S_{sc}(\omega) \approx \delta(\omega)$ that corresponds to the condition $R_{sc}(\tau) \equiv 1$ obtained from the conditions in Equation (5.55).

Because the normalized correlation function $R_{sc}(\tau)$ is different from zero if the conditions $|\tau| \leq \tau_c$ and $\tau_c << T_a$ are satisfied, we can write

$$\sin 0.5\Omega_a \tau \cong 0.5\Omega_a \tau \tag{5.60}$$

and, using Equation (5.54), we obtain

$$\Delta\beta_0 \cong \beta_m \Omega_a \tau \cos \Omega_a t = \Omega_a \sqrt{\beta_m^2 - \beta_0^2(t)}\;\tau\;, \tag{5.61}$$

where $\beta_m \Omega_a \cos \Omega_a t$ is the instantaneous velocity of radar antenna scanning. Exchanging the function $\Delta\beta_0(t)$ given by Equation (5.53) within the limits of the interval that is of interest to us by the segment of the tangent given by Equation (5.61) implies that we can neglect the variations in the radar antenna angular velocity within the limits of the correlation interval if the condition $\Delta_a << \beta_m$ is satisfied. In this case, the target return signal is the quasistationary process.

Substituting Equation (5.61) in Equation (5.14) or Equation (5.15) under the condition $\Delta\gamma_0 = 0$, we obtain the formulae coinciding with Equation (5.34) or Equation (5.39), in which

$$\tau_c = \frac{2\,k_a}{\Omega_a \sqrt{L(t)}}\;, \tag{5.62}$$

where $k_a = \frac{1}{\sqrt{2}}$ in Equation (5.34) or $k_a = \frac{2}{3}$ in Equation (5.39). Naturally, Equation (5.62) takes the same form as Equation (5.58).

Hereafter, we can use all results discussed in Section 5.2.1 for the case of one-line radar antenna scanning and, in particular, we can define the instantaneous power spectral densities of the fluctuations, which coincide with the formulae in Equation (5.37), Equation (5.41), Equation (5.48), and Equation (5.49), but the effective bandwidth of which is equal to $\Delta F_{sc} = \tau_c^{-1}$, where the correlation interval τ_c given by Equation (5.62) is a function of the angle $\beta_0(t)$. For example, in the case of the Gaussian directional diagram, it follows from

Equation (5.48) that the power spectral density $S_{sc}(\omega)$ of the fluctuations can be written in the following form:

$$S_{sc}(\omega) \approx e^{-\frac{\omega^2}{2\pi\Omega_a^2 L(t)}} \cdot \sum_{k=0}^{\infty} \delta(\omega - k\Omega_a) = \sum_{k=0}^{\infty} \delta(\omega - k\Omega_a) \cdot e^{-\frac{k^2}{2\pi L(t)}}. \tag{5.63}$$

We can prove without any problem that under the condition $L(t) >> 1$, the power spectral densities of the fluctuations given by Equation (5.59) and Equation (5.63) are close in shape. Using the discussed technique in this section, we can investigate other laws of radar antenna scanning that are different from the law given by Equation (5.52).

5.2.4 Line Circular Scanning with Various Directional Diagrams under Transmitting and Receiving Conditions

Consider the case when the directional diagrams are different under transmitting and receiving conditions. Then, the function $g(\varphi, \psi)$ can be replaced with the product of the directional diagrams given by Equation (2.73). First, consider a situation when one of the directional diagrams, for instance, under the transmitting condition, is nondirected, i.e., the condition $g_u^t(\varphi, \psi) \equiv 1$ is satisfied. In this case, the normalized correlation function $R_{sc}^h(\Delta\beta_0)$ of the target return signal fluctuations can be determined in the following form:

$$R_{sc}^h(\Delta\beta_0) = N \cdot \int g_{uh}^r(\varphi) \cdot g_{uh}^r(\varphi + \Delta\beta_0 \cos\gamma_0)\, d\varphi \tag{5.64}$$

instead of Equation (5.7). Assume that the distribution law of electromagnetic field within the limits of the radar antenna area is uniform, so that

$$g_{uh}^r(\varphi) = \mathbf{sinc}\left(\tfrac{\pi\varphi}{\Delta_r}\right), \tag{5.65}$$

where Δ_r is the horizontal-coverage directional diagram width under the receiving condition. In the case of the **sinc**-directional diagram, the effective horizontal-coverage directional diagram width is the same both in power and in voltage despite the width of the main lobe being different.

Taking into consideration the condition $\Delta\beta_0 = \Omega_{sc}\tau$, the normalized correlation function $R_{sc}^h(\tau)$ can be determined in the following form:

$$R_{sc}^h(\tau) = \mathbf{sinc}\left(\tfrac{\pi\Omega_{sc}\cos\gamma_0\tau}{\Delta_r}\right). \tag{5.66}$$

The power spectral density of fluctuations corresponding to $R^h_{sc}(\tau)$, given by Equation (5.66), is the square waveform power spectral density within the limits of the interval $\left[-\frac{\pi\,\Omega_{sc}\cos\gamma_0}{\Delta_r},\ \frac{\pi\,\Omega_{sc}\cos\gamma_0}{\Delta_r}\right]$. The power spectral density bandwidth of the fluctuations is determined in the following form:

$$\Delta F = \frac{\Omega_{sc}}{\Delta_r} \cdot \cos\gamma_0 \,. \tag{5.67}$$

It is easy to define how the form of the directional diagram acts on the shape of the power spectral density $S_{sc}(\omega)$ under the transmitting condition. If the distribution law of electromagnetic field within the limits of the transmitting radar antenna area is uniform, $S_{sc}(\omega)$ has the shape of a trapezium (see Figure 5.12). The width of the bottom base is given by Equation (5.67) and the width of the top base is determined in the following form:

$$\Delta F' = \left[\frac{\Omega_{sc}}{\Delta_r} - \frac{\Omega_{sc}}{\Delta_t}\right] \cdot \cos\gamma_0 \,, \tag{5.68}$$

where Δ_t is the width of the directional diagram by power under the transmitting condition.

If the condition $\Delta_t >> \Delta_r$ is satisfied, we can neglect the stimulus of the transmitting radar antenna area and consider the previous case. If the condition $\Delta_t << \Delta_r$ is true and the directional diagrams are the same under the transmitting and receiving conditions according to Equation (5.63), we can assume that $\Delta F' = 0$ and the trapezoidal power spectral density $S_{sc}(\omega)$ is

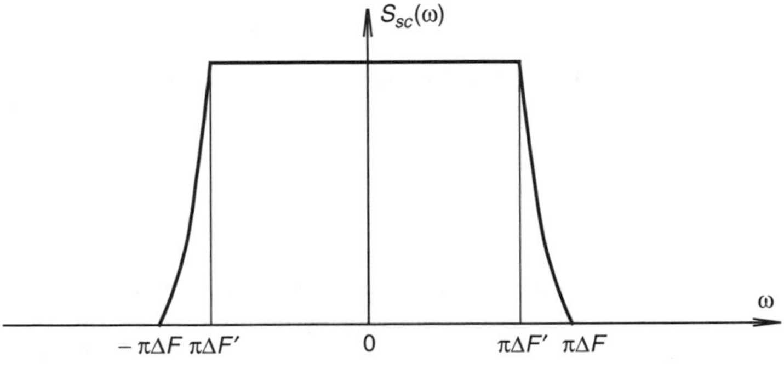

FIGURE 5.12
The power spectral density with line scanning by the radar antenna for the different transmitting and receiving directional diagrams of the radar antenna.

converted into the triangular one [see Equation (5.41)]. If under transmitting and receiving conditions the directional diagrams are defined by the Gaussian law and have the different effective widths by power Δ_t and Δ_r, respectively, then the normalized correlation function $R_{sc}(\tau)$ and the power spectral density $S_{sc}(\omega)$ are determined by Equation (5.34) and Equation (5.37), respectively, in which the parameter Δ_r must be replaced with the parameter $\frac{\sqrt{2}\,\Delta_r\Delta_t}{\sqrt{\Delta_r^{(2)}+\Delta_t^{(2)}}}$.

5.3 Conical Scanning

Consider a classical one-beam one-channel navigational system with the conical radar antenna scanning of open and hidden forms from a large number of various navigational systems with the target tracking by direction.[13,14] For navigational systems with conical radar antenna scanning of the open type, the directional diagrams under transmitting and receiving conditions are the same and matched, and move in space continuously. In this case, an essential disadvantage is the transmitting radar antenna scanning because information regarding frequency and phase of conical radar antenna scanning is known and can be used to generate deliberate interference. The radar with a hidden frequency of antenna scanning is used with the purpose of hiding the law under which radar antenna scanning is only for the receiving directional diagram. If the dimensions of the transmitting and receiving radar antennas are comparable, the radar with a hidden frequency of antenna scanning has a lesser slope of bearing characteristic and requires formation of two independent directional diagrams — stationary for the transmitting radar antenna and scanning for the receiving radar antenna. As a rule, navigational systems with target tracking use the pulsed searching signals that solve the problem of definition of distance to the target. Under the stationary condition, there is range tracking by navigational systems, and only the interperiod fluctuations exist at the point of input of the receiver or detector.

5.3.1 Three-Dimensional (Space) Target Tracking

Let the radar antenna scanning be the open conical scanning with the angle θ_{sc} and frequency Ω_{sc} (see Figure 5.13). Then we can write

$$\gamma_0 = \gamma_{eq} + \theta_{sc}\sin\alpha_0 \qquad \text{and} \qquad \tilde{\beta}_0 = \beta_0\cos\gamma_0 = \theta_{sc}\cos\alpha_0\,, \tag{5.69}$$

where γ_{eq} is the aspect angle of equisignal direction. Under this law of radar antenna scanning, based on Equation (5.17), we can write

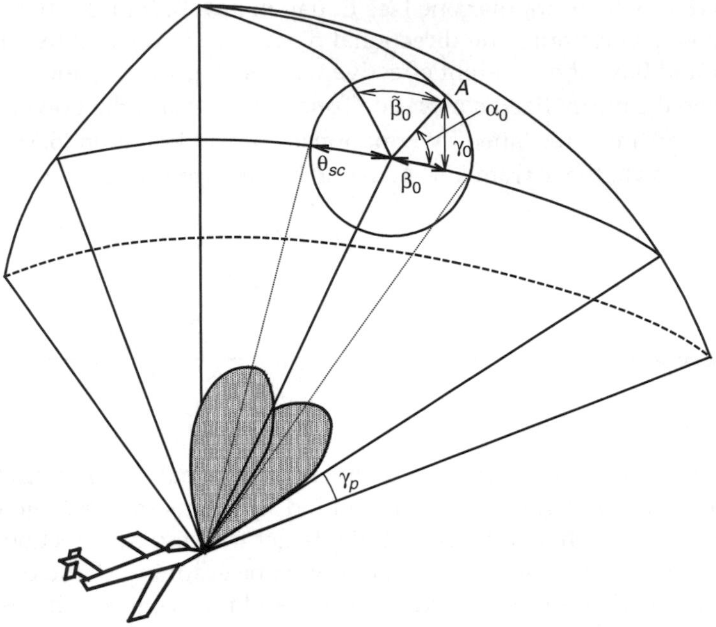

FIGURE 5.13
The coordinate system with conical scanning by the radar antenna.

$$\Delta\gamma_0 = 2\theta_{sc}\sin 0.5\Delta\alpha_0 \cos\alpha_0 \quad \text{and} \quad \Delta\tilde{\beta}_0 = \Delta\beta_0 \cos\gamma_0 = 2\theta_{sc}\sin 0.5\Delta\alpha_0 \sin\alpha_0 \,, \tag{5.70}$$

where

$$\alpha_0 = \Omega_{sc}\cdot t \qquad \text{and} \qquad \Delta\alpha_0 = \Omega_{sc}\cdot\tau \,. \tag{5.71}$$

Substituting Equation (5.70) and Equation (5.71) in Equation (5.14) and assuming that the directional diagram is defined by the two-dimensional Gaussian function, we can write the normalized correlation function $R_{sc}[\tau, \alpha_0(t)]$ of the target return signal fluctuations in the following form:

$$R_{sc}[\tau,\alpha_0(t)] = e^{-2\pi\theta_{sc}^2\cdot\left(\frac{\sin^2\Omega_{sc}t}{\Delta_h^{(2)}} + \frac{\cos^2\Omega_{sc}t}{\Delta_v^{(2)}}\right)\sin^2 0.5\Omega_{sc}\tau} \,. \tag{5.72}$$

Under the condition $\Delta_h \neq \Delta_v$, $R_{sc}[\tau, \alpha_0(t)]$ given by Equation (5.72) depends on the parameter $\alpha_0(t)$. The target return signal is the nonstationary stochastic process because the directional diagram has no axial symmetry and takes different positions with various values of α_0 with respect to the direction of the moving radar producing a periodic change in the power spectral density bandwidth of the fluctuations. The formula in Equation (5.72) is similar to that in Equation (5.55) and allows us to define the instantaneous power spectral density of the fluctuations. However, unlike in Section 5.2.3, we are here interested in the power spectral density that is averaged during the period of radar antenna scanning because the target return signal is filtered, and the frequency of radar antenna scanning is defined.

After averaging with respect to the parameter $\alpha_0(t)$, we can define the averaged normalized correlation function $\overline{R_{sc}[\tau,\alpha_0(t)]}$ in the following form:

$$\overline{R_{sc}[\tau,\alpha_0(t)]} = I_0[\pi\,\theta_{sc}^2 b \sin^2 0.5\Omega_{sc}\tau]\cdot e^{-\pi\,\theta_{sc}^2 a \sin^2 0.5\Omega_{sc}\tau}, \tag{5.73}$$

where

$$a = \Delta_h^{-(2)} + \Delta_v^{-(2)} \quad \text{and} \quad b = \Delta_h^{-(2)} - \Delta_v^{-(2)} . \tag{5.74}$$

Rigorously speaking, it is necessary to average the correlation function of the fluctuations, not the normalized correlation function. However, in this case, it is of no importance because the power of the target return signal under the scanning of the uniform three-dimensional (space) target is independent of the parameter α_0.

If the directional diagram has axial symmetry, i.e., the condition $\Delta_h = \Delta_v = \Delta_a$ is satisfied, $R_{sc}(t, \tau)$, based on Equation (5.72), can be written in the following form:

$$R_{sc}(t,\tau) = R_{sc}(\tau) = e^{-\frac{2\pi\,\theta_{sc}^2}{\Delta_a^{(2)}}\sin^2 0.5\Omega_{sc}\tau} . \tag{5.75}$$

In this case, the target return signal is the stationary stochastic process. The normalized correlation functions $\overline{R_{sc}(t,\tau)}$ and $R_{sc}(t,\tau)$ given by Equation (5.73) and Equation (5.75), respectively, are periodic functions with respect to the variable τ that indicates the periodic character of the averaged target return signal. For this reason, the power spectral density of the fluctuations of the averaged target return signal is the regulated function with distance between harmonics equal to Ω_{sc}.

However, the Fourier transform of the averaged normalized correlation function $\overline{R_{sc}(t,\tau)}$ of the fluctuations, which is given by Equation (5.73), is not determined exactly. We can determine it with some accuracy if we employ a series expansion for the Bessel function $I_0(x)$ used in Equation (5.73), assuming that the argument is very low in value. Because the value of the parameter $\pi\,\theta_{sc}^2 b$ is low, even in the case of the high axial asymmetric directional diagram, it is sufficient to be limited by one or two terms of the series expansion. For instance,

$$S_{sc}(\omega) \approx \sum_{k=0}^{\infty} N_k \cdot \delta(\omega - k\Omega_{sc}), \qquad (5.76)$$

where

$$N_k = I_k(x) - (\tfrac{y}{4})^2[I_{k-1}(x) + I_{k+1}(x)] + (\tfrac{y}{8})^2[I_{k-2}(x) + I_{k+2}(x)]\,; \qquad (5.77)$$

$$x = 0.5\pi\,\theta_{sc}^2 a \qquad \text{and} \qquad y = 0.5\pi\,\theta_{sc}^2 b\,. \qquad (5.78)$$

As a rule,

$$\frac{\theta_{sc}}{\Delta_a} \in \left[\left(\sqrt{4\pi}\right)^{-1}, \left(\sqrt{2\pi}\right)^{-1}\right] \qquad \text{and} \qquad \Delta_h \cong \Delta_v\,. \qquad (5.79)$$

If the directional diagram is axial symmetric, i.e., the conditions

$$a = 2\Delta_v^{-(2)} \qquad \text{and} \qquad b = 0 \qquad (5.80)$$

are satisfied, then the power spectral density $S_{sc}(\omega)$ of the fluctuations can be written in the following form:

$$S_{sc}(\omega) \approx \sum_{k=0}^{\infty} I_k\left[\tfrac{\pi\,\theta_{sc}^2}{\Delta_a^{(2)}}\right] \cdot \delta(\omega - k\Omega_{sc})\,. \qquad (5.81)$$

Under the condition $\frac{\theta_{sc}}{\Delta_a} = 0.3$ that is characteristic for a conical scanning of the radar antenna, the power of the first harmonic of the power spectral density of the fluctuations is for about 13% from the zero harmonic, and the power of the second harmonic is approximately 1% from the zero harmonic. For this case, the power spectral density of the fluctuations caused by radar antenna scanning consists approximately of three components — a main component and two side components. Therefore, we can neglect the power of the other components of the power spectral density of fluctuations of the target return signal. Axial asymmetry of the directional diagram leads to some increase in the power of harmonics, but this increase becomes significant only if the axial asymmetry of the directional diagram is high.

Now consider the navigational system with the hidden scanning of the radar antenna. Let us assume that the directional diagram of the transmitting radar antenna is stationary and the directional diagram of the receiving radar antenna performs a conical scanning with the frequency Ω_{sc}. For this case, the two-dimensional directional diagram $g(\varphi, \psi)$ can be expressed as the product of the directional diagram $g_u^t(\varphi, \psi)$ of the transmitting radar antenna

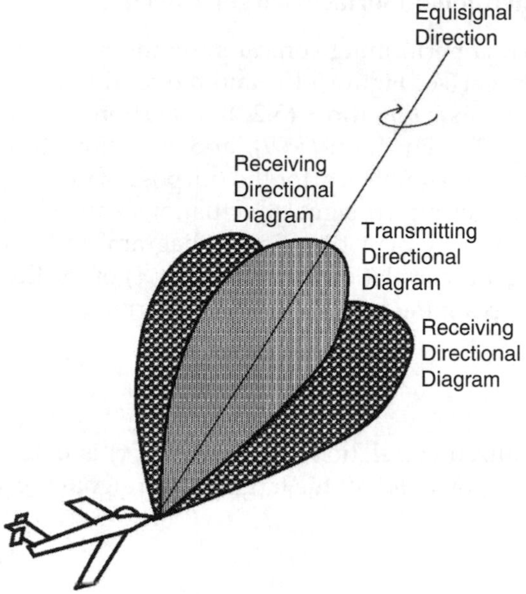

FIGURE 5.14
Radar antenna directional diagrams with hidden conical scanning.

and the directional diagram $g_u^r(\varphi, \psi)$ of the receiving radar antenna shifted by the angles $\tilde{\beta}_0$ and γ_0 [see Equation (5.69) and Figure 5.14]

$$g(\varphi, \psi) = g_u^t(\varphi, \psi) \cdot g_u^r(\varphi + \tilde{\beta}_0, \psi + \gamma_0) , \tag{5.82}$$

where the angles φ and ψ are reckoned from the directional diagram axis of the transmitting radar antenna. Let us substitute Equation (5.82) in Equation (5.3), and take into consideration the fact that the transmitting radar antenna directional diagram is stationary. Then, the normalized correlation function $R_{sc}(\Delta\beta_0, \Delta\gamma_0)$ of the fluctuations takes the following form:

$$\begin{aligned} R_{sc}(\Delta\beta_0, \Delta\gamma_0) = N \cdot \iint & [g_u^t(\varphi, \psi)]^2 \cdot g_u^r(\varphi + \tilde{\beta}_0, \psi + \gamma_0) \\ & \times g_u^r(\varphi + \tilde{\beta}_0 + \Delta\beta_0 \cos\gamma_0, \ \psi + \gamma_0 + \Delta\gamma_0)\, d\varphi\, d\psi \end{aligned} \quad . \tag{5.83}$$

The formula in Equation (5.83) is true for any arbitrary shape of the transmitting and receiving radar antenna directional diagrams. If the one-dimensional directional diagram is defined by the axial symmetric Gaussian functions, then the finite result can be obtained immediately by replacing the parameter θ_{sc} in Equation (5.72), Equation (5.73), Equation (5.75), Equation (5.76), and Equation (5.81) with the parameter $0.5\theta_{sc}$.[12]

5.3.2 Two-Dimensional (Surface) Target Tracking

If the radar antenna performing conical scanning searches the two-dimensional (surface) target (see Figure 5.15), and the condition $\Delta_p << \Delta_v$ is satisfied, it is necessary to use Equation (5.20), Equation (5.21), and Equation (5.23)–Equation (5.27), Equation (5.69), and Equation (5.70), replacing the parameter γ_0 with the parameter γ_* for the purpose of defining the correlation function of the target return signal fluctuations caused by radar antenna scanning. With the Gaussian directional diagram and under open radar antenna scanning, the correlation function $R_{sc}^{en}(t,\tau)$ of the fluctuations caused by radar antenna scanning has the following form:

$$R_{sc}^{en}(t,\tau) = p(t)\cdot R_{sc}^{en}(t,\tau)\ , \tag{5.84}$$

where the normalized correlation function $R_{sc}^{en}(t,\tau)$ is determined by Equation (5.72), and the power $p(t)$ of the target return signal is given by Equation (2.138), in which

$$g_v(\gamma_0-\gamma_*) = e^{-\pi\frac{(\gamma_p+\theta_{sc}\cos\Omega_{sc}t-\gamma_*)^2}{\Delta_v^{(2)}}}\ . \tag{5.85}$$

Target return signal fluctuations are a nonstationary and nonseparable stochastic process. This is true because both the normalized correlation function $R_{sc}[\tau, \alpha_0(t)]$ of the fluctuations caused by radar antenna scanning [which is determined by Equation (5.72)] and the power $p(t)$ of the target return signal depend on the parameter t, due to the amplitude modulation of the

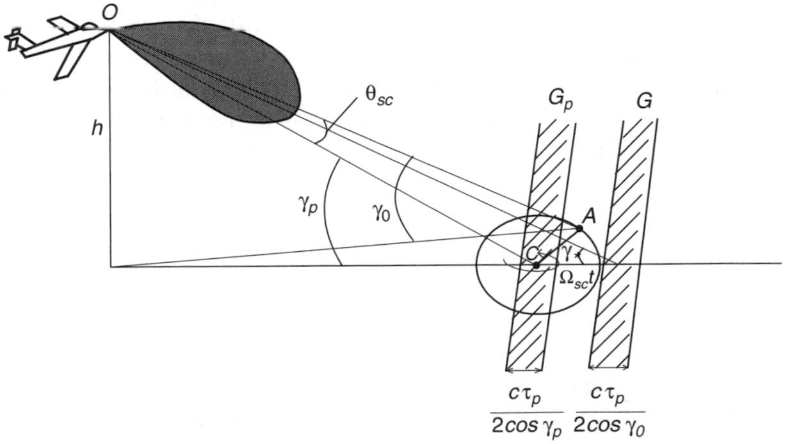

FIGURE 5.15
Two-dimensional (surface) target tracking: OC — the equisignal direction; OA — the axis of the radar antenna directional diagram; G_p and G — the gated range element.

received target return signal gated in the radar range, which is scattered from the two-dimensional (surface) target.[13,14] In the case of the axial symmetric directional diagram, when $R_{sc}[\tau, \alpha_0(t)]$ caused by radar antenna scanning is independent of the parameter t [see Equation (5.75)], the target return signal is a nonstationary but separable stochastic process. In this case, the averaged power spectral density of the fluctuations caused by radar antenna scanning coincides with the instantaneous one if we do not consider the power of the target return signal and it has the same shape as in the case of scanning the three-dimensional (space) target [see Equation (5.81)].

5.4 Conical Scanning with Simultaneous Rotation of Polarization Plane

If under scanning the scatterers oriented chaotically in space, the position of the polarization plane is changed simultaneously with the radar antenna moving, we can assume that $\Delta\xi = const$ and $\Delta\zeta = 0$. Let us define the normalized correlation function $R_q(\Delta\xi)$ of the target return signal space fluctuations caused by rotation of the polarization plane of the radar antenna, which is determined by Equation (5.4), for the case when half-wave dipoles are scatterers. In this case, with consistent polarization at the transmitting and receiving conditions, we can write[15]

$$q(\xi,\zeta) = \cos^2\xi \sin^2\zeta \,. \tag{5.86}$$

Then, the normalized correlation function $R_q(\Delta\xi, \Delta\zeta)$ of the space fluctuations caused by rotation of the polarization plane simultaneously with the moving radar antenna can be written in the following form:

$$R_q(\Delta\xi,\Delta\zeta) = R_q(\Delta\xi) \approx 0.33(2+\cos 2\Delta\xi) \,. \tag{5.87}$$

If the polarization plane is rotated simultaneously with the angular velocity Ω_{sc}, then $\Delta\xi = \Omega_p\tau$, and the normalized correlation function $R_q(\tau)$ of the space fluctuations caused by rotation of the polarization plane, simultaneously with radar antenna scanning, takes the following form:

$$R_q(\tau) \approx 0.33(2+\cos 2\Omega_p\tau) \,. \tag{5.88}$$

The corresponding power spectral density $S_q(\omega)$ of the space fluctuations caused by rotation of the polarization plane simultaneously with radar antenna scanning has the following form:

$$S_q(\omega) \cong \delta(\omega) + 0.25[\delta(\omega - 2\Omega_p) + \delta(\omega + 2\Omega_p)] \tag{5.89}$$

and consists of three components:[16,17] the main (central) component and two subcomponents offset by $\pm 2\Omega_p$ from the central component. The power of each side component is four times less than the power of the central component. A presence of $\pm 2\Omega_p$ is the consequence of the two-lobe polarization characteristic of scatterers.

If a radar antenna with axial symmetry Gaussian directional diagram performs conical scanning, then the normalized correlation function of the target return signal fluctuations, which takes into account the radar antenna scanning and rotation of the polarization plane, can be written under the condition $\Omega_p = \Omega_{sc}$ as the product of Equation (5.75) and Equation (5.88):

$$R(\tau) \approx 0.33(2 + \cos 2\Omega_{sc}\tau) \cdot e^{-2\pi \frac{\theta_{sc}^2}{\Delta_a^{(2)}} \sin^2 0.5\Omega_{sc}\tau}. \tag{5.90}$$

As before, the power spectral density is the regulated function. The power of harmonics of this power spectral density is determined by

$$P_{N_k} \cong I_k\left(\tfrac{\pi\, \theta_{sc}^2}{\Delta_a^{(2)}}\right) + 0.25\left[I_{k-2}\left(\tfrac{\pi\, \theta_{sc}^2}{\Delta_a^{(2)}}\right) + I_{k+2}\left(\tfrac{\pi\, \theta_{sc}^2}{\Delta_a^{(2)}}\right)\right]. \tag{5.91}$$

Comparing Equation (5.91) with Equation (5.81), one can see an essential increase in the number of harmonics of the power spectral density $S_q(\omega)$ of the fluctuations caused by the rotation of the polarization plane of the radar antenna. The second harmonic differs from zero and is equal to 0.25 from the zero harmonic by power due to the rotation of the polarization plane, even if the condition $\theta_{sc} = 0°$ is satisfied.

In the case of the cross-polarized signal given by

$$q(\xi, \zeta) = \sin 2\xi \sin^2 \zeta, \tag{5.92}$$

we can write

$$R_q(\Delta\xi, \Delta\zeta) = R_q(\Delta\xi) = \cos 2\Delta\xi; \tag{5.93}$$

$$R_q(\tau) = \cos 2\Omega_p\tau. \tag{5.94}$$

The power spectral density with the stationary radar antenna can be written in the following form:

$$S_q(\omega) \cong \delta(\omega - 2\Omega_p) + \delta(\omega + 2\Omega_p) \tag{5.95}$$

and consists only of two subfrequencies without the central component.

Under simultaneous conical scanning, i.e., when the condition $\Omega_p = \Omega_{sc}$ is true, the normalized correlation function $R(\tau)$ of the fluctuations, which takes into consideration the radar antenna scanning and rotation of the polarization plane, has the following form:

$$R(\tau) = \cos 2\Omega_{sc}\tau \cdot e^{-2\pi\frac{\theta_{sc}^2}{\Delta_a^{(2)}}\sin^2 0.5\Omega_{sc}\tau}. \tag{5.96}$$

The power of harmonics of the power spectral density is determined by the following form:

$$P_{N_k} \cong I_{k-2}\left(\frac{\pi\,\theta_{sc}^2}{\Delta_a^{(2)}}\right) + I_{k+2}\left(\frac{\pi\,\theta_{sc}^2}{\Delta_a^{(2)}}\right). \tag{5.97}$$

5.5 Conclusions

If for any navigational system the radar is stationary, the radar antenna scanning is the main source of the slow target return signal fluctuations. In this section we investigate the correlation function and power spectral density of the target return signal fluctuations caused by the moving (rotation) of the radar antenna under scanning of the three-dimensional (space) and two-dimensional (surface) targets. Two of the most universally adopted forms of the radar antenna moving (rotation) are considered: the line scanning used, as a rule, in the detection of targets and the conical scanning used in tracking moving targets. The case of the radar antenna conical scanning with simultaneous rotation of the polarization plane of the radar antenna is also investigated.

In the general case, under scanning of the three-dimensional (space) target, the space–time normalized correlation function of fluctuations under conditions discussed in this section is defined by the product of three normalized correlation functions: the normalized correlation function of time fluctuations in the radar range, which is caused by the propagation of the pulsed searching signal; the normalized correlation function of space fluctuations which is caused by the radar antenna moving (rotation); and the normalized correlation function of space fluctuations which is caused by the rotation of the radar antenna polarization plane.

In scanning the two-dimensional (surface) target by the pulsed searching signal, the normalized correlation function of the fluctuations can be expressed as the product of the periodic normalized correlation function of the fluctuations in the radar range and the normalized correlation function of the slow fluctuations caused by radar antenna moving (rotation). The power spectral density of the target return signal fluctuations with line scanning of both the three-dimensional (space) target and the two-dimensional (surface) target is the square of the power spectral density of the amplitude of the target return signal, which is the Fourier transform of the individual elementary signal, under covering of all elementary scatterers by the directional diagram during the radar antenna scanning.

With line scanning by radar antenna and in the case of the continuous searching signal, the normalized correlation function of the fluctuations in terms of periodic iteration of the same elementary signals with each revolution of the radar antenna is determined by Equation (5.47). The power spectral density is the regulated function given by Equation (5.48). In the case of the pulsed searching signal, the normalized correlation function of the fluctuations caused by the radar antenna line scanning is defined by the product of the normalized correlation function given by Equation (5.47) and the normalized correlation function of the time fluctuations in the radar range, which is determined by Equation (5.2). The power spectral density is the convolution between the power spectral densities given by Equation (3.15) and Equation (5.48).

In conical scanning by the radar antenna, the normalized correlation function of the target return signal fluctuations is nonstationary, in the general case, and can be determined by Equation (5.72) that allows us to define the instantaneous power spectral density of the target return signal fluctuations. If the directional diagram is in axial symmetry, the normalized correlation function of the target return signal fluctuations caused by conical scanning is stationary and the power spectral density is the regulated function. Under radar antenna conical scanning with simultaneous rotation of the polarization plane, the normalized correlation function and power spectral density of the target return signal fluctuations are determined by Equation (5.88) and Equation (5.89) in the case of the polarization plane rotation with uniform angular velocity. The power spectral density of the target return signal fluctuations consists only of three components: the main component and two subcomponents. In the case of the axial symmetry Gaussian directional diagram, the normalized correlation function and power spectral density of the fluctuations caused by conical scanning and simultaneous rotation of the polarization plane of the radar antenna are determined by Equation (5.90) and Equation (5.91), respectively. In this case, there is an essential increase in the number of harmonics of the power spectral density of the target return signal fluctuations.

References

1. Farina, A., *Antenna-Based Signal Processing Techniques for Radar Systems*, Artech House, Norwood, MA, 1992.
2. Schleher, D., *MTI and Pulsed Doppler Radar*, Artech House, Norwood, MA, 1991.

3. Compton, R., *Adaptive Antennas,* Prentice Hall, Englewood Cliffs, NJ, 1988.
4. Johnson, D. and Dudgeon, D., *Array Signal Processing: Concepts and Techniques,* Prentice Hall, Englewood Cliffs, NJ, 1993.
5. Hudson, J., *Adaptive Array Principles,* Peter Peregrinus, London, 1981.
6. Feldman, Yu and Reznikov, V. Reduction of reflections under scanning the Earth surface, *Problems in Radio Electronics,* Vol. OT, No. 1, 1968, pp. 55–66 (in Russian).
7. Rappaport, T., *Wireless Communications: Principle and Practice,* Prentice Hall, Englewood Cliffs, NJ, 1996.
8. Jablou, N., Adaptive beamforming with generalized side-lobe canceller in the presence of array imperfections, *IEEE Trans.,* Vol. AP-34, No. 8, 1986, pp. 996–1012.
9. Buckley, K. and Griffiths, L., An adaptive generalized side-lobe canceller with derivative constraints, *IEEE Trans.,* Vol. AP-34, No. 3, 1986, pp. 311–319.
10. Arnold, H., Cox, D., and Murray, R., Macroscopic diversity performance measured in the 800-MHz portable radio communications environment, *IEEE Trans.,* Vol. AP-36, No. 2, 1988, pp. 277–280.
11. Graziano, V., Propagation correlation at 900 MHz, *IEEE Trans.,* Vol. VT-27, No. 5, 1978, pp. 182–189.
12. Monzingo, R. and Miller, T., *Introduction to Adaptive Arrays,* John Wiley & Sons, New York, 1980.
13. Feldman, Yu, Gidaspov, Yu, and Gomzin, V., *Moving Target Tracking,* Soviet Radio, Moscow, 1978 (in Russian).
14. Blackman, S. and Popoli, R., *Design and Analysis of Modern Tracking Systems,* Artech House, Norwood, MA, 1999.
15. Bogomolov, A., *Foundations of Radar,* Soviet Radio, Moscow, 1954 (in Russian).
16. Appelbaum, S. and Chapman, D., Adaptive arrays with main beam constraints, *IEEE Trans.,* Vol. AP-24, No. 9, 1976, pp. 650–662.
17. Li, Q., Rothwell, E., Chen, K., and Nyquist, D., Scattering center analysis of radar targets using fitting scheme and genetic algorithm, *IEEE Trans.,* Vol. AP-44, No. 1, 1996, pp. 198–207.

6

Fluctuations Caused by the Moving Radar with Simultaneous Radar Antenna Scanning

6.1 General Statements

Target return signal fluctuations caused by the moving radar with simultaneous radar antenna scanning are of prime interest. These fluctuations are encountered in navigational systems, for example, when the target is searched against the background of clouds moving under stimulus of the wind.[1,2] Fluctuations caused by the moving radar with simultaneous radar antenna scanning are a nonstationary stochastic process.[3] This nonstationary stochastic process possesses a set of peculiarities that cannot be investigated by individually studying the fluctuations caused only by the moving radar or those caused only by radar antenna scanning. Problems caused by the moving radar with simultaneous radar antenna scanning are numerous. In this chapter, we study only the most important problems: moving radar with simultaneous radar antenna line scanning and moving radar with simultaneous radar antenna conical scanning.

6.1.1 The Correlation Function in the Scanning of the Three-Dimensional (Space) Target

In the general case, with the pulsed searching signal, the correlation function of the target return signal fluctuations caused by the moving radar with simultaneous radar antenna scanning can be determined by Equation (2.93)–Equation (2.95). The intraperiod and interperiod target return signal fluctuations can be separated because the total normalized correlation function of the target return signal fluctuations is defined by the product of the normalized correlation functions of the intraperiod and interperiod fluctuations. However, because the shifts $\Delta\varphi$, $\Delta\psi$, and $\Delta\rho$ are functions of the angles β_0 and γ_0 — see Equation (2.84)–Equation (2.87) and Equation (2.92), where

$$\Delta\rho = \Delta\ell \cos\beta_0 \cos\gamma_0 \tag{6.1}$$

is the radial displacement along the radar antenna directional diagram axis — and the angles β_0 and γ_0, in turn, are functions of time during the radar antenna scanning, the fluctuations become a nonstationary stochastic process. The character of the nonstationary state of the normalized correlation functions given by Equation (2.94) and Equation (2.95) is different.

Dependence on the angles $\beta_0(t)$ and $\gamma_0(t)$ in the normalized correlation function of the fluctuations in the radar range, which is determined by Equation (2.95), manifests itself by the radial displacement $\Delta\rho_0$ along the directional diagram axis [see Equation (6.1)]. As the radar moves uniformly with the velocity V, this dependence acts on the coefficient of scale μ given by Equation (3.11) and, as a result, the coefficient μ becomes a function of the time

$$\mu = 1 + 2\, V_r c^{-1} = 1 + 2\, V c^{-1} \cos\beta_0(t) \cos\gamma_0(t). \tag{6.2}$$

This means that the comb structure of the normalized correlation function $R_p(t)$ (see Figure 3.1a) is kept unchanged, but the scale with respect to the variable τ depends on the time t. The instantaneous power spectral density $S_p(\omega)$ of the fluctuations in the radar range is the same as in the case of the stationary radar antenna, i.e., it is a regulated function (see Figure 3.1b). However, the scale of $S_p(\omega)$ with respect to the axis of frequency is a function of the time. As a rule, we can neglect these changes. The main effect is exhibited by the characteristics of slow (interperiod) fluctuations. First, with the moving radar and stationary radar antenna, the shape, bandwidth, and average frequency of the power spectral density depends on the radar antenna orientation, i.e., the angles β_0 and γ_0. Second, under radar antenna scanning, the variation of the angles β_0 and γ_0 shows the regular deformation of the power spectral density, i.e., the nonstationary state. For this reason, in this chapter, the slow fluctuations are the main subject of study. The instantaneous normalized correlation function, which is determined by Equation (2.94), can be written in the following symmetric form:

$$R_g(\Delta\ell, \Delta\beta_0, \Delta\gamma_0) = R_{mov,sc}(\Delta\ell, \Delta\beta_0, \Delta\gamma_0) = N \iint g(\varphi - 0.5\Delta\varphi, \psi - 0.5\Delta\psi) \times g(\varphi - 0.5\Delta\varphi, \psi - 0.5\Delta\psi) \cdot e^{4j\pi \frac{\Delta\rho(\varphi,\psi)}{\lambda}}\, d\varphi\, d\psi. \tag{6.3}$$

The interperiod fluctuations with the pulsed searching signal and the slow fluctuations with the simple harmonic searching signal, in regard to the long-range area, are the same under scanning of the three-dimensional (space) target.

The formula in Equation (6.2) contains solutions for both: in the case when the radar is moving and when the radar antenna scanning is absent, i.e., when the radar antenna is stationary and the condition $\Delta\varphi = \Delta\psi = 0°$ is satisfied, and when the radar is stationary and the condition $\Delta\rho = 0$ is true, i.e., there is radar antenna scanning. It can easily be seen that, in the general case, we cannot express the correlation function of the fluctuations caused by the moving radar with simultaneous radar antenna scanning as the product (or another combination) of the correlation function caused by the moving radar and that caused by radar antenna scanning.

However, there is at least one exception if the directional diagram is Gaussian.[4] There, the total normalized correlation function of the fluctuations caused by the moving radar with simultaneous radar antenna scanning can be written in the following form:

$$R_{mov,sc}(\Delta\ell, \Delta\beta_0, \Delta\gamma_0) = R_{mov}(\Delta\ell) \cdot R_{sc}(\Delta\beta_0, \Delta\gamma_0), \tag{6.4}$$

where

$$R_{mov}(\Delta\ell) = N \iint e^{-2\pi\left(\frac{\varphi^2}{\Delta_h^{(2)}} + \frac{\psi^2}{\Delta_v^{(2)}}\right) + 4j\pi\frac{\Delta\rho(\varphi,\psi)}{\lambda}} d\varphi\, d\psi \tag{6.5}$$

is the normalized correlation function caused only by the moving radar. Here, $\Delta\rho[\beta_0(t), \gamma_0(t)]$ is a function of time;

$$R_{sc}(\Delta\beta_0, \Delta\gamma_0) = e^{-0.5\pi\left(\frac{\varphi^2}{\Delta_h^{(2)}} + \frac{\psi^2}{\Delta_v^{(2)}}\right)} \tag{6.6}$$

is the normalized correlation function caused only by radar antenna scanning. Naturally, the total power spectral density of the fluctuations is the convolution between the power spectral density caused only by the moving radar and that caused only by radar antenna scanning.

If the directional diagram is not Gaussian, but the variables φ and ψ are separable in the function $g(\varphi, \psi)$, then, for a linear approximation of the radar displacement $\Delta\rho_0$ along the directional diagram axis, which is determined by Equation (3.47), based on Equation (6.3), we can write

$$R_{mov,sc}(\Delta\ell, \Delta\beta_0, \Delta\gamma_0) = R^h_{mov,sc}(\Delta\ell, \Delta\beta_0) \cdot R^v_{mov,sc}(\Delta\ell, \Delta\gamma_0) \cdot e^{-4j\pi\frac{\Delta\rho_0}{\lambda}}, \tag{6.7}$$

where

$$R^h_{mov,sc}(\Delta\ell, \Delta\beta_0) = N \int g_h(\varphi - 0.5\Delta\varphi) \cdot g_h(\varphi + 0.5\Delta\varphi) \cdot e^{-4j\pi\varphi\frac{\Delta\rho_h}{\lambda}} d\varphi; \tag{6.8}$$

$$R^v_{mov,sc}(\Delta\ell, \Delta\gamma_0) = N \int g_v(\psi - 0.5\Delta\psi) \cdot g_v(\psi + 0.5\Delta\psi) \cdot e^{-4j\pi\psi\frac{\Delta\rho_v}{\lambda}} d\psi; \tag{6.9}$$

$\Delta\rho_0$ is given by Equation (6.1);

$$\Delta\rho_h = \Delta\ell \cdot \sin\beta_0 \; ; \qquad \text{and} \qquad \Delta\rho_v = \Delta\ell \cdot \cos\gamma_0 \sin\gamma_0 . \tag{6.10}$$

6.1.2 The Correlation Function in the Scanning of the Two-Dimensional (Surface) Target

In scanning the two-dimensional (surface) target, the correlation function of the target return signal fluctuations [see Equation (2.122)] is seen as more complex.[5,6] With the continuous searching signal, we need to consider that the target return signal power depends on the aspect angle and radar range. With the pulsed searching signal, we need to consider that the interperiod and intraperiod fluctuations are not separable (see Chapter 4). If and only if the directional diagram is Gaussian can we define the normalized correlation function of the fluctuations caused by radar antenna scanning, which is given by Equation (6.6), as the individual cofactor using the definition of the total correlation function [see Equation (2.122)]. The second cofactor is the normalized correlation function of fluctuations caused by the moving radar, which has been discussed in Chapter 4. If the variables φ and ψ are separable in the directional diagram $g(\varphi, \psi)$, then, with the pulsed searching signal, the total correlation function can be expressed as the product of the azimuth-normalized and the aspect-angle normalized correlation functions [see Equation (2.133)–Equation (2.135)]. Under radar antenna scanning, the main changes are seen as arising in the interperiod fluctuations, as in the case of scanning the three-dimensional (space) target. Intraperiod fluctuations have the same character, but only the coefficient of scale μ given by Equation (6.2) becomes a function of $\beta_0(t)$ and $\gamma_0(t)$.

In the representation of the pulsed searching signal with short duration, i.e., when the condition $\Delta_p << \Delta_v$ is satisfied in the form of the delta function assuming $\tau = nT_p'$, and using Equation (2.142) and Equation (5.22), based on Equation (2.122), we can write the normalized correlation function $R_{mov,sc}(\Delta\ell, \Delta\beta_0, \Delta\gamma_0)$ of fluctuations caused by the moving radar with simultaneous radar antenna scanning in the following form:

$$\begin{aligned} R_{mov,sc}(\Delta\ell, \Delta\beta_0, \Delta\gamma_0) = N \int g(\varphi - 0.5\Delta\varphi, \psi_* - 0.5\Delta\psi) \\ \times g(\varphi + 0.5\Delta\varphi, \psi_* + 0.5\Delta\psi) \cdot e^{4j\pi \frac{\Delta\rho(\varphi,\psi_*)}{\lambda}} d\varphi \, , \end{aligned} \tag{6.11}$$

where

$$\Delta\rho(\varphi, \psi_*) = \Delta\ell \cdot \cos\left[\beta_0(t) + \tfrac{\varphi}{\cos\gamma_*}\right] \cos\gamma_* \, , \tag{6.12}$$

$$\gamma_* = \psi_* + \gamma_0 . \tag{6.13}$$

The normalized correlation function $R_{mov,sc}(\Delta\ell, \Delta\beta_0, \Delta\gamma_0)$, given by Equation (6.11), is similar to the instantaneous normalized correlation function $R_g(\Delta\ell, \Delta\beta_0, \Delta\gamma_0)$ of slow fluctuations given by Equation (6.3). The difference is that an integration is carried out only with respect to the variable φ, not the two variables φ and ψ. In the general case, $R_{mov,sc}(\Delta\ell, \Delta\beta_0, \Delta\gamma_0)$ is not divided into the normalized correlation function $R_{mov}(\Delta\ell)$ of fluctuations caused only by the moving radar and the normalized correlation function $R_{sc}(\Delta\beta_0, \Delta\gamma_0)$ of fluctuations caused only by radar antenna scanning. In the case of the Gaussian directional diagram, the formula in Equation (6.4) is true, where $R_{mov}(\Delta\ell)$ can be written in the following form:

$$R_{mov}(\Delta\ell) = N \int_{\varphi_1}^{\varphi_2} e^{-2\pi \frac{\varphi^2}{\Delta_h^{(2)}} + 4j\pi \frac{\Delta\rho(\varphi,\psi_*)}{\lambda}} d\varphi . \tag{6.14}$$

In this case, $R_{sc}(\Delta\beta_0, \Delta\gamma_0)$, caused only by radar antenna scanning is completely determined by Equation (6.6). In the case of the Gaussian directional diagram, and if variables φ and ψ are separable in the function $g(\varphi, \psi)$, Equation (6.7) and Equation (6.8) are true for a linear approximation of the parameter $\Delta\rho$. Therefore, Equation (6.9) takes the following form:

$$R^{v}_{mov,sc}(\Delta\ell, \Delta\gamma_0) = N \int g_v(\psi_* - 0.5\Delta\psi) \cdot g_v(\psi_* + 0.5\Delta\psi) \cdot e^{-4j\pi\,\psi_* \frac{\Delta\rho_v}{\lambda}} d\psi, \tag{6.15}$$

where

$$\Delta\rho_v = \Delta\ell \cos\beta_0 \sin\gamma_* . \tag{6.16}$$

6.2 The Moving Radar with Simultaneous Radar Antenna Line Scanning

6.2.1 Scanning of the Three-Dimensional (Space) Target: The Gaussian Directional Diagram

Let us assume that the radar moves uniformly and linearly with the velocity V. The radar antenna rotates about its own vertical axis, i.e., the condition $\gamma_0 = const$ is satisfied, with the angular velocity $\Omega_{sc}(t)$ making the line (circular or segment) scanning the three-dimensional (space) target. First, we consider the uniform line circular scanning by radar antenna with the constant angular velocity $\Omega_{sc} = const$. In this case, the formula in Equation (6.4) is true. The normalized correlation function $R_{sc}(\Delta\beta_0, \Delta\gamma_0)$ of fluctuations caused only by radar antenna scanning is Gaussian [see Equation (5.34)]. The normalized correlation function $R_{mov}(\Delta\ell)$ of fluctuations caused only by the moving radar depends on orientation of the radar antenna with respect to the vector of velocity V of the moving radar and, in the case of the high-deflected radar antenna, is also Gaussian [see Equation (3.79)]. Because of this, the total normalized correlation function $R_{mov,sc}(\Delta\ell, \Delta\beta_0, \Delta\gamma_0)$ of fluctuations caused by the moving radar with simultaneous radar antenna scanning, given by Equation (6.4), can be written in the following form:

$$R_{mov,sc}(t,\tau) = e^{-\pi\,(\Delta F_{mov}^2 + \Delta F_{sc}^2) + j\Omega_0\tau}, \tag{6.17}$$

where ΔF_{mov} is the effective bandwidth of the power spectral density of the fluctuations caused only by the moving radar [see Equation (3.81)], in which $\beta_0 = \Omega_{sc}t$ is a function of time, ΔF_{sc} is the effective bandwidth of the power spectral density of the fluctuations caused only by radar antenna scanning [see Equation (5.38)], Ω_0 is the averaged Doppler frequency given by Equation (3.48) and Equation (3.49) and is also a function of $\beta_0(t)$. The instantaneous power spectral density of fluctuations caused by the moving radar with simultaneous radar antenna scanning is Gaussian with the effective bandwidth determined in the following form:

$$\Delta F(t) = \sqrt{\Delta F_{mov}^2(t) + \Delta F_{sc}^2}\,. \tag{6.18}$$

More rigorously speaking, to define the resulting normalized correlation function $R_{mov,sc}(t, \tau)$ of fluctuations caused by the moving radar with simultaneous radar antenna scanning, we must use the normalized correlation function $R_{sc}^h(\tau, nT_{sc})$ of fluctuations caused only by radar antenna scanning, which is determined by Equation (5.47), instead of the normalized correlation function $R_{sc}(\tau)$ given by Equation (5.34). To define the total instantaneous power spectral density of fluctuations caused by the moving radar with simultaneous radar antenna scanning, we must carry out the convolution between the continuous power spectral density $S_{mov}(\omega, t)$ of fluctuations caused only by the moving radar, which is determined by Equation (3.80), and the regulated power spectral density $S_{sc}(\omega)$ of fluctuations caused only by radar antenna scanning, which is given by Equation (5.49). It is not difficult to show that when the effective bandwidth of $S_{mov}(\omega, t)$ of fluctuations caused only by the moving radar is significantly greater than the frequency of radar antenna scanning, i.e., the condition $\Delta F_{mov} >> F_{sc}$ or $\tau_c^{mov} << T_{sc}$ is satisfied, the result is the same.

However, the effective bandwidth ΔF_{sc} of the power spectral density $S_{sc}(\omega)$ of fluctuations caused only by radar antenna line circular scanning can change substantially in value during the radar antenna line circular scanning, especially if the aspect angle γ_0 of the radar antenna is not so high in value.[7] Under the condition $\beta_0 \approx 0°$, the effective bandwidth of the moving radar ΔF_{mov} of the power spectral density $S_{mov}(\omega, t)$ is very low in value and can be less than the frequency of the radar antenna scanning F_{sc}. Moreover, the low-deflected radar antenna, $S_{mov}(\omega, t)$, caused only by the moving radar, is the more complex function, which is not Gaussian (see Section 3.3). In particular, under the condition $\beta_0 = 0°$ and if the radar antenna has axial symmetry, $S_{mov}(\omega, t)$ takes the exponential form with the effective bandwidth ΔF_{mov} that is very low in value.

With the Gaussian directional diagram, the total normalized correlation function of the fluctuations $R_{mov,sc}(\Delta\ell, \Delta\beta_0, \Delta\gamma_0)$ caused by the moving radar with simultaneous radar antenna scanning can be expressed as the product of the normalized correlation functions $R_{mov}(\Delta\ell)$ of fluctuations caused only by the moving radar and $R_{sc}(\Delta\beta_0, \Delta\gamma_0)$ caused only by radar antenna scanning. This is despite the fact that with some positions of the radar antenna, the power spectral density $S_{mov}(\omega, t)$ of fluctuations caused only by the moving radar does not obey the Gaussian law. Because of this, for all cases, the resulting power spectral density $S_{mov,sc}(\omega, t)$ of fluctuations caused by the moving radar with simultaneous radar antenna scanning is defined by the convolution between $S_{mov}(\omega, t)$ and the regulated power spectral density $S_{sc}(\omega)$ of fluctuations caused only by radar antenna scanning, which is given by Equation (5.49) and can be written in the following form:

$$S_{mov,sc}(\omega, t) = \sum_{k=0}^{\infty} S_{mov}(\omega - k\Omega_{sc}) \cdot e^{-\pi \frac{k^2\Omega_{sc}^2}{\Delta\Omega_{mov}^2}} . \tag{6.19}$$

The formula in Equation (6.19) is true for any relationships between $\Delta\Omega_{mov} = 2\pi\Delta F_{mov}$ and Ω_{sc}. The power spectral density $S_{mov}(\omega, t)$ of fluctuations caused by the moving radar with simultaneous radar antenna scanning is the sum of the power spectral densities $S_{mov}(\omega, t)$ of fluctuations caused only by the moving radar, which are shifted to the frequencies $k\Omega_{sc}$, and multiplied by the coefficients $S_{sc}(k\Omega_{sc})$. Under the condition $\Delta\Omega_{mov} >> \Omega_{sc}$, we obtain the integral convolution of the already discussed power spectral densities $S_{mov}(\omega, t)$ and $S_{sc}(\omega)$ and the already discussed result (see Figure 6.1). Under the condition $\Delta\Omega_{mov} < \Omega_{sc}$, the power spectral density $S_{mov}(\omega, t)$ has a comb structure. In doing so, the shape of the teeth of the comb power spectral density $S_{mov}(\omega, t)$ can be changed from the Gaussian form — the radar antenna is high deflected and the velocity of the moving radar is low in value (see Figure 6.2) — to the exponential form — the radar antenna is not deflected (see Figure 6.3).

Thus, during the radar antenna line scanning, the instantaneous power spectral density $S_{mov,sc}(\omega, t)$ is deformed continuously taking various shapes in accordance with the position and width of the directional diagram, velocity

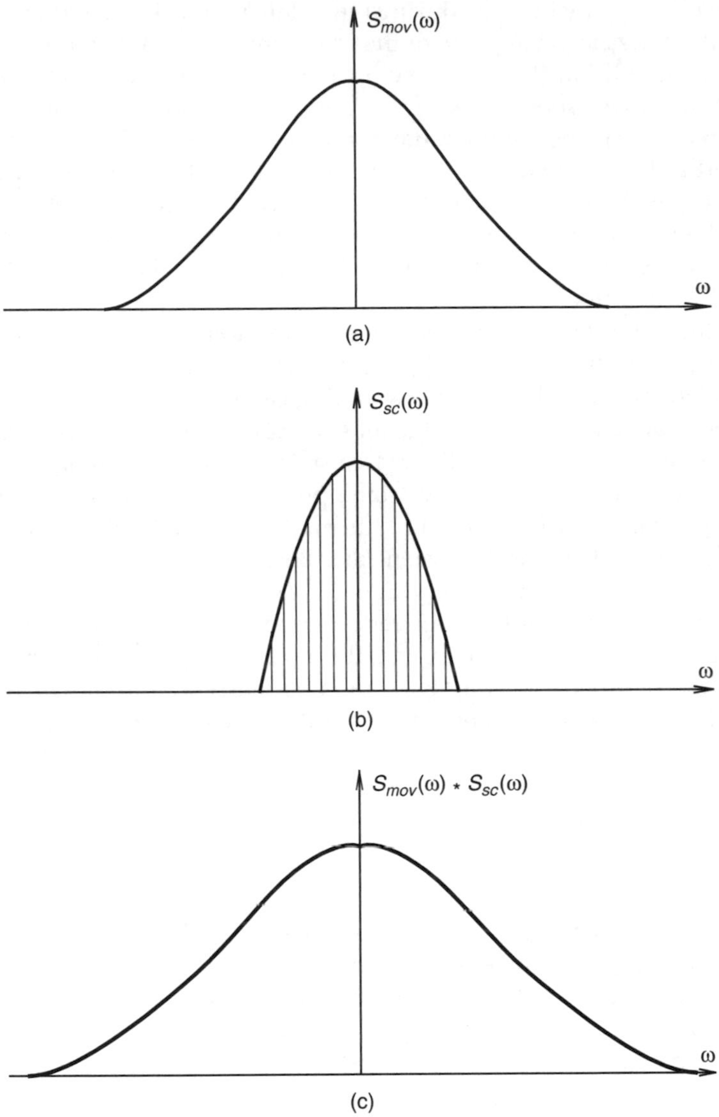

FIGURE 6.1
Moving radar with simultaneous radar antenna scanning, $\Delta F_{mov} >> F_{sc}$. The power spectral density of interperiod fluctuations of the target return signal caused by (a) only moving radar, (b) only radar antenna scanning, and (c) moving radar with simultaneous radar antenna scanning.

of the moving radar, and frequency of radar antenna line scanning. The Doppler shift in the frequency of the partial power spectral densities $S_{mov}(\omega, t)$ with regard to the frequencies $k\Omega_{sc}$ is also continuously varied within the limits of the interval $[-\Omega_{\max} \cos \gamma_0, \Omega_{\max} \cos \gamma_0]$.

With radar antenna segment hunting, the instantaneous angular velocity Ω_{sc} becomes a function of time but Equation (6.4) is kept true. The normalized

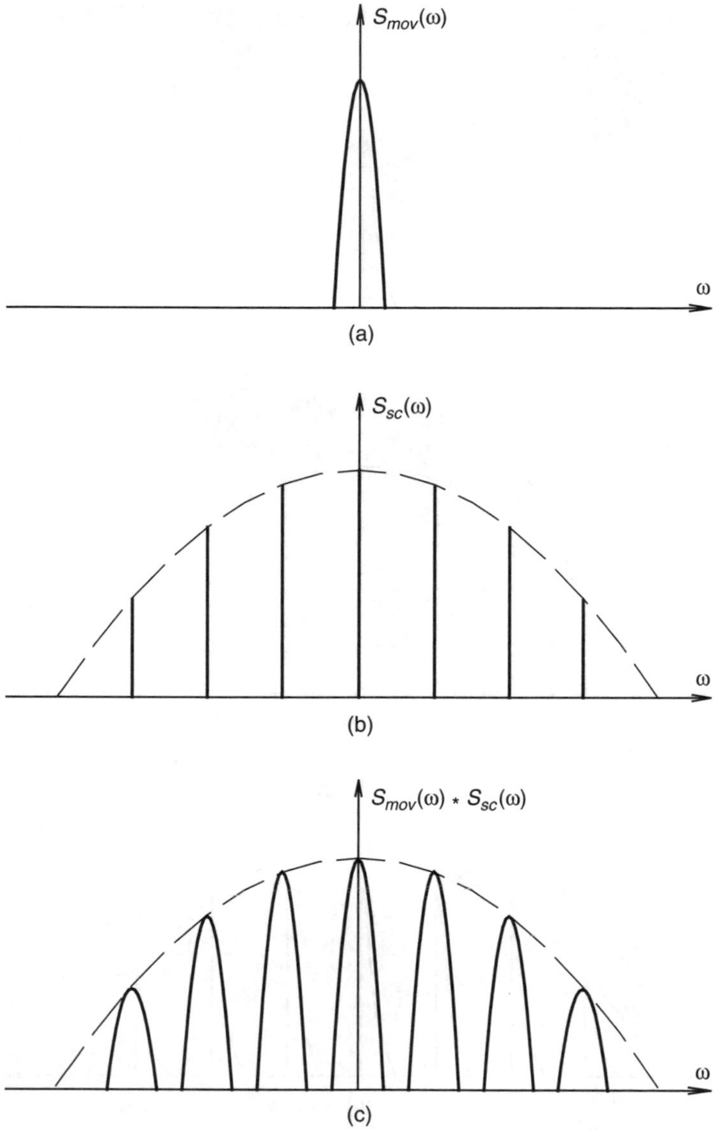

FIGURE 6.2
Moving radar with simultaneous radar antenna scanning. Radar antenna is high deflected. Velocity of the moving radar is low by value, $\Delta F_{mov} << F_{sc}$. Power spectral density of interperiod fluctuations of the target return signal caused by (a) only moving radar, (b) only radar antenna scanning, and (c) moving radar with simultaneous radar antenna scanning.

correlation function $R_{sc}(\Delta\beta_0, \Delta\gamma_0)$ of the fluctuations caused only by the radar antenna scanning becomes more complex. If the condition $\Delta_a << \beta_m$ (see Section 5.2.3) is satisfied, so that we can neglect a variation of the angular velocity during the time of the radar antenna rotation by the angle equal to the directional diagram width — this is called quasi-stationary radar antenna scanning

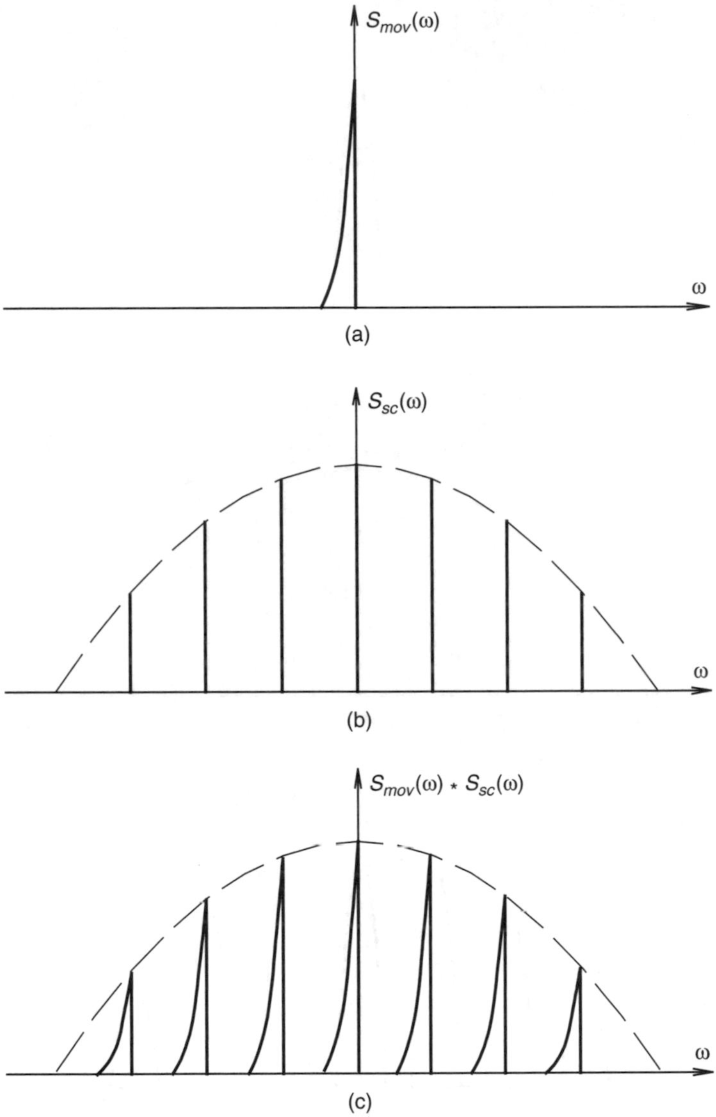

FIGURE 6.3
Moving radar with simultaneous radar antenna scanning. Radar antenna is not deflected, $\Delta F_{mov} << F_{sc}$. The power spectral density of interperiod fluctuations of the target return signal caused by (a) only moving radar, (b) only radar antenna scanning, and (c) moving radar with simultaneous radar antenna scanning.

— Equation (6.17) and Equation (6.18) are true. In this case, the effective bandwidth ΔF_{sc} of the power spectral density $S_{sc}(\omega)$ of fluctuations caused only by radar antenna scanning becomes a function of time, as the effective bandwidth ΔF_{mov} of the power spectral density $S_{mov}(\omega, t)$ of fluctuations caused only by the moving radar. If the radar antenna segment hunting is rapid, $R_{sc}(\Delta\beta_0, \Delta\gamma_0)$ is given by Equation (5.55). $S_{mov,sc}(\omega, t)$ of fluctuations caused by the moving radar with simultaneous radar antenna scanning is defined by the corresponding convolution between the power spectral densities $S_{mov}(\omega, t)$ and $S_{sc}(\omega)$.

6.2.2 Scanning of the Two-Dimensional (Surface) Target: The Gaussian Directional Diagram

Consider the normalized correlation function $R_{mov,sc}(\Delta\ell, \Delta\beta_0, \Delta\gamma_0)$ of fluctuations of the target return signal caused by the moving radar with simultaneous radar antenna scanning, which is given by Equation (6.11). Using the Gaussian directional diagram, we obtain the product of the two normalized correlation functions: $R_{sc}(\Delta\beta_0, \Delta\gamma_0)$ of fluctuations caused only by radar antenna scanning, which is determined by Equation (6.6), and $R_{mov}(\Delta\ell)$ of the Doppler fluctuations caused by the difference in the radial velocities in the azimuth plane, which is given by Equation (6.14). Because with the high-deflected radar antenna, both normalized correlation functions $R_{mov}(\Delta\ell)$ and $R_{sc}(\Delta\beta_0, \Delta\gamma_0)$ obey Gaussian law, the formulae in Section 6.2.1 are true with the only difference that with the pulsed searching signal, the effective bandwidth ΔF_{mov} of the power spectral density $S_{mov}(\omega, t)$ is given by Equation (4.141) instead of Equation (3.81). Also, the effective bandwidth ΔF_{sc} of the power spectral density $S_{sc}(\omega)$ is given by Equation (5.38), in which the variable γ_0 is replaced with the variable γ_*. In the case of the continuous searching signal, the formulae obtained in Section 6.2.1 can be used without any changes.

In the case of the low-deflected radar antenna, we can use the same statements and conclusions as in Section 6.2.1. If it is necessary to obtain rigorous formulae, we must use the results discussed in Section 4.6 to obtain the total power spectral density $S_{mov,sc}(\omega, t)$ of the fluctuations caused by the moving radar with simultaneous radar antenna scanning. The main differences rise in the determination of the Fourier transform for the total normalized correlation function $R_{mov,sc}(\Delta\ell, \Delta\beta_0, \Delta\gamma_0)$. $R_{mov,sc}(\Delta\ell, \Delta\beta_0, \Delta\gamma_0)$ consists of three cofactors: the normalized correlation function $R_{sc}^h(\tau, nT_{sc})$ of the fluctuations caused only by radar antenna scanning, which is given by Equation (5.47); the aspect-angle normalized correlation function $R_\gamma(t, \tau)$ of the interperiod fluctuations caused by the glancing radar range, which is given by Equation (4.67); and the azimuth-normalized correlation function $R_\beta(\tau)$ of the Doppler fluctuations caused by various values of the azimuth angles within the horizontal-coverage directional diagram width. However, these differences are the same as determined for the power spectral density $S_{mov}(\omega, t)$ of the fluctuations caused only by the moving radar in Chapter 4.

6.2.3 Short-Range Area: The Gaussian Directional Diagram

Using an example of scanning the underlying surface,[8,9] we consider the short-range area. We need to add the angle shift $\Delta\varphi_{sc}$ given by Equation (2.84) and Equation (2.85) caused by radar antenna scanning (naturally, we must replace the variable γ_0 with the variable γ_*) to the previous angle shift $\Delta\varphi_{mov}$ given by Equation (2.86) and Equation (2.87) caused by the moving radar. When the radar moves uniformly and the radar antenna rotates uniformly, we can write

$$\Delta\varphi = \Delta\varphi_{mov} + \Delta\varphi_{sc} = [V\rho_*^{-1}\sin\beta_0 \cos\gamma_* + \Omega_{sc}\cos\gamma_*]\,\tau. \tag{6.20}$$

Based on Equation (6.20) we can obtain that the total normalized correlation function $R_{mov,sc}(\Delta\ell, \Delta\beta_0, \Delta\gamma_0)$ is Gaussian in the case of the Gaussian directional diagram [see Equation (6.17)]. The effective bandwidth ΔF of the total power spectral density $S_{mov,sc}(\omega, t)$ of fluctuations caused by the moving radar with simultaneous radar antenna scanning can be determined in the following form:

$$\Delta F = \sqrt{\Delta F_{mov}^2 + \Delta\tilde{F}_{sc}^2} = \sqrt{\Delta F_{mov}^2 + (\Delta F_{sc} + \Delta\tilde{F}_{mov})^2} \tag{6.21}$$

instead of Equation (6.18), where

$$\Delta F_{mov} = \sqrt{2}\, V\lambda^{-1}\Delta_h \sin\beta_0 \tag{6.22}$$

is the effective bandwidth of the power spectral density $S_\beta(\omega)$ of the Doppler fluctuations caused by the difference in the Doppler frequencies in the azimuth plane;

$$\Delta\tilde{F}_{sc} = \Delta F_{sc} + \Delta\tilde{F}_{mov} = \frac{\Omega_{sc}}{\sqrt{2}\,\Delta_h}\cdot\cos\gamma_* + \frac{V}{\sqrt{2}\,\rho_*\Delta_h}\cdot\sin\beta_0\cos\gamma_* \tag{6.23}$$

is the effective bandwidth of the power spectral density $S_\gamma(\omega)$ of fluctuations caused by the angle displacement of scatterers within the directional diagram width due to radar antenna scanning (the term ΔF_{sc}) and the moving radar (the term $\Delta\tilde{F}_{mov}$). Because

$$\Omega_\varphi = V\rho_*^{-1}\sin\beta_0 \tag{6.24}$$

is the angular velocity of displacement of scatterers due to the moving radar, we can write

$$\Delta\tilde{F}_{sc} = \frac{\Omega_{sc} + \Omega_\varphi}{\sqrt{2}\,\Delta_h}\cdot\cos\gamma_*\,. \tag{6.25}$$

Here, the following fact is of interest. The angular velocity Ω_φ changes its sign while changing the sign of the azimuth deviation β_0. In other words, taking into account the angle shift $\Delta\varphi_{mov}$ in the short-range area leads to the extension of the effective bandwidth ΔF (the condition $\beta_0 > 0$ is satisfied) and to the contraction of the effective bandwidth ΔF (the condition $\beta_0 < 0$ is true) of the total power spectral density $S_{mov,sc}(\omega, t)$ of the fluctuations caused by the moving radar with simultaneous radar antenna scanning.

Thus, the effective bandwidth ΔF of the total power spectral density $S_{mov,sc}(\omega, t)$ of fluctuations in the quadrants I and II differs from the effective bandwidth ΔF of the total power spectral density $S_{mov,sc}(\omega, t)$ of fluctuations caused by the moving radar with simultaneous radar antenna scanning in the quadrants IV and III, respectively, under the same absolute deviation of the radar antenna beam in the azimuth plane — the phenomenon of asymmetry (see Figure 6.4 and Figure 6.5). This phenomenon can be easily explained based on the physical meaning. The relative displacements of two scatterers spaced by the distance ρ from the radar under the angles $\pm\beta$ with respect to the direction of the moving radar are shown in Figure 6.4. These displacements lead us to the angle shifts $\Delta\varphi_{mov}$ with various signs depending on the sign of the angle β. At the same time, the radar antenna scanning gives rise to the angle shifts $\Delta\varphi_{sc}$ with respect to the axis of the directional diagram. The sign of the angle shifts $\Delta\varphi_{sc}$ is independent of the sign of the angle β and depends only on the direction of the radar antenna scanning. If the radar antenna rotates clockwise, the angle shift $\Delta\varphi_{mov}$ is subtracted from the angle shift $\Delta\varphi_{sc}$ in quadrant I and is added to the angle shift $\Delta\varphi_{sc}$ in quadrant IV.

At the surface points with the coordinates satisfying the equality $\Omega_\varphi = \Omega_{sc}$, the angle shifts $\Delta\varphi_{mov}$ and $\Delta\varphi_{sc}$ compensate each other, which leads to the radar antenna scanning stopping from the viewpoint of the target return signal fluctuations.[10,11] The geometric locus of these surface points forms a circle shown in Figure 6.5 by the solid line. The circle spaced symmetrically in the quadrants III and IV is shown in Figure 6.5 by the dotted line. This circle corresponds to the double effective bandwidth of the power spectral density of the fluctuations caused by the angle displacements of scatterers in comparison with the effective bandwidth $\Delta\tilde{F}_{sc}$ of the power spectral density $S_{sc}(\omega)$ of the fluctuations caused only by radar antenna scanning.

6.2.4 The Sinc²-Directional Diagram

Let us assume, as before, that V, Ω_{sc}, and γ_0 are the constant values. If the radar antenna searches the three-dimensional (space) target, the total normalized correlation function $R_{mov,sc}(\Delta\ell, \Delta\beta_0, \Delta\gamma_0)$ of fluctuations of the target return signal caused by the moving radar with simultaneous radar antenna scanning can be given by Equation (6.7) — Equation (6.9). In doing so, we must assume that $\Delta\psi = 0°$ in Equation (6.9). The normalized correlation function $R^h_{mov,sc}(\Delta\ell, \Delta\beta_0)$ is of prime interest because it contains all the information regarding the moving radar with simultaneous radar antenna scanning.

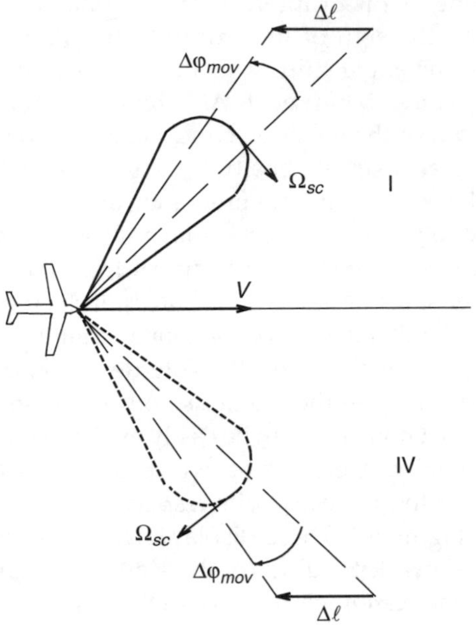

FIGURE 6.4
The power spectral density in the short-range area. The sum of velocities under moving radar with simultaneous radar antenna scanning.

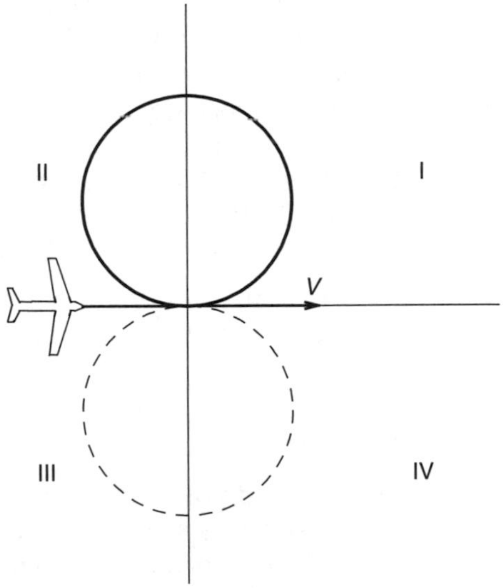

FIGURE 6.5
The power spectral density in the short-range area. Geometric locus of compensation (solid line) and the double effective bandwidth (dotted line) of the power spectral density of fluctuations of the target return signal.

After very complicated mathematical transformations, $R^h_{mov,sc}(\Delta\ell, \Delta\beta_0)$ can be determined in the following forms:

$$R^h_{mov,sc}(t,\tau) = 6a^{-3} \cdot [2\sin ab - \sin(a - ab) + a(1 - 0.5b)\cos ab - ab\cos(a - ab)] \times e^{j\Omega_0\tau} \quad \text{at} \quad 0 \le b \le 0.5; \tag{6.26}$$

$$R^h_{mov,sc}(t,\tau) = 6a^{-3} \cdot [\sin ab - a(1 - b)\cos(a - ab)] \cdot e^{j\Omega_0\tau} \quad \text{at} \quad 0.5 \le b \le 1; \tag{6.27}$$

$$R^h_{mov,sc}(t,\tau) = 0 \quad \text{at} \quad b > 1, \tag{6.28}$$

where

$$a = 3\pi \Delta F_{sc} \cdot |\tau|; \tag{6.29}$$

$$b = 0.75\Delta F_{mov} \cdot |\tau|; \tag{6.30}$$

ΔF_{sc} is the effective bandwidth of the power spectral density $S_{sc}(\omega)$ of the fluctuations caused only by radar antenna scanning, which is given by Equation (5.42), at the **sinc**-directional diagram; ΔF_{mov} is the effective bandwidth of the power spectral density $S_{mov}(\omega, t)$ of the fluctuations caused only by the moving radar, which is given by Equation (3.87), at the **sinc**-directional diagram (see Figure 6.6). Here the effective bandwidth ΔF_{mov} is not the total effective bandwidth of the power spectral density $S_{mov}(\omega, t)$. It is only that part of the power spectral density $S_{mov}(\omega, t)$ that corresponds to the power spectral density of the Doppler fluctuations caused by the difference in the Doppler frequencies in the azimuth plane. The total normalized correlation function $R_{mov,sc}(\Delta\ell, \Delta\beta_0, \Delta\gamma_0)$ is defined according to Equation (6.7) by the product of $R^h_{mov,sc}(\Delta\ell, \Delta\beta_0)$ given by Equation (6.8) and $R^v_{mov,sc}(\Delta\ell, \Delta\gamma_0)$ given by Equation (6.9). For the case considered here, $R^h_{mov,sc}(\Delta\ell, \Delta\beta_0)$ has been determined only by Equation (6.26) — Equation (6.30) and $R^v_{mov,sc}(\Delta\ell, \Delta\gamma_0)$ is given by Equation (3.88).

As one can see from Figure 6.6, under the condition $q = \frac{\Delta F_{sc}}{\Delta F_{mov}} = 0$, i.e., when the radar antenna scanning is stopped, the total normalized correlation function $R_{mov,sc}(t, \tau)$ of the fluctuations caused by the moving radar with simultaneous radar antenna scanning is smooth and coincides with the normalized correlation function $R^v_{mov,sc}(t,\tau)$ given by Equation (3.88). With an increase in the value of q, we can observe the side lobes, the level of which increases and at $q = 2$ reaches the maximum (for about 5%). As $q \to \infty$, i.e., the radar is stationary, the number of side lobes increases, but their level decreases and $R_{mov,sc}(t, \tau)$ becomes a smooth function again and coincides with $R^h_{mov,sc}(\tau)$ given by Equation (5.39).

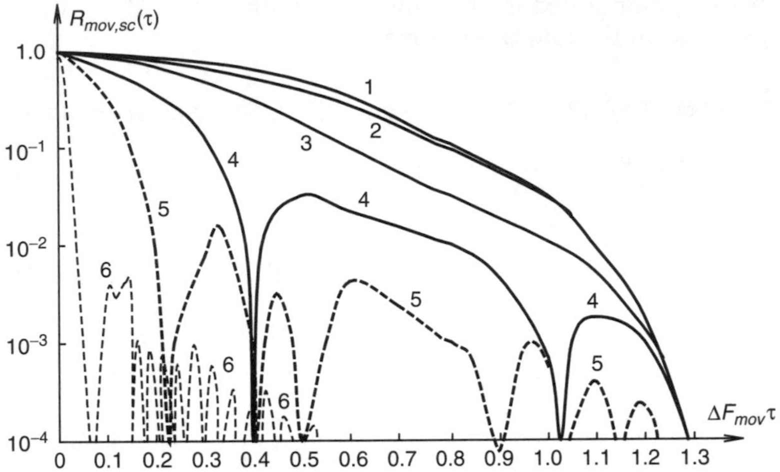

FIGURE 6.6
The normalized correlation function of fluctuations of the target return signal caused by the moving radar with simultaneous radar antenna line scanning for the **sinc**-directional diagram of radar antenna at various values of $q = \frac{\Delta F_{sc}}{\Delta F_{mov}}$: (1) $q = 0$; (2) $q = 0.5$; (3) $q = 1$; (4) $q = 2$; (5) $q = 5$; (6) $q = 40$.

Under scanning the two-dimensional (surface) target with the pulsed searching signals and radar antenna line scanning to define the total normalized correlation function $R_{mov,sc}(t, \tau)$ it is necessary, according to Section 6.1.2, to take the product of $R^{v}_{mov,sc}(t,\tau)$ given by Equation (6.26) and Equation (6.30), in which the variable γ_0 is replaced with the variable γ_*, and $R_\gamma(t, \tau)$ given by Equation (2.135). If we use the short pulsed searching signal, i.e., the duration of the pulsed searching signal is low in value so that the correlation interval given by Equation (4.70) is much more than that of $R^{v}_{mov,sc}(t,\tau)$, given by Equation (6.26)–Equation (6.30), it characterizes completely the interperiod fluctuations caused by the moving radar with simultaneous radar antenna scanning.

6.3 The Moving Radar with Simultaneous Radar Antenna Conical Scanning

6.3.1 The Instantaneous Power Spectral Density

Let us assume that the radar moving uniformly with the velocity V and searching the three-dimensional (space) target makes the open conical scanning by radar antenna with the angular velocity Ω_{sc}, and the angle between the directional diagram axis and the equisignal direction is equal to θ_{sc}.[12] The equisignal

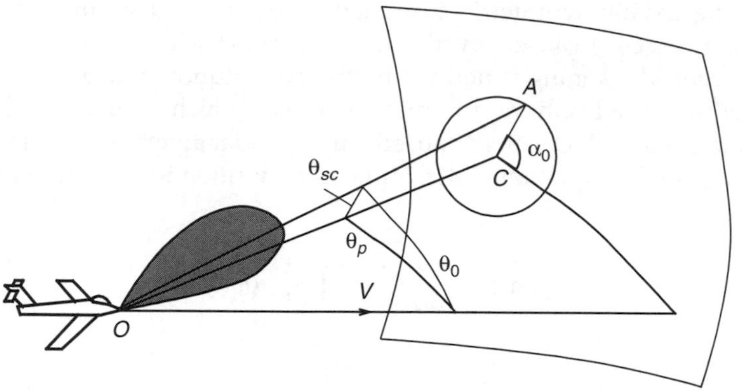

FIGURE 6.7
The coordinate system under the moving radar with simultaneous radar antenna conical scanning: *OC* — equisignal direction; *OA* — the axis of the radar antenna directional diagram.

direction is deflected on the arbitrary angle θ_p from the vector of velocity V (see Figure 6.7). Then, the directional diagram axis at the arbitrary instant of time is set under the angle

$$\theta_0 = \sqrt{\theta_p^2 + \theta_{sc}^2 + 2\theta_p\theta_{sc} \cdot \cos\alpha_0} \tag{6.31}$$

with respect to the vector of velocity V, where $\alpha_0 = \Omega_{sc}t$ is the angle characterizing the moving position of the radar antenna. The angle α_0 is reckoned from a plane passing through the equisignal direction and the vector of velocity V (see Figure 6.7). If the condition $\theta_p >> \theta_{sc}$ is satisfied, we can write

$$\theta_0 \approx \theta_p + \theta_{sc} \cdot \cos\alpha_0 \, . \tag{6.32}$$

If the condition $\theta_p = 0°$ is true, then $\theta_p = \theta_{sc} = const.$

The shape, bandwidth, and average frequency of the power spectral density $S_{mov}(\omega, t)$ of the fluctuations caused only by the moving radar, as was shown in Chapter 3, depend essentially on the angle θ_0. In other words, during the radar antenna conical scanning, the power spectral density of the fluctuations $S_{mov}(\omega, t)$ caused only by the moving radar is continuously deformed, i.e., $S_{mov}(\omega, t)$ becomes a function of time.

With the Gaussian directional diagram, the instantaneous normalized correlation function $R_{mov,sc}(\omega, t)$ of the fluctuations caused by the moving radar with simultaneous radar antenna scanning can be determined by Equation (6.4). The total instantaneous power spectral density of fluctuations of $S_{mov,sc}(\omega, t)$ caused by the moving radar with simultaneous radar antenna conical scanning is defined by the convolution between $S_{mov}(\omega, t)$ caused only by the moving radar and $S_{sc}(\omega)$ caused only by radar antenna conical scanning.

With the axially symmetric Gaussian directional diagram of the radar antenna, $S_{mov,sc}(\omega, t)$ caused by the moving radar with simultaneous radar antenna conical scanning is defined by the convolution between $S_{mov}(\omega, t)$ of fluctuations caused only by the moving radar, which is given by Equation (3.80), and $S_{sc}(\omega)$ of fluctuations caused only by radar antenna conical scanning, which is given by Equation (5.81), and can be written in the following form:

$$S_{mov,sc}(\omega, t) \cong \sum_{k=0}^{\infty} I_k\left[\tfrac{\pi\,\theta_{sc}^2}{\Delta_a^{(2)}}\right] \cdot S_{mov}(\omega - k\Omega_{sc} t)\ . \tag{6.33}$$

The coefficients of the series $I_k(x)$ with the used values of the parameter $\frac{\pi\theta_{sc}^2}{\Delta_a^{(2)}}$ decrease fast, so that in practice the process can be limited only by three terms: $k = -1, 0, 1$.[13]

If the effective bandwidth ΔF_{mov} of the power spectral density $S_{mov}(\omega, t)$ of fluctuations caused only by the moving radar is satisfied by the condition $\Delta F_{mov} < F_{sc}$, where F_{sc} is the frequency of radar antenna conical scanning, the total instantaneous power spectral density of fluctuations $S_{mov,sc}(\omega, t)$ caused by the moving radar with simultaneous radar antenna conical scanning, which is given by Equation (6.33), consists of three individual parts spaced near the frequencies: $\omega_0 + \Omega_0$, $\omega_0 + \Omega_0 + \Omega_{sc}$, and $\omega_0 + \Omega_0 - \Omega_{sc}$. The shape of these individual parts of $S_{mov,sc}(\omega, t)$ depends on a value of the angle θ_0 (see Figure 6.8 and Figure 6.9).

Under the high-deflected equisignal direction, every individual part of $S_{mov,sc}(\omega, t)$ has the Gaussian form (see Figure 6.8). Under the nondeflected equisignal direction, every individual part of $S_{mov,sc}(\omega, t)$ has the shape of the function given by Equation (3.141) (see Figure 6.9). If the condition $\Delta F_{mov} > F_{sc}$ is satisfied, these individual parts of $S_{mov,sc}(\omega, t)$ are overlapped. Under the

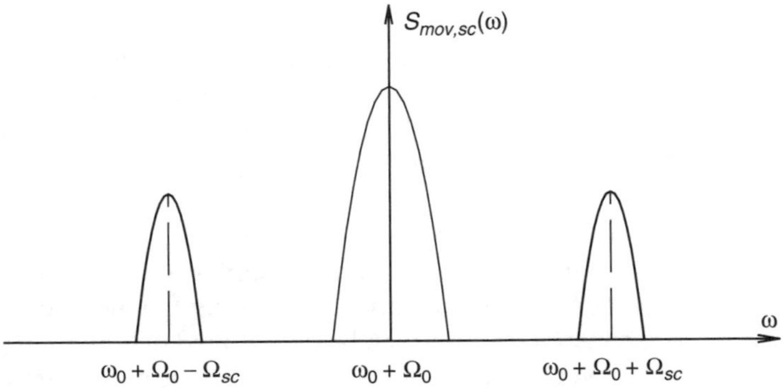

FIGURE 6.8
The instantaneous power spectral density of fluctuations of the target return signal under the moving radar with simultaneous radar antenna conical scanning. Equisignal direction is high deflected, and $\Delta F_{mov} << F_{sc}$.

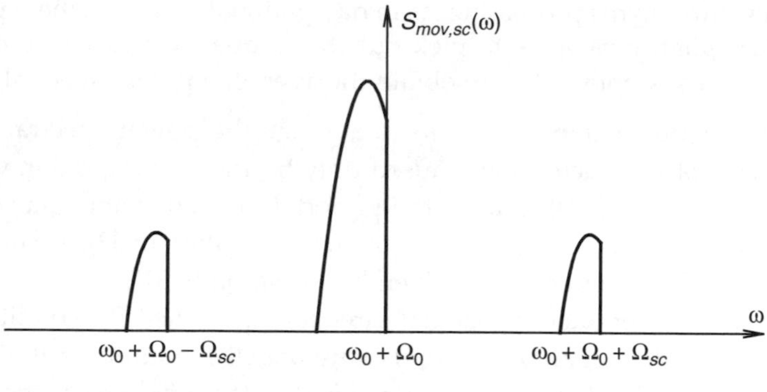

FIGURE 6.9
The instantaneous power spectral density of fluctuations of the target return signal under the moving radar with simultaneous radar antenna conical scanning. Equisignal direction is not deflected, and $\Delta F_{mov} << F_{sc}$.

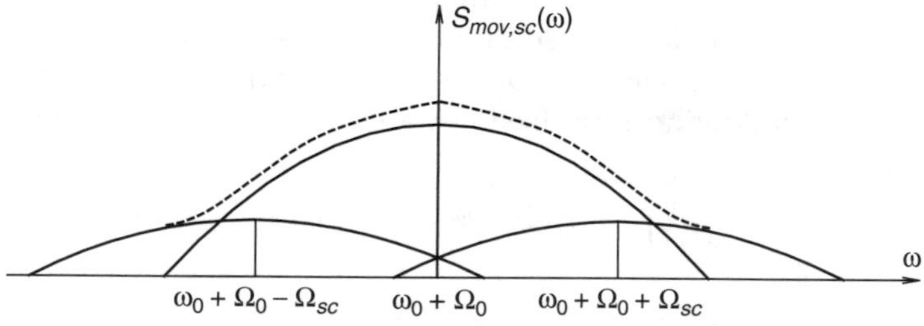

FIGURE 6.10
The instantaneous power spectral density of fluctuations of the target return signal under the moving radar with simultaneous radar antenna conical scanning. Equisignal direction is high deflected, and $\Delta F_{mov} >> F_{sc}$.

condition $\Delta F_{mov} >> F_{sc}$, the effective bandwidth and shape of $S_{mov,sc}(\omega, t)$ are defined by the effective bandwidth and shape of the central part (see Figure 6.10). In practice, the condition $\Delta F_{mov} >> F_{sc}$ is true when the radar antenna is high deflected, and the condition $\Delta F_{mov} < F_{sc}$ is true with the nondeflected radar antenna.[14,15]

6.3.2 The Averaged Power Spectral Density

In target tracking in navigational systems, the averaged power spectral density $\overline{S_{mov,sc}(\omega, t)}$ of fluctuations of the target return signal caused by the moving radar with simultaneous radar antenna conical scanning is more important compared to the total instantaneous power spectral density $S_{mov,sc}(\omega, t)$ of fluctuations of the target return signal caused by the moving radar with simultaneous radar antenna conical scanning. This is because the target return signal is subjected to only frequency signal processing, not frequency–time signal processing, as in navigational systems where purpose of the surveillance radar is to pick out the frequency and phase of radar antenna conical scanning.[16,17] To obtain the averaged power spectral density $\overline{S_{mov,sc}(\omega,t)}$ of fluctuations, we need to average the power spectral density $S_{mov}[\omega, \theta_0(t)]$ of the fluctuations caused only by the moving radar, which is given by Equation (3.80), during the period T_{sc} of radar antenna scanning for every term of $S_{mov,sc}(\omega, t)$, which is given by Equation (6.33). We must bear in mind that the parameter θ_0 is given by Equation (6.31).

Under the nondeflected equisignal direction, i.e., when the condition $\theta_p = 0°$ is satisfied, the equality $\theta_0 = \theta_{sc}$ is true and the parameter θ_0 does not depend on time. The target return signal is the stationary stochastic process and there is no need to average it. $\overline{S_{mov,sc}(\omega,t)}$, which is given by Equation (6.33), coincides with the averaged $\overline{S_{mov,sc}(\omega,t)}$. Under the high-deflected equisignal direction, this problem can be solved only approximately.[18]

Let us consider the total instantaneous normalized correlation function $R_{mov,sc}(t, \tau)$ of the fluctuations caused by the moving radar with simultaneous radar antenna conical scanning. For the case considered here, we can write $R_{mov,sc}(t, \tau)$ in the following form [see Equation (3.79) and Equation (3.83)]:

$$R_{mov,sc}(t,\tau) = e^{-\pi \Delta F_{mov}^2 \tau^2 - 2\delta_{sc}\sin^2(0.5\Omega_{sc}\tau) + j\Omega_{\max}\cos\theta_0(t)\tau}, \tag{6.34}$$

where

$$\Delta F_{mov} = \sqrt{2}\, V\lambda^{-1}\Delta_a \sin\theta_0; \tag{6.35}$$

$$\delta_{sc} = \frac{\pi\,\theta_{sc}^2}{\Delta_a^{(2)}} \tag{6.36}$$

is the parameter of radar antenna conical scanning; the function $\theta_0(t)$ is given by Equation (6.32). Thus, $R_{mov,sc}(t, \tau)$ is a function of time, due to the effective bandwidth ΔF_{mov} of the power spectral density $S_{mov}(\omega, t)$ of the fluctuations caused only by the moving radar and the averaged Doppler frequency $\Omega_{\max} \times \cos\theta_0$ being functions of the parameter $\theta_0(t)$.

With the high-deflected radar antenna, i.e., with the condition $\theta_p >> \theta_{sc}$ being satisfied, changes in the effective bandwidth ΔF_{mov} of the power spectral density $S_{mov}(\omega, t)$ are very low in value, and so can be neglected. We must use only the angle θ_p in Equation (6.35) instead of the angle θ_0. Changes in frequency are very important. Because of this, we can write

$$\cos\theta_0 \approx \cos\theta_p - \theta_{sc}\sin\theta_p \cdot \cos\Omega_{sc} t. \tag{6.37}$$

Taking into consideration Equation (6.37) and averaging $R_{mov,sc}(t, \tau)$, which is given by Equation (6.34), with respect to the variable t, the averaged normalized correlation function $\overline{R_{mov,sc}(t,\tau)}$ of fluctuations caused by the moving radar with simultaneous radar antenna conical scanning can be written in the following form:

$$\overline{R_{mov,sc}(t,\tau)} = I_0\left[\sqrt{8\pi\delta_{sc}}\ \Delta F_{mov,\theta_p}\tau\right]\cdot e^{-\pi\,\Delta F^2_{mov,\theta_p}\tau^2-2\delta_{sc}\cdot\sin^2(0.5\Omega_{sc}\tau)\,+\,j\Omega_p\tau}, \tag{6.38}$$

where

$$\Omega_p = \Omega_{\max}\cdot\ \cos\theta_p \tag{6.39}$$

is the Doppler shift in the frequency corresponding to the equisignal direction; and

$$\Delta F_{mov,\theta_p} = \sqrt{2}\ V\lambda^{-1}\Delta_a \sin\theta_p \tag{6.40}$$

is the effective bandwidth of the power spectral density $S_{mov}(\omega, t)$ of fluctuations caused only by the radar moving when the directional diagram axis is deflected by the angle θ_p.

As one can see from Figure 6.11, under the condition $F_{sc} \le \Delta F_{mov,\theta_p}$, the envelope of the averaged normalized correlation function $\overline{R_{mov,sc}(t,\tau)}$, which is given by Equation (6.38), is smoothed. Under $F_{sc} > \Delta F_{mov,\theta_p}$, the envelope of $\overline{R_{mov,sc}(t,\tau)}$ is wavy. Therefore, the number of waves is high, and the value of the frequency F_{sc} of radar antenna conical scanning is also high.

The Fourier transform of $\overline{R_{mov,sc}(t,\tau)}$ which is given by Equation (6.38), is not determined exactly. The computer-calculated averaged power spectral densities $\overline{S_{mov,sc}(\omega,t)}$ of fluctuations caused by the moving radar with simultaneous radar antenna conical scanning are shown in Figure 6.12. Under the condition $F_{sc} \le \Delta F_{mov,\theta_p}$, $\overline{S_{mov,sc}(\omega,t)}$ is smoothed. Under the condition $F_{sc} > \Delta F_{mov,\theta_p}$, $\overline{S_{mov,sc}(\omega,t)}$ have the waves corresponding to the frequency F_{sc} of radar antenna conical scanning.

The averaged normalized correlation function $\overline{R_{mov,sc}(t,\tau)}$ of fluctuations caused by the moving radar with simultaneous radar antenna conical scanning can be determined in the following form based on results discussed in Feldman et al.:[19]

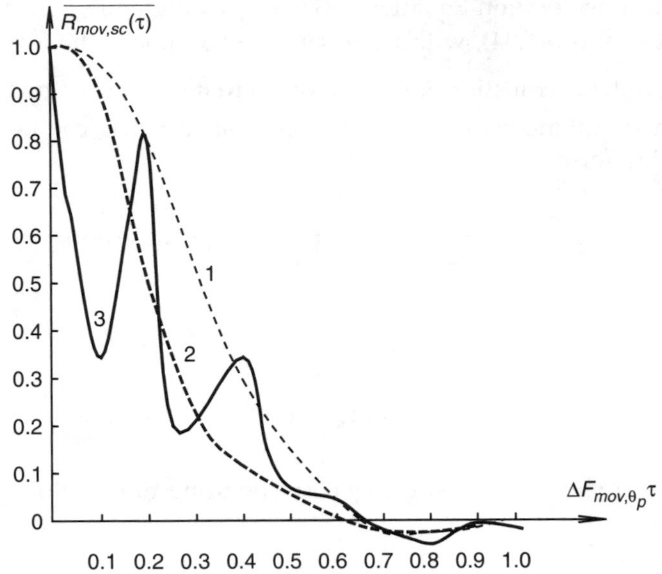

FIGURE 6.11
The envelope of the averaged normalized correlation function of fluctuations of the target return signal under the moving radar with simultaneous radar antenna conical scanning, $\frac{\Delta\theta_{sc}}{\Delta_a} = 4$: (1) $\frac{F_{sc}}{\Delta F_{mov}} = 0.2$; (2) $\frac{F_{sc}}{\Delta F_{mov}} = 1$; (3) $\frac{F_{sc}}{\Delta F_{mov}} = 5$.

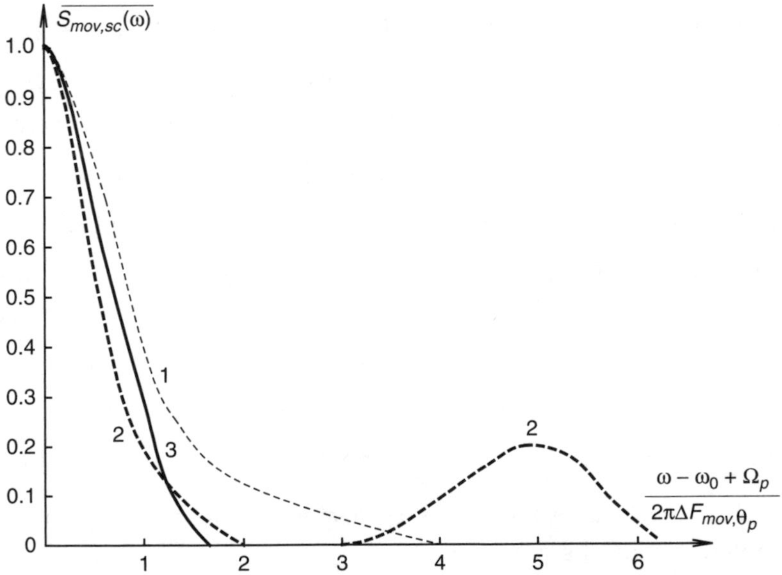

FIGURE 6.12
The envelope of the averaged power spectral density of fluctuations of the target return signal under the moving radar with simultaneous radar antenna conical scanning, $\frac{\Delta\theta_{sc}}{\Delta_a} = 4$: (1) $\frac{F_{sc}}{\Delta F_{mov}} = 0.2$; (2) $\frac{F_{sc}}{\Delta F_{mov}} = 1$; (3) $\frac{F_{sc}}{\Delta F_{mov}} = 5$.

$$\overline{R_{mov,sc}(t,\tau)} \approx e^{-\pi\,\Delta F_1^2\tau^2 - 2\delta_{sc}\cdot\sin^2(0.5\Omega_{sc}\tau) + j\Omega_p\tau}, \tag{6.41}$$

where

$$\Delta F_1 = \frac{e^{\delta_{sc}}}{I_0(\delta_{sc})}\cdot \Delta F_{mov,\theta_p}. \tag{6.42}$$

The resulting instantaneous power spectral density $S_{mov,sc}(\omega, t)$ of fluctuations is defined by the convolution between the Gaussian power spectral density $S_{mov}(\omega, t)$ of fluctuations caused only by the moving radar and the power spectral density $S_{sc}(\omega)$ of fluctuations caused only by radar antenna conical scanning, given by Equation (5.81) that corresponds to Figure 6.12. All formulae obtained and discussed in this section, results, and conclusions are true for navigational systems employing the radar with the hidden radar antenna conical scanning. In this case, we must replace the parameter θ_{sc} with the parameter $0.5\theta_{sc}$.

6.4 Conclusions

Target return signal fluctuations caused by the moving radar with simultaneous radar antenna scanning are the nonstationary stochastic process, possessing a set of peculiarities that cannot be investigated by individually studying fluctuations caused only by the moving radar or caused only by radar antenna conical scanning. Here, the moving radar with simultaneous radar antenna line scanning and with simultaneous radar antenna conical scanning are discussed.

When scanning both the three-dimensional (space) target and the two-dimensional (surface) target, in general, the normalized correlation function of the fluctuations caused by the moving radar with simultaneous radar antenna conical scanning cannot be expressed as the product of the normalized correlation function of fluctuations caused only by the moving radar and those caused only by radar antenna scanning. An exception is the case where the directional diagram is Gaussian. For this, the total instantaneous power spectral density of the fluctuations caused by the moving radar with simultaneous radar antenna scanning is defined by the convolution between the power spectral densities of fluctuations caused only by the moving radar and only by radar antenna scanning.

Under the moving radar with simultaneous radar antenna uniform line circular scanning, the normalized correlation function of the target return signal fluctuations can be expressed as the product of the normalized correlation function of the fluctuations caused only by the moving radar and that caused only by radar antenna line scanning. In this case, the normalized correlation function of the fluctuations caused only by the moving radar depends on the orientation of the radar antenna with respect to the velocity vector of the moving radar and obeys the Gaussian law with the high-deflected radar antenna. The normalized correlation function of fluctuations by radar antenna line scanning is also Gaussian. The total instantaneous power spectral density of fluctuations caused by the moving radar with simultaneous radar antenna line scanning obeys the Gaussian law. During the radar antenna line scanning, the instantaneous power spectral density of fluctuations caused by the moving radar with simultaneous radar antenna line scanning is deformed, continuously taking various shapes in accordance with the position and width of the directional diagram, velocity of moving radar, and frequency of radar antenna line scanning.

In the case of the moving radar with simultaneous radar antenna conical scanning, and when the directional diagram is Gaussian, the instantaneous normalized correlation function of fluctuations is expressed as the product of the normalized correlation functions of fluctuations caused only by the moving radar and those caused only by radar antenna conical scanning. The total instantaneous power spectral density of fluctuations caused by the moving radar with simultaneous radar antenna conical scanning is defined by the convolution between the instantaneous power spectral densities of fluctuations caused only by the moving radar and those caused only by radar antenna conical scanning.

References

1. Blackman, S., *Multiple-Target Tracking with Radar Applications,* Artech House, Norwood, MA, 1986.
2. Alexander, S., *Adaptive Signal Processing: Theory and Applications,* Springer-Verlag, New York, 1986.
3. Haykin, S., *Communication Systems,* 3rd ed., John Wiley & Sons, New York, 1994.
4. Rappaport, T., *Wireless Communications: Principles and Practice,* Prentice Hall, Englewood Cliffs, NJ, 1996.
5. Godard, D., Self-recovering equalization and carrier tracking in two-dimensional data communication systems, *IEEE Trans.,* Vol. COM-28, No. 11, 1980, pp. 1867–1875.
6. Headrick, J. and Skolnik, M., Over-the-horizon radar in the HF band, in *Proceedings of the IEEE,* Vol. 6, No. 6, 1974, pp. 664–673.
7. Jao, J., A matched array beamforming technique for low angle radar tracking in multipath, in *Proceedings of the IEEE National Radar Conference,* 1994, pp. 171–176.
8. Krolik, J. and Andersen, R., Maximum likelihood coordinate registration for over-the-horizon radar, *IEEE Trans.,* Vol. SP-45, No. 4, 1997, pp. 945–959.
9. Papazoglou, M. and Krolik, J., Matched-field estimation of aircraft altitude from multiple over-the-horizon radar revisits, *IEEE Trans.,* Vol. SP-47, No. 4, 1999, pp. 966–976.

10. Kohno, R. et al., Combination on adaptive array antenna and a canceller of interference for direct-sequence spread-spectrum multiple-access systems, *IEEE J. Select. Areas Commun.*, Vol. 8. No. 5, 1990, pp. 641–649.
11. Widrow, B. and Stearus, S., *Adaptive Signal Processing*, Prentice Hall, Englewood Cliffs, NJ, 1985.
12. Vaidyanathau, C. and Buckley, K., An adaptive decision feedback equalizer antenna array for multiuser CDMA wireless communications, in *Proceedings of the 30th Asilomar Conference on Circuits, Systems, Computers*, Pacific Grove, CA, November 1996, pp. 340–344.
13. Machi, O., *Adaptive Processing*, John Wiley & Sons, New York, 1995.
14. Schmidt, R., Multiple emitter location and signal parameter estimation, *IEEE Trans.*, Vol. AP-34, No. 3, 1986, pp. 276–280.
15. McNamara, L., *The Ionosphere: Communications, Surveillance, and Direction Finding*, Krieger, Malabar, FL, 1991.
16. Monzingo, R. and Miller, T., *Introduction to Adaptive Arrays*, John Wiley & Sons, New York, 1980.
17. Benedetto, S. and Biglieri, E., Non-linear equalization of digital satellite channels, *IEEE J. Select. Areas Commun.*, Vol. SAC-1, No. 1, 1983, pp. 57–62.
18. Liu, H. and Zoltowski, M., Blind equalization in antenna array CDMA systems, *IEEE Trans.*, Vol. SP-45, No. 1, 1997, pp. 161–172.
19. Feldman, Yu, Gidaspov, Yu, and Gomzin, V., *Moving Target Tracking*, Soviet Radio, Moscow, 1978 (in Russian).

7

Fluctuations Caused by Scatterers Moving under the Stimulus of the Wind

7.1 Deterministic Displacements of Scatterers under the Stimulus of the Layered Wind

Fluctuations of the target return signal caused by displacements of scatterers under the stimulus of the wind are of great interest.[1–7] Let us consider the slow fluctuations of the target return signal caused, for example, by a moving cloud of scatterers under the stimulus of the wind, gravity, stream of reactive aircraft, and other sources. As regards the rapid fluctuations of the target return signal in the radar range with the pulsed searching signal, all statements and conclusions made in Section 3.1.2 are true for the case considered there. With the continuous searching signal, the slow fluctuations completely cover the whole spectrum of fluctuation sources. The moving radar can be considered as the particular case of the consistent motion of scatterers with the velocity equal in value to the velocity of the moving radar with the opposite sign. In the general case, the motion of the cloud of scatterers is conveniently split into two components:[8,9] the deterministic motion of scatterers with various velocities, which can be defined by the nonstochastic function of coordinates and time (for example, variations in the velocity of the wind as a function of altitude [the layered wind], i.e., the motion of cloud of scatterers as a whole) and the stochastic motion of scatterers with the velocity at random and varied in time, in the general case.

Let us consider the fluctuations caused by the deterministic displacements of scatterers, in particular, the simultaneous stimulus of the layered wind and moving radar. In principle, this problem is analogous to the problem of the moving radar only and was investigated in more detail in Chapter 3. We can solve this problem and obtain a solution if at the given law of motion of scatterers under the stimulus of the wind we can define the radial displacements $\Delta\rho_w$ of scatterers as the function of the angle coordinates β and γ. Obviously, the radial displacements $\Delta\rho_w$ of scatterers must be added to the displacements $\Delta\rho_r$ of scatterers caused by the moving radar.

Let us assume that the velocity V_w of the wind and the velocity of the moving scatterers vary linearly as a function of altitude

$$V_w = V_{w_0} + |\vec{g}|h, \tag{7.1}$$

where V_{w_0} is the velocity of the wind at the altitude h at the point of cloud of the scatterers; $\vec{g}$ is the gradient of velocity of the wind; and h is the altitude reckoned from the plane of the moving radar (see Figure 7.1). Let us suppose that the velocity of the wind is directed under the angle β_w with respect to the velocity of the moving radar. As the following equality

$$h = \rho \sin\gamma \tag{7.2}$$

is true, where ρ is the distance between the radar and the corresponding pulsed volume, the radial components of the velocity of scatterers can be determined in the following form:

$$V_{w_r} = V_w \cos(\beta - \beta_w)\cos\gamma = (V_{w_0} + |\vec{g}|\rho\sin\gamma)\cos(\beta - \beta_w)\cos\gamma. \tag{7.3}$$

Let us take into consideration the moving radar. The velocity of the moving radar V_a relative to the Earth's surface is equal to the vector sum of the

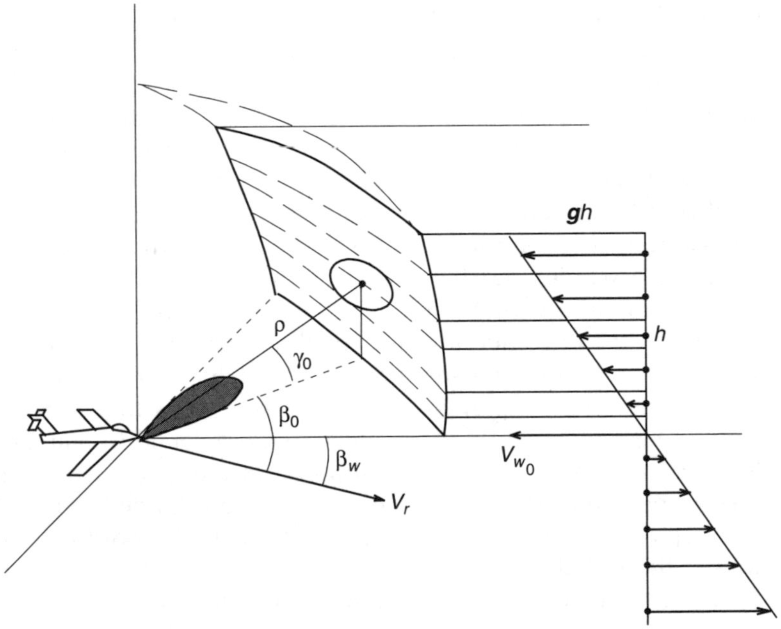

FIGURE 7.1
The power spectral density formation with the layered wind.

velocity of the moving radar in air V_r and the velocity V'_{w_0} of the wind at the point of the radar in space. If the distance between the radar and the cloud of scatterers, which forms the target return signal, is high in value, then the velocities V_{w_0} and V'_{w_0} cannot be equal. The projection of the velocity V_a of the moving radar relative to the Earth's surface in the direction defined by the angles β and γ can be written in the following form:

$$V_{a_r} = V_r \cos\beta \cos\gamma + V'_{w_0} \cos(\beta - \beta_w)\cos\gamma. \tag{7.4}$$

The radial component of the velocity of scatterers relative to the radar is determined by

$$V^r_{scat} = V_{a_r} - V_{w_r} = \{V_r \cos\beta + (\Delta V_{w_0} - |\vec{g}|\rho\sin\gamma)\}\cos\gamma, \tag{7.5}$$

where

$$\Delta V_{w_0} = V'_{w_0} - V_{w_0}\,. \tag{7.6}$$

Using Equation (2.79), we can define the angles β and γ as the function of the variables φ and ψ. We use the Taylor-series expansion for the radial component V^r_{scat}, limiting to linear terms

$$V^r_{scat} = V'_0 - \varphi \cdot V'_h - \psi \cdot V'_v\,, \tag{7.7}$$

where

$$V'_0 = V_r \cos\beta_0 \cos\gamma_0 + (\Delta V_{w_0} - |\vec{g}|\rho\sin\gamma_0)\cos(\beta_0 - \beta_w)\cos\gamma_0\,; \tag{7.8}$$

$$V'_h = V_r \sin\beta_0 + (\Delta V_{w_0} - |\vec{g}|\rho\sin\gamma_0)\sin(\beta_0 - \beta_w)\,; \tag{7.9}$$

$$V'_v = V_r \cos\beta_0 \sin\gamma_0 + (\Delta V_{w_0} \sin\gamma_0 + |\vec{g}|\rho\cos 2\gamma_0)\cos(\beta_0 - \beta_w). \tag{7.10}$$

Further investigation can be carried out as well for the high-deflected radar antenna using the formulae in Section 3.2, in which the variables Ω_0, Ω_h, Ω_v must be replaced with the following variables:

$$\Omega'_0 = 4\,\pi\, V'_0\lambda^{-1};\quad \Omega'_h = 4\,\pi\, V'_h\lambda^{-1};\quad \text{and}\quad \Omega'_v = 4\,\pi\, V'_v\lambda^{-1}; \tag{7.11}$$

respectively. If the wind is absent, Equation (7.11) coincides with Equation (3.48)–Equation (3.51). As a result, the power spectral density of the fluctuations is given by Equation (3.70), in which the variables Ω_0' and

$$A = \sqrt{\Omega_h'^2 + \Omega_v'^2} \tag{7.12}$$

determined by Equation (3.48)–Equation (3.51) must be replaced with the variables Ω_0', Ω_h', and Ω_v' given by Equation (7.11).

As before, the variable Ω_0' defines the Doppler shift in the averaged frequency of the power spectral density of the fluctuations. Because the velocity of the wind can have any sign, the presence of the wind leads to both increase and decrease in the averaged frequency. According to Equation (3.70), the power spectral density of the fluctuations coincides in shape with the squared directional diagram by power in the plane passing through the plane φ (the horizon, see Figure 2.7) under the angle

$$\kappa' = \text{arctg}\,\frac{\Omega_v'}{\Omega_h'}\,. \tag{7.13}$$

However, now this plane does not pass through the direction of the moving radar and the directional diagram axis, neither does it occur in the absence of the wind. If the condition in Equation (3.73) is satisfied, Equation (3.74) should be used to define the power spectral density of the fluctuations. In this case, it takes the following form:

$$S(\omega) = \bar{g}_h^2\left(\frac{\omega - \omega_0 - \Omega_0'}{A}\right), \tag{7.14}$$

where $\bar{g}_h(\ldots)$ is the directional diagram in the plane κ'. The effective bandwidth of the power spectral density of the fluctuations is defined, as in Equation (3.76), by the squared directional diagram width by power in the plane κ':

$$\Delta F = 0.5\pi^{-1} A\, \Delta_\kappa^{(2)}. \tag{7.15}$$

The value of ΔF, as in Equation (3.76), is independent of the parameter λ, but unlike Equation (3.76), it is a complex function of the angles β_0, γ_0, and β_w.

Now let us assume that the directional diagram is Gaussian. Then, the normalized correlation function and the power spectral density of the fluctuations are given by Equation (3.79) and Equation (3.80), respectively, in which the effective bandwidth is determined by the following form:

$$\Delta F = \sqrt{2}\lambda^{-1}\{\Delta_h^{(2)}[V_r \sin\beta_0 + (\Delta V_{w_0} - |\vec{g}|\rho \sin\gamma_0)\sin(\beta_0 - \beta_w)]^2$$
$$+ \Delta_v^{(2)}[V_r \cos\beta_0 \sin\gamma_0 + (\Delta V_{w_0} \sin\gamma_0 + |\vec{g}|\rho\cos 2\gamma_0)\cos(\beta_0 - \beta_w)]^2\}^{0.5}. \tag{7.16}$$

The formula in Equation (7.16) is the generalization of Equation (3.81) for the case of the proper deterministic motion of scatterers. In doing so, under the conditions $\Delta V_{w_0} = 0$ and $|\vec{g}| = 0$ — i.e., the wind is absent — Equation (3.81) follows from Equation (7.16). Because in the general case the values of ΔV_{w_0} and $|\vec{g}|$ can have any sign, the presence of the wind leads to both extension and narrowing of the effective bandwidth of the power spectral density of the target return signal fluctuations caused by the moving radar. Dependence of the effective bandwidth ΔF of the power spectral density of the fluctuations on the parameter ρ — the distance between the radar and the target — is a very interesting peculiarity of Equation (7.16). This can be explained by the fact that, in accordance with the parameter ρ, the directional diagram illuminates an area of the cloud of scatterers, which have various velocities as a function of altitude. Let us consider some particular cases.

7.1.1 The Radar Antenna Is Deflected in the Horizontal Plane

In this case, the condition $\gamma_0 = 0$ is true. Let $\Delta V_{w_0} = 0$. Then,

$$\Omega_0' = 4\pi V_r \lambda^{-1} \cos\beta_0 ; \tag{7.17}$$

$$\Delta F = \sqrt{2}\lambda^{-1}\sqrt{\Delta_h^{(2)} V_r^2 \sin^2\beta_0 + \Delta_v^{(2)} |\vec{g}|^2 \rho^2 \cos^2(\beta_0 - \beta_w)} = \sqrt{\Delta F_{mov}^2 + \Delta F_w^2}, \tag{7.18}$$

where ΔF_{mov} is the effective bandwidth of the power spectral density of the target return signal fluctuations caused only by the moving radar, which is given by Equation (3.81); ΔF_w is the effective bandwidth of the power spectral density of fluctuations of the target return signal caused only by the stimulus of the wind (the radar is stationary). Reference to Equation (7.18) shows that if the radar antenna is deflected in the azimuth plane and the aspect angle is low in value, the stimulus of the layered wind leads always to extension of the power spectral density of the fluctuations caused by the moving radar.

7.1.2 The Radar Antenna Is Deflected in the Vertical Plane

In this case, the condition $\beta_0 = 0°$ is true. Let us also assume that the direction of the wind coincides with the direction of the moving radar, i.e., the conditions $\Delta V_{w_0} = 0$ and $\beta_w = 0°$ are satisfied. Then we can write

$$\Omega_0' = 4\pi\lambda^{-1}(V_r - |\vec{g}|\rho\sin\gamma_0)\cos\gamma_0\,; \tag{7.19}$$

$$\Delta F = \sqrt{2}\,\Delta_h\lambda^{-1}\,|\,V_r\sin\gamma_0 + |\vec{g}|\rho\cos 2\gamma_0\,| = |\,\Delta F'_{mov} + \Delta F'_w\,|, \tag{7.20}$$

where $\Delta F'_{mov}$ is the effective bandwidth of the power spectral density of fluctuations caused only by the moving radar; $\Delta F'_w$ is the effective bandwidth of the power spectral density of fluctuations caused only by the stimulus of the wind.

If the radar antenna is deflected in the vertical plane and the direction of the wind is matched with the direction of the moving radar, the effective bandwidth of the resulting power spectral density of the fluctuations is defined by the algebraic sum of the effective bandwidths of the power spectral density of the fluctuations caused only by the moving radar and that caused only by the stimulus of the wind. In this case, both the extension and the narrowing of the power spectral density are possible because components can be compensated by each other. With some values of the angle γ_0 and the distance ρ, the effective bandwidth of the power spectral density of the fluctuations is equal to zero for the considered approximation.

Narrowing of the power spectral density can be easily understood if we take into consideration the fact that with an increase in the value of the aspect angle γ, the radial component of the velocity of the moving radar decreases and the radial component of the velocity of the wind increases with the corresponding sign of the gradient of velocity of the wind. Due to this fact, their sum can be approximately constant within the limits of the narrow interval of the aspect angle γ. This phenomenon is illustrated in Figure 7.2, where the segments AB and $A'B'$ are the projections of the velocity V_r on two different directions within the directional diagram. The segments BC and $B'C'$ are the projections of the velocity $|\vec{g}|h$ of the wind as functions of altitude. As one can see from Figure 7.2, under definite conditions, the sums of the segments $AB + BC$ and $A'B' + B'C'$ can be the same for various values of the aspect angle γ. This equality of projections of the resulting velocity is kept constant with a given accuracy within the limits of some interval of values of the aspect angle γ covering the directional diagram width. By virtue of this fact, the total power spectral density of the target return signal fluctuations contracts to a discrete line.

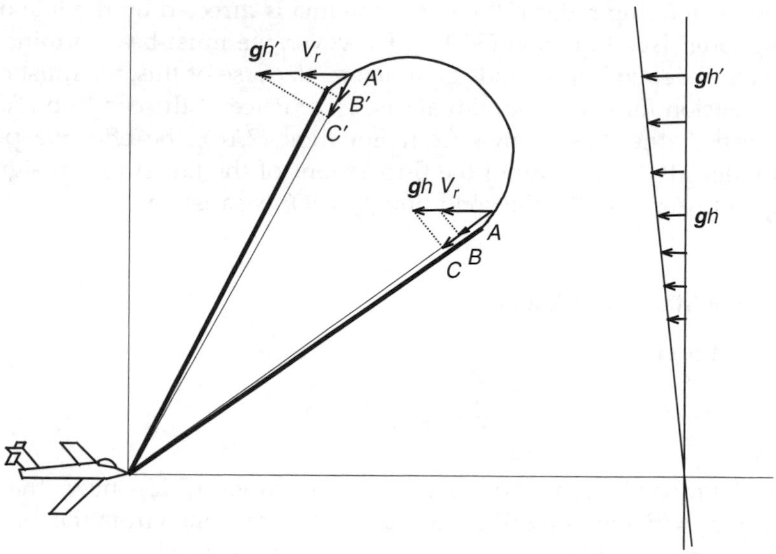

FIGURE 7.2
The power spectral density narrowing with the layered wind.

7.1.3 The Radar Antenna Is Directed along the Line of the Moving Radar

In this case, the conditions $\beta_0 = 0°$ and $\gamma_0 = 0°$ are satisfied. Assume that $\Delta V_{w_0} = 0$. Then, we can write

$$\Omega'_0 = 4\pi\lambda^{-1}V_r\ ; \tag{7.21}$$

$$\Delta F = \sqrt{2}\lambda^{-1}\Delta_h\,|\vec{g}\,|\,\rho\cos\beta_w\ . \tag{7.22}$$

Reference to Equation (7.21) and Equation (7.22) shows that if the condition $\beta_w \neq 90°$ is satisfied, the effective bandwidth of the power spectral density of the target return signal fluctuations is not equal to zero. Unlike in Equation (3.81), the formula in Equation (7.22) is true in the case when the radar antenna is directed exactly along the line of the moving radar. This phenomenon can be explained by the fact that with a given position of the radar antenna, the effective power spectral density bandwidth of the fluctuations caused by stimulus of the layered wind is much more than that caused only by the moving radar. Because of this, we can neglect the target return signal fluctuations caused by the moving radar.

Let us consider an example. Let $\Delta_h = 2°$, $\lambda = 3$ cm, $\rho = 10$ km, $|\vec{g}\,| = 2\,\frac{\text{m}}{\text{sec}}$ km; this value is universally adopted for the middle latitudes. Then, we obtain that $\Delta F \approx 30$ Hz. This value is a whole order of magnitude greater than the effective bandwidth of the power spectral density of the fluctuations caused only by the moving radar if the radar antenna is directed on the line of the moving radar [see Equation (3.141)]. However, we must bear in mind that these values depend on various parameters. Because of this, we must check this conclusion for every individual case in practice. If the condition $\Delta V_{w_0} \neq 0$ is satisfied, then, as follows from Equation (7.16), the effective power spectral density bandwidth of the fluctuations of the target return signal is not equal to zero even if the condition $\beta_w = 90°$ is satisfied.

7.1.4 The Stationary Radar

If the condition

$$\vec{V}_a = \vec{V}_r + \vec{V}_{w_0} \equiv 0 \tag{7.23}$$

is true, the radar is stationary. Reference to Equation (7.23) shows that $V_r = -V'_{w_0}$ and $\beta_w = 0°$. Let us recall that the angle β_w is reckoned from the direction of the vector $\vec{V}_r$. Under these conditions, we obtain that

$$\Omega'_0 = 4\pi\lambda^{-1}\left[V_{w_0} + |\vec{g}|\rho \sin\gamma_0\right]\cos\beta_0 \cos\gamma_0\,; \tag{7.24}$$

$$\Delta F = \sqrt{2}\lambda^{-1} \times \sqrt{\Delta_h^{(2)}(V_{w_0} + |\vec{g}|\rho\sin\gamma_0)^2 \sin^2\beta_0 + \Delta_v^{(2)}(V_{w_0}\sin\gamma_0 - |\vec{g}|\rho\cos 2\gamma_0)^2\cos^2\beta_0} \tag{7.25}$$

Formulae in Equation (7.24) and Equation (7.25) are the main ones that consider the deterministic wind in navigational systems. In particular, taking into consideration Equation (7.24) and Equation (7.25), we can show that if the directional diagram axis is deployed along the direction of the wind, i.e., the condition $\beta_0 = 0°$ is satisfied, the first term in Equation (7.25) is equal to zero. Then, with low values of the aspect angle γ_0, the layered wind plays the main role to form the power spectral density of the target return signal fluctuations. If the directional diagram axis is perpendicular to the direction of the wind, the second term in Equation (7.25) is equal to zero. Then, with low values of the aspect angle γ_0, the velocity V_{w_0} of the wind plays the main role. In some cases, Equation (7.25) can be used to measure the gradient $\vec{g}$ of the wind velocity. Under the conditions $\beta_0 = 0°$ or $\beta_0 = \pm 0.5\pi$, we can choose such values of the aspect angle γ_0 and the parameter ρ that satisfy the condition $\Delta F = 0$ — the target return signal must be gated by the radar range. Then, knowing the values of the aspect angle γ_0, the parameter ρ, and the velocity V_{w_0}, we can easily determine the magnitude of the gradient $\vec{g}$ of the wind velocity.

7.2 Scatterers Moving Chaotically (Displacement and Rotation)

Let us consider the target return signal fluctuations with the chaotic motion of scatterers, taking into account both changes in the phase of elementary signals caused by the radial displacements of scatterers and changes in the amplitudes of elementary signals caused by the rotation of scatterers. We neglect changes in phases of the elementary signals caused by the rotation of scatterers and changes in the amplitudes of elementary signals caused by the radial displacements of scatterers.[10–13]

We assume that the radial displacements $\Delta\rho$ and the angle shifts $\Delta\xi$ and $\Delta\zeta$ of scatterers are random variables and that the amplitude S of the target return signal is also a random variable. We assume that the joint probability distribution density $f(\Delta\rho, \Delta\xi, \Delta\zeta, S)$ is known. In the general case, these values can be dependent. Let us then assume that the joint probability distribution density $f(\Delta\rho, \Delta\xi, \Delta\zeta, S)$ is the same for all scatterers and is independent of the coordinates ρ, φ, and ψ. Because the number of scatterers is very high, the magnitude of $f(\Delta\rho, \Delta\xi, \Delta\zeta, S)d(\Delta\rho)d(\Delta\xi)d(\Delta\zeta)dS$ characterizes the number of scatterers, in which the radial displacements, angle shifts, and amplitudes of the target return signal are within the limits of the corresponding intervals $[\Delta\rho, (\Delta\rho + d(\Delta\rho)]$, $[\Delta\xi, \Delta\xi + d(\Delta\xi)]$, $[\Delta\zeta, \Delta\zeta + d(\Delta\zeta)]$, and $[S, S + dS]$.

At first, we consider the totality of scatterers, in which the radial displacements, angle shifts, and amplitudes of the target return signal are within the limits of the intervals just mentioned. Then, we can use Equation (2.74), in which it is necessary to assume $\tau = nT_p$ because we consider only the slow fluctuations, $\Delta\varphi = 0°$, $\Delta\psi = 0°$, and $\Delta\rho = 0$ for the argument of the pulse function P, and the radial displacements $\Delta\rho$ in exponent are independent of the angles φ and ψ. Then, without determining exactly the power of the target return signal, the correlation function of the fluctuations for the considered totality can be written in the following form:

$$R'(\Delta\rho, \Delta\xi, \Delta\zeta) \cong S^2 R'(\Delta\rho, \Delta\xi, \Delta\zeta) \times f(\Delta\rho, \Delta\xi, \Delta\zeta, S)\, d(\Delta\rho)\, d(\Delta\xi)\, d(\Delta\zeta)\, dS, \tag{7.26}$$

where

$$R'(\Delta\rho, \Delta\xi, \Delta\zeta) = N \cdot \int\int q(\xi, \zeta) \cdot q(\xi + \Delta\xi, \zeta + \Delta\zeta) \cdot e^{-2j\omega_0 \frac{\Delta\rho}{c}} \sin\zeta\, d\xi\, d\zeta\,; \tag{7.27}$$

S^2 is the value that is proportional to the power of the target return signal;

$$f(\Delta\rho, \Delta\xi, \Delta\zeta, S)d(\Delta\rho)d(\Delta\xi)d(\Delta\zeta)dS \tag{7.28}$$

is the averaged number of scatterers for the given totality.

The resulting correlation function of the fluctuations from all scatterers for arbitrary values of $\Delta\rho$, $\Delta\xi$, $\Delta\zeta$, and S in accordance with Equation (2.17) can be written in the following form:

$$R^{en}_{\Delta\rho,\Delta\xi,\Delta\zeta}(\Delta\rho, \Delta\xi, \Delta\zeta, S) \cong \int_{-\infty}^{\infty}\int_{-\infty}^{\infty}\int_{-\infty}^{\infty}\int_{0}^{\infty} S^2 R'(\Delta\rho, \Delta\xi, \Delta\zeta) \times f(\Delta\rho, \Delta\xi, \Delta\zeta, S)\, d(\Delta\rho)\, d(\Delta\xi)\, d(\Delta\zeta)\, dS, \tag{7.29}$$

where integration with respect to the variables $\Delta\xi$ and $\Delta\zeta$ is carried out within the limits of the interval $[-\infty, \infty]$ because there are no limitations for these variables. Based on Equation (7.29), the normalized correlation function of the fluctuations has the following form:

$$R^{en}_{\Delta\rho,\Delta\xi,\Delta\zeta}(\Delta\rho, \Delta\xi, \Delta\zeta, S) = \frac{\int_{-\infty}^{\infty}\int_{-\infty}^{\infty}\int_{-\infty}^{\infty}\int_{0}^{\infty} S^2 R'(\Delta\rho, \Delta\xi, \Delta\zeta)\cdot f(\Delta\rho, \Delta\xi, \Delta\zeta, S)\, d(\Delta\rho)\, d(\Delta\xi)\, d(\Delta\zeta)\, dS}{\int_{0}^{\infty} S^2 f_S(S)\, dS}, \tag{7.30}$$

where $f_s(S)$ is the probability distribution density of the amplitude of the target return signal.

Hereafter, we assume that the angle shifts $\Delta\xi$ and $\Delta\zeta$ are independent of the radial displacements $\Delta\rho$ and the amplitude S of the target return signal. Because of this, we can write

$$f(\Delta\rho, \Delta\xi, \Delta\zeta, S) = f_{\Delta\rho,S}(\Delta\rho, S)\cdot f_{\Delta\xi,\Delta\zeta}(\Delta\xi, \Delta\zeta). \tag{7.31}$$

Reference to Equation (7.30) shows that

$$R^{en}_{\Delta\rho,\Delta\xi,\Delta\zeta}(\Delta\rho, \Delta\xi, \Delta\zeta, S) = R_{\Delta\rho}(\Delta\rho)\cdot R_{\Delta\xi,\Delta\zeta}(\Delta\xi, \Delta\zeta), \tag{7.32}$$

where

$$R_{\Delta\rho}(\Delta\rho) = \frac{\int_{-\infty}^{\infty}\int_{0}^{\infty} S^2 f_{\Delta\rho,S}(\Delta\rho, S)\cdot e^{4j\pi\frac{\Delta\rho}{\lambda}} d(\Delta\rho)\, dS}{\int_{0}^{\infty} S^2 f_S(S)\, dS}; \tag{7.33}$$

$$R_{\Delta\xi,\Delta\zeta}(\Delta\xi, \Delta\zeta) = \frac{\int_{0}^{\pi}\int_{-\pi}^{\pi}\int_{-\infty}^{\infty}\int_{-\infty}^{\infty} q(\xi,\zeta)\cdot q(\xi+\Delta\xi, \zeta+\Delta\zeta)\cdot f_{\Delta\xi,\Delta\zeta}(\Delta\xi,\Delta\zeta)\sin\zeta\, d(\Delta\xi)\, d(\Delta\zeta)\, d\xi\, d\zeta}{\int_{0}^{\pi}\int_{-\pi}^{\pi} q^2(\xi,\zeta)\sin\zeta\, d\xi\, d\zeta}. \tag{7.34}$$

Thus, under the condition that the chaotic radial displacements and rotations of scatterers are the independent random variables, the normalized correlation function of the fluctuations caused by simultaneous chaotic radial displacements and rotations of scatterers is defined by the product of the normalized correlation function of the fluctuations caused only by the radial displacements of scatterers and that caused only by the rotation of scatterers. Let us consider particular cases in the use of Equation (7.33) and Equation (7.34).

7.2.1 Amplitudes of Elementary Signals Are Independent of the Displacements of Scatterers

If the amplitude S of the target return signal — the target return signal is the sum of elementary signals, as before — and the radial displacements $\Delta\rho$ of scatterers are independent random variables, we can write

$$f_{\Delta\rho,S}(\Delta\rho, S) = f_{\Delta\rho}(\Delta\rho)\cdot f_S(S). \tag{7.35}$$

Then, instead of Equation (7.25), the normalized correlation function of the target return signal fluctuations caused only by the radial displacements of scatterers can be written in the following form:

$$R_{\Delta\rho}(\Delta\rho) = \int_{-\infty}^{\infty} f_{\Delta\rho}(\Delta\rho)\cdot e^{4j\pi\frac{\Delta\rho}{\lambda}} d(\Delta\rho). \tag{7.36}$$

This formula in Equation (7.36) coincides with the results discussed in Vanshtein and Zubakov.[14] Therefore, we will consider it briefly here.

Let us assume that the random radial displacements $\Delta\rho$ of scatterers obey the Gaussian law, i.e.,

$$f_{\Delta\rho}(\Delta\rho) = \frac{1}{\sqrt{2\pi}\ \sigma_{\Delta\rho}} \cdot e^{-\frac{\Delta\rho^2}{2\sigma^2_{\Delta\rho}}}, \tag{7.37}$$

where $\sigma^2_{\Delta\rho}$ is the variance of the radial displacements $\Delta\rho$ of scatterers. Under this assumption, the normalized correlation function of the fluctuations caused only by the radial displacements $\Delta\rho$ of scatterers has the following form:

$$R_{\Delta\rho}(\Delta\rho) = e^{-8\pi^2\frac{\sigma^2_{\Delta\rho}}{\lambda^2}}. \tag{7.38}$$

To define the normalized correlation function $R_{\Delta\rho}(\Delta\rho)$ of the fluctuations caused only by the radial displacements $\Delta\rho$ of scatterers, which is given by Equation (7.12), as a function of the shift τ in time, it is necessary to know the variance $\sigma^2_{\Delta\rho}$ as a function of τ.

It would appear reasonable that changes in the velocities of scatterers are low in value and can be considered approximately as constant values within short time intervals. Because of this, with low values of τ, we have

$$\Delta\rho = V^r_{scat} \cdot \tau \qquad \text{and} \qquad \sigma^2_{\Delta\rho} = \sigma^2_{V^r_{scat}} \cdot \tau^2, \tag{7.39}$$

where V^r_{scat} is the radial component of the velocity of scatterers; $\sigma^2_{V^r_{scat}}$ is the variance of the fluctuations of the radial component V^r_{scat} of the velocity of scatterers. Substituting Equation (7.39) in Equation (7.38), we can easily define the conditions in which we conclude that the velocities of scatterers are constant in the determination of the normalized correlation function of the fluctuations $R_{\Delta\rho}(\Delta\rho)$ caused only by the radial displacements $\Delta\rho$ of scatterers, i.e., we can consider this stochastic process as a quasi-stationary process. If the condition in Equation (7.39) is true even for values of τ satisfying the condition

$$8\,\pi^2\lambda^{-2}\sigma^2_{V^r_{scat}}\tau^2 > 4 \qquad \text{or} \qquad \sigma_{V^r_{scat}} \cdot \tau = \sigma_{\Delta\rho} > 0.25\lambda, \tag{7.40}$$

in which $R_{\Delta\rho}(\Delta\rho)$, given by Equation (7.38) tends to approach zero, the velocities of scatterers can be considered constant if their variation is low in value within the limits of the time interval in which the root mean square deviation $\sigma_{\Delta\rho}$ of the radial displacements of scatterers becomes greater than the ratio 0.25λ. It seems likely that this condition is true for scatterers moving under the stimulus of the wind if the value λ is very low.

In the case of constant or approximately constant velocities of the moving scatterer, the power spectral density of the fluctuations caused only by the radial displacements ($\Delta\rho$) of scatterers takes the Gaussian shape after multiplying Equation (7.38) with the exponent $e^{j\omega_0\tau}$ and can be written in the following form:

$$S(\omega) \cong e^{-\pi \frac{(\omega-\omega_0)^2}{(2\pi\,\Delta F_{\Delta\rho})^2}}. \tag{7.41}$$

The effective bandwidth of the power spectral density $S(\omega)$ of the fluctuations caused only by the radial displacements ($\Delta\rho$) of scatterers, which is given by Equation (7.41), can be determined in Hz by

$$\Delta F_{\Delta\rho} = \sqrt{8\pi}\sigma_{V^r_{scat}}\lambda^{-1} \approx 5\sigma_{V^r_{scat}}\lambda^{-1}. \tag{7.42}$$

The effective bandwidth $\Delta F_{\Delta\rho}$ of the power spectral density $S(\omega)$ depends on the variance of the radial velocities of scatterers, which is the function of the characteristics of the wind and the aerodynamics of scatterers, and on the length of the wave λ.[6,15,16] Under the same conditions of observation, $\Delta F_{\Delta\rho}$ is higher and the wavelength λ is shorter, unlike the effective bandwidth of the power spectral density of the Doppler fluctuations of the target return signal, which is independent of the wavelength λ with the deflected directional diagram.

7.2.2 The Velocity of Moving Scatterers Is Random but Constant

Let us assume that scatterers move with approximately constant (in the sense elaborated in the previous section) but random velocities so that $\Delta\rho = V^r_{scat} \cdot \tau$. We do not make any simple assumptions regarding the probability distribution density of the radial displacements or velocities of moving scatterers and the amplitude of elementary signals. Then, based on Equation (7.33), the normalized correlation function of the fluctuations can be determined in the following form:

$$R(\tau) = \frac{\int\limits_{-\infty}^{\infty}\int\limits_{0}^{\infty} S^2 f_{V^r_{scat},S}(V^r_{scat}, S) \cdot e^{4j\pi\frac{V^r_{scat}\cdot\tau}{\lambda}}\, dV^r_{scat}\, dS}{\int\limits_{0}^{\infty} S^2 f_S(S)\, dS}. \tag{7.43}$$

Multiplying the normalized correlation function $R(\tau)$ of the fluctuations by the exponent $e^{j\omega_0\tau}$ and using the Fourier transform, we can write the power spectral density of the fluctuations in the following form:

$$S(\omega) \cong \frac{\lambda}{4\pi}\cdot\int_0^{\infty} S^2 f_{V^r_{scat},S}\left[\tfrac{\lambda}{4\pi}(\omega-\omega_0), S\right] dS\,. \tag{7.44}$$

In particular, if the amplitude S of the target return signal and the radial velocity V^r_{scat} of the radial displacements $\Delta\rho$ of scatterers are mutually independent, the normalized correlation function $R(\tau)$ and the power spectral density $S(\omega)$ of the fluctuations can be determined in the following form:

$$R(\tau) = \int_{-\infty}^{\infty} f_{V^r_{scat}}(V^r_{scat})\cdot e^{4j\pi\frac{V^r_{scat}\tau}{\lambda}}\, dV^r_{scat}\,, \tag{7.45}$$

$$S(\omega) \cong \frac{\lambda}{4\pi}\cdot f_{V^r_{scat}}\left[\tfrac{\lambda(\omega-\omega_0)}{4\pi}\right] \tag{7.46}$$

instead of Equation (7.43) and Equation (7.44).

The formulae in Equation (7.44) and Equation (7.46) have a simple physical meaning. Reference to Equation (7.46) shows that if the scatterers move with random but constant velocities and the amplitude S of the target return signal is independent of the velocity of moving scatterers, the power spectral density of the fluctuations is defined by the probability distribution density of radial velocities of moving scatterers if the condition

$$V^r_{scat} = 0.25\pi^{-1}\,\lambda(\omega - \omega_0) \tag{7.47}$$

is satisfied in Equation (7.46). It is, naturally, because

$$\omega - \omega_0 = 4\,\pi\, V^r_{scat}\lambda^{-1} \tag{7.48}$$

is the Doppler shift in frequency with the velocity V^r_{scat} of moving scatterers. If the velocities V^r_{scat} of moving scatterers and the amplitude S of the target return signal are functionally related, the power spectral density of the fluctuations is defined by the joint probability distribution density of the power of the target return signal and the radial velocities V^r_{scat} of moving scatterers.

The formulae in Equation (7.44) and Equation (7.46) can be used successfully to determine the power spectral density of the fluctuations if these are caused by moving scatterers and the radial velocities V^r_{scat} of moving scatterers are constant or vary slowly. For example, the power spectral density of the fluctuations, which is given by Equation (7.41), follows from Equation (7.46) if we use the probability distribution density of V^r_{scat} based on the probability distribution density of the radial displacements $\Delta\rho$ of scatterers [see Equation (7.37)] using Equation (7.39). We will discuss further that the formulae in Equation (7.44) and Equation (7.46) can be used in the cases where the radial velocities V^r_{scat} of moving scatterers are not random variables but are definite functions of space coordinates.

7.2.3 The Amplitude of the Target Return Signal Is Functionally Related to Radial Displacements of Scatterers

This case, in a certain sense, is the opposite of the case discussed in Section 7.2.1. Let us assume that function $S = U(\Delta\rho)$ exists between the radial displacements $\Delta\rho$ of scatterers and the amplitude S of the target return signal. The joint probability distribution density takes the following form:

$$f_{\Delta\rho,S}(\Delta\rho, S) = f_{\Delta\rho}(\Delta\rho)\cdot f_S(S\,|\,\Delta\rho), \tag{7.49}$$

where $f_s(S\,|\,\Delta\rho)$ is the conditional probability distribution density of the amplitude S of the target return signal under the condition that the radial displacements $\Delta\rho$ of scatterers takes a given value. Owing to the strong functional relationship between S and $\Delta\rho$, we can write

$$f_S(S\,|\,\Delta\rho) = \delta[S - U(\Delta\rho)]\,. \tag{7.50}$$

Then, based on Equation (7.33) the normalized correlation function $R_{\Delta\rho}(\Delta\rho)$ of the fluctuations caused only by the radial displacements $\Delta\rho$ of scatterers can be determined by the following form:

$$R_{\Delta\rho}(\Delta\rho) = \frac{\int\limits_{-\infty}^{\infty} U^2(\Delta\rho) f_{\Delta\rho}(\Delta\rho)\cdot e^{4j\pi\frac{\Delta\rho}{\lambda}}\, d(\Delta\rho)}{\int\limits_{-\infty}^{\infty} U^2(\Delta\rho) f_{\Delta\rho}(\Delta\rho)\, d(\Delta\rho)}\,. \tag{7.51}$$

The formula in Equation (7.51) can be applied in the case of the deterministic motion of scatterers when the power of the searching signal and the radial displacements $\Delta\rho$ are hardly related by the function. This function can appear, for example, when scatterers making radial displacements $\Delta\rho$ are localized within the definite volume of space and illuminated by the definite part of the directional diagram. It occurs if scatterers with the various effective scattering areas move with the different radial velocities V^r_{scat}. For example, large raindrops have a more effective scattering area and fall faster in comparison with small rain drops; passive scatterers of various dimensions and forms move with different velocities.[17]

The shape of the function $U(\Delta\rho)$ depends on the shape of the directional diagram and on the law of moving scatterers. Using Equation (7.51), we can solve the problem considered here before dealing with problems concerning the moving radar and scatterers under the stimulus of the regular wind. The formula in Equation (3.8) was used to solve the problem of the deterministic motion of scatterers, and it has the same physical content as the formula in Equation (7.51), and both can be transformed by each other. These formulae are consequences of the general formula given by Equation (2.74). For this reason, we do not further investigate the formula in Equation (7.51) in this section.

7.2.4 Chaotic Rotation of Scatterers

Let us consider the normalized correlation function $R_{\Delta\xi,\Delta\zeta}(\Delta\xi, \Delta\zeta)$ of the fluctuations caused only by the rotation of scatterers under the assumption that the half-wave dipoles are scatterers. As was discussed in Section 5.4, the following definition $q(\xi, \zeta) = \cos^2\xi\sin^2\zeta$ is true for the component of the target return signal matched by polarization with the searching signal. Then, under the condition that the variables $\Delta\xi$ and $\Delta\zeta$ are independent of the variables ξ and ζ, based on Equation (7.34), $R_{\Delta\xi,\Delta\zeta}(\Delta\xi, \Delta\zeta)$ takes the following form:

$$R_{\Delta\xi,\Delta\zeta}(\Delta\xi,\Delta\zeta) = R_{\Delta\xi}(\Delta\xi)R_{\Delta\zeta}(\Delta\zeta) = \frac{1}{9}\Big[2+\int_{-\infty}^{\infty} f_{\Delta\xi}(\Delta\xi)\cos 2\Delta\xi\, d(\Delta\xi)\Big] \times\Big[2+\int_{-\infty}^{\infty} f_{\Delta\zeta}(\Delta\zeta)\cos 2\Delta\zeta\, d(\Delta\zeta)\Big] \tag{7.52}$$

If the variables $\Delta\xi$ and $\Delta\zeta$ (rotation of scatterers) obey the Gaussian law with zero mean and the variances $\sigma^2_{\Delta\xi}$ and $\sigma^2_{\Delta\zeta}$, respectively, the normalized correlation function $R_{\Delta\xi,\Delta\zeta}(\Delta\xi, \Delta\zeta)$ is determined by

$$R_{\Delta\xi,\Delta\zeta}(\Delta\xi,\Delta\zeta) \cong 0.11\cdot\big[2+e^{-2\sigma^2_{\Delta\xi}}\big]\big[2+e^{-2\sigma^2_{\Delta\zeta}}\big]. \tag{7.53}$$

If the random angular velocities Ω_ξ and Ω_ζ of scatterer rotations are slowly varied, we can define

$$\Delta\xi = \Omega_\xi\cdot\tau \quad\text{and}\quad \Delta\zeta = \Omega_\zeta\cdot\tau; \tag{7.54}$$

$$\sigma^2_{\Delta\xi} = \sigma^2_{\Omega_\xi} \cdot \tau^2 \quad \text{and} \quad \sigma^2_{\Delta\zeta} = \sigma^2_{\Omega_\zeta} \cdot \tau^2\,, \tag{7.55}$$

where $\sigma^2_{\Omega_\xi}$ is the variance of the angular velocity Ω_ξ of the rotation $\Delta\xi$ of scatterers; $\sigma^2_{\Omega_\zeta}$ is the variance of the angular velocity Ω_ζ of the rotation $\Delta\zeta$ of scatterers. As in the previous section, we define the conditions in which the angular velocities Ω_ξ and Ω_ζ can be considered the constant values. In this case, we obtain

$$\sigma^2_{\Delta\xi} = \sigma^2_{\Omega_\xi} \cdot \tau^2 > 2 \quad \text{and} \quad \sigma^2_{\Delta\zeta} = \sigma^2_{\Omega_\zeta} \cdot \tau^2 > 2 \tag{7.56}$$

instead of Equation (7.40).

This means that the angular velocities Ω_ξ and Ω_ζ of scatterers can be considered constant if we can neglect their variation within the limits of the time interval, in which the root mean square deviations $\sigma_{\Delta\xi}$ and $\sigma_{\Delta\zeta}$ of the rotations $\Delta\xi$ and $\Delta\zeta$ of scatterers become more than 90°, i.e., $\sqrt{2}$ radian. Substituting Equation (7.55) and Equation (7.56) in Equation (7.53), and multiplying after substitution on the exponent $e^{j\omega_0\tau}$, and using the Fourier transform, the power spectral density of the fluctuations caused only by the chaotic rotations $\Delta\xi$ and $\Delta\zeta$ of scatterers can be written in the following form:

$$S(\omega) \cong 4\delta(\omega - \omega_0) + \frac{e^{-\pi\frac{(\omega-\omega_0)^2}{\Delta\Omega^2_{\Delta\xi} + \Delta\Omega^2_{\Delta\zeta}}}}{\sqrt{\Delta\Omega^2_{\Delta\xi} + \Delta\Omega^2_{\Delta\zeta}}} + 2 \cdot \frac{e^{-\pi\frac{(\omega-\omega_0)^2}{\Delta\Omega^2_{\Delta\xi}}}}{\Delta\Omega_{\Delta\xi}} + 2 \cdot \frac{e^{-\pi\frac{(\omega-\omega_0)^2}{\Delta\Omega^2_{\Delta\zeta}}}}{\Delta\Omega_{\Delta\zeta}}\,, \tag{7.57}$$

where

$$\Delta\Omega_{\Delta\xi} = 2\,\pi\,\Delta F_{\Delta\xi} = \sqrt{8\pi}\,\sigma_{\Omega_\xi} \quad \text{and} \quad \Delta\Omega_{\Delta\zeta} = 2\,\pi\,\Delta F_{\Delta\zeta} = \sqrt{8\pi}\,\sigma_{\Omega_\zeta}\,. \tag{7.58}$$

Thus, the power spectral density $S(\omega)$ of the fluctuations caused only by the chaotic rotations $\Delta\xi$ and $\Delta\zeta$ of scatterers consists of the one discrete line at the frequency ω_0 of the searching signal and the continuous power spectral density in the form of three Gaussian power spectral densities (see Figure 7.3). The effective bandwidth of each Gaussian power spectral density is equal to $\Delta F_{\Delta\xi}$, $\Delta F_{\Delta\zeta}$, and $\sqrt{\Delta F^2_{\Delta\xi} + \Delta F^2_{\Delta\zeta}}$, respectively.

Reference to Equation (7.53) shows that 4/9 of the total power of the target return signal is concentrated in the discrete component of the power spectral density $S(\omega)$ of the fluctuations caused only by the chaotic rotations $\Delta\xi$ and $\Delta\zeta$ of scatterers at the frequency ω_0; 5/9 is concentrated in the continuous

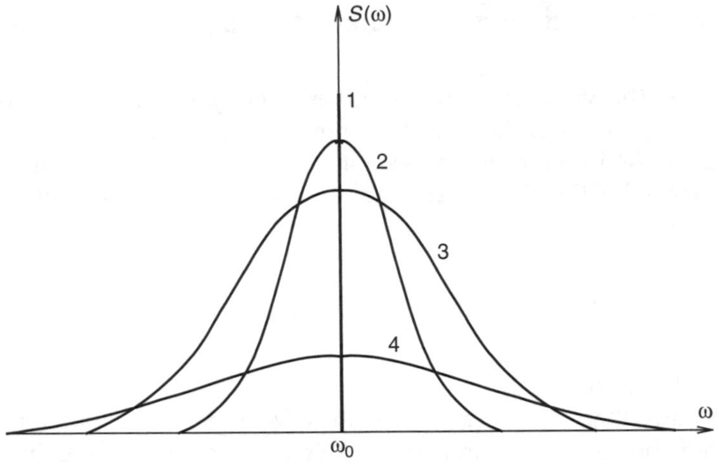

FIGURE 7.3
The power spectral density. Chaotic rotation of scatterers: (1) $\delta(\omega - \omega_0)$; (2) $S_{\Delta\xi}(\omega)$; (3) $S_{\Delta\zeta}(\omega)$; (4) $S_{\Delta\xi,\Delta\zeta}(\omega)$.

part. If the scatterers are stationary, the total power of the target return signal is concentrated near the discrete line of $S(\omega)$ caused by the radial displacements $\Delta\rho$ of scatterers and $\Delta\xi$ and $\Delta\zeta$ at the frequency ω_0. The chaotic rotations $\Delta\xi$ and $\Delta\zeta$ lead us to the generation of the continuous part of $S(\omega)$ caused by $\Delta\rho$ and $\Delta\xi$ and $\Delta\zeta$ at the frequency ω_0, which possesses 5/9 of the total power of the target return signal independent of the shape of the probability distribution density. Unlike this, the chaotic motion (the radial displacements $\Delta\rho$ and chaotic rotations $\Delta\xi$ and $\Delta\zeta$) of scatterers leads to the spread power spectral density of the fluctuations in which all discrete components are absent and the total power of the target return signal is concentrated in the continuous power spectral density of fluctuations.

7.2.5 Simultaneous Chaotic Displacements and Rotations of Scatterers

Comparing Equation (7.42) and Equation (7.58), we can see that if the scatterers moving with approximately the same linear and angular velocities make one rotation on average within the limits of the segment with length equal to the length of the wave λ, i.e., when the condition $\frac{\sigma_{V^r_{scat}}}{\lambda} \approx \frac{\sigma_\Omega}{2\pi}$ is satisfied, then the chaotic displacements $\Delta\rho$ and rotations $\Delta\xi$ and $\Delta\zeta$ of scatterers form the power spectral densities of the fluctuations with the same effective bandwidths.

Because with the simultaneous chaotic displacements $\Delta\rho$ and rotations $\Delta\xi$ and $\Delta\zeta$ of scatterers the normalized correlation function $R_{\Delta\rho}(\Delta\rho)$ of the fluctuations caused only by the radial displacements $\Delta\rho$ of scatterers, which is given by Equation (7.33) and the normalized correlation function $R_{\Delta\xi,\Delta\zeta}(\Delta\xi,$

$\Delta\zeta$) of the fluctuations caused only by the rotations $\Delta\xi$ and $\Delta\zeta$ of scatterers, which is given by Equation (7.34), are multiplied by each other, then, with slowly varying linear and angular velocities, the resulting normalized correlation function $R_{\Delta\rho,\Delta\xi,\Delta\zeta}(\Delta\rho, \Delta\xi, \Delta\zeta)$ of the fluctuations caused by simultaneous chaotic radial displacements $\Delta\rho$ and rotations $\Delta\xi$ and $\Delta\zeta$ is defined by the product of $R_{\Delta\rho}(\Delta\rho)$, which is given by Equation (7.38), and $R_{\Delta\xi,\Delta\zeta}(\Delta\xi, \Delta\zeta)$, which is given by Equation (7.53). Thus, Equation (7.39), Equation (7.54), and Equation (7.55) must be satisfied in Equation (7.38) and Equation (7.53), respectively. The resulting normalized correlation function $R_{\Delta\rho,\Delta\xi,\Delta\zeta}(\Delta\rho, \Delta\xi, \Delta\zeta)$ can be determined by the following form:

$$R_{\Delta\rho,\Delta\xi,\Delta\zeta}(\Delta\rho,\Delta\xi,\Delta\zeta) \cong 0.11 \cdot \left[2 + e^{-2\sigma^2_{\Delta\xi}\cdot\tau^2}\right]\left[2 + e^{-2\sigma^2_{\Delta\zeta}\cdot\tau^2}\right] \cdot e^{-8\pi\frac{\sigma^2_{V^r_{scat}}\tau^2}{\lambda^2}} . \tag{7.59}$$

The resulting power spectral density of the fluctuations caused by simultaneous chaotic radial displacements $\Delta\rho$ and rotations $\Delta\xi$ and $\Delta\zeta$ of scatterers, which is shifted by the frequency ω_0, has the following form:

$$S(\omega) = S_1(\omega) + S_2(\omega) + S_3(\omega) + S_4(\omega) , \tag{7.60}$$

where

$$S_1(\omega) \cong \frac{4}{\Delta\Omega_{\Delta\rho}} \cdot e^{-\pi\frac{(\omega-\omega_0)^2}{\Delta\Omega^2_{\Delta\rho}}} ; \tag{7.61}$$

$$S_2(\omega) \cong \frac{2}{\sqrt{\Delta\Omega^2_{\Delta\rho} + \Delta\Omega^2_{\Delta\xi}}} \cdot e^{-\pi\frac{(\omega-\omega_0)^2}{\Delta\Omega^2_{\Delta\rho}+\Delta\Omega^2_{\Delta\xi}}} ; \tag{7.62}$$

$$S_3(\omega) \cong \frac{2}{\sqrt{\Delta\Omega^2_{\Delta\rho} + \Delta\Omega^2_{\Delta\zeta}}} \cdot e^{-\pi\frac{(\omega-\omega_0)^2}{\Delta\Omega^2_{\Delta\rho}+\Delta\Omega^2_{\Delta\zeta}}} ; \tag{7.63}$$

$$S_4(\omega) \cong \frac{1}{\sqrt{\Delta\Omega^2_{\Delta\rho} + \Delta\Omega^2_{\Delta\xi} + \Delta\Omega^2_{\Delta\zeta}}} \cdot e^{-\pi\frac{(\omega-\omega_0)^2}{\Delta\Omega^2_{\Delta\rho}+\Delta\Omega^2_{\Delta\xi}+\Delta\Omega^2_{\Delta\zeta}}} ; \tag{7.64}$$

$$\Delta\Omega_{\Delta\rho} = 2\,\pi\,\Delta F_{\Delta\rho} \tag{7.65}$$

is determined by Equation (7.42); $\Delta\Omega_{\Delta\xi}$ and $\Delta\Omega_{\Delta\zeta}$ are given by Equation (7.58).

The resulting power spectral density $S(\omega)$, which is given by Equation (7.60), consists of four overlapping continuous power spectral densities containing 4/9, 2/9, 2/9, and 1/9 of the total target return signal power, respectively, and takes the Gaussian shape. The effective bandwidths of each component of $S(\omega)$ are determined in Hz by $\Delta F_{\Delta\rho}$, $\sqrt{\Delta F^2_{\Delta\rho} + \Delta F^2_{\Delta\xi}}$, $\sqrt{\Delta F^2_{\Delta\rho} + \Delta F^2_{\Delta\zeta}}$, and $\sqrt{\Delta F^2_{\Delta\rho} + \Delta F^2_{\Delta\xi} + \Delta F^2_{\Delta\zeta}}$, respectively (see Figure 7.4).

Thus, we can conclude as follows. If the chaotic rotations $\Delta\xi$ and $\Delta\zeta$ of scatterers are slow, i.e., their rotations on average are much less than 2π with the radial displacements $\Delta\rho$ on the wavelength λ, then the following condition $\Delta F_{\Delta\xi}, \Delta F_{\Delta\zeta} \ll \Delta F_{\Delta\rho}$ is true, and we can neglect the chaotic rotations $\Delta\xi$ and $\Delta\zeta$ of scatterers. In this case, the power spectral density $S(\omega)$ of the target return signal fluctuations is Gaussian and can be determined by Equation (7.41). If $\Delta\xi$ and $\Delta\zeta$ are fast, i.e., their rotations on average are much more than 2π with the radial displacements $\Delta\rho$ on the wavelength λ, then the following condition $\Delta F_{\Delta\xi}, \Delta F_{\Delta\zeta} \gg \Delta F_{\Delta\rho}$ is true. In this case, $S(\omega)$ is defined by the sum of four Gaussian power spectral densities: the narrow power spectral density $S_1(\omega)$ with the effective bandwidth equal to $\Delta F_{\Delta\rho}$, and three wide power spectral densities $S_1(\omega)$, $S_2(\omega)$, and $S_3(\omega)$. In the limiting case, as $\Delta F_{\Delta\rho} \to 0$ and $S_1(\omega) \to \delta(\omega - \omega_0)$, the resulting $S(\omega)$ of the target return signal fluctuations caused by simultaneous chaotic radial displacements $\Delta\rho$ and rotations $\Delta\xi$ and $\Delta\zeta$ of scatterers tends to approach Equation (7.57).

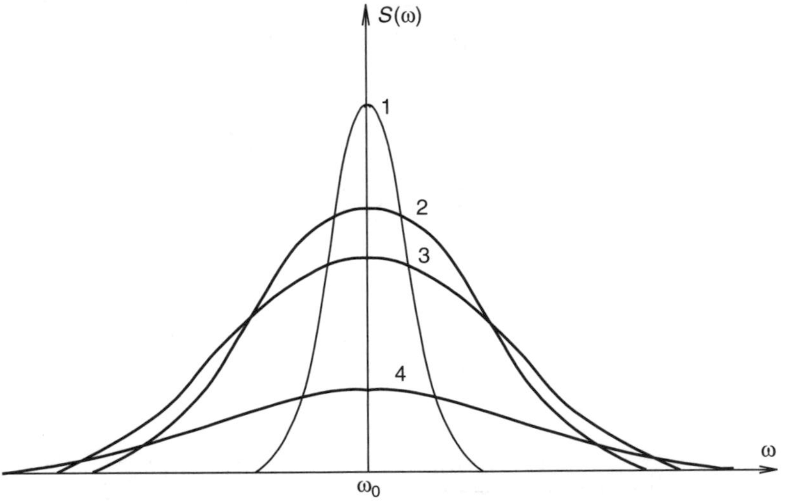

FIGURE 7.4
The power spectral density. Simultaneous chaotic displacements and rotations of scatterers: (1) $S_1(\omega)$; (2) $S_2(\omega)$; (3) $S_3(\omega)$; (4) $S_4(\omega)$.

7.3 Simultaneous Deterministic and Chaotic Motion of Scatterers

7.3.1 Deterministic and Chaotic Displacements of Scatterers

In the study of the simultaneous deterministic and chaotic displacements $\Delta\rho$ of scatterers without rotation, we should follow from the fact that Equation (7.33) is true for both cases. If scatterers take part in both forms of displacements, we can write[18–20]

$$\Delta\rho = \Delta\rho_d + \Delta\rho_{ch}\,, \tag{7.66}$$

where $\Delta\rho_d$ and $\Delta\rho_{ch}$ are the displacements of scatterers caused by the deterministic motion and the chaotic motion, respectively. In the general case, the amplitude S of the target return signal is functionally related to both displacements. The character of this function has a pronounced effect on the shape of the resulting normalized correlation function of the fluctuations.

Let us consider the simplest particular case. Let us assume that the amplitude S of the target return signal is independent of both $\Delta\rho_d$ and $\Delta\rho_{ch}$. This case can occur with the rectangle directional diagram or if the width of the directional diagram is much more than the angle dimensions of the cloud of scatterers. It can also occur if scatterers with definite displacements $\Delta\rho$ have equiprobable distribution in space.

Reference to Equation (7.36), which is true with the independent amplitudes S of the target return signal and displacements $\Delta\rho$ of scatterers, shows that to determine the normalized correlation function of the fluctuations, it is necessary to know the probability distribution density of the displacements $\Delta\rho$ of scatterers given by Equation (7.66). Because the deterministic displacements $\Delta\rho_d$ and the chaotic displacements $\Delta\rho_{ch}$ of scatterers are mutually independent the probability distribution density of simultaneous deterministic and chaotic displacements $\Delta\rho$ of scatterers can be defined by the convolution between the probability distribution density of $\Delta\rho_d$ and $\Delta\rho_{ch}$ in the following form:

$$f_{\Delta\rho}(\Delta\rho) = f_{\Delta\rho_d}(\Delta\rho_d) * f_{\Delta\rho_{ch}}(\Delta\rho_{ch})\,. \tag{7.67}$$

Substituting Equation (7.67) in Equation (7.36), we can prove that with the independent values of S and $\Delta\rho$, the normalized correlation function of the fluctuations $R_{\Delta\rho}(\Delta\rho)$ caused simultaneously by the deterministic $\Delta\rho_d$ and chaotic $\Delta\rho_{ch}$ displacements is defined by the product of two normalized correlation functions of the target return signal: that of $R_{\Delta\rho_d}(\Delta\rho_d)$ caused only by $\Delta\rho_d$ and that of $R_{\Delta\rho_{ch}}(\Delta\rho_{ch})$ caused only by $\Delta\rho_{ch}$. $R_{\Delta\rho}(\Delta\rho)$ can be written in the following form:

$$R_{\Delta\rho}(\Delta\rho) = R_{\Delta\rho_d}(\Delta\rho_d)\cdot R_{\Delta\rho_{ch}}(\Delta\rho_{ch})\,. \tag{7.68}$$

The normalized correlation function of the fluctuations $R_{\Delta\rho_d}(\Delta\rho_d)$ caused only by the deterministic displacements $\Delta\rho_d$ and the normalized correlation function $R_{\Delta\rho_{ch}}(\Delta\rho_{ch})$ caused only by the chaotic displacements $\Delta\rho_{ch}$ can be determined based on Equation (7.33) if we change the displacements $\Delta\rho$ in $\Delta\rho_d$ and $\Delta\rho_{ch}$, respectively.

Consider the general case, when the amplitude S of the target return signal is functionally related to the deterministic $\Delta\rho_d$ and chaotic $\Delta\rho_{ch}$ displacements. To determine the normalized correlation function $R_{\Delta\rho}(\Delta\rho)$ caused simultaneously by the deterministic $\Delta\rho_d$ and chaotic $\Delta\rho_{ch}$, which is given by Equation (7.33), it is necessary to first define the joint probability distribution density of the variables S and $\Delta\rho$, where the displacements $\Delta\rho$ are given by Equation (7.66). Naturally, we assume that the functional relation between the amplitude S and the displacements $\Delta\rho_d$ and $\Delta\rho_{ch}$ is known. In other words, we assume that the probability distribution densities $f_{\Delta\rho_d,S}(\Delta\rho_d, S)$ and $f_{\Delta\rho_{ch},S}(\Delta\rho_{ch}, S)$ are known. Then, we can write

$$f_{\Delta\rho,S}(\Delta\rho, S) = f_S(S)\cdot f_{\Delta\rho}(\Delta\rho\,|\,S)\,, \tag{7.69}$$

where $f_{\Delta\rho}(\Delta\rho\,|\,S)$ is the conditional probability distribution density of the displacements $\Delta\rho$ of scatterers.

As is well known, the conditional probability distribution density is subject to the same laws as the usual (unconditional) probability distribution density. Because of this, we can write

$$\begin{aligned} f_{\Delta\rho}(\Delta\rho\,|\,S) &= f_{\Delta\rho}[(\Delta\rho_d + \Delta\rho_{ch})\,|\,S] = f_{\Delta\rho_d}(\Delta\rho_d\,|\,S) * f_{\Delta\rho_{ch}}(\Delta\rho_{ch}\,|\,S) \\ &= \int_{-\infty}^{\infty} f_{\Delta\rho_d}(\Delta\rho_d\,|\,S)\cdot f_{\Delta\rho_{ch}}[(\Delta\rho - \Delta\rho_d)\,|\,S]\, d(\Delta\rho_d) \end{aligned}\,. \tag{7.70}$$

Substituting Equation (7.70) in Equation (7.33) and Equation (7.69),

$$R_{\Delta\rho}(\Delta\rho) = \frac{\int_0^{\infty} S^2 f_S(S) \int_{-\infty}^{\infty} f_{\Delta\rho}(\Delta\rho\,|\,S)\cdot e^{4j\pi\frac{\Delta\rho}{\lambda}} d(\Delta\rho)\, dS}{\int_0^{\infty} S^2 f_S(S)\, dS}\,. \tag{7.71}$$

Note that the inside integral in the numerator of the formula in Equation (7.71) coincides with Equation (7.36). The difference is only that the conditional probability distribution density $f_{\Delta\rho}(\Delta\rho\,|\,S)$ is subjected to the Fourier transform, and not the unconditional probability distribution density. Let us consider that the conditional probability distribution density is defined by the convolution given by Equation (7.70) between the two conditional probability distribution densities, as well as the unconditional probability distribution density given by Equation (7.67). Let us introduce a designation

$$R_{\Delta\rho\,|\,S}(\Delta\rho\,|\,S) = \int\limits_{-\infty}^{\infty} f_{\Delta\rho}(\Delta\rho\,|\,S)\cdot e^{4j\pi\frac{\Delta\rho}{\lambda}}\,d(\Delta\rho) \tag{7.72}$$

that is similar to Equation (7.36). We call the function $R_{\Delta\rho|S}(\Delta\rho\,|\,S)$ the conditional normalized correlation function of the fluctuations caused simultaneously by the deterministic $\Delta\rho_d$ and chaotic $\Delta\rho_{ch}$ displacements of scatterers. Then we can write

$$R_{\Delta\rho\,|\,S}(\Delta\rho\,|\,S) = R_{\Delta\rho_d|S}(\Delta\rho_d\,|\,S)\cdot R_{\Delta\rho_{ch}|S}(\Delta\rho_{ch}\,|\,S)\,. \tag{7.73}$$

Substituting Equation (7.73) in Equation (7.71), in the general case, we see that the normalized correlation function of the fluctuations $R_{\Delta\rho}(\Delta\rho)$ caused simultaneously by the deterministic $\Delta\rho_d$ and chaotic $\Delta\rho_{ch}$ displacements of scatterers can be determined by the following form:

$$R_{\Delta\rho}(\Delta\rho) = \frac{\int\limits_0^{\infty} S^2 f_S(S)\cdot R_{\Delta\rho_d|S}(\Delta\rho_d\,|\,S)\cdot R_{\Delta\rho_{ch}|S}(\Delta\rho_{ch}\,|\,S)\,dS}{\int\limits_0^{\infty} S^2 f_S(S)\,dS}\,, \tag{7.74}$$

i.e., Equation (7.68) is not true. However, if we assume that $\Delta\rho_d$ or $\Delta\rho_{ch}$ are independent of the amplitude S, Equation (7.68) becomes true. From the preceding discussion particularly, it follows that Equation (7.68) is true if S is functionally related, for example, with $\Delta\rho_d$ and is independent of $\Delta\rho_{ch}$. This circumstance allows us to define $R_{\Delta\rho}(\Delta\rho)$ caused simultaneously by $\Delta\rho_d$ and $\Delta\rho_{ch}$ based on the results discussed in the previous sections.

Consider an example where the scatterers make simultaneously the deterministic motion caused by the moving radar and the layered wind and chaotic motion with velocities, which are distributed by the Gaussian law and varied slowly. The resulting normalized correlation function of the fluctuations is defined by the product of the normalized correlation functions given by Equation (3.79) and Equation (7.36). Therefore, Equation (7.37) must be substituted in Equation (7.36).

The resulting power spectral density of the fluctuations caused simultaneously by the deterministic $\Delta\rho_d$ and chaotic $\Delta\rho_{ch}$ displacements of scatterers is defined by the convolution between the power spectral densities given by

Equation (3.80) and Equation (7.41). Because both these power spectral densities obey the Gaussian law, the resulting power spectral density caused simultaneously by $\Delta\rho_d$ and $\Delta\rho_{ch}$ takes the Gaussian shape with the effective bandwidth determined by

$$\Delta F = \sqrt{\Delta F^2_{\Delta\rho_d} + \Delta F^2_{\Delta\rho_{ch}}}\,, \tag{7.75}$$

where $\Delta F_{\Delta\rho_d}$ is the effective bandwidth of the fluctuations caused only by $\Delta\rho_d$ which is given by Equation (7.16), and $\Delta F_{\Delta\rho_{ch}}$ caused only by $\Delta\rho_{ch}$.

7.3.2 Chaotic Rotation of Scatterers and Rotation of the Polarization Plane

First, let us prove that by using Equation (7.34) and Equation (7.52) we can define the normalized correlation function of the fluctuations caused by the rotation of the radar antenna polarization plane. Because with rotation of the radar antenna polarization plane the angles $\Delta\xi$ are the same for all scatterers, the probability distribution density of the variable $\Delta\xi$ can be written in the following form:[21,22]

$$f_{\Delta\xi}(\Delta\xi) = \delta(\Delta\xi - \Delta\xi_d)\,, \tag{7.76}$$

where $\Delta\xi_d$ is the angle of rotation of the radar antenna polarization plane during the time τ. Assuming that scatterers are the half-dipoles, we can substitute Equation (7.76) in Equation (7.52). After integration, we obtain the normalized correlation function of the fluctuations caused by rotation of the radar antenna polarization plane in the following form:

$$R_q(\Delta\xi_d) \cong 0.33\cdot(2 + \cos 2\Delta\xi_d)\,. \tag{7.77}$$

As we can see, Equation (7.77) coincides with Equation (5.88).

Now, consider the simultaneous chaotic rotation of scatterers and the rotation of the polarization plane of the radar antenna.[23] For this case, we can write

$$\Delta\xi = \Delta\xi_d + \Delta\xi_{ch}\,, \tag{7.78}$$

where $\Delta\xi_{ch}$ and $\Delta\xi_d$ are the angles of rotation due to the simultaneous chaotic rotation of scatterers and rotation of the polarization plane of the radar antenna, respectively. Because the random variable $\Delta\xi_{ch}$ is independent of the nonrandom variable $\Delta\xi_d$, the probability distribution density $f_{\Delta\xi}(\Delta\xi)$ of the variable $\Delta\xi$ is defined by the convolution between the probability distribution density of the chaotic rotation $\Delta\xi_{ch}$ of scatterers and that of the rotation $\Delta\xi_d$ of the radar antenna polarization plane, and can be written in the following form:

$$f_{\Delta\xi}(\Delta\xi) = f_{\Delta\xi_{ch}}(\Delta\xi_{ch}) * f_{\Delta\xi_d}(\Delta\xi_d) = f_{\Delta\xi_{ch}}(\Delta\xi - \Delta\xi_d) . \tag{7.79}$$

In this particular case, when the random variable $\Delta\xi_{ch}$ obeys the Gaussian law, we can write

$$f_{\Delta\xi}(\Delta\xi) = \frac{1}{\sqrt{2\pi}\, \sigma_{\Delta\xi_{ch}}} \cdot e^{-\frac{(\Delta\xi - \Delta\xi_d)^2}{2\sigma^2_{\Delta\xi_{ch}}}} . \tag{7.80}$$

Substituting Equation (7.80) in Equation (7.34) and assuming that, as before, the random variables $\Delta\xi$ and $\Delta\zeta$ are mutually independent, we can write the resulting normalized correlation function of the fluctuations $R_{\Delta\xi,\Delta\zeta}(\Delta\xi, \Delta\zeta)$ in the following form:

$$R_{\Delta\xi,\Delta\zeta}(\Delta\xi, \Delta\zeta) \cong 0.11 \cdot [2 + \cos 2\Delta\xi_d \cdot e^{-2\sigma^2_{\Delta\xi_{ch}}}][2 + e^{-2\sigma^2_{\Delta\zeta}}] . \tag{7.81}$$

Comparing Equation (7.81) with Equation (7.53) and Equation (7.77), we can see that the resulting $R_{\Delta\xi,\Delta\zeta}(\Delta\xi, \Delta\zeta)$ given by Equation (7.81) is not defined by the product of the normalized correlation functions $R_{\Delta\xi,\Delta\zeta}(\Delta\xi, \Delta\zeta)$ and $R_q(\Delta\xi_d)$ of the fluctuations, which are given by Equation (7.53) and Equation (7.77), respectively.

If scatterers rotate with slow varied velocities, i.e., the conditions

$$\sigma_{\Delta\xi_{ch}} = \sigma_{\Omega_\xi} \cdot \tau \quad \text{and} \quad \sigma_{\Delta\zeta} = \sigma_{\Omega_\zeta} \cdot \tau \tag{7.82}$$

are satisfied, and the polarization plane of the radar antenna rotates uniformly with the angular velocity Ω_d, i.e., the condition $\Delta\xi_d = \Omega_d \cdot \tau$ is satisfied, then the normalized correlation function $R(\tau)$ of the fluctuations caused simultaneously by the chaotic rotation of scatterers and the rotation of the polarization plane of the radar antenna can be written in the following form:

$$R(\tau) \cong 0.11 \cdot [2 + \cos 2\Omega_d \tau \cdot e^{-2\sigma^2_{\Omega_\xi}\tau^2}][2 + e^{-2\sigma^2_{\Omega_\zeta}\tau^2}] . \tag{7.83}$$

Multiplying Equation (7.83) by the exponent $e^{j\omega_0\tau}$ and using the Fourier transform, the power spectral density of the fluctuations caused simultaneously by the chaotic rotation of scatterers and the rotation of the polarization plane takes the following form:

$$S(\omega) \cong 8\delta(\omega - \omega_0) + S_1(\omega) + S_2(\omega) + S_3(\omega), \tag{7.84}$$

where

$$S_1(\omega) \cong \frac{4}{\Delta\Omega_{\Delta\zeta}} \cdot e^{-\pi \frac{(\omega-\omega_0)^2}{\Delta\Omega^2_{\Delta\zeta}}}; \tag{7.85}$$

$$S_2(\omega) \cong \frac{2}{\Delta\Omega_{\Delta\xi}} \cdot \left[e^{-\pi \frac{(\omega-\omega_0-2\Omega_d)^2}{\Delta\Omega^2_{\Delta\xi}}} + e^{-\pi \frac{(\omega-\omega_0+2\Omega_d)^2}{\Delta\Omega^2_{\Delta\xi}}} \right]; \tag{7.86}$$

$$S_3(\omega) \cong \frac{1}{\sqrt{\Delta\Omega^2_{\Delta\xi} + \Delta\Omega^2_{\Delta\zeta}}} \cdot \left[e^{-\pi \frac{(\omega-\omega_0-2\Omega_d)^2}{\Delta\Omega^2_{\Delta\xi}+\Delta\Omega^2_{\Delta\zeta}}} + e^{-\pi \frac{(\omega-\omega_0+2\Omega_d)^2}{\Delta\Omega^2_{\Delta\xi}+\Delta\Omega^2_{\Delta\zeta}}} \right]. \tag{7.87}$$

The power spectral density $S(\omega)$ of the fluctuations caused simultaneously by the chaotic rotation of scatterers and the rotation of the radar antenna polarization plane, which is given by Equation (7.84), consists of the one discrete line and the continuous power spectral density $S_1(\omega)$ at the frequency ω_0 and two continuous power spectral densities $S_2(\omega)$ and $S_3(\omega)$ at the frequencies $\omega_0 \pm 2\Omega_d$, respectively (see Figure 7.5): $S_2'(\omega)$ and $S_3'(\omega)$ at the frequency $\omega_0 + 2\Omega_d$, and $S_2''(\omega)$ and $S_3''(\omega)$ at the frequency $\omega_0 - 2\Omega_d$. The discrete component $8\delta(\omega - \omega_0)$ of $S(\omega)$ possesses 8/18 of the total power of the target return signal; 4/18 of the total power is concentrated in $S_1(\omega)$ at the frequency ω_0 with the effective bandwidth equal to $\Delta\Omega_{\Delta\zeta}$, and 2/18 in $S_2'(\omega)$ at the frequency $\omega_0 - 2\Omega_d$ with the effective bandwidth equal to $\Delta\Omega_{\Delta\xi}$; 2/18 is concentrated in $S_2''(\omega)$ at the frequency $\omega_0 - 2\Omega_d$ with the effective bandwidth equal to $\Delta\Omega_{\Delta\xi}$; 1/18 in $S_3'(\omega)$ at the frequency $\omega_0 - 2\Omega_d$ with the effective bandwidth equal to $\sqrt{\Delta\Omega^2_{\Delta\xi} + \Delta\Omega^2_{\Delta\zeta}}$; 1/18 in $S_3''(\omega)$ at the frequency $\omega_0 - 2\Omega_d$ with the effective bandwidth equal to $\sqrt{\Delta\Omega^2_{\Delta\xi} + \Delta\Omega^2_{\Delta\zeta}}$.

7.3.3 Chaotic Displacements of Scatterers and Rotation of the Polarization Plane

Because the normalized correlation function of the fluctuations caused only by the rotation of the radar antenna polarization plane, as well as those caused only by the chaotic rotation of scatterers, can be determined by Equation (7.34),

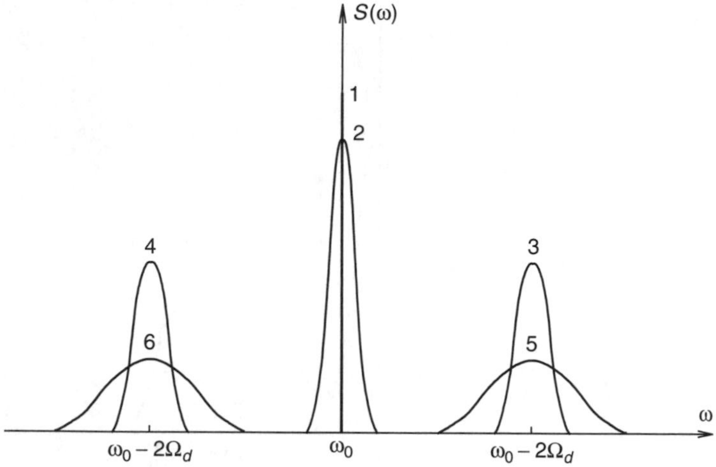

FIGURE 7.5
The power spectral density. Simultaneous chaotic rotation of scatterers and rotation of the polarization plane: (1) $\delta(\omega - \omega_0)$; (2) $S_1(\omega)$; (3) $S_2'(\omega)$; (4) $S_2''(\omega)$; (5) $S_3'(\omega)$; (6) $S_3''(\omega)$.

and the normalized correlation function of the fluctuations caused only by chaotic displacements of scatterers is given by Equation (7.33), in accordance with Equation (7.32), the resulting normalized correlation function is defined by the product of the normalized correlation functions.

Let us consider the simplest example. Let us assume that the chaotic displacements of scatterers are distributed according to the Gaussian law. Scatterers move with approximately constant velocities, and the polarization plane of the radar antenna rotates uniformly with the angular velocity Ω_d.[24,25] Using Equation (7.38), Equation (7.39), and Equation (7.77), and taking into consideration the condition $\Delta\xi_d = \Omega_d \cdot \tau$, the resulting normalized correlation function $R(\tau)$ caused simultaneously by the chaotic displacements of scatterers and the rotation of the radar antenna polarization plane can be written in the following form:

$$R(\tau) \cong 0.33 \cdot (2 + \cos 2\Omega_d\tau) \cdot e^{-8\pi \frac{\sigma^2_{V^r_{scat}} \tau^2}{\lambda^2}} . \tag{7.88}$$

The corresponding power spectral density $S(\omega)$ (see Figure 7.6) can be determined by the following form:

$$S(\omega) \cong 4S_{\Delta\rho}(\omega) + S_{\Delta\rho}(\omega - 2\Omega_d) + S_{\Delta\rho}(\omega + 2\Omega_d) , \tag{7.89}$$

where the power spectral density $S_{\Delta\rho}(\omega)$ is given by Equation (7.41). In an analogous way, we can define the normalized correlation function and power spectral density of the fluctuations with simultaneous deterministic

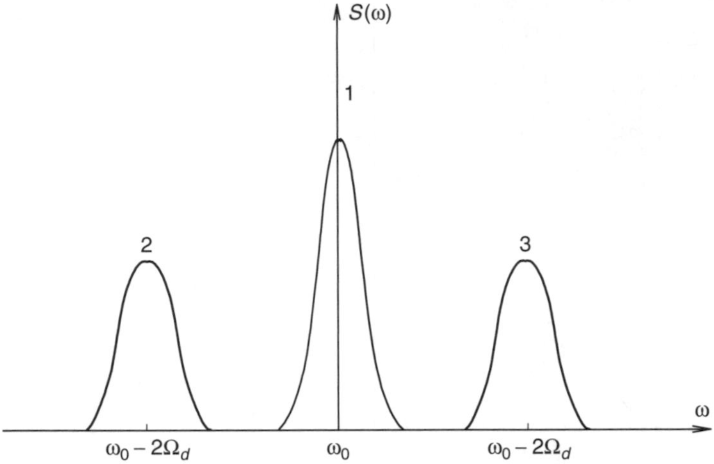

FIGURE 7.6
The power spectral density. Simultaneous chaotic displacements of scatterers and rotations of the polarization plane: (1) $4S_{\Delta\rho}(\omega)$; (2) $S_{\Delta\rho}(\omega - 2\Omega_d)$; (3) $S_{\Delta\rho}(\omega + 2\Omega_d)$.

displacements and chaotic rotations of scatterers, as in this case, and the formula in Equation (7.32) is also true.

7.4 Conclusions

In the course of investigation of the target return signal fluctuations caused by moving scatterers under the stimulus of the wind, we can conveniently consider two motion components of the cloud of scatterers: deterministic motion with different velocities, which can be defined by the nonstochastic function of coordinates and time (for example, variation of the velocity of the wind as a function of altitude [the layered wind] and the motion of the cloud of scatterers as a whole) and stochastic motion with the random velocity and velocity that is varied in time, in the general case.

With the target return signal fluctuations caused by deterministic displacements of scatterers, in particular by the simultaneous stimulus of the layered wind and moving radar, the power spectral density coincides in shape with the squared directional diagram by power in the plane passing through the plane φ (the horizon) under the angle κ' given by Equation (7.13). However, this plane does not pass through the direction of the moving radar and the directional diagram axis, neither does it occur in the absence of the wind. The effective bandwidth depends on distance between the radar and the target. This phenomenon exists because, in accordance with the distance between the radar and the target, the directional diagram illuminates areas of the cloud of scatterers having various velocities as a function of altitude.

With simultaneous chaotic radial displacements and rotations of scatterers, if these are the independent random variables, the normalized correlation function of the target return signal fluctuations is defined by the product of that caused only by the chaotic radial displacements of scatterers and that caused only by the rotation of scatterers.

In the general case, because with simultaneous chaotic displacements and rotations of scatterers the normalized correlation function $R_{\Delta\rho}(\Delta\rho)$ of the fluctuations caused only by the chaotic displacements of scatterers and $R_{\Delta\xi,\Delta\zeta}(\Delta\xi, \Delta\zeta)$ of the fluctuations caused only by the rotation of scatterers are multiplied with each other, then, with linear and angular velocities that vary very slowly, the resulting normalized correlation function $R_{\Delta\rho,\Delta\xi,\Delta\zeta}(\Delta\rho, \Delta\xi, \Delta\zeta)$ of the fluctuations caused simultaneously by chaotic displacements and rotations of scatterers is defined by the product of $R_{\Delta\rho}(\Delta\rho)$, which is given by Equation (7.38), and $R_{\Delta\xi,\Delta\zeta}(\Delta\xi, \Delta\zeta)$,which is given by Equation (7.53).

The power spectral density of the fluctuations caused simultaneously by chaotic displacements and rotations of scatterers consists of four overlapping power spectral densities possessing 4/9, 2/9, 2/9, and 1/9 of the total power of the target return signal, respectively (see Figure 7.4).

With the simultaneous deterministic and chaotic displacements (without rotations) of scatterers, in the general case, the resulting normalized correlation function of the fluctuations takes a more complex form [see Equation (7.74)]. The resulting power spectral density is defined by the convolution between the power spectral densities caused only by the deterministic displacements of scatterers and chaotic displacements (without rotation) of scatterers.

With the simultaneous chaotic rotation of scatterers and rotation of the polarization plane of the radar antenna, the power spectral density consists of (see Figure 7.5) the one discrete line at the frequency ω_0 containing 8/18 of the total power of the target return signal; the continuous power spectral density at the frequency ω_0 containing 4/18 of the total power of the target return signal; two continuous power spectral densities at the frequencies $\omega_0 \pm 2\Omega_d$, with each possessing 2/18 of the total power of the target return signal; and two continuous power spectral densities at the frequencies $\omega_0 \pm 2\Omega_d$, with each possessing 1/18 of the total power of the target return signal.

With chaotic displacements of scatterers and the rotation of the radar antenna polarization plane, the power spectral density of the target return signal fluctuations consists of three continuous power spectral densities (see Figure 7.6): with the center at the frequency ω_0, possessing 2/3 of the total power of the target return signal; with the center at the frequency $\omega_0 + 2\Omega_d$, possessing 1/6 of the total power of the target return signal; and with the center at the frequency $\omega_0 - 2\Omega_d$, possessing 1/6 of the total power of the target return signal.

References

1. Melnik, A., Zubkovich, C., Stepanenko, V. et al., *Radar Methods of Investigation of the Earth Surface*, Soviet Radio, Moscow, 1980 (in Russian).
2. Lourtie, I. and Carter, G., Signal detectors for random ocean media, *J. Acoust. Soc. Amer.*, Vol. 92, No. 3, 1992, pp. 1420–1427.
3. Takao, K., Fujita, H., and Nishi, T., An adaptive array under directional constraint, *IEEE Trans.*, Vol. AP-21, No. 9, 1976, pp. 662–669.
4. Stepanenko, V., *Radar in Meteorology*, Hydrometizdat, Leningrad, 1973 (in Russian).
5. Hall, H., A new model for impulsive phenomena: application to atmospheric-noise communication channel, *Technical Report 3412-8*, Stanford University, Stanford, CA, August 1966.
6. Dulevich, V., *Theoretical Foundations of Radar*, Soviet Radio, Moscow, 1978 (in Russian).
7. Borkus, M., Energy spectrum of signals scattered by atmosphere aerosol particles caused by moving radar, *Problems in Radio Electronics*, Vol. OT, No. 1, 1977, pp. 43–50 (in Russian).
8. Ursin, B. and Bertenssen, K., Comparison of some inverse methods for wave propagation: layered media, in *Proceedings of the IEEE*, Vol. 74, 1986, pp. 389–400.
9. Giordano, A. and Haber, F., Modeling of atmospheric noise, *Radio Sci.*, Vol. 7, No. 8, 1972, pp. 1101–1123.
10. Levanon, N., *Radar Principles*, John Wiley & Sons, New York, 1988.
11. Andersh, D., Lee, S., and Ling, H., XPATCH: a high frequency electromagnetic scattering prediction code and environment for complex three-dimensional objects, *IEEE Antennas Propagat. Mag.*, Vol. 36, No. 1, 1994, pp. 65–69.
12. Yagle, A. and Frolik, J., On the feasibility of impulse reflection response data for the two-dimensional inverse scattering problem, *IEEE Trans.*, Vol. AP-44, No. 8, 1996, pp. 1551–1564.
13. Rendas, M. and Monra, J., Ambiguity in radar and sonar, *IEEE Trans.*, Vol. SP-46, No. 2, 1998, pp. 294–305.
14. Vanshtein, L. and Zubakov, V., *Picking out the Signals in Noise*, Soviet Radio, Moscow, 1960 (in Russian).
15. Potter, L. and Moses, R., Attributed scattering centers for SAR ATR, *IEEE Trans.*, Vol. IP-6, No. 1, 1997, pp. 79–91.
16. Foschini, G., Layered space-time architecture for wireless communication in a fading environment when using multi-element antennas, *Bell Labs Tech. J.*, Vol. 1, No. 2, 1996, pp. 41–59.
17. Habibi-Ashrafi, F. and Mendel, J., Estimation of parameters in loss-less layered media systems, *IEEE Trans.*, Vol. AC-27, No. 1, 1982, pp. 31–48.
18. Widrow, B., Mantley, P., Griffiths, L., and Goode, B., Adaptive antennas systems, in *Proceedings of the IEEE*, Vol. 55, No. 12, 1967, pp. 2143–2159.
19. Compton, R., *Adaptive Antennas*, Prentice Hall, Englewood Cliffs, NJ, 1988.
20. Pillai, S., *Array Signal Processing*, Springer-Verlag, New York, 1989.
21. Holm, W., Polarimetric fundamentals and techniques, in *Principles of Modern Radar*, J. Eaves and E. Reedy, Eds., Van Nostrand Reinhold, New York, 1987.

22. Ulaby, F. and Elachi, C., *Radar Polarimetry for Geoscience Applications,* Artech House, Norwood, MA, 1990.
23. Drane, C., *Positioning Systems: A Unified Approach,* Springer-Verlag, New York, 1992.
24. Cloude, S., Polarimetric techniques in radar signal processing, *Microwave J.,* Vol. 26, No. 7, 1983, pp. 119–127.
25. Guili, D., Polarization diversity in radar, in *Proceedings of the IEEE,* Vol. 74, No. 2, 1986, pp. 245–269.

8

Fluctuations under Scanning of the Two-Dimensional (Surface) Target with the Continuous Frequency-Modulated Signal

8.1 General Statements

The continuous frequency-modulated searching signal, in accordance with Equation (2.54), can be written in the following, form:

$$W(t) = W_0(t) \cdot e^{-j[\omega_0 t + \Psi(t)]} \,. \tag{8.1}$$

The instantaneous frequency of the searching signal given by Equation (8.1) has the following form:

$$\omega(t) = \omega_0 + \frac{d\Psi(t)}{dt} \,. \tag{8.2}$$

In accordance with the general formula in Equation (2.55), the correlation function of target return signal fluctuations reflected by the two-dimensional (surface) target, which is modulated by frequency (or phase), has the following form:

$$\begin{aligned} R(t,\tau) = p_0 \cdot e^{j\omega_0\tau} \iint \tilde{g}^2(\varphi,\psi) \cdot e^{-2j\omega_0 \frac{\Delta\rho}{c}} \\ \times e^{j\left[\Psi\left(t - \frac{2\rho}{c} - \frac{\Delta\rho}{c} + 0.5\tau\right) - \Psi\left(t - \frac{2\rho}{c} - \frac{\Delta\rho}{c} - 0.5\tau\right)\right]} d\varphi \, d\psi \end{aligned} \,, \tag{8.3}$$

where $\tilde{g}^2(\varphi,\psi)$ is the generalized radar antenna directional diagram given by Equation (4.14);

$$p_0 = \frac{PG_0^2\lambda^2}{64\pi^3 h^2} \,. \tag{8.4}$$

Representation of the searching signal in the form given by Equation (8.1) and the correlation function of the fluctuations in the form determined by Equation (8.3) is convenient if the function $\Psi(\mathrm{t})$ is continuous for all values of t, as is the case with harmonic frequency modulation. If the function $\Psi(\mathrm{t})$ or its derivative is discontinuous, which is characteristically the case in linear frequency modulation, it is worthwhile to express the frequency-modulated searching signal as a time periodic sequence of signals that is coherent from period to period,[1] i.e., the value of $\omega_0 T_M$ is multiplied by 2π:

$$W(t) = W_0(t)\sum_{n=-\infty}^{\infty} e^{-j[\omega_0 t+\Psi(t-nT_M)]} \,, \tag{8.5}$$

where $0 < |t - nT_M| < T_M$ and T_M is the period of modulation. In this case, we can write

$$R(t,\tau) = p_0 \cdot e^{j\omega_0\tau}\iint \tilde{g}^2(\varphi,\psi)\cdot e^{-2j\omega_0\frac{\Delta\rho}{c}}$$

$$\times \sum_{k=-\infty}^{\infty}\sum_{n=-\infty}^{\infty} e^{j\left\{\Psi\left[t-kT_M-\frac{2\rho}{c}-\frac{\Delta\rho}{c}+0.5(\tau-nT_M)\right]-\Psi\left[t-kT_M-\frac{2\rho}{c}-\frac{\Delta\rho}{c}-0.5(\tau-nT_M)\right]\right\}}\, d\varphi\, d\psi \,. \tag{8.6}$$

The correlation function of the fluctuations given in Equation (8.6) is analogous to the correlation function of pulsed target return signal sequence fluctuations and is periodic, both with respect to t and with respect to τ. This correlation function defines the fluctuations from period to period, within the limits of a period. If we introduce the function $\Pi(t)$ into the integrand, the correlation function of the fluctuations given by Equation (8.6) allows us to define the correlation function of pulsed target return signal fluctuations under intrapulsed and interpulsed frequency modulation.

In most cases of radar with frequency-modulated searching signals, the target return signal is multiplied by the searching signal, which is a heterodyne signal.[2–5] Let us assume that the heterodyne signal with unit amplitude is shifted by the intermediate frequency ω_{im}, which is much larger than the bandwidth of the searching signal. Henceforth, the correlation function and power spectral density of the target return signal fluctuations are defined under the condition $\omega_{im} \neq 0$. For $\omega_{im} = 0$, we must use $|\Omega_0 - n\omega_M|$ instead of frequencies $\omega_{im} + \Omega_0 - n\omega_M$ in all formulae, as negative frequencies have no physical meaning.

When we use the searching signal given by Equation (8.1), the heterodyne signal takes the following form:

$$W_h(t) = e^{-j\,[(\omega_0 - \omega_{im})t + \Psi(t)]}\,. \tag{8.7}$$

In this case, the transformed target return signal from the *i*-th scatterer located at a distance ρ from the radar can be written in the following form: at the time instant $t - 0.5\tau$, we obtain

$$W_{tr_i}(t - 0.5\tau) = W_{0_i}(t)\cdot W_h(t - 0.5\tau)\cdot W^*_{r_i}\left(t - 0.5\tau - \tfrac{2\rho}{c} + \tfrac{\Delta\rho}{c}\right) \tag{8.8}$$

and at the time instant $t + 0.5\tau$, we obtain

$$W^*_{tr_i}(t + 0.5\tau) = W_{0_i}(t)\left[W_h(t + 0.5\tau)\cdot W^*_{r_i}\left(t + 0.5\tau - \tfrac{2\rho}{c} - \tfrac{\Delta\rho}{c}\right)\right]^*, \tag{8.9}$$

where $W_{tr_i}(t)$ is the transformed target return signal from the *i*-th scatterer; $W_{r_i}(t)$ is the target return signal from the *i*-th scatterer; and $W^*(t)$ is the signal conjugated with the signal $W(t)$.

Multiplying the target return signal from the *i*-th scatterer by the heterodyne signal $W_h(t)$ does not change the random character of the target return signal phase. Because of this, the transformed target return signal $W_{tr_i}(t)$ can be considered as a stochastic process, and we can use the general formula in Equation (2.55) to determine the correlation function of the fluctuations

$$R(t,\tau) = p_0 \cdot e^{j\omega_{im}\tau} \iint \tilde{g}^2(\varphi,\psi)\cdot e^{-2j\omega_0\frac{\Delta\rho}{c}}$$
$$\times\, e^{j\left[\Psi(t-0.5\tau) - \Psi\left(t - 0.5\tau - \frac{2\rho}{c} + \frac{\Delta\rho}{c}\right)\right]}\cdot e^{-j\left[\Psi(t+0.5\tau) - \Psi\left(t + 0.5\tau - \frac{2\rho}{c} - \frac{\Delta\rho}{c}\right)\right]} d\varphi\, d\psi. \tag{8.10}$$

In an analogous way, if the signal $W(t)$ is given by Equation (8.5), we can write the correlation function of the fluctuations in the following form:

$$R(t,\tau) = p_0 \cdot e^{j\omega_{im}\tau} \iint \tilde{g}^2(\varphi,\psi)\cdot e^{-2j\omega_0\frac{\Delta\rho}{c}}$$
$$\times \sum_{k=-\infty}^{\infty}\sum_{n=-\infty}^{\infty} e^{j\left\{\Psi\left[t - kT_M - 0.5(\tau - nT_M)\right] - \Psi\left[t - kT_M - 0.5(\tau - nT_M) - \frac{2\rho}{c} + \frac{\Delta\rho}{c}\right]\right\}} \tag{8.11}$$
$$\times\, e^{-j\left\{\Psi\left[t - kT_M + 0.5(\tau - nT_M)\right] - \Psi\left[t - kT_M + 0.5(\tau - nT_M) - \frac{2\rho}{c} - \frac{\Delta\rho}{c}\right]\right\}} d\varphi\, d\psi$$

The function $\Psi(\tau)$ of the heterodyne signal given by Equation (8.8) is taken within the limits of the same period *k*, as a shift in *t* by an integer does not change the target return signal due to the fact that the heterodyne signal is coherent. The formulae in Equation (8.10) and Equation (8.11), as well as those in Equation (8.3) and Equation (8.6), define transformed target return signal fluctuations and depend on time. This fact indicates a nonstationary state of target return signal fluctuations.[6–8]

Dependence of the instantaneous power spectral densities of pulsed target return signal fluctuations on time allows us to isolate information regarding pulsed target return signal delay, which defines the distance between the radar and the scanned target. As is well known, in the use of radar with frequency-modulated searching signals, information about the distance between the radar and the two-dimensional (surface) target is extracted from the frequency of the transformed target return signal. To obtain the frequency of the transformed target return signal from different targets, we use a set of filters. The bandwidth of each filter is approximately equal to the modulation frequency Ω_M. The comb characteristics of these filters are the analogs of pulses gated in time under the use of radar with pulsed searching signals. For this reason, the correlation function of transformed target return signal fluctuations is of prime interest to us, not the instantaneous correlation function. Thus, the correlation function of transformed target return signal fluctuations must be averaged over the time period of frequency modulation.[9,10]

In principle, the correlation function of transformed target return signal fluctuations averaged over time can be obtained based on Equation (8.10) and Equation (8.11), averaging over the period T_M of the predetermined correlation function $R(t, \tau)$. Here we use the technique discussed in Section 2.3.4, which does not require determination of the instantaneous correlation function of the fluctuations. For this purpose, each signal component given by Equation (8.8) and Equation (8.9)

$$e^{j\left[\Psi(t-0.5\tau)-\Psi\left(t-0.5\tau-\frac{2\rho}{c}+\frac{\Delta\rho}{c}\right)\right]} \tag{8.12}$$

and

$$e^{-j\left[\Psi(t+0.5\tau)-\Psi\left(t+0.5\tau-\frac{2\rho}{c}-\frac{\Delta\rho}{c}\right)\right]} \tag{8.13}$$

is replaced in the integrand of Equation (8.10) with the equivalent Fourier series. In the integrand, we average the product of the Fourier series over the period T_M of frequency modulation

$$R(\tau) = p_0 \cdot e^{j\omega_{im}\tau} \iint \tilde{g}^2(\varphi,\psi)\cdot e^{-2j\omega_{av}\frac{\Delta\rho}{c}} \cdot \sum_{n=-\infty}^{\infty} C^n_{\rho-0.5\Delta\rho} C^n_{\rho+0.5\Delta\rho} \cdot e^{-jn\omega_M\left(\tau-\frac{\Delta\rho}{c}\right)} d\varphi\, d\psi\,, \tag{8.14}$$

where $C^n_{\rho-0.5\Delta\rho}$ and $C^n_{\rho+0.5\Delta\rho}$ are the coefficients of the Fourier-series expansion of signal components given by Equation (8.12) and Equation (8.13), and ω_{av} is the high frequency averaged over the modulation period.

In an analogous manner, we can define the correlation function $R(\tau)$ if the searching signal has the form given by Equation (8.5). Unlike in Equation (8.10), differences between the function $\Psi(t)$ and the coefficients $C^n_{\rho-0.5\Delta\rho}$ and $C^n_{\rho+0.5\Delta\rho}$ of the Fourier-series expansion, respectively, should be estimated within the limits of several periods, as the arguments of the function $\Psi(t)$ in Equation (8.8) and Equation (8.9) are shifted in each exponent. The correlation function averaged over time is the sum of the correlation functions given by Equation (8.14) with coefficients $C^n_{\rho\pm0.5\Delta\rho}$ for each period. An example of a specific determination of $R(\tau)$ is discussed in Section 8.3. It should be noted that the general formula in Equation (8.14) coincides with the formula discussed in Jukovsky et al.,[11] which is based on solving the electromagnetic problem of back scattering of radio waves by a rough surface under Gaussian statistics. Let us define the correlation function of the target return signal fluctuations for the main forms of frequency-modulated searching signals.

8.2 The Linear Frequency-Modulated Searching Signal

Let us write the functions $\Psi(t)$ and $\omega(t)$ in the following form:

$$\Psi(t) = 0.5k_\omega t^2 \quad \text{and} \quad \omega(t) = \omega_0 + k_\omega t\,, \tag{8.15}$$

where k_ω is the velocity of frequency variation, which can be both positive (see Figure 8.1a) and negative (see Figure 8.1b). Assume that a linear variation in frequency is unlimited in time. In the case of the periodic searching signal, this really corresponds to the condition that the period of frequency modulation is very large in value so that $T_d << T_M$ and $\tau_c << T_M$, where τ_c is the correlation length of the stochastic process. The difference $\Delta\Psi$ of the modulating functions given by Equation (8.3) can be written for the case of Equation (8.15) in the following form:

$$\Delta\Psi = k_\omega(\tau - 2\Delta\rho c^{-1})(t - 2\rho c^{-1})\,. \tag{8.16}$$

Substituting Equation (8.15) and Equation (8.16) in Equation (8.3), we can write the correlation function $R(t, \tau)$ in the following form:

$$R(t,\tau) = p_0 \cdot e^{j\omega(t)\tau} \iint \tilde{g}^2(\varphi,\psi)\cdot e^{-2j\omega(t)\frac{\Delta\rho}{c}} \cdot e^{-jk_\omega\left(\tau-\frac{2\Delta\rho}{c}\right)\frac{2\rho}{c}}\, d\varphi\, d\psi\,. \tag{8.17}$$

Reference to Equation (8.17) shows that the target return signal is a nonstationary frequency-modulated stochastic process (the cofactor $e^{j\omega(t)\tau}$). The Doppler frequency is a function of time, too (the cofactor $e^{-2j\omega(t)\frac{\Delta\rho}{c}}$). In addition

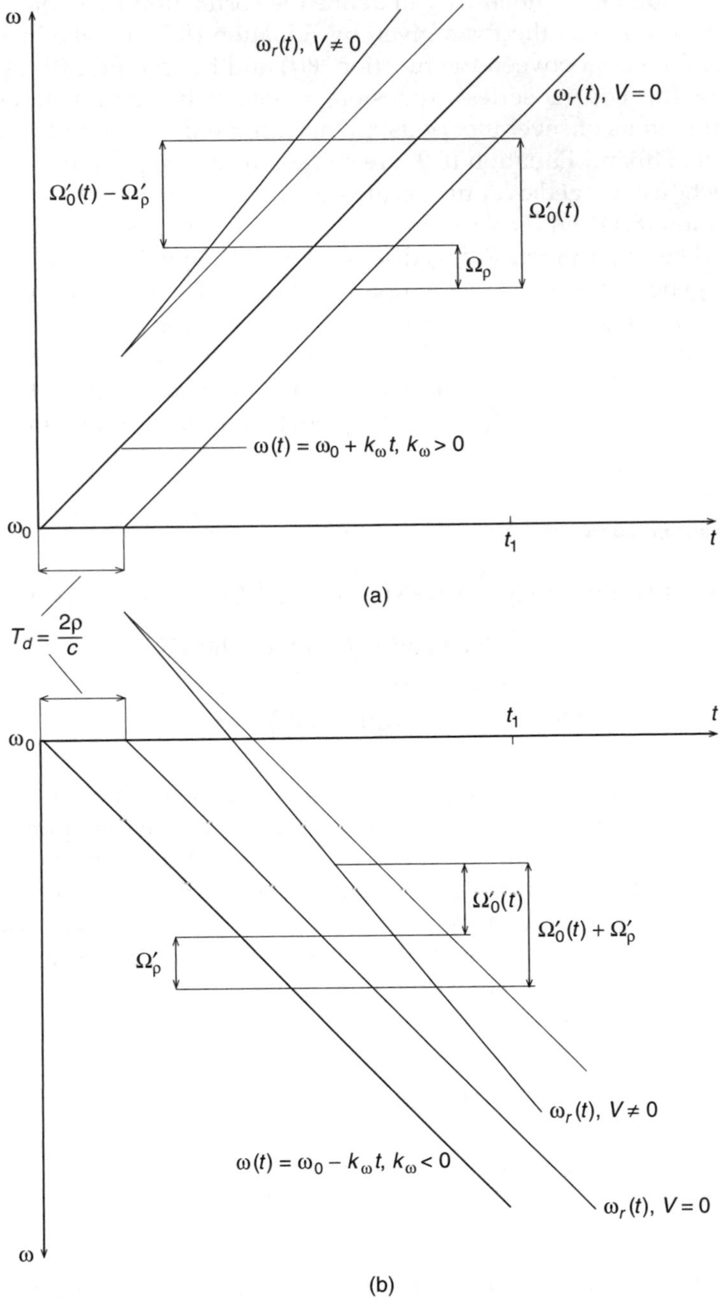

FIGURE 8.1
Variation of instantaneous frequency of searching and target return signals under saw-tooth frequency modulation when the radar is stationary and is moving: (a) $k_\omega > 0$ and (b) $k_\omega < 0$.

to the Doppler shift in frequency, the target return signal is shifted in frequency by the so-called range-finder frequency Ω_ρ (the cofactor $e^{-2jk_\omega \tau\rho c^{-1}}$) determined by

$$\Omega_\rho = k_\omega 2\rho c^{-1}. \tag{8.18}$$

The range-finder frequency Ω_ρ is shifted by the Doppler frequency (the cofactor $e^{2j\Omega_\rho \frac{\Delta\rho}{c}}$). Using Equation (8.10), we can define the correlation function of transformed target return signal fluctuations, which is different from the correlation function given by Equation (8.17) only in that the cofactor $e^{j\omega(t)\tau}$ is replaced with the cofactor $e^{j\omega_{im}\tau}$, i.e., frequency modulation is absent in the case of high frequency.

Consider slope scanning of the two-dimensional (surface) target when the velocity vector of moving radar is outside the directional diagram; the directional diagram is Gaussian; and the specific effective scattering area $S°(\psi)$ is approximated by the exponent [see Equation (2.150)]. Using Equation (2.142), Equation (4.15), Equation (4.16), and Equation (8.17), we can define the instantaneous power spectral density of the fluctuations in the following form:

$$S(\omega, t) = \frac{p(t)}{\Delta\Omega_{FM}} \cdot e^{-\pi \frac{[\omega - \omega(t) - \Omega'(t)]^2}{\Delta\Omega_{FM}^2}}, \tag{8.19}$$

where $p(t)$ is the power of the target return signal in the absence of frequency modulation [see Equation (2.167)];

$$\begin{aligned}
&\Omega'(t) = \Omega'_0(t) - \Omega'_\rho; \quad \Omega'_0(t) = \Omega_0(t)\cdot(1 - \delta_1 - \delta_2);\\
&\Omega_0(t) = \Omega_{\max}(t)\cdot \cos\theta_0; \quad \Omega_{\max}(t) = 2Vc^{-1}\omega(t);\\
&\Omega'_\rho = \omega_{\rho_0}[1 + 2Vc^{-1}\cos\theta_0 - \delta_3] \approx \Omega_{\rho_0}(1 - \delta_3);\\
&\Omega_{\rho_0} = k_\omega \cdot \frac{2h}{c\sin\gamma_0} = 2k_\omega \rho_0 c^{-1};
\end{aligned} \tag{8.20}$$

$$\begin{aligned}
&\delta_1 = -\frac{k_1 \Delta_v^{(2)} a_1}{4\pi\cos\theta_0}; \quad \delta_2 = -0.25\pi^{-1}\Delta_v^{(2)} a_1 \operatorname{ctg}\gamma_0;\\
&\delta_3 = 0.25\pi^{-1}\Delta_v^{(2)}(k_1 + \operatorname{ctg}\gamma_0)\operatorname{ctg}\gamma_0;
\end{aligned} \tag{8.21}$$

$$\Delta\Omega_{FM} = \sqrt{\Delta\Omega_h^2 + (\Delta\Omega_v + \Delta\Omega_\rho)^2}\ ;\ \Delta\Omega_h = \Delta_h^{(2)} b_1 \Omega_{\max}(t)\ ;$$
$$\Delta\Omega_v = \Delta_v^{(2)} a_1 \Omega_{\max}(t)\ ;\ \Delta\Omega_\rho = \Delta_v^{(2)} \Omega_{\rho_0}\ \mathrm{ctg}\,\gamma_0\ ; \tag{8.22}$$

$$\cos\theta_0 = \cos\varepsilon_0 \cos\gamma_0 \cos(\alpha + \beta_0) - \sin\varepsilon_0 \sin\gamma_0\ ; \tag{8.23}$$

a_1 and b_1 are determined by Equation (II.26) (see Appendix II).

As follows from Equation (8.19), the shape of the instantaneous power spectral density of the transformed fluctuations at a fixed t coincides with the shape of the power spectral density of the fluctuations with the nonmodulated (in frequency) searching signal. The values of $\Omega'(t)$ and $\Delta\Omega_{FM}$ depend on the time t, too. The variation of the Doppler frequency $\Omega_0(t)$ as a function of time is defined by the values of k_ω and t. For instance, at $f_0 = 10$ GHz, $F_0 = 10^4$ Hz, and $k_f = 2000$ GHz/sec for $t_1 = 10^{-2}$ sec, the change in the Doppler frequency $\Omega_0(t)$ is 20 Hz. The frequency $\omega_r(t)$ of the target return signal is plotted in Figure 8.1a and Figure 8.1b as a function of time at $V = 0$ and $V \neq 0$. The frequencies $\Omega_0'(t)$ and $\Omega_\rho'(t)$ are subtracted ($k_\omega > 0$, $V > 0$, see Figure 8.1a) or added ($k_\omega < 0$, $V > 0$, see Figure 8.1b), depending on the sign of k_ω. In addition, the instantaneous Doppler frequency as a function of carrier frequency at $V \neq 0$ is shown in Figure 8.1. Under these conditions and with the ratio $\frac{T_{d0}}{t_1} = 0.5 \times 10^{-3}$, the range-finder frequency Ω_ρ given by Equation (8.20) is equal to 10 kHz. The Doppler shift Ω_ρ' in the range-finder frequency given by Equation (8.20) is equal to 0.01 Hz under the same conditions, i.e., it is infinitesimal. The shifts δ_1 and δ_2 are the same as the shifts in the case of the nonmodulated (in frequency) searching signal with the specific effective scattering area $S°(\gamma)$ and distance ρ.

The shift δ_3 in the range-finder frequency given by Equation (8.21) coincides with the shift $\delta_{y_1} + \delta_{\rho_y}$ [see Equation (II.19) and Equation (II.20), Appendix II] due to the specific effective scattering area $S°(\gamma)$ and is equal to a time-shift of the pulsed target return signal center [see Equation (2.151)]. The bandwidth $\Delta\Omega_{FM}$ given by Equation (8.22) depends on time, too. However, this dependence is different for $\Delta\Omega_v$, $\Delta\Omega_h$, and $\Delta\Omega_\rho$. The values of $\Delta\Omega_h$ and $\Delta\Omega_v$ are independent of frequency [see Equation (3.76)], whereas the value of $\Delta\Omega_\rho$ depends on frequency due to Δ_v. At $\frac{\Delta\Omega_\rho}{\Omega_\rho} = 0.1$ and with the data given previously, the change in $\Delta\Omega_\rho$ is equal to 2 Hz, i.e., we can neglect this value. Because of this, we can neglect changes in the power spectral density bandwidth of the fluctuations caused by the function $\omega(t)$, the only exceptions being specific cases.

Reference to Equation (8.22) shows that under the condition $\alpha = \beta_0 = \varepsilon_0 = 0$, we can write

$$\Delta\Omega_{FM} = \Delta_v^{(2)}[\Omega_\rho \mathrm{ctg}\,\gamma_0 - 2Vc^{-1}\omega(t)\sin\gamma_0]\,. \tag{8.24}$$

As one can see from Equation (8.24), at $t = 0$ and if the conditions

$$\Delta\Omega_{v} = \Delta\Omega_{\rho} \qquad \text{and} \qquad \Omega_{0} = \Omega_{\rho}\,\text{ctg}^{2}\gamma_{0} \tag{8.25}$$

are satisfied, the bandwidth $\Delta\Omega_{FM}$ is equal to zero. For the considered case, the value of $\Delta\Omega_{FM}$ varies from 0 at $t = 0$ to 2 Hz at $t_1 = 10^{-2}$ sec. In accordance with Equation (8.20), the values of Ω_0 and Ω_ρ have different signs. Reference to Equation (8.25) shows that with power spectral density compression of the fluctuations, we can write

$$\frac{V}{h} = \frac{k_{\omega}}{f_{0}} \cdot \frac{\cos\gamma_{0}}{\sin^{3}\gamma_{0}}. \tag{8.26}$$

Consequently, varying the value of k_ω with known values of f_0 and γ_0 so that the bandwidth becomes minimum, we can define a navigational parameter ratio $\frac{V}{h}$.

The effect of power spectral density compression of the fluctuations can be explained based on the physical meaning, using Figure 8.2. Doppler $\Omega_0 = const$ and range-finder $\Omega_\rho = const$ isofrequency lines in the direction of moving radar touch one another if the condition $\beta_0 = 0°$ is satisfied. With a decrease in the angle γ_0 by $\frac{\Delta V}{2\sqrt{2}}$, the Doppler frequency increases by $0.5\Delta\Omega$, and vice versa. In accordance with Equation (8.20), the range-finder frequency must be negative and is equal to $-\Omega_{\rho_0}$ at the angle γ_0. However, it is equal to $-\Omega_{\rho_0} - 0.5\Delta\Omega_\rho$ at the angle $\gamma_0 + \frac{\Delta V}{2\sqrt{2}}$; and at the angle $\gamma_0 - \frac{\Delta V}{2\sqrt{2}}$, we obtain $-\Omega_{\rho_0} + 0.5\Delta\Omega_\rho$. Consequently, at the angle $\gamma_0 + \frac{\Delta V}{2\sqrt{2}}$, the sum of the range-finder frequency shift and the Doppler shift is given by

$$\Omega_{0} + 0.5\Delta\Omega - \Omega\rho_{0} - 0.5\Delta\Omega_{\rho} = \Omega_{0} - \Omega_{\rho_{0}}; \tag{8.27}$$

at the angle $\gamma_0 - \frac{\Delta V}{2\sqrt{2}}$, we have

$$\Omega_{0} + 0.5\Delta\Omega - \Omega\rho_{0} + 0.5\Delta\Omega_{\rho} = \Omega_{0} - \Omega_{\rho_{0}}; \tag{8.28}$$

and at the angle γ_0, we obtain $\Omega_0 - \Omega_{\rho_0}$.

Thus, relative shifts in the radar-range frequency Ω_{ρ_0} and Doppler frequency Ω_0 with respect to the difference $\Omega_0 - \Omega_{\rho_0}$ within the directional diagram compensate each other at the angle γ_0. If the Doppler frequency Ω_0 is negative, the range-finder frequency Ω_{ρ_0} has to be positive to ensure power spectral density compression of the fluctuations under sloping scanning of the underlying surface of a two-dimensional target. It should be noted that complete compression can take place not only in the horizontal direction of

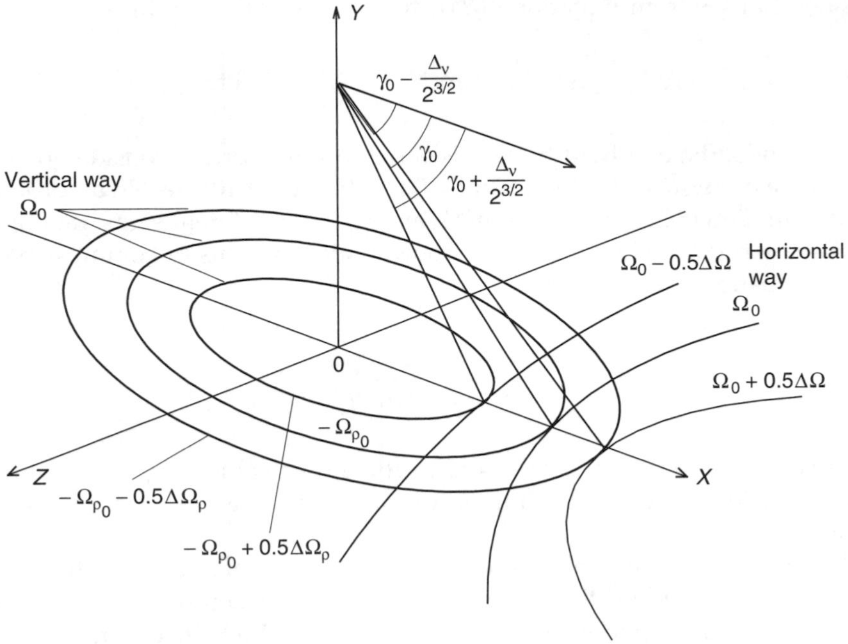

FIGURE 8.2
The Ω_0 and Ω_ρ isofrequency lines on the scattering surface.

moving radar, but in all cases when the velocity vector of moving radar and the directional diagram axis are in the same vertical plane. However, it should be noted that the average [see Equation (8.19)] power spectral density frequency of the fluctuations varies in time according to the saw-tooth law due to variation of the functions $\omega(t)$ and $\Omega'(t)$ both in the presence and absence of compression. The Doppler frequency as a function of time is present in the transformed target return signal and can hide the effect of compression.[12,13]

Using Equation (2.66) and Equation (8.19), we can easily define the power spectral density $\overline{S(\omega)}$ of the nonperiodic fluctuations averaged over the time interval $[0, t_1]$. If we do not take into consideration the effect of the term $k_\omega t$ on the value of $\Delta\Omega_{FM}$ [see Equation (8.22)], the power spectral density $\overline{S(\omega,t)}$ of the fluctuations averaged over time takes the following form:

$$\overline{S(\omega,t)} = S(\omega) = 0.5p(t)\cdot\left\{\Phi\left[\tfrac{\sqrt{\pi}}{\Delta\Omega_{FM}}(\omega - \Omega_0' - \omega_0 + \Omega_\rho' + 0.5\Delta\omega_M)\right]\right.$$
$$\left. - \Phi\left[\tfrac{\sqrt{\pi}}{\Delta\Omega_{FM}}(\omega - \Omega_0' - \omega_0 + \Omega_\rho' - 0.5\Delta\omega_M)\right]\right\}, \tag{8.29}$$

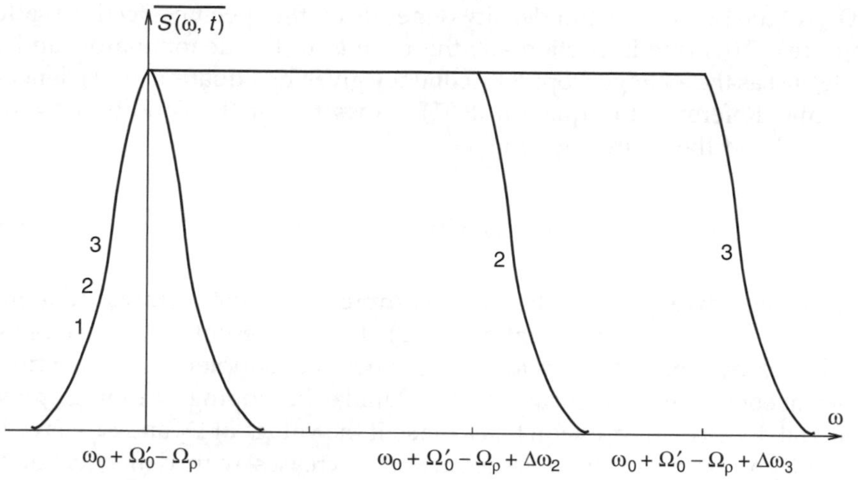

FIGURE 8.3
The power spectral density (averaged over time) of the saw-tooth frequency-modulated target return signal: (1) $\Delta\omega_1 = 0$; (2) $\Delta\omega_2 \neq 0$, $\Delta\omega_2 < \Delta\omega_3$; (3) $\Delta\omega_3 \neq 0$, $\Delta\omega_2 < \Delta\omega_3$.

where $\Delta\omega_M = K_\omega t_1$ and $\Phi(x)$ is given by Equation (1.27). The power spectral density of the fluctuations given by Equation (8.29) is shown in Figure 8.3. The power spectral density bandwidth at the level 0.5 is equal to the deviation of the frequency $\Delta\omega_M$; the width of front and cut is equal to $\Delta\Omega_{FM}$.

Omitting the intermediate mathematics, we consider the power spectral density of the fluctuations for the case when the directional diagram axis is directed vertically down and the velocity vector of moving radar is located on this axis. We assume that the directional diagram is axial symmetric and has width Δ; and the specific effective scattering area $S^\circ(\psi)$ is approximated by the Gaussian law [see Equation (2.187) and Equation (II.8), Appendix II]. Using the same technique as before, we obtain

$$S(\omega) = \frac{p(t)}{\Delta\Omega_{FM}} \cdot e^{-\pi \frac{\omega - \omega(t) - \Omega(t)}{\Delta\Omega_{FM}}}, \tag{8.30}$$

where $\omega < \Omega(t) + \omega(t)$ at $\Omega_0 \geq \Omega_{\rho_0}$ and $\omega > \Omega(t) + \omega(t)$ at $\Omega_0 \leq \Omega_{\rho_0}$; $p(t)$ is the target return signal power given by Equation (2.203);

$$\begin{gathered} \Omega(t) = \Omega_0(t) - \Delta\Omega_{\rho_0}; \; \Omega_0(t) = -\frac{2V}{c} \cdot \omega(t) \sin\varepsilon_0, \\ \varepsilon_0 = \pm 90^\circ; \; \Delta\Omega_{FM} = \frac{\Delta^2 |\, \Omega_0(t) + \Omega_{\rho_0} \,|}{4\pi a^2}; \; \Delta\Omega_{\rho_0} = \frac{2h}{c} \cdot k_\omega . \end{gathered} \tag{8.31}$$

The power spectral density of the fluctuations has an exponential shape, as in the case of the non-frequency-modulated searching signal. The bandwidth

$\Delta\Omega_{FM}$ of the power spectral density depends on the specific effective scattering area $S°(\psi)$ (see in Section 4.7, the term a^2 in the denominator) and on time, just as the average Doppler frequency given by Equation (8.31) depends on time. Reference to Equation (8.31) shows that if the condition $t = 0$ is satisfied and the following equality

$$2Vc^{-1}\omega_0 \sin\varepsilon_0 = 2hc^{-1}k_\omega \tag{8.32}$$

is true, the power spectral density is compressed to the bandwidth defined by the function $\omega(t)$ in Equation (8.31). In the present case, this can be explained by complete coincidence between the Doppler and range-finder isofrequency lines, which are circles. Unlike in sloping scanning, power spectral density compression takes place if the values of Ω_0 and Ω_{ρ_0} have the same sign, as the Doppler frequency Ω_0 increases with deflection of the directional diagram axis from $\varepsilon_0 = 90°$.

8.3 The Asymmetric Saw-Tooth Frequency-Modulated Searching Signal

In this case, the time-periodic sequence of searching signals coherent from period to period is given by Equation (8.5), where the functions $\Psi(t)$ and $\omega(t)$ are determined by Equation (8.15) and $k_\omega = \frac{\Delta\omega_M}{T_M}$. Changes in the frequency of this searching signal are shown in Figure 8.4. Figure 8.5 shows the frequency ω_h of the heterodyne signal and the frequency ω_r of the target return signal as a function of the distance ρ. It is assumed in Figure 8.5 that

$$\omega_{im} = 0 \quad \text{and} \quad V = 0 \ ; \tag{8.33}$$

$$t_0 = 0.5\tau, \quad t_1 = 0.5\tau + 2\rho c^{-1}, \quad t_2 = 0.5\tau + T_M, \quad \text{and} \quad t_3 = 0.5\tau + T_M + 2\rho c^{-1}. \tag{8.34}$$

The transformed target return signal $W_r(t + 0.5\tau)$ shifted left by τ with respect to the signal $W_r(t - 0.5\tau)$ on the time axis presents an analogous picture. In the general case, the signals $W_r(t - 0.5\tau)$ and $W_r(t + 0.5\tau)$ may not overlap within the limits of the period $n = 0$. Reference to Figure 8.6 shows that the signal $W_r(t - 0.5\tau)$ or $W_r(t + 0.5\tau)$ consists of two closing pulses within the limits of each period T_M: the first pulse with frequency $k_\omega T_d$ and duration $T_M - T_d$ and the second pulse with frequency $k_\omega(T_M - T_d)$ and duration T_d. For

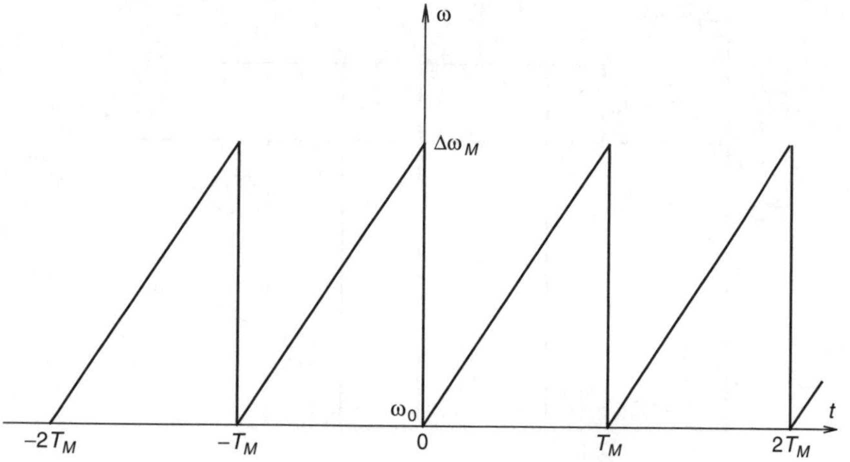

FIGURE 8.4
The frequency of the searching signal under asymmetric saw-tooth frequency modulation.

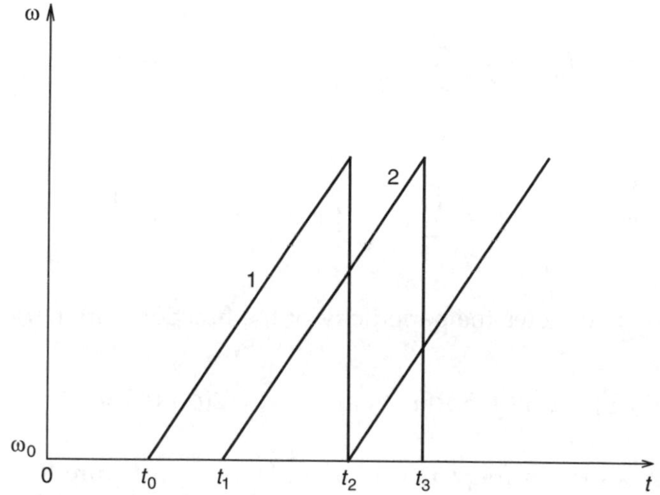

FIGURE 8.5
The frequency of the heterodyne (1) and target return (2) signals under asymmetric saw-tooth frequency modulation.

further analysis, define differences $\Delta\Psi_1(t)$ in the exponents of Equation (8.10), which depend on $t - 0.5\tau$ and $t + 0.5\tau$:

$$\Delta\Psi_1(t \mp 0.5\tau) = \pm k_\omega (t \mp 0.5\tau)\cdot\left(\tfrac{2\rho}{c} \mp \tfrac{\Delta\rho}{c}\right). \tag{8.35}$$

We can define the functions $\Delta\Psi_1(t)$ within the limits of the period T_M. For all $z_{1,2} = t \mp 0.5\tau$ we can use the Fourier-series expansion

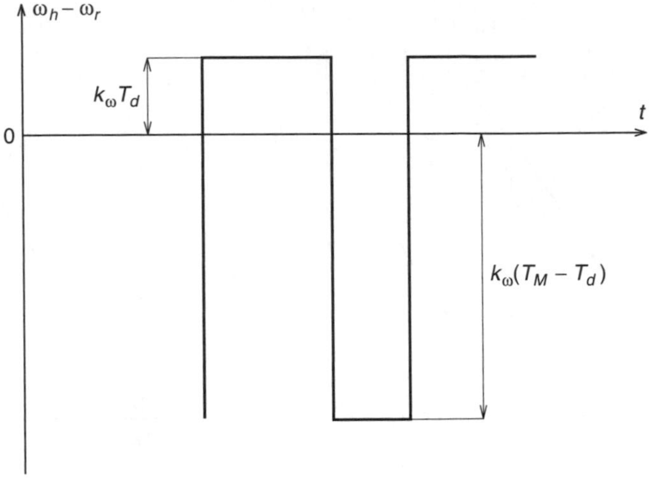

FIGURE 8.6
The frequency of the transformed target return signal under asymmetric saw-tooth frequency modulation.

$$\boldsymbol{Per}_{T_M}\left(e^{jk_\omega z_1 T_d'}\right) = \sum_{n=-\infty}^{\infty}\left[\frac{1}{T_M}\int_0^{T_M} e^{jk_\omega z_1 T_d'}\cdot e^{-jn\omega_M z_1}dz_1\right]\cdot e^{jn\omega_M z_1}\ ; \tag{8.36}$$

$$\boldsymbol{Per}_{T_M}\left(e^{jk_\omega z_2 T_d''}\right)^* = \sum_{n=-\infty}^{\infty}\left[\frac{1}{T_M}\int_0^{T_M} e^{jk_\omega z_2 T_d''}\cdot e^{-jn\omega_M z_2}dz_2\right]\cdot e^{-jn\omega_M z_2}\ , \tag{8.37}$$

where $\boldsymbol{Per}_{T_M}(x)$ denotes the periodicity of the function with period T_M;

$$T_d' = 2(\rho - 0.5\Delta\rho)c^{-1} \quad \text{and} \quad T_d'' = 2(\rho + 0.5\Delta\rho)c^{-1}; \tag{8.38}$$

ω_M is the modulation frequency. In accordance with Figure 8.4–Figure 8.6, the coefficients of the Fourier-series expansion in Equation (8.36) and Equation (8.37) should be defined for the function given by Equation (8.36) within the limits of the intervals $[t_1, t_2]$ and $[t_2, t_3]$ independently of the function given by Equation (8.37). This is also true for the function given by Equation (8.37). It is necessary to substitute Equation (8.36) and Equation (8.37) in Equation (8.10) and to average over time within the limits of the period T_M.

Carrying out the mathematics, we can define the correlation function of the fluctuations averaged over time for arbitrary positions of the velocity vector of moving radar and radar antenna beam with respect to the two-dimensional (surface) target

$$R(\tau) = p_0 \cdot e^{j\omega_{im}\tau} \iint \tilde{g}^2(\varphi, \psi) \cdot e^{-2j(\omega_0 + 0.5\Delta\omega_M)\frac{\Delta\rho}{c}}$$

$$\times \sum_{n=-\infty}^{\infty} \left[(1 - \tfrac{T_d'}{T_M})(1 - \tfrac{T_d''}{T_M}) \cdot \textbf{sinc}\, x' \,\textbf{sinc}\, x'' \right. \quad , \tag{8.39}$$

$$\left. + \tfrac{T_d'}{T_M} \cdot \tfrac{T_d''}{T_M} \cdot \textbf{sinc}\, y' \,\textbf{sinc}\, y'' \right] \cdot e^{-jn\omega_M(\tau - \frac{\Delta\rho}{c})} d\varphi\, d\psi$$

where

$$x' = 0.5(k_\omega T_d' - n\omega_M) \cdot \left(1 - \tfrac{T_d'}{T_M}\right) \cdot T_M$$
$$\text{and} \quad x'' = 0.5(k_\omega T_d'' - n\omega_M) \cdot \left(1 - \tfrac{T_d''}{T_M}\right) \cdot T_M \, ; \tag{8.40}$$

$$y' = 0.5 \cdot \left[k_\omega (T_M - T_d') + n\omega_M \right] \cdot T'$$
$$\text{and} \quad y'' = 0.5 \cdot \left[k_\omega (T_M - T_d'') + n\omega_M \right] \cdot T_d'' ; \tag{8.41}$$

$$\textbf{sinc}\, x = \frac{\sin x}{x}. \tag{8.42}$$

Comparing Equation (8.14) and Equation (8.39), one can see that the coefficients C_ρ'' have the form of the function **sinc** x and $\omega_{av} = \omega_0 + 0.5\Delta\omega_M$. The power spectral density of the fluctuations consists of a set of partial power spectral densities defined by harmonics of the modulation frequency ω_M. Unlike the case of the nonperiodic searching signal given by Equation (8.17), the correlation function given by Equation (8.39) possesses two range-finder frequencies

$$\Omega_\rho' = k_\omega T_d' \quad \text{and} \quad \Omega_\rho'' = k_\omega (T_M - T_d') \tag{8.43}$$

(see Figure 8.6), which are defined by different intervals of the modulation period. Harmonics of the modulation frequency ω_M are shifted by the one-half Doppler frequency $\frac{\omega_M \Delta\rho}{c}$. Because all components of the target return signal should be shifted by their Doppler frequency, not the one-half Doppler frequency, we can reason that another one-half Doppler shift in frequency is given by Equation (8.39) using the arguments $k_\omega T_d'$ and $k_\omega T_d''$ of the **sinc** x functions. As the condition $n\omega_M << \omega_0$ is satisfied, as a rule, we can neglect the Doppler shift in the frequency ω_M.

Reference to Equation (8.39) shows that the target return signal power at the range-finder frequency $k_\omega T_d$ is proportional to

$$\left(1-\tfrac{T'_d}{T_M}\right)\cdot\left(1-\tfrac{T''_d}{T_M}\right)\approx\left(1-\tfrac{T_{d_0}}{T_M}\right)^2, \tag{8.44}$$

i.e., to the square of the relative decrease in the period T_M, as we can suppose here that

$$T'_d = T''_d = T_{d_0} = 2\rho_0 c^{-1}, \tag{8.45}$$

where ρ_0 corresponds to the distance in the direction along the directional diagram axis. The second term in Equation (8.39) defines the part of the power spectral density that lies within another modulation period. The power of this part is proportional to

$$\frac{T'_d}{T_M}\cdot\frac{T''_d}{T_M}\approx\frac{T^2_{d_0}}{T^2_M}. \tag{8.46}$$

If the first target return signal with the range-finder frequency $k_\omega T_{d_0}$ becomes maximum at $n_0\omega_M$, the second target return signal with the range-finder frequency $k_\omega(T_M - T_{d_0})$ becomes maximum under the condition

$$k_\omega(T_M - T_{d_0}) = -n_1\omega_M. \tag{8.47}$$

If the condition $T_M >> T_{d_0}$ is satisfied, we can write

$$k_\omega(T_M - T_{d_0}) \approx \Delta\omega_M, \tag{8.48}$$

i.e., the frequency given by Equation (8.48) is approximately equal to the frequency deviation. The maximum of the power spectral density of the fluctuations of the first target return signal takes place at the frequency $\omega_{im} + \Omega_0 - n_0\omega_M$. The maximum of the power spectral density of the fluctuations of the second target return signal takes place at the frequency $\omega_{im} + \Omega_0 + n_1\omega_M$. The power ratio of these components is equal to $\frac{T_M^2}{T_{d_0}^2}$ and is independent of the value of $\Delta\omega_M$. For instance, at $\rho_0 = 3$ km, $T_{d_0} = 2 \cdot 10^{-5}$ sec, and $T_M = 2 \times 10^{-2}$ sec, we get $\frac{T_M}{T_{d_0}} = 10^3$. For this reason, here, we can neglect the second term.

If the condition $T_M < T_{d_0}$ is satisfied, Equation (8.39) is true, too. Then, the parameter T_d must be expressed in the form $lT_M + t_d$, where $t_d \leq T_M$, and we need to replace the parameter T_d with the parameter t_d in Equation (8.39). The frequencies $k_\omega T_{d_0}$ and $k_\omega(T_M - T_{d_0})$, as in the case of radar with pulsed searching signals, do not contain any information regarding lT_M if the period is less than T_d and will be proportional to $k_\omega t_d$ and $k_\omega(T_M - t_d)$, respectively. Henceforth, we assume that $T_d' << T_M$, $T_d'' << T_M$, and $T_d' = T_d'' = T_d$, as we can neglect the effect of the parameter $\Delta\rho$ on the parameters T_d' and T_d''. In this case, we can write

$$x' = x'' = \frac{\pi(k_\omega T_d - n\omega_M)}{\omega_M} = \pi(\Delta f_M T_d - n)\,. \tag{8.49}$$

Denote

$$X_\rho = \Delta f_M T_d = 2\Delta f_M \frac{\rho}{c} = \frac{\Omega_\rho}{\omega_M}\,. \tag{8.50}$$

The pure parameter X_ρ is normalized with respect to the modulation frequency ω_M by the range-finder frequency Ω_ρ. The sign of the parameter X_ρ depends on the sign of the parameter Δf_M. Because

$$\rho = \frac{h}{\sin(\psi + \gamma_0)}\,, \tag{8.51}$$

the pure parameter X_ρ can be thought of as a function of the angle ψ. For simplicity and convenience, in the subsequent discussion, we will use the frequency ω_0 instead of the frequency $\omega_0 + 0.5\Delta\omega_M$. Taking into consideration the previously mentioned statements, the correlation function of the fluctuations given by Equation (8.39) can be written in the following form:

$$R(\tau) = p_0 \cdot e^{j\omega_{im}\tau} \iint \tilde{g}^2(\varphi,\psi) \cdot e^{-2j\omega_0 \frac{\Delta\rho}{c}} \sum_{n=-\infty}^{\infty} \mathbf{sinc}^2[\pi(X_\rho - n)]\, d\varphi\, d\psi\,. \tag{8.52}$$

Let us consider some specific cases.

8.3.1 Sloping Scanning

As follows from Section 8.2, the power spectral density characteristics of the target return signal fluctuations with the range-finder frequency depend only on characteristics of the vertical-coverage directional diagram. Because of this, consider first the case of sloping scanning (i.e., the condition $\gamma_0 + \varepsilon_0 < 90°$ should be satisfied), when the velocity vector of moving radar lies in the same vertical plane as the vertical-coverage directional diagram axis, is outside the directional diagram, and is toward directed, i.e., $\alpha = \beta_0 = 0°$.

Let us assume that the variables φ and ψ in the function $\tilde{g}(\varphi, \psi)$ are separable and that the specific effective scattering area $\mathbf{S}°$ is a function of the angle γ only. Under these conditions, Equation (8.50) can be written in the following form:

$$X_\rho = X_{\rho_0}(1-\psi \operatorname{ctg} \gamma_0) \quad \text{and} \quad X_{\rho_0} = 2\Delta f_M h c^{-1} \sin\gamma_0 = \Delta f_M T_{d_0} \,. \tag{8.53}$$

Using the Fourier transform for Equation (8.52) and taking into consideration the dependence between X_ρ and ψ, we can write the power spectral density of the fluctuations in the following form:

$$S(\omega) = \frac{p_1}{\Delta\Omega_v} \cdot \sum_{n=-\infty}^{\infty} \tilde{g}^2\left[\tfrac{\Omega_n-\omega}{\Omega_v}\right] \cdot \operatorname{sinc}^2\pi\left[X_{\rho_0} - n + \tfrac{\omega-\Omega_n}{\Delta\Omega_v} \cdot d_\rho\right], \tag{8.54}$$

where

$$\Omega_n = \omega_{im} + \Omega_0 - n\omega_M ; \tag{8.55}$$

$$d_\rho = \frac{\Delta\Omega_\rho}{\omega_M} = \Delta f_M T_r = X_{\rho_0} \Delta_v^{(2)} \operatorname{ctg} \gamma_0 ; \tag{8.56}$$

the parameters p_1, Ω_0, Ω_v, and $\Delta\Omega_v$ are the same as in the absence of frequency modulation:

$$\Omega_0 = 4\pi\, V c^{-1} \cos(\gamma_0 + \varepsilon_0) ; \quad \Omega_v = 4\pi\, V c^{-1} \sin(\gamma_0 + \varepsilon_0) ;$$
$$\text{and} \quad \Delta\Omega_v = \Omega_v \Delta_v^{(2)} ; \tag{8.57}$$

$$p_1 = \frac{P G_0^2 \lambda^2 \Delta_v^{(2)} S^\circ(\gamma_0) \sin\gamma_0 \int g_h^2(\varphi)\, d\varphi}{64\pi^3 h^2} \,. \tag{8.58}$$

The parameter d_ρ is the bandwidth of the range-finder spectrum $\Delta\Omega_\rho$ normalized with respect to the modulation frequency ω_M [see Equation (8.22)].

If the condition $\Delta f_M = 0$ is satisfied, the power spectral density given by Equation (8.54) defines the power spectral density of the target return signal Doppler fluctuations in the absence of frequency modulation, which is shifted by the frequency ω_{im}:

$$S(\omega) = \frac{p_1}{\Delta\Omega_v} \cdot \tilde{g}_v^2\left(\tfrac{\Omega_0 + \omega_{im} - \omega}{\Omega_v}\right) . \tag{8.59}$$

Assume that the directional diagram is Gaussian and that the specific effective scattering area $\mathbf{S}^\circ(\psi)$ is defined by exponent.[14,15] As is well known, the main lobe of the function $\mathbf{sinc}^2\, x$ can be approximated by the Gaussian law:

$$\mathbf{sinc}^2\pi(X_\rho - n) = e^{-\pi(X_\rho - n)^2}. \tag{8.60}$$

Using the approximation given by Equation (8.60), both functions have the same effective bandwidth 1 (or ω_M in the nonnormalized case). In this case, based on Equation (8.54) we can write

$$S(\omega) = \frac{p_{FM}}{|\Delta\Omega_{v_{FM}}|} \cdot \sum_{n=-\infty}^{\infty} e^{-\pi\frac{(X'_{\rho_0} - n)^2}{1+d_\rho^2}} \cdot e^{-\pi\frac{(\omega-\Omega_n)^2}{\Delta\Omega^2_{v_{FM}}}}, \tag{8.61}$$

where

$$p_{FM} = \frac{PG_0^2\lambda^2 S^\circ(\gamma_0)\Delta_v\Delta_h \sin\gamma_0}{128\pi^3 h^2\sqrt{1+d_\rho^2}} \cdot e^{\frac{\Delta_v^{(2)}(k_1 + \mathrm{ctg}\,\gamma_0)}{8\pi(1+d_\rho^2)}}; \tag{8.62}$$

$$\begin{aligned} \Omega_n &= \omega_{im} + \Omega'_0 - n\omega_M - \frac{\Delta\Omega_v d_\rho (X_{\rho_0} - n)}{1+d_\rho^2} \\ &= \omega_{im} + \Omega'_0 - n\omega_M - \frac{\Delta\Omega_v d_\rho X_{\rho_0}}{1+d_\rho^2} + n\left[\frac{\Delta\Omega_v d_\rho}{1+d_\rho^2} - \omega_M\right] \end{aligned}; \tag{8.63}$$

$$\Omega'_0 = \Omega_0(1 - \delta_{1_{FM}} - \delta_{2_{FM}}); \tag{8.64}$$

$$\delta_{1_{FM}} = \frac{k_1\Delta_v^{(2)}\,\mathrm{tg}\,(\gamma_0 + \varepsilon_0)}{4\pi(1+d_\rho^2)}; \tag{8.65}$$

$$\delta_{2_{FM}} = \frac{\Delta_v^{(2)}\mathrm{ctg}\,\gamma_0 \cdot \mathrm{tg}\,(\gamma_0 + \varepsilon_0)}{4\pi(1+d_\rho^2)}; \tag{8.66}$$

$$\Delta\Omega_{v_{FM}} = \frac{\Delta\Omega_v}{\sqrt{1+d_\rho^2}}; \tag{8.67}$$

$$X'_{\rho_0} = X_{\rho_0}(1 - \delta_3). \tag{8.68}$$

Reference to Equation (8.62), Equation (8.64)–Equation (8.66), and Equation (8.67) shows that the parameters p_{FM}, δ_{1FM}, δ_{2FM}, and $\Delta\Omega_{vFM}$ are equal to the corresponding parameters of the power spectral density of the fluctuations in the absence of frequency modulation, replacing the parameter Δ_v with the parameter $\frac{\Delta_v}{\sqrt{1+d_\rho^2}}$. The shift δ_3 in the normalized range-finder frequency [see Equation (8.68)] is equal to the range-finder frequency shift of the nontransformed target return signal [see Equation (8.21)].

First, consider the stationary radar, i.e., the condition $V = 0$ is true. In this case, under the condition $\Delta\Omega_{vFM} \to 0$, based on Equation (8.61) we can determine the power spectral density of the fluctuations as follows:

$$S(\omega) = p_{FM} \sum_{n=-\infty}^{\infty} \delta[\omega - (\omega_{im} - n\omega_M)] \cdot e^{-\pi \frac{(X'_{\rho_0} - n)^2}{1+d_\rho^2}}. \tag{8.69}$$

Reference to Equation (8.69) shows that the power spectral density is a set of discrete components at the frequencies $\omega_{im} - n\omega_M$. The power of each component is defined by the modulating function given by Equation (8.60) and by the parameter d_ρ, which depends on the parameters of the vertical-coverage directional diagram. Under the condition $X'_{\rho_0} > 0$, there are only harmonics at frequencies $\omega_{im} - n\omega_M$ in the power spectral density and under the condition $X'_{\rho_0} < 0$, only harmonics at frequencies $\omega_{im} + n\omega_M$. If the conditions $d_\rho << 1$ and $\Omega_{\rho_0} < \omega_M$ are satisfied, based on Equation (8.69) we can write

$$S(\omega) = p_{FM} \sum_{n=-\infty}^{\infty} \delta[\omega - (\omega_{im} - n\omega_M)] \cdot e^{-\pi(X'_{\rho_0} - n)^2} \tag{8.70}$$

and the power spectral density is independent of the directional diagram and is defined only by the modulating function given by Equation (8.60). The effective bandwidth of this power spectral density envelope is equal to 1 or ω_M for nonnormalized values. The only harmonic $n_0\omega_M$ takes place under the condition $X'_{\rho_0} = n_0$ within the limits of the effective power spectral density bandwidth given by Equation (8.70). In spite of the fact that in the present case, the target return signal consists of the individual target return signals from a set of random scatterers spaced by the distance ρ_0, the shape of the power spectral density of the fluctuations corresponds to reflections from the point target.

As can be easily seen, the condition $d_\rho << 1$ corresponds to the condition $\delta\rho_0 >> \delta\rho$, where

$$\delta\rho_0 = \frac{c}{2\Delta f_M} \quad \text{and} \quad \delta\rho = \frac{h\Delta_v \operatorname{ctg}\gamma_0}{\sqrt{2}\,\sin\gamma_0}. \tag{8.71}$$

As is well known, the parameter $\delta\rho_0$ is the radar resolution with the frequency-modulated searching signal for point targets; $\delta\rho$ is the length of the scanned area of the two-dimensional (surface) target (see Figure 8.7 and Figure 8.8). The radar resolution with the frequency-modulated searching signal depends only on frequency deviation and corresponds to the resolution $0.5c\tau_p$ of radar with the pulsed searching signal under the condition $\tau_p = \frac{1}{\Delta f_M}$. If the condition $\delta\rho_0 >> \delta\rho$ is satisfied, we can say that reflection is caused by an area no larger than one zone of the radar range (see Figure 8.7).

Under the condition $d_\rho >> 1$, we can assume that $\Omega_{\rho_0} >> \omega_M$ and the power spectral density takes the following form:

$$S(\omega) = p_{FM} \sum_{n=-\infty}^{\infty} \delta[\omega - (\omega_{im} - n\omega_M)] \cdot e^{-\pi \frac{(X'_{\rho_0} - n)^2}{d_\rho^2}} . \tag{8.72}$$

In this case, the power spectral density envelope of the fluctuations in non-normalized units with respect to the modulation frequency ω_M coincides with the power spectral density envelope of the fluctuations with the range-finder frequency, the bandwidth of which is equal to $\Delta\Omega_\rho$, and is defined completely by the shape of the vertical-coverage directional diagram. Under

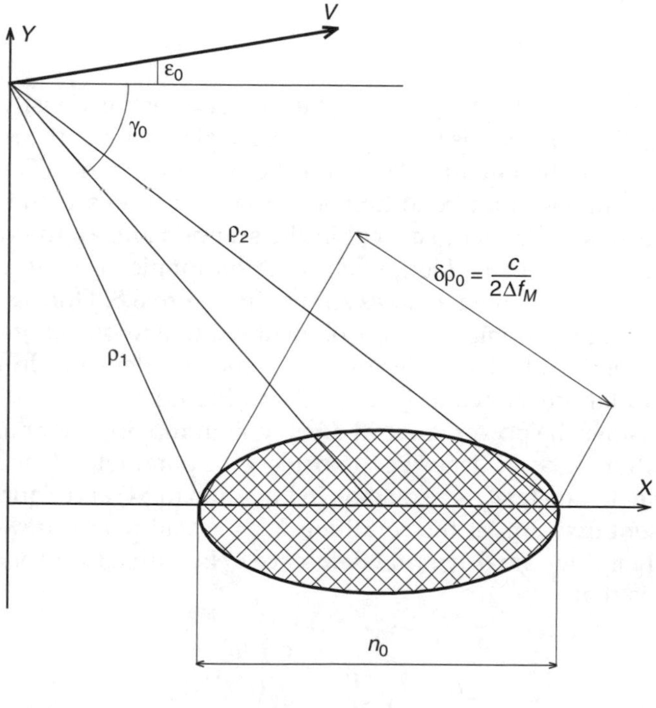

FIGURE 8.7
A single radar range zone.

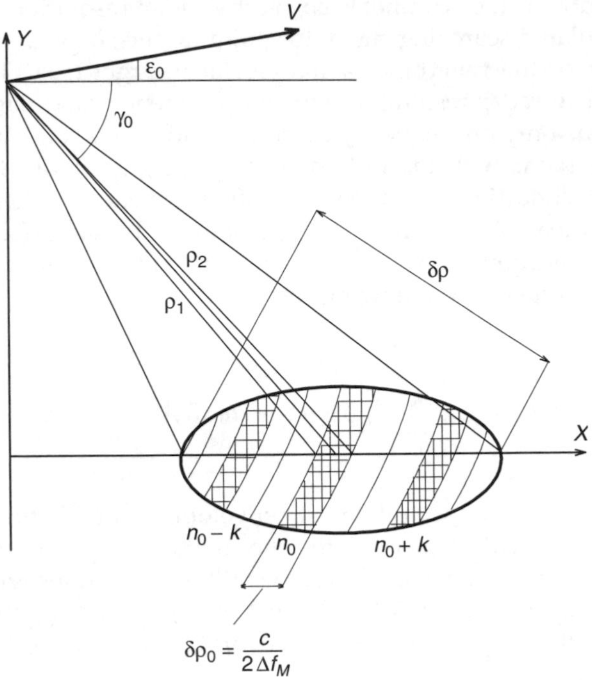

FIGURE 8.8
Several radar range zones ($\frac{\delta\rho}{\delta\rho_0} = 11$) .

the condition $\omega_M \to 0$ or $T_M \to \infty$, the power spectral density given by Equation (8.72) corresponds to the power spectral density of the nonperiodic fluctuations given by Equation (8.19) for the stationary radar. The condition $d_\rho >> 1$ is analogous to the condition $\delta\rho_0 << \delta\rho$. For this case, the number of radar range zones is equal to d_ρ within the scanned area of the two-dimensional (surface) target (see Figure 8.8). Each harmonic $\omega_{im} - (n \pm k) \times \omega_M$ is formed in its own scanned area, as shown in Figure 8.8. During formation, the neighboring harmonics are noncoherent relative to one another and are formed independently. Examples of power spectral densities in the case of stationary radar are shown in Figure 8.9 and Figure 8.10.

Now, consider the power spectral density with moving radar, i.e., $V \neq 0$.[16] If the condition $d_\rho << 1$ is true, we can neglect the parameter d_ρ and the terms in which the parameter d_ρ is a cofactor, in Equation (8.54) and Equation (8.61). In the present case, under the condition $X'_{\rho_0} = n_0$, all partial power spectral densities are infinitesimal at $n \neq n_0$. Reference to Equation (8.54) and Equation (8.61) shows that

$$S(\omega) = \frac{p}{|\Delta\Omega_v|} \cdot \tilde{g}^2\left(\tfrac{\Omega_{n_0} - \omega}{\Omega_v}\right), \tag{8.73}$$

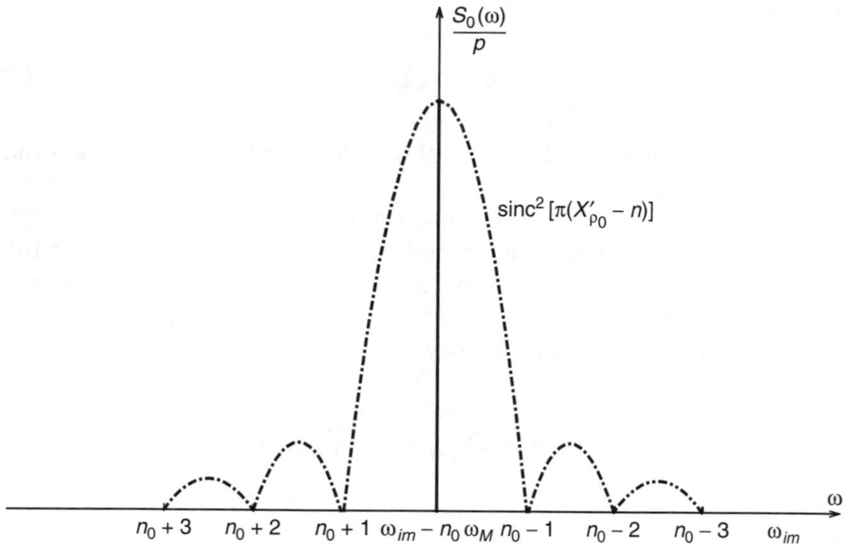

FIGURE 8.9
The power spectral density of target return signal fluctuations with stationary radar and the asymmetric saw-tooth frequency-modulated searching signal: $d_p << 1$.

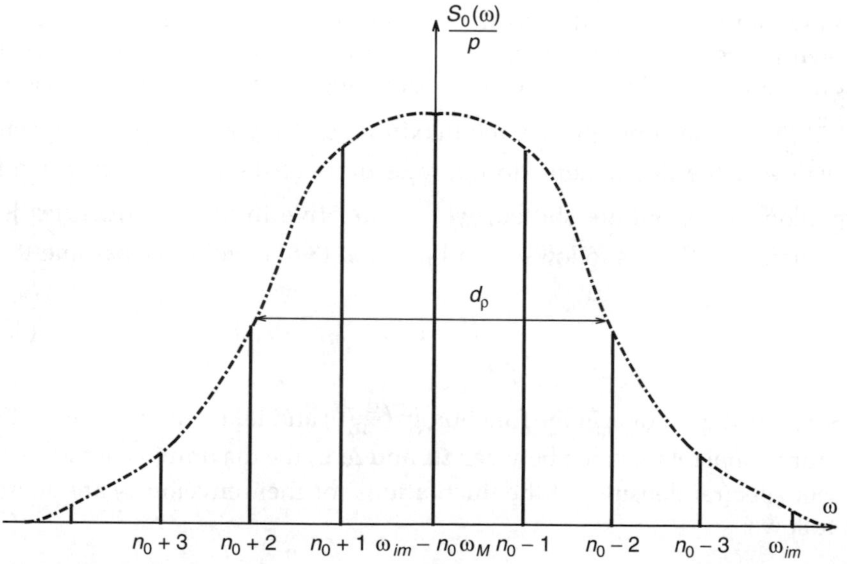

FIGURE 8.10
The power spectral density of target return signal fluctuations with stationary radar and the asymmetric saw-tooth frequency-modulated searching signal: $d_p >> 1$.

where

$$\Omega_{n_0} = \omega_{im} + \Omega'_0 - n_0\omega_M , \tag{8.74}$$

i.e., only one power spectral density of the Doppler fluctuations takes place at the frequency Ω_{n_0}.

Under the condition $d_\rho >> 1$, the maximums of all partial power spectral densities of the fluctuations are defined by the "equality to zero" condition of the argument of the function $\mathbf{sinc}^2\, x$ in Equation (8.54). For instance, the maximum partial power spectral density corresponding to the $(n_0 + k)$-th harmonic takes place at the frequency

$$\omega_k = \Omega_{n_0+k} + k \cdot \frac{\Delta\Omega_v}{d_\rho} , \tag{8.75}$$

where

$$\Omega_{n_0+k} = \omega_{im} + \Omega'_0 - (n_0 + k)\omega_M . \tag{8.76}$$

Based on the physical meaning we can explain this in the following manner. The vertical-coverage directional diagram is divided into d_ρ partial directional diagrams due to the frequency-modulated searching radar signal as shown in Figure 8.7 and Figure 8.8. The Doppler shift in frequency corresponding to the middle of the k-th radar range zone relative to n_0 is equal to $\frac{\Omega'_0 + k\Delta\Omega_v}{d_\rho}$. As a consequence, the maximums of the partial power spectral densities of the fluctuations do not take place at the frequencies given by Equation (8.76), but are shifted by $\frac{k\Delta\Omega_v}{d_\rho}$ relative to these frequencies [see Equation (8.75)]. This follows from Equation (8.61), too, if we assume that

$$d_\rho^2 \approx 1 + d_\rho^2 . \tag{8.77}$$

Substituting ω_k for ω in the function $\tilde{g}_v^2(\frac{\Omega_n - \omega}{\Omega_v})$ and taking into consideration the functional relationship between Ω_v and $\Delta\Omega_v$, the maximums of the partial power spectral densities of the fluctuations (or their envelopes) are defined as follows:

$$S(\omega_k) = \frac{p_1}{\Delta\Omega_v} \cdot \tilde{g}_v^2(-\tfrac{k\Delta_v^{(2)}}{d_\rho}) . \tag{8.78}$$

The shape of the partial power spectral density is determined by

$$S_{n_0+k}(\omega) = \frac{p_1}{\Delta\Omega_v} \cdot \tilde{g}_v^2\left(\frac{\omega_k - \frac{k\Delta\Omega_v}{d_\rho} - \omega}{\Omega_v}\right) \cdot \mathbf{sinc}^2\left[\frac{\pi(\omega-\omega_k)d_\rho}{\Delta\Omega_v}\right]. \tag{8.79}$$

The function $\mathbf{sinc}^2\left[\frac{\pi(\omega-\omega_k)d_\rho}{\Delta\Omega_v}\right]$ will "cut" the power spectral density with an effective bandwidth equal to $\frac{\Delta\Omega_v}{d_\rho}$ from the function $\tilde{g}_v^2(\psi)$ under tuning on the frequency $\frac{k\Delta\Omega_v}{d_\rho}$. In the general case, the power spectral density is not symmetrical, because the "cutting" is carried out on the slope of the function $\tilde{g}_v^2(\psi)$. Only in the case of the Gaussian vertical-coverage directional diagram and if the condition in Equation (8.60) is satisfied, as shown from Equation (8.61), is the power spectral density symmetric, with a bandwidth of $\frac{\Delta\Omega_v}{d_\rho}$.

Reference to Equation (8.75) shows that in accordance with the sign of the parameters $\Delta\Omega_v$ and d_ρ, the frequency ω_k can be greater than the frequency Ω_{n_0+k} (if $\Delta\Omega_v$ and d_ρ have the same sign) or less (if $\Delta\Omega_v$ and d_ρ have different signs). In the present case, the bandwidth $\Delta\Omega_v$ is positive, i.e., the condition $\alpha = \beta_0 = 0°$ is satisfied. Let us define a difference between the frequencies

$$\omega_{k+1} - \omega_k = \omega_M\left(\frac{\Delta\Omega_v}{d_\rho\omega_M} - 1\right). \tag{8.80}$$

One can see from Equation (8.80) that if the condition

$$\Delta\Omega_v = d_\rho\omega_M \tag{8.81}$$

is true, the partial power spectral densities are superimposed on one another. In this case, the power spectral density given by Equation (8.61) has the following form:

$$S(\omega) = \frac{p_1}{\omega_M} \cdot e^{-\pi\frac{(\omega-\Omega'_{n_0})^2}{\omega_M^2}}, \tag{8.82}$$

where

$$\Omega'_{n_0} = \omega_{im} + \Omega'_0 - n_0\omega_M = \omega_{im} + \Omega'_0 - X'_{\rho_0}\omega_M. \tag{8.83}$$

The effective compressed power spectral density bandwidth corresponds to reflection from a single radar range zone [see Equation (8.71)] and is equal to ω_M. To compress the power spectral density, it is necessary to satisfy the condition given by Equation (8.81), and the Doppler and range-finder frequencies [see Equation (8.83)] must have different signs.

If the condition $\alpha + \beta_0 = \pi$ is true, i.e., if the radar antenna beam is directed back with respect to the line of radar motion, the parameter d_p must be negative to compress the power spectral density. In this case,

$$\Omega'_{n_0} = \omega_{im} - \Omega'_0 + n_0\omega_M . \tag{8.84}$$

The physical significance of power spectral density compression is discussed in Section 8.2. However, in the present case, the searching signal is periodic, and power spectral density compression is possible only up to the modulation frequency ω_M. In the case of the periodic searching signal, a complete power spectral density compression of the fluctuations takes place only under the condition $t = 0$, and changes in time of the Doppler frequency are retained due to carrier frequency variation with the frequency-modulated searching radar signal. This variation is averaged in the correlation function (averaged over time) of the transformed fluctuations and a power spectral density compression of the fluctuations takes place at the frequency $\omega_0 + 0.5\omega_M$. To define the power spectral density under the condition $\alpha + \beta_0 \neq 0°$ or π, it is necessary to carry out a convolution of the power spectral density given by Equation (8.61) and the power spectral density given by Equation (4.22) for the continuous nonmodulated searching signal or carry out the mathematics discussed in Section 8.2.

Omitting the mathematics used in accordance with Baker,[17] we can say that the power spectral density of the fluctuations will coincide with the power spectral density given by Equation (8.61) on replacing the bandwidth $\Delta\Omega_{v_{FM}}$ with the bandwidth $\Delta\Omega_v$ in Equation (8.61), where

$$\Delta\Omega_{FM} = \sqrt{\Delta\Omega^2_{v_{FM}} + \Delta\Omega^2_h} \quad \text{and} \quad \Delta\Omega_h = 4\sqrt{2}\,\pi\, V\lambda^{-1} b_1 \Delta_h , \tag{8.85}$$

and the frequencies Ω_0 and Ω_v with the frequencies

$$\Omega_0 = 4\pi\, V\lambda^{-1}\cos\theta_0 \tag{8.86}$$

and

$$\Omega_v = -4\pi\, V\lambda^{-1} a_1 , \tag{8.87}$$

respectively. If the condition $\alpha + \beta_0 = 0°$ is not satisfied, a compression of the bandwidth $\Delta\Omega_{v_{FM}}$ is only possible for an arbitrary position of the sloped radar antenna beam. If the condition given by Equation (8.81) is satisfied, the power spectral density bandwidth is determined by

$$\Delta\Omega_{v_{FM}} = \sqrt{\omega^2_M + \Delta\Omega^2_h} . \tag{8.88}$$

The partial power spectral densities with the use of the Gaussian approximation and approximation by the function **sinc**^{2} x for the vertical-coverage directional diagram are shown in Figure 8.11 and Figure 8.12, respectively, at the same and different signs of the frequencies Ω_0' and $n\omega_M$ for the frequency Ω_n. In the second case, as one can easily see, partial compression of the symmetric spectral density takes place. To define the total power spectral densities of target return signal fluctuations shown in Figure 8.11 and Figure 8.12, it is necessary to add the ordinates of all partial power spectral densities.

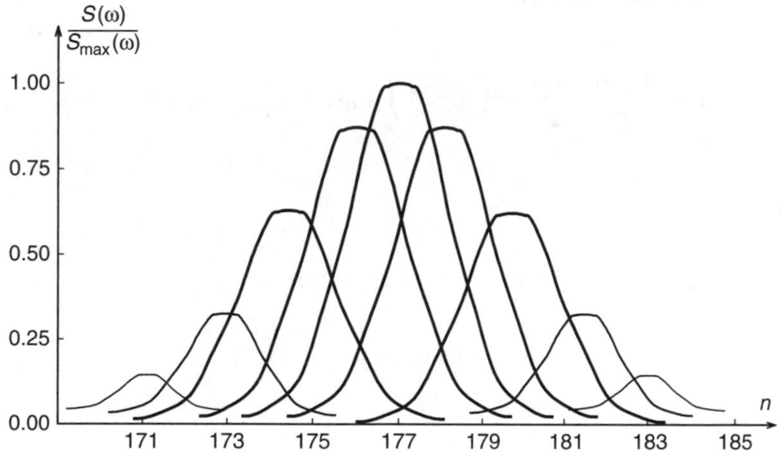

FIGURE 8.11
Partial power spectral densities of target return signal fluctuations with an asymmetric sawtooth frequency-modulated searching signal at the total frequency: $d_\rho >> 1$.

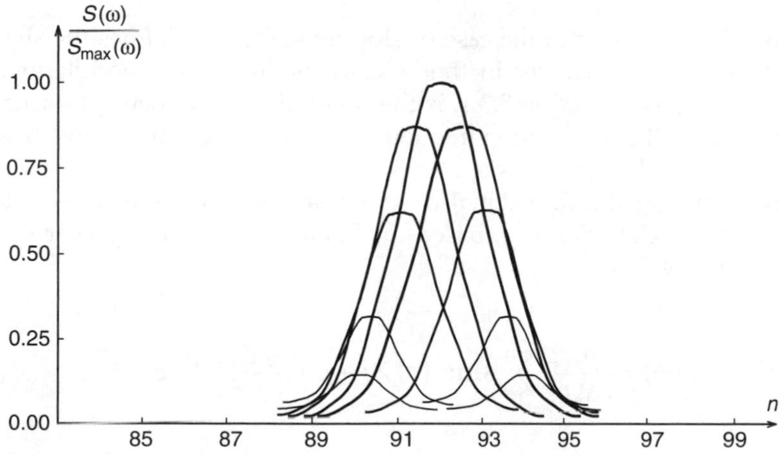

FIGURE 8.12
Partial power spectral densities of target return signal fluctuations with the asymmetric sawtooth frequency-modulated searching signal at the difference frequency: $d_\rho >> 1$.

8.3.2 Vertical Scanning and Motion

Under vertical scanning, the condition $\gamma_0 = 90°$ is true. Assume that the velocity vector of moving radar is directed along the directional diagram axis and is equal to $V \cdot \sin \varepsilon_0$.[18,19] The Doppler frequency Ω_0 is negative under the condition $\varepsilon_0 = 90°$ and is positive under the condition $\varepsilon_0 = -90°$. Suppose directional diagram has axial symmetry and depends on both the specific effective scattering area $S°$ and the angle in the vertical plane. Omitting the mathematics, we can write the power spectral density of target return signal fluctuations in the following form:

$$S(\omega) = \frac{2\pi p_0}{|\Omega_0|} \cdot \sum_{n=-\infty}^{\infty} \tilde{g}^2\left(\sqrt{\tfrac{\Omega_n - \omega}{0.5\Omega_0}}\right) \cdot \mathbf{sinc}^2\left\{\pi\left[X_h - n + \tfrac{(\Omega_n - \omega)X_h}{\Omega_0}\right]\right\}, \quad (8.89)$$

where

$$\Omega_n = \omega_{im} + \Omega_0 - n\omega_M; \quad (8.90)$$

$$\frac{\Omega_n - \omega}{\Omega_0} \geq 0; \quad (8.91)$$

$$X_h = 2h\Delta f_M c^{-1}; \quad (8.92)$$

$$\tilde{g}^2\left(\sqrt{\tfrac{\Omega_n - \omega}{0.5\Omega_0}}\right) = g^2\left(\sqrt{\tfrac{\Omega_n - \omega}{0.5\Omega_0}}\right) \cdot S°\left(\sqrt{\tfrac{\Omega_n - \omega}{0.5\Omega_0}}\right). \quad (8.93)$$

The function $\tilde{g}^2(\psi)$, as in the case of sloping scanning, defines the shape of the power spectral density in the absence of frequency modulation. The parameter X_h, as in Section 8.3.1, is the range-finder frequency (normalized with respect to the modulation frequency ω_M) corresponding to the direction of the directional diagram axis.

Approximating the directional diagram and specific effective scattering area $S°$ by the Gaussian law, based on Equation (8.89) the power spectral density has the form:

$$S(\omega) = \frac{p_0'}{|\Delta\Omega|} \cdot \sum_{n=-\infty}^{\infty} \mathbf{sinc}^2\left\{\pi\left[X_h - n + \tfrac{(\Omega_n - \omega)d_h}{2\Delta\Omega}\right]\right\} \cdot e^{-\frac{\Omega_n - \omega}{\Delta\Omega}}, \quad (8.94)$$

where

$$\omega < \Omega_n \text{ at } \Omega_0 > 0, \quad \text{and} \quad \omega > \Omega_n \text{ at } \Omega_0 < 0; \quad (8.95)$$

$$d_h = 0.5\pi^{-1}\Delta^{(2)}X_h a^{-2}\ ;\ a^2 = 1 + 0.5\pi^{-1}k_2\Delta^{(2)}\ ;\ \text{and}$$
$$\Delta\Omega = 0.25\pi^{-1}\Omega_0\Delta^{(2)}a^{-2}\ ; \tag{8.96}$$

p'_0 is the target return signal power in the absence of frequency modulation. The parameter d_h, as before, defines the number of radar range zones.

In the case of stationary radar and vertical scanning, the power spectral density takes the following form:[20]

$$S(\omega) = \frac{p'_0}{|d_h|} \cdot e^{-\pi d_h^2} \sum_{n=-\infty}^{\infty} \left\{1 - \Phi\left[\tfrac{(1+\pi d_h)(X_h - n)}{\sqrt{\pi}\, d_h}\right]\right\} \cdot \delta[\omega - (\omega_{im} - n\omega_M)] \cdot e^{\frac{X_h - n}{0.5 d_h}}, \tag{8.97}$$

where $\Phi(x)$ is the error integral given by Equation (1.27). Reference to Equation (8.97) shows that if the condition $d_h << 1$ is satisfied, the argument in the function $\Phi(x)$ is high in value and we can use the following approximation:

$$1 - \Phi(x) = \frac{e^{-x}}{x\sqrt{\pi}}. \tag{8.98}$$

Thus, based on Equation (8.97) the power spectral density can be written in the following form:

$$S(\omega) = p'_0 \cdot \sum_{n=-\infty}^{\infty} \delta[\omega - (\omega_{im} - n\omega_M)] \cdot e^{-\pi(X_h - n)^2}. \tag{8.99}$$

The formula in Equation (8.99) coincides with Equation (8.70). In other words, under the condition $d_h << 1$, the power spectral density is independent of the characteristics of the directional diagram and is only defined by the modulating function.

Reference to Equation (8.97) shows that with an increase in the value of d_h, i.e., in the number of radar range zones, the power spectral density becomes asymmetric with an increase in n relative to n_0. In the opposite direction, the power spectral density decreases according to the function $\mathbf{sinc}^2\, \pi(X_h - n)$ in Equation (8.94), which can be explained in the following manner. The radar range ρ can only increase with respect to the altitude h within the directional diagram. Consider some examples of the power spectral densities under the condition $d_h > 1$ in the case of the symmetric sawtooth law of frequency modulation.

Consider the case of moving radar. Reference to Equation (8.89) and Equation (8.94) shows that if the conditions $d_h << 1$ and $|\Delta\Omega| << \left|\frac{\Omega_0}{X_h}\right|$ are satisfied, we can neglect the second term in the function $\mathbf{sinc}^2\, x$, and the power spectral density of the Doppler fluctuations at the frequency Ω_{n_0} takes place. Under the condition $d_h >> 1$, the bandwidth of the partial power spectral density of the fluctuations is defined by the function

$$\mathbf{sinc}^2\left[\frac{\pi(\Omega_n-\omega)X_h}{\Omega_0}\right] = e^{-\frac{\pi(\Omega_n-\omega)^2 X_h^2}{\Omega_0^2}} = e^{-\frac{\pi(\Omega_n-\omega)^2 d_h^2}{2\Delta\Omega^2}}. \tag{8.100}$$

The bandwidth of this function is much less than the bandwidth of the power spectral density of the Doppler fluctuations. Because the power spectral density of the Doppler fluctuations is one-sided under the condition $X_h = n_0$, the product of the function $\tilde{g}^2(\psi)$ and Equation (8.100) defines the one-sided power spectral density of the fluctuations.

Under the condition $d_h >> 1$, the maximums of other partial power spectral densities of the fluctuations take place at the frequencies

$$\omega_k = \Omega_{n_0+k} - k\cdot\frac{\Omega_0}{X_h} = \Omega_{n_0} - k\omega_M\left[\frac{\Omega_0}{X_h\omega_M}+1\right] \tag{8.101}$$

when the function $\mathbf{sinc}^2\, x = 1$. The frequency ω_k, as with sloping scanning, differs from the frequency Ω_{n_0+k}. This can be explained, as before, by the fact that each n-th harmonic is formed by an area (in the present case by a ring) corresponding to the resolved area of radar with the frequency-modulated searching signal. In the present case, the Doppler isofrequency lines are circles (see Figure 8.2). The k-th ring resolved with respect to the circle has a Doppler shift equal to $\frac{k\Omega_0}{X_h}$ at n_0, which is subtracted from the frequency Ω_{n_0+k} in Equation (8.101).

Substituting Equation (8.101) in Equation (8.89) or Equation (8.94), we can write the power spectral density of the fluctuations in the following form:

$$S(\omega) = p_0\cdot\frac{2\pi}{|\Omega_0|}\sum_{n=-\infty}^{\infty}\tilde{g}^2\left(\sqrt{\frac{\omega_k + k\frac{\Omega_0}{X_h} - \omega}{0.5\Omega_0}}\right)\cdot\mathbf{sinc}^2\left[(\omega_k-\omega)\frac{X_h}{\Omega_0}\right]. \tag{8.102}$$

As noted above, the partial power spectral density at the frequency $\Omega_{n_0}(k=0)$ is one-sided and its shape is defined by the function $\mathbf{sinc}^2\, x$ in Equation (8.102) under the condition $d_h >> 1$, i.e., the shape is the one-sided Gaussian curve. Under the condition $k \neq 0$, the functions $\tilde{g}^2(\varphi, \psi)$ and $\mathbf{sinc}^2\, x$ are shifted in argument relative to the frequency Ω_{n_0+k} by the value $\frac{k\Omega_0}{X_h}$. At the point $\omega = \omega_k$, the functions $\tilde{g}^2(\varphi, \psi)$ and $\mathbf{sinc}^2\, x$ are not one-sided and the partial power spectral density takes the shape of the two-sided Gaussian curve with average frequency ω_k.

Substituting Equation (8.101) in Equation (8.94), we can define the envelope of partial power spectral densities in the following form:

$$S(\omega_k) = p_0 \cdot \frac{2\pi}{|\Omega_0|} \tilde{g}^2\left(\sqrt{\tfrac{2k}{X_h}}\right). \tag{8.103}$$

Thus, the value of $\frac{k}{X_h}$ is always positive due to the one-sided power spectral density of the Doppler fluctuations. When X_h is negative, k must be negative, too, and only frequencies $(n_0 - k)w_M$ are in the power spectral density. If the condition $X_h > 0$ is satisfied, we can observe the same picture, but at frequencies $(n_0 + k)w_M$. The difference in average frequencies of two neighboring partial power spectral densities is determined by

$$\omega_k - \omega_{k+1} = \omega_M\left(\tfrac{\Omega_0}{\omega_M X_h} + 1\right). \tag{8.104}$$

The condition of compression of all partial power spectral densities has the following form:

$$\Omega_0 = -\omega_M X_h = -n_0\omega_M , \tag{8.105}$$

i.e., the frequencies Ω_0 and $n\omega_M$ must have the same sign as the frequency Ω_n. The compressed power spectral density is the one-sided power spectral density at the frequency Ω_n and the sum of two-sided power spectral densities at the same frequency.

In the absence of frequency modulation under the condition $\Omega_0 > 0$, i.e., as the radar moves closer to the Earth's surface, the maximum of the power spectral density of the Doppler fluctuations is at the frequency $\omega_{im} + \Omega_0$ (see Figure 8.13b). The power spectral density of the fluctuations with the frequency-modulated searching signal is shown in Figure 8.13a under the condition $X_h > 0$, i.e.,

$$\omega_k = \omega_{im} + \Omega_0 - (n_0 + k)\omega_M . \tag{8.106}$$

The bandwidth of the partial power spectral densities is equal to $\frac{2\Delta\Omega}{d_h}$. The compressed power spectral density is shown in Figure 8.14 on a decreased scale, under the condition $X_h < 0$. This power spectral density is obtained by superposition of all partial power spectral densities in the frequency region

$$\omega_{im} + 2\Omega_0 = \omega_{im} + 2n_0\omega_M . \tag{8.107}$$

It should be noted that under vertical scanning and with $\varepsilon_0 \pm 90°$, the values of d_h and $\Delta\Omega$ are much less in comparison with the case of sloping scanning if the velocity vector of moving radar is outside the directional diagram. For instance, at $\Delta f_M = 20$ MHz, $\Delta = 0.1$ (6°), and $h = 750$ m, we get $X_h = 100$ m and $d_h = 0.15$. In the present case, the power spectral density

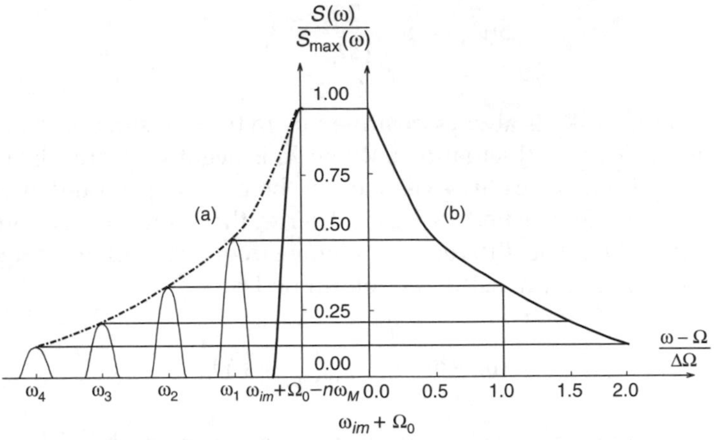

FIGURE 8.13
Power spectral densities of target return signal fluctuations with vertically moving radar and vertical scanning: (a) the asymmetric saw-tooth frequency-modulated searching signal; (b) frequency modulation is absent: $d_\rho >> 1$.

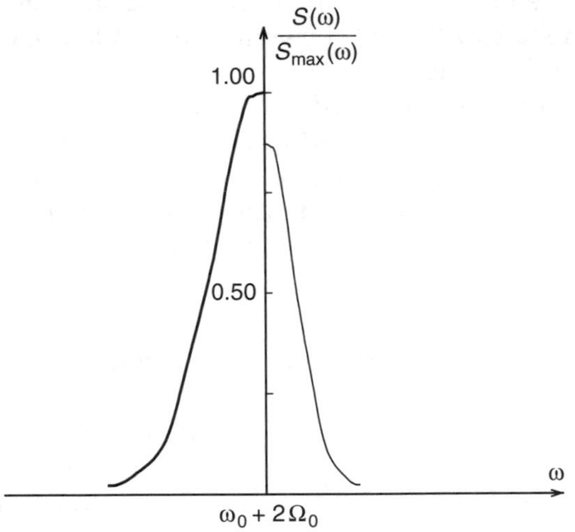

FIGURE 8.14
Power spectral densities of target return signal fluctuations with vertically moving radar and vertical scanning, with the asymmetric saw-tooth frequency-modulated searching signal. Condition of compression of the power spectral density is satisfied.

corresponds to the condition $d_h << 1$ and consists only of a single power spectral density of the Doppler fluctuations at the frequency Ω_{n0}; definition of the average frequency of the target return signal is not difficult. Consider now another case that is very important in practice.

8.3.3 Vertical Scanning: The Velocity Vector Is Outside the Directional Diagram

Assume that the directional diagram is directed vertically down. The velocity vector of moving radar is close to the horizontal direction. Suppose that the axial symmetric directional diagram and the specific scattering area **S°** are defined by Gaussian surfaces. The power spectral density of the fluctuations can be written in the following form:

$$S(\omega) = \frac{p_0'}{\pi\,|\,d_h \Delta\Omega\,|} \sum_{n=-\infty}^{\infty} \sqrt{1 + \pi d_h \left[X_h - n + \pi d_h \cdot \frac{(\omega-\Omega_n)^2}{2\Delta\Omega^2}\right]} \cdot e^{-\pi \frac{(\omega-\Omega_n)^2}{\Delta\Omega^2}},$$

$$\times\, e^{-\pi\left[X_h - n + \pi d_h \frac{(\omega-\Omega_n)^2}{2\Delta\Omega^2}\right]} K_{0.25}(z^2) \cdot e^{z^2} \tag{8.108}$$

where

$$z^2 = \frac{\left\{1 + \pi d_h \left[X_h - n + \pi d_h \cdot \frac{(\omega-\Omega_n)^2}{2\Delta\Omega^2}\right]\right\}^2}{2\pi d_h^2}; \tag{8.109}$$

p_0' is the target return signal power in the absence of frequency modulation [see Equation (2.203)];

$$\Omega_0 = 4\pi V \lambda^{-1} \sin \varepsilon_0 ; \tag{8.110}$$

$$\Delta\Omega = 2\sqrt{2}\,\pi V \lambda^{-1} a^{-2} \cos \varepsilon_0 ; \tag{8.111}$$

$$K_{0.25}(x) = (\sqrt{2})^{-1}\pi \cdot \left[I_{-0.25}(x) - I_{0.25}(x)\right] \tag{8.112}$$

is the modified Bessel function of the second kind; and $I_{\pm 0.25}(x)$ is the modified Bessel function of the first kind. The functions $I_{\pm 0.25}(x)$ are tabulated. Under the condition $x < 0.1$, we can use a very close approximation for the functions $I_{\pm 0.25}(x)$. Thus, we can write

$$K_{0.25}(x) = 2.158 \cdot \left(x^{-0.25} - 0.9596 \cdot x^{0.25}\right). \tag{8.113}$$

If the required accuracy is not more than 15%, we can use an approximation given by Equation (8.113) under the condition $x \le 0.5$. With high values of x, we can use the following approximation:

$$K_{0.25}(x) = \sqrt{0.5\pi x^{-1}} \cdot e^{-x}. \tag{8.114}$$

Reference to Equation (8.108) shows that the maximum of the partial power spectral density under the conditions

$$\omega = \Omega_{n_0} \qquad \text{and} \qquad X_h = n_0 \tag{8.115}$$

has the following form:

$$S(\Omega_{n_0}) = \frac{p'_0}{\pi | d_h \Delta\Omega |} \cdot K_{0.25}\left(\tfrac{1}{2\,\pi\, d_h^2}\right) \cdot e^{\frac{1}{2\,\pi\, d_h^2}}. \tag{8.116}$$

At the frequency Ω_{n_0+k}, we can write

$$S(\Omega_{n_0+k}) = \frac{p'_0}{\pi | d_h \Delta\Omega |} \cdot K_{0.25}\left[\tfrac{(1-\pi\, d_h k)^2}{2\,\pi\, d_h^2}\right] \cdot \sqrt{|1 - \pi d_h k|} \cdot e^{-\pi k^2} \cdot e^{\frac{(1-\pi\, d_h k)^2}{2\,\pi\, d_h^2}}. \tag{8.117}$$

The maximums of the partial power spectral densities are at the frequencies given by the condition

$$X_h - n + \pi \cdot \frac{d_h(\omega - \Omega_n)^2}{2\Delta\Omega^2} = 0. \tag{8.118}$$

Reference to Equation (8.118) shows that there are two maximums at the frequencies equal to

$$\omega'_k = \Omega_{n_0+k} + \sqrt{2\pi^{-1} k d_h^{-1}} \qquad \text{and} \qquad \omega''_k = \Omega_{n_0+k} - \sqrt{2\pi^{-1} k d_h^{-1}} \tag{8.119}$$

for each partial power spectral density. As one can see from Equation (8.119), the frequencies ω'_k and ω''_k are symmetric with respect to the frequency Ω_{n0+k}. The parameters k and d_h in Equation (8.119) must have the same sign because the power spectral density, as in Section 8.3.1, is one-sided. Substituting Equation (8.119) in Equation (8.103), we can define these maximums

$$S(\omega_k) = \frac{p'_0}{\pi | d_h \Delta\Omega |} \cdot K_{0.25}\left(\tfrac{1}{2\pi\, d_h^2}\right) \cdot e^{-\frac{2k}{d_h}} \cdot e^{\frac{1}{2\pi\, d_h^2}} = S(\Omega_{n_0}) \cdot e^{-\frac{2k}{d_h}}. \tag{8.120}$$

The partial power spectral densities are shown in Figure 8.15–Figure 8.19 for various values of k and d_h. At $x \le 0.1$, the approximation given by Equation (8.113) is used.

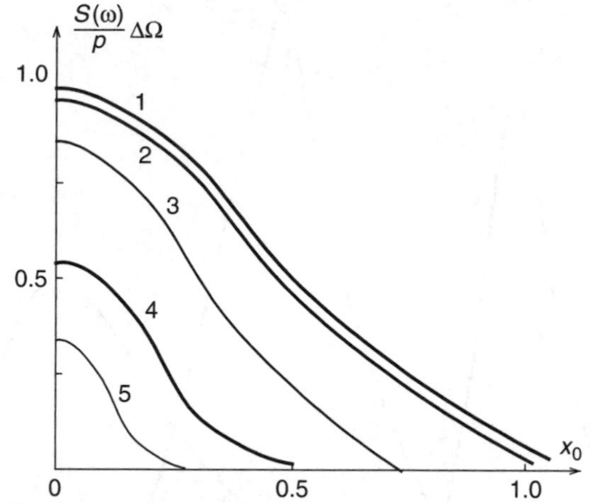

FIGURE 8.15
Partial power spectral densities of target return signal fluctuations with the frequency-modulated searching signal and vertical scanning. The velocity vector of moving radar is outside the radar antenna directional diagram, $k = 0$, $x_0 = \frac{\omega - (\omega_{im} + \Omega'_0 - n_0\omega_M)}{\Delta\Omega}$: (1) $d_h = 0.1$; (2) $d_h = 0.3$; (3) $d_h = 1$; (4) $d_h = 3$; (5) $d_h = 10$.

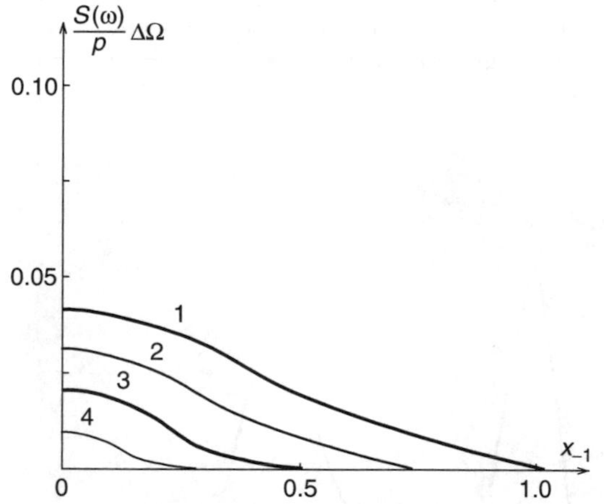

FIGURE 8.16
Partial power spectral densities of target return signal fluctuations with the frequency-modulated searching signal and vertical scanning. The velocity vector of moving radar is outside the radar antenna directional diagram, $k = -1$, $x_{-1} = \frac{\omega - [\omega_{im} + \Omega'_0 - (n_0 - 1)\omega_M]}{\Delta\Omega}$: (1) $d_h = 0.1$; (2) $d_h = 0.3$; (3) $d_h = 1$; (4) $d_h = 3$.

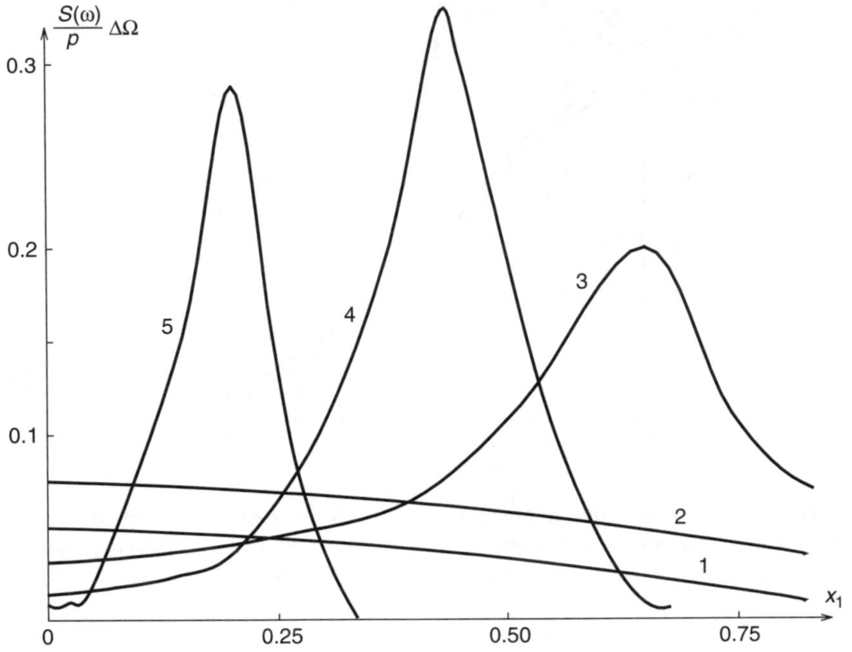

FIGURE 8.17
Partial power spectral densities of target return signal fluctuations with the frequency-modulated searching signal and vertical scanning. The velocity vector of moving radar is outside the radar antenna directional diagram, $k = 1$, $x_1 = \frac{\omega - [\omega_{im} + \Omega_0' - (n_0 + 1)\omega_M]}{\Delta\Omega}$: (1) $d_h = 0.1$; (2) $d_h = 0.3$; (3) $d_h = 1$; (4) $d_h = 3$; (5) $d_h = 10$.

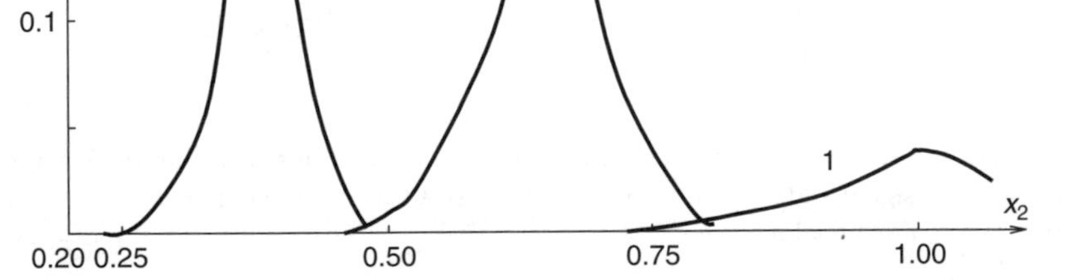

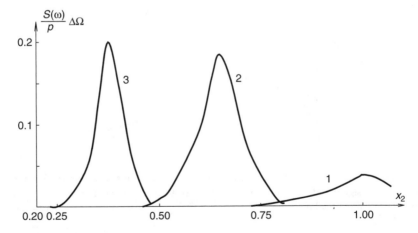

FIGURE 8.18
Partial power spectral densities of target return signal fluctuations with the frequency-modulated searching signal and vertical scanning. The velocity vector of moving radar is outside the radar antenna directional diagram, $k = 2$, $x_2 = \frac{\omega - [\omega_{im} + \Omega_0' - (n_0 + 2)\omega_M]}{\Delta\Omega}$: (1) $d_h = 1$; (2) $d_h = 3$; (3) $d_h = 10$.

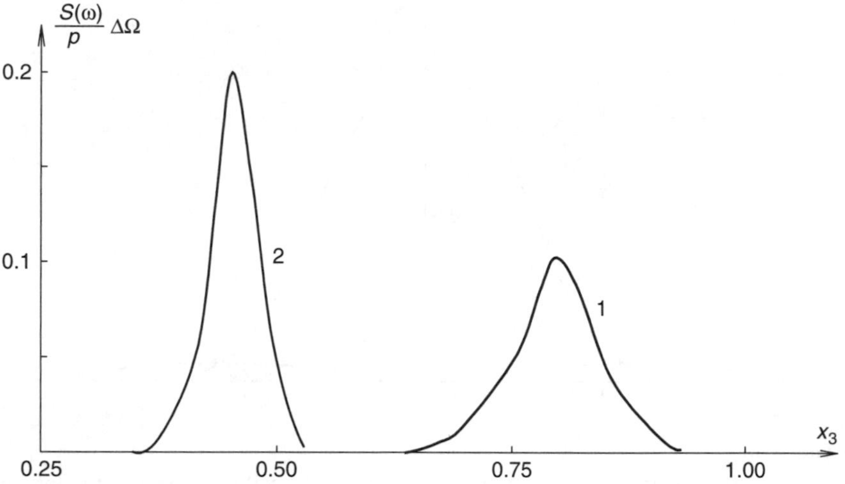

FIGURE 8.19
Partial power spectral densities of target return signal fluctuations with the frequency-modulated searching signal and vertical scanning. The velocity vector of moving radar is outside the radar antenna directional diagram, $k = 3, x_3 = \frac{\omega - [\omega_{im} + \Omega'_0 - (n_0+3)\omega_M]}{\Delta\Omega}$: (1) $d_h = 3$; (2) $d_h = 10$.

Only positive branches of the power spectral densities are presented in Figure 8.15–Figure 8.19 at $x_k > 0$. At $d_h > 1$, the shape of the one-half partial power spectral densities is not symmetric: in the case $\omega < \omega_k$, they are more slanting in comparison with the case $\omega > \omega_k$. With an increase in the value of d_h, this asymmetry decreases. At $d_h = 0.1$ or 0.3, the power spectral density bandwidth approximately coincides with the power spectral density bandwidth of the Doppler fluctuations under the condition $k = 0$, which is defined by the Gaussian law. At high values of d_h, the power spectral density bandwidth is defined by $\frac{1}{\sqrt{|d_h|}}$. At $k \geq 1$ and $d_h \geq 1$, the bandwidth of the power spectral densities is determined by

$$\Delta\omega_k = \frac{\Delta\Omega}{\sqrt{2\pi k d_h}}. \tag{8.121}$$

The power spectral densities under the condition $\Delta\Omega = \omega_M$ are shown in Figure 8.20 and Figure 8.21. At $d_h = 0.3$, the partial power spectral densities under the conditions $k = 0$ and $k = 1$ are symmetric and are at the frequency Ω_{n_0+k}. At $d_h = 3$, each partial power spectral density is divided into two parts in accordance with Equation (8.119). Neighboring partial power spectral densities partially overlap. The normalized envelope $S(\omega_k)$ given by Equation (8.120) is shown in Figure 8.21 by the dotted line.

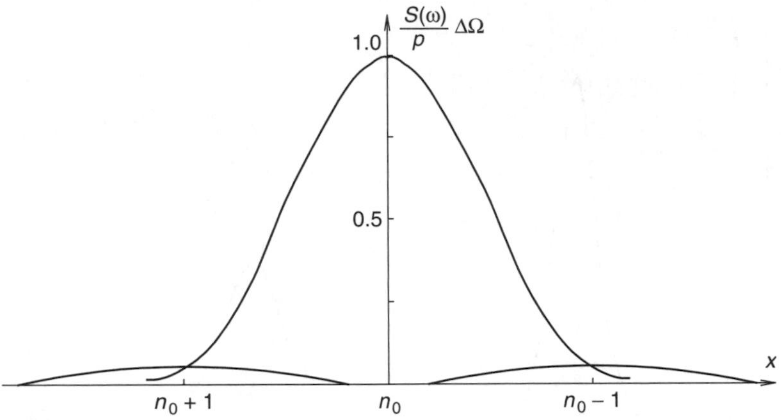

FIGURE 8.20
Power spectral densities of target return signal fluctuations with the asymmetric saw-tooth frequency-modulated searching signal and vertical scanning. The velocity vector of moving radar is outside the radar antenna directional diagram: $\omega_M = \Delta\Omega$ and $d_h = 0.3$.

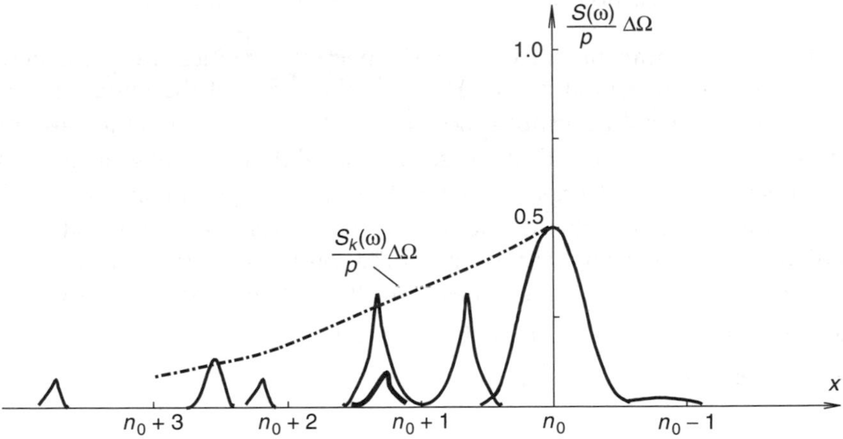

FIGURE 8.21
Power spectral densities of target return signal fluctuations with the asymmetric saw-tooth frequency-modulated searching signal and vertical scanning. The velocity vector of moving radar is outside the radar antenna directional diagram: $\omega_M = \Delta\Omega$ and $d_h = 3$.

8.4 The Symmetric Saw-Tooth Frequency-Modulated Searching Signal

In the case of symmetric saw-tooth frequency modulation, the coherent searching signal can be determined by Equation (8.5), where

$$\begin{cases} \Psi(t-nT_M)=0.5k_\omega(t-nT_M)^2 & \text{at} \quad nT_M<t<(n+0.5)T_M \ ; \\ \Psi[t-(n+1)T_M]=-0.5k_\omega[t-(n+1)T_M]^2 & \text{at} \quad (n+0.5)T_M<t<(n+1)T_M \ . \end{cases} \tag{8.122}$$

We have, respectively (see Figure 8.22),

$$\begin{aligned} &\omega_1(t-nT_M)=\omega_0+k_\omega(t-nT_M) \quad \text{and} \\ &\omega_2[t-(n+1)T_M]=\omega_0-k_\omega[t-(n+1)T_M] \end{aligned} . \tag{8.123}$$

With the same frequency deviation $\Delta\omega_M$ and the same period T_M, the parameter k_ω, in the present case, is twice that in the case of asymmetric saw-tooth frequency modulation:

$$k_\omega=2\Delta\omega_M T_M^{-1}. \tag{8.124}$$

To estimate the correlation function of the fluctuations it is necessary, first, to define the transformed elementary signals given by Equation (8.8) and Equation (8.9) within the limits of the various intervals $t_1, \ldots, t_5$ (see Figure 8.23) belonging to the period T_M; second, to define the differences $\Delta\Psi$ given by Equation (8.35); third, to use the Fourier-series expansion for the function $e^{j\Delta\Psi}$ in accordance with Equation (8.37); fourth, to substitute results from the Fourier-series expansion into the correlation function of the fluctuations, and to carry out averaging over the period T_M as in Section 8.3. As one can see from Figure 8.24, the frequency of the transformed target return signal within the limits of the intervals $[t_2, t_3]$ and $[t_4, t_5]$ varies linearly in time. To keep the sign of the frequency shift $k_\omega T_d$ constant, the parameter ω_{im} is introduced in Figure 8.24. In principle, we can define the correlation function of the fluctuations taking these intervals into consideration. However, if the condition $T_d << T_M$ is true, we can neglect these intervals (as in Section 8.3), as the power they contribute is proportional to $\frac{4T_d^2}{T_M^2}$. In the case of radar with the frequency-modulated searching signal, the receiver does not operate at this instant of time and the target return signals are not processed.

Using the same mathematics as in Section 8.3, we can write the correlation function of the fluctuations in the following form:

$$\begin{aligned} R(\tau)=0.25p_0'\cdot e^{j\omega_{im}\tau}\iint \tilde{g}^2(\varphi,\psi)\cdot e^{-2j(\omega_0+0.5\Delta\omega_M)\frac{\Delta\rho}{c}} \\ \times\sum_{n=-\infty}^{\infty}[\mathbf{sinc}^2\pi(X_\rho-0.5n)+\mathbf{sinc}^2\pi(X_\rho+0.5n)]\cdot e^{-jn\omega_M\tau}d\varphi\, d\psi, \end{aligned} \tag{8.125}$$

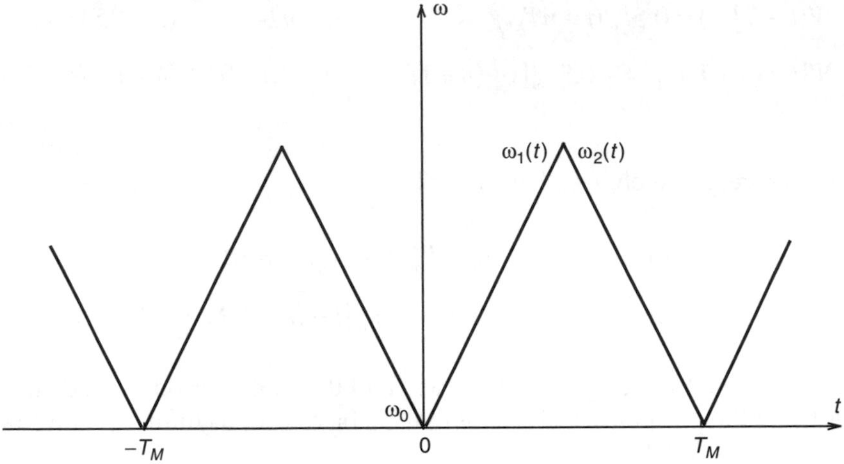

FIGURE 8.22
The frequency of the searching signal under symmetric saw-tooth frequency modulation.

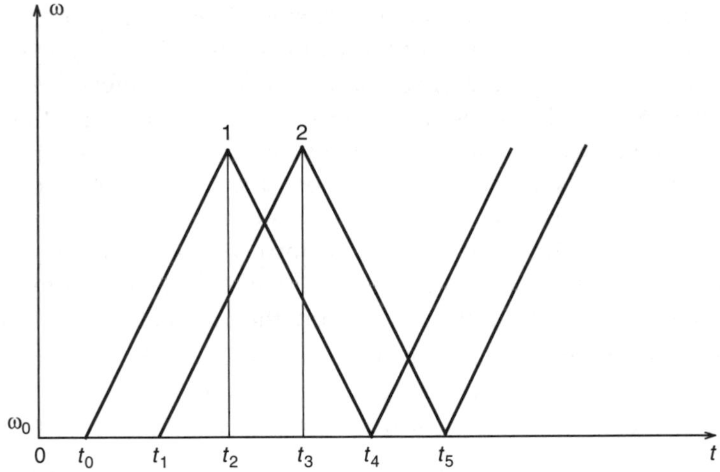

FIGURE 8.23
The frequency of the heterodyne (1) and target return (2) signals under symmetric saw-tooth frequency modulation.

where p_0' and X_ρ are the same as in Equation (2.203) and Equation (3.12), but the value of X_ρ is positive. Unlike in Equation (8.52), we have two $\mathbf{sinc}^2\, x$ functions. The function $\mathbf{sinc}^2\,(X_\rho - 0.5n)$ is different from zero at positive values of n; the function $\mathbf{sinc}^2\,(X_\rho + 0.5n)$ is different from zero at negative values of n. Furthermore, unlike in Equation (8.52), the function $\mathbf{sinc}^2\, x$ has as an argument $0.5n$ instead of n. This is because the value of k_ω is twice that in the case of asymmetric saw-tooth frequency modulation.

The range-finder frequency at the radar range ρ is double too:

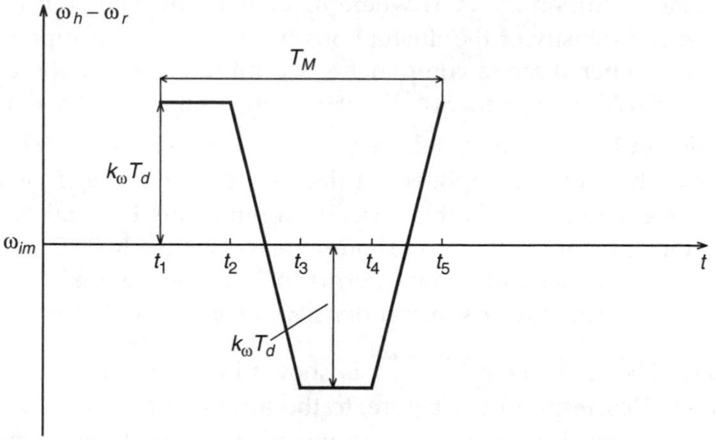

FIGURE 8.24
The frequency of the transformed target return signal under symmetric saw-tooth frequency modulation.

$$\Omega_\rho = 2X_\rho\omega_M = n_0\omega_M\,. \tag{8.126}$$

The functions **sinc**$^2\,x$ in Equation (8.125) are twice as wide as those in Equation (8.52). The bandwidth of the functions **sinc**$^2\,x$ in Equation (8.125) is equal to two in normalized units. Taking the previously mentioned peculiarities into consideration, based on Equation (8.52) and Equation (8.125) the power spectral density of the fluctuations takes the following form:

$$S(\omega) = 0.25[S_h(\omega - \Omega'_n, X_{\rho_0} - 0.5n) + S_h(\omega - \Omega''_n, X_{\rho_0} + 0.5n)]\,, \tag{8.127}$$

where

$$\Omega'_n = \omega_{im} + \Omega'_0 - n\omega_M \quad \text{and} \quad \Omega''_n = \omega_{im} + \Omega'_0 + n\omega_M\,. \tag{8.128}$$

Using Equation (8.127), we can extend all the results discussed in Section 8.3 to the case of symmetric saw-tooth frequency modulation.

Let us consider the main characteristics of the power spectral density of the fluctuations under sloping scanning. In the case of stationary radar and with the use of the approximation given by Equation (8.60), the power spectral density of the fluctuations takes the following form:

$$S(\omega) = 0.25p_{FM}\sum_{n=-\infty}^{\infty}\delta[\omega - (\omega_{im} - n\omega_M)]\cdot e^{-\pi\frac{(X'_{\rho_0} - 0.5n)^2}{1+d_\rho^2}} + \delta[\omega - (\omega_{im} + n\omega_M)]\cdot e^{-\pi\frac{(X'_{\rho_0} + 0.5n)^2}{1+d_\rho^2}}\,. \tag{8.129}$$

Under the condition $d_\rho \ll 1$, where d_ρ is given by Equation (8.56), the power spectral density of the fluctuations has six discrete components (see Figure 8.25); other discrete components are infinitesimal. If the condition $2X_{\rho_0} = n_0$ is true, the frequencies of these discrete components are determined by Equation (8.128), i.e., at $\Omega_0 = 0$, we have $n_0 - 1$, n_0, and $n_0 + 1$. The relative power of each discrete component at the frequency $\omega_{im} + n_0\omega_M$ is equal to one. The relative power of other discrete components is equal to 0.456 or less. The total power of all discrete components is equal to 0.956, and not 1, as we take into consideration only terms included in the main lobe of the function $\mathbf{sinc}^2\, x$.The power spectral density is shown in Figure 8.25 for the given case. The envelope $e^{-\pi(X'_{\rho_0} \mp 0.5n)^2}$ is shown by the dotted line. The condition $d_\rho \ll 1$ corresponds, as before, to the condition $\delta\rho_0 \gg \delta\rho$ (see Figure 8.7), as radar resolution with a frequency-modulated searching signal depends only on the frequency deviation Δf_M.

Under the condition $d_\rho \gg 1$, the power spectral densities correspond to the condition $\delta\rho_0 \ll \delta\rho$ (see Figure 8.26). Because the target return signals from each radar range zone are noncoherent, all side components in Figure 8.25 are summarized power. The effective power spectral density bandwidth with the range-finder frequency is equal to $2d_\rho\omega_M$. Each radar range zone corresponds to two harmonics. Because the total bandwidth is equal to $2d_\rho\omega_M$, the number of radar range zones is equal to d_ρ, as in the case of asymmetric saw-tooth frequency modulation.

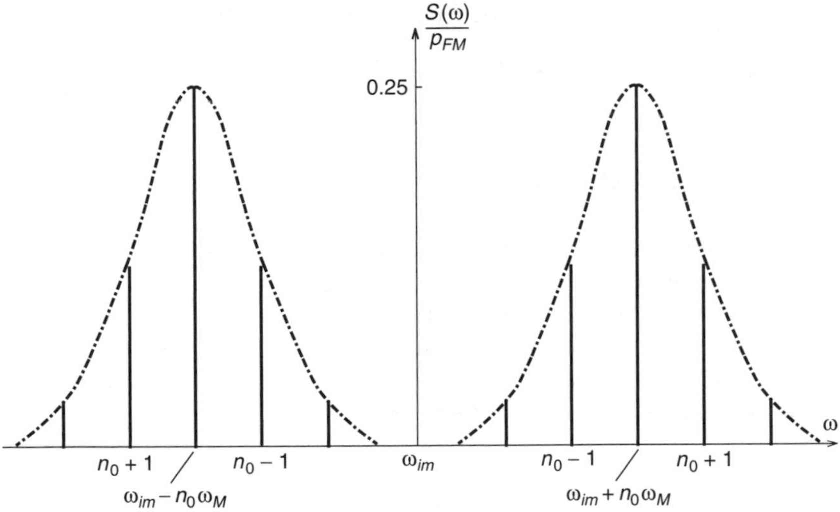

FIGURE 8.25
The power spectral density of target return signal fluctuations with the symmetric saw-tooth frequency-modulated searching signal: $d_\rho \ll 1$.

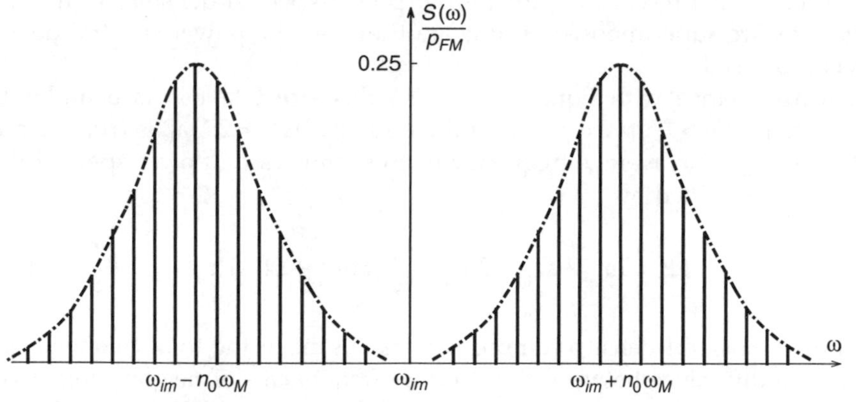

FIGURE 8.26
The power spectral density of target return signal fluctuations with the symmetric saw-tooth frequency-modulated searching signal: $d_\rho >> 1$.

Let us define the power spectral density of the fluctuations with moving radar. We should take into consideration that the envelope $S(\omega_k)$ for an arbitrary shape of the function $\tilde{g}^2(\varphi, \psi)$ is determined by Equation (8.78), replacing d_ρ with $2d_\rho$. The formula in Equation (8.61) for the case of symmetrical saw-tooth frequency modulation has the following form

$$S(\omega) = \frac{p_{FM}}{4|\Delta\Omega_{v_{FM}}|}\sum_{n=-\infty}^{\infty}\left[e^{-\pi\frac{(\omega-\Omega'_n)^2}{\Delta\Omega^2_{v_{FM}}}} + e^{-\pi\frac{(\omega-\Omega''_n)^2}{\Delta\Omega^2_{v_{FM}}}}\right]\cdot e^{-\pi\frac{(X'_{\rho_0}-0.5|n|)^2}{1+d_\rho^2}}, \quad (8.130)$$

where the parameters p_{FM}, $\Delta\Omega_{v_{FM}}$, d_ρ, and $X'_{\rho 0}$ are the same as in Equation (8.61), but the values of d_ρ and X'_{ρ_0} are always positive. The values of Ω'_n and Ω''_n are determined by

$$\Omega_n = \omega_{im} + \Omega'_0 - \frac{\Delta\Omega_v d_\rho X_{\rho 0}}{1+d_\rho^2} + n\left[\frac{\Delta\Omega_v d_\rho}{2(1+d_\rho^2)} \mp \omega_M\right], \quad (8.131)$$

where the sign "–" at the frequency ω_M corresponds to Ω'_n, i.e., to the frequency $\omega_{im} + \Omega'_0 - n\omega_M$; the sign "+" corresponds to the frequency $\omega_{im} + \Omega'_0 + n\omega_M$.

Reference to Equation (8.131) shows that the frequency Ω'_n is independent of n if the condition

$$\omega_M = \frac{\Delta\Omega_v d_\rho}{2(1+d_\rho^2)} \quad (8.132)$$

is true. In the present case, the partial power spectral densities of the fluctuations are superimposed on one another, i.e., the power spectral density is compressed.

In accordance with Equation (8.132), this effect takes place under the condition $\Delta\Omega_v > 0$ or $\alpha = \beta_0 = 0°$ if the equality $\Delta\Omega_v = 2d_\rho\omega_M$ is true, because $d_\rho^2 \approx 1 + d_\rho^2$. The average frequency of the compressed power spectral density is determined by

$$\Omega'_n = \omega_{im} + \Omega'_0 - 2\omega_M X_{\rho_0} = \omega_{im} + \Omega'_0 - n_0\omega_M . \tag{8.133}$$

Power spectral density compression is absent in the total frequency Ω''_n and the difference between the average frequencies of neighboring partial power spectral densities is equal to $2\omega_M$. If the directional diagram is directed back, in accordance with Equation (8.58) the frequency Ω'_0, as well as Ω_v and $\Delta\Omega_v$, would be negative. In the present case, power spectral density compression is possible within the limits of the other one-half modulation period, as ω_M and $\Delta\Omega_v$ in Equation (8.131) have different signs. With a further increase in the value of d_ρ, the number of partial power spectral densities increases and the difference of their average frequencies tends to approach the frequency ω_M. The effective bandwidth of the compressed power spectral density is equal to $2\omega_M$, i.e., it corresponds to the effective power spectral density bandwidth in the case of stationary radar.

It is not difficult to define the power spectral density $S(\omega)$ under the condition $\beta_0 \neq 0°$.[21,22] For this purpose, it is necessary to use Equation (8.61) and Equation (8.127), and to replace $\Delta\Omega_{FM}$ with the value of $\Delta\Omega_{FM}$ given by Equation (8.67). If the condition of power spectral density compression of the fluctuations is satisfied, the bandwidth $\Delta\Omega_{FM}$ is determined by

$$\Delta\Omega_{FM} = \sqrt{4\omega_M^2 + \Delta\Omega_h^2} . \tag{8.134}$$

The partial power spectral densities are shown in Figure 8.27. The total power spectral density is shown in Figure 8.28. The power spectral densities at the total and difference frequencies have approximately the same bandwidth, which is close to $\Delta\Omega$.

In an analogous way, we can consider the power spectral density under vertical scanning and vertically moving radar [see Equation (8.89) and Equation (8.90)]. The main peculiarity of this power spectral density is that the partial power spectral densities are asymmetrical at $n_0 - 1$ and $n_0 + 1$.

The power spectral densities for the cases of stationary radar and vertical scanning in the presence of the frequency-modulated searching signal are shown in Figure 8.29–Figure 8.31 for various values of d_h. The power spectral densities are one-sided in the direction of increasing n. The left part of the power spectral density corresponds to a decrease in the function $\mathbf{sinc}^2\, x$. Figure 8.29 corresponds to reflection from one radar range zone, and Figure

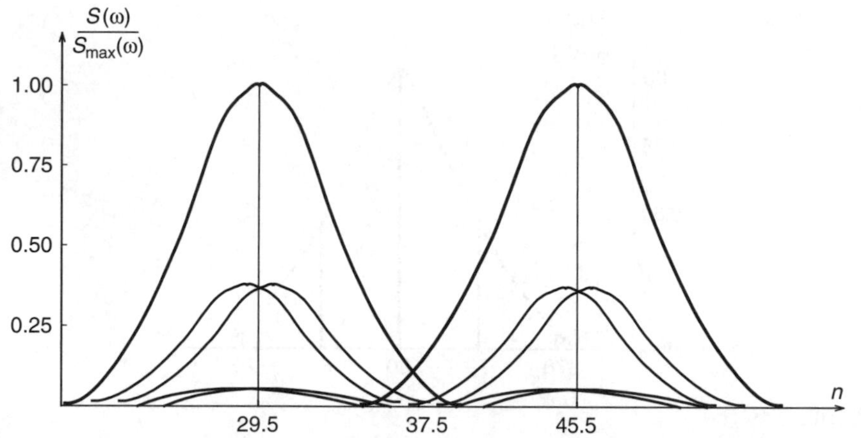

FIGURE 8.27
Partial power spectral densities of target return signal fluctuations with the symmetric sawtooth frequency-modulated searching signal at total and difference frequencies: $X_{\rho_0}\omega_M < \Omega_0$; $d_\rho < 1$; $\Omega_0 = 37.5\ \omega_M$; $X_{\rho_0} = 4$; $\Delta_v = \Delta_h = 0.1$; $\gamma_0 = 65°$; and $\beta_0 = 45°$.

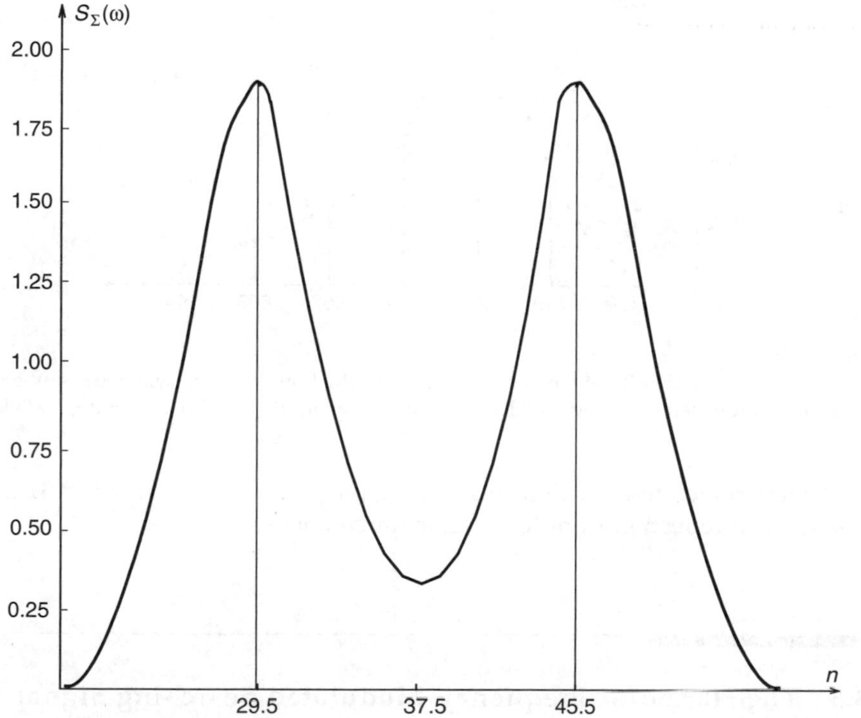

FIGURE 8.28
The total power spectral density of target return signal fluctuations with the symmetric sawtooth frequency-modulated searching signal at total and difference frequencies: $X_{\rho_0}\omega_M < \Omega_0$; $d_\rho < 1$; $\Omega_0 = 37.5\ \omega_M$; $X_{\rho_0} = 4$; $\Delta_v = \Delta_h = 0.1$; $\gamma_0 = 65°$; and $\beta_0 = 45°$.

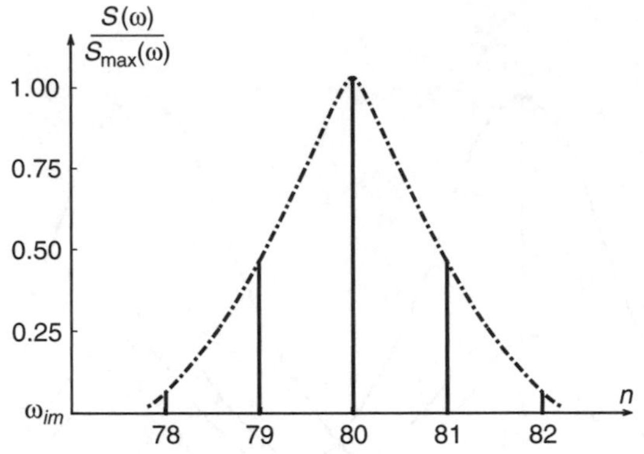

FIGURE 8.29
The power spectral density of target return signal fluctuations with the symmetric saw-tooth frequency-modulated searching signal and stationary radar: $\Delta f_M = 20$ MHz; $\Delta_v = 0.1$; $h = 300$ m; $X_h = 40$.

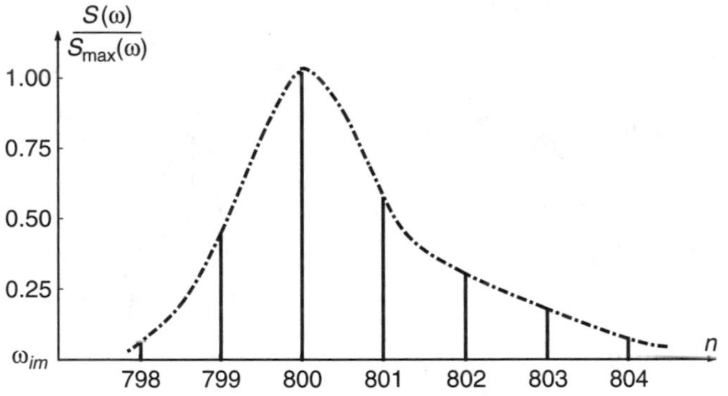

FIGURE 8.30
The power spectral density of target return signal fluctuations with the symmetric saw-tooth frequency-modulated searching signal and stationary radar: $\Delta f_M = 20$ MHz; $\Delta_v = 0.1$; $h = 3$ km; $X_h = 400$.

8.30 corresponds to reflection from two radar range zones. Figure 8.31 corresponds to reflection from four radar range zones.

8.5 The Harmonic Frequency-Modulated Searching Signal

In the case of the harmonic frequency-modulated searching signal, we can write

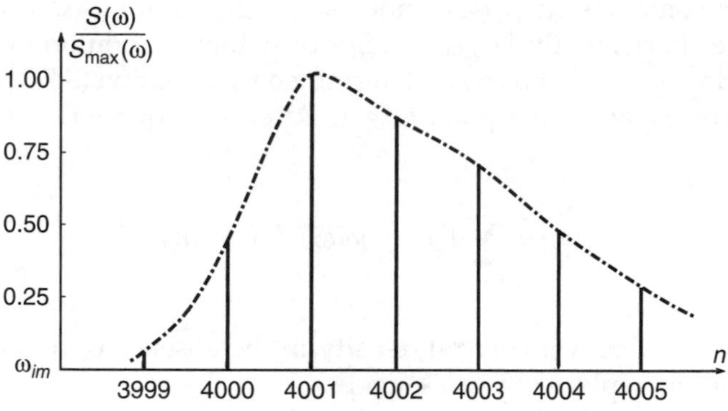

FIGURE 8.31
The power spectral density of target return signal fluctuations with the symmetric saw-tooth frequency-modulated searching signal and stationary radar: $\Delta f_M = 20$ MHz; $\Delta_v = 0.1$; $h = 15$ km; $X_h = 2000$.

$$\Psi(t) = \frac{\Delta\omega_M}{\omega_M} \cdot \cos\omega_M t\,. \tag{8.135}$$

The power spectral density of the nontransformed fluctuations (averaged over time) with harmonic frequency modulation is determined by Equation (2.70), where

$$|N_n|^2 = J_n^2\left(\tfrac{\Delta\omega_M}{\omega_M}\right). \tag{8.136}$$

The correlation function of the nontransformed fluctuations (averaged over time) is given by Equation (8.14) where

$$C^n_{\rho-0.5\Delta\rho} = J_n\left[\tfrac{2\Delta\omega_M}{\omega_M} \cdot \sin\tfrac{\omega_M(\rho-0.5\Delta\rho)}{c}\right] \quad \text{and} \quad C^n_{\rho+0.5\Delta\rho} = J_n\left[\tfrac{2\Delta\omega_M}{\omega_M} \cdot \sin\tfrac{\omega_M(\rho+0.5\Delta\rho)}{c}\right]. \tag{8.137}$$

Neglecting the Doppler shift in frequency ω_M, based on Equation (8.14) and in terms of Equation (8.135), the correlation function can be written in the following form

$$R(\tau) = p_0 \sum_{n=-\infty}^{\infty} e^{j(\omega_{im}+n\omega_M)\tau} \iint \tilde{g}^2(\varphi,\psi) J_n^2(M_\rho) \cdot e^{-2j\omega_0\frac{\Delta\rho}{c}}\, d\varphi\, d\psi\,, \tag{8.138}$$

where

$$M_\rho = \frac{2\Delta\omega_M}{\omega_M} \cdot \sin\frac{\omega_M \rho}{c}\,. \tag{8.139}$$

Consider some specific cases under sloping scanning. Assume that the radar moves horizontally, i.e., $\varepsilon_0 = 0$. Suppose that the condition $\omega_M\ \delta\rho << 1$ [the condition $\frac{T_r}{T_M} << 1$ and $\delta\rho$ are determined by Equation (8.71)] is satisfied. This case corresponds to a point target. Based on Equation (8.138) we can write

$$S_h(\omega) = \sum_{n=-\infty}^{\infty} J_n^2(M_{\rho_0})\delta[\omega - (\omega_{im} + n\omega_M)], \tag{8.140}$$

where $S(\omega)$ is the power spectral density in the absence of frequency modulation, with ω_0 replaced by $\omega_{im} + n\omega_M$.

We use Equation (8.140) to define allowable spurious harmonic frequency modulation caused, for example, by low-frequency pulsed power supply voltage in the super high-frequency radar generator with continuous non-modulated searching signals. Consider two cases: $\Delta\Omega > \omega_M$ and $\Delta\Omega < \omega_M$. In the first case, with an increase in M_{ρ_0}, the power spectral density initially expands; its maximum decreases and after that it becomes two-mode and further, multimode, but it remains continuous (see Figure 8.32). The computer-modeling ratio $\frac{\Delta\Omega_{FM}}{\Delta\Omega}$ as a function of $\frac{\omega_M}{\Delta\Omega}$ with one-mode power spectral density and with an allowable extension of 50% is shown in Figure 8.33. For instance, assume that $h = 15$ km, $\gamma_0 = 65°$, $F_0 = 10$ kHz, $\Delta F = 1.5$ kHz, and $f_M = 400$ Hz. At $\frac{\omega_M}{\Delta\Omega} = 0.27$, based on Figure 8.33 we can find that $M_{\rho_0} = 2$ and the allowable value of frequency deviation Δf_M is equal to 3.0 kHz. In the case $\Delta\Omega < \omega_M$, the power spectral densities $S[\omega - (\omega_{im} + n\omega_M)]$ are not close to one another. Given the ratio $\frac{J_0^2}{J_n^2}$, we can find the allowable frequency deviation. For instance, at $h = 3$ km, $\gamma_0 = 65°$, $f_M = 400$ Hz, and $\frac{J_0^2}{J_n^2} = 30$, the allowable frequency deviation is equal to 5.3 kHz.

Assume that $\frac{\omega_M\rho}{c} << 1$ and $2\pi X_{\rho_0} << 1$, i.e., that there is a low radar range or low frequency modulation at a not so high frequency deviation. Then,

$$J_n^2\left[\tfrac{2\Delta\omega_M}{\omega_M} \cdot \sin\tfrac{\omega_M\rho}{c}\right] \approx \frac{(2\pi X_{\rho_0})^{2|n|}}{2^{2|n|}(|n|!)^2} \cdot \left(1 + 2|n|\psi \operatorname{ctg}\gamma_0\right). \tag{8.141}$$

Substituting Equation (8.141) in Equation (8.138), we can define the correlation function and power spectral density. With the Gaussian directional diagram, the power spectral density takes the following form:[23]

$$S(\omega) = \sum_{n=-\infty}^{\infty} \frac{p_0'}{\Delta\Omega} \cdot e^{-\pi\frac{(\omega - \Omega_n')^2}{\Delta\Omega^2}}, \tag{8.142}$$

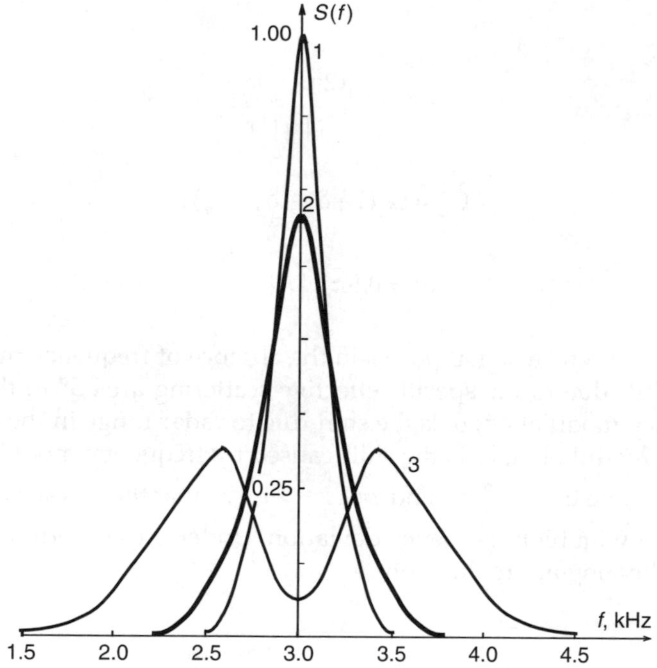

FIGURE 8.32
Power spectral densities of target return signal fluctuations with the harmonic frequency-modulated searching signal, f_M = 400 kHz: (1) Δf_M = 0 kHz; (2) Δf_M = 4 kHz; (3) Δf_M = 8 kHz.

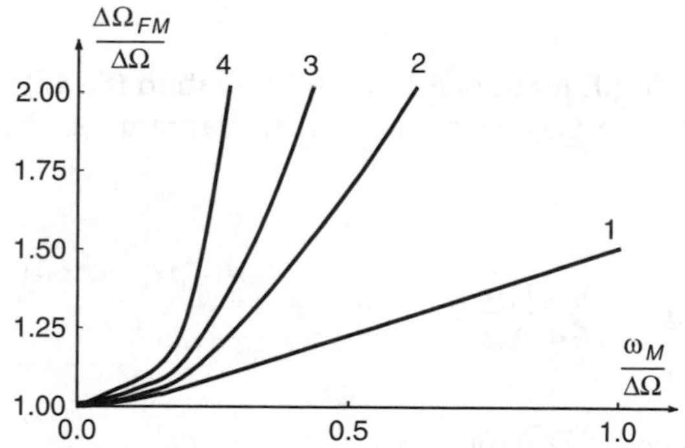

FIGURE 8.33
Relative expansion $\frac{\Delta\Omega_{FM}}{\Delta\Omega}$ of power spectral density of target return signal Doppler fluctuations with the frequency-modulated searching signal and various values of M_{ρ_0}; $\Delta\Omega$ is the power spectral density bandwidth of the target return signal Doppler fluctuations in the absence of frequency modulation : (1) M_{ρ_0} = 1; (2) M_{ρ_0} = 1.5; (3) M_{ρ_0} = 2; (4) M_{ρ_0} = 5.

where

$$p'_{0_n} = \frac{p'_0(2\pi X_{\rho_0})^{2|n|}}{2^{2|n|}(|n|!)^2};\tag{8.143}$$

$$\Omega'_0 = \Omega_0(1-\delta_1-\delta_2+\delta_n);\tag{8.144}$$

$$\delta_n = 0.5\pi^{-1}|n|\Delta_v^{(2)};\tag{8.145}$$

p'_0 is the target return signal power in the absence of frequency modulation; δ_1 is the shift due to the specific effective scattering area S° in the absence of frequency modulation; δ_2 is the shift due to radar range in the absence of frequency modulation; δ_n is the shift caused by frequency modulation.

Let us assume that $\frac{\omega_M \rho}{c} < 1$ and $2\pi X_{\rho_0} >> 1$, i.e., that there is slow frequency modulation with high frequency deviation. Under these conditions, we can use the following approximation:

$$J_n^2(2\pi X_\rho) = \frac{1}{\pi^2 X_\rho}\cdot e^{-\frac{\pi^2}{\alpha_1^2}[2X_\rho - 0.25(2n+1)]},\tag{8.146}$$

where

$$-0.5 \le 2X_\rho - 0.25(2n+1) \le 0.5,\tag{8.147}$$

and $\alpha_1 = 0.78$. Using Equation (8.146) and Equation (8.147), the power spectral density with the Gaussian directional diagram can be determined as follows:

$$S(\omega) = \sum_{n=-\infty}^{\infty} \frac{p_{FM}}{\Delta\Omega}\cdot e^{-\pi\frac{(\omega-\Omega'_n)^2}{\Delta\Omega^2}}\cdot e^{-\frac{\pi^2}{\alpha_1^2(1+d_\rho^2)}[2X'_{\rho_0} - 0.25(2n+1)]^2},\tag{8.148}$$

where p_{FM} is given by Equation (8.62);

$$\Omega_n = \omega_{im} + \Omega'_0 + \frac{\Delta\Omega_v\sqrt{\pi}d_\rho[2X'_{\rho_0} - 0.25(2n+1)]}{\alpha_1^2(1+d_\rho^2)} + n\omega_M;\tag{8.149}$$

Ω'_0 and X'_{ρ_0} are the same as in the case of the saw-tooth frequency-modulated searching signal.

The partial power spectral densities with the harmonic frequency-modulated searching signal, as well as with the symmetric saw-tooth frequency-modulated searching signal, are at the total and difference frequencies Ω_n. Thus, there is a narrowing of the power spectral density at the difference frequency and an expansion at the total frequency. We can define the condition of power spectral density compression as follows: The sum of the terms with index n in Equation (8.149) should be zero.

For the case $\alpha = \beta_0 = \varepsilon_0 = 0°$, the power spectral density can be defined in an explicit form without additional simplifications. If the velocity vector **V** of moving radar is outside the directional diagram and the variables φ and ψ are separable in the function $\tilde{g}^2(\varphi, \psi)$, the power spectral density can be written in the following form:

$$S(\omega) = \frac{p_1}{|\Omega_v|} \cdot \sum_{n=-\infty}^{\infty} \tilde{g}^2\left(\tfrac{\Omega_n - \omega}{\Omega_v}\right) J_n^2\left\{\tfrac{2\Delta\omega_M}{\omega_M} \cdot \sin\left[\frac{\omega_M h}{c\sin\left(\gamma_0 + \frac{\Omega_n - \omega}{\Omega_v}\right)}\right]\right\}, \tag{8.150}$$

where the parameters p_1, Ω_v, and Ω_n are determined by Equation (8.58).

The formula in Equation (8.150) is analogous to Equation (8.54). The partial power spectral densities are distorted due to the function $J_n^2(x)$, which is the analog of the function $\mathbf{sinc}^2\, x$ in Equation (8.54). For instance, at the frequency $\omega = \Omega_{n_0}$ and under the condition $\sin \frac{\omega_M h}{c \sin\gamma_0} = 0$, $J_n = 0$ $(n \neq 0)$ and the power spectral density is zero. The target return signal power with "zero" harmonic at the frequency $\omega_{im} + \Omega_0'$ is maximal. Due to the directional diagram length, the target return signal power can be different from zero at other Doppler frequencies.

As one can see from Figure 8.34, the power spectral density of the Doppler fluctuations is highly distorted at the condition $\beta_0 = 0°$ in comparison with the power spectral density under the condition $\beta_0 = 45°$. If the radar antenna is deflected in the direction of moving radar, i.e., if the condition $\beta_0 \neq 0°$ is satisfied, the lines for constant values of ρ and isofrequency Doppler lines on the surface of a two-dimensional target are mismatched, as it takes place under the condition $\beta_0 = 0°$. In this case, we cannot carry out the integration in Equation (8.138) and it is necessary to use numerical integration techniques.

Because the Bessel function depends only on the angle ψ, we can use it together with the function $\tilde{g}_v^2(\psi)$ in the following way:

$$\tilde{g}_1^2(\psi) = \tilde{g}_v^2(\psi) \cdot J_n^2(\psi). \tag{8.151}$$

Using the function $\tilde{g}_1^2(\psi)$ under the condition $T_r < T_M$, we can define a deterioration of the target return signal due to frequency modulation at the n-th harmonic in comparison with the nonmodulated searching signal. In addition, we can define the shift $\Delta\gamma$ in the average angle γ_0 with the Gaussian directional diagram $(n > 0)$

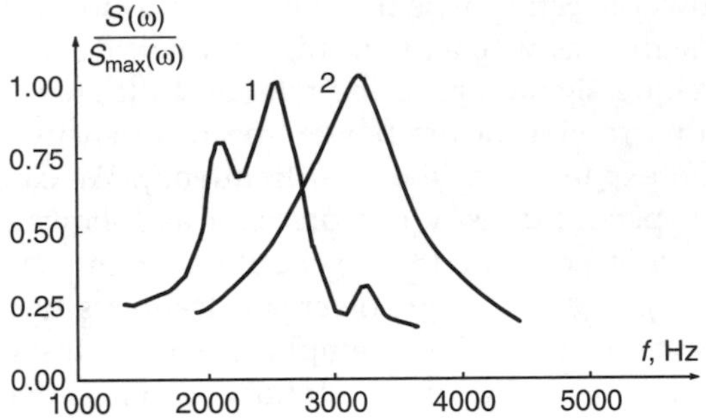

FIGURE 8.34
Experimental power spectral densities of target return signal Doppler fluctuations with the harmonic frequency-modulated searching signal, $f_M = 1$ MHz, $\Delta f_M = 2.4$ MHz, $n = 3$: (1) $\beta_0 = 0°$; (2) $\beta_0 = 45°$.

$$d_n\left(\tfrac{h}{T_M}\right) = J_n^2(M_\rho) + \frac{\Delta_v^{(2n)}[J_2^2(M_\rho)]^{(2n)}}{2^{3n} n!\, \pi^n} \;; \qquad (8.152)$$

$$\Delta\gamma = -\frac{\Delta_v^{(2)}[J_n^2(M_\rho)]^{(1)}}{4\pi d_n\left(\tfrac{h}{T_M}\right)}, \qquad (8.153)$$

where $[J_n^2(M_\rho)]^{(2n)}$ is a derivative of the second order with respect to γ. We need use only the term of the derivative that contains the Bessel function $J_0^2(M_\rho)$ of zero order at the neighboring point $J_n^2(M_\rho) = 0$, because other terms are negligible. Under the condition $n = 1$, we can write

$$\begin{cases} [J_1^2(M_\rho)]^{(2)} = 0.5 J_0^2(M_\rho) \cdot M_\rho'^2 \;; \\ [J_1^2(M_\rho)]^{(1)} = J_1(M_\rho)[J_0(M_\rho) - J_2(M_\rho)] \cdot M_\rho'^2 \;; \\ M_\rho' = -2\pi \frac{\Delta\omega_M}{\omega_M} \cdot \frac{T_d}{T_M} \cdot \operatorname{ctg} \gamma_0 \operatorname{ctg} \frac{\omega_M \rho}{c} \;. \end{cases} \qquad (8.154)$$

An example of the determination of the value of $d_1(\frac{h}{T_M})$ is shown in Figure 8.35. The computer-modeled and experimental values of the power P_r of the transformed target return signal, which are normalized to the power P_n of noise, are shown in Figure 8.36. The segment of curve 2 in Figure 8.36 corresponding to low values of h, in which the value of P_r is approximately independent of h in accordance with Equation (8.143)–Equation (8.145), can be seen. The experimental dependence of the ratio $\frac{P_r}{P_n}$ as a function of h is plotted in Figure 8.37.

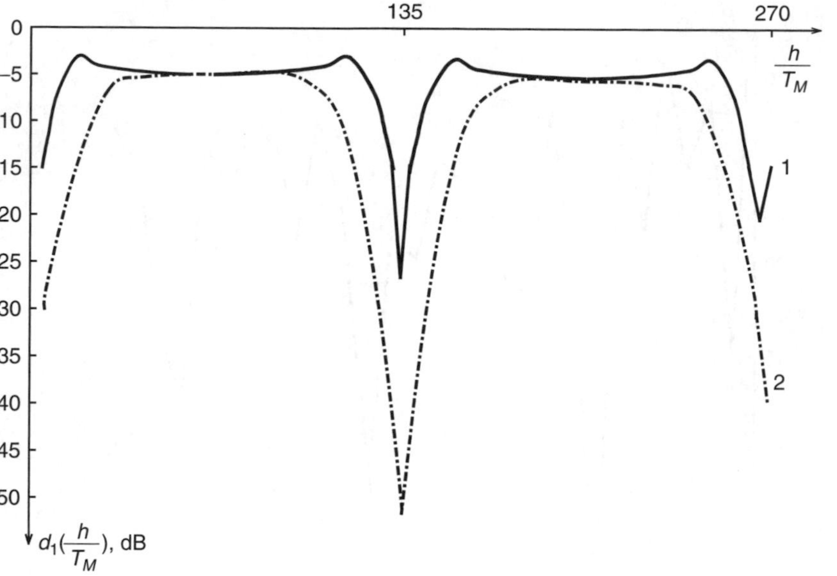

FIGURE 8.35
Deterioration of the target return signal $d_1(\frac{h}{T_M})$ with the harmonic frequency-modulated searching signal at the first harmonic of frequency modulation, $\gamma_0 = 65°$, $\Delta_v = 5°$, $\frac{\Delta\omega_M}{\omega_M} = 1.2$: (1) one-sided frequency; (2) two-sided convoluted frequencies.

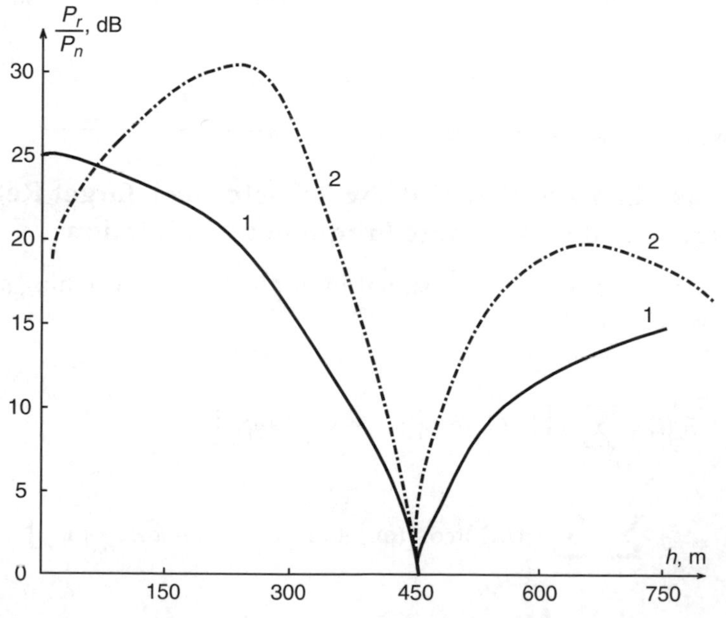

FIGURE 8.36
Experimental and computer-modeling ratio $\frac{P_r}{P_n}$ as a function of altitude h, $\gamma_0 = 65°$, $\Delta_v = 5°$, $\frac{\Delta\omega_M}{\omega_M} = 1.2$, $n = 1$, $f_M = 300$ kHz: (1) one-sided frequency; (2) two-sided convoluted frequencies.

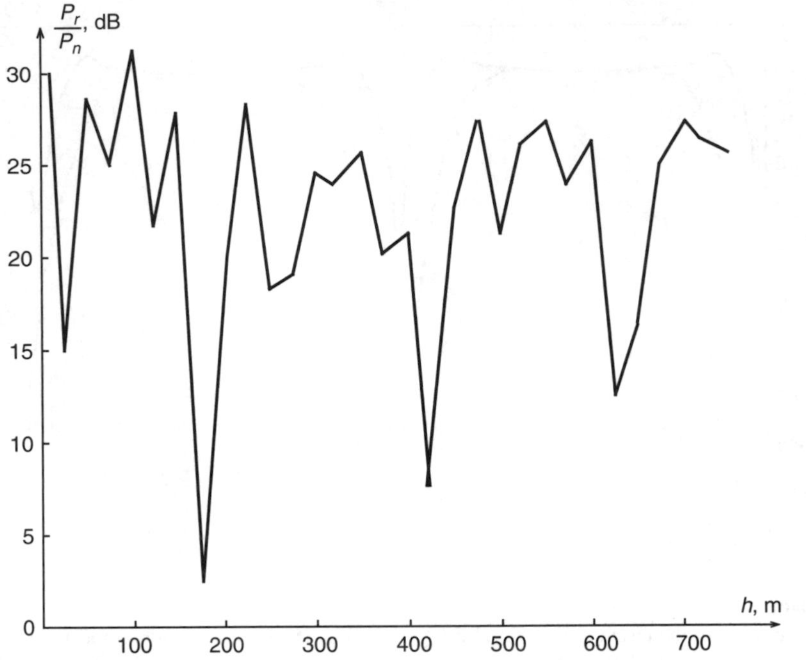

FIGURE 8.37
Experimental ratio $\frac{P_r}{P_n}$ with the harmonic frequency-modulated searching signal at the third harmonic of frequency modulation (two-sided convoluted): $\gamma_0 = 65°$, $\Delta_v = 5°$, $f_M = 1$ MHz, $\frac{\Delta\omega_M}{\omega_M} = 2.4$.

8.6 Phase Characteristics of the Transformed Target Return Signal under Harmonic Frequency Modulation

Let us express the target return signal multiplied by the searching signal in the following form:[24]

$$
\begin{aligned}
S_r(t) = \sum_{i=-\infty}^{\infty} S_i \Big\{ J_0(M_\rho)\cos[(\omega_{im} + \Omega_{0_i})t + \varphi_{0_i}] \\
+ \sum_{n=-\infty}^{\infty}\sum_{i=-\infty}^{\infty} J_n(M_{\rho_i})\Big\{\cos\Big[(\omega_{im} + \Omega_{0_i} + n\omega_M)t - \pi n \tfrac{T_{d_i}}{T_M} + \varphi_{0_i}\Big], \quad (8.155) \\
+ (-1)^n \cos\Big[(\omega_{im} + \Omega_{0_i} - n\omega_M)t + \pi n \tfrac{T_{d_i}}{T_M} + \varphi_{0_i}\Big]\Big\}\Big\}
\end{aligned}
$$

where φ_{0_i} is the phase of an elementary signal distributed randomly within the limits of the interval [0, 2π]. Due to the random character of the phase φ_{0_i}, the phase of each term in the sum is random, too. At the same time, the phase $\pi n \frac{T_d}{T_M}$ varies within the limits of the narrow region defined by the directional diagram under the condition $T_r << T_M$. The average phase $\pi n \frac{T_d}{T_M}$ of the target return signal given by Equation (8.155) defines some average distance between the radar and the two-dimensional (surface) target. To except the random phase φ_{0_i}, which does not allow us to measure the phase $\pi n \frac{T_d}{T_M}$, we should take the product of the signals with frequencies $\omega_{im} + \Omega_{0_i} + n\omega_M$ and $\omega_{im} + \Omega_{0_i} - n\omega_M$, as a rule. For this purpose, the signal $S_r(t)$ must be squared. As a result, we can write

$$S_{r_{sq}}(t) = \sum_{i=-\infty}^{\infty} S_i^2(t) J_n^2(M_{\rho_i}) \cos 2\left(n\omega_M t - \pi n \tfrac{T_{d_i}}{T_M}\right). \tag{8.156}$$

Denote

$$\boldsymbol{\varphi}_i = 2\pi n \cdot \frac{T_{d_i}}{T_M} = 4\pi n \cdot \frac{h}{cT_M \sin(\psi + \gamma_0)}. \tag{8.157}$$

Based on Equation (8.157) we can write Equation (8.156) in the following form:

$$S_{r_{sq}}(t) = E\cos(2n\omega_M + \boldsymbol{\varphi}) = X\cos(2n\omega_M t) + Y\sin(2n\omega_M t), \tag{8.158}$$

where

$$X = \sum_{i=-\infty}^{\infty} S_i^2(t) J_n^2(M_{\rho_i}) \cos\boldsymbol{\varphi}_i = E\cos\boldsymbol{\varphi} \ ; \tag{8.159}$$

$$Y = \sum_{i=-\infty}^{\infty} S_i^2(t) J_n^2(M_{\rho_i}) \sin\boldsymbol{\varphi}_i = E\sin\boldsymbol{\varphi} \ ; \tag{8.160}$$

$$\mathrm{tg}\,\boldsymbol{\varphi} = \frac{Y}{X}. \tag{8.161}$$

The average value $\boldsymbol{\varphi}_{av}$ of the phase can be determined by

$$\overline{\text{tg}\,\boldsymbol{\varphi}} = \overline{\text{tg}\,(\boldsymbol{\varphi}_0 + \Delta\boldsymbol{\varphi}_i)} = \frac{\overline{\sum_{i=-\infty}^{\infty} S_i^2(t) J_n^2(M_{\rho_i}) \sin \boldsymbol{\varphi}_i}}{\overline{\sum_{i=-\infty}^{\infty} S_i^2(t) J_n^2(M_{\rho_i}) \cos \boldsymbol{\varphi}_i}}, \tag{8.162}$$

where $\Delta\boldsymbol{\varphi}_i = \boldsymbol{\varphi}_i - \boldsymbol{\varphi}_0$. Averaging is carried out over an ensemble. Assuming that the phase deviation $\nu = \boldsymbol{\varphi}_i - \boldsymbol{\varphi}_0$ is not high in value within the directional diagram, i.e., that the condition $\gamma_0 < 90°$ is satisfied, based on Equation (8.162) we can write

$$\boldsymbol{\varphi}_{av} = \boldsymbol{\varphi}_0 + \overline{\Delta\boldsymbol{\varphi}}, \tag{8.163}$$

where

$$\overline{\Delta\boldsymbol{\varphi}} = \frac{\sum_{i=-\infty}^{\infty} S_i^2(t) J_n^2(M_{\rho_i}) \Delta\boldsymbol{\varphi}_i}{\sum_{i=-\infty}^{\infty} S_i^2(t) J_n^2(M_{\rho_i})}. \tag{8.164}$$

Reference to Equation (8.157) shows that

$$\varphi = \frac{\boldsymbol{\varphi}_i - \boldsymbol{\varphi}_0}{\boldsymbol{\varphi}_0 \text{ctg}\,\gamma_0} = -\frac{\nu}{\boldsymbol{\varphi}_0 \text{ctg}\,\gamma_0}; \tag{8.165}$$

$$\boldsymbol{\varphi}_0 = \frac{4\pi nh}{cT_M \sin \gamma_0}, \tag{8.166}$$

where the angle γ_0 corresponds to the direction of the directional diagram axis for transmission and reception. In terms of Equation (8.166), we can express Equation (8.164) in the following form:

$$\overline{\Delta\boldsymbol{\varphi}} = \frac{\int_{\nu_1}^{\nu_2} \nu A(\nu)\, d\nu}{\int_{\nu_1}^{\nu_2} A(\nu)\, d\nu}, \tag{8.167}$$

where $A(\nu)$ is the integrand function $\psi(\nu)$ in Equation (8.138) under the condition $\tau = 0$:

$$A(\nu) = \frac{1}{\varphi_0 \text{ctg}\,\gamma_0}\tilde{g}_v^2\left(-\tfrac{\nu}{\varphi_0 \text{ctg}\,\gamma_0}\right)\cdot J_n^2\left\{\frac{\frac{2\Delta\omega_M}{\omega_M}}{\sin\left[\frac{2\pi h}{cT_M \sin\left(\gamma_0 - \frac{\nu}{\varphi_0 \text{ctg}\,\gamma_0}\right)}\right]}\right\}. \tag{8.168}$$

The value of $\overline{\Delta\varphi}$ given by Equation (8.167) defines a shift in the average value of the transformed target return signal phase with respect to the directional diagram axis for transmission and reception caused by various factors: the asymmetrical directional diagram; effect of the specific effective scattering area of the underlying surface of a two-dimensional target; influence of the function $J_n^2(x)$, and changes in the radar range within the directional diagram. If we assume that the function $\tilde{g}_v^2(\psi)$ can be approximated by the Gaussian law with an effective width Δ_v and average angle γ_0,[25,26] and that the specific effective scattering area $S°(\psi)$ can be approximated by the exponent,[27] and if we take into consideration the condition $T_r << T_M$, as in the opposite case it is possible to measure the phase φ, based on Equation (8.167) and in terms of Equation (8.168), we can write

$$\Delta\varphi_{sh} = \overline{\Delta\varphi} = \varphi_{av} - \varphi_0 = -\varphi_0 \cdot \frac{\Delta_v^{(2)}}{4\pi}\left\{k_1 + \text{ctg}\,\gamma_0 + \frac{[J_n^2(M_\rho)]^{(1)}}{d_n\left(\frac{h}{T_M}\right)}\right\}\cdot \text{ctg}\,\gamma_0 . \tag{8.169}$$

Because $\varphi \cong \rho$, we can replace $\Delta\varphi_{sh}$ by $\Delta\rho_{sh}$ in Equation (8.169).

Reference to Equation (8.169) shows that we can define the relative error $\frac{\Delta\rho_{sh}}{\rho_0}$ in measuring the distance between the radar and the scanned two-dimensional (surface) target. For instance, at $\Delta_v = 5°$, $\gamma_0 = 65°$, $n = 1$, $k_1 = 12.9$ (the weak rough sea with the wind velocity within the limits of the interval 0, … ,11 km/h), and $\frac{\Delta\omega_M}{\omega_M} = 1.2$, the component of the error $\frac{\Delta\rho_{sh}}{\rho_0}$ due to the specific effective scattering area $S°(\psi)$ is equal to 0.47%; the component of the error $\frac{\Delta\rho_{sh}}{\rho_0}$ due to changes in the radar range ρ within the directional diagram is equal to 0.02%; and the component of the error $\frac{\Delta\rho_{sh}}{\rho_0}$ due to the function $J_n^2(M_\rho)$ is equal to 0.22%. The component of the error $\frac{\Delta\rho_{sh}}{\rho_0}$ due to the function $J_n^2(M_\rho)$ is decreased, approximately by the same law as is the modulation index $\frac{\Delta\omega_M}{\omega_M}$. Because all cofactors of the function $A(\nu)$ can be considered in Equation (8.168) to be linear functions at low values of ν, except for the term $g_v^2\left(-\frac{\nu}{\varphi_0 \text{ctg}\,\gamma_0}\right)$, the limits of variations of the phase φ_i are defined only by the function $g_v^2(\psi)$. If the function $g_v^2(\psi)$ takes the shape of the Gaussian curve, we can write

$$g_v^2\left(-\frac{v}{\varphi_0 \operatorname{ctg} \gamma_0}\right) \approx e^{-\pi \frac{(\varphi-\varphi_0)^2}{\Delta_\varphi^{(2)}}}, \tag{8.170}$$

where

$$\Delta_\varphi = n\omega_M T_r = 2\sqrt{2}\pi n \cdot \frac{h\Delta_v \cos\gamma_0}{cT_M \sin^2\gamma_0} \tag{8.171}$$

is the effective bandwidth of the phase characteristic of radar at the frequency $2n\omega_M$. The value of Δ_φ can be used to estimate the variance of the fluctuations of the phase φ (or radar range ρ) caused by the limited width of the directional diagram

$$\sigma_\varphi^2 = \frac{\Delta_\varphi^{(2)}}{2\pi} \quad \text{or} \quad \sigma_\rho^2 = \frac{c^2 T_r^2}{8\pi}. \tag{8.172}$$

For instance, at $\Delta_v = 1°$ and $\gamma_0 = 65°$, we get $\frac{\sigma_\rho}{\rho_0} = 1.27\%$.

8.7 Conclusions

The fact that the instantaneous power spectral density of the target return signal fluctuations is a function of time allows us to isolate information regarding pulsed target return signal delay, which defines the distance between the radar and the scanned two-dimensional (surface) target. As is well known, in the use of radar with frequency-modulated searching signals, information about the distance between the radar and the two-dimensional (surface) target is extracted from the frequency of the transformed target return signal. To obtain the frequency of the transformed target return signal from different targets, we use a set of filters. The bandwidth of each filter is approximately equal to the modulation frequency. The comb characteristics of these filters are analogous to pulses gated in time under the use of radar with pulsed searching signals. Because of this, the correlation function of the fluctuations is of prime interest to us and there is no need to consider the instantaneous correlation function. Thus, the correlation function of the transformed fluctuations must be averaged over the period of frequency modulation. The general formula for definition of the correlation function of the fluctuations is given by Equation (8.14). The correlation function averaged over time is the sum of the correlation functions given by Equation (8.14) with specific coefficients for each interval.

With linear variation of the searching signal frequency, the correlation function of the fluctuations is a nonstationary frequency-modulated process. Thus, the Doppler frequency is a function of time. In addition to the Doppler shift in frequency, the target return signal is shifted in frequency by the range-finder frequency. The range-finder frequency is shifted by its own Doppler frequency, too.

Relative shifts in the range-finder- and Doppler frequencies with respect to their differences within the directional diagram compensate each other under a fixed aspect angle. If the Doppler frequency is negative, the range-finder frequency must be positive to ensure power spectral density compression of the fluctuations under sloping scanning of the underlying surface of a two-dimensional target. It should be noted that complete compression of the power spectral density can take place not only with horizontally moving radar, but in all cases when the velocity vector of moving radar and the directional diagram axis are in the same vertical plane. The Doppler frequency as a function of time is present in the transformed target return signal and can hide the effect of power spectral density compression.

With the asymmetric saw-tooth frequency-modulated searching signal, the coefficients of the Fourier-series expansion of the correlation function of the fluctuations take the form of the function $\mathbf{sinc}^2\, x$ and the average frequency $\omega_{av} = \omega_0 + 0.5\Delta\omega_M$. The power spectral density consists of a set of partial power spectral densities defined by harmonics of the modulation frequency ω_M. Unlike the case of the nonperiodic searching signal, the correlation function of the fluctuations possesses two range-finder frequencies given by Equation (8.43), which are defined by different intervals of the modulation period. Harmonics of the modulation frequency ω_M are shifted by the one-half Doppler frequency. Because all components of the target return signal must be shifted by their Doppler frequency, not the one-half Doppler frequency, we can reason that another one-half Doppler shift in frequency is given by Equation (8.39). As the condition $n\omega_M << \omega_0$ is satisfied, as a rule, we can neglect the Doppler shift in modulation frequency ω_M. The same conclusion can be reached in the case of the symmetric saw-tooth frequency-modulated searching signal.

The partial power spectral densities of the fluctuations with the harmonic frequency-modulated searching signal, as well as with the symmetric saw-tooth frequency-modulated searching signal, are at the total and difference frequencies Ω_n. There is a narrowing of the power spectral density at the difference frequency and an expansion at the total frequency.

References

1. Scharf, L., *Statistical Signal Processing, Detection, Estimation, and Time Series Analysis,* Addison-Wesley, Reading, MA, 1991.
2. Li, J., Liu, G., Jiang, N., and Stoica, P., Moving target feature extraction for airborne high-range resolution phased-array radar, *IEEE Trans.*, Vol. SP-49, No. 2, 2001, pp. 277–289.
3. Schleher, D., *MTI and Pulsed Doppler Radar,* Artech House, Norwood, MA, 1991.
4. Ward, K., Baker, C., and Watts, S., Maritime surveillance radar. Part 1: Radar scattering from the ocean surface, *Proceedings of the IEE F,* Vol. 137, No. 2, 1990, pp. 51–62.
5. Dong, X., Beaulieu, N., and Wittke, P., Signaling constellations for fading channels, *IEEE Trans.,* Vol. COM-47, No. 5, 1999, pp. 703–714.
6. Rappaport, T., *Wireless Communications: Principles and Practice,* Prentice Hall, Englewood Cliffs, NJ, 1996.
7. Stuber, G., *Principles of Mobile Communications,* Kluwer, Boston, MA, 1996.
8. Hudson, J., *Adaptive Array Principles,* Peter Peregrinus, London, 1981.
9. Muqnet, B., Courville, M., Duhamel, P., and Buzenac, V., A subspace based blind and semi-blind channel identification method for OFDM systems, in *Proceedings of the IEEE 2nd Workshop on Signal Processing Advances in Wireless Communications,* Annapolis, MD, May 9–12, 1999, pp. 170–173.
10. Graziano, V., Propagation correlation at 900 MHz, *IEEE Trans.,* Vol. VT-27, No. 11, 1978, pp. 1182–1189.
11. Jukovsky, A., Onoprienko, E., and Chijov, V., *Theoretical Foundations of Radar Altimetry,* Soviet Radio, Moscow, 1979. (In Russian.)
12. Jacobs, S. and O'Sullivan, J., Automatic target recognition using sequences of high resolution radar range profiles, *IEEE Trans.,* Vol. AES-36, No. 4, 2000, pp. 364–382.
13. Gerlach, K. and Steiner, M., Fast converging adaptive detection of Doppler-shifted range-distributed targets, *IEEE Trans.,* Vol. SP-48, No. 9, 2000, pp. 2538–2541.
14. Melvin, W., Wicks, M., Autonik, P., Salama, X., Li, P., and Schuman, H., Knowledge-based space–time adaptive processing for airborne early warning radar, *IEEE Aerosp. Electron. Syst. Mag.,* April 1998, pp. 37–42.
15. Barndoff–Nielsen, O. and Cox, D., *Asymptotic Techniques for Use in Statistics,* Chapman & Hall, London, 1989.
16. Arnold, H., Cox, D., and Murray, R., Macroscopic diversity performance measured in the 800-MHz portable radio communications environment, *IEEE Trans.,* Vol. AP-36, No. 2, 1988, pp. 277–280.
17. Baker, C., *The Numerical Treatment of Integral Equations,* Oxford University Press, Oxford, U.K., 1977.
18. Besson, O., Stoica, P., and Gershman, A., Simple and accurate direction of arrival estimator in the case of imperfect spatial coherence, *IEEE Trans.,* Vol. SP-49, No. 4, 2001, pp. 730–737.
19. Amin, M., Time-frequency spectrum analysis and estimation for nonstationary random processes, in *Time-Frequency Signal Analysis: Methods and Applications,* Longman-Cheshire, London, 1992.

20. Tyapkin, L., Spectral characteristics of frequency-modulated target return signals, *Problems in Radio Electronics,* Vol. OT, No. 4, 1976, pp. 3–23. (In Russian.)
21. Ho, P. and Lane, P., Spectrum, distance, and receiver complexity of encoded continuous phase modulation, *IEEE Trans.,* Vol. IT-34, No. 9, 1988, pp. 1021–1032.
22. Sari, H., Orthogonal frequency-division multiple access with frequency-hopping and diversity, in *Multi-Carrier Spread Spectrum,* K. Fazel and G. Fettweis, Eds., Kluwer, Norwell, MA, 1997, pp. 57–68.
23. Tyapkin, L. and Mandurovsky, I., Spectral characteristics of frequency-modulated target return signals, *Problems in Radio Electronics,* Vol. OT, No. 2, 1970, pp. 35–58. (In Russian.)
24. Kolchinsky, V., Mandurovsky, L., and Konstantinovsky, M., *Doppler Devices and Navigational Systems,* Soviet Radio, Moscow, 1979. (In Russian.)
25. Narayanan, K. and Stuber, G., Performance of trellis-coded CPM with iterative demodulation and decoding, *IEEE Trans.,* Vol. COM-49, No. 4, 2001, pp. 676–687.
26. Buzzi, S., Lops, M., and Tulino, A., Time-varying narrowband interference rejection in asynchronous multiuser DS/CDMA systems over frequency-selective fading channels, *IEEE Trans.,* Vol. COM-47, No. 10, 1999, pp. 1523–1536.
27. Buzzi, S., Lops, M., and Tulino, A., MMSE RAKE reception for asynchronous DS/CDMA overlay systems and frequency-selective fading channels, in *Wireless Personal Communications,* Kluwer, Norwell, MA, Vol. 13, June 2000, pp. 295–318.

9

Fluctuations under Scanning of the Three-Dimensional (Space) Target by the Continuous Signal with a Frequency that Varies with Time

9.1 General Statements

Let us assume that the radar illuminates the three-dimensional (space) target with the continuous searching signal whose amplitude and frequency vary with time [see Equation (2.54)]. At first, let us suppose that the three-dimensional (space) target has a limited length in the radar range interval $[\rho_1, \rho_2]$ and that the boundary values ρ_1 and ρ_2 are constants. In other words, the boundary values ρ_1 and ρ_2 are independent of the variables φ and ψ within the radar antenna directional diagram. This assumption does not change the physical essence of the considered problem but makes it simple to solve.[1–3]

If the frequency of the searching signal is a periodic function of time with period T_M, then a product of the shifted signals, $W(t - 0.5\tau) \cdot W^*(t + 0.5\tau)$, similar to Equation (2.62), is a periodic function both of the time t and of the shift in time τ. Thus, the correlation function of the fluctuations in the target return signal frequency given by Equation (2.55) can be written in the following form

$$R_{\Delta\rho,\Delta\omega}(t,\tau) = p\sum_{n=-\infty}^{\infty}\int_{-\pi}^{\pi}\int_{-0.5\pi}^{0.5\pi}\int_{\rho_1}^{\rho_2} g^2(\varphi,\psi)\rho^{-2}\cdot e^{-j\Delta\tilde{\Psi}_n}d\rho\, d\varphi\, d\psi \tag{9.1}$$

if the amplitude of the target return signal is given by Equation (2.72) and the variables ρ, φ, and ψ are independent, where

$$\Delta\tilde{\Psi}_n = \omega_0(\tau - nT_M) - 2\omega_0\Delta\rho c^{-1} + \Delta\Psi_n\,; \tag{9.2}$$

$$\Delta\Psi_n = \Psi\left[t - \tfrac{2\rho}{c} + 0.5(\tau - nT_M) - \tfrac{\Delta\rho}{c}\right] - \Psi\left[t - \tfrac{2\rho}{c} - 0.5(\tau - nT_M) + \tfrac{\Delta\rho}{c}\right]; \tag{9.3}$$

$$p = \frac{PG_0^2\lambda^2\mathbf{S}^\circ}{64\pi^3} . \tag{9.4}$$

Here we retain the cofactor $e^{j\omega_0 nT_M}$ for the sake of convenience. In the case of the coherent searching signal, this cofactor is equal to one.

We assume that the velocity of variations in frequency $k_\omega(t)$ is a very slow function of the time t so that frequency changes can be considered as a linear function within time intervals that can be compared in value with the correlation length τ_c. Therefore, we can write

$$\Psi'(t) = \Omega(t) = \Omega(t_0 + t - t_0) \cong \Omega(t_0) + k_\omega(t - t_0) \tag{9.5}$$

and

$$\omega(t) = \omega_0 + \Omega(t) = \omega_0 + \Omega(t_0) + k_\omega(t - t_0) , \tag{9.6}$$

where $k_\omega(t)$ can be a slow function of time, in particular, a constant (but not necessarily) low in value function [see Equation (2.54)].

Without any loss of generality, we can assume that $t_0 = 0$ and $\Omega(t_0) = 0$. Then we can write

$$\Psi'(t) = \Omega(t) = k_\omega t \qquad \text{and} \qquad \omega(t) = \omega_0 + k_\omega t . \tag{9.7}$$

In the case of the moving radar with constant velocity, i.e., $\Delta\rho = -V_r \cdot \tau$, reference to Equation (9.2) and Equation (9.3) shows that

$$\Delta\Psi_n = (\mu\tau - nT_M)\cdot\Psi'(t - 2\rho c^{-1}) = (\mu\tau - nT_M)\cdot k_\omega(t - 2\rho c^{-1}) ; \tag{9.8}$$

$$\Delta\tilde{\Psi}_n = \omega(t)\cdot(\mu\tau - nT_M) + \Omega_D\tau - \Omega_\rho(\mu\tau - nT_M) , \tag{9.9}$$

where

$$\mu = 1 + 2\, V_r(\varphi, \psi)c = \mu(\varphi, \psi) \tag{9.10}$$

is a coefficient that takes into consideration the Doppler distortion in the target return signal frequency — here, $\mu(\varphi, \psi)$ is a function of the coordinates φ and ψ, so far we have been using only the average value $\mu = 1 + \frac{2V_{r0}}{c}$;

$$\Omega_D = 2\omega(t)V_r(\varphi, \psi)c^{-1} = 4\pi V_r(\varphi, \psi)\lambda^{-1}(t) = \Omega_D(t, \varphi, \psi) \tag{9.11}$$

is the Doppler shift in the target return signal frequency that is proportional to the frequency $\omega(t)$ and depends on time;

$$\Omega_\rho = 2k_\omega c^{-1}\rho = k_\omega T_d = \Omega_\rho(t,\rho) \tag{9.12}$$

is the range-finder frequency that is equal to the searching signal frequency differential within the limits of the period $T_d = \frac{2\rho}{c}$.

The frequencies $\omega(t)$ and $\Omega_D(t)$ are periodic functions of time with the period T_M. Neglecting the Doppler shift in the range-finder frequency $\frac{2\rho\,\Omega_\rho}{c}$ in Equation (9.9), which is very low in value, we can write

$$\Delta\tilde{\Psi}_n = [\omega(t) - \Omega_\rho]\cdot(\tau - nT_M) + \Omega_D\tau\ . \tag{9.13}$$

The formula in Equation (9.13) shows that there are two components. The first component depends on the periodic difference $\omega(t) - \Omega_\rho$; the second component, caused by the Doppler shift, monotonically increases from period to period together with the shift $\tau = nT_M$. In the case of the periodic frequency-modulated searching signal, the condition of slow variation of the velocity of changes in frequency $k_\omega(t)$ is not always true. For instance, in the case of the linear frequency-modulated searching signal at definite instants of time, the velocity of variations in frequency $k_\omega(t)$ is a jump-like function. As a rule, these instants of time can be excluded from consideration because $k_\omega(t) = const$ within the main part of the period.[4] In the case of the harmonic frequency-modulated searching signal, we can write

$$\Psi(t) = M\cdot\sin\omega_M t \quad \text{and} \quad \Omega(t) = \Psi'(t) = \Delta\omega_M\cdot\cos\omega_M t; \tag{9.14}$$

$$\omega(t) = \omega_0 + \Delta\omega_M\cdot\cos\omega_M t; \tag{9.15}$$

$$M = \frac{\Delta\omega_M}{\omega_M}; \tag{9.16}$$

$$\Delta\tilde{\Psi}_n = \mu\omega_0\tau + 2M\cdot\sin 0.5\mu\omega_M\tau\cdot\cos[\omega_M(t - 2\rho c^{-1})]. \tag{9.17}$$

We can easily show that if the conditions $T_d = \frac{2\rho}{c} << T_M$ and $\tau_c << T_M$ are satisfied — the last condition is true at high deviation — Equation (9.17) can be rewritten in the form of Equation (9.13), where $\omega(t)$ is determined by Equation (9.15) and Ω_ρ is given by Equation (9.12), and

$$\Omega_D = 2\ V_r(\varphi,\psi)c^{-1}(\omega_0 + \Delta\omega_M\cos\omega_M t); \tag{9.18}$$

$$k_\omega(t) = -\Delta\omega_M \cdot \omega_M \sin \omega_M t. \quad (9.19)$$

If the target return signal was transformed by the product of the heterodyne signal with varying frequency, we must replace the parameter $\Delta\Psi_n$ with $\Delta\Psi_{n_0}$ in Equation (9.1)–Equation (9.3), where

$$\Delta\Psi_{n_0} = \Psi[t + 0.5(\tau - nT_M)] - \Psi[t - 0.5(\tau - nT_M)]. \quad (9.20)$$

Then, with the previously mentioned assumptions and based on the results discussed in Section 8.3, we obtain

$$\Delta\tilde{\Psi}_n = (\omega_{im} + \Omega_D)\tau - \Omega_\rho(\tau - nT_M) \quad (9.21)$$

instead of Equation (9.13). In the case of the harmonic frequency-modulated searching signal, we can find an exact formula to define $A\tilde{\Psi}_n$, method is cumbersome. Therefore, we use the approximation

$$\Delta\tilde{\Psi}_n = [\omega_{im} + \Omega_D(t)]\tau + 4M \cdot \sin 0.5\omega_M\tau \cdot \sin\frac{\omega_M\rho}{c} \cdot \sin\left[\omega_M\left(t - \tfrac{\rho}{c}\right)\right] \quad (9.22)$$

when the conditions $T_d << T_M$, $\tau_c << T_M$, and $\frac{2V_r}{c} << 1$ are true. We can also use Equation (9.21) when Ω_D is determined by Equation (9.18) and Equation (9.19), Ω_ρ by Equation (9.12), and $k_\omega(t)$ by Equation (9.18) and Equation (9.19).

9.2 The Nontransformed Target Return Signal

9.2.1 The Searching Signal with Varying Nonperiodic Frequency

In the case of the nonperiodic searching signal, we can use the formulae in Section 9.1 under the condition $n = 0$. From Equation (9.1) and Equation (9.13) it follows that the normalized correlation function of the target return signal fluctuations has the following form:

$$R_{\Delta\rho,\Delta\omega}(t,\tau) = R_g(t,\tau) \cdot R_\omega(t,\tau), \quad (9.23)$$

where

$$R_g(t,\tau) = N \iint g^2(\varphi,\psi) \cdot e^{j\Omega_D(t,\varphi,\psi)\tau} d\varphi\, d\psi \quad (9.24)$$

is the normalized correlation function of the Doppler fluctuations in the target return signal frequency;

$$R_{\omega}(t,\tau) = R_{\omega}^{\circ}(t,\tau)\cdot e^{j\omega(t)\tau}, \tag{9.25}$$

where

$$R_{\omega}^{\circ}(t,\tau) = N\int_{\rho_1}^{\rho_2}\rho^{-2}\cdot e^{-j\Omega_{\rho}(t,\rho)\tau}d\rho \tag{9.26}$$

is the normalized correlation function of the fluctuations in the target return signal frequency. The target return signal given by the normalized correlation function $R_{\Delta\rho,\Delta\omega}(t,\tau)$ is a nonstationary and nonseparable stochastic process because the variables t and τ are not separable.

Consider the Doppler fluctuations in the target return signal frequency. Comparing Equation (9.24) with Equation (3.8) and taking into consideration Equation (9.11), we can see the distinction is that the Doppler frequency Ω_D in Equation (9.24) uses the variable frequency $\omega(t)$ instead of the constant frequency ω_0. Because of this, we can use all the formulae of Chapter 3 in the definition of the normalized correlation function $R_g(t,\tau)$ if the frequency $\Omega_{\max} = \frac{2\omega_0 V}{c} = \frac{4\pi V}{\lambda}$ is replaced with

$$\Omega_{\max}(t) = 2\omega(t)Vc^{-1} = 4\pi V\lambda^{-1}(t). \tag{9.27}$$

Thus, the average Doppler frequency of the power spectral density of the fluctuations in the target return signal frequency depends on the time t. If the velocity vector of moving radar is outside the directional diagram, the bandwidth of the power spectral density of the Doppler fluctuations in the target return signal frequency is independent of the wavelength or frequency [for example, see Equation (3.76)] and so does not vary with time. If the velocity vector of moving radar is within the directional diagram, the bandwidth is inversely proportional to frequency [for example, see Equation (3.123)]. When relative variations in the target return signal frequency are low in value, i.e., if the condition $\Delta\omega_M << \omega_0$ is satisfied, we can neglect these variations.[5]

When the radar is stationary, the Doppler fluctuations in the target return signal frequency are absent, i.e., $R_g(t,\tau) \equiv 1$, and there are only the fluctuations in frequency that are characteristic of the considered case. The cofactor $e^{j\omega(t)\tau}$ in the normalized correlation function $R_{\omega}(t,\tau)$ defines the average frequency of the power spectral density of the fluctuations in the target return signal frequency as a function of the time t and leads us to the nonstationary state of the target return signal — a regular dependence on the time t. Another source of the nonstationary state of the target return signal is the dependence of the velocity of variations in frequency $k_{\omega}(t)$ on the time t.

Under the condition $k_\omega(t) = const$, the normalized correlation function $R^\circ_\omega(t, \tau)$ of the fluctuations in the target return signal frequency is independent of the time t. This means that the shape and bandwidth of the power spectral density of the fluctuations in frequency do not vary with time in spite of the fact that the frequency of the power spectral density is variable.[6]

Reference to Equation (9.26) shows that the instantaneous power spectral density of the fluctuations in frequency takes the following form:

$$S_\omega(\omega, t) \cong \int_{\rho_1}^{\rho_2} \rho^{-2} \cdot \delta\left[\omega(t) - k_\omega \cdot \tfrac{2\rho}{c} - \omega\right] d\rho \cong \frac{k_\omega}{[\omega - \omega(t)]^2}\,, \tag{9.28}$$

where

$$\omega_1(t) \le \omega \le \omega_2(t) \qquad \text{and} \qquad \omega_{1,2}(t) = \omega_0 + k_\omega\left(t - \tfrac{2\rho_{1,2}}{c}\right) = \omega(t) - k_\omega \cdot \tfrac{2\rho_{1,2}}{c}\,. \tag{9.29}$$

The power spectral density given by Equation (9.29) is shown in Figure 9.1 and Figure 9.2. At the boundary frequencies, the power spectral density $S_\omega(\omega_{1,2}) \cong \rho_{1,2}^{-2}$ and shape are independent of the time t. The bandwidth of the power spectral density does not vary with time, too, if the condition $k_\omega(t) = const$ is satisfied:

$$\Delta\Omega_\rho = \omega_2(t) - \omega_1(t) = 2k_\omega(\rho_2 - \rho_1)c^{-1} = k_\omega \Delta T_d\,. \tag{9.30}$$

The average frequency of the power spectral density depends on time:

$$\omega_{av} = 0.5[\omega_1(t) + \omega_2(t)] = \omega(t) - 2k_\omega\rho_0 c^{-1} = \omega(t) - \Omega_{\rho_0}\,, \tag{9.31}$$

where

$$\Delta T_d = 2(\rho_2 - \rho_1)c^{-1}; \tag{9.32}$$

$$\rho_0 = 0.5(\rho_1 + \rho_2); \tag{9.33}$$

$$\Omega_{\rho_0} = 2k_\omega \rho c^{-1}. \tag{9.34}$$

If $\rho_1 < \rho_2$, then $S(\omega_1) > S(\omega_2)$. At $k_\omega > 0$, we have $\omega_1 > \omega_2$ and at $k_\omega < 0$, we get $\omega_1 < \omega_2$ (see Figure 9.1 and Figure 9.2).

The correlation lengths in time τ_c and in frequency $\Delta\omega_c$ have the following form:

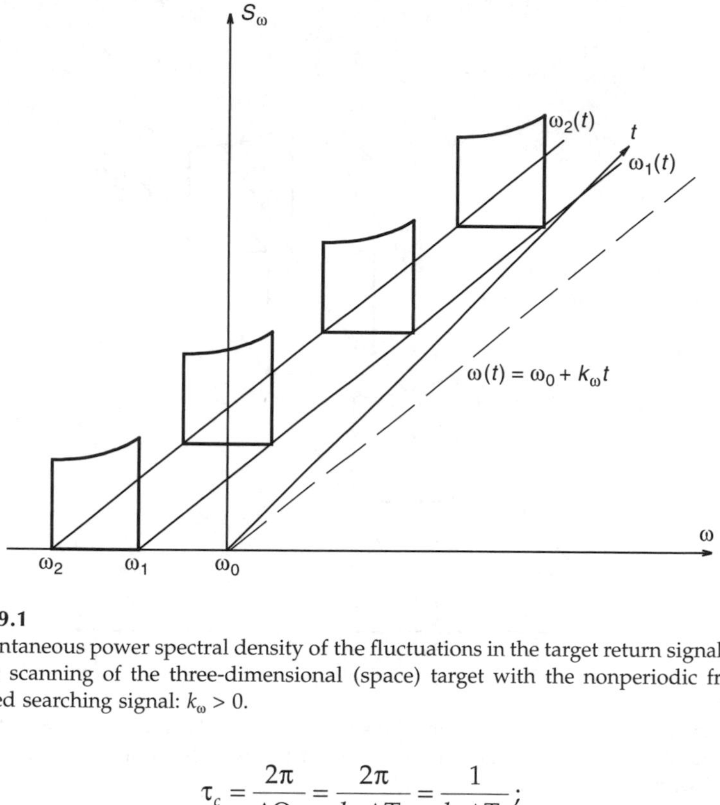

FIGURE 9.1
The instantaneous power spectral density of the fluctuations in the target return signal frequency under scanning of the three-dimensional (space) target with the nonperiodic frequency-modulated searching signal: $k_\omega > 0$.

$$\tau_c = \frac{2\pi}{\Delta\Omega_\rho} = \frac{2\pi}{k_\omega \Delta T_d} = \frac{1}{k_f \Delta T_d}; \tag{9.35}$$

$$\Delta\omega_c = k_\omega \tau_c = 2\pi(\Delta T_d)^{-1} \quad \text{or} \quad \Delta f_c = (\Delta T_d)^{-1}. \tag{9.36}$$

Thus, the greater the length $\rho_1 - \rho_2$ of the three-dimensional (space) target, the less is the correlation length in time and frequency. Thus, the correlation length in frequency is independent of k_ω and is defined by the value of ΔT_d as a whole. However, it is possible to draw an incorrect conclusion based on Equation (9.35) and Equation (9.36), i.e., as $\rho \to \infty$ the values of τ_c and $\Delta\omega_c$ tend to approach zero. This can be explained by an insufficiently rigorous definition of the bandwidth of the power spectral density of the fluctuations in frequency, given by Equation (9.30) as a difference of the boundary frequencies. For a rigorous estimation it is necessary to introduce an effective bandwidth of the power spectral density.[7]

Here, it is appropriate to recall that τ_c, which is inversely proportional to the bandwidth of the power spectral density, when multiplied by 0.5τ corresponds to the radar range resolution that can be obtained with the searching signal having the same bandwidth. Reference to Equation (9.35) shows that under the condition $k_f > \Delta T_d^{-2}$, we get $\tau_c < \Delta T_d$, i.e., the radar range resolution is less than the length of the three-dimensional (space) target.

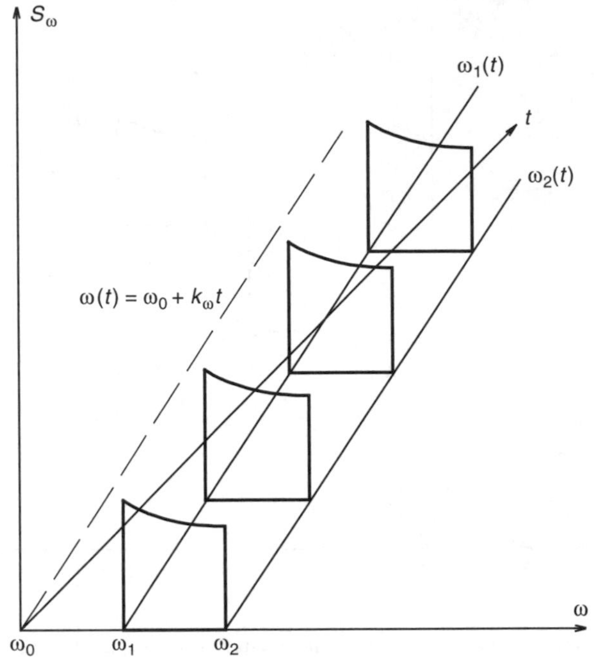

FIGURE 9.2
The instantaneous power spectral density of the fluctuations in the target return signal frequency under scanning of the three-dimensional (space) target with the nonperiodic frequency-modulated searching signal: $k_\omega < 0$.

In the case of stationary radar, the fluctuations in the target return signal frequency are completely defined by the normalized correlation function given by Equation (9.25) and Equation (9.26) and the power spectral density determined by Equation (9.28). When the radar is moving, it is necessary to define a convolution between the power spectral density given by Equation (9.28) and the power spectral density of the Doppler fluctuations in the target return signal frequency with the purpose of defining the total power spectral density. This leads to a shift in the average frequency of the instantaneous power spectral density of the fluctuations in frequency by the average Doppler frequency $\Omega_{D_0}(t) = \frac{2V_{r_0}\omega(t)}{c}$ and to an increase in the bandwidth of the power spectral density of the Doppler fluctuations in frequency. If at some instant of time we have

$$\omega(t) = k_\omega \rho_0 V_{r_0}^{-1}, \tag{9.37}$$

the total shift in frequency is determined by $\Omega_{D_0} - \Omega_{\rho_0} = 0$. The total power spectral density of the Doppler fluctuations and the fluctuations in the target return signal frequency obtained as a result of convolution of the power spectral density of the Doppler fluctuations and the power spectral density of the fluctuations in the target return signal frequency is always wider than the widest of that. The total power spectral density compression of the fluctuations in the target return signal frequency is absent, as in the case of scanning the two-dimensional (surface) target. This can be explained by the independence between the variable ρ and the coordinates φ and ψ. In the case of the fixed Doppler frequency, i.e., with fixed coordinates φ and ψ, there is a scattering over ρ — the power spectral density of the fluctuations in the range-finder frequency is formed and vice versa, i.e., with ρ fixed, there is scattering at the coordinates φ and ψ — and the power spectral density of the Doppler fluctuations in frequency of the target return signal is formed.[8]

9.2.2 The Periodic Frequency-Modulated Searching Signal

In this case, Equation (9.23) and Equation (9.24) are true, as before, but instead of Equation (9.25) and Equation (9.26) we can write the normalized correlation function of the fluctuations in frequency in the following form:

$$R_\omega(t,\tau) = N\sum_{n=-\infty}^{\infty}\int_{\rho_1}^{\rho_2}\rho^{-2}\cdot e^{j\left[\omega(t)-k_\omega\cdot\frac{2\rho}{c}\right](\tau-nT_M)}d\rho\,. \tag{9.38}$$

The normalized correlation function $R_\omega(t, \tau)$ given by Equation (9.38) is a periodic function of τ as opposed to the normalized correlation function $R^{\circ}_{\omega}(t, \tau)$ given by Equation (9.25) and Equation (9.26). The normalized correlation function $R_\omega(t, \tau)$ given by Equation (9.38) is shown in Figure 9.3 with the assumption that changes in the target return signal power are low in value within the limits of the interval $[\rho_1, \rho_2]$ so that we can replace the value ρ^{-2} with the value ρ_0^2. Because of this, the instantaneous power spectral density of the fluctuations in frequency is linear, with the distance between harmonics equal to ω_M. As before, the average frequency of the instantaneous power spectral density is shifted by the range-finder frequency $\Omega_{\rho_0} = \frac{2\rho_0 k_\omega}{c}$ with respect to the frequency $\omega(t)$,which is a periodic function of the time t. As a rule, if the conditions $\frac{2\rho_0}{c} = T_{d_0} \le T'_M$, $\frac{2\rho_0}{c} = T_{d_0} \le T''_M$, and $|k'_\omega T'_M| = |k''_\omega T''_M| = \Delta\omega_M$ are satisfied, then the condition $\Omega_{\rho_0} \ll \Delta\omega_M$ is true. The power spectral density bandwidth is determined by Equation (9.30), as before. The target return signal is a periodic nonstationary stochastic process.

However, periodic frequency modulation is impossible without periodic variations in the velocity of changes in frequency k_ω. The bandwidth and shape of the power spectral density periodically vary together with variations in the value of k_ω. The instantaneous power spectral density in the case of the saw-tooth asymmetric linear frequency-modulated searching signal is

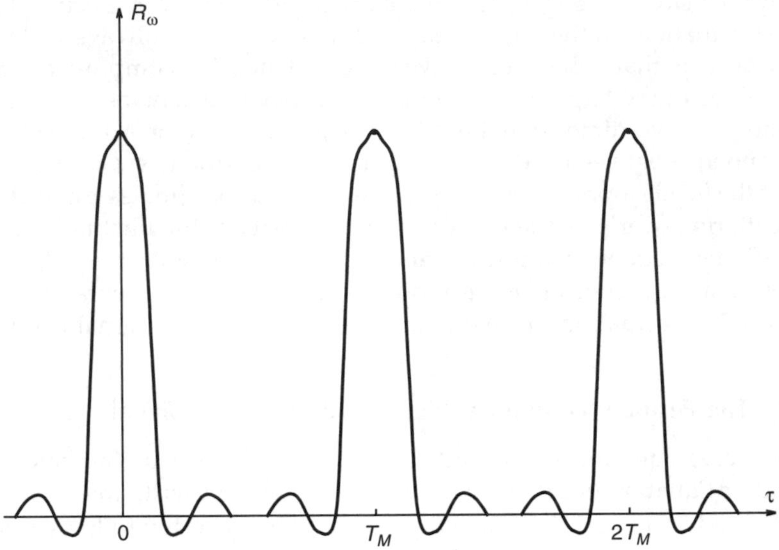

FIGURE 9.3
The correlation function of the fluctuations in the target return signal frequency under scanning of the three-dimensional (space) target with the periodic frequency-modulated searching signal when the radar is stationary.

shown in Figure 9.5. The bandwidth and shape of the power spectral density are constant within linear intervals. When the sign of the parameter k_ω is changed, we get a jump-like shift in the power spectral density with respect to the frequency axis.[9] Based on Figure 9.5 we can construct the shape of the instantaneous power spectral density both in the cases of the symmetric and one-sided saw-tooth linear frequency-modulated searching signal and also in the case of the harmonic frequency-modulated searching signal (see Figure 9.6), if the frequency $\omega(t)$ is determined by Equation (9.16) and the velocity of variations in frequency k_ω is given by Equation (9.18) and Equation (9.19). The transition processes with duration ΔT_d arise in time T_d after the jump-like variation of the velocity of changes in frequency k_ω, which is accompanied by jump-like variations in shape and in the average frequency of the power spectral density. These processes are not investigated here because we assume that the condition $\Delta T_d << T_M$ is satisfied.

Due to the periodic behavior of the fluctuations in the target return signal frequency, we can define the intra- and interperiod fluctuations as in the case of the pulsed searching signal (see Chapter 3). Under the condition $\tau = nT_M$, the normalized correlation function $R_\omega(t, \tau)$ is equal to one, i.e., the fluctuations caused by the frequency-modulated searching signal are absent, and the interperiod fluctuations in frequency are reduced to the Doppler fluctuations in frequency, as in the case without frequency modulation of the searching signal. The normalized correlation function $R_g(t, \tau)$ of the Doppler fluctuations plays the same role of envelope with respect to the normalized correlation function $R_\omega(t, \tau)$ (see Figure 9.4) as in the case of the pulsed

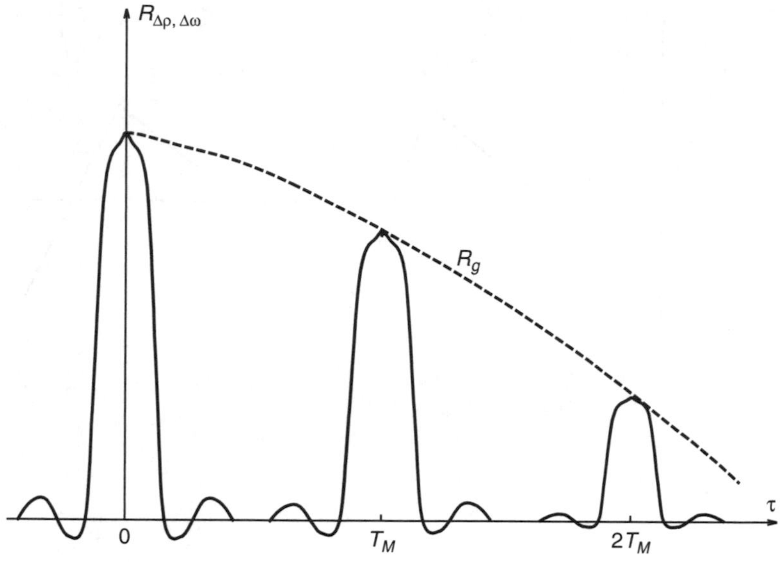

FIGURE 9.4
The correlation function of the fluctuations in the target return signal frequency under scanning of the three-dimensional (space) target with the periodic frequency-modulated searching signal when the radar is moving.

searching signal. The interperiod Doppler fluctuations can be observed as the fluctuations of the envelope both at the instants of time fixed with respect to the origin of the modulation period (as in the case of the pulsed searching signal) and at the output of the comb filter, the central frequency of which is within the limits of the interval $[\omega_1(t), \omega_2(t)]$.

Under the condition $n = 0$, the normalized correlation function $R_\omega(t, \tau)$ given by Equation (9.38) defines the intraperiod fluctuations. Thus, the normalized correlation function $R_\omega(t, \tau)$ coincides with that given by Equation (9.25) and Equation (9.26). The power spectral density of the intraperiod fluctuations is matched with the power spectral density given by Equation (9.28). The intraperiod fluctuations are similar to the previously mentioned fluctuations in the target return signal frequency when the frequency of the searching signal is not periodic. The correlation length of the intraperiod fluctuations given by Equation (9.35) depends on the parameter $k_f(\rho_1 - \rho_1)$. Intraperiod fluctuations are analogous to the rapid fluctuations in the case of the pulsed searching signal, but a distinction is that their power does not vary within the modulation period and the intraperiod fluctuations cannot be caused by propagation of the pulsed searching signal along the three-dimensional (space) or two-dimensional (surface) target — the updated set of scatterers in the pulse volume — because at each instant of the time, the target return signal contains elementary signals from all scatterers within the limits of the interval $[\rho_1, \rho_2]$. In this case, the intraperiod fluctuations are caused by an updated set of target return signal frequencies, and there is an analogy with the stochastic process observed at the output of the filter having

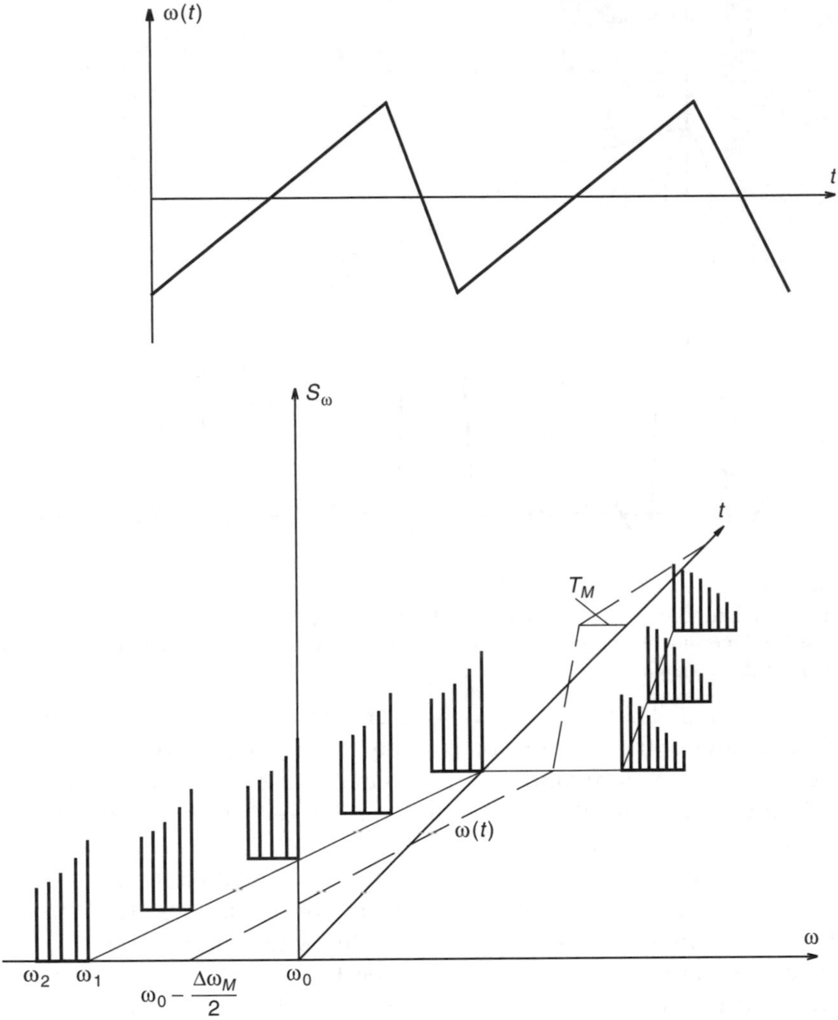

FIGURE 9.5
The instantaneous power spectral density of the fluctuations in the target return signal frequency under scanning of the three-dimensional (space) target with the asymmetric linear frequency-modulated searching signal when the radar is stationary.

the frequency response given by Equation (9.28) if the process at the filter input is white Gaussian noise.[10]

In the case of moving radar, the instantaneous power spectral density is a convolution of the power spectral densities of the Doppler fluctuations and the fluctuations in the target return signal frequency. As a result, each harmonic of the regular power spectral density $S_\omega(\omega, t)$ has the power spectral density $S_g(\omega, t)$ and the total power spectral density shifted by the average Doppler frequency

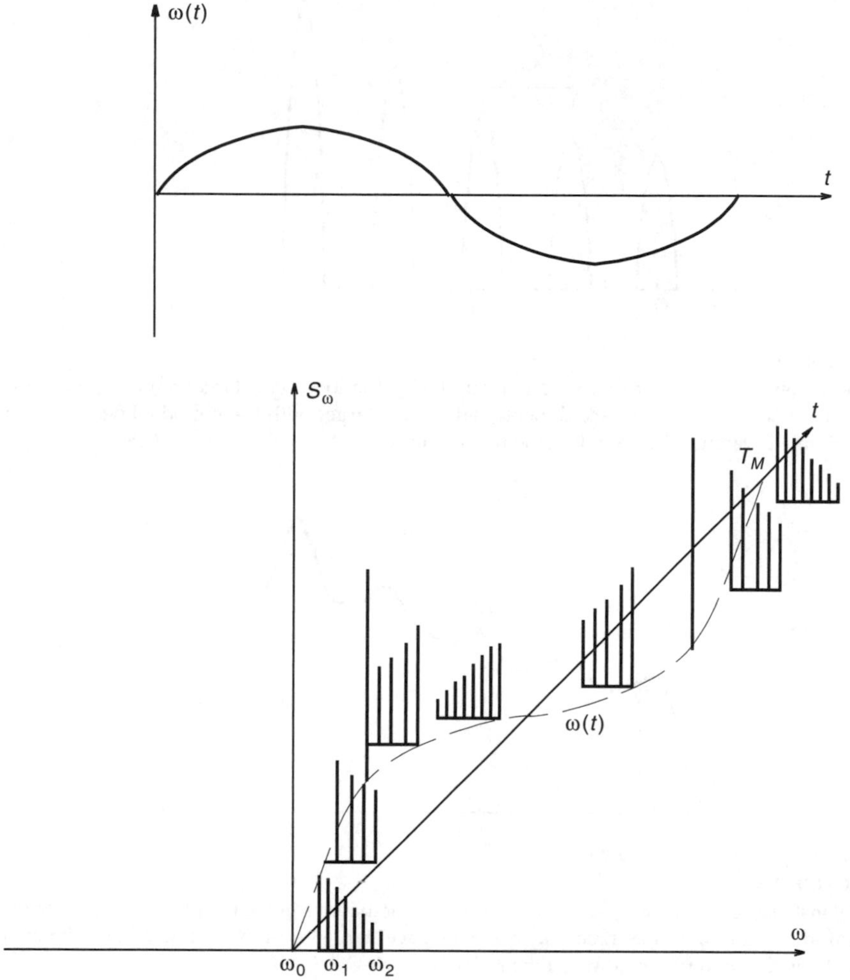

FIGURE 9.6
The instantaneous power spectral density of the fluctuations in the target return signal frequency under scanning of the three-dimensional (space) target with the harmonic frequency-modulated searching signal when the radar is stationary.

$$\Omega_{D_0} = 2\omega(t)V_{r_0}c^{-1} \; . \tag{9.39}$$

If the bandwidth of the power spectral density of the Doppler fluctuations is less than the modulation frequency, i.e., if the condition $\Delta F_D < f_M$ is satisfied, the instantaneous power spectral density becomes a comb, with clearly detected teeth and with envelope $S_\omega(\omega, t)$ (see Figure 9.7). Under the condition $\Delta F_D > f_M$, the power spectral density of the fluctuations in the target return signal frequency becomes continuous (see Figure 9.8).

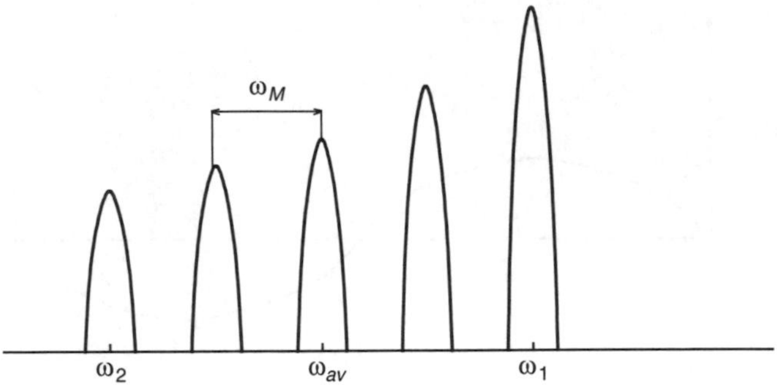

FIGURE 9.7
The instantaneous power spectral density of the fluctuations in the target return signal frequency under scanning of the three-dimensional (space) target with the periodic linear frequency-modulated searching signal when the radar is moving: $\Delta F_D < f_M$.

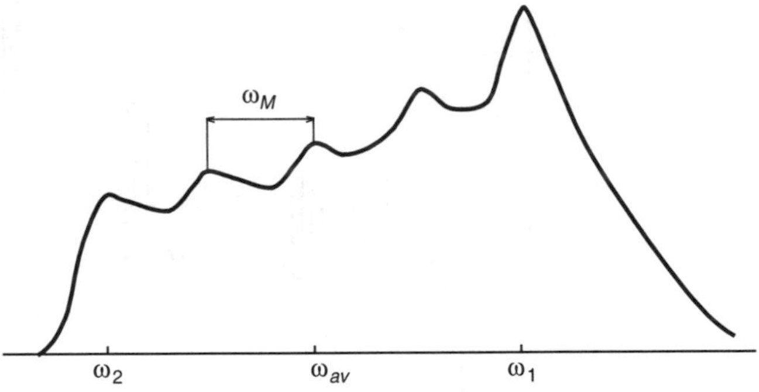

FIGURE 9.8
The instantaneous power spectral density of the fluctuations in the target return signal frequency under scanning of the three-dimensional (space) target with the periodic linear frequency-modulated searching signal when the radar is moving: $\Delta F_D > f_M$.

9.2.3 The Average Power Spectral Density with the Periodic Frequency-Modulated Searching Signal

We can average the correlation function of the fluctuations in the target return signal frequency using the Fourier-series expansion for the target return signals and multiplying them, and averaging over time without determining the instantaneous correlation function (see Chapter 8). With this approach, we unfortunately lose very important information regarding the instantaneous correlation function and power spectral density. In this section, we define the average normalized correlation function of the fluctuations in the target return signal frequency by averaging over time the instantaneous correlation function.

Let us consider the case of an asymmetric linear frequency-modulated searching signal and assume that k'_ω and T'_M are the velocity of frequency variations and duration of the modulation function front, and that k''_ω and T''_M are the velocity of frequency variations and duration of the modulation function trailing edge. Integrating the total normalized correlation function $R_{\Delta\rho,\Delta\omega}(t,\tau)$, which is defined by the product of $R_g(t,\tau)$ given by Equation (9.24) and $R_\omega(t,\tau)$ given by Equation (9.38), with respect to the variable t within the limits of the period $T_M = T'_M + T''_M$, we obtain the average correlation function of the fluctuations in the target return signal frequency.[11]

In the case of stationary radar $R_g(t,\tau) \equiv 1$, we average only the cofactor $e^{j\omega(t)\tau}$ in Equation (9.38):

$$\overline{R_\omega(t,\tau)} = N \cdot e^{j\omega_0\tau} \sum_{n=-\infty}^{\infty} \left[\frac{T'_M}{T_M} \cdot R'_\omega(\tau - nT_M) + \frac{T''_M}{T_M} \cdot R''_\omega(\tau - nT_M) \right], \tag{9.40}$$

where

$$\left.\begin{aligned} R'_\omega(\tau) &= \mathbf{sinc}\,\frac{\Delta\omega_M \tau}{2} \cdot \int_{\rho_1}^{\rho_2} \rho^{-2} \cdot e^{-jk'_\omega \cdot \frac{2\rho}{c}\tau} d\rho \quad \text{and} \\ R''_\omega(\tau) &= \mathbf{sinc}\,\frac{\Delta\omega_M \tau}{2} \cdot \int_{\rho_1}^{\rho_2} \rho^{-2} \cdot e^{-jk''_\omega \cdot \frac{2\rho}{c}\tau} d\rho \end{aligned}\right\} ; \tag{9.41}$$

$$\Delta\omega_M = |k'_\omega T'_M| = |k''_\omega T''_M| ; \tag{9.42}$$

$$T_M = T'_M + T''_M . \tag{9.43}$$

Thus, in the case of stationary radar, $\overline{R_\omega(t,\tau)}$ is periodic. The average power spectral density is regular. An envelope of the average power spectral density is obtained by convolution of the power spectral density $S_\omega(\omega, t)$ given by Equation (9.28) with the bandwidth given by Equation (9.30) and the rectangle power spectral density with the bandwidth equal to $\Delta\omega_M$.

In the case of the two-sided linear frequency-modulated searching signal, the power spectral density consists of two overlapping regions that are within the limits of the intervals T'_M and T''_M . Their average frequencies are shifted to the right or left with respect to the frequency ω_0:

$$\Omega'_{\rho_0} = 2k'_\omega \rho c^{-1} \tag{9.44}$$

and

$$\Omega''_{\rho_0} = 2k''_\omega \rho c^{-1} , \tag{9.45}$$

where k'_ω and k''_ω have different signs. As the condition $\Delta\omega_M >> \Delta\Omega_\rho$ is satisfied [see Equation (9.30)], an envelope of the power spectral density is very close to the shape of a trapezium with edges having width $\Delta\omega_M$ and length $\Delta\Omega_\rho$. In the case of the one-sided linear frequency-modulated searching signal, there is only a single region corresponding to $R'_\omega(\tau)$ or $R''_\omega(\tau)$.

If the radar is moving, we can assume that the bandwidth of the power spectral density of the Doppler fluctuations is independent of the time t. Then, the average power spectral density is shifted by the average Doppler frequency and the harmonics spaced at frequencies $\omega_0 + n\omega_M$ are subjected to convolution with the power spectral density of the Doppler fluctuations, and the regular power spectral density is transformed to the comb power spectral density.

Let us consider the case of the harmonic frequency-modulated searching signal. Substituting Equation (9.17) in Equation (9.1) and averaging over time t within the limits of the period, the average normalized correlation function of the Doppler fluctuations and the fluctuations in the target return signal frequency can be written in the following form:

$$\overline{R_{\Delta\rho,\Delta\omega}(t,\tau)} = N\iint g^2(\varphi,\psi)\cdot J_0[2M\cdot\sin(0.5\mu\omega_M\tau)]\cdot e^{j\mu\omega_0\tau}d\varphi\, d\psi. \quad (9.46)$$

Using the well-known expansion

$$J_0(z\cdot\sin\alpha) = J_0^2(0.5z) + 2\sum_{n=1}^{\infty} J_n^2(0.5z)\cdot\cos 2n\alpha \quad (9.47)$$

and taking into consideration that

$$\cos x = 0.5(e^{jx} + e^{-jx}), \quad (9.48)$$

we can write

$$\overline{R_{\Delta\rho,\Delta\omega}(t,\tau)} = N\sum_{n=-\infty}^{\infty} J_n^2(M)\cdot R_{g_n}(\tau), \quad (9.49)$$

where

$$R_{g_n}(\tau) = e^{j(\omega_0+n\omega_M)\tau}\iint g^2(\varphi,\psi)\cdot e^{j(\omega_0+n\omega_M)\tau\frac{2V_r(\varphi,\psi)}{c}}d\varphi\, d\psi \quad (9.50)$$

is the normalized correlation function of the Doppler fluctuations for the case of the continuous searching signal [see Equation (3.8)] at the frequency $\omega = \omega_0 + n\omega_M$.

Then, the average power spectral density can be determined as follows:

$$\overline{S(\omega,t)} = \sum_{n=\infty-\infty}^{\infty} J_n^2(M)\cdot S_n(\omega-\omega_0-n\omega_M), \tag{9.51}$$

where $S_n(\omega - \omega_0 - n\omega_M)$ is the power spectral density in the case of moving radar and the simple harmonic searching signal with the frequency $\omega_0 + n\omega_M$, discussed in more detail in Section 3.2 and Section 3.3. In the case of stationary radar, the average power spectral density given by Equation (9.51) is regular, as in the case of a point target, as geometric characteristics of the three-dimensional (space) target are absent in Equation (9.51) (see Figure 9.9):

$$\overline{S(\omega,t)} = \sum_{n=-\infty}^{\infty} J_n^2(M)\cdot \delta(\omega-\omega_0-n\omega_M)\,. \tag{9.52}$$

If the radar is moving, each harmonic of the average power spectral density $\overline{S(\omega,t)}$ given by Equation (9.52) has a corresponding Doppler shift. The bandwidth and shape of the power spectral density of the Doppler fluctuations depend on the shape and position of the directional diagram with respect to the velocity vector of moving radar.

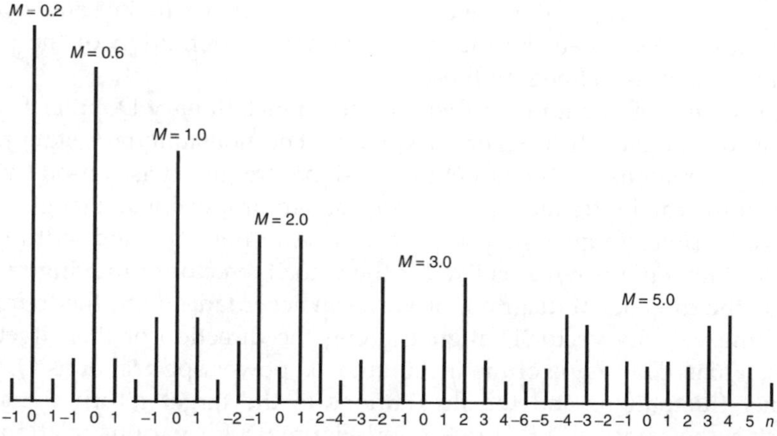

FIGURE 9.9
The average power spectral density of the fluctuations in the target return signal frequency under scanning of the three-dimensional (space) target with the harmonic frequency-modulated searching signal with various modulation indexes when the radar is stationary.

9.3 The Transformed Target Return Signal

9.3.1 Nonperiodic and Periodic Frequency-Modulated Searching Signals

In the case of the transformed target return signal, Equation (9.23) is true, in which $R_g(t, \tau)$ as given by Equation (9.24) is the same, and the normalized correlation function $R_\omega(t, \tau)$ given by Equation (9.25) and Equation (9.26) is modified in the following manner: If the searching signal is nonperiodic we replace the frequency $\omega(t)$ with the parameter ω_{im}, and if the searching signal is periodic we use Equation (9.38) instead of Equation (9.25) and Equation (9.26), into which we replace the frequency $\omega(t)$ with the frequency ω_{im}, too.

The main distinction between periodic and nonperiodic searching signals is the following: As long as the velocity of variations in frequency k_ω does not vary with time, i.e., as long as $k_\omega = const$, the average frequency of the power spectral density does not vary in time, too, if the radar is stationary. When the radar is moving, the average frequency of the power spectral density varies moderately, as does the average Doppler frequency Ω_{D0}. The nonstationary state of the fluctuations in target return signal frequency is caused only by changes in the value and sign of k_ω within the period. In the case of the one-sided saw-tooth linear frequency-modulated searching signal, the instantaneous power spectral density is independent of time within the main part of the period if the condition $k_\omega = const$ is satisfied (see Figure 9.10).

In the case of the two-sided saw-tooth linear frequency-modulated searching signal, the value and sign of k_ω and the range-finder frequency Ω_{P_0} exhibit a jump-like variation within the period. The bandwidth and average frequency $\omega_{av} = \omega_{im} - \Omega_{\mathrm{P}_0}$ of the power spectral density vary in the same manner (see Figure 9.11). When the radar is stationary, a definition of the power spectral density is defined by these data.

In the case of moving radar, there are the nonstationary Doppler fluctuations in the target return signal frequency. The nonstationary state of the Doppler fluctuations in the target return signal frequency is caused by variations of the carrier frequency $\omega(\tau)$ and leads to proportional changes in the average Doppler frequency. The power spectral density bandwidth of the Doppler fluctuations remains fixed if the velocity vector of moving radar is outside the directional diagram, or varies in accordance with the frequency $\omega(t)$ if the velocity vector is aligned along the direction of the directional diagram axis. The shape of the instantaneous power spectral density of the Doppler fluctuations and the fluctuations in the target return signal frequency is the same as in Figure 9.7 and Figure 9.8 for various relationships between the power spectral density bandwidth of the Doppler fluctuations and the modulation frequency ω_M.

In the case of the harmonic frequency-modulated signal, the normalized correlation function $R_{\Delta\rho,\Delta\omega}(t, \tau)$ is approximately determined by Equation (9.23)–Equation (9.26) under the conditions $T_d << T_M$ and $\tau_c << T_M$ when we

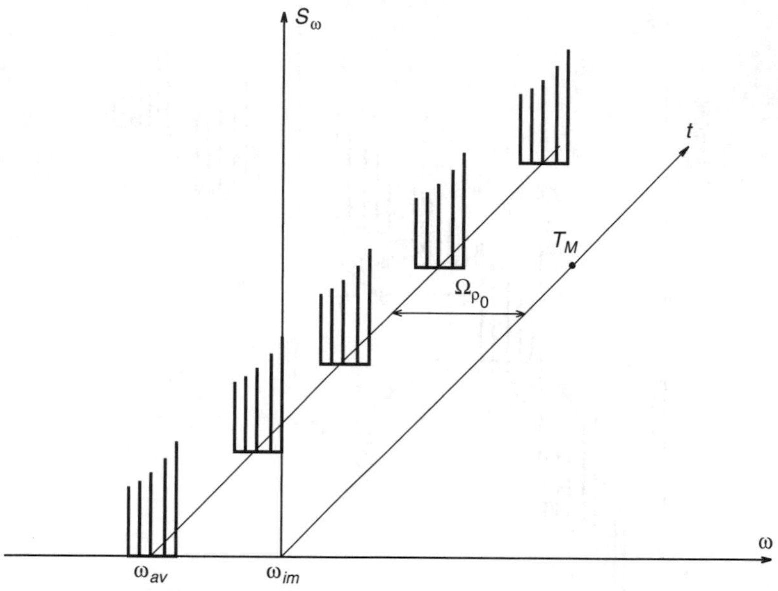

FIGURE 9.10
The instantaneous power spectral density of the fluctuations in the transformed target return signal frequency with the one-sided linear frequency-modulated searching signal when the radar is stationary.

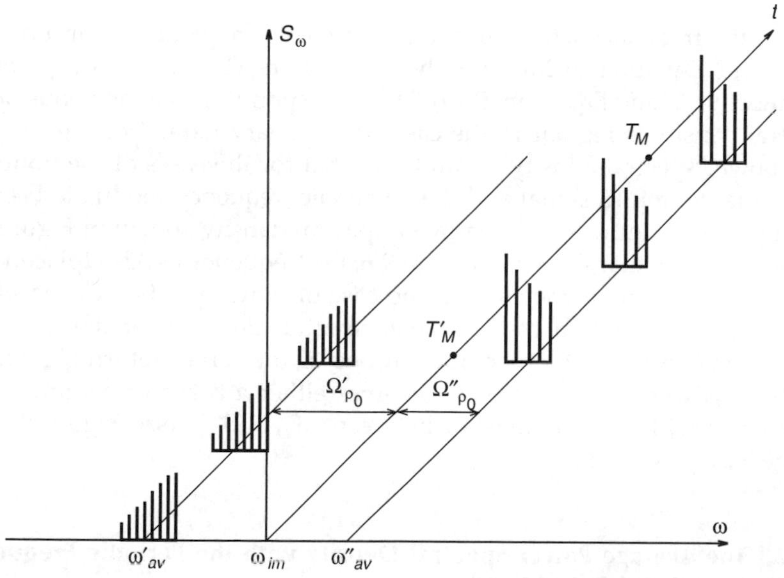

FIGURE 9.11
The instantaneous power spectral density of the fluctuations in the transformed target return signal frequency with the two-sided linear frequency-modulated searching signal when the radar is stationary.

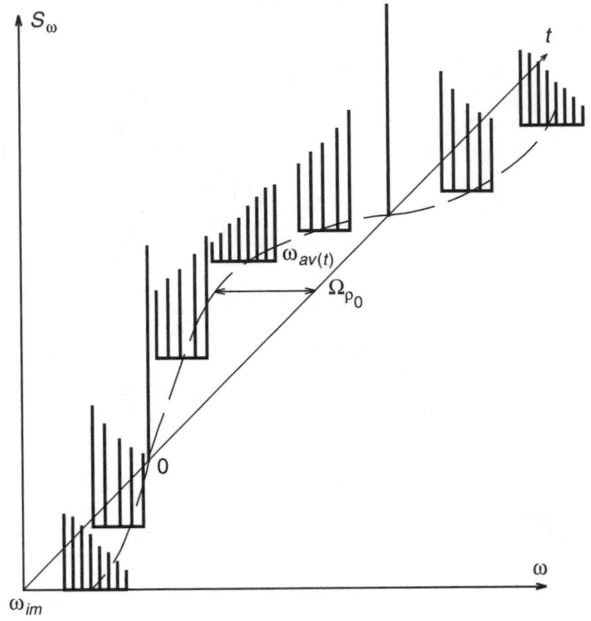

FIGURE 9.12
Instantaneous power spectral density of the fluctuations in the transformed target return signal frequency with the harmonic frequency-modulated searching signal when the radar is stationary.

replace the frequency $\omega(t)$ with the intermediate frequency ω_{im} in Equation (9.25) and Equation (9.26), and the parameters Ω_D and k_ω are given by Equation (9.18) and Equation (9.19). The corresponding instantaneous power spectral density is regular in the case of stationary radar (see Figure 9.12). This power spectral density is similar to that for the cases of the nontransformed target return signal and the harmonic frequency-modulated searching signal (see Figure 9.6). The power spectral density shown in Figure 9.12 is different from that shown in Figure 9.6; the frequency $\omega(t)$ is replaced with the intermediate frequency ω_{im}. In the case of moving radar, it is necessary to define a convolution of the power spectral densities of the frequency fluctuations and the Doppler fluctuations of the target return signal. The resulting power spectral density becomes either a comb or continuous, in accordance with the relationship between ΔF_D and f_M (see Figure 9.7 and Figure 9.8).

9.3.2 The Average Power Spectral Density with the Periodic Frequency-Modulated Searching Signal

As was shown in Section 9.3.1, when the radar is stationary, i.e., when the condition k_ω = *const* is satisfied, and with the linear frequency-modulated searching signal, the instantaneous power spectral density of the fluctuations

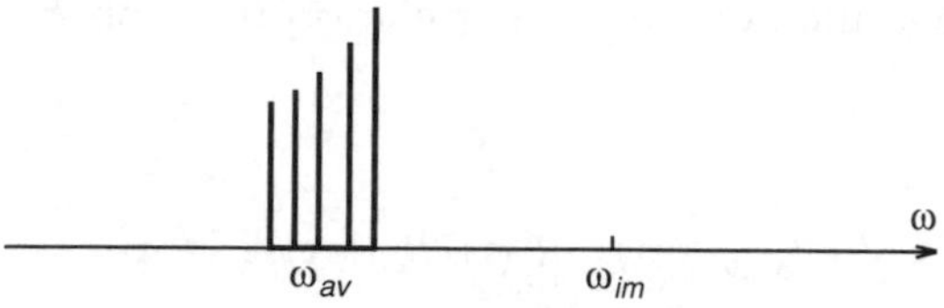

FIGURE 9.13
The average power spectral density of the fluctuations in the transformed target return signal frequency with the one-sided linear frequency-modulated searching signal when the radar is stationary.

in the target return signal frequency is independent of the time t. For this reason, with the one-sided linear frequency-modulated searching signal, the power spectral density averaged over the modulation period coincides with the instantaneous power spectral density and remains regular (see Figure 9.13). In the case of the two-sided linear frequency-modulated searching signal, the average power spectral density (regular, too) is defined by a weighted sum of the instantaneous power spectral densities $S'_\omega(\omega)$ and $S''_\omega(\omega)$ (see Figure 9.14):

$$S_\omega(\omega,t) = \frac{T'_M}{T_M}\cdot S'_\omega(\omega) + \frac{T''_M}{T_M}\cdot S''_\omega(\omega)\,, \tag{9.53}$$

where T_M is given by Equation (9.43).

If the radar is moving and the regular power spectral density is transformed to a comb, as shown in Figure 9.7 and Figure 9.8, the average frequency of the power spectral density becomes a weak periodic function in time due to the Doppler shift in frequency. An envelope of the power spectral density averaged over the modulation period is additionally fuzzified by the value of $\frac{2V_{r_0}\Delta\omega_M}{c}$. If $\frac{2V_{r_0}\Delta\omega_M}{c}$ is considerably less than the power spectral density bandwidth of the Doppler fluctuations in the target return signal frequency, we can neglect it.

Define the average power spectral density in the case of the harmonic frequency-modulated searching signal. For this purpose, we use Equation

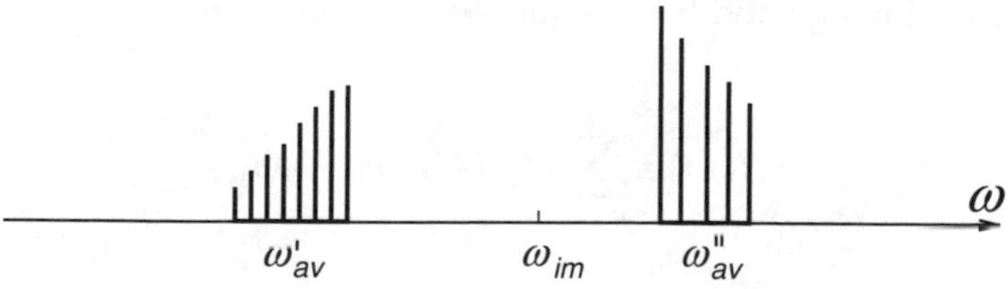

FIGURE 9.14
The average power spectral density of the fluctuations in the transformed target return signal frequency with the two-sided linear frequency-modulated searching signal when the radar is stationary.

(9.22) and average the normalized correlation function $R_{\Delta\rho,\Delta\omega}(t, \tau)$ over the time t:

$$\overline{R_{\Delta\rho,\Delta\omega}(t,\tau)} = R_g(\tau) \cdot \overline{R_\omega^\circ(t,\tau)} \cdot e^{j\omega_{im}\tau} \text{ p}, \tag{9.54}$$

where

$$R_g(\tau) = N \iint g^2(\varphi,\psi) \cdot e^{-2j\omega_0\tau \frac{V_r(\varphi,\psi)}{c}} \, d\varphi \, d\psi \, ; \tag{9.55}$$

$$\overline{R_\omega^\circ(t,\tau)} = N \int_{\rho_1}^{\rho_2} J_0\left[4M \cdot \sin\tfrac{\omega_M \rho}{c} \sin(0.5\omega_M \tau)\right] \cdot \rho^{-2} d\rho \, . \tag{9.56}$$

In this approximation, the Doppler fluctuations and the fluctuations in the target return signal frequency are separable. In the case of stationary radar, the normalized correlation function $\overline{R_\omega^\circ(t,\tau)}$ given by Equation (9.56) defines exactly and completely the fluctuations in the target return signal frequency. Using Equation (9.47), we can write

$$\overline{R_\omega^\circ(t,\tau)} = \sum_{n=-\infty}^{\infty} N_n(\rho_1,\rho_2) \cdot e^{jn\omega_M\tau} \, , \tag{9.57}$$

where

$$N_n = \int_{\rho_1}^{\rho_2} J_n^2\left[2M \cdot \sin\tfrac{\omega_M \rho}{c}\right] \cdot \rho^{-2} d\rho \, . \tag{9.58}$$

Thus, the average power spectral density is regular (the distance between harmonics being ω_M) and has the following form:

$$\overline{S_\omega(\omega,t)} \cong \sum_{n=-\infty}^{\infty} N_n \cdot \delta(\omega - \omega_0 - n\omega_M) \, . \tag{9.59}$$

The amplitudes of harmonics depend on the modulation index M and the interval $[\rho_1, \rho_2]$. Note that amplitudes of harmonics change with n as the distance between the radar and the three-dimensional (space) target varies.

We can define this function depending on the determination of the coefficients N. For instance, if the condition $\frac{2\rho_0}{c} << T_M$ is true, we can write

$$N_n \cong \int_{\rho_1}^{\rho_2} J_n^2\left[\Delta\omega_M \cdot \tfrac{2\rho}{c}\right] \cdot \rho^{-2} d\rho \,. \tag{9.60}$$

In the case of moving radar, the resulting power spectral density is defined by convolution of the regular power spectral density $\overline{S_\omega(\omega, t)}$ given by Equation (9.60) and the power spectral density of the Doppler fluctuations. Each harmonic is transformed into the power spectral density of the Doppler fluctuations in frequency of the target return signal corresponding to the normalized correlation function $R_g(t, \tau)$.[12]

9.4 Conclusions

The average Doppler frequency of the power spectral density of the fluctuations in the target return signal frequency is a function of time. If the velocity vector of moving radar is outside the directional diagram, the power spectral density bandwidth of the Doppler fluctuations is independent of wavelength or frequency and, for this reason, is not a function of time. If the velocity vector of moving radar is within the directional diagram, the power spectral density bandwidth of the Doppler fluctuations is inversely proportional to frequency. As relative variations in frequency are very low in value, i.e., as the condition $\Delta\omega_M << \omega_0$ is satisfied, we can neglect frequency variations.

In the case of stationary radar, the Doppler fluctuations are absent and there are only the fluctuations in the target return signal frequency that are characteristic of the considered case. The cofactor $e^{j\omega(t)\tau}$ in the normalized correlation function $R_\omega(t, \tau)$ defines the average frequency of the power spectral density as a function of time and leads to nonstationary target return signals — a regular dependence on time. Another source of the nonstationary state of target return signals is the dependence of the velocity of variations in frequency k_ω on time. If the parameter k_ω does not vary with time, the normalized correlation function $R_\omega^\circ(t, \tau)$ is independent of time. This means that the shape and bandwidth of the power spectral density are independent of time in spite of the fact that the frequency of the power spectral density is floating.

The total power spectral density of the Doppler fluctuations and the fluctuations in the target return signal frequency obtained as a result of convolution of the power spectral density of the Doppler fluctuations and the power spectral density of the fluctuations in the target return signal frequency is always wider than the widest of that. The power spectral density compression of the fluctuations in the target return signal frequency is absent, as in the case of scanning the two-dimensional (surface) target. This can be explained by independence between the variable ρ and the coordinates φ and ψ. With a fixed Doppler frequency (the fixed coordinates φ and ψ), there is scattering over the radar range ρ, i.e., the power spectral density of the fluctuations in the range-finder frequency of the target return signal is formed. With a fixed radar range ρ, there is scattering at the coordinates φ and ψ, i.e., the power spectral density of the Doppler fluctuations is formed.

In the case of periodic behavior of the fluctuations in the target return signal frequency, we can define the intra- and interperiod fluctuations as in the case of the pulsed searching signal. The interperiod Doppler fluctuations of the target return signal can be observed as the fluctuations of the envelope, both at the instants of time fixed with respect to the origin of the modulation period, as in the case of the pulsed searching signal, and at the output of the comb filter, the central frequency of which is within the limits of the interval $[\omega_1(t), \omega_2(t)]$. The intraperiod fluctuations in the target return signal frequency are analogous to the rapid fluctuations in frequency with the pulsed searching signal, but a distinction is that their power does not vary within the modulation period, and the intraperiod fluctuations cannot be caused by propagation of the pulsed searching signal along the three-dimensional (space) or two-dimensional (surface) target — the updated set of scatterers in the pulse volume — because at each instant of time, the target return signal contains elementary signals from all scatterers within the limits of the interval $[\rho_1, \rho_2]$.

In the case of the transformed target return signal, when the radar is moving there are the nonstationary Doppler fluctuations in the target return signal frequency. The nonstationary state of the Doppler fluctuations is caused by variations of the carrier frequency and leads to proportional variations in the average Doppler frequency. The power spectral density bandwidth of the Doppler fluctuations in the target return signal frequency remains fixed if the velocity vector of moving radar is outside the directional diagram, or varies in accordance with the frequency $\omega(t)$ if the velocity vector is matched with the directional diagram axis. The shape of the instantaneous power spectral density of the Doppler fluctuations and the fluctuations in the target return signal frequency is the same as with the nontransformed target return signal for various relationships between the power spectral density bandwidth of the Doppler fluctuations in the target return signal frequency and the modulation frequency.

References

1. Blackman, S. and Popoli, R., *Design and Analysis of Modern Tracking Systems,* Artech House, Boston, MA, 1999.
2. Farina, A., *Antenna-Based Signal Processing Techniques for Radar Systems,* Artech House, Norwood, MA, 1992.
3. Hughes, P., A high resolution range radar detection strategy, *IEEE Trans.,* Vol. AES-19, No. 5, 1983, pp. 663–667.
4. Li, J., Liu, G., Jiang, N., and Stoica, P., Moving target feature extraction for airborne high-range resolution phased-array radar, *IEEE Trans.,* Vol. SP-49, No. 2, 2001, pp. 277–289.
5. Schleher, D., *MTI and Pulsed Doppler Radar,* Artech House, Norwood, MA, 1991.
6. Dickey, F., Labitt, M., and Standaher, F., Development of airborne moving target-radar for long range surveillance, *IEEE Trans.,* Vol. AES-27, No. 6, 1991, pp. 959–971.
7. Muirhead, R., *Aspects of Multivariate Statistical Theory,* John Wiley & Sons, New York, 1982.
8. Farina, A., Scannabieco, F., and Vinelli, F., Target detection and classification with very high range resolution radar, in *Proceedings of the International Conference on Radar,* Versailles, France, April 1989, pp. 20–25.
9. Winitzky, A., *Basis of Radar under Continuous Generation of Radio Waves,* Soviet Radio, Moscow, 1961 (in Russian).
10. Wehner, D., *High Resolution Radar,* Artech House, Norwood, MA, 1987.
11. Jacobs, S. and O'Sullivan, J., High resolution radar models for joint tracking and recognition, in *Proceedings of the IEEE National Conference on Radar,* Syracuse, NY, May 1997, pp. 99–104.
12. Mensa, D., *High Resolution Radar Cross-Section Imaging,* Artech House, Norwood, MA, 1991.

10

Fluctuations Caused by Variations in Frequency from Pulse to Pulse

10.1 Three-Dimensional (Space) Target Scanning

10.1.1 Nonperiodic Variations in the Frequency of the Searching Signal

Let us consider the fluctuations of the target return signals from the three-dimensional (space) target caused by nonperiodic variations in the frequency of the pulsed searching signal from pulse to pulse. A frequency jump $\Delta\omega$ is accompanied by variations in the phase of elementary signals $\frac{2\rho\Delta\omega}{c}$ that are random because the distance ρ is a random variable. Variations in the phase of elementary signals can be reproduced exactly by navigational systems if the searching signal possesses a constant frequency, but scatterers have to be moving according to a definite law, i.e., radial displacements (velocities) would be proportional to the distance between the radar and the scatterer.

Using Equation (2.65) and Equation (2.72), the total instantaneous correlation function of frequency fluctuations of the target return signal is defined by the product of the normalized correlation functions of the Doppler frequency fluctuations and the target return signal frequency fluctuations.

$$R_{\Delta\rho,\Delta\omega}(t,\tau) = R_g(t,\tau)\cdot R_\omega(t,\tau), \tag{10.1}$$

where the normalized correlation function $R_g(t, \tau)$ of the Doppler frequency fluctuations coincides with Equation (2.94) or, in the case of absence of scanning, with Equation (3.8). $R_\omega(t, \tau)$ is determined by the following form:

$$R_\omega(t,\tau) = N\cdot e^{j\omega(t)\tau}$$
$$\times\sum_{n=-\infty}^{\infty}\int \boldsymbol{P}\left[z-0.5(\tau-nT_M)+\tfrac{\Delta\rho_0}{c}\right]\cdot\boldsymbol{P}^*\left[z+0.5(\tau-nT_M)-\tfrac{\Delta\rho_0}{c}\right]\cdot e^{j\Delta\omega z}dz. \tag{10.2}$$

Let us assume that the intrapulse frequency modulation of the searching signal is absent, i.e., the conditions $\boldsymbol{P}(t) \equiv \Pi(t)$ and $V = const$ or $\Delta\rho_0 = -V_{r_0}\tau$ are satisfied, and variations in frequency are defined by the regular function of time

$$\omega(t) = \omega_0 + \Omega(t), \tag{10.3}$$

for example, instability in the frequency of the searching signal caused by supply voltages. Then, $R_\omega(t, \tau)$ can be represented as a function of the time shift τ. After this we can define the power spectral density of fluctuations in frequency of the target return signal. Hereafter, we assume that the frequency of the searching signal is a function that varies slowly. We can consider this function as a linear function within the limits of time intervals, the duration of which is much more then the period T_p so that

$$\Delta\omega = k_\omega(t)\cdot\tau \quad \text{or} \quad \Delta f = k_f(t)\cdot\tau\,. \tag{10.4}$$

Then

$$\begin{aligned} R_\omega(t,\tau) &= N\cdot e^{j\omega(t)\tau} \\ &\times \sum_{n=-\infty}^{\infty} \int \boldsymbol{P}[z-0.5(\mu\tau - nT_M)]\cdot \boldsymbol{P}^*[z+0.5(\mu\tau - nT_M)]\cdot e^{jk_\omega(t)z\tau}dz\,. \end{aligned} \tag{10.5}$$

The normalized correlation function $R_\omega(t, \tau)$ of fluctuations in the frequency of the target return signal given by Equation (10.5) is similar to that determined by Equation (4.66) in the case of the wide vertical-coverage radar antenna directional diagram, i.e., $g_v \equiv 1$. For this reason, we can use here some results discussed in Section 4.4. In particular, in the case of the square waveform pulsed searching signals, the results discussed in Section 4.4.4 are true if we replace the parameter $c_*\Omega_\gamma$ with the parameter $-k_\omega$, or the parameter ΔF_τ with the parameter $-k_f\tau_p$.

However, we do not study here the normalized correlation function $R_\omega(t, \tau)$ of the target return signal frequency fluctuations given by Equation (10.5) because the most important features — the interperiod frequency fluctuations with frequency variations of the searching signal from period to period — are obtained immediately, and the rapid target return signal frequency fluctuations within the limits of the period when $\tau < T_p$ and $n = 0$ are the same as in the case of the stable searching signal frequency. We carry out a complete analysis of $R_\omega(t, \tau)$ only in the case of the Gaussian searching signal, taking into consideration the intra- and interperiod fluctuations in the target return signal frequency:

$$R_\omega(t,\tau) = e^{j\omega(t)\tau - 0.5\pi k_f^2 \tau_p^2 \tau} \cdot \sum_{n=-\infty}^{\infty} e^{-\frac{\pi(\mu\tau - nT_p)}{2\tau_p^2}} . \tag{10.6}$$

$R_\omega(t, \tau)$ is a comb function with teeth having the period $T_p' = \frac{T_p}{\mu}$ (see Figure 3.5). The teeth have a Gaussian shape with the effective bandwidth equal to $\sqrt{2}\tau_p$, as in the case of the pulsed searching signal with stable frequency. The envelope of the teeth also possesses the Gaussian shape, with the effective width equal to

$$\tau_c = \frac{\sqrt{2}}{k_f \tau_p} \tag{10.7}$$

that defines the correlation length in time of the interperiod fluctuations of the target return signal. The correlation length in frequency is determined by the following form:

$$\Delta f_c = k_f \tau_c = \sqrt{2}\,\tau_p^{-1}. \tag{10.8}$$

These values coincide with analogous values in the case of the continuous searching signal if we suppose that $\tau_p \equiv T_r$ and have a similar physical meaning.

The instantaneous power spectral density of fluctuations in the target return signal frequency is a comb function and similar to that given by Equation (3.150) (see Figure 10.1). The teeth are Gaussian with the effective bandwidth equal to

$$\Delta F_\omega = (\sqrt{2})^{-1} k_f \tau_p \tag{10.9}$$

and characterize the power spectral density of the target return signal frequency fluctuations. Speaking rigorously, we have to use the absolute value $|k_\omega|$ instead of the value k_ω because the power spectral density bandwidth is independent of the sign of the parameter k_ω. The envelope of the teeth is also Gaussian. The bandwidth is defined by the pulse duration and is equal to $\Delta F_p = \frac{1}{\sqrt{2}\tau_p}$. The average frequency of the power spectral density $\omega(t) = \omega_0 + \Omega_0(t)$ is a function of the time:

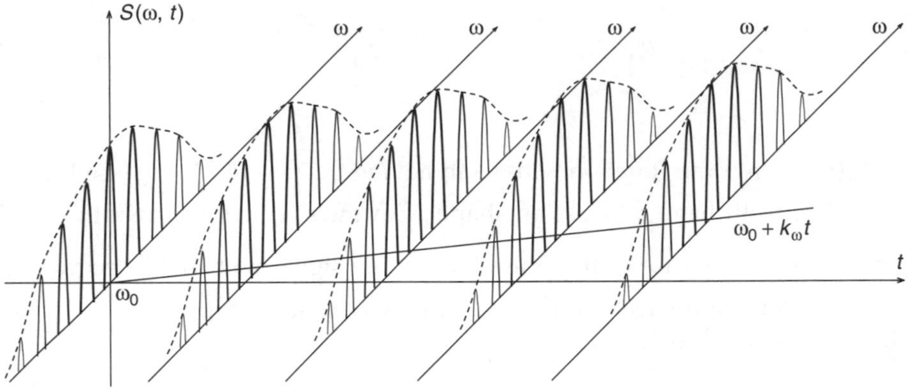

FIGURE 10.1
The instantaneous power spectral density of fluctuations in the target return signal frequency with linear variations in frequency from pulse to pulse.

$$S_{\omega}(\omega, t) \cong e^{-\frac{\pi[\omega-\omega_0-\Omega(t)]^2}{\Delta\Omega_p^2}} \cdot \sum_{n=-\infty}^{\infty} e^{-\frac{\pi[\omega-\omega_0-\Omega(t)-n\Omega'_p]^2}{\Delta\Omega_{\omega}^2}}. \tag{10.10}$$

The total normalized correlation function $R_{\Delta\rho,\Delta\omega}(t, \tau)$ of the target return signal frequency fluctuations given by Equation (10.1) is defined by the product of $R_{\omega}(t, \tau)$ frequency fluctuations determined by Equation (10.6) and $R_g(t, \tau)$ of the Doppler fluctuations, which depends on the shape of the directional diagram and the orientation of its axis relative to the velocity vector of the moving radar. The total power spectral density is defined by the convolution between the power spectral densities frequency fluctuations [see Equation (10.10)] and the Doppler frequency fluctuations. Thus, the total power spectral density is shifted by the average Doppler frequency

$$\Omega_0 = 2\omega(t)V_{r_0}c^{-1} \tag{10.11}$$

that is proportional to the carrier frequency $\omega(t)$ and, for this reason, depends on the time t by the same law as the carrier frequency.

When the directional diagram is high deflected and Gaussian, the power spectral density of the target return signal frequency fluctuations is also Gaussian, but the effective bandwidth of the teeth increases and becomes equal to

$$\Delta F = \sqrt{\Delta F_{\omega}^2 + \Delta F_D^2}, \tag{10.12}$$

where ΔF_D is the effective bandwidth of the power spectral density of Doppler fluctuations in the target return signal frequency [see Equation (3.81)].

It should be pointed out that the effective bandwidth of the power spectral density of Doppler fluctuations in the target return signal frequency is independent of the frequency in the case of the high-deflected directional diagram (see Section 3.2.1). Because of this, the effective bandwidth is not varied in time, as the dependence on the wavelength λ has disappeared in Equation (3.81): $\Delta F_D \cong \Delta_a \lambda^{-1} \cong d_a^{-1}$. When the directional diagram is not deflected, we get a linear function of the wavelength λ if Equation (3.123) is true for ΔF_D and we get a function of the time t.

10.1.2 The Interperiod Fluctuations

In the case of the glancing radar range, assuming $\tau = \frac{nT_p}{\mu}$ in Equation (10.5), we can write that

$$R_\omega(t,\tau) = N \cdot e^{j\omega(t)\tau} \int \Pi^2(z) \cdot e^{jk_\omega(t)z\tau} dz\,. \tag{10.13}$$

Based on Equation (10.13) after using the Fourier transform, we can write the instantaneous power spectral density of fluctuations in the target return signal frequency in the following form:

$$S_\omega(\omega,t) \cong \Pi^2\left[\tfrac{\omega-\omega(t)}{k_\omega(t)}\right]. \tag{10.14}$$

Reference to Equation (10.14) shows that the power spectral density $S_\omega(\omega, t)$ coincides in shape with the squared pulsed searching signal. The main parameters of $S_\omega(\omega, t)$ — the average frequency $\omega(t)$ and the effective bandwidth

$$\Delta F = 0.5\pi^{-1} k_\omega(t)\tau_p^{(2)} \tag{10.15}$$

— vary slowly in time. If $k_\omega = const$, the effective bandwidth of $S_\omega(\omega, t)$ is independent of time. It follows from Equation (10.14) that in the case of the Gaussian pulsed searching signal, the $S_\omega(\omega, t)$ can be written in the following form:

$$S_\omega(\omega,t) \cong e^{-\pi\frac{[\omega-\omega(t)]^2}{[2\pi\Delta F(t)]^2}}, \tag{10.16}$$

where

$$\Delta F(t) = (\sqrt{2})^{-1} k_f(t)\tau_p \quad \text{and} \quad k_f = 0.5\pi^{-1} k_\omega\,. \tag{10.17}$$

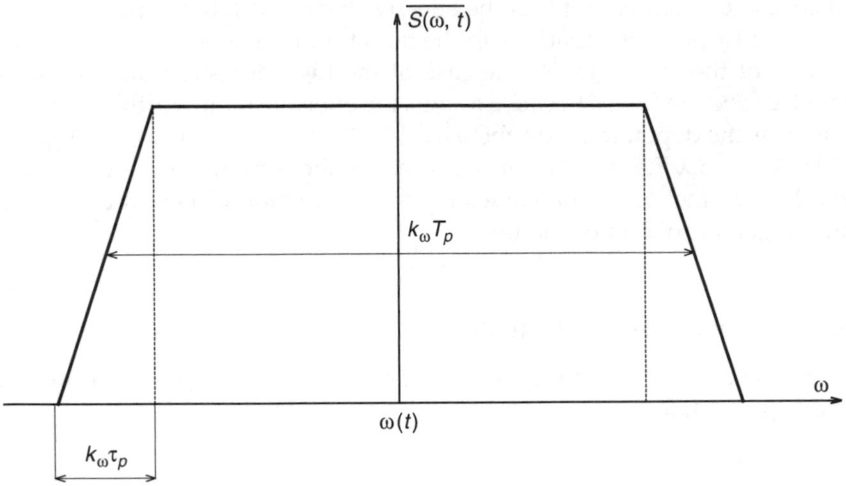

FIGURE 10.2
The average power spectral density of interperiod fluctuations in the target return signal frequency in the case of the square waveform pulsed linear frequency-modulated searching signal.

In the case of the square waveform pulsed searching signal, the power spectral density of fluctuations in the target return signal frequency takes the square waveform and has the effective bandwidth equal to (Figure 10.2):

$$\Delta F(t) = k_f(t)\tau_p . \tag{10.18}$$

The normalized correlation function $R_\omega(t, \tau)$ takes the following form:

$$R_\omega(t,\tau) = \mathbf{sinc}\ [0.5k_\omega(t)\tau_p\tau] \cdot e^{j\omega(t)\tau} . \tag{10.19}$$

The correlation length in time is equal to $\tau_c = \frac{1}{k_f \tau_p}$. The correlation length in frequency is equal to $\Delta f_c = k_f \tau_c = \frac{1}{\tau_p}$, i.e., it is $\sqrt{2}$ times less than in the case of the Gaussian pulsed searching signal.

In the fixed radar range $\tau = nT_p$, based on Equation (10.5) we can write

$$R_\omega(t,\tau) = N \cdot e^{j\omega(t)\tau} \int \Pi(z - 0.5\Delta\rho_0) \cdot \Pi(z + 0.5\Delta\rho_0) \cdot e^{jk_\omega(t)z\tau} dz . \tag{10.20}$$

In the case of the Gaussian pulsed searching signal, we can write

$$R_\omega(t,\tau) = e^{-\pi(\Delta F_\omega^2 + \Delta F'^2)\tau^2 + j\omega(t)\tau} , \tag{10.21}$$

where

$$\Delta F_{\omega} = (\sqrt{2})^{-1} k_f \tau_p \tag{10.22}$$

is the effective bandwidth of the power spectral density of fluctuations in the target return signal frequency caused by the exchange of scatterers [compare with Equation (3.163) at $D = 0$].

The duration of the pulsed searching signal is long, the fluctuations in the target return signal frequency caused by the exchange of scatterers are weak, and the impact of variations in frequency is strong. As the duration of the pulsed searching signal is decreased, the stimulus of the unstable state of frequency is decreased, and the stimulus of exchange of scatterers is increased.[1] In the case of the square waveform searching signal, we can write

$$R_{\omega}(t,\tau) = e^{j\omega(t)\tau} \cdot \frac{\sin\left[0.5k_{\omega}\tau_p\tau\left(1 - \frac{2V_{r_0}|\tau|}{c\tau_p}\right)\right]}{0.5k_{\omega}\tau_p\tau}. \tag{10.23}$$

Comparing Equation (10.23) and Equation (3.31), we can see that $R_{\omega}(t, \tau)$ given by Equation (10.23) defines the fluctuations of the stochastic process formed by superposition of the square waveform pulsed searching signals with the duration determined by

$$\frac{c\tau_p}{2V_{r_0}} = \frac{c\tau_p}{2\,V\cos\beta_0\cos\gamma_0} \tag{10.24}$$

with the linear frequency modulation with deviation $k_{\omega}(t)\tau_p$. The normalized correlation function $R_{\omega}(t, \tau)$ of fluctuations in the target return signal frequency is shown in Figure 3.5 and Figure 3.6. The instantaneous power spectral density of fluctuations in the target return signal frequency is determined by Equation (3.32), in which we have to replace the parameter $\Delta\omega'_M$ with the parameter $k_{\omega}(t)\tau_p$, and the parameter τ'_p with the parameter $\frac{c\tau_p}{2V_{r_0}}$.

10.1.3 The Average Power Spectral Density

Let us consider the power spectral density of fluctuations of the target return signal averaged within the limits of the interval $[0, T_p]$. If the velocity of variations in the frequency $k_{\omega}(t)$ is not varied in time, or if we can neglect it within the limits of the interval $[0, T_p]$, then the averaging of the power spectral density of the target return signal fluctuations is not a difficult problem. Reference to Equation (10.13) shows that the only cofactor $e^{j\omega(t)\tau}$ in

Equation (10.13) is averaged within the limits of the interval $[t - 0.5T_p, t + 0.5T_p]$. Under the linear frequency modulation law for the searching signal, we can write

$$\overline{e^{j\omega(t)\tau}} = e^{j\omega(t)\tau} \cdot \mathbf{sinc}\,(0.5k_\omega T_p \tau)\,. \tag{10.25}$$

The corresponding power spectral density of fluctuations in the frequency of the target return signal has a square waveform with the effective bandwidth equal to $\Delta F = k_f T_p$.

The average normalized correlation function of interperiod fluctuations in the target return signal frequency is defined based on Equation (10.13), Equations (10.19)–(10.21), and Equation (10.23) after replacing the cofactor $e^{j\omega(t)\tau}$ with Equation (10.25). The average power spectral density of interperiod fluctuations in the target return signal frequency is defined by the convolution of the instantaneous power spectral densities with a square waveform and the effective bandwidth equal to $\Delta F = k_f T_p$. Let us consider some examples.

In the case of the Gaussian pulsed searching signal, the average normalized correlation function $\overline{R_\omega(t,\tau)}$ takes the following form:

$$\overline{R_\omega(t,\tau)} = \mathbf{sinc}[0.5k_\omega T_p \tau] \cdot e^{\pi\, \Delta F_\omega^2 \tau^2 + j\omega(t)\tau}\,, \tag{10.26}$$

where $\Delta F_\omega = \frac{k_f \tau_p}{\sqrt{2}}$. In this case, the power spectral density of fluctuations in the frequency of the target return signal is shown in Figure 10.3. As before, the bandwidth of the power spectral density is equal to $\Delta F = k_f T_p$. The difference from the previous case is in the shape of the front and trailing edges of the power spectral density, the bandwidth of which is determined by $\Delta F = k_f T_p$. If the frequency is different by the value $\pm\frac{k_f \tau_p}{4}$ from the frequency corresponding to the middle or front, the power spectral density value is reduced from 0.8 to 0.2. If the condition $T_p >> \tau_p$ is satisfied, as before, the width of the front and trailing edges is very low by value in comparison with the bandwidth of the power spectral density.

Thus, under the condition $T_p >> \tau_p$, the target return signal having fluctuating parameters does not impact the shape of the average power spectral density of fluctuations in the target return signal frequency. This can be explained by the following fact. Regular variations in the frequency of the searching signal lead to high extension of the power spectral density of fluctuations in the target return signal frequency, and therefore, the power spectral density of slow fluctuations in the target return signal amplitude and phase is very narrow and does not contribute to the total shape of the power spectral density. The regular frequency modulation of the searching

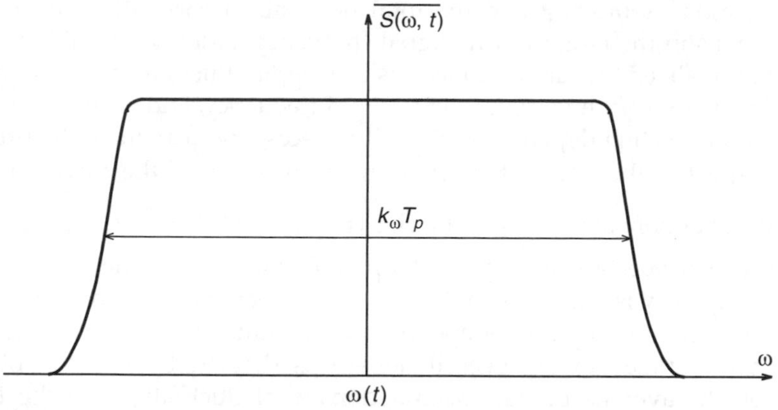

FIGURE 10.3
The average power spectral density of interperiod fluctuations in the target return signal frequency in the case of the Gaussian pulsed searching signal.

signal leads us to find that power spectral density, in which the phases of its components depend on the frequency by a definite law. The phases of the components of the power spectral density of fluctuations in the target return signal frequency are random. Due to simultaneous regular and random changes in the frequency of the searching signal, which is characteristic of the nonstationary stochastic process, there are correlation relationships between phases of the individual components. Because the instantaneous and average normalized correlation functions and power spectral densities do not contain the information regarding phase relationships of the target return signals, the nonstationary stochastic processes are not characterized completely. The complete information can be found in the two-dimensional power spectral densities of the target return signal fluctuations (see Section 2.2.2).

Fluctuating changes in parameters of the target return signal are defined by the slowly varying cofactors **sinc** $[0.5k_\omega \tau_p \tau]$ and $e^{-0.5\pi^{-1}k_\omega^2 \tau_p^2 \tau^2}$ in the average normalized correlation functions of fluctuations in the target return signal frequency given by Equation (10.26). The rapid varying cofactor **sinc** $[0.5k_\omega \tau_p \tau]$ defines regular changes in parameters of the target return signal. This cofactor plays the same role as the cofactor $e^{j\omega_0 \tau}$, which always exists in the high-frequency correlation function of fluctuations in the target return signal frequency and defines rapid regular (sinusoidal) changes in parameters of the target return signal. The cofactor $e^{j\omega\tau}$ does not define the peculiarities of the target return signal as a stochastic process. In the case considered here, the cofactors **sinc** $[0.5k_\omega \tau_p \tau]$ and $e^{j\omega(t)\tau}$ are considered as functions characterizing the regular changes in parameters of the target return signal and do not define the peculiarities of the target return signal as a stochastic process.

If the radar is moving and the total normalized correlation function of fluctuations in the target return signal frequency is defined by the product of the normalized correlation functions of Doppler fluctuations in frequency and fluctuations in the target return signal frequency, and each normalized correlation function depends on time, it is necessary to average the product with respect to the time t. However, in the majority of the cases,[2] we can consider that only the average Doppler frequency $\Omega_0(t) = \frac{2\omega(t)V}{c} \cdot \cos\beta_0 \cos\gamma_0$. of the power spectral density of Doppler fluctuations in the target return signal frequency is a function of the time t. The bandwidth and shape of the power spectral density are independent of the time t. Under this condition, the total average power spectral density is defined by the convolution between the average power spectral density of fluctuations in the target return signal frequency and the power spectral density of Doppler fluctuations in the target return signal frequency, and by the shift in the average frequency on the average Doppler frequency. In particular, with the Gaussian pulsed searching signal, the average normalized correlation function taking into consideration Doppler fluctuations in the target return signal frequency is determined by the following form:

$$\overline{R_\omega(t,\tau)} = \mathbf{sinc}[0.5k_\omega T_p \tau] \cdot e^{-\pi(\Delta F_\omega^2 + \Delta F_D^2)\tau^2 + j[\omega(t)+\Omega_0(t)]\tau} \tag{10.27}$$

instead of Equation (10.26), where ΔF_D is given by Equation (3.81).

10.1.4 Periodic Frequency Modulation

Let us consider the periodic one-sided linear frequency-modulated searching signal with the period $T_M >> T_p$. Instead of Equation (10.5) we can write

$$R_\omega(t,\tau) = N \cdot e^{j\omega(t)\tau} \sum_{n=-\infty}^{\infty} \sum_{m=-\infty}^{\infty} \int \Pi[z - 0.5(\mu\tau - nT_p)] \cdot \Pi[z + 0.5(\mu\tau - nT_p)] \times e^{jk_\omega(\tau - mT_M)z} dz. \tag{10.28}$$

With the Gaussian pulsed searching signal, we can write the normalized correlation function $R_\omega(t, \tau)$ of fluctuations in the target return signal frequency in the following form:

$$R_\omega(t,\tau) = e^{j\omega(t)\tau} \cdot \sum_{n=-\infty}^{\infty} \sum_{m=-\infty}^{\infty} e^{-0.5\tau\left[\frac{(\mu\tau - nT_p)^2}{\tau_p^2} + k_f^2\tau_p^2(\tau - mT_M)^2\right]}. \tag{10.29}$$

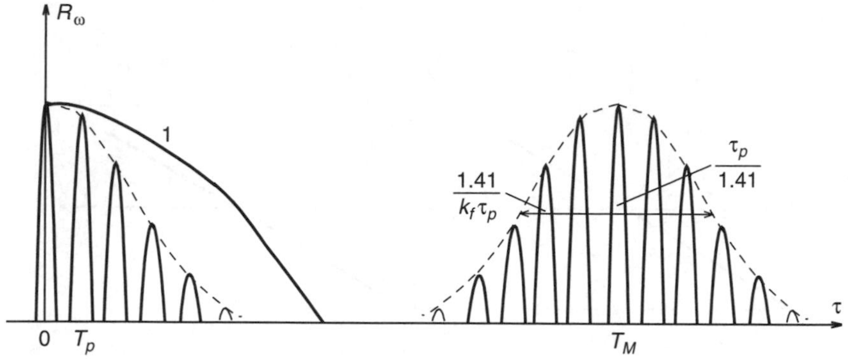

FIGURE 10.4
The instantaneous normalized correlation function of fluctuations in the target return signal frequency in the case of the one-sided linear frequency-modulated searching signal.

$R_\omega(t, \tau)$ is shown in Figure 10.4 without the cofactor $e^{j\omega(t)\tau}$. This function has double periodicity with the periods T_M and T_p. The width and shape of the teeth of $R_\omega(t, \tau)$ are the same as in the case of the nonperiodic frequency-modulated searching signal.

When the radar is moving, $R_\omega(t, \tau)$ given by Equation (10.28) must be multiplied by the normalized correlation function of Doppler fluctuations in the target return signal frequency given by Equation (3.8). The correlation length of Doppler fluctuations in frequency can be both greater and less than the period T_M. If the correlation length is less than the period T_M (see Figure 10.4, curve 1), the periodic peculiarities of the target return signal caused by the frequency-modulated searching signal disappear and are hidden completely by random Doppler fluctuations in frequency.

The normalized correlation function $R_\omega(t, \tau)$ given by Equation (10.29) is similar to the normalized correlation function of the target return signal caused by the pulsed searching signal and radar antenna scanning (see Figure 5.5b) if there is a double periodicity with the periods T_p and T_{sc}. The power spectral density of fluctuations in the target return signal frequency corresponding to $R_\omega(t, \tau)$ given by Equation (10.29) is regular with discreteness F_M (see Figure 10.5a). The width of the teeth is equal to $\frac{k_f\tau_p}{\sqrt{2}}$; the width of the envelope is equal to $\frac{\sqrt{2}}{T_p}$. If the radar is moving, it is necessary to carry out a convolution with the power spectral density of Doppler fluctuations in the target return signal frequency, and thus, each harmonic is transformed into the power spectral density of Doppler fluctuations in the target return signal frequency. When the power spectral density is narrow, i.e., the condition $\Delta F_D < F_M$ is true, the total power spectral density takes the shape of a double comb. When it is wide, i.e., the condition $\Delta F_D > F_M$ is satisfied, the double comb structure disappears.[3]

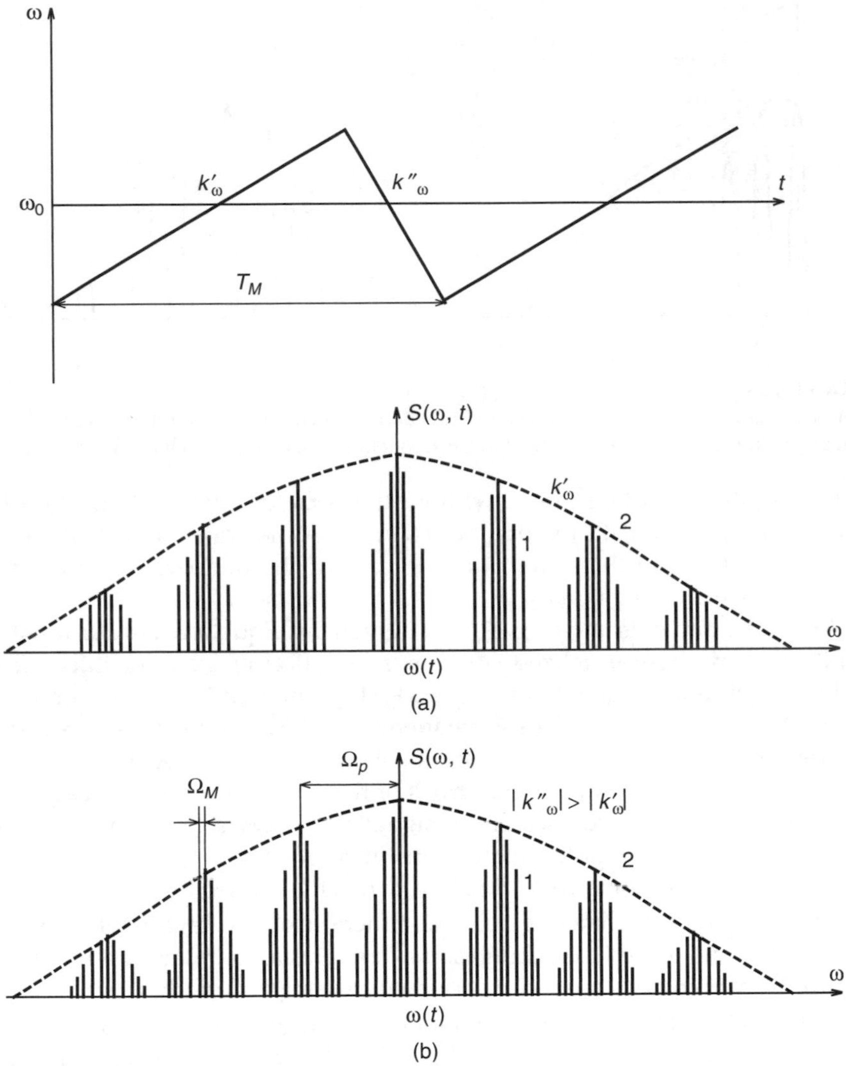

FIGURE 10.5
The instantaneous power spectral density of fluctuations in the target return signal frequency in the case of the two-sided asymmetric linear frequency-modulated searching signal: a) stationary radar, b) moving radar.

The nonstationary state of the target return signal manifests itself as the dependence of the average frequency and the Doppler shift in frequency of the power spectral density of fluctuations in the target return signal frequency on time and as a dependence of the bandwidth of the power spectral density of fluctuations in the target return signal frequency on the parameter k_ω if the parameter k_ω is a function of the time. Comparing Figure 10.5a and Figure 10.5b, we can see that the widths of the teeth are different with variations in the absolute value of the parameter k_ω. With the symmetric saw-tooth frequency-modulated searching signal, i.e., when the condition $k'_\omega = k''_\omega$ is satisfied, the nonstationary state of the target return signal appears as periodic changes in frequency. The shape and bandwidth of the power spectral density are not varied in time within the limits of the modulation period because they are independent of sign of the parameter k_ω.

As $k_\omega \to \infty$, the two-sided saw-tooth linear frequency-modulated searching signal is transformed into the one-sided one and the instantaneous regular power spectral density of fluctuations in the target return signal frequency with unlimited bandwidth (the instantaneous regular white noise) is formed at the instant of the jump in frequency. We can define the instantaneous power spectral density of fluctuations in the target return signal frequency in the case of the harmonic frequency-modulated searching signal. In this case, we have to use the following formulae:

$$\omega(t) = \omega_0 + \Delta\omega_M \cdot \sin \omega_M t \qquad \text{and} \qquad k_\omega = \omega_M \cdot \Delta\omega_M \cos \omega_M t\,. \quad (10.30)$$

The corresponding instantaneous power spectral density is shown in Figure 10.6 at various instants of time. At $k_\omega = 0$, each tooth of the power spectral density consisting from some δ-functions under the condition $k_\omega \neq 0$ is contracted into a single δ-function.

In conclusion, we pay attention to the following circumstance. With the pulsed searching signal, the radar heterodyne frequency is not varied, the range finder frequency is absent, and the average frequency of the power spectral density of fluctuations in the target return signal frequency is matched with the transmitter frequency shifted on the Doppler frequency.

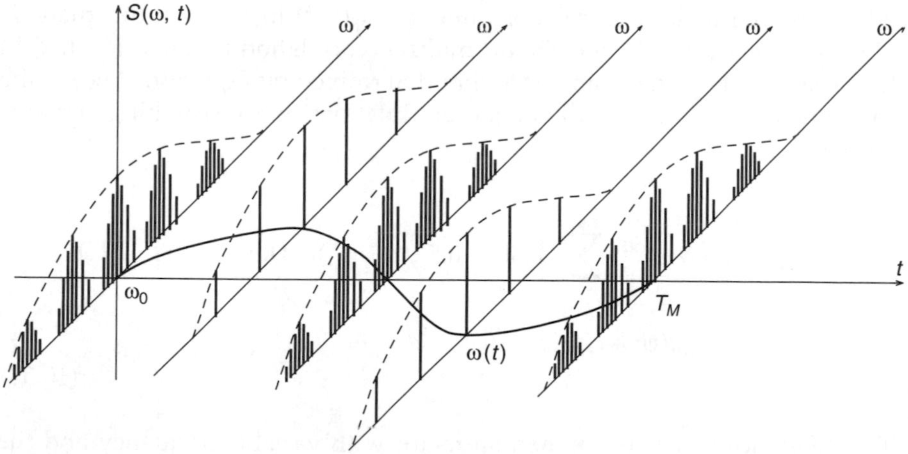

FIGURE 10.6
The instantaneous power spectral density of fluctuations in the target return signal frequency in the case of the harmonic frequency-modulated searching signal.

In some cases, the heterodyne frequency can vary in a jump-like manner simultaneously with the variation in frequency of the frequency-modulated searching signal. Then the intermediate frequency of the transformed target return signal is constant for all elementary signals and we have to assume that $\omega(t) = \omega_{im} = const$, but the Doppler shift in the frequency is proportional to $\omega(t)$, as before. In this case, if the Doppler shift in the frequency is absent and if the condition $k_\omega = const$ is satisfied, the instantaneous power spectral density of fluctuations in the target return signal frequency is independent of the time t and the average power spectral density coincides with the instantaneous power spectral density.

10.2 Two-Dimensional (Surface) Target Scanning

Let us consider the fluctuations of target return signals from the two-dimensional (surface) target caused by variations in frequency from the pulsed searching signal to the pulsed searching signal and the moving radar. In this case, using Equation (2.65) and Equation (4.6)–Equation (4.10), we can find that the total normalized correlation function of fluctuations in the target return signal frequency, as in the case of Equation (4.6), is determined by the product of the normalized correlation functions in the azimuth and aspect-angle planes

$$R_{\Delta\rho,\Delta\omega}(t,\tau) = R_\beta(t,\tau)\cdot R_{\gamma,\omega}(t,\tau)\ , \tag{10.31}$$

where the normalized correlation function $R_\beta(t, \tau)$ in the azimuth plane is given by Equation (4.7) and the normalized correlation function $R_{\gamma,\omega}(t, \tau)$ in the aspect-angle plane is similar to that determined by Equation (4.66) in the case where the intrapulse frequency modulation for the searching signal is absent:

$$R_{\gamma,\,\omega}(t,\tau) = N\cdot e^{j\omega(t)\tau}\sum_{n=-\infty}^{\infty}\int \Pi[z-0.5(\mu\tau-nT_p)]\cdot \Pi[z+0.5(\mu\tau-nT_p)]$$
$$\times\, g_v^2(\psi_* + c_* z)\cdot e^{j[k_\omega(t)-c_*\Omega_\gamma]z\tau}dz\,. \tag{10.32}$$

The difference is in the exponential factor with variable frequency and the exponent in the integrand: in Equation (10.32) we can see the sum $k_\omega(t) - c_*\Omega_\gamma$ instead of the parameter $-c_*\Omega_\gamma$ in Equation (4.66).

The normalized correlation function $R_{\gamma,\omega}(t, \tau)$ in the aspect-angle plane given by Equation (10.32), as in Equation (4.66), takes into consideration the interperiod target return signal fluctuations caused by variations in the searching signal frequency in addition to those caused by the propagation of the searching signal and those caused by the difference in the Doppler frequencies at the aspect-angle plane. $R_{\gamma,\omega}(t,\tau)$ can be investigated in the same way as the normalized correlation function given by Equation (4.66). Here we discuss only some peculiarities.

Let us consider only the interperiod fluctuations of the target return signal in the glancing radar range when the condition $\tau = \frac{nT_p}{\mu}$ is satisfied. The intra-period fluctuations of the target return signal are the same as in the case of the stable frequency of the searching signal. Then, the normalized correlation function $R_{\gamma,\omega}(t,\tau)$ of fluctuations in the target return signal frequency in the aspect-angle plane can be written in the following form:

$$R_{\gamma,\omega}(t,\tau) = N \cdot e^{j\omega(t)\tau} \sum_{n=-\infty}^{\infty} \int \Pi^2(z) \cdot g_v^2(\psi_* + c_* z) \cdot e^{j[k_\omega(t) - c_*\Omega_\gamma]z\tau} dz \,. \quad (10.33)$$

Using Equation (10.33), we can define the power spectral density of the interperiod target return signal fluctuations, which is similar to that given by Equation (4.68). The difference is the following: we use the sum $k_\omega(t) - c_*\Omega_\gamma$ in Equation (10.33) instead of the parameter $-c_*\Omega_\gamma$ in Equation (4.68). In the case of the wide vertical-coverage directional diagram, i.e., the condition $\Delta_v >> \Delta_p$ is satisfied, the shape of the power spectral density of fluctuations in the target return signal frequency, as in Equation (4.69), is defined only by the shape of pulsed searching signals:

$$S_{\gamma,\omega}(\omega,t) \cong \Pi^2\left[\frac{\omega - \omega(t)}{k_\omega(t) - c_*\Omega_\gamma}\right]. \quad (10.34)$$

The average frequency of the power spectral density $S_{\gamma,\omega}(\omega,t)$ given by Equation (10.34) is a function of the time. The effective bandwidth of $S_{\gamma,\omega}(\omega,t)$ is determined by the following form:

$$\Delta\Omega_{\omega,\tau}(t) = k_p \tau_p [k_\omega(t) - c_*\Omega_\gamma] = \Delta\Omega_\omega(t) - \Delta\Omega_\tau \quad (10.35)$$

where

$$\Delta\Omega_\omega(t) = k_p \tau_p k_\omega(t) \quad (10.36)$$

and $\Delta\Omega_\tau$ is given by Equation (4.70).

Thus, linear variations in frequency from the pulsed searching signal to the pulsed searching signal as a function of the sign of the parameter $k_\omega(t)$ can lead us to both expansion and narrowing of the bandwidth $\Delta\Omega_\tau$ of the power spectral density of target return signal fluctuations in the aspect-angle plane, as in the case of the continuous linear frequency-modulated searching signal when the power spectral density of the target return signal fluctuations in the range finder frequency can be added to or subtracted from that of the Doppler fluctuations (see Section 8.2).

If the condition $k_\omega(t) = const$ is satisfied, the condition $\Delta\Omega_\omega(t) = const$ is also true. The bandwidth of the power spectral density of the target return signal fluctuations in the aspect-angle plane is a function of the value of the aspect angle γ_* or the time-cross-section **t**. Dependence on the aspect angle γ_* is shown in Figure 4.10 by the solid line. Because of this, the bandwidth $\Delta\Omega_{\omega,\tau}$ given by Equation (10.18) is also a function of the aspect angle γ_* or the time cross section **t**. Addition of the term $\Delta\Omega_\omega$ to the term $\Delta\Omega_\tau$ can lift ($k_\omega(t) < 0$) or lower ($k_\omega(t) > 0$) the bandwidth $\Delta\Omega_\tau(\gamma_*)$.

Under the condition $\Delta\Omega_\omega = \Delta\Omega_\tau$ or $k_\omega = c_*\Omega_\gamma$, the bandwidth $\Delta\Omega_{\omega,\tau} = 0$, i.e., the Doppler fluctuations of the target return signal in the aspect-angle plane are completely compensated by fluctuations in the target return signal frequency. This phenomenon takes place when

$$k_f = \frac{f V \cdot \cos\beta_0 \sin^3\gamma_*}{h \cdot \cos\gamma_*} . \tag{10.37}$$

Because the bandwidth of the power spectral density of the target return signal fluctuations is a nonnegative value, then, under the condition $\gamma_* = \overline{\gamma_*}$, in which Equation (10.37) is true, the bandwidth $\Delta\Omega_{\omega,\tau}$ shows a bend (Figure 4.10, dotted line). For example, at $\beta_0 = 0°$, $V = 200\ \frac{\text{m}}{\text{sec}}$, $h = 10\,000$ m, $f = 10{,}000$ MHz, the equality $\Delta\Omega_{\omega,\tau} = 0$ is true if the conditions $\gamma_* = 45°$ and $k_f = 100\ \frac{\text{MHz}}{\text{sec}}$ are satisfied.

Consider the total power spectral density of the target return signal fluctuations taking into consideration the fluctuations both in the aspect-angle plane and in frequency, and the interperiod target return signal fluctuations in the azimuth plane for the simplest case when the pulsed searching signal and the directional diagram are Gaussian and the directional diagram is wide. Then the power spectral density is also Gaussian, and is determined by Equation (4.140). The bandwidth of the power spectral density of the target return signal fluctuations is determined by

$$\Delta\Omega = \sqrt{\Delta\Omega_h^2 + (\Delta\Omega_\tau - \Delta\Omega_\omega)^2} \tag{10.38}$$

instead of Equation (4.141). When $\beta_0 = 0°$, at the first approximation we can consider that $\Delta\Omega_h = 0$ [see Equation (4.119)]. Then Equation (10.38) is transformed into Equation (10.35). For this approximation, the interperiod fluctuations of the target return signal are absent in the cross section

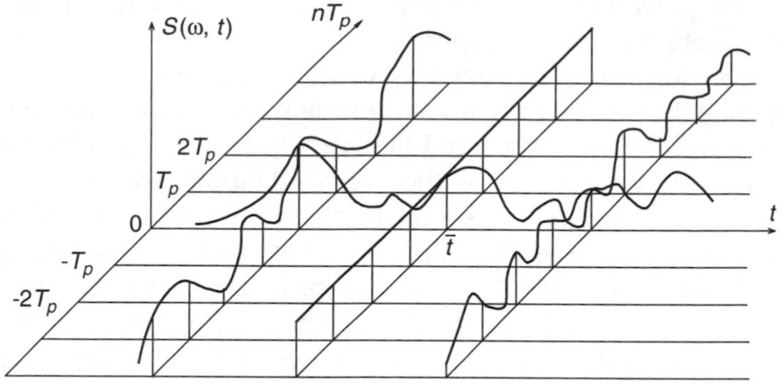

FIGURE 10.7
The Doppler fluctuations and interperiod fluctuations in the target return signal frequency for different time-cross-sections at $\beta_0 = 0°$.

$\overline{\gamma_*} = \overline{\gamma_*} \; (t = \overline{t})$ (see Figure 10.7). If $\beta_0 \neq 0°$, the bandwidth of the power spectral density of the target return signal fluctuations given by Equation (10.38) is not equal to zero, but reaches a maximum in value equal to $\Delta\Omega_h$.

10.3 Conclusions

Under three-dimensional (space) target scanning, in the case of target return signal fluctuations caused by nonperiodic variations in frequency from pulse to pulse, the total normalized correlation function of the target return signal frequency fluctuations is defined by the product of the normalized correlation functions of the target return signal frequency fluctuations and the Doppler fluctuations, which depends on the shape of the directional diagram and the orientation of its axis relative to the vector of velocity of the moving radar. The total power spectral density of the target return signal frequency fluctuations is defined by the convolution between the power spectral densities frequency fluctuations and the Doppler frequency fluctuations of the target return signal. Thus, the total power spectral density of the target return signal fluctuations is shifted on the average Doppler frequency that is proportional to the carrier frequency of the target return signal and, for this reason, the total power spectral density depends on time by the same law as the carrier frequency.

If the radar is moving and the total normalized correlation function of the target return signal frequency fluctuations is defined by the product of the normalized correlation functions of Doppler frequency fluctuations and the target return signal frequency fluctuations, and each normalized correlation function depends on time, it is necessary to average the product with respect to the time. However, in the majority of cases, we can consider that only the average Doppler frequency is a function of the time. The bandwidth and shape of the power spectral density of the Doppler target return signal frequency fluctuations are independent of the time. In this condition, the total average power spectral density of the target return signal frequency fluctuations is defined by the convolution between the average power spectral density and the power spectral density of Doppler fluctuations and by shift of the average frequency by the average Doppler frequency.

In the case of target return signal fluctuations caused by variations in frequency from the pulsed searching signal to the pulsed searching signal and the moving radar, under the two-dimensional (surface) target scanning, the total normalized correlation function of the target return signal frequency fluctuations is defined by the product of the normalized correlation functions in the azimuth and aspect-angle planes. That in the aspect-angle plane takes into consideration the interperiod target return signal frequency fluctuations caused by variations in the frequency of the searching signal in addition to the intraperiod target return signal frequency fluctuations caused by the propagation of the searching signal and by a difference in the Doppler frequencies in the aspect-angle plane.

References

1. Wehner, D., *High Resolution Radar,* Artech House, Norwood, MA, 1987.
2. Mensa, D., *High Resolution Radar Cross-Section Imaging,* Artech House, Norwood, MA, 1991.
3. Schleher, D., *MIT and Pulsed Doppler Radar,* Artech House, Norwood, MA, 1991.

Part II

Generalized Approach to Space–Time Signal and Image Processing in Navigational Systems

11

Foundations of the Generalized Approach to Signal Processing in Noise

The generalized approach to signal processing in the presence of noise is based on a seemingly abstract idea: the introduction of an additional noise source, which does not carry any information about the signal, for the purpose of improving the detection performance and noise immunity of complex navigational systems. The proposed generalized approach in the presence of noise allows us to formulate decision-making rules based on the determination of the jointly sufficient statistics of the likelihood function or functional mean and variance. Classical and modern signal processing theories allow us to define only the sufficient statistic of the likelihood function or functional mean. The presence of additional information about the statistical likelihood function or functional parameters leads to better qualitative characteristics of signal detection in comparison to the optimal signal processing algorithms of classical and modern theories.

11.1 Basic Concepts

The simplest signal detection problem is that of binary detection in the presence of additive Gaussian noise with zero mean and the power spectral density 0.5 N_0. The optimal detector can be realized as the matched filter or the correlation receiver. Detection quality depends on the normalized distance between two signal points of the decision-making space. This distance is characterized by signal energies, the coefficient of correlation between the signals, and the spectral power density of the additive Gaussian noise. If the signal energies are equal, then the optimal coefficient of correlation is equal to -1. Further, the signal waveform is of no consequence. Despite the classical signal processing theory being very orderly and smooth, it cannot provide the most complete answer to the questions posed in the following text. Let us consider briefly the results discussed in References 1 to 49.

The hypothesis H_0 must be chosen so that the input stochastic process is normal Gaussian and has a zero mean, vs. the alternative H_1, which is also normal Gaussian but has a mean that varies according the known law $a(t)$. In a statistical context, this problem is solved as follows. Let $X(t)$ be the input stochastic process, which is observed within the limits of the time interval $[0, T]$; $a(t)$, the signal; and $\xi(t)$, the additive Gaussian noise with a zero mean and the known variance σ_n^2:

$$X(t) = a(t) + \xi(t) \Rightarrow H_1 \quad \text{and} \quad X(t) = \xi(t) \Rightarrow H_0 . \tag{11.1}$$

As elements of the observed input stochastic sample, we take the uncorrelated coordinates

$$X_i = \sqrt{\lambda_i} \int_0^T X(t)\Xi_i(t)\, dt, \tag{11.2}$$

where $X(t)$ is the realization of the input stochastic process within the limits of the time interval $[0, T]$, and λ_i and Ξ_i are the eigenvalues and eigenfunctions of the integral equation

$$F(t) = \lambda \int_0^T R(y-t)\Xi(y)\, dy, \quad 0 < t < T, \tag{11.3}$$

where $R(t)$ is the known correlation function of the additive Gaussian noise.

As a rule, we take only the first N coordinates. Thus, for the hypothesis H_0 the likelihood function of the observed input stochastic sample $X_1, \ldots, X_N$ has the following form (note: for simplicity, we set the noise variance to be equal to unity):

$$f_{X|H_0}(X \mid H_0) = \frac{1}{(2\pi)^{0.5N}} \cdot \exp\left\{-0.5\sum_{i=1}^{N} X_i^2\right\}. \tag{11.4}$$

This notation corresponds to a "no" signal in the observed input stochastic sample $X_1, \ldots, X_N$. For the observed input stochastic sample with a nonzero mean $a(t)$, for example, when considering the hypothesis H_1, we take for observed coordinates the values

$$X_i = a_i + \xi_i = \sqrt{\lambda_i} \int_0^T [a(t) + \xi(t)]\Xi_i(t)\, dt, \tag{11.5}$$

where $\xi(t)$ is the additive Gaussian noise with a zero mean and known variance.

Then the likelihood function in the presence of a signal in the observed input stochastic sample $X_1, \ldots, X_N$ has the following form:

$$f_{X|H_1}(X \mid H_1) = \frac{1}{(2\pi)^{0.5N}} \cdot \exp\left\{-0.5\sum_{i=1}^{N}\{X_i - a_i\}^2\right\}. \tag{11.6}$$

This notation corresponds to a "yes" signal in the observed input stochastic sample $X_1, \ldots, X_N$. Using Equation (11.4) and Equation (11.6), we can write the likelihood function ratio in the following form:

$$\begin{aligned}\frac{f_{X|H_1}(X \mid H_1)}{f_{X|H_0}(X \mid H_0)} &= \frac{\exp\left\{-0.5\sum_{i=1}^{N}\{X_i - a_i\}^2\right\}}{\exp\left\{-0.5\sum_{i=1}^{N}X_i^2\right\}} \\ &= \exp\left\{\sum_{i=1}^{N}X_i a_i - 0.5\sum_{i=1}^{N}a_i^2\right\} = \mathcal{L}(X_1, \ldots, X_N) = C,\end{aligned} \tag{11.7}$$

where C is the constant, which is determined by the performance criterion of the decision-making rule. Taking the logarithm in Equation (11.7), we can write

$$\sum_{i=1}^{N} X_i a_i > K_{op} \Rightarrow H_1 \quad \text{or} \quad \sum_{i=1}^{N} X_i a_i \le K_{op} \Rightarrow H_0\,, \quad K_{op} = \ln C + 0.5\sum_{i=1}^{N} a_i^2, \tag{11.8}$$

where $\sum_{i=1}^{N} a_i^2$ is the signal energy and K_{op} is the threshold. Letting $N \to \infty$ and transitioning to the integral form, and using the Parseval theorem,[8] we can maintain generality and write

$$\begin{aligned}&\int_0^T X(t)a(t)\,dt > K_{op} \Rightarrow H_1 \quad \text{or} \\ &\int_0^T X(t)a(t)\,dt \le K_{op} \Rightarrow H_0\,, \quad K_{op} = \ln\, C + 0.5\int_0^T a^2(t)\,dt,\end{aligned} \tag{11.9}$$

where $\int_0^T a^2(t)\,dt$ is the signal energy, and $[0, T]$ is the time interval, within the limits of which the input stochastic process is observed. It is asserted that the signal detection algorithm given by Equation (11.8) and Equation (11.9) reduces to the calculation of the value $\sum_{i=1}^{N} X_i a_i$ or $\int_0^T X(t)a(t)\,dt$ and comparison to the threshold K_{op}.

This signal processing algorithm is optimal for any of the following chosen performance criteria: the Bayesian criterion (including, as particular cases, the *a posteriori* probability distribution density maximum and maximal likelihood), the Neyman–Pearson criterion, and the mini–max criterion, and is called the correlation signal processing algorithm because the mutual correlation function between the input stochastic process $X(t)$ and signal $a(t)$ is defined. Analysis of the signal processing algorithm given by Equation (11.8) and Equation (11.9) yields a property that, in comparison with other factors, defines the noise immunity. The essence of the analysis reduces to substituting the actual values $X_i = a_i + \xi_i$ and $X(t) = a(t) + \xi(t)$ (the hypothesis H_1) or $X_i = \xi_i$ and $X(t) = \xi(t)$ (the hypothesis H_0) into Equation (11.9):

$$\sum_{i=1}^{N} X_i a_i = \sum_{i=1}^{N} a_i^2 + \sum_{i=1}^{N} a_i \xi_i \Rightarrow H_1\ ; \quad \sum_{i=1}^{N} X_i a_i = \sum_{i=1}^{N} a_i \xi_i \Rightarrow H_0 \tag{11.10}$$

or

$$\begin{aligned} &\int_0^T X(t)a(t)\,dt = \int_0^T a^2(t)\,dt + \int_0^T a(t)\xi(t)\,dt \Rightarrow H_1\ ; \\ &\int_0^T X(t)a(t)\,dt = \int_0^T a(t)\xi(t)\,dt \Rightarrow H_0, \end{aligned} \tag{11.11}$$

where the terms $\sum_{i=1}^{N} a_i^2$ and $\int_0^T a^2(t)\,dt$ are the signal energy, and the terms $\sum_{i=1}^{N} a_i \xi_i$ and $\int_0^T a(t)\xi(t)\,dt$ are the noise component with a zero mean and the finite variance defined

$$\lim_{N\to\infty} \overline{\left\{\sum_{i=1}^{N} a_i \xi_i\right\}^2} = 0.5 E_a N_0 \quad \text{as} \quad N \to \infty \tag{11.12}$$

or as $T \to \infty$

$$\overline{\left[\int_0^\infty a(t)\xi(t)\,dt\right]^2} = \int_0^\infty dt \int_0^\infty a(t)a(s)\overline{\xi(t)\xi(s)}\,ds = 0.5E_a N_0. \qquad (11.13)$$

The detection parameter $q = \sqrt{\frac{2E_a}{N_0}}$ is taken as a qualitative characteristic of the signal detection algorithm given by Equation (11.8) and Equation (11.9). This parameter may also be called the voltage signal-to-noise ratio (SNR). This parameter is very important and, together with other factors, defines the noise immunity of any signal processing system and also, in particular, of navigational systems.

11.2 Criticism

Let us consider those factors generating questions in the synthesis of the signal processing algorithm given by Equation (11.8) and Equation (11.9). It is known that $\sum_{i=1}^{N} X_i$, which is the sufficient statistic of the mean, and $\sum_{i=1}^{N} X_i^2$, which is the sufficient statistic of the variance, are the jointly sufficient statistics characterizing the distribution law of the random variable X_i.[8,19] The sufficient statistics $\sum_{i=1}^{N} X_i^2$ of the likelihood functions $f_{X|H_1}(X\,|\,H_1)$ and $f_{X|H_0}(X\,|\,H_0)$ are reduced in the synthesis of the signal processing algorithm given by Equation (11.8) and Equation (11.9). This is indeed the case in regard to the form of the expressions and assumptions of the statistical decision theory of decision-making. However, in the physical form, it causes a specific perplexity.

The point is that a "yes" signal — the mean a_i of the observed input stochastic sample $X_1, \ldots, X_N$ is not zero — is indicated in the numerator of Equation (11.7), and a "no" signal is indicated in its denominator with the same coordinates. It would be difficult to imagine another approach to the same input stochastic sample $X_1, \ldots, X_N$ in both the numerator and denominator of the likelihood function ratio. The first question that arises is: Might a signal processing algorithm be constructed without loss of sufficient statistic of variance, which is one of the characteristics of the probability distribution density? Another factor generating questions is that the signal processing is performed against the background of the noise component

$\sum_{i=1}^{N} a_i\xi_i$ or $\int_0^T a(t)\xi(t)\,dt$ caused by the interaction between the signal and noise.

The variance of the noise component is proportional to the signal energy, which follows from Equation (11.12) and Equation (11.13). The fact that the voltage signal-to-noise ratio for the signal processing algorithm, which is given by Equation (11.8) and Equation (11.9), defined by Equation (11.12), is not proportional to $\frac{2E_a}{N_0}$ but rather to the square root of the value $\frac{2E_a}{N_0}$ is a consequence of this. The resulting question would be: Is this good or bad? We would believe it is good if the condition $\frac{2E_a}{N_0} < 1$ is satisfied. However, if the condition is $q < 1$ and the probability of false alarm P_F is equal to 10^{-3}, for example, the probability of detection P_D does not exceed 0.1, which is a practically inoperative region for signal detection. If the conditions $\frac{2E_a}{N_0} > 1$ and $q = \sqrt{\frac{2E_a}{N_0}}$ are satisfied, then the probability of detection P_D is smaller in comparison to the proportional dependence $q = \frac{E_a}{N_0}$. This conclusion seems unusual, but it is real and is shown in References 50 and 51.

Analyzing Equation (11.8) and Equation (11.9), we may note that this signal processing algorithm is considered to be optimal during the following conditions. The likelihood function or functional ratio is formed using the same input stochastic sample where the numerator assumes a "yes" signal and the denominator assumes a "no" signal in the input stochastic process. In this case, the standard quadratic statistic is reduced and the additional information is lost — the sufficient statistic of the variance of the likelihood function or functional. The expression obtained ensures the calculation of the sufficient statistic of the likelihood function or functional mean only: $\sum_{i=1}^{N} X_i a_i$ or $\int_0^T X(t)a(t)\,dt$, where a_i or $a(t)$ is the known signal. Theoretically speaking, the signal processing algorithm given by Equation (11.8) and Equation (11.9) is not realizable for the following reasons: (1) the mutual correlation function between the input stochastic process X_i or $X(t)$ and the signal a_i or $a(t)$ is defined by the left side of the Equation (11.8) and Equation (11.9), respectively; (2) the left side of Equation (11.8) and Equation (11.9) vanishes given a "no" signal in the input stochastic process X_i or $X(t)$: $\sum_{i=1}^{N} X_i a_i$ or $\int_0^T X(t)a(t)\,dt$, where $a_i = 0$ or $a(t) = 0$, *and any physical meaning is lost*. In practice, the signal processing algorithm given by Equation (11.8) and Equation (11.9) is realized if the signal structure a_i or $a(t)$ is replaced by its model a_i^* or $a^*(t)$ at the receiver, as a_i or $a(t)$ is the completely known signal $-a_i^* = ka_i$ or $a^*(t) = ka(t)$, where k is the coefficient of proportionality

$\sum_{i=1}^{N} X_i a_i^*$ or $\int_0^T X(t)a^*(t)\,dt$. In this case, the left side of Equation (11.8) and Equation (11.9) has a specific physical form. If the signal structure a_i or $a(t)$ is replaced by its model a_i^* or $a^*(t)$, then the noise component $\sum_{i=1}^{N} a_i^* \xi_i$ or $\int_0^T a^*(t)\xi(t)\,dt$ arises, caused by the interaction between the model signal and noise, and always exists independently of what hypothesis (H_0 or H_1) is considered. The variance of the noise component noted here is proportional to the energy of the model signal, i.e., $0.5E_a^* N_0$, where E_a^* is the energy of the model signal and $0.5N_0$ is the power spectral density of the noise. The signal processing algorithm given by Equation (11.8) and Equation (11.9) does not allow us to obtain the ratio between the energy characteristics of the signal and noise in the pure form, for example, in the form $\frac{2E_a}{N_0}$. It causes the probability of detection P_D to be a function of the square root of the signal and noise energy characteristics ratio, i.e., the voltage signal-to-noise ratio is proportional to $\sqrt{\frac{2E_a}{N_0}}$. The signal processing algorithm given by Equation (11.8) and Equation (11.9) does not afford detection of the signal and definition of the signal parameters, whose structure does not correspond to that of the model signal at the receiver. In general, a receiver constructed according to the signal processing algorithm given by Equation (11.8) and Equation (11.9) must be a tracker, not a true detector, because the instant of signal's appearance on the time axis is unknown relative to the origin.

Considering the conditions of optimality of the signal processing algorithm given by Equation (11.8) and Equation (11.9) as briefly outlined above, if the same input stochastic sample is observed in the numerator and denominator of the likelihood function or functional ratio, it is the author's opinion that it is necessary to undertake a critical review of the initial premises constituting the basis of the classical and modern signal processing theories.

11.3 Initial Premises

The signal processing algorithm given by Equation (11.8) and Equation (11.9) is based on the assumption that the frequency–time region Z of the noise exists where a signal may be present; for example, there is an observed stochastic sample from this region, relative to which it is necessary to make the decision a "yes" signal (the hypothesis H_1) or a "no" signal (the hypothesis H_0). We now proceed to modify the initial premises of the classical and modern signal processing theories. Let us suppose there are two independent frequency–time regions Z and Z^* belonging to the space A (see Figure 11.1). Noise from these regions obeys the same probability distribution densities with the same statistical parameters. For the sake of simplicity of this analysis, the same probability distribution density and equality of the statistical parameters have been chosen. However, generally, these parameters may not be equal. A "yes" signal is possible in the noise region Z as before. *It is known* a priori *that a "no" signal is obtained in the noise region* Z^*. In the following text we will call the noise region Z^* the reference region, and consequently, the observed sample from this region is called the reference sample.

It is necessary to make the decision a "yes" signal (the hypothesis H_1) or a "no" signal (the hypothesis H_0) in the observed stochastic sample from the region Z, by comparing the probability distribution density statistical parameters of this sample with those of the sample from the reference region Z^*. The problem posed in the preceding sections of this chapter must be solved using the statistical decision-making theory. Thus, it is necessary to accumulate and compare statistical data defining the probability distribution density statistical parameters of the observed input stochastic samples from two independent frequency–time regions Z and Z^*.

If the probability distribution density statistical parameters for the two samples are equal or agree with each other within the limits of a given accuracy, then the decision of a "no" signal in the observed input stochastic sample from the region Z is made — the hypothesis H_0. If the probability distribution density statistical parameters of the observed input stochastic sample from the region Z differ from those of the reference sample from the region Z^* by a value that exceeds the prescribed error limit, then the decision of a "yes" signal in the region Z is made — the hypothesis H_1.

11.4 Likelihood Ratio

Now the problem is to obtain jointly sufficient statistics to define the statistical parameters of the probability distribution densities. For this purpose, let us avail ourselves of one of the well-known results.[7,23,28,30,44,52–54] It is known that the sufficient statistic is determined from the condition that the likelihood function has an extremum. In general, the condition of an extremum of the likelihood function, relative to the parameter to be determined with the prescribed accuracy, is determined in the following form:

$$\frac{\partial f_X(X_1,\ldots,X_N \mid \vartheta)}{\partial \vartheta} = 0, \qquad (11.14)$$

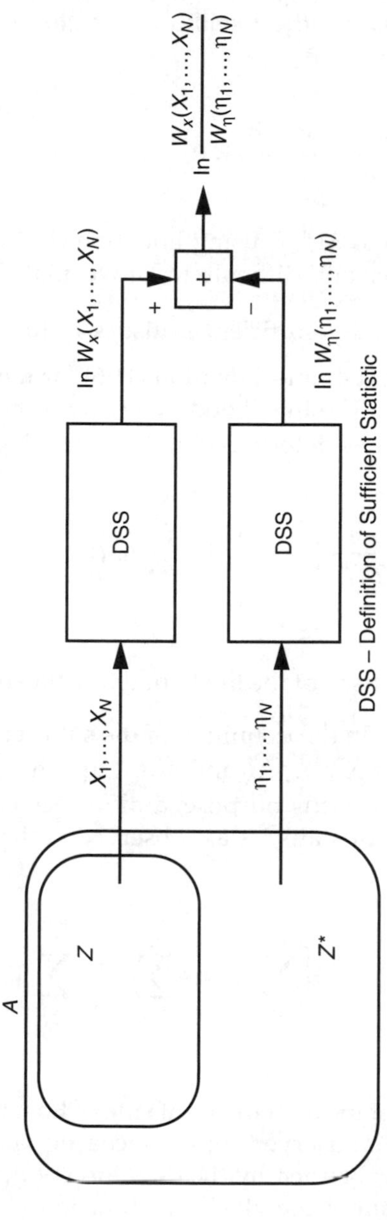

FIGURE 11.1
The definition of jointly sufficient statistics.

where N is the sample size determining the prescribed accuracy, and ϑ is the parameter to be determined. However, this equation is not used in practice. A simple mathematical procedure simplifies the representation of this equation. Because the logarithm is a monotonic function, the extrema of the functions $f_X(X_1, \ldots, X_N)$ and $\ln f_X(X_1, \ldots, X_N)$ are reached at the same values of the parameter ϑ. Therefore, the likelihood function equation is usually written in the following form:

$$\frac{\partial \ln f_X(X_1, \ldots, X_N \mid \vartheta)}{\partial \vartheta} = 0 . \tag{11.15}$$

As was shown in References 19 and 31,[19,31] using Equation (11.4), Equation (11.6), and Equation (11.15), it is not difficult to prove that the values $\sum_{i=1}^{N} X_i a_i$ and $\sum_{i=1}^{N} X_i^2$ are the jointly sufficient statistics of the likelihood function parameters in Equation (11.4) and Equation (11.6) for the observed input stochastic sample $X_1, \ldots, X_N$. The likelihood function for the reference sample $\eta_1, \ldots, \eta_N$ for unit variance is determined in the following form:

$$f_\eta(\eta_1, \ldots, \eta_N) = \frac{1}{(2\pi)^{0.5N}} \cdot \exp\left\{-0.5\sum_{i=1}^{N} \eta_i^2\right\} , \tag{11.16}$$

where $\sum_{i=1}^{N} \eta_i^2$ is the sufficient statistic of the likelihood function parameters for the reference sample $\eta_1, \ldots, \eta_N$. In the definition of the sufficient statistics using the input stochastic samples $X_1, \ldots, X_N$ and $\eta_1, \ldots, \eta_N$, the problem of their comparison arises. Usually, for this purpose, a difference is used (see Figure 11.1). The resulting sufficient statistics are observed at the output of the difference device:

$$\ln f_X(X_1, \ldots, X_N) - \ln f_\eta(\eta_1, \ldots, \eta_N) = 0.5\left\{\sum_{i=1}^{N} 2X_i a_i - \sum_{i=1}^{N} X_i^2 + \sum_{i=1}^{N} \eta_i^2 - \sum_{i=1}^{N} a_i^2\right\} . \tag{11.17}$$

It is customary to reference the last term on the right side of Equation (11.17) to a threshold independent of the observed input stochastic sample, as in Equation (11.8). Equation (11.17), obtained by the definition of the resulting sufficient statistics, is the logarithm of the likelihood function.

The signal processing algorithm based on the two independent observed input stochastic samples, one of which is the reference sample with *a priori* information of a "no" signal, follows from Equation (11.17)

$$\ln f_X(X_1,\ldots,X_N) - \ln f_\eta(\eta_1,\ldots,\eta_N) = \ln\left\{\frac{f_X(X_1,\ldots,X_N)}{f_\eta(\eta_1,\ldots,\eta_N)}\right\}$$
$$= 0.5\left\{\sum_{i=1}^{N} 2X_i a_i - \sum_{i=1}^{N} X_i^2 + \sum_{i=1}^{N} \eta_i^2 - \sum_{i=1}^{N} a_i^2\right\} = \ln C \tag{11.18}$$

or

$$\sum_{i=1}^{N} 2X_i a_i - \sum_{i=1}^{N} X_i^2 + \sum_{i=1}^{N} \eta_i^2 = K_g\,, \tag{11.19}$$

where K_g is the threshold.

Proceeding from generally accepted concepts, it follows that the hypothesis H_1 — a "yes" signal in the observed input stochastic sample $X_1, \ldots, X_N$ — is assumed if the following inequality is satisfied:

$$\sum_{i=1}^{N} 2X_i a_i - \sum_{i=1}^{N} X_i^2 + \sum_{i=1}^{N} \eta_i^2 > K_g\,, \tag{11.20}$$

and the hypothesis H_0 — a "no" signal in the observed input stochastic sample $X_1, \ldots, X_N$ — is assumed if the opposite inequality is satisfied. The first term on the left side of Equation (11.20) is the signal processing algorithm given by Equation (11.8) and Equation (11.9) with the factor 2. The more rigorous form of Equation (11.20), based on the analysis performed in Sections 11.1 and 11.2, is the following:

$$\sum_{i=1}^{N} 2X_i a_i^* - \sum_{i=1}^{N} X_i^2 + \sum_{i=1}^{N} \eta_i^2 > K_g\,, \tag{11.21}$$

where a_i^* is the model signal. Letting $N \to \infty$ and transitioning to the integral form, and using the Parseval theorem,[8] we maintain generality and can write

$$2\int_0^T X(t)a_i^*(t)\,dt - \int_0^T X^2(t)\,dt + \int_0^T \eta^2(t)\,dt > K_g, \tag{11.22}$$

where $[0, T]$ is the time interval, within the limits of which the input stochastic process is observed. Analysis of the signal processing algorithm in Equation (11.21) and Equation (11.22), performed by the same procedure as in Sections 11.1 and 11.2, shows that when considering the hypothesis H_1:

$X_i = a_i + \xi_i$ or $X(t) = a(t) + \xi(t)$ given $a_i = a_i^*$ or $a^*(t)$, the left side of Equation (11.21) and Equation (11.22) has the following form:

$$2\sum_{i=1}^{N}[a_i + \xi_i]a_i^* - \sum_{i=1}^{N}[a_i + \xi_i]^2 + \sum_{i=1}^{N}\eta_i^2 = \sum_{i=1}^{N}a_i^2 + \sum_{i=1}^{N}\eta_i^2 - \sum_{i=1}^{N}\xi_i^2 \quad (11.23)$$

or

$$\begin{aligned} &2\int_0^T [a(t)+\xi(t)]a^*(t)\,dt - \int_0^T [a(t)+\xi(t)]^2\,dt + \int_0^T \eta^2(t)\,dt \\ &= \int_0^T a^2(t)\,dt + \int_0^T \eta^2(t)\,dt - \int_0^T \xi^2(t)\,dt \end{aligned}, \quad (11.24)$$

respectively, where the terms $\sum_{i=1}^{N} a_i^2$ and $\int_0^T a^2(t)\,dt$ are the signal energy and the terms $\sum_{i=1}^{N}\eta_i^2 - \sum_{i=1}^{N}\xi_i^2$ and $\int_0^T \eta^2(t)\,dt - \int_0^T \xi^2(t)\,dt$ are the background noise. When considering the hypothesis H_0: $X_i = \xi_i$ or $X(t) = \xi(t)$ and the conditions $a_i = 0$ or $a(t) = 0$ given $a_i^* = a_i$ or $a(t) = a^*(t)$, the left side of Equation (11.20) and Equation (11.21) has the following form: $\sum_{i=1}^{N}\eta_i^2 - \sum_{i=1}^{N}\xi_i^2$, or $\int_0^T \eta^2(t)\,dt - \int_0^T \xi^2(t)\,dt$.

Subsequent analysis of the signal processing algorithm given by Equation (11.21) and Equation (11.22) will only be performed under the conditions $a_i^* = a_i$ or $a(t) = a^*(t)$. This statement is very important for further understanding of the generalized approach to signal processing in the presence of noise. How we do this becomes clear in References 50 and 51. It must be emphasized that $\sum_{i=1}^{N}\eta_i^2 - \sum_{i=1}^{N}\xi_i^2 \to 0$ as $N \to \infty$ or $\int_0^T \eta^2(t)\,dt - \int_0^T \xi^2(t)\,dt \to 0$ as $T \to \infty$ in the statistical sense because the processes ξ_i and η_i, or $\xi(t)$ and $\eta(t)$, are uncorrelated and have the same power spectral density of the additive Gaussian noise $0.5N_0$ according to the initial conditions. Thus, it has been shown that both the signal processing algorithms based on the observed input stochastic sample $X_1, \ldots, X_N$ and the two independently observed input stochastic samples $X_1, \ldots, X_N$ and $\eta_1, \ldots, \eta_N$ have the same approach and are defined by the likelihood function using the statistical theory of decision-making. The difference is that the numerator and denominator of the likelihood function used for the synthesis of the algorithms given by Equation (11.8) and Equation (11.9) involve the same observed input stochastic sample [see Equation (11.4) and Equation (11.6)], but a "yes" signal is assumed in the numerator and a "no" signal in the denominator. The numerator of the likelihood function used for synthesis of the algorithm given by Equation (11.21) and Equation (11.22) involves the observed input stochastic sample, where a "yes" signal may be present and the denominator involves the reference sample, which is known *a priori* to contain a "no" signal.

On this basis, it may be stated that only the sufficient statistic $\sum_{i=1}^{N} X_i a_i$ or $\int_0^T X(t)a(t)\,dt$ has been applied to define the mean of the likelihood function in the signal processing algorithm given by Equation (11.8) and Equation (11.9), respectively. In the algorithm given by Equation (11.21) and Equation (11.22), the jointly sufficient statistics $\sum_{i=1}^{N} 2X_i a_i$ and $\sum_{i=1}^{N} (\eta_i^2 - X_i^2)$ or $2\int_0^T X(t)\,a(t)\,dt$ and $\int_0^T [\eta^2(t) - X^2(t)]\,dt$ are used to define the mean and variance of the likelihood function. This fact permits us to obtain more complete information in the decision-making process in comparison to the algorithm given by Equation (11.8) and Equation (11.9). The algorithm given by Equation (11.21) and Equation (11.22) is free from a number of conditions unique to the algorithm given by Equation (11.8) and Equation (11.9). As the algorithm given by Equation (11.8) and Equation (11.9) is a component of the algorithm given by Equation (11.21) and Equation (11.22), the last has been called the generalized signal processing algorithm.

11.5 The Engineering Interpretation

The technical realization of independent sampling from the regions Z and Z^* obeying the same probability distribution density with the same statistical parameters is not difficult. The solution of the problem of detecting the signal $a(t)$ with the additive Gaussian noise $n(t)$ is well known.[1–49,55] The observed input stochastic process $X(t)$ is examined at the output of the linear section of the receiver, which has an ideal amplitude-frequency response and the bandwidth ΔF. The noise at the linear section input of the receiver is the additive "white" Gaussian noise, having the correlation function $0.5N_0\delta(t_2 - t_1)$, where $\delta(x)$ is the delta function.

The signal $a(t)$ is assumed to be completely known, and the signal energy is taken to be equal to 1. The power spectral density $0.5N_0$ is considered an *a priori* indeterminate parameter. The gain of the linear section of the receiver is equal to 1. Through analysis, the problem is reduced to testing the complex hypothesis with the decision function

$$\mathrm{Re}\int_0^T \dot{X}(t)\dot{a}^*(t)\,dt > K(P_F)\sqrt{\int_0^T |\dot{X}(t)|^2\,dt}, \tag{11.25}$$

where $\dot{a}^*(t)$ is the filter matched with the signal, P_F is the threshold defined by the probability of false alarm $K(P_F)$, and $\int_0^T |\dot{X}(t)|^2\,dt$ is the statistic defining the decision function.

It turns out that the signal detector constructed in accordance with the above decision function renders the probability of false alarm P_F stable, given an unknown noise power, and has the greatest probability of detection P_D for any signal-to-noise ratio.[28] Let us interpret this problem. We use two linear sections of the receiver (instead of one section) for our set of statistics. These linear sections will be called the preliminary (PF) and additional (AF) filters. The amplitude-frequency responses of PF and AF must obey the same law. The resonant frequencies of PF and AF must be detuned relative to each other by a value determined from the well-known results in References 56 and 57, for the purpose of providing uncorrelated statistics at the outputs of PF and AF. The detuning value between the resonant frequencies of PF and AF exceeds the effective signal bandwidth ΔF_a. As is well known,[56,57] if this value reaches more than $4\Delta F_a$, the coefficient of correlation between the statistics at the outputs of PF and AF tends to approach zero. In practice, these statistics may be regarded as uncorrelated. The effective bandwidth of PF is equal to that of the signal frequency spectrum and can be even greater, but this is undesirable because the noise power at the output of PF is proportional to the effective bandwidth. The effective bandwidth of AF may be smaller than that of PF; however, for simplicity of analysis, in this chapter the effective bandwidth of AF is assumed to be the same as that of PF.

Thus, we can assume that uncorrelated samples of the observed input stochastic process are formed at the outputs of PF and AF. These samples obey the same probability distribution density with the same statistical parameters given that the same process is present at the inputs of PF and AF, even if this process is the additive "white" Gaussian noise having the correlation function $0.5N_0\delta(t_2 - t_1)$. The physicotechnical interpretation of the signal processing algorithm given by Equation (11.21) and Equation (11.22) is the following. AF may serve as the source of the observed reference sample $\eta_1, \ldots, \eta_N$ from the interference region Z^*. The AF resonant frequency is detuned relative to the carrier frequency of the signal by a value that can be determined on the basis of well-known results,[19,56,57] depending on the specific practical situation. PF serves as the source of the sample $X_1, \ldots, X_N$ of the observed input stochastic process from the interference region Z. The bandwidth of PF is matched with the effective bandwidth of the signal. The bandwidth value of PF is matched with that of AF. The first term of the generalized signal processing algorithm, given by Equation (11.21) and Equation (11.22) corresponds to the synthesis of the correlation channel with twice the gain. The second term, given by Equation (11.21) and Equation (11.22), corresponds to the synthesis of the autocorrelation channel coupled with PF. The third term, given by Equation (11.21) and Equation (11.22), corresponds to the synthesis of the autocorrelation channel coupled with AF. The statistic of the autocorrelation channel coupled with PF is subtracted from the statistic of the autocorrelation channel coupled with AF. As a result, $\sum_{i=1}^{N} \eta_i^2 - \sum_{i=1}^{N} \xi_i^2 \to 0$ as $N \to \infty$ or $\int_0^T \eta^2(t)\,dt - \int_0^T \xi^2(t)\,dt \to 0$ as $N \to \infty$ in the statistical sense. The statistic of the autocorrelation channel coupled with PF is subtracted from that of the correlation channel. As a result, a complete compensation of the noise component $\sum_{i=1}^{N} a_i^* \xi_i$ or $\int_0^T a^*(t)\xi(t)\,dt$ of the signal processing algorithm given by Equation (11.8) and Equation (11.9) is achieved in the statistical sense if the conditions $a_i^* = a_i$ or $a(t) = a^*(t)$ are satisfied, where a_i^* or $a^*(t)$ is the model signal, and a_{1i}, or a_1 is the signal at the PF output. The detector shown in Figure 11.2 is based on the physicotechnical interpretation of the generalized approach to signal processing in noise[50,51,58–61] stated in the preceding text.

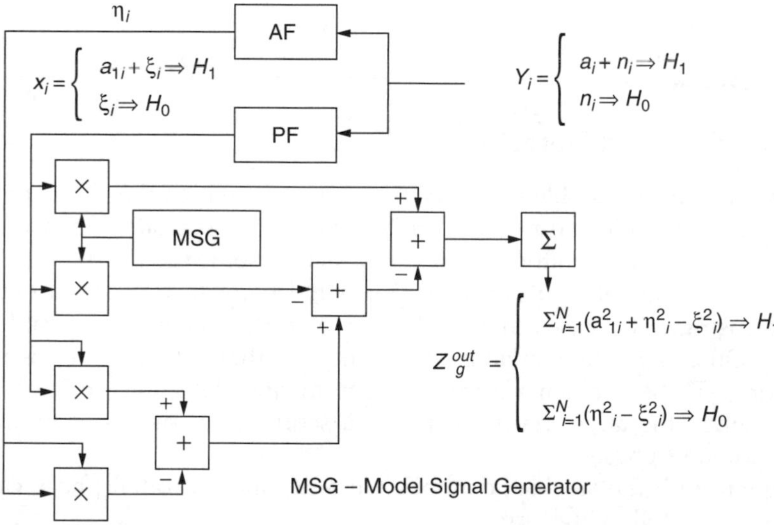

FIGURE 11.2
The physicotechnical interpretation of the generalized detector.

It is of special interest to compare these statements to the statements and analysis in Bacut et al.[29] Some opponents of the generalized approach to signal processing in the presence of noise erroneously believe that this approach is the same as the one-input two-sample signal processing approach discussed in Bacut et al.[29] For this purpose, we briefly recall the main statements of the one-input two-sample signal processing algorithm. First, the signal sample is generated only at the time instants that correspond to the expected signal. In other words, the signal sample is generated only at the time instants when the expected signal may appear in the time frequency space. Second, the noise channel is formed within the limits of the time intervals in which it is *a priori* known that the signal is absent.

The generalized approach to signal processing in the presence of noise is based on the following statements. The first sample is generated independently of the time instants that correspond to the emergence of the signal in the time frequency space. The reference sample — the second sample — is formed simultaneously with the first at the same time intervals as the first and exists without any limitations in time or readings, but it is known *a priori* that a "no" signal exists in the reference sample due to the conditions for generating the reference sample. The sample sizes of the first (signal) and second (reference) samples are the same.

These differences between the generalized approach in the presence of noise and the one-input two-sample approach are very important. In addition, we can see that the engineering interpretation of the generalized approach in the presence of noise differs greatly from the one-input two-sample approach in Bacut et al.[29]

11.6 Generalized Detector

Let us consider the problem of specific interest, in which the signal has the stochastic amplitude and random initial phase. The necessity of considering this problem stems from the fact that, in practice, some satellite navigational systems use channels with an ionosphere mechanism of propagation and operate using frequencies that are higher than the maximal allowable frequency. Other satellite navigational systems use channels with tropospheric scattering. These problems arise in radar during detection of fluctuating targets when the target return signal is a sequence of pulses of unknown amplitude and phase.

A signal with the stochastic amplitude and random initial phase can be written in the following form:

$$a(t, A, \varphi_0) = A(t)S(t)\cos[\omega_0 t + \Psi_a(t) - \varphi_0], \tag{11.26}$$

where ω_0 is the carrier frequency of the signal $a(t, A, \varphi_0)$; $S(t)$ is the known modulation law of the amplitude of the signal $a(t, A, \varphi_0)$; $\Psi_a(t)$ is the known modulation law of the phase of the signal $a(t, A, \varphi_0)$; φ_0 is the random initial phase of the signal $a(t, A, \varphi_0)$ which is uniformly distributed within the limits of the interval $[-\pi, \pi]$ and is time variant within the limits of the time interval $[0, T]$; and $A(t)$ is the amplitude factor, which is a random value and a function of time in the general case. Let us consider the generalized approach to signal processing in the presence of noise for the signals with the stochastic amplitude and random initial phase.[63–66] According to the generalized approach, it is necessary to form the following: (1) the reference sample with *a priori* information of a "no" signal in the input stochastic process; (2) the autocorrelation channel for the purpose of compensating for the noise component of the correlation channel, which is caused by the interaction between the model signal and noise.

The main principles of construction of the generalized detector for the signals with the stochastic amplitude and random initial phase using the Neyman–Pearson criterion are in References 66 to 70. The input stochastic process $Y(t)$ must pass through the preliminary filter (PF). The effective bandwidth of PF is equal to ΔF_a, where ΔF_a is the effective spectrum bandwidth of the signal.

$$X(t) = a_1(t, A, \varphi_0) + \xi(t) \tag{11.27}$$

is the process at the output of PF, if a "yes" signal exists in the input stochastic process — the hypothesis H_1.

$$X(t) = \xi_1(t) \tag{11.28}$$

is the process at the output of PF if a "no" signal exists in the input stochastic process — the hypothesis H_0; $a_1(t, A, \varphi_0)$ is the signal at the output of PF given by Equation (11.26); and $\xi(t)$ is the noise at the output of PF.

We need to form the reference sample for the generation of the jointly sufficient statistics of the likelihood function mean and variance. For this purpose, the additional filter (AF) is formed parallel to PF. The amplitude-frequency response of AF is analogous over the entire range of parameters to that of PF, but it is detuned in the resonant frequency relative to PF for the purpose of providing uncorrelated statistics at the outputs of PF and AF. The detuning value must be larger than the effective spectrum bandwidth of the signal so that the processes at the outputs of PF and AF will be uncorrelated. As was shown in References 56 and 57, if this detuning value reaches the value between $4\Delta F_a$ and $5\Delta F_a$, the processes at the outputs of PF and AF are not correlated practically. In this condition, the correlation coefficient between the statistics at the outputs of PF and AF is not more than 0.05 for all practical purposes. It may be considered as a value tending to approach zero. Thus, the process $\eta(t)$ is formed at the output of AF:

$$\eta(t) = \xi_2(t)\cos[\omega'_n t + \upsilon'(t)], \tag{11.29}$$

where $\xi_2(t)$ is the random envelope of the amplitude of the noise at the output of AF; $\upsilon'(t)$ is the random phase of the noise at the output of AF; and ω'_n is the medium frequency of the noise at the output of AF.

The generalized detector for the signals with the stochastic amplitude and random initial phase is shown in Figure 11.3. There are two filters with the nonoverlapping amplitude-frequency responses that must obey the same law: PF and the AF. PF is matched with the effective spectrum bandwidth of the signal with the carrier frequency ω_0. AF does not pass the frequency ω_0. By this means, there is the following requirement for AF: the resonant frequency of AF must be detuned with respect to that of PF to ensure the uncorrelated statistics at the outputs of both AF and PF.[66–72] This requirement is necessary to ensure the complete compensation of the background noise constant component at the output of the generalized detector in the statistical sense. For the generation of the jointly sufficient statistics of the likelihood function mean and variance, in accordance with the generalized approach to signal processing in the presence of noise, it is necessary to form the autocorrelation channel. These actions allow us to compensate for the total noise component in the statistical sense. We now proceed to show this.

The generalized detector for the signals with the stochastic amplitude and random initial phase shown in Figure 11.3 consists of the correlation channel (the multipliers *1* and *2*, the integrators *1* and *2*, the square-law function generators *5* and *6*, the summator 2, the model signal generator MSG); the autocorrelation channel [the multipliers *3* and *4*, the summator *1*, the integrator *3*, the square-law function generator *7*, the amplifier (>)]; the compensating channel (the summators 3 and 4, the compensation of the total noise component in the statistical sense is carried out by the summators *3* and *4*); the delay blocks *1–5* are only used for specific technical problems, and are not taken into consideration during the analysis of the theoretical principles of functionality. The compensating channel of the generalized detector allows us to compensate the correlation channel noise component and the autocorrelation channel random component both of the generalized detector, in the statistical sense. The noise component is created by the interaction between the model signal and noise. The random component, which will be described below, is caused by the interaction between the signal and noise.

Let us analyze the generalized detector shown in Figure 11.3 for the condition

$$S(t) = S^*(t), \tag{11.30}$$

i.e., the model signal $a^*(t)$ is completely matched with the signal $a_1(t, A, \varphi_0)$ at the output of PF, and $\tau = 0$. How we are able to do this becomes clear in the discussion of the experimental and application results presented in References 50 and 51. We must take into consideration that the frequencies $2\omega_0$, $2\omega_n$, or $2\omega'_n$ cannot pass through PF and AF, respectively. The analysis is

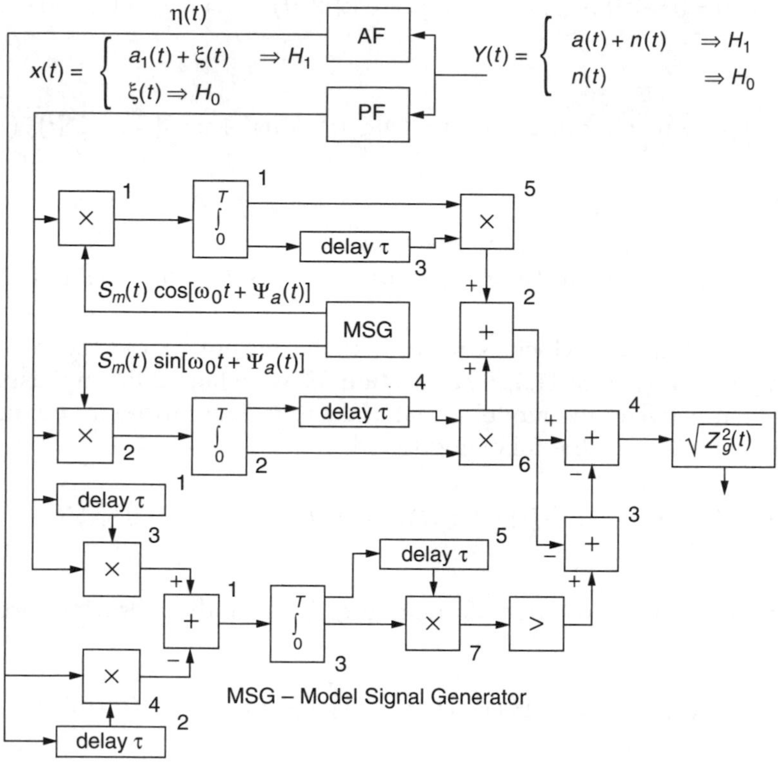

FIGURE 11.3
The generalized detector for the signal with stochastic parameters.

based on the results discussed in References 62 to 67[8] and the hypothesis H_1 — a "yes" signal in the input stochastic process.

In terms of Equation (11.26), the processes at the outputs of the multipliers *1–4* take the following form:

$$\begin{aligned} y_1(t) &= [a_1(t, A, \varphi_0) + \xi(t)]a_1^*(t) \\ &= \{A(t)S(t)\cos[\omega_0 t + \Psi_a(t) - \varphi_0] + \xi_1(t)\cos[\omega_n t + \upsilon(t)]\}S^*(t)\cos[\omega_0 t + \Psi_a(t)] \\ &= 0.5\cos\varphi_0 A(t)S(t)S^*(t) + 0.5S^*(t)\xi_1(t)\cos[\Psi_a(t) - \upsilon(t)]; \end{aligned} \tag{11.31}$$

$$\begin{aligned} y_2(t) &= [a_1(t, A, \varphi_0) + \xi(t)]a_2^*(t) \\ &= \{A(t)S(t)\cos[\omega_0 t + \Psi_a(t) - \varphi_0] + \xi_1(t)\cos[\omega_n t + \upsilon(t)]\}S^*(t)\sin[\omega_0 t + \Psi_a(t)] \\ &= 0.5\sin\varphi_0 A(t)S(t)S^*(t) + 0.5S^*(t)\xi_1(t)\sin[\Psi_a(t) - \upsilon(t)]; \end{aligned} \tag{11.32}$$

$$y_3(t) = 0.5A^2(t)S^2(t) + A(t)S(t)\xi_1(t)\cos[\Psi_a(t) - \upsilon(t) - \varphi_0] + 0.5\xi_1^2(t)\,; \tag{11.33}$$

$$y_4(t) = \eta^2(t) = \xi_2(t)\cos[\omega_n' + \upsilon'(t)]\xi_2(t)\cos[\omega_n' + \upsilon'(t)] = 0.5\xi_2^2(t)\,, \tag{11.34}$$

where

$$a_1^*(t) = S^*(t)\cos[\omega_0 t + \Psi_a(t)] \quad \text{and} \quad a_2^*(t) = S^*(t)\sin[\omega_0 t + \Psi_a(t)] \tag{11.35}$$

is the model signal, which is formed at the output of the model signal generator MSG (Figure 11.3). The notation $S^*(t)$ is retained to emphasize the interaction between the model signal and noise. The process at the output of summator *1* takes the following form:

$$y_5(t) = 0.5A^2(t)S^2(t) + A(t)S(t)\xi_1(t)\cos[\Psi_a(t) - \upsilon(t) - \varphi_0] - 0.5[\xi_2^2(t) - \xi_1^2(t)]. \tag{11.36}$$

The processes at the outputs of integrators *1–3* take the following form:

$$Z_1(t) = 0.5\cos\varphi_0 \int_0^T A(t)S(t)S^*(t)\,dt + 0.5\int_0^T S^*(t)\xi_1(t)\cos[\Psi_a(t) - \upsilon(t)]\,dt\,; \tag{11.37}$$

$$Z_2(t) = 0.5\sin\varphi_0 \int_0^T A(t)S(t)S^*(t)\,dt + 0.5\int_0^T S^*(t)\xi_1(t)\sin[\Psi_a(t) - \upsilon(t)]\,dt\,; \tag{11.38}$$

$$Z_3(t) = 0.5\int_0^T A^2(t)S^2(t)\,dt + \int_0^T A(t)S(t)\xi_1(t)\cos[\Psi_a(t) - \upsilon(t) - \varphi_0]\,dt - 0.5\int_0^T [\xi_2^2(t) - \xi_1^2(t)]\,dt. \tag{11.39}$$

It should be particularly emphasized that all integration operations must be read in the statistical sense.

Let us consider the process $Z_3(t)$ at the output of integrator *3* given by Equation (11.39). The first term of the process $Z_3(t)$ is proportional to the energy of the signal within the limits of the time interval [0, *T*]. The second term is the random component of the autocorrelation channel of the generalized detector, which is caused by the interaction between the signal and noise. The third term is the difference between the powers of the noise, which are formed at the outputs of PF and AF, respectively. The third term does not participate in compensating the noise components of the generalized detector correlation channel $\int_0^T S(t)S^*(t)\,dt \int_0^T A(t)S^*(t)\xi_1(t)\cos[\Psi_a(t) - \upsilon(t) - \varphi_0]dt$ and $\left\{\int_0^T S^*(t)\xi_1(t)\,dt\right\}^2$, which are caused by the interaction between the model signal and noise. The third term has the following physical sense. The integral $\int_0^T [\xi_2^2(t) - \xi_1^2(t)]\,dt$ is the background noise at the output of the generalized detector and tends to approach zero as $T \to \infty$ in the statistical sense (see Section 11.4). The background noise is only used for definition of the threshold K_g during decision making. Based on this statement, the third term of the process $Z_3(t)$ given by Equation (11.39) can be discarded in the following analysis, but we will take it into account in the end result.

The processes at the outputs of the square-law function generators *1* and *2* take the following form:

$$\begin{aligned}
Z_1^2(t) &= 0.25\cos^2\varphi_0 \int_0^T A^2(t)S(t)S^*(t)\,dt \int_0^T S(t)S^*(t)\,dt \\
&+ 0.25\left\{\int_0^T S^*(t)\xi_1(t)\cos[\Psi_a(t) - \upsilon(t)]\,dt\right\}^2 \\
&+ 0.5\cos\varphi_0 \int_0^T S(t)S^*(t)\,dt \int_0^T A(t)S^*(t)\xi_1(t)\cos[\Psi_a(t) - \upsilon(t)]\,dt,
\end{aligned} \tag{11.40}$$

$$\begin{aligned}
Z_2^2(t) &= 0.25\sin^2\varphi_0 \int_0^T A^2(t)S(t)S^*(t)\,dt \int_0^T S(t)S^*(t)\,dt \\
&+ 0.25\left\{\int_0^T S^*(t)\xi_1(t)\sin[\Psi_a(t) - \upsilon(t)]\,dt\right\}^2 \\
&+ 0.5\sin\varphi_0 \int_0^T S(t)S^*(t)\,dt \int_0^T A(t)S^*(t)\xi_1(t)\sin[\Psi_a(t) - \upsilon(t)]\,dt.
\end{aligned} \tag{11.41}$$

Using the straightforward mathematical transformations in References 25 and 33, it is not difficult to show that

$$0.25\left\{\int_0^T S^*(t)\xi_1(t)\cos[\Psi_a(t)-\upsilon(t)]\,dt\right\}^2$$
$$=0.25\left\{\int_0^T S^*(t)\xi_1(t)\sin[\Psi_a(t)-\upsilon(t)]\,dt\right\}^2 \quad (11.42)$$
$$=0.25\left\{\int_0^T S^*(t)\xi_1(t)\,dt\right\}^2.$$

The process at the output of summator 2 in terms of Equation (11.42) has the following form:

$$Z_\Sigma^2(t)=0.25\int_0^T A^2(t)S(t)S^*(t)\,dt\int_0^T S(t)S^*(t)\,dt+0.25\left\{\int_0^T S^*(t)\xi_1(t)\,dt\right\}^2$$
$$+0.5\int_0^T S(t)S^*(t)\,dt\int_0^T A(t)S^*(t)\xi_1(t)\cos[\Psi_a(t)-\upsilon(t)-\varphi_0]\,dt. \quad (11.43)$$

The process at the output of the square-law function generator 7 in terms of Equation (11.42) takes the following form:

$$Z_3^2(t)=0.25\int_0^T A^2(t)S(t)^2\,dt\int_0^T A^2(t)S^2(t)\,dt$$
$$+0.25\left\{\int_0^T A(t)S^*(t)\xi_1(t)\,dt\right\}^2 \quad (11.44)$$
$$+\int_0^T A^2(t)S^2(t)\,dt\int_0^T A(t)S(t)\xi_1(t)\cos[\Psi_a(t)-\upsilon(t)-\varphi_0]\,dt.$$

Considering the second and third terms in Equation (11.43) and Equation (11.44), respectively, we can see that they differ by the factor $A^2(t)$ under the condition given by Equation (11.30), and the second and third terms in Equation (11.42) and Equation (11.43) agree within the factor of 2.

Before proceeding to questions of compensation of these terms in the statistical sense, let us consider two plausible cases. The first case implies that the amplitude envelope of the signal is not the stochastic function of time within the limits of the time interval $[0, T]$ for a single sample and can be stochastic from sample to sample — the case of the slow fluctuations. This case is very important for certain radar navigational systems where the signal may occur over and over. The second case is based on the fact that the envelope of the amplitude of the signal is the stochastic function of time within the limits of the time interval $[0, T]$ for a single sample, or in other words, the case of the rapid fluctuations of the envelope of amplitude of the signal. Let us consider both cases in detail.

11.6.1 The Case of the Slow Fluctuations

In this case, the compensation between the second and third terms in Equation (11.43) and Equation (11.44) in the statistical sense is conceivable by averaging on a set M of realizations of the input stochastic process $X(t)$. All statistical characteristics of the input stochastic process $X(t)$ are invariant within the limits of the time interval $[0, T]$. Then, Equation (11.44) may be written in the following form:

$$\sum_{j=1}^{M} Z_{3_j}^2(t) = 0.25\sum_{j=1}^{M}\left\{\int_0^T A_j^2(t)S_j^2(t)\,dt\int_0^T A_j^2(t)S_j^2(t)\,dt\right\}$$

$$+0.5\sum_{j=1}^{M}\left\{\int_0^T A_j(t)S_j(t)\xi_{1_j}(t)\,dt\right\}^2$$

$$+\sum_{j=1}^{M}\left\{\int_0^T A_j^2(t)S_j^2(t)\,dt\int_0^T A_j(t)S_j(t)\xi_{1_j}(t)\cos[\Psi_{a_j}(t)-\upsilon_j(t)-\varphi_{0_j}]\,dt\right\}. \tag{11.45}$$

Equation (11.43) may be written in identical form:

$$\sum_{j=1}^{M} Z_{\Sigma_j}^2(t) = 0.25\sum_{j=1}^{M}\left\{\int_0^T A_j^2(t)S_j(t)S_j^*\,dt\int_0^T S_j(t)S_j^*(t)\,dt\right\}$$

$$+0.25\sum_{j=1}^{M}\left\{\int_0^T S_j^*(t)\xi_{1_j}(t)\,dt\right\}^2$$

$$+0.5\sum_{j=1}^{M}\left\{\int_0^T S_j(t)S_j^*(t)\,dt\int_0^T A_j(t)S_j^*(t)\xi_{1_j}(t)\cos[\Psi_{a_j}(t)-\upsilon_j(t)-\varphi_{0_j}]\,dt\right\}. \tag{11.46}$$

We introduce the designations for the condition given by Equation (11.30)

$$E_{a_{1j}} = \int_0^T S_j(t)S_j^*(t)\,dt \quad \text{and} \quad \sigma_{A_j}^2 E_{a_{1j}} = \int_0^T A_j^2(t)S_j(t)S_j^*(t)\,dt, \tag{11.47}$$

where $E_{a_{1j}}$ is the energy of the signal within the limits of the time interval in the j-th realization of the input stochastic process, and $\sigma_{A_j}^2$ is the variance of the amplitude envelope factor $A_j(t)$ of the signal within the limits of the time interval $[0, T]$ in the j-th realization of the input stochastic process. An analogous representation of the variance $\sigma_{A_j}^2$ is used in References 8 and 16.[8,16]

Let us consider the second term in Equation (11.45). This term can be represented as a product of two integrals:

$$\sum_{j=1}^{M}\left\{\int_0^T A_j(t)S_j(t)\xi_{1_j}(t)\,dt\right\}^2 = \sum_{j=1}^{M}\left\{\int_0^T dt\int_0^T A_j^2(t)S_j(t)S_j(\tau)\xi_{1_j}(t)\xi_{1_j}(\tau)\,d\tau\right\}.$$

Averaging the integrand with respect to the amplitude envelope factor $A_j(t)$ in the j-th realization for a set M of realizations of the input stochastic process, we can write

$$\int_0^T dt\int_0^T \overline{A_j^2(t)S_j(t)S_j(\tau)\xi_{1_j}(t)\xi_{1_j}(\tau)}\,d\tau = \sigma_{A_j}^2\int_0^T dt\int_0^T S_j(t)S_j(\tau)\xi_{1_j}(t)\xi_{1_j}(\tau)\,d\tau. \tag{11.48}$$

In terms of Equation (11.47) and Equation (11.48), Equation (11.45) and Equation (11.46) can take the following form:

$$\begin{aligned}\sum_{j=1}^{M} Z_{3_j}^2(t) &= 0.25\sum_{j=1}^{M}\sigma_{A_j}^2 E_{a_{1j}}\int_0^T A_j^2(t)S_j^2(t)\,dt \\ &+ 0.5\sum_{j=1}^{M}\sigma_{A_j}^2\left\{\int_0^T S_j(t)\xi_{1_j}(t)\,dt\right\}^2 \\ &+ \sum_{j=1}^{M}\sigma_{A_j}^2 E_{a_{1j}}\int_0^T A_j(t)S_j(t)\xi_{1_j}(t)\cos[\Psi_{a_j}(t) - \upsilon_j(t) - \varphi_{0_j}]\,dt\,;\end{aligned} \tag{11.49}$$

$$\sum_{j=1}^{M} Z_{\Sigma_j}^2(t) = 0.25\sum_{j=1}^{M} E_{a_{1j}} \int_0^T A_j^2(t) S^*(t) S_j(t)\, dt$$
$$+\,0.25\sum_{j=1}^{M} \sigma_{A_j}^2 \left\{ \int_0^T S_j^*(t)\xi_{1_j}(t)\, dt \right\}^2 \tag{11.50}$$
$$+\,0.5\sum_{j=1}^{M} E_{a_{1j}} \int_0^T A_j(t) S_j^*(t)\xi_{1_j}(t)\cos[\Psi_{a_j}(t) - \upsilon_j(t) - \varphi_{0_j}]\, dt.$$

Comparing Equation (11.49) with Equation (11.50), we can see that the second and third terms differ by value $\sum_{j=1}^{M} \sigma_{A_j}^2$ and by the factor 2. Because of this, the amplifier (>) of the autocorrelation channel of the generalized detector has the amplification factor $\frac{1}{\sigma_A^2}$ and is connected to the input of the compensating channel of the generalized detector. Taking this fact into account, we may write Equation (11.49) in the following form:

$$\sum_{j=1}^{M} Z_{3_j}^2(t) = 0.25\sum_{j=1}^{M} \sigma_{A_j}^2 E_{a_{1j}} \int_0^T A_j^2(t) S_j^2(t)\, dt$$
$$+\,0.5\sum_{j=1}^{M} \sigma_{A_j}^2 \left\{ \int_0^T S_j(t)\xi_{1_j}(t)\, dt \right\}^2 \tag{11.51}$$
$$+\sum_{j=1}^{M} \sigma_{A_j}^2 E_{a_{1j}} \int_0^T A_j(t) S_j(t)\xi_{1_j}(t)\cos[\Psi_{a_j}(t) - \upsilon_j(t) - \varphi_{0_j}]\, dt.$$

Taking into consideration the condition given by Equation (11.30) and the discarded third term of the process $Z_3(t)$ in Equation (11.39), the process at the output of the compensating channel of the generalized detector — the output of the summator *4* — takes the following form:

$$[Z_g^{out}(t)]^2 = 0.25\sum_{j=1}^{M} E_{a_{1j}} \int_0^T A_j^2(t) S_j^2(t)\, dt + 0.25\sum_{j=1}^{M} \left\{ \int_0^T [\xi_{2_j}^2(t) - \xi_{1_j}^2(t)]\, dt \right\}^2. \tag{11.52}$$

Reference to Equation (11.52) shows that the compensation between the second and third terms in Equation (11.43) and Equation (11.44), in the statistical sense, is performed at the output of the compensating channel of the generalized detector — the output of summator 4 — during averaging on a set M of realizations of the input stochastic process $X(t)$. It should be considered that all integration operations are implied in the statistical sense.

In this conjunction, the process at the output of the generalized detector takes the following form:

$$Z_g^{out}(t) = 0.5\sqrt{\sum_{j=1}^{M} E_{a_{1j}} \int_0^T A_j^2(t) S_j^2(t)\, dt + \sum_{j=1}^{M} \left\{ \int_0^T [\xi_{2_j}^2(t) - \xi_{1_j}^2(t)]\, dt \right\}^2} . \quad (11.53)$$

The first term in Equation (11.53), at the output of the generalized detector, is the energy of the signal, and the second term is the background noise that tends to approach zero in the statistical sense.

11.6.2 The Case of the Rapid Fluctuations

In this case, the envelope of the amplitude of the signal is the stochastic function of the time within the limits of the time interval $[0, T]$ for a single realization of the input stochastic process $X(t)$. The process at the output of summator 2 — the output of the correlation channel of the generalized detector — takes the following form:

$$\begin{aligned} Z_\Sigma^2(t) &= 0.25 \int_0^T S^*(t) S(t)\, dt \int_0^T A^2(t) S^*(t) S(t)\, dt \\ &+ 0.25 \left\{ \int_0^T S^*(t) \xi_1(t)\, dt \right\}^2 \\ &+ 0.5 \int_0^T S^*(t) S(t)\, dt \int_0^T A(t) S^*(t) \xi_1(t) cvos[\Psi_a(t) - \upsilon(t) - \varphi_0]\, dt. \end{aligned} \quad (11.54)$$

The process at the output of the autocorrelation channel of the generalized detector — the output of square-law function generator 7 — takes the following form:

$$\begin{aligned} Z_3^2(t) &= 0.25 \int_0^T A^2(t) S^2(t)\, dt \int_0^T A^2(t) S^2(t)\, dt \\ &+ 0.5 \left\{ \int_0^T A(t) S(t) \xi_1(t)\, dt \right\}^2 \\ &+ \int_0^T A^2(t) S^2(t)\, dt \int_0^T A(t) S(t) \xi_1(t) \cos[\Psi_a(t) - \upsilon(t) - \varphi_0]\, dt. \end{aligned} \quad (11.55)$$

We proceed to introduce the designations in accordance with the results discussed in References 71 to 74 for the condition given by Equation (11.30):

$$E_{a_1} = \int_0^T S^*(t)S(t)\,dt \quad \text{and} \quad \sigma_A^2 E_{a_1} = \int_0^T A^2(t)S^*(t)S(t)\,dt\,, \tag{11.56}$$

where E_{a_1} is the energy of the signal within the limits of the time interval [0, T], and σ_A^2 is the variance of the amplitude envelope factor $A(t)$ within the limits of the time interval [0, T].

Let us consider the second term in Equation (11.55) that can be expressed as the product of two integrals.[25] Let us consider the double integral $\left\{\int_0^T A(t)S(t)\xi_1(t)\,dt\right\}^2 = \int_0^T dt \int_0^T A(t)A(\tau)S(t)S(\tau)\xi_1(t)\xi_1(\tau)\,d\tau$. Averaging the integrand with respect to the amplitude envelope factor $A(t)$ within the limits of the time interval [0, T], we can write

$$\int_0^T dt \int_0^T \overline{A(t)A(\tau)}S(t)S(\tau)\xi_1(t)\xi_1(\tau)\,d\tau = \sigma_A^2 \int_0^T dt \int_0^T S(t)S(\tau)\xi_1(t)\xi_1(\tau)\,d\tau\,. \tag{11.57}$$

Equation (11.54) and Equation (11.55) take the following form in terms of Equation (11.56) and Equation (11.57):

$$\begin{aligned} Z_\Sigma^2(t) &= 0.25E_{a_1}\int_0^T A^2(t)S^*(t)S(t)\,dt + 0.25\left\{\int_0^T S^*(t)\xi_1(t)\,dt\right\}^2 \\ &+ 0.5E_{a_1}\int_0^T A(t)S^*(t)\xi_1(t)\cos[\Psi_a(t) - \upsilon(t) - \varphi_0]dt; \end{aligned} \tag{11.58}$$

$$\begin{aligned} Z_3^2(t) &= 0.25\sigma_A^2 E_{a_1}\int_0^T A^2(t)S^2(t)\,dt + 0.5\sigma_A^2\left\{\int_0^T S(t)\xi_1(t)\,dt\right\}^2 \\ &+ \sigma_A^2 E_{a_1}\int_0^T A(t)S(t)\xi_1(t)\cos[\Psi_a(t) - \upsilon(t) - \varphi_0]dt\,. \end{aligned} \tag{11.59}$$

Comparing Equation (11.58) with Equation (11.59), we can see that the second and third terms differ by the variance σ_A^2 of the amplitude envelope factor and a factor of 2. Because of this, the amplifier (>) of the generalized detector autocorrelation channel has the amplification factor $\frac{1}{\sigma_A^2}$ and is connected to the input of the generalized detector compensating channel. Taking this fact into consideration, Equation (11.59) can be written in the following form:

$$Z_3^2(t) = 0.25E_{a_1}\int_0^T A^2(t)S^2(t)\,dt + 0.5\left\{\int_0^T S(t)\xi_1(t)\,dt\right\}^2 + E_{a_1}\int_0^T A(t)S(t)\xi_1(t)\cos[\Psi_a(t) - \upsilon(t) - \varphi_0]\,dt. \tag{11.60}$$

For the condition given by Equation (11.30) and in terms of the discarded third term of the process $Z_3(t)$ in Equation (11.39), the process at the output of the generalized detector compensating channel — the output of summator *4* — takes the following form:

$$\left[Z_g^{out}(t)\right]^2 = 2Z_\Sigma^2(t) - Z_3^2(t) = 0.25E_{a_1}\int_0^T A^2(t)S^2(t)\,dt + 0.25\left\{\int_0^T [\xi_2^2(t) - \xi_1^2(t)]\,dt\right\}^2. \tag{11.61}$$

Equation (11.61) shows that the compensation between the second and third terms in Equation (11.43) and Equation (11.44), in the statistical sense, is performed at the output of the generalized detector compensating channel — the output of summator *4* — during averaging of the input stochastic process $X(t)$ within the limits of the time interval $[0, T]$. It should be noted that all integration operations are implied in the statistical sense. The process at the output of the generalized detector takes the following form:

$$Z_g^{out}(t) = 0.5\sqrt{E_{a_1}\int_0^T A^2(t)S^2(t)\,dt + \left\{\int_0^T [\xi_2^2(t) - \xi_1^2(t)]\,dt\right\}^2}. \tag{11.62}$$

The first term in Equation (11.62) is the energy of the signal at the output of the generalized detector, and the second term is the background noise at the output of the generalized detector that tends to approach zero in the statistical sense.

Thus, if the envelope of amplitude of the signal is the stochastic function of the time within the limits of the time interval $[0, T]$, it is possible to perform the compensation between the correlation channel noise component of the generalized detector [which is created by the interaction between the model signal and noise — the second and third terms in Equation (11.54)] and the random component of the autocorrelation channel of the generalized detector [which is caused by the interaction between the signal and noise — the second and third terms in Equation (11.55)], without averaging of the input stochastic process $X(t)$ on a set of realizations. This effect of compensation takes place within the limits of the time interval $[0, T]$. To attain these ends, the amplification factor of the generalized detector's autocorrelation channel amplifier (>) must differ by $\frac{1}{\sigma_A^2}$ in comparison with the correlation channel amplification factor.

Generation of the jointly sufficient statistics of the likelihood function mean and variance during the use of the generalized detector for the signals with the stochastic amplitude and random initial phase allows us, in principle, to compensate the generalized detector's correlation channel noise component (which is created by the interaction between the model signal and noise) and its autocorrelation channel random component (which is created by the interaction between the signal and noise), using its compensating channel. The generalized detector allows us to increase the signal-to-noise ratio at the output of the detector in comparison to the optimal detectors of the classical and modern signal processing theories during the same input conditions.

11.7 Conclusions

Let us summarize briefly the main results discussed in this chapter. The proposed modification of the initial premises of the classical and modern signal processing theories assumes that the frequency–time regions of the noise exist where a "yes" signal may be found, and where it is known *a priori* that a "no" signal exists. This modification allows us to perform the theoretical synthesis of the generalized signal processing algorithm. Two uncorrelated samples are used, one of which is the reference sample because it is known *a priori* that a "no" signal is found in this sample. This fact allows us to obtain the jointly sufficient statistics of the likelihood function mean and variance. The optimal signal processing algorithms of the classical and modern theories for the signals with known and unknown amplitude-phase-frequency structure allow us to obtain only the sufficient statistic of the likelihood function mean and are components of the generalized signal processing algorithm.

The physicotechnical interpretation of the generalized approach to signal processing in the presence of noise is a combination of the optimal approaches of the classical and modern theories for signals with both known and unknown amplitude-phase-frequency structure. The additional filter (AF) is the source of the reference sample. The resonant frequency of AF is detuned relative to that of the preliminary filter (PF). The value of the detuning is greater than the effective spectral bandwidth of the signal. The use of AF jointly with PF forms the background noise at the output of the generalized detector. The background noise is the difference between the energy characteristics of the noise at the outputs of PF and AF and tends to approach zero in the statistical sense. In other words, the background noise at the output of the generalized detector is formed as a result of the generation of the jointly sufficient statistics of the likelihood function mean and variance for the generalized approach to signal processing in the presence of noise. The background noise is caused by both the noise at the output of the PF and the noise at the output of the AF. The background noise at the output of the generalized detector is independent of both the signal and the model signal.

The correlation between the noise component $\sum_{i=1}^{N} a_i^* \xi_i$ or $\int_0^T a^*(t)\xi(t)\,dt$ of the generalized detector's correlation channel and the autocorrelation channel random component $\sum_{i=1}^{N} a_{1_i}\xi_i$ or $\int_0^T a_1(t)\xi(t)\,dt$ allows us to generate the jointly sufficient statistics of the likelihood function mean and variance. The noise component of the correlation channel is caused by the interaction between the model signal and noise. The random component of the autocorrelation channel is caused by the interaction between the signal and noise. The effect of compensation between these two components is caused by the generation of the jointly sufficient statistics of the likelihood function mean and variance for the generalized approach to signal processing in the presence of noise and employing generalized detectors in various complex navigational systems. The effect of this compensation is carried out within the limits of the sample size [1, N] or of the time interval [0, T], for which the input stochastic process is observed.

The use of the generalized detector for the signals with the stochastic amplitude and random initial phase in various complex navigational systems has the following peculiarity. If the envelope of the amplitude of the signal is not a stochastic function of time within the limits of the time interval [0, T] for a single realization of the input stochastic process and can be stochastic from realization to realization, the generation of the jointly sufficient statistics of the likelihood function mean and variance is carried out by averaging on a set M of realizations of the input stochastic process. If the envelope of the amplitude of the signal is a stochastic function of time within the limits of the time interval [0, T], then the generation of the jointly sufficient statistics of the likelihood function mean and variance is possible within the limits of that time interval without averaging the input stochastic processes on a set of realizations.

References

1. Kotelnikov, V., *Potential Noise Immunity Theory*, Soviet Radio, Moscow, 1956 (in Russian).
2. Wiener, N., *Non-Linear Problems in Stochastic Process Theory*, McGraw-Hill, New York, 1959.
3. Middleton, D., *An Introduction to Statistical Communication Theory*, McGraw-Hill, New York, 1961.
4. Shannon, K., *Research on Information Theory and Cybernetics*, McGraw-Hill, New York, 1961.
5. Wiener, N., *Cybernetics or Control and Communication in the Animal and the Machine*, 2nd ed., John Wiley & Sons, New York, 1961.
6. Selin, I., *Detection Theory*, Princeton University Press, Princeton, NJ, 1965.
7. Miller, R., *Simultaneous Statistical Inference*, McGraw-Hill, New York, 1966.
8. Van Trees, H., *Detection, Estimation, and Modulation Theory. Part 1: Detection, Estimation, and Linear Modulation Theory*, John Wiley & Sons, New York, 1968.
9. Helstrom, C., *Statistical Theory of Signal Detection*, 2nd ed., Pergamon Press, Oxford, 1968.
10. Gallager, R., *Information Theory and Reliable Communication*, John Wiley & Sons, New York, 1968.
11. Thomas, J., *An Introduction to Statistical Communication Theory*, John Wiley & Sons, New York, 1969.
12. Jazwinski, A., *Stochastic Processes and Filtering Theory*, Academic Press, New York, 1970.
13. Van Trees, H., *Detection, Estimation, and Modulation Theory. Part 2: Non-Linear Modulation Theory*, John Wiley & Sons, New York, 1970.
14. Schwartz, M., *Information, Transmission, Modulation, and Noise*, 2nd ed., McGraw-Hill, New York, 1970.
15. Wong, E., *Stochastic Processes in Information and Dynamical Systems*, McGraw-Hill, New York, 1971.
16. Van Trees, H., *Detection, Estimation, and Modulation Theory. Part 3: Radar-Sonar Signal Processing and Gaussian Signals in Noise*, John Wiley & Sons, New York, 1972.
17. Box, G. and Tiao, G., *Byesian Inference in Statistical Analysis*, Addison-Wesley, Cambridge, MA, 1973.
18. Stratonovich, R., *Principles of Adaptive Processing*, Soviet Radio, Moscow, 1973 (in Russian).
19. Levin, B., *Theoretical Foundations of Statistical Radio Engineering, Parts 1–3*, Soviet Radio, Moscow, 1974–1976 (in Russian).
20. Tikhonov, V. and Kulman, N., *Non-Linear Filtering and Quasideterministic Signal Processing*, Soviet Radio, Moscow, 1975 (in Russian).
21. Repin, V. and Tartakovsky, G., *Statistical Synthesis under a priori Uncertainty and Adaptation of Information Systems*, Soviet Radio, Moscow, 1977 (in Russian).

22. Kulikov, E. and Trifonov, A., *Estimation of Signal Parameters in Noise,* Soviet Radio, Moscow, 1978 (in Russian).
23. Sosulin, Yu, *Detection and Estimation Theory of Stochastic Signals,* Soviet Radio, Moscow, 1978 (in Russian).
24. Ibragimov, I. and Rozanov, Y., *Gaussian Random Processes,* Springer-Verlag, New York, 1978.
25. Anderson, B. and Moore, J., *Optimal Filtering,* Prentice Hall, Englewood Cliffs, NJ, 1979.
26. Shirman, Y. and Manjos, V., *Theory and Methods in Radar Signal Processing,* Radio and Svyaz, Moscow, 1981 (in Russian).
27. Huber, P., *Robust Statistics,* John Wiley & Sons, New York, 1981.
28. Blachman, N., *Noise and Its Effect in Communications,* 2nd ed., Krieger, Malabar, FL, 1982.
29. Bacut, P. et al., *Signal Detection Theory,* Radio and Svayz, Moscow, 1984 (in Russian).
30. Anderson, T., *An Introduction to Multivariate Statistical Analysis,* 2nd ed., John Wiley & Sons, New York, 1984.
31. Lehmann, E., *Testing Statistical Hypothesis,* 2nd ed., John Wiley & Sons, New York, 1986.
32. Silverman, B., *Density Estimation for Statistics and Data Analysis,* Chapman & Hall, London, 1986.
33. Bassevillee, M. and Benveniste, A., *Detection of Abrupt Changes in Signals and Dynamical Systems,* Springer-Verlag, Berlin, 1986.
34. Trifonov, A. and Shinakov, Yu, *Joint Signal Differentiation and Estimation of Signal Parameters in Noise,* Radio and Svayz, Moscow, 1986 (in Russian).
35. Thomas, A., *Adaptive Signal Processing: Theory and Applications,* John Wiley & Sons, New York, 1986.
36. Blahut, R., *Principles of Information Theory,* Addison-Wesley, Reading, MA, 1987.
37. Weber, C., *Elements of Detection and Signal Design,* Springer-Verlag, New York, 1987.
38. Skolnik, M., *Radar Applications,* IEEE Press, New York, 1988.
39. Kassam, S., *Signal Detection in Non-Gaussian Noise,* Springer-Verlag, Berlin, 1988.
40. Poor, V., *Introduction to Signal Detection and Estimation,* Springer-Verlag, New York, 1988.
41. Brook, D. and Wynne, R., *Signal Processing: Principles and Applications,* Pentech Press, London, 1988.
42. Porter, W. and Kak, S., *Advances in Communications and Signal Processing,* Springer-Verlag, Berlin, 1989.
43. Adrian, C., *Adaptive Detectors for Digital Modems,* Pentech Press, London, 1989.
44. Scharf, L., *Statistical Signal Processing, Detection, Estimation, and Time Series Analysis,* Addison-Wesley, Reading, MA, 1991.
45. Cover, T. and Thomas, J., *Elements of Information Theory,* John Wiley & Sons, New York, 1991.
46. Basseville, M. and Nikiforov, I., *Detection of Abrupt Changes,* Prentice Hall, Englewood Cliffs, NJ, 1993.
47. Dudgeon, D. and Johnson, D., *Array Signal Processing: Concepts and Techniques,* Prentice Hall, Englewood Cliffs, NJ, 1994.
48. Porat, B., *Digital Processing of Random Signals: Theory and Methods,* Prentice Hall, Englewood Cliffs, NJ, 1994.

49. Helstrom, C., *Elements of Signal Detection and Estimation,* Prentice Hall, Englewood Cliffs, NJ, 1995.
50. McDonough, R. and Whallen, A., *Detection of Signals in Noise,* 2nd ed., Academic Press, New York, 1995.
51. Tuzlukov, V., *Signal Detection Theory,* Springer-Verlag, New York, 2001.
52. Tuzlukov, V., *Signal Processing Noise,* CRC Press, Boca Raton, FL, 2002.
53. Crammer, H., *Mathematical Methods of Statistics,* Princeton University Press, Princeton, NJ, 1946.
54. Grenander, U., *Stochastic Processes and Statistical Inference,* Arkiv Mat, Uppsala, Sweden, 1950.
55. Rao, C., *Advanced Statistical Methods in Biometric Research,* John Wiley & Sons, New York, 1952.
56. Van Trees, H., *Detection, Estimation, and Modulation Theory. Parts 1–4,* John Wiley & Sons, New York, 2002–2003.
57. Maximov, M., Joint correlation of fluctuating noise at the outputs of frequency filters, *Radio Eng.,* No. 9, 1956, pp. 28–38.
58. Chernyak, Y., Joint correlation of noise voltage at the outputs of amplifiers with non-overlapping responses, *Radio Phys. Electron.,* No. 4, 1960, pp. 551–561 (in Russian).
59. Tuzlukov, P. and Tuzlukov, V., Reliability increasing in signal processing in noise in communications, *Automatized Systems in Signal Processing,* Vol. 7, 1983, pp. 80–87 (in Russian).
60. Tuzlukov, V., Detection of deterministic signal in noise, *Radio Eng.,* No. 9, 1986, pp. 57–60.
61. Tuzlukov, V., Signal detection in noise in communications, *Radio Phys. Electron.,* Vol. 15, 1986, pp. 6–12.
62. Tuzlukov, V., Detection of deterministic signal in noise, *Telecomm. Radio Eng.,* Vol. 41, No. 10, 1987, pp. 128–131.
63. Tuzlukov, V., Detection of signals with stochastic parameters by employment of generalized algorithm, in *Proceedings of SPIE's 1997 International Symposium on AeroSense: Aerospace/Defense Sensing, Simulation, and Controls,* Orlando, FL, April 20–25, 1997, Vol. 3079, pp. 302–313.
64. Tuzlukov, V., Noise reduction by employment of generalized algorithm, in *Proceedings of the 13th IEEE International Conference on Digital Signal Processing (DSP97),* Santorini, Greece, July 2–4, 1997, pp. 617–620.
65. Tuzlukov, V., Detection of signals with random initial phase by employment of generalized algorithm, in *Proceedings of SPIE's 1997 International Symposium on Optical Science, Engineering, and Instrumentation,* San Diego, CA, July 27–August 1, 1997, Vol. 3162, pp. 61–72.
66. Tuzlukov, V., Generalized detection algorithm for signals with stochastic parameters, in *Proceedings of the 1997 IEEE International Geoscience and Remote Sensing Symposium (IGARSS'97),* August 4–8, 1997, Singapore, pp. 139–141.
67. Tuzlukov, V., Tracking systems for stochastic signal processing by employment of generalized algorithm, in *Proceedings of the 1st IEEE International Conference on Information Communications and Signal Processing (ICICS'97),* Singapore, September 9–12, 1997, pp. 311–315.
68. Tuzlukov, V., A new approach to signal detection theory, *Digital Signal Process. Rev. J.,* Vol. 8, No. 3, 1998, pp. 166–184.

69. Tuzlukov, V., Signal-to-noise ratio improvement under detection of stochastic signal using generalized detector, in *Proceedings of the 1998 International Conference on Applications of Photonics Technology (ICAPT'98)*, Ottawa, Canada, July 27–30, 1988.
70. Tuzlukov, V., Signal processing in noise in communications: a new approach. Tutorial No. 3, in *Proceedings of the 6th IEEE International Conference on Electronics, Circuits, and Systems,* Paphos, Cyprus, September 5–8, 1999, pp. 5–128.
71. Tuzlukov, V., Detection of stochastic signals using generalized detector, in *Proceedings of the 1999 IASTED International Conference on Signal and Image Processing,* Nassau, Bahamas, October 18–21, 1999, pp. 95–99.
72. Tuzlukov, V., New remote sensing algorithms under detection of minefields in littoral waters, in *Proceedings of the 3rd International Conference on Remote Sensing Technologies for Minefield Detection and Monitoring,* Easton, Washington, D.C., May 17–20, 1999, pp. 182–241.
73. Tuzlukov, V., New remote sensing algorithms on the basis of generalized approach to signal processing in noise. Tutorial No. 2, in *Proceedings of the 2nd International ICSC Symposium on Engineering of Intelligent Systems (EIS 2000),* University of Paisley, Paisley, Scotland, June 27–30, 2000.
74. Tuzlukov, V., Detection of quasideterministic signals in additive Gaussian noise, *News of the Belarusian Academy of Sciences, Ser. Phys.-Tech. Sci.,* No. 4, 1985, pp. 98–104 (in Russian).
75. Tuzlukov, V., Detection of signals with stochastic amplitude and random initial phase in additive Gaussian noise, *Radio Eng.,* No. 9, 1988, pp. 59–61 (in Russian).
76. Tuzlukov, V., Interference compensation in signal detection for a signal of arbitrary amplitude and initial phase, *Telecomm. Radio Eng.,* Vol. 44, No. 10, 1989, pp. 131–132.
77. Tuzlukov, V., The generalized algorithm of detection in statistical pattern recognition, *Pattern Recognition and Image Analysis,* Vol. 3, No. 4, 1993, pp. 474–485.
78. Tuzlukov, V., Signal-to-noise improvement in video signal processing, in *Proceedings of SPIE's 1993 International Symposium on High-Definition Video,* Berlin, Germany, April 5–9, 1993, Vol. 1976, pp. 346–358.
79. Tuzlukov, V., Probability distribution density of background noise at the output of generalized detector, *News of the Belarusian Academy of Sciences. Ser. Phys.-Tech. Sci.,* No. 4, 1993, pp. 63–70 (in Russian).
80. Tuzlukov, V., Distribution law at the generalized detector output, in *Proceedings PRIA'95,* Minsk, Belarus, September 19–21, 1995, pp. 145–150.
81. Tuzlukov, V., Statistical characteristics of process at the generalized detector output, in *Proceedings PRIA'95,* Minsk, Belarus, September 19–21, 1995, pp. 151–156.
82. Tuzlukov, V., Signal fidelity in radar processing by employment of generalized algorithm under detection of mines and mine-like targets, in *Proceedings SPIE's 1998 International Symposium on AeroSense: Aerospace/Defense Sensing, Simulations and Controls,* Orlando, FL, April 13–17, 1998, Vol. 3392, pp. 1206–1217.
83. Tuzlukov, V., Employment of the generalized detector for noise signals in radar systems, in *Proceedings of SPIE's 2000 International Symposium on AeroSense: Aerospace/Defense Sensing, Simulations and Controls,* Orlando, FL, April 24–28, 2000, Vol. 4048.
84. Tuzlukov, V., Adaptive generalized detector for unknown power spectral densities of noise, in *Proceedings of the IEEE International Signal Processing Conference (ISPC'03),* Dallas, TX, March 31–April 4, 2003.

85. Tuzlukov, V., Signal detection in compound Gaussian noise: generalized detector, in *Proceedings of the 3rd International Symposium on Image and Signal Processing and Analysis (ISPA'03)*, Rome, Italy, September 18–20, 2003.
86. Tuzlukov, V., Adaptive beam-former generalized detector, in *Proceedings of IASTED International Conference on Signal Processing, Pattern Recognition, Applications (SPPRA 2003)*, Rhodes, Greece, June 30–July 2, 2003.
87. Kim, Y., Yoon, W-S., and Tuzlukov, V., Generalized approach to distributed signal processing with randomized data selection in wireless sensor networks, in *Proceedings of the 1st International Conference on Information and Technology*, Suwon, Korea, November 28–30, 2003, pp. 54–67.
88. Tuzlukov, V., Yoon, W.-S., and Kim, Y., Adaptive beam-former generalized detector in wireless sensor networks, in *Proceedings of the IASTED International Conference on Parallel and Distributed Computing and Networks (PDCN 2004)*, Innsbruck, Austria, 2004, pp. 195–200.

12

Theory of Space–Time Signal and Image Processing in Navigational Systems

12.1 Basic Concepts of Navigational System Functioning

The main principle of navigational system functioning is the following. To define the location of the navigational object, for example, an aircraft, a moving image of the Earth's surface or the totality of landmarks, relative to which the target moves, is formed on the board of the navigational object. An image of the same Earth's surface or the totality of landmarks, which was obtained under the assumption that the target moves according to a predetermined trajectory, is formed in advance and is called the model image. The moving image is compared with the model image, and in doing so, a noncoincidence in the location of the totality of landmarks on the observation plane both for the moving image and for the model image is a measure of deviation of the aircraft flight track from the predetermined or the required aircraft flight track is characterized by mutual noncoincidence between the moving image and the model image relative to each other. Let us recall that a certain region of the given coordinate system is called the observation plane. In some cases, the region called the observation plane has a relative character. For example, in optical signal processing in navigational systems, the focus plane of the optical receiver is called the observation plane. Then, during the formation of the moving and model images as the reference totality, the domain of possible values of these readings related to the predetermined coordinate system is called the observation plane.[1,2]

The totality of landmarks used in solving navigational problems such as the Earth's surface during navigation of aircraft flight track, the image of the starry sky, and so on is observed, as a rule, in the background of interferences and noise generated by various sources. For this reason, the effectiveness of solving navigational problems is defined by the quality of the moving image, i.e., by the quality of signal and image processing of information components of the moving image containing useful information regarding the totality of landmarks. It is obvious that the methods and techniques of space–time signal and image processing in navigational systems are defined by the nature of the signals used. Because of this, it is worthwhile to carry out a brief investigation of the space–time fields and interference and noise generated by these fields before discussing the methods and techniques of signal and image processing in navigational systems.

Generally speaking, any fields obtained by measurement of the characteristics of landmarks, the Earth's surface, or the totality of other references, for example, the stars, can be considered as a source of information regarding the position of the navigational target. These characteristics, systematized in a fixed way, can be presented as the field of both the moving image and the model image. However, it is worthwhile to pay attention to the definition of those space–time signal processing methods and techniques that are widely used.

Physical fields generating the space–time signals can be divided into natural and manmade fields. Anomalous magnetic fields, gravitation fields, the totality of readings of the Earth's surface relief, optical fields, and heat and radio heat fields that are the consequence of the lighting of the Earth's surface by the sun, moon, and stars, tectonic activity of the Earth, nonidentical radiant emittance of various nonhomogeneities of the Earth's surface and fields, which are generated by human activity, are called the natural fields. In the case of human activity, the "natural field" statement is conditional because this field is not the result of purposeful human activity and can more rigorously be called the second-organized heat field. The field of radio signals caused by the radiation of radar, communications, and other signal processing systems can be considered the nonorganized field. This field is the secondary consequence of their functioning. Incidentally, the anomalous magnetic field can be considered the secondary nonorganized field because its character can be defined by manmade metallic constructions, for example, a big bridge, dam, etc. The fields formed during the reflection of radiation energy directed to the navigational object are called manmade fields. They are formed by the radiation of the Earth's surface by radar, laser, sonar, etc.[3,4]

The types of fields mentioned in the preceding text have various characteristics that can define the method of space–time signal and image processing. Moreover, the type of receiver or detector also influences the chosen method of the space–time signal and image processing. Let us explain this statement in more detail. As well known, the scanning of landmarks by, for example, the radar antenna of an aircraft navigational system can be both sequential and parallel. In the case of radar aircraft navigational systems, scanning and forming of an image of the two-dimensional (surface) target are carried out by the radar antenna, i.e., by the sequential displacement of the radar antenna. Thus, at each instant of time, the area of the scanned two-dimensional (surface) target is defined by the width of the directional diagram. This sweep can be partially realized due to the moving radar and antenna scanning in the direction that is orthogonal to the trajectory of the aircraft's flight (see Figure 12.1). An example of parallel space–time signal and image processing is the multielement optical receiver (matrix detector) in which an image of the scanned two-dimensional (surface) target is formed

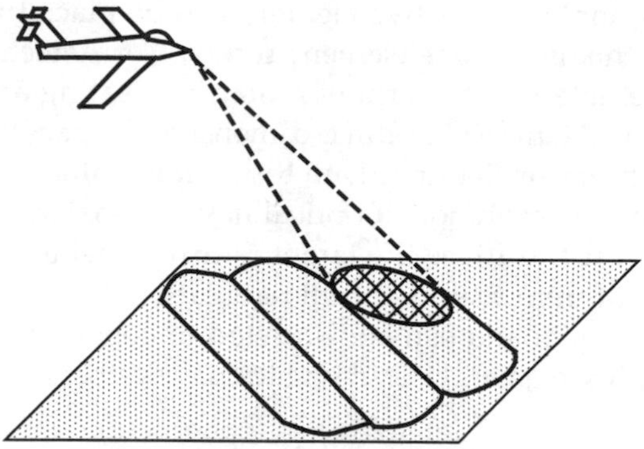

FIGURE 12.1
Orthogonal radar antenna scanning.

instantaneously. Evidently, in the first case, the spatial image is transformed into the signal as a function of time and, in the second case, we have a spatial image of the two-dimensional (surface) target. It is clear that signal and image processing methods based on the generalized approach in the presence of noise[5–8] are different for each case.

Let us consider the sensor type stimulus of the considered field on the signal and image processing algorithm. For instance, let us consider the radar navigational system. In the radar, the target return signal, as a rule, is a narrow-band stochastic process, though we would expect that it could be a high wideband signal process. The target return signal comes in at the input of the narrow-band linear tract or antenna of the navigational system. Resolution or accuracy of reproduction of the two-dimensional (surface) target is defined by the wavelength of the searching signal and by the parameters and characteristics of the receiver or detector, first and foremost by the width of the directional diagram. For the wavelength range in centimeter, the resolution of the aircraft radar navigational systems is a few hundred meters, and for the millimeter range, the resolution is a few tens of meters.[9]

In sonar navigational systems, the target return signal is characterized by mechanical oscillations. However, signal processing techniques in sonar navigational systems are very close to those in the radar because a sensor in the sonar field transforms mechanical oscillations into electromagnetic oscillations similar to the radar video signal. In optical navigational systems, except in the case of laser and heat optic radiation of gases, the target return signal is a set of simple harmonic oscillations within the limits of the interval from 0.001 to 10 μm of electromagnetic waves and, therefore, signal and image processing under the use of resonance techniques and devices is a very complex problem in practice.

The main type of receiver or detector in navigational systems operating in the optical range is the integral detector.[10] Its basic principles are founded on the ability to transform optical signals into electromagnetic waves. Various techniques have been employed in the design of optical receivers and detectors. For example, a sensitive element can be placed in the focus of a concave mirror; this is the one-element receiver or detector. Sensitive elements can be presented in the form of a matrix, creating an image plane of the two-dimensional (surface) or three-dimensional (space) target; this is the multielement receiver or detector. Due to the low value of the electromagnetic wavelength, the resolution of optical navigational system is within the limits of the interval 1 to 10 m. The target return signal in laser navigational systems is different from that in optical navigational systems because it is a simple harmonic signal and can be subjected to coherent signal and image processing. For this reason, laser navigational systems have higher noise immunity.[10]

The correlation radius of the geophysical field, which consists of magnetic and gravity fields, depends on the altitude, using which we can observe these fields. Even for low values of altitude, the correlation radius is within the limits of a few hundred meters.[11] Moreover, there are no current methods and techniques to transform the geomagnetic field into the electromagnetic wave in radar navigational systems, similar to the case in optical navigational systems where the target return signal is a function of the optical image in the coordinates of the observation plane. The existing sensors in geophysical fields allow measurement of a field at the point corresponding to the location of the navigational object at the instant of measurement. A set of discrete readings obtained with the moving radar navigational system allows us to create a function, which corresponds to the cross section of the geomagnetic field by the plane coinciding with the trajectory of the moving radar navigational system (see Figure 12.2). Comparing the observed moving image with the model images corresponding to possible variants of airborne flight tracks and related to a given coordinate system, we can define the location of the navigational object based on analysis of the target return signals.[12–15] It is obvious that in this case the signal and image processing of the moving image should be different from those of the space–time target return signals in the optical or radar fields.

The moving image of the Earth's surface relief is formed due to periodic changes in the aircraft's flight altitude. As an example, to solve the navigational problem of the moving ship, the sea depth is measured. In both the first and second cases, the moving image is a function of time because of the moving radar navigational system. In this case, the signal and image processing is similar to the signal and image processing in time, as in the observation of the geophysical field. Thus, methods and techniques of spatial target return signal processing can be classified into space signal and image processing, space–time signal and image processing, and signal and image processing in time.

In space signal and image processing, each operation in defining the elements composing the moving image is carried out at the same instants of time. In other words, elements of the moving image are processed in parallel. Therefore, space signal and image processing used in this case is a particular case of parallel processing of the target return signal containing all the

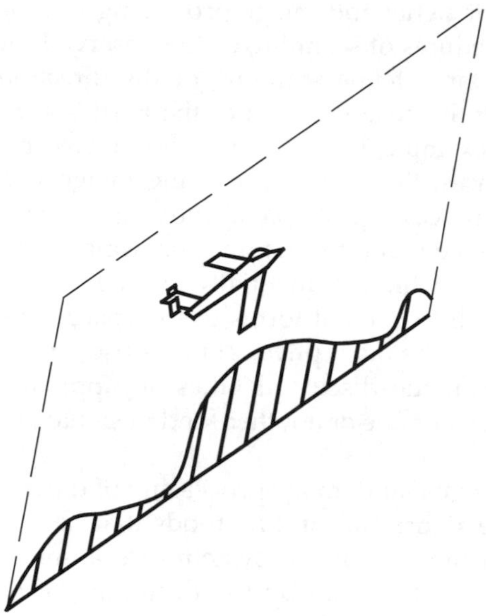

FIGURE 12.2
Geomagnetic field cross section.

necessary information. It is used when the sensor is the matrix receiver that is widely used in optics. In practice, these navigational systems are called raster or screen systems.

The space–time signal and image processing is different in that one part of the moving image is subjected to parallel processing and the other part to sequential processing. An example is shown in Figure 12.3. An optical image of the observed area of the Earth's surface is obtained by the mosaic linear sensor that is perpendicular to the axis of an aircraft. Thus, the moving image is obtained due to a shift of screen string formed at the receiver or detector output in navigational systems.

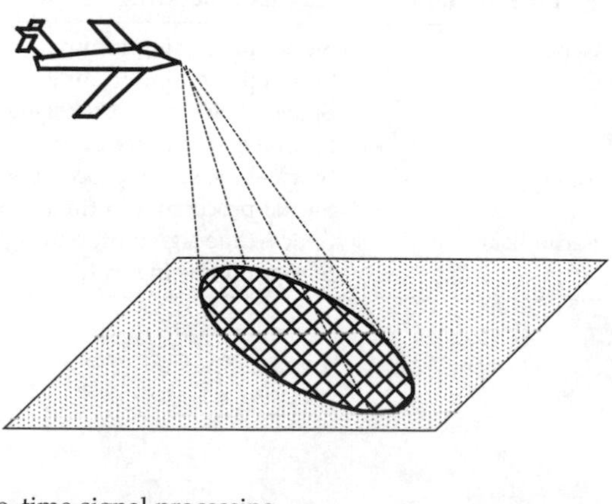

FIGURE 12.3
An example of space–time signal processing.

The techniques of signal and image processing in time are characterized by sequential procedures of scanning of the observed moving image. They are realized by radar antenna scanning or the directional diagram of the sensor. Scanning of the observed area of the Earth's surface can be done in various ways, for example, by spiral, row-frame sweep, moving radar, etc. However, in all cases, the observed moving image is a function of time. Signal and image processing of spatial moving images in time belongs in the technique of sequential information processing because each element of the moving image is defined sequentially in time.

Table 12.1 shows the different forms of the space–time signal and image processing applied to the main physical fields used in practice. It should be noted that the technique discussed takes an approximate character and shows the superiority of one or another kind of signal and image processing technique.

The progress in signal and image processing of the moving image allows us to realize those algorithms and methods that have not been described previously. For instance, a mixed system of scanning the observed area of the Earth's surface followed by the designing of the matrix image is widely used in practice. The element of the moving image in this navigational system is viewed on the exterior plane of the mirror by the infrared lens. The exterior plane of the mirror reflects the optical signal on the matrix element of the sensor receiver. The output signal of each element of the sensor receiver controls the photodiode of the indicator after amplification. The exterior plane of the mirror is used for scanning a heat image, and the interior plane of the mirror is used to form a visible image in the plane of the photodiodes. In fast scanning by the mirror, this image can be subjected to parallel signal and image processing due to the inertness of photodiodes.[16,17]

TABLE 12.1

Forms of Space-Time Signal Processing

Type of Field	Signal Processing Method
Geophysical	Signal processing in time
Optical	Signal processing in time
	Space–time signal processing
	Spatial signal processing
Radar	Space–time signal processing
	Signal processing in time
Radio heat	Space–time signal processing
	Signal processing in time

12.2 Basics of the Generalized Approach to Signal and Image Processing in Time

As noted earlier, the signal and image processing in time or the sequential signal and image processing of spatial signals is based on the transformation of the navigational field into the signal as a function of time with further signal and image processing. The solution of navigational problems such as defining the location of the navigational object is carried out under ambiguous conditions. The uncertainty is due to the following main factors: external interference, sensor noise, interference that can arise with the moving radar navigational system, indefinite information of landmark coordinates, etc. These interferences caused by external noise and internal sensor noise are called signal interference or signal noise. Interferences acting on navigational systems are called direct interferences.[18] Incorrect knowledge of the location of landmarks, i.e., the relative position of the navigational system in the coordinate system, is known as *a priori* uncertainty in the definition of the navigational object coordinates.

For the reasons already noted, we can discuss only the estimation of the navigational object coordinates. We can consider the estimation to be optimal when it is the best in a definite sense. Thus, the main problem in navigational systems is defining the optimal estimation of the navigational object coordinates, for example, defining the true aircraft flight track, using information including the signals generated by various physical sources.

The definition of the optimal estimation of the navigational object coordinates is carried out in the following manner. Let us assume that there is information regarding the intensity and character of direct and signal interferences and we have information about the moving navigational object. Then, based on measurements that are available at this instant, we are able to truly estimate the navigational object coordinates. The process of defining these estimations is signal and image processing.

Let us consider the basic statements of classical signal processing theory within the limits of the Bayes approach to decision-making rules in the presence of uncertainty. The observed signal at the navigational system receiver input, in the general case, can be written in the following form:

$$\mathbf{X} = (\mathbf{\Lambda}, \mathbf{n}). \tag{12.1}$$

Here $\mathbf{X}$ is the m-dimensional vector of the observed signal at the receiver or detector input of the navigational system; $\mathbf{\Lambda}$ is the n-dimensional vector of parameters of coordinates of the navigational object; and $\mathbf{n}$ is the ℓ-dimensional noise vector ($\ell \leq m$).

We assume that the signal-to-noise ratio is known and already given. With the Bayesian approach, we consider that the joint probability distribution density $f(\mathbf{\Lambda}, \mathbf{n})$ is also already given. All information regarding the components of the vector of parameters $\mathbf{\Lambda}$ of the navigational object coordinates is included in the conditional probability distribution density $f(\mathbf{\Lambda}/\mathbf{X})$ that is the *a posteriori* probability distribution density of the vector $\mathbf{\Lambda}$. In other words, $f(\mathbf{\Lambda}/\mathbf{X})$ contains all the information that may be used in the definition of the estimation of the navigational object coordinates. According to the Bayes theorem, we can write

$$f(\mathbf{\Lambda}/\mathbf{X}) = f_{ps}(\mathbf{\Lambda}) = \frac{f_{pr}(\mathbf{\Lambda}) \cdot f(\mathbf{X}/\mathbf{\Lambda})}{f(\mathbf{X})}, \tag{12.2}$$

where $f_{ps}(\mathbf{\Lambda})$ is the *a posteriori* probability distribution density; $f(\mathbf{X}/\mathbf{\Lambda})$ is the conditional probability distribution density defining a statistic of the observed signal at the receiver or detector input of the navigational system; and $f_{pr}(\mathbf{\Lambda})$ is the *a priori* probability distribution density.

The conditional probability distribution density $f(\mathbf{X}/\mathbf{\Lambda})$, considered as a function of $\mathbf{\Lambda}$, is called the likelihood function $\mathscr{L}(\mathbf{\Lambda})$. It contains all the information about the statistic of the observed signal samples at the input of the navigational system. These samples are processed to form the vector estimation $\mathbf{\Lambda}^*$ of the vector of parameters of the navigational object coordinates. With fixed values of the observed signal at the navigational system input, the likelihood function makes it possible for us to define the most probable value of $\mathbf{\Lambda}$. In the definition of the vector estimation $\mathbf{\Lambda}^*$, its value tends to approach the optimal. The optimality of estimation is considered in accordance with the chosen criteria; let us recall the three widely used criteria.[19]

The first criterion is that of the maximum *a posteriori* probability distribution density $f_{ps}(\mathbf{\Lambda})$. With this criterion, the probability of the event that $\mathbf{\Lambda} = \mathbf{\Lambda}^*$ becomes maximal. The solution $\mathbf{\Lambda}^* = f_{ps}(\mathbf{\Lambda})$ is called the most probable or unconditional estimation of the maximum probability distribution density $f_{ps}(\mathbf{\Lambda})$. The second criterion is the minimum of the mean square error, i.e., $\min\{\langle(\mathbf{\Lambda} - \mathbf{\Lambda}^*)^T \mathbf{O}(\mathbf{\Lambda} - \mathbf{\Lambda}^*)\rangle\}$ where $\langle\ldots\rangle$ denotes the average in the statistical sense, and $\mathbf{O}$ is the positively defined matrix introduced with the purpose of defining the weight of the components of the error vector $\Delta\mathbf{\Lambda} = \mathbf{\Lambda} - \mathbf{\Lambda}^*$. The maximum of the *a posteriori* probability distribution density coincides with the minimum of the mean square error in the case of some probability distribution densities, particularly the normal Gaussian probability distribution density. The solution $\mathbf{\Lambda}^* = \langle\mathbf{\Lambda}/\mathbf{X}\rangle$ is the estimation of the minimum variance. This estimation ensures the minimal width of the *a posteriori* probability distribution density. The third criterion is that in which the mean of the absolute error is minimized. The solution corresponding to the criterion can be written in the form $\mathbf{\Lambda}^* = Me\, f_{ps}(\mathbf{\Lambda}/\mathbf{X})$.

Let us consider the particular case of the linear equation [see Equation (12.1)] as an example of the optimal estimation definition. For simplicity, we assume that the variables considered are not varied in time, i.e., we can write

$$\mathbf{X} = \mathbf{H}\mathbf{\Lambda} + \mathbf{n}; \tag{12.3}$$

$f(\mathbf{\Lambda}, \mathbf{n})$ is the Gaussian probability distribution density that can be determined in the following form:

$$f(\mathbf{\Lambda},\mathbf{n}) = f(\mathbf{\Lambda}) \cdot f(\mathbf{n})\,; \tag{12.4}$$

$\mathbf{A}_0$ is the covariance matrix of $\mathbf{\Lambda}$; $\langle\mathbf{n}\rangle = 0$; $\langle\mathbf{n}\mathbf{n}^{\mathrm{T}}\rangle = \mathbf{N}$, where

$$\mathbf{N} = \begin{Vmatrix} N_1 \ldots 0 \\ \ldots\ldots\ldots \\ 0 \ldots N_m \end{Vmatrix}. \tag{12.5}$$

Due to the stimulus of the additive noise, the probability distribution density $f(\mathbf{X})$ is Gaussian with the mean $\langle\mathbf{X}\rangle = \mathbf{H}\mathbf{\Lambda}_0$ and the covariance matrix equal to $\mathbf{H}\mathbf{A}_0\mathbf{H}^{\mathrm{T}} + \mathbf{N}$.

Let us define the *a posteriori* probability distribution density using the generalized approach to signal processing in noise.[5–8] For this purpose, we will use the main statements of the generalized approach to signal processing in noise discussed in Chapter 11. In accordance with this, we have to form the additional reference signal $\boldsymbol{\eta}$ at the input of the navigational system. It is known *a priori* that the "no" information signal obtains in the additional reference signal $\boldsymbol{\eta}$. The additional reference signal $\boldsymbol{\eta}$, by its nature, is the noise or interference. Thus, we can write

$$\boldsymbol{\eta} = \mathbf{Q}^*(\mathbf{n}^*)\,, \tag{12.6}$$

where $\langle\mathbf{n}^*\rangle = 0$; $\langle\mathbf{n}^*\mathbf{n}^{*\mathrm{T}}\rangle = \mathbf{N}^*$, and

$$\mathbf{N}^* = \begin{Vmatrix} N_1^* \ldots 0 \\ \ldots\ldots\ldots \\ 0 \ldots N_m^* \end{Vmatrix}. \tag{12.7}$$

The additional reference signal $\boldsymbol{\eta}$ is the m-dimensional vector, and $\mathbf{n}^*$ is the ℓ-dimensional noise vector ($\ell \leq m$). At the input of the navigational system signal $\mathbf{X}$ and additional reference signal $\boldsymbol{\eta}$ are observed and are uncorrelated processes by the initial premises of the generalized approach to signal processing in noise (see Chapter 11). For simplicity, we assume that the statistics $\mathbf{N}$ and $\mathbf{N}^*$ are the same. Then, the *a posteriori* probability distribution density takes the following form:

$$\ln \mathcal{L}(\mathbf{\Lambda}^*) = \ln\,[f_{ps}(\mathbf{\Lambda}^*)] = \frac{2\mathbf{X}^{\mathrm{T}}\mathbf{N}^{-1}(\mathbf{\Lambda}-\mathbf{\Lambda}_0) - \mathbf{X}^{\mathrm{T}}\mathbf{N}^{-1}\mathbf{X} + \boldsymbol{\eta}^{\mathrm{T}}\mathbf{N}^{-1}\boldsymbol{\eta}}{\sqrt{(\mathbf{\Lambda}-\mathbf{\Lambda}_0)^{\mathrm{T}}\mathbf{N}^{-1}(\mathbf{\Lambda}-\mathbf{\Lambda}_0)}}\,. \tag{12.8}$$

If the *a priori* probability distribution density $f_{pr}(\mathbf{\Lambda})$ is unknown, we can consider that it is uniform. In this case, the maximization of the *a posteriori* probability distribution density reduces to maximization of the likelihood function, i.e., to the definition of estimations that is optimal by the criterion of the likelihood function maximum. Let the signal $\mathbf{X}$ observed at the input of navigational system be the function of time defined at the time instants reckoned over the same interval Δt. Then the *a posteriori* probability distribution density can be written in the following form:

$$f_{ps}(\mathbf{\Lambda}) = f(\mathbf{\Lambda} / \mathbf{X}), \tag{12.9}$$

where $\mathbf{X} = (\mathbf{X}_1, \mathbf{X}_2, \ldots, \mathbf{X}_p)$ is the sample of the signal vector at the input of the navigational system at the time instants $t_i (i = 1, 2, \ldots, p)$. The definition of the optimal procedure of estimation of the vector $\mathbf{\Lambda}$ in the case of the continuous signal is carried out using the same rules as $\Delta t \to 0$. In doing so, a totality of samples of the signal observed at the input of the navigational system within the limits of the time interval $[0, T]$ is replaced by the continuous process $\mathbf{X}(t)$. The likelihood function is replaced by the likelihood functional

$$L(\mathbf{\Lambda}) = \lim_{\Delta t \to 0} \mathcal{L}(\mathbf{\Lambda}). \tag{12.10}$$

Let us consider some examples. We assume that it is necessary to define the delay in the harmonic signal observed at the input of the navigational system. For simplicity, we assume that this delay can be considered as not varying in the time function during the time measurement interval and that the initial value of signal delay is equal to zero.

12.2.1 The Signal with Random Initial Phase

Let us consider the working principles of the generalized detector shown in Figure 12.4. The initial conditions are the following. The signal at the input of the navigational system is observed within the limits of the time interval $[0,T]$. The signal is the narrow-band process with the given amplitude modulation law $S(t)$ and phase modulation law $\Psi_a(t)$. The random initial phase φ_0 of the signal observed at the input of navigational system is distributed uniformly within the limits of the interval $[-\pi, \pi]$ and is not time variant. The signal observed at the input of the navigational system has the following form:

$$a(t) = S(t)\cos[\omega_0 t - \Psi_a(t) + \varphi_0], \tag{12.11}$$

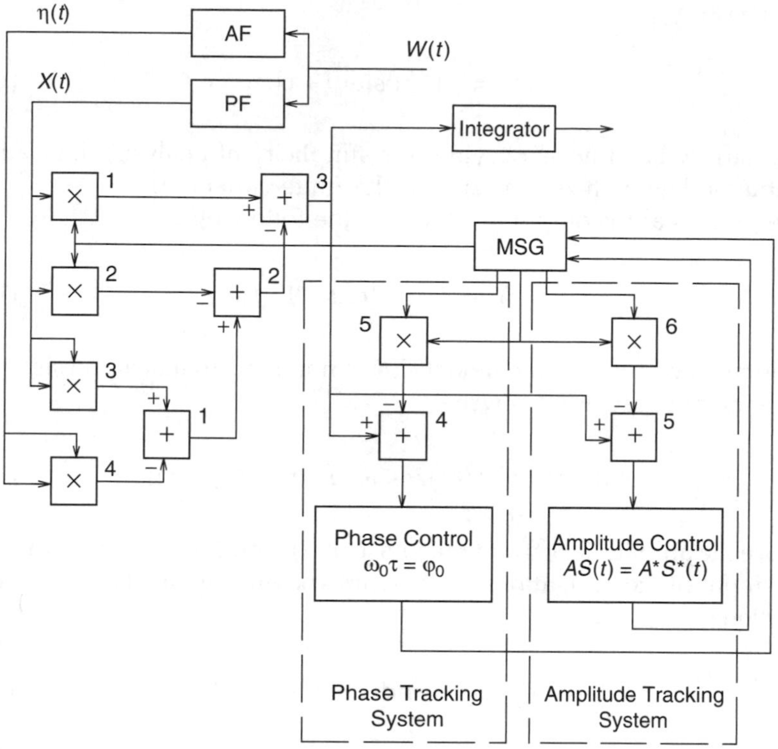

FIGURE 12.4
The generalized detector with tracking systems.

where ω_0 is the carrier frequency of the signal $a(t)$. The realizations of the input stochastic process correspond to variations of the random initial phase φ_0 within the limits of the interval $[-\pi, \pi]$.

Using the peculiarities of the generalized detector and the known amplitude and phase modulations of the signal at the input of the navigational system, we can construct a detector that is able to solve the problem of signal detection in the input stochastic process and that of definition and estimation of the random initial phase of the signal and, consequently simultaneously, signal delay. Because the model signal generated in the generalized detector is accurate with regard to the initial phase, it is not particularly a problem to design a device to measure and control the initial phase of the signal at the input of navigational system using the variation in the initial phase of the model signal at the output of the model signal generator (MSG) of the generalized detector.

The process at the output of the preliminary filter (PF) of the generalized detector shown in Figure 12.4 takes the following form:

$$X(t) = S(t)\cos[\omega_0 t + \Psi_a(t) - \varphi_0] + \xi_1(t)\cos[\omega_0 t + \upsilon(t)]\,. \qquad (12.12)$$

The process

$$\xi(t) = \xi_1(t)\cos[\omega_0 t + \upsilon(t)] \tag{12.13}$$

is the narrow-band noise obeying, for simplicity of analysis, the Gaussian distribution law with zero mean and the finite variance σ_n^2.

The process at the output of MSG has the following form:

$$a^*(t) = S^*(t)\cos[\omega_0 t + \Psi_a^*(t)]. \tag{12.14}$$

Taking into account that the model $a^*(t)$ is not liable to match the signal $a_1(t)$ by the initial phase, we can write

$$a^*(t-\tau) = S^*(t-\tau)\cos[\omega_0(t-\tau) + \Psi_a^*(t-\tau)]. \tag{12.15}$$

Because the amplitude $S(t)$ and the phase $\Psi_a(t)$ modulation laws of the signal at the input of the navigational system are known, we can always satisfy the conditions

$$S^*(t-\tau) = S(t) \quad \text{and} \quad \Psi_a^*(t-\tau) = \Psi_a(t). \tag{12.16}$$

Only under these conditions should the following analysis be carried out. Therefore, we drop the intermediate mathematics and consider only the main end results.

The process at the output of the summator *1* takes the following form:

$$Z_1(t) = 0.5S^2(t) + S(t)\xi_1(t)\cos[\Psi_a(t) - \upsilon(t) + \varphi_0] + 0.5[\xi_1^2(t) - \xi_2^2(t)]. \tag{12.17}$$

The processes at the outputs of the multipliers *1* and *2* have the following form:

$$Z_2(t) = 0.5S^2(t)\cos[\omega_0\tau - \varphi_0] + 0.5S(t)\xi_1(t)\cos[\omega_0\tau - \Psi_a(t) + \upsilon(t)]. \tag{12.18}$$

The process at the output of the summator *2* takes the following form:

$$\begin{aligned} Z_3(t) &= 0.5S^2(t)\{1 - \cos[\omega_0\tau - \varphi_0]\} \\ &+ 0.5S(t)\xi_1(t)\{2\cos[\Psi_a(t) - \upsilon(t) - \varphi_0] - \cos[\omega_0\tau - \Psi_a(t) + \upsilon(t)]\} \\ &+ 0.5[\xi_1^2(t) - \xi_2^2(t)]. \end{aligned} \tag{12.19}$$

The process at the output of the summator *3* takes the following form:

$$\begin{aligned} Z_4(t) &= S^2(t)\cos[\omega_0\tau - \varphi_0] \\ &- 0.5\{S^2(t) - S(t)\xi_1(t)\{2\cos[\omega_0\tau - \Psi_a(t) + \upsilon(t)] - 2\cos[\Psi_a(t) - \upsilon(t) - \varphi_0]\}\} \\ &+ 0.5[\xi_1^2(t) - \xi_2^2(t)]. \end{aligned} \tag{12.20}$$

In the condition $\omega_0\tau = \varphi_0$, the process at the output of the generalized detector has the following form:

$$Z_g^{out}(t) = 0.5\left\{\int_0^T S^2(t)\,dt + \int_0^T [\xi_2^2(t) - \xi_1^2(t)]\,dt\right\}. \tag{12.21}$$

The condition $\omega_0\tau = \varphi_0$ is satisfied during the use of the phase tracking system (see Figure 12.4). The phase tracking system has the circuit that controls the initial phase of the model signal at the output of MSG of the generalized detector and fulfills the required condition.

12.2.2 The Signal with Stochastic Amplitude and Random Initial Phase

The signal with stochastic amplitude and random initial phase can be written in the following form:

$$a(t, \varphi_0, A) = A(t)S(t)\cos[\omega_0\tau + \Psi_a(t) - \varphi_0], \tag{12.22}$$

where ω_0 is the carrier frequency of the signal $a(t, \varphi_0, A)$; $S(t)$ is the known modulation law of the amplitude of the signal $a(t, \varphi_0, A)$; $\Psi_a(t)$ is the known modulation law of phase of the signal $a(t, \varphi_0, A)$; φ_0 is the random initial phase of the signal $a(t, \varphi_0, A)$, which is uniformly distributed within the limits of the interval $[-\pi, \pi]$ and is not time variant within the limits of the time interval $[0, T]$; and $A(t)$ is the amplitude factor, which is the random value and the function of time in the general case.

The signal with stochastic amplitude and random initial phase at the output of PF of the generalized detector shown in Figure 12.4 for the case of the rapid fluctuations of the amplitude factor $A(t)$ within the limits of the time interval $[0, T]$ takes the following form:

$$a_1(t) = A(t)S(t)\cos[\omega_0\tau + \Psi_a(t) - \varphi_0]. \tag{12.23}$$

The model signal at the output of MSG of the generalized detector takes the form:

$$a^*(t) = A^*(t-\tau)S^*(t-\tau)\cos[\omega_0(t-\tau) + \Psi_a^*(t-\tau)]. \quad (12.24)$$

This written form is correct because the amplitude and phase modulation laws of the signal at the input of navigational system are known. Therefore, we can fulfill the conditions

$$S^*(t-\tau) = S(t) \quad \text{and} \quad \Psi_a^*(t-\tau) = \Psi_a(t). \quad (12.25)$$

With the arguments of Subsection 12.2.1, we can be assured that it is necessary to construct a detector that can solve the detection problem, define the time delay of the signals with stochastic parameters, and possess the amplitude and phase tracking systems. This detector is shown in Figure 12.4.

In light of the main statements of Subsection 12.2.1, it is reasonably safe to suggest that the process at the output of the summator *1* takes the following form:

$$Z_1(t) = 0.5A^2(t)S^2(t) + A(t)S(t)\xi_1(t)\cos[\Psi_a(t) - \upsilon(t) - \varphi_0] + 0.5[\xi_1^2(t) - \xi_2^2(t)]. \quad (12.26)$$

The process at the outputs of the multipliers *1* and *2* takes the following form:

$$\begin{aligned} Z_2(t) &= 0.5A(t)A^*(t)S^2(t)\cos[\omega_0\tau - \varphi_0] \\ &+ 0.5A^*(t)S(t)\xi_1(t)\cos[\omega_0\tau - \Psi_a(t) + \upsilon(t)]. \end{aligned} \quad (12.27)$$

The process at the output of the summator *2* takes the following form:

$$\begin{aligned} Z_3(t) &= 0.5S^2(t)\{A^2(t) - A(t)A^*(t)\cos[\omega_0\tau - \varphi_0]\} \\ &+ 0.5S(t)\xi_1(t)\{2A(t)\cos[\Psi_a(t) - \upsilon(t) - \varphi_0] \\ &- A^*(t)\cos[\omega_0\tau - \Psi_a(t) + \upsilon(t)]\} + 0.5[\xi_1^2(t) - \xi_2^2(t)]. \end{aligned} \quad (12.28)$$

The process at the output of the summator *3* takes the following form:

$$\begin{aligned} Z_4(t) &= A(t)A^*(t)S^2(t)\cos[\omega_0\tau - \varphi_0] - 0.5\{A^2(t)S^2(t) \\ &- S(t)\xi_1(t)\{2A^*(t)\cos[\omega_0\tau - \Psi_a(t) + \upsilon(t)] \\ &- 2A(t)\cos[\Psi_a(t) - \upsilon(t) - \varphi_0]\}\} + 0.5[\xi_1^2(t) - \xi_2^2(t)]. \end{aligned} \quad (12.29)$$

Given $\omega_0\tau = \varphi_0$, this process takes the following form:

$$\begin{aligned} Z_4(t) &= A(t)A^*(t)S^2(t) - 0.5A^2(t)S^2(t) \\ &+ \{A^*(t) - A(t)\}S(t)\xi_1(t)\cos[\Psi_a(t) - \upsilon(t) - \varphi_0] \\ &+ 0.5[\xi_1^2(t) - \xi_2^2(t)] \end{aligned} \tag{12.30}$$

and under the additional condition $A(t) = A^*(t)$, the process at the output of the generalized detector has the form

$$Z_g^{out}(t) = 0.5\left\{\int_0^T A^2(t)S^2(t)\,dt + \int_0^T [\xi_2^2(t) - \xi_1^2(t)]\,dt\right\}. \tag{12.31}$$

Thus, with fulfillment of the conditions

$$\omega_0\tau = \varphi_0 \qquad \text{and} \qquad A(t) = A^*(t), \tag{12.32}$$

we are able to define the signal delay with high accuracy of estimation and minimum error; incidentally, the variance of error tends to approach zero under the condition $T >> \tau_c$, where τ_c is the correlation length of the signal. The conditions in Equation (12.32) are fulfilled using the amplitude and phase tracking systems (see Figure 12.4). There are circuits that control both the amplitude and the initial phase of the model signal at the output of the model signal generator and provide fulfillment of the conditions in Equation (12.32).

The results obtained indicate that we have a good chance of using the generalized detector in navigational systems in the signal processing of the target return signal with stochastic parameters, and that the estimation of the definition of the navigational object coordinates have a high accuracy. Let us proceed as follows: we construct amplitude and phase tracking systems that control the appropriate parameters of the model signal generated at the output of the model signal generator of the generalized detector. In principle, the construction of the amplitude and phase tracking systems is made possible by using the condition $T >> \tau_c$. The amplitude and phase tracking systems may take various forms. One of the variants is shown in Figure 12.4. The detection performances of the generalized detector are shown in Figure 12.5 in comparison with the potential detection performances of the classical and modern signal processing theories.

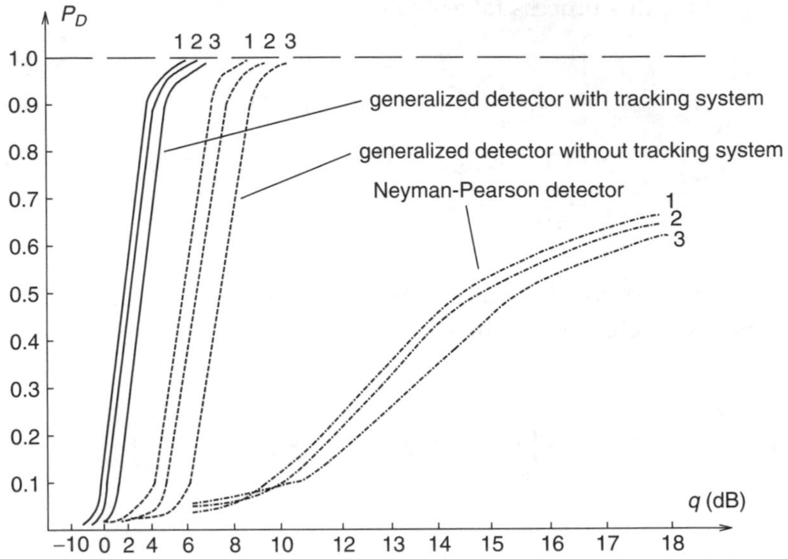

FIGURE 12.5
Detection performances of the generalized detector: (1) $T\beta$ = 1000; (2) $T\beta$ = 100; (3) $T\beta$ = 10; $P_F = 10^{-5}$.

12.3 Basics of the Generalized Approach to Space–Time Signal and Image Processing

In practice, we deal with the spatial signals, not just time-dependent stochastic processes. The spatial signals are the radar or optical image of the Earth's surface relief, functions defining the magnetic or gravity fields of the Earth,[12,15,18,20] and radio signals from various manmade sources, etc. The starry sky also forms a field of spatial signals. The definition of real space–time stochastic processes is related to their presentation by the space–time signals.[18,21] The application area of the space–time signals and images is very extensive, for instance, navigation, guidance, pattern recognition, pattern analysis, etc. Because space–time signal and image processing algorithms are similar to signal and image processing in time algorithms, and because space–time noise and interference also occur, the problem of optimal space–time signal and image processing arises in this case, too. Let us first consider the widely used models of the space–time signals and images before considering the main results of the use of the generalized approach to space–time signal and image processing in noise.

The analytical model of the space–time signals is defined by the physical nature of these signals and problems, for solving which an observation is

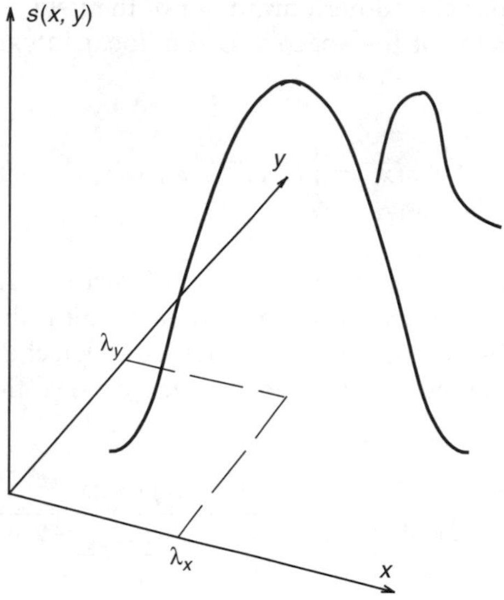

FIGURE 12.6
The parameter vector Λ defines the position of the signal.

carried out. One of the widely used models of the space–time signals corresponds to the surface potential of the space–time signal and image field[22,23]

$$a(x, y, \mathbf{\Lambda}, t) = a(x - \lambda_x, y - \lambda_y, t) \quad \text{and} \quad \mathbf{\Lambda} = \left\| \lambda_x, \lambda_y \right\|^{\mathrm{T}}. \tag{12.33}$$

The position of the signal $a(x, y, \mathbf{\Lambda}, t)$ in the Cartesian coordinate system on the plane XOY is defined by the parameter vector $\mathbf{\Lambda}$ (see Figure 12.6). The projections λ_x and λ_y of $\mathbf{\Lambda}$ define the shift of the signal along the corresponding axes. The field model given by Equation (12.33) is called the first-kind model. Its application area is too much varied. It is used in various navigational problems that are solved using optical and radar images of the Earth's surface or individual landmarks; it is also used successfully in pattern recognition theory and image analysis. In a particular case, a totality of reference points, for example, a region of the starry sky, can be determined in the following form:

$$a(x, y, \mathbf{\Lambda}, t) = \sum_{i=1}^{M} A_i \delta(x - \lambda_{x_i}, y - \lambda_{y_i}, t), \tag{12.34}$$

where $\delta(x)$ is the delta function.

In some cases, it is necessary to know amplitude characteristics of the signal at a given point of 3-D space. This is characteristic of the air magnetic survey of the Earth's anomalous magnetic gravity field when it is necessary to estimate the potential or other parameter of the field.[24–27] The field potential $a(\bar{x}_z)$ for each point of the space $\bar{x}_z$ is the linear integral operator of the surface potential

$$a(\bar{x}_z) = \int_X G(\bar{x}, \bar{x}_z) \cdot a(\bar{x})\, d\bar{x}, \tag{12.35}$$

where $G(\bar{x}, \bar{x}_z)$ is the known scalar function of vector arguments (the Green function) and $\bar{x}_z$ is the function of time characterizing the moving position of the sensor and the law of scanning. If we can neglect the curvature of the Earth's surface in the measurement of the geophysical field, we can write

$$a(\bar{x}_z) = \frac{x_{3z}}{2\pi} \int_{-\infty}^{\infty} \int_{-\infty}^{\infty} \frac{a(x_1, x_2)\, dx_1 dx_2}{\sqrt{[(x_1 - x_{1z})^2 + (x_2 - x_{2z})^2 + x_{3z}]^2}}. \tag{12.36}$$

where

$$\bar{x}_z = (x_{1z}, x_{2z}, x_{3z}) \quad \text{and} \quad \bar{x} = (x_1, x_2) \tag{12.37}$$

are the Cartesian coordinate system shown in Figure 12.7. If we take into account the curvature of the Earth's surface in spherical approximation, the spherical coordinates take the following form:

$$\bar{x}_z = (\varphi_z, \lambda_z, h_z) \qquad \text{and} \qquad \bar{x} = (\varphi, \lambda), \tag{12.38}$$

where φ_z is the latitude, λ_z is the longitude, and h_z is the altitude of sensor with respect to the Earth's surface (see Figure 12.8). Then the potential can be written in the following form:

$$a(\bar{x}_z) = \frac{(1 + h_1^2) - 1}{4\pi} \iint_S \frac{a(\varphi, \lambda) \cos\varphi \, d\varphi \, d\lambda}{\sqrt{[1 + (1 + h_1^2) - 2(1 + h_1)\cos\Psi]^3}}, \tag{12.39}$$

$$\cos\Psi = \sin\varphi \sin\varphi_z + \cos\varphi \cos\varphi_z \cos(\lambda - \lambda_z), \tag{12.40}$$

$h_1 = hR^{-1}$, where R is the radius of the Earth.

In both cases, we believe that the field characteristics are not varied in time to be true both for gravity and anomalous magnetic fields. This field model, called the second-kind model, was designed by A. Krasovsky.[12,28] To use the second-kind field model it is necessary to define the Green function for each measurement. In particular, in the measurement of a single component of the anomalous magnetic field strength or a single "gradient," i.e., the

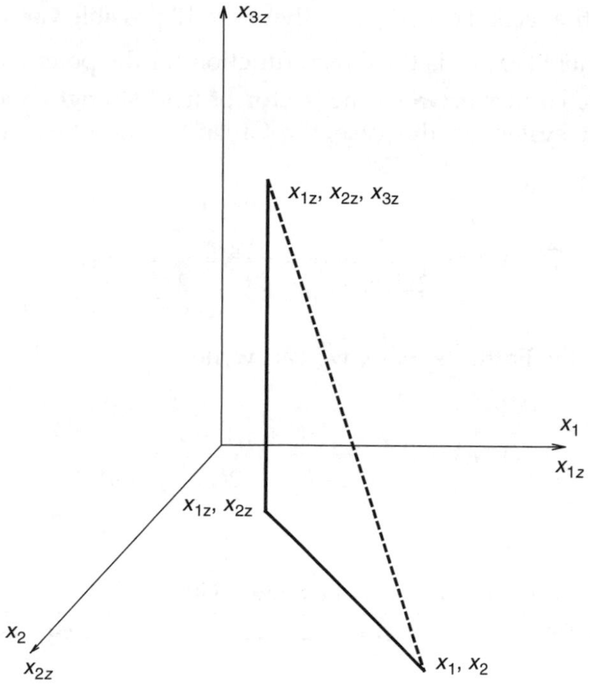

FIGURE 12.7
The Cartesian coordinate system.

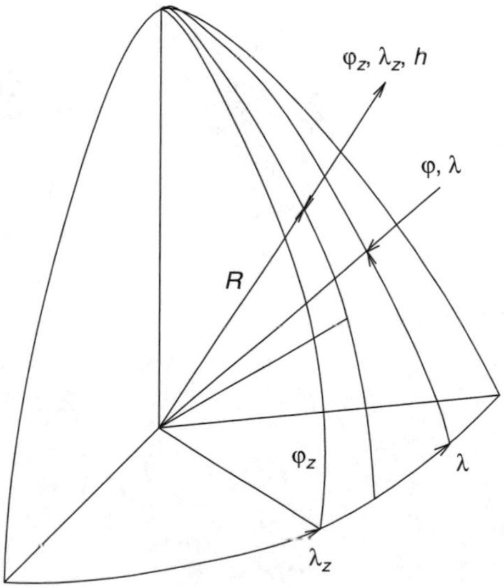

FIGURE 12.8
The spherical coordinate system.

derivative of the second order $\frac{\partial^2 a}{\partial x_{zi}\partial x_{zj}}$, there are 10 possible variants shown in Table 12.2, where $G(x, x_z)$ is the Green function for the potential, and $\varepsilon_{T_{xi}}(i = 1, 2, 3)$ are the cosines between the vector of field strength and the axes of the coordinate system. In this case, the Green function takes the following form:

$$G(\bar{x},\bar{x}_z) = \frac{x_{3z}}{\sqrt{2\pi[(x_1 - x_{1z})^2 + (x_2 - 2x_{2z})^2 + x_{3z}^2]^3}}. \tag{12.41}$$

In the case of the Earth's sphere, we can write

$$G(\bar{x},\bar{x}_z) = \frac{(1+h_1)^2 - 1}{\sqrt{[1+(1+h_1)^2 - 2(1+h_1)\cos\Psi]^3}}. \tag{12.42}$$

TABLE 12.2

Possible Variants of Measurement of a Single "Gradient"

Measured Value	Type of Operator	Resulting Kerner
Horizontal strength	$\frac{\partial}{\partial x_{1z}}$	$\frac{\partial}{\partial x_{1z}} G(x,x_z)$
Horizontal strength	$\frac{\partial}{\partial x_{2z}}$	$\frac{\partial}{\partial x_{2z}} G(x,x_z)$
Vertical strength	$\frac{\partial}{\partial x_{3z}}$	$\frac{\partial}{\partial x_{3z}} G(x,x_z)$
Total absolute value	$\sum_1^3 \varepsilon_{T_{xi}} \frac{\partial}{\partial x_{iz}}$	$\sum_1^3 \varepsilon_{T_{xi}} \frac{\partial}{\partial x_{iz}} G(x,x_z)$
x_1x_1	$\frac{\partial^2}{\partial x_{1z}^2}$	$\frac{\partial^2}{\partial x_{1z}^2} G(x,x_z)$
x_1x_2	$\frac{\partial^2}{\partial x_{1z}\partial x_{2z}}$	$\frac{\partial^2}{\partial x_{1z}\partial x_{2z}} G(x,x_z)$
x_1x_3	$\frac{\partial^2}{\partial x_{1z}\partial x_{3z}}$	$\frac{\partial^2}{\partial x_{1z}\partial x_{3z}} G(x,x_z)$
x_2x_2	$\frac{\partial^2}{\partial x_{2z}^2}$	$\frac{\partial^2}{\partial x_{2z}^2} G(x,x_z)$
x_2x_3	$\frac{\partial^2}{\partial x_{2z}\partial x_{3z}}$	$\frac{\partial^2}{\partial x_{2z}\partial x_{3z}} G(x,x_z)$
x_3x_3	$\frac{\partial^2}{\partial x_{3z}^2}$	$\frac{\partial^2}{\partial x_{3z}^2} G(x,x_z)$

The space–time signal $a(x, y, t)$ can be often expressed as the sum of the harmonic components. Each harmonic component can be considered as a simple harmonic field, for which the inverse Fourier transform with respect to all frequencies is true:[18,29]

$$a(x,y,t)=\frac{1}{(2\pi)^3}\int_{-\infty}^{\infty}\int_{-\infty}^{\infty}\int_{-\infty}^{\infty}F(\omega,\omega_x,\omega_y)\cdot e^{j(\omega_x x+\omega_y y-\omega t)}d\omega_x d\omega_y d\omega=\mathcal{F}^{-1}(\omega,\omega_x,\omega_y), \tag{12.43}$$

where $\mathcal{F}^{-1}(\omega, \omega_x, \omega_y)$ is the inverse Fourier transform;

$$F(\omega,\omega_x,\omega_y)=\int_{-\infty}^{\infty}\int_{-\infty}^{\infty}\int_{-\infty}^{\infty}A_\omega(x,y)\cdot e^{-j(\omega_x x+\omega_y y-\omega t)}dx\,dy\,dt, \tag{12.44}$$

where $A_\omega(x, y)$ is the complex amplitude of the field component at the given frequency ω; ω_x is the spatial frequency along the axis x; and ω_y is the spatial frequency along the axis y. Under these conditions, the space–time signal can be determined in the following form:

$$a(x,y)=\frac{1}{4\pi^2}\int_{-\infty}^{\infty}\int_{-\infty}^{\infty}F(\omega_x,\omega_y)\cdot e^{j(\omega_x x+\omega_y y)}d\omega_x d\omega_y=\mathcal{F}^{-1}\{F(\omega_x,\omega_y)\}, \tag{12.45}$$

$$F(\omega_x,\omega_y)=\int_{-\infty}^{\infty}\int_{-\infty}^{\infty}a(x,y)\cdot e^{-j(\omega_x x+\omega_y y)}dx\,dy. \tag{12.46}$$

Let us recall that for a couple of the Fourier transforms to exist, the function $a(x, y)$ must satisfy the Dirichlet conditions. The first condition is: the observation interval can be divided on a finite number of intervals, within the limits of which the function $a(x, y)$ is continuous and monotonic. The second condition is: the function $a(x, y)$ should have limited numbers of finite breaks. Thus, there is one exponent with the weight coefficient $G(\omega_x, \omega_y)$ for each couple of values of the spatial frequencies ω_x and ω_y in the generalized sum corresponding to Equation (12.45).

Let us consider this form of exponent for the given values of spatial frequencies. This exponent is the complex function similar to the Fourier transform of function in time. It is impossible to show this function graphically due to the double dependence on x and y. Therefore, the concept of zero-phase regions is introduced to guess the form of this function. Zero-phase regions can be determined in the following form:

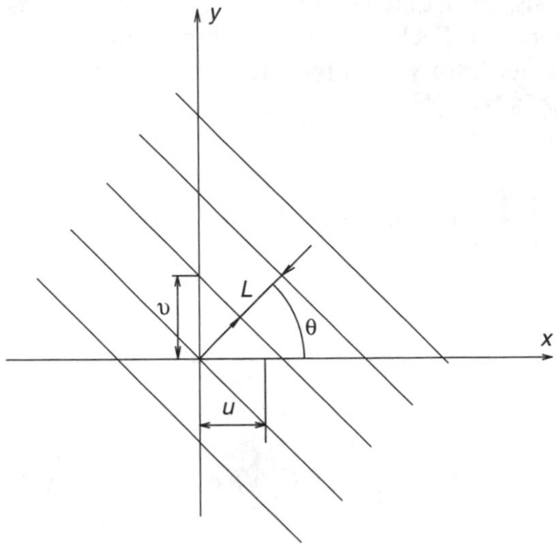

FIGURE 12.9
The nonzero phase region.

$$y = -(ux + 2\pi n)\upsilon^{-1}; \quad u = 0.5\pi^{-1}\omega_x; \quad \text{and} \quad \upsilon = 0.5\pi^{-1}\omega_y. \tag{12.47}$$

In accordance with Equation (12.47), the zero-phase regions can be represented by parallel lines with the distance between the lines equal to $L = \sqrt{u^2 + \upsilon^2}$. The slope of these lines is defined by the angle of orientation equal to $\vartheta = \text{arctg}\,\frac{u}{v}$ (see Figure 12.9). When the spatial frequencies are high, the zero phase lines are located frequently. To understand this clearly, we should assume that the absolute value of the energy spectrum of spatial frequencies is equal to zero anywhere, excepting at the points (u_1, υ_1) and ($-u_2$, υ_1). In this case, the spatial signal can be approximated by the sum

$$a(x, y) = 0.5A \cdot \left[e^{2j\pi(u_1x+\upsilon_1y)} + e^{2j\pi(-u_1x-\upsilon_1y)}\right], \tag{12.48}$$

where $A = const$ and is the real value that can be presented in the form of sine curve with the unit amplitude (see Figure 12.10). Peaks of this surface are the parallel lines similar to zero phase lines.

Because, in the general case, the signal $a(x, y, t)$ is a stochastic process, we should use the statistical characteristics, the mean, the variance, and the correlation function to define it. The correlation function for uniform — stationary by space — and stationary signals can be determined in the following form:

$$R(\Delta x, \Delta y, \Delta t) = \langle a(x + \Delta x, y + \Delta y, t) \cdot a(x, y, t) \rangle. \tag{12.49}$$

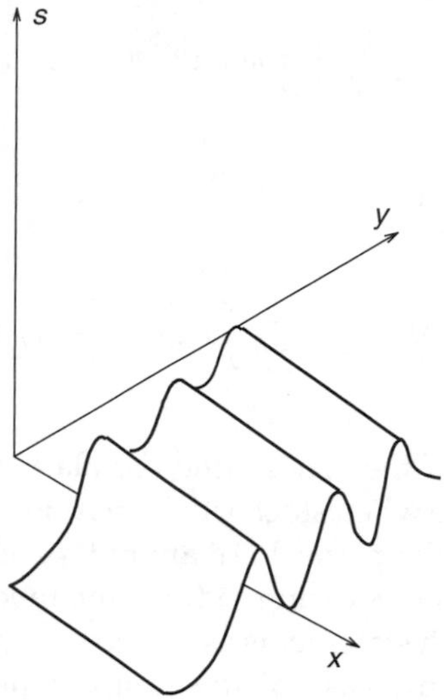

FIGURE 12.10
Representation of the spatial signal.

In the case of the complex signals $a(x, y, t)$, we can write

$$R(\Delta x, \Delta y, \Delta t) = < a(x + \Delta x, y + \Delta y, t + \Delta t) \cdot \ddot{a}(x, y, t) > . \qquad (12.50)$$

Here, the symbol <.> denotes the mean in the statistical sense and the symbol $\ddot{a}$ denotes a complex conjugated value. The correlation functions given by Equation (12.49) and Equation (12.50) are called the space–time correlation functions.

If we have a stationary space–time and uniform signal, which can be expressed as the product of the components of individual radiation sources, then we can write

$$a(x, y, t) = \sum_{i=1}^{n} a_{1i}(x, y) \cdot a_{2i}(t), \qquad (12.51)$$

where $a_{1i}(x, y)$ is the modulation law of the signal amplitude from the i-th source. If the modulation law of the signal amplitude from the i-th source is varied as a function of time in accordance with $a_{2i}(t)$, the space–time correlation function can be determined in the form of the product of the space and time components:

$$R(\Delta x, \Delta y, \Delta t) = R(\Delta x, \Delta y) \cdot R(\Delta t). \qquad (12.52)$$

Here

$$R(\Delta x, \Delta y) = \lim_{\substack{X\to\infty \\ Y\to\infty}} \frac{1}{XY} \int_{-X}^{X} \int_{-Y}^{Y} a(x, y) \cdot a(x + \Delta x, y + \Delta y)\, dx\, dy \tag{12.53}$$

is the space component and

$$R(\Delta t) = \lim_{T\to\infty} \frac{1}{2T} \int_{-T}^{T} a(t) \cdot a(t + \Delta t)\, dt \tag{12.54}$$

is the time component of the space–time correlation function. The formula in Equation (12.52) is true when stochastic variations of the space–time signal $a(x, y, t)$ in time and on the plane XOY are independent.

As in the case of signal processing in time, the theory of space–time signal processing possesses such statements as the power spectral density of spatial frequency $S(\omega_x, \omega_y)$. In the case of the uniform field, the power spectral density $S(\omega_x, \omega_y)$ is defined by the energy spectrum of the electromagnetic field:[30,31]

$$S(\omega_x, \omega_y) = \lim_{\substack{X\to\infty \\ Y\to\infty}} |F(\omega_x, \omega_y)|^2 . \tag{12.55}$$

The power spectral density of spatial frequencies of the uniform field is related with the space–time correlation function by the Wiener–Heanchen theorem:

$$S(\omega_x, \omega_y) = \int_{-\infty}^{\infty} \int_{-\infty}^{\infty} R(\Delta x, \Delta y) \cdot e^{-j(\omega_x \Delta x + \omega_y \Delta y)} d(\Delta x)\, d(\Delta y), \tag{12.56}$$

$$R(\Delta x, \Delta y) = \frac{1}{4\pi^2} \int_{-\infty}^{\infty} \int_{-\infty}^{\infty} S(\omega_x, \omega_y) \cdot e^{j(\omega_x \Delta x + \omega_y \Delta y)} d\omega_x d\omega_y = \mathcal{F}^{-1}[S(\omega_x, \omega_y)]. \tag{12.57}$$

For the correlation function given by Equation (12.52), the power spectral density $S(\omega_x, \omega_y, \omega)$ is defined by the product of the angular power spectral density $S(\omega_x, \omega_y)$ and the frequency power spectral density $S(\omega)$:

$$S(\omega_x, \omega_y, \omega) = S(\omega_x, \omega_y) \cdot S(\omega) . \tag{12.58}$$

Optimization of the measurement of the navigational object parameter vector $\mathbf{\Lambda}$ for the space-time signals model of the first kind observed in the presence of additive noise can be carried out based on the definition of the likelihood functional for the space–time signal and image processing.[18,32] The

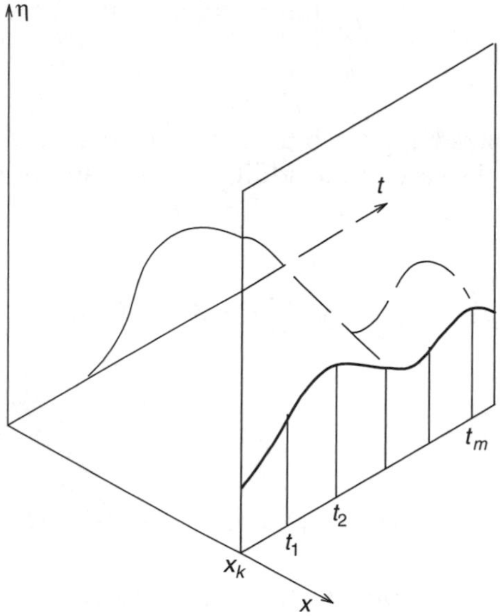

FIGURE 12.11
Space–time stochastic process $\eta(x, y, t)$.

algorithm to solve this problem is as follows. First, we consider the multidimensional probability distribution density of the stochastic process $\zeta(x, y, t)$ that is discrete in space and time. The totality of the real values $\zeta(x_k, y_l, t_m)$ where $k, l, m = 1, 2, \ldots, M$, of the stochastic process $\zeta(x, y, t)$, which are called the samples, can be represented by readings of the specific realization obtained under the cross section of the stochastic process $\zeta(x, y, t)$ in the plane $x = x_k$ at the instant of the time t_m (see Figure 12.11). For all values of x_k, y_l, and t_m, the discrete readings $\zeta(x_k, y_l, t_m)$ are the possible values of the stochastic process $\zeta(x, y, t)$ reckoned over the intervals Δx, Δy, and Δt. The totalities of the probability distribution densities of these values for all M form the multidimensional probability distribution density $f^{(M)}[\zeta(x, y, t)]$. The limit of the multidimensional probability distribution density $f^{(M)}[\zeta(x, y, t)]$ is defined as the functional of the probability distribution density of the space–time stochastic process:[18]

$$
\begin{aligned}
&f[\zeta(x = x_1, y = y_1, t = t_1); \zeta(x = x_1, y = y_1, t = t_2); \ldots;\\
&\quad \zeta(x = x_1, y = y_1, t = t_M)];
\end{aligned}
\tag{12.59}
$$

$$
\begin{aligned}
f^{(M)}[\zeta(x, y, t)] &= f[\zeta(x = x_1, y = y_1, t = t_1)\ldots\\
&\quad \zeta(x = x_1, y = y_1, t = t_M)\ldots\zeta(x = x_2, y = y_1, t = t_1)\ldots\\
&\quad \zeta(x = x_2, y = y_1, t = t_M)\ldots\zeta(x = x_M, y = y_M, t = t_M)];
\end{aligned}
\tag{12.60}
$$

$$L(\mathbf{X}) = \lim_{\substack{\Delta x \to 0 \\ \Delta y \to 0 \\ \Delta t \to 0}} f^{(M)}[\zeta(x, y, t)]. \quad (12.61)$$

Using the generalized approach to signal processing in noise,[5–8] the likelihood functional in the case of the additive Gaussian noise can be written in the following form:

$$L(\mathbf{X}) = C \cdot \exp\left\{-0.5 \int_X \int_Y \left[2\mathbf{X}^{\mathrm{T}}\mathbf{S}_0^{-1}\mathbf{a}^* - \mathbf{X}^{\mathrm{T}}\mathbf{S}_0^{-1}\mathbf{X} + \boldsymbol{\eta}^{\mathrm{T}}\mathbf{S}_1^{-1}\boldsymbol{\eta}\right] dx\, dy\, dt\right\}, \quad (12.62)$$

where

$$\mathbf{X} = \mathbf{X}(x, y, t) = \left\| X_1(x, y, t) \ldots X_m(x, y, t) \right\|^{\mathrm{T}} \quad (12.63)$$

is the moving image or input stochastic process at the input of the navigational system that can contain the information signal;

$$\mathbf{a}^* = \mathbf{a}^*(x, y, t) = \left\| a_1^*(x, y, t) \ldots a_m^*(x, y, t) \right\|^{\mathrm{T}} \quad (12.64)$$

is the model image; and

$$\boldsymbol{\eta} = \boldsymbol{\eta}(x, y, t) = \left\| \eta_1(x, y, t) \ldots \eta_m(x, y, t) \right\|^{\mathrm{T}} \quad (12.65)$$

is the reference additional stochastic process at the input of the navigational system and it is the known *a priori* "no" information signal in the reference additional stochastic process; X is the observed region along the axis x; Y is the observed region along the axis y; $\mathbf{S}_0$ and $\mathbf{S}_1$ are the matrices of the power spectral densities in the sense of ordinary and spatial frequencies with the same statistics; and C is the normalized factor. Recall that the "white" Gaussian space–time noise is the Gaussian process with the non-time-variant power spectral density of components S_0 or S_1 in the sense of ordinary and spatial frequencies. The correlation function of the "white" Gaussian space–time noise can be written in the following form:

$$R(\Delta x, \Delta y, \Delta t) = \mathbf{S}_0 \delta(\Delta x, \Delta y, \Delta t). \quad (12.66)$$

In the case of the one-dimensional (scalar) signal, we can write

$$L(X) = C \cdot \exp\Big\{-0.5S_0^{-1} \times \int_x \int_y \int_t \big[2X(x,y,t) \cdot a^*(x,y,t) - X^2(x,y,t) + \eta^2(x,y,t)\big]\, dx\, dy\, dt\Big\}. \quad (12.67)$$

Thus, the space–time signal and image processing using the generalized approach to signal processing in the presence of noise is to define the maximum of the integral

$$I = \int_X \int_Y \big[2X(x,y,t) \cdot a^*(x,y,\mathbf{\Lambda}^*,t) - X^2(x,y,t) + \eta^2(x,y,t)\big]\, dx\, dy, \quad (12.68)$$

where $X(x, y, t)$ is the moving image and $a^*(x, y, \mathbf{\Lambda}^*, t)$ is the model image. If the values X and Y are much more than the correlation length of the signal and the signal is the stationary stochastic process, the integral in Equation (12.68) matches the maximum of the likelihood functional. The definition of this maximum is carried out with the use of the generalized detector tracking systems discussed in Section 12.2. Thus, the discrimination characteristics are defined along the corresponding axes:

$$I_x = \frac{\partial}{\partial \lambda_x^*} I \quad \text{and} \quad I_y = \frac{\partial}{\partial \lambda_y^*} I. \quad (12.69)$$

Navigational systems realizing the generalized approach to signal processing in noise are called the correlation-extremal systems because Equation (12.68) allows us to define the maximum of the likelihood functional only under the main conditions of functioning of the generalized detector. The functions given by Equation (12.68) and Equation (12.69) are often called the criterial functions.

The criterial functions given by Equation (12.69) are proportional to the measurement errors, i.e.,

$$I_x \approx \frac{\partial^2 R(\Delta x, \Delta y, \Delta t)}{\partial(\Delta x)^2}\bigg|_{\Delta x=0} \cdot \Delta x \quad \text{and} \quad I_y \approx \frac{\partial^2 R(\Delta x, \Delta y, \Delta t)}{\partial(\Delta y)^2}\bigg|_{\Delta y=0} \cdot \Delta y, \quad (12.70)$$

because the background noise in using the generalized approach to signal processing tends to approach zero in the statistical sense, and the noise component $2\int_X \int_Y a^*(x,y,\mathbf{\Lambda}^*,t) n(x,y,t) dx dy$ of the generalized detector correlation channel caused by the interaction between the model image and noise and the random component $2\int_X\int_Y a(x,y,\boldsymbol{\Lambda},t)n(x,y,t)dxdy$ of the generalized detector autocorrelation channel caused by the interaction between the moving image and noise are compensated by each other in the statistical sense (see Chapter 11).

Correlation extremal receivers or detectors constructed according to the generalized approach to signal processing in the presence of noise may be used in navigational systems without tracking systems. The use of the generalized receivers or detectors of this form implies the employment of the signals with known amplitude-phase-frequency structure. Comparison of the incoming moving image at the input of the navigational system with the model image defined by the form of the signal and *a priori* knowledge regarding the location of the signal on the observation plane allows us to estimate the true value of the parameter vector of $\boldsymbol{\Lambda}$. In doing so, this value of the parameter vector $\boldsymbol{\Lambda}$ is automatically controlled. The main condition of the navigational system functioning without tracking devices is the correlation between the moving image $\mathbf{X}(x, y, t)$ and the model image $\mathbf{a}^*(x, y, \boldsymbol{\Lambda}^*, t)$. In other words, we cannot recognize the information signal or image $\mathbf{a}(x, y, \boldsymbol{\Lambda}, t)$ in the input moving image $\mathbf{X}(x, y, t)$. The use of the generalized detector with tracking systems (see Section 12.2) allows us to solve this problem and to recognize the type of the signal $\mathbf{a}(x, y, \boldsymbol{\Lambda}, t)$ and to define and measure the parameter vector $\boldsymbol{\Lambda}$.

The criterial functions may be defined in another manner. For example, in Equation (12.56) and Equation (12.57), which define the relationship between the space–time correlation function and the power spectral density of spatial frequencies, the criterial function takes the following form:

$$I = \frac{1}{4\pi^2}\int_{-\infty}^{\infty}\int_{-\infty}^{\infty} S(\omega_x,\omega_y,t)\cdot e^{j(\omega_x\Delta x+\omega_y\Delta y)}d\omega_x d\omega_y, \tag{12.71}$$

where

$$S(\omega_x,\omega_y,t) = \int_{-\infty}^{\infty}\int_{-\infty}^{\infty} R(\Delta x,\Delta y,t)\cdot e^{-j(\omega_x\Delta x+\omega_y\Delta y)}d(\Delta x)\, d(\Delta y). \tag{12.72}$$

For simplicity, we assume that the signal and noise are uncorrelated between each other in Equation (12.71) and the differentiation with respect to the parameters λ_x^* and λ_y^* is not taken into consideration.

The discussion in the preceding text defines the so-called spectral technique of space–time signals processing. In accordance with this technique, the moving and model images are processed only in the spatial frequencies region. In this case, both the moving and model images are transformed in the power spectral densities of spatial frequencies according to Equation

(12.46). Therefore, the model signal $a^*(x - \lambda_x^*, y - \lambda_y^*, t)$ is defined by $F(\omega_x, \omega_y) \cdot e^{-j(\omega_x \lambda_x^* + \omega_y \lambda_y^*)}$ in accordance with the theorem of shift. The power spectral density of the information signal in the moving image is defined by $F(\omega_x, \omega_y) \cdot e^{-j(\omega_x \lambda_x + \omega_y \lambda_y)}$. Under these conditions, the criterial function given by Equation (12.71) has the following form:

$$I = \int_{-\infty}^{\infty} \int_{-\ldots\infty}^{\infty} F(\omega_x, \omega_y) \cdot F^*(\omega_x, \omega_y) \cdot e^{-j(\omega_x \Delta x + \omega_y \Delta y)} d\omega_x d\omega_y, \qquad (12.73)$$

$$\Delta x = \lambda_x - \lambda_x^* \quad \text{and} \quad \Delta y = \lambda_y - \lambda_y^*, \qquad (12.74)$$

where $F^*(\omega_x, \omega_y)$ denotes the complex conjugate Fourier transform. In other words, the signal processing given by Equation (12.71) and Equation (12.73) specifies the transformation of the moving image into the power spectral density of spatial frequencies, the generation of the product of the complex conjugate power spectral densities of the moving image and model image, and fulfillment of the inverse Fourier transform to define the mutual correlation function between the information signal $a(x, y, t)$ and the model signal $a^*(x, y, \mathbf{\Lambda}^*, t)$ with further determination of its derivative at the point $\Delta\mathbf{\Lambda} = \mathbf{\Lambda} - \mathbf{\Lambda}^*$. This technique is used successfully in optical navigational systems. We do not take into consideration the additive noise $n(x, y, t)$ in Equation (12.71) and Equation (12.73). This was made with the purpose of stressing the external character of the criterial function. In practice, naturally, all spectral transformations are carried out with the moving image $\mathbf{X}(x, y, t)$.

Because we are able to represent the moving image in the form of a set of harmonic components, the problem of the possible employment of various spatial frequency filters arises. It is well known[5–8] that the generalized approach to signal processing in noise provides the definition of the mutual correlation function derivative of the moving and model images, which is equivalent to passing the moving image through the filter of the spatial frequencies with the pulse characteristic

$$h(x, y) = \frac{\partial a(x - \lambda_x^*, y - \lambda_y^*)}{\partial \lambda_{x(y)}^*}. \qquad (12.75)$$

Actually, the criterial functions I_x and I_y are similar to the Duhamel integral, which is characteristic of the response of any system to external action. In this case, the derivative of the model image is considered as the pulse response of the preliminary filter of the generalized detector. We can use the preliminary filter of the generalized detector spatial frequencies with the pulse response $h(x, y) = a^*(x - \lambda_x^*, y - \lambda_y^*)$, but in accordance with the generalized approach to signal processing in the presence of noise, it is necessary to differentiate the signal at the output of the preliminary filter with respect to the variables λ_x^* and λ_y^* for each channel measuring the parameters λ_x and λ_y, respectively. For this purpose, we can use the preliminary filter of high spatial frequencies of the generalized detector. The stimulus of the preliminary filter of the generalized detector high spatial frequencies is analogous to spatial differentiation.

Let us consider this problem in more detail using the example of the shift definition of the signal $a(t)$ along the axis x, i.e., the example of the definition of the parameter λ_x. In accordance with Equation (12.69), the model image $a^*(x - \lambda_x^*)$ is the totality of the spatial waves of various lengths. The definite power spectral density corresponds to this representation. In the simplest case, when the stochastic process at the input of the navigational system has a single spatial harmonic in the condition $\omega_x = \omega_x'$, we can write

$$a^*(x - \lambda_x^*) = 0.5A \cdot \left[e^{j(\omega_x' x - \omega_y' \lambda_x^*)} + e^{-j(\omega_x' x - \omega_x' \lambda_x^*)}\right]. \tag{12.76}$$

In this case, the zero phase regions shown in Figure 12.9 have the form of an unlimited number of vertical parallel lines at the condition $\theta = 0°$ and are spaced by the distance $x_0 = (\omega_x')^{-1}$. The energy spectrum of spatial frequencies can be presented using the delta function $A\delta(\omega_x - \omega_x')$. In the general case, the space–time signal is formed by the sum of unlimited numbers of spatial harmonics, for example

$$a^*(x - \lambda_x^*) = \sum_{i=0}^{\infty} 0.5A\left[e^{j(\omega_{xi} x - \omega_{xi}\lambda_x^*)} + e^{-j(\omega_{xi} x - \omega_{xi}\lambda_x^*)}\right] = \sum_{i=-\infty}^{\infty} c_i e^{j\omega_{xi}\lambda_x^*}, \tag{12.77}$$

where $c_i = 0.5A \cdot e^{j\omega_{xi} x}$. Trigonometrically, we can write

$$a^*(x - \lambda_x^*) = \sum_{i=0}^{\infty} A_i \cos(\omega_x x - \omega_x \lambda_x^*). \tag{12.78}$$

The pulse response of the spatial frequency preliminary filter of the generalized detector has the following form:

$$h(x) = \sum_{i=-\infty}^{\infty} \hat{c}_i \cdot e^{-j\omega_{xi}\lambda_{xi}} \quad \text{and} \quad \hat{c}_i = 0.5A_i \cdot e^{-j\omega_{xi} x}. \tag{12.79}$$

Taking into consideration the averaging within the limits of the interval $X >> x_{0i} = \frac{2\pi}{\omega_{xi}}$, in terms of Equation (12.68) the criterial function has the following form:

$$I_X = X^{-1}\int_X \sum_{i=-\infty}^{\infty}\sum_{k=-\infty}^{\infty} 0.25A_iA_k \cdot e^{-j\omega_{x_{ik}}\Delta x}dx$$
$$+ 2X^{-1}\int_X \sum_{i=-\infty}^{\infty} 0.5A_i \cos(\omega_{x_i}x - \omega_{x_i}\lambda_x^*)\xi(x)dx$$
$$- 2X^{-1}\int_X \sum_{i=-\infty}^{\infty} 0.5A_i \cos(\omega_{x_i}x - \omega_{x_i}\lambda_x^*)\xi(x)dx + \int_X \eta^2(x)\,dx - \int_X \xi^2(x)\,dx, \tag{12.80}$$

where the second term in Equation (12.80) is the noise component of the generalized detector correlation channel caused by the interaction between the model image and noise; the third term in Equation (12.80) is the random component of the generalized detector autocorrelation channel caused by the interaction between the information signal of the moving image and noise. The fourth and fifth terms in Equation (12.80) are the background noise of the generalized detector. As discussed in Chapter 11, the second and third terms are compensated between each other in the statistical sense and the background noise [the fourth and fifth terms in Equation (12.80)] tends to approach zero in the statistical sense. Based on the results discussed in Chapter 11, Equation (12.80) can be written in the following form:

$$I_X \cong \sum_{i=0}^{\infty} 0.25A_i^2 \cdot e^{-j\omega_{x_i}\Delta x} \cong \sum_{i-0}^{\infty} 0.5A_i^2 \cos(\omega_{x_i}\Delta x), \tag{12.81}$$

where $\Delta x = \lambda - \lambda_x^*$ and $\omega_{x_{ik}} = \omega_{x_i} - w_{x_k}$.

Thus, to obtain the criterial function in the form given by Equation (12.68), the preliminary filter of the generalized detector should be matched with the model image. The amplitude response of the amplitude-frequency characteristics of the generalized detector preliminary filter is the totality of the delta functions $A_i\delta(\omega_x - \omega_{x_i})$ (see Figure 12.12). Obviously, in practice we can only explain more close approximation of the amplitude-frequency response, for example, by the bell-shaped function (see Figure 12.13)

$$\sum_{i=1}^{\infty} 0.5A_i \cdot e^{-j\frac{(\omega_{x_i}x - \omega_{x_i}\lambda_x^2)^2}{2c}}, \tag{12.82}$$

where $c = const$ and the value of c is chosen as soon as it is low. Note that the use of spatial differentiation is equivalent to the procedure for emphasizing the contour lines that have been widely used in practice to process the optical images.

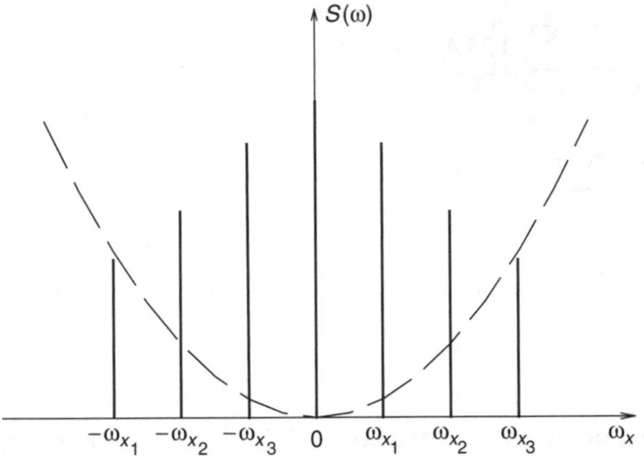

FIGURE 12.12
Spectral characteristic of the high-frequency preliminary filter of the generalized detector.

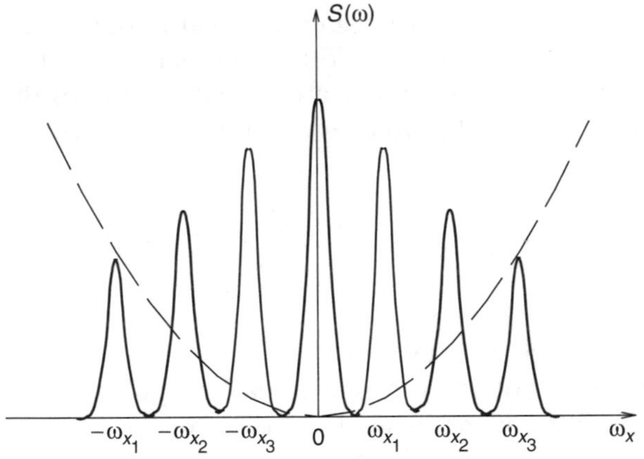

FIGURE 12.13
Spectral characteristic of the bell-shaped preliminary filter of the generalized detector.

In moving navigational systems, the zero-phase regions are shifted in parallel relative to the observed image of the Earth's surface. In other words, the image of the Earth's surface is formed due to the propagation of spatial harmonics with various spatial periods. The output voltage of the comb preliminary filter of the generalized detector is independent of the parallel shifts of the observed Earth's surface image. For this reason, the preliminary filter of the generalized detector is kept matched with the model image. However, this preliminary filter is very critical with respect to the angle of

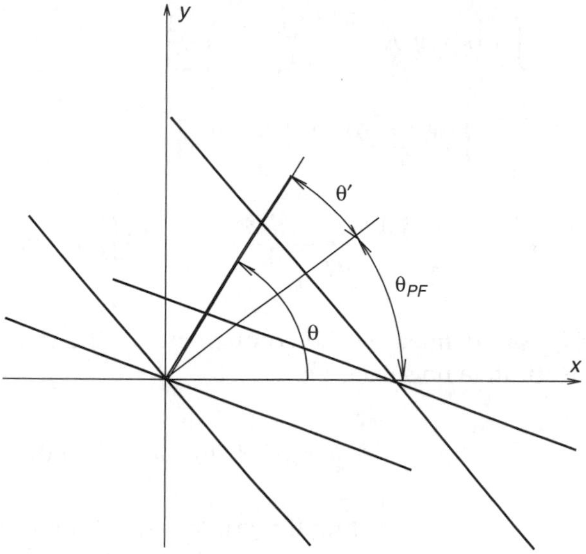

FIGURE 12.14
Angles of mutual orientation.

mutual orientation $\theta' = \theta - \theta_{PF}$ between the direction of the spatial wave normal θ and the axis θ_{PF} of the generalized detector preliminary filter (see Figure 12.14). If the pulse-transient function of the generalized detector preliminary filter can be considered as the model image or its derivative, the previous limitations tell us about the inadmissibility of the rotations between the information signal of the moving image and the model image. One way to cushion this requirement is to make the transfer function of the filtering elements of the generalized detector preliminary filter more blurred, which ensures the invariance of the preliminary filter with respect to the angle of mutual orientation. In particular, in the one-dimensional case, i.e., $\theta = 0°$, using n one-dimensional filtering elements, the condition of invariance with respect to the allowable value of the mutual orientation angle can be defined as[32–35] $n \geq 0.5\pi\theta'_{allow}$.

According to the character of the additive noise $n(x, y)$ at the input of the navigational system, we can use the generalized detector preliminary filter of various forms. For simplicity, we assume that the noise $n_1(x, y)$ is the Gaussian narrow-band process. Then, the criterial function for each estimation channel λ_x and λ_y is defined by the derivative of the space–time correlation function of signal derivatives with the weight coefficients that are inversely proportional to the spectral bandwidth of the noise $n_1(x, y)$ in the region of spatial and ordinary frequencies. In measuring the value of λ_x, this criterial function can be determined in the following form:

$$I = \int_X \int_Y \Big\{ a(x,y,\mathbf{\Lambda}) \cdot \frac{\partial a^*(x,y,\mathbf{\Lambda}^*)}{\partial \lambda_x^*} + \frac{1}{2\alpha_x^2} \times \frac{\partial}{\partial \lambda_x^*} \Big[\frac{\partial a(x,y,\mathbf{\Lambda})}{\partial x} \cdot \frac{\partial a^*(x,y,\mathbf{\Lambda}^*)}{\partial x} \Big] + \frac{1}{2\alpha_y^2} \cdot \frac{\partial}{\partial \lambda_y^*} \Big[\frac{\partial a(x,y,\mathbf{\Lambda})}{\partial y} \cdot \frac{\partial a^*(x,y,\mathbf{\Lambda}^*)}{\partial y} \Big] \Big\} \, dx\, dy, \tag{12.83}$$

where α_x^{-1} and α_y^{-1} are defined by the components of the spatial correlation function of the additive noise $n(x, y)$:

$$R_n(\Delta x, 0) = R_n(0,0) \cdot e^{-\alpha_x \Delta x} \quad \text{and} \quad R_n(0, \Delta y) = R_n(0,0) \cdot e^{-\alpha_y \Delta y}. \tag{12.84}$$

In an analogous way we can define the criterial function in measuring the value of λ_y. In estimating the one-dimensional parameter λ, Equation (12.84) can be written in the simplest form:

$$I = \int_X \Big\{ a(x-\lambda_x) \cdot \frac{\partial a^*(x-\lambda_x^*)}{\partial \lambda_x^*} + \frac{1}{2\alpha_x^2} \cdot \frac{\partial}{\partial \lambda_x^*} \Big[\frac{\partial a(x-\lambda_x)}{\partial x} \cdot \frac{\partial a^*(x-\lambda_x^*)}{\partial x} \Big] \Big\} dx. \tag{12.85}$$

The physical form of Equation (12.83) and Equation (12.85) is, as noted, a differentiation that corresponds to the filtering of high frequencies. Therefore, the spectral bandwidth of the additive noise is wider, the components of the high spatial frequencies of the moving image information signal are lower in value, and *vice versa,* i.e., if the main energy part of the additive noise $n(x, y)$ is concentrated in the region of low frequencies, the most true information is contained in the high frequencies of the moving image information signal. If the space–time additive noise $n(x, y, t)$is a wide-band process so that $\alpha_x^{-1} \to 0$ and $\alpha_y^{-1} \to 0$, Equation (12.83) and Equation (12.85) have the form that is analogous to the "white" Gaussian noise considered earlier.

In solving the navigational problems, the generalized approach to signal processing in the presence of noise allows us to use the following procedure. Let us assume that the intensity of the considered field can be defined as the totality of the values $\mathbf{a}(t, \mathbf{\Lambda}_i)$. The additive noise is a stochastic process depending only on time, not on the observed space coordinates. The *a priori* information regarding the measured parameter vector $\mathbf{\Lambda}_i$ is determined by

$$\frac{d\mathbf{\Lambda}}{dt} = \mathbf{Q}(t, \mathbf{\Lambda}) + \mathbf{m}(t), \tag{12.86}$$

where

$$\mathbf{Q}(t,\mathbf{\Lambda}) = \left\| q_1(t,\mathbf{\Lambda}),\ldots,q_n(t,\mathbf{\Lambda})\right\|^{\mathrm{T}} \tag{12.87}$$

is the deterministic vector function of the parameter vector $\mathbf{\Lambda}$ and time; $\mathbf{m}(t)$ is the vector Gaussian noise with zero mean and given power spectral density matrix $\mathbf{M}_i$; and

$$\mathbf{\Lambda} = \left\|\lambda_1,\lambda_2,\ldots,\lambda_n\right\|^{\mathrm{T}}. \tag{12.88}$$

In this case, the additive noise $\mathbf{n}(t)$ is a time stochastic process. Therefore, the likelihood functional in the case of the space–time Gaussian noise $\mathbf{n}(x, y, t)$ with zero mean and power spectral density $\mathbf{C}_0$ can be written in the following form:

$$L(\mathbf{X}) = c\cdot\exp\Big\{-0.5\int_X\int_Y\int_T\big[2\mathbf{X}^{\mathrm{T}}(x,y,t)\mathbf{C}_0^{-1}\mathbf{a}^*(x,y,\mathbf{\Lambda},t) - \mathbf{X}^{\mathrm{T}}(x,y,t)\mathbf{C}_0^{-1}\mathbf{X}(x,y,t)$$
$$+\,\boldsymbol{\eta}^{\mathrm{T}}(x,y,t)\mathbf{C}_1^{-1}\boldsymbol{\eta}(x,y,t)\big]dxdydt\Big\}. \tag{12.89}$$

In the case of the additive noise $\mathbf{n}(t)$ as a function of time only with zero mean and the power spectral density matrix $\mathbf{N}_0$, we can write

$$L(\mathbf{X}) = c\cdot\exp\Big\{-0.5\int_T\big[2\mathbf{X}^{\mathrm{T}}(t)\mathbf{N}_0^{-1}\mathbf{a}^*(t,\mathbf{\Lambda}) - \mathbf{X}^{\mathrm{T}}(t)\mathbf{N}_0^{-1}\mathbf{X}(t) + \boldsymbol{\eta}^{\mathrm{T}}(t)\mathbf{N}_1^{-1}\boldsymbol{\eta}(t)\big]\,dt\Big\}. \tag{12.90}$$

To realize in practice this space–time signal and image processing algorithm based on the generalized approach to signal processing in the presence of noise for solving navigational problems, we should have a special memory block to store both the map of the field (the model image) and the map of the first partial derivatives of the map of the field (the first partial derivatives of the model image).

12.4 Space–Time Signal Processing and Pattern Recognition Based on the Generalized Approach to Signal Processing

The pattern recognition process can be represented in the form of two steps. Measurements of definite estimations jointly with values of parameters belonging to the predetermined model image are carried out during the first step. The extremal value of the criterion, which is chosen with the purpose of comparison, allows us to define with some probability the membership of the observed sample image in a definite group. Thus, two cases are possible. In the first case, *a priori* knowledge about the observed image is very poor. Therefore, in the observation of the navigational object, we should compare the observed image (the moving image) with the great number of model images defining the power of similarity. In the second case, we use the only model image, comparing it with the moving image using the totality of parameters. In this case, we are able to give the answer to the following question: Does the observed navigational object or the moving image belong to the same group as the model image or not? The pattern recognition problem is a more complex problem, i.e., in observing the moving image and using the only model image, we should answer the following question: Is there the expected image in the observed totality of the moving images or not?

Thus, in the pattern recognition problem solving, we should carry out first, the measurements, and second, the comparison between the moving and model images by analysis, and comparison of their parameters and features. Thus, the first step precedes the second. In accordance with the generalized approach to signal processing in the presence of noise,[5–8] the estimation of the moving image parameters is related to the use of the model image, which is *a priori* identical to the moving image. Actually, to measure the moving image parameters using the generalized approach to signal processing in the presence of noise, it is necessary to know the shape of the moving image (the observed object) and, consequently, we should define to which group the observed object (the moving image) belongs. To solve this problem, it is necessary to estimate the parameters and features of the moving image under the conditions of *a priori* uncertainty. Let us try to solve this illusory contradiction.

The totality of estimations and distinctive features of the moving image (the recognized object) characterizes a certain point in the q-dimensional space. For example, the estimations of the two features x_1 and x_2 can be considered as coordinates on the plane XOY. The region of possible or allowable estimations of distinctive features forms the subspace of the moving image belonging to the definite group. The totality of other values of estimations forms the next region, and so on.

The widely used estimation of identity of images is the minimum of the difference between the vector **X** (the moving image) and the vector $\mathbf{a}^*$ (the model image). If the vector components are dimensionless or have the same dimensions, we can use two nonweighted sum of difference squares between components of these vectors:

$$\Delta\mathbf{X} = (\mathbf{X} - \mathbf{a}^*)^{\mathrm{T}}(\mathbf{X} - \mathbf{a}^*). \tag{12.91}$$

The observed image (the moving image) belongs to the *i*-th group if the quadratic measure given by Equation (12.91) for the *i*-th model image is less than the same measure for any other model image, i.e.,

$$(\mathbf{X} - \mathbf{a}_i^*)^{\mathrm{T}}(\mathbf{X} - \mathbf{a}_i^*) < (\mathbf{X} - \mathbf{a}_j^*)^{\mathrm{T}}(\mathbf{X} - \mathbf{a}_j^*) \quad \text{for all} \quad i \neq j. \tag{12.92}$$

Each model image is on the q-dimensional sphere with the radius ρ, i.e., $\rho^2 = \mathbf{X}^{\mathrm{T}}\mathbf{a}_j^*$ is constant for all values of j. Therefore, the technique considered for the minimization of the distance between the vectors **X** and $\mathbf{a}_j^*$ is that the image characterized by the set of features $x_1, x_2, \ldots, x_q$ belongs to the *i*-th group if

$$\mathbf{X}^{\mathrm{T}}\mathbf{a}_i^* > \mathbf{X}^{\mathrm{T}}\mathbf{a}_j^*, \quad i \neq j. \tag{12.93}$$

The boundaries of the groups are fuzzy because we carry out measurements in the background of noise and interference, and allowable estimations for the given group of images are not exact due to *a priori* uncertainty. By this we mean that the end result of the pattern recognition problem has a probabilistic character as an estimation of distinctive features. Therefore, the statistical characteristics are used as the measure of convergence between the moving and model images. Under definite conditions, the correlation function between the vectors **X** and $\mathbf{a}^*$, i.e., $R = \langle\mathbf{X}^{\mathrm{T}}\mathbf{a}_i^*\rangle$, is the most convenient in this sense.

Let us assume that there is a device measuring components of the vector **X**. Each component is a parameter of the observed object (the moving image). Let M be the possible image groups. The conditional probability distribution density $f_{\mathbf{x}|H_i}(\mathbf{X}\,|\,H_i)$ and *a priori* probability distribution density $f(H_i)$ of the event H_i corresponding to the *i*-th image group are related to each group belonging to M. The pattern recognition problem is to define the technique of processing the observed vector **X** with the purpose of defining the moving image group.

As an example, let us consider the problem of detection of the signal with known shape $AS(t)$ with unknown amplitude A in the background of the Gaussian noise. The input stochastic process has the form $X(t) = AS(t) + n(t)$. The input stochastic process $X(t)$ is limited by the spectral bandwidth ΔF of the generalized detector preliminary filter. The input stochastic process $X(t)$

is observed within the limits of the time interval [0, *T*]. We should make the decision a "yes" (hypothesis H_1) or "no" (hypothesis H_0) signal in the input stochastic process $X(t)$. The conditional probability distribution density under the hypothesis H_1 can be defined by the likelihood functional

$$f_{X|H_1}(X \mid H_1) = c \cdot \exp\left\{-\frac{1}{4\sigma_n^4}\int_0^T [2X(t)a^*(t) - X(t)X(t) + \eta(t)\eta(t)]\,dt\right\}, \tag{12.94}$$

where σ_n^2 is the variance of the noise at the input of the generalized detector preliminary filter, and c is the normalized cofactor.

If we know the *a priori* probability distribution density $f_{pr}(A)$ of the amplitude of the signal, then the generalized detector determines the conditional probability distribution density $f(X \mid H_1)$ in the following form:

$$f_{X|H_1}(X \mid H_1) = \int_A f_{X|H_1}(X \mid H_1, A) \cdot f_{ps}(A)\,dA\,. \tag{12.95}$$

The statistic $f_{X|H_1}(X \mid H_1)$ at the output of the generalized detector is compared with the threshold. If the statistic exceeds the threshold, we make the decision a "yes" information signal in the moving image and *vice versa*. Thus, the pattern recognition problem solving requires the preliminary estimation of the amplitude of the signal that should be taken into consideration in determining the probability distribution density $f_{X|H_1}(X \mid H_1)$ given by Equation (12.95). If the preliminary filter of the generalized detector has the spectral bandwidth ΔF, then the input stochastic process can be represented as the vector consisting of $2T\Delta F$ readings:[5–8]

$$\mathbf{X} = \left\| \begin{matrix} X(\frac{1}{2\Delta F}) \\ X(\frac{1}{\Delta F}) \\ \cdots \\ X(T) \end{matrix} \right\|. \tag{12.96}$$

The conditional probability distribution density under the hypothesis H_1 takes the following form:

$$f_{X|H_1}(X \mid H_1, A) = \frac{1}{(8\sigma_n^4 \Delta F)^{T\Delta F}} \cdot \exp\left\{-\frac{2\mathbf{X}^{\mathrm{T}}\mathbf{C}_0^{-1}A\mathbf{S} - \mathbf{X}^{\mathrm{T}}\mathbf{C}_0^{-1}\mathbf{X} + \boldsymbol{\eta}^{\mathrm{T}}\mathbf{C}_1^{-1}\boldsymbol{\eta}}{8\sigma_n^4 \Delta F}\right\}, \tag{12.97}$$

where

$$\mathbf{S} = \begin{Vmatrix} S(\frac{1}{2\Delta F}) \\ S(\frac{1}{\Delta F}) \\ \cdots \\ S(T) \end{Vmatrix}. \tag{12.98}$$

For simplicity, we assume that the additional filter has the same spectral bandwidth ΔF in value.

In observing the spatial signals $a(x, y)$ and space–time signals $a(x, y, t)$, the pattern recognition principles are the following.[36,37] We should find the region on the observed plane XOY in which the function of intensity is similar to the predetermined function of intensity called the model function. In this case, the criterion of similarity is the quadratic measure given by Equation (12.91), which can be reduced to the correlation function between the moving and model images

$$R = < a(x,y) \cdot a^*(x-\lambda_x, y-\lambda_y) > \tag{12.99}$$

in the case of the Gaussian probability distribution density of the noise where the parameters λ_x and λ_y define the shift between the model image relative to the moving image belonging to the same group of images.

The maximum of the correlation function given by Equation (12.99) corresponds to the complete matching between the model image and the moving image. A position of the model image in the coordinate system XOY is known *a priori* (see Figure 12.15). Thus, the definition of the maximum of the correlation function given by Equation (12.99) allows us to solve both the pattern recognition problem and to define and measure the position of the moving image in the coordinate system XOY. Let us demonstrate this. Let the information space–time signal $a_i(x - \lambda, t)$ belonging to the i-th group be observed in the background of the space–time noise $n(x, t)$ with zero mean and power spectral density C_0. For simplicity, we do not take into consideration the coordinate y of the coordinate system XOY. We need to define what is the group p, to which the moving image belongs. In the general case, the pattern recognition procedure can be carried out using the p-channel generalized receiver or detector. The conditional probability distribution density of the statistic at the generalized detector output under the hypothesis H_i corresponding to the presence of the i-th signal ($i = 1, 2, \ldots, p$) can be determined in the following form:[5–8]

$$f_{X|H_i}(X \mid H_i) = \int f_{X|H_i}[(X \mid H_i) \mid \lambda_i] \cdot f(\lambda_i)\, d\lambda_i \tag{12.100}$$

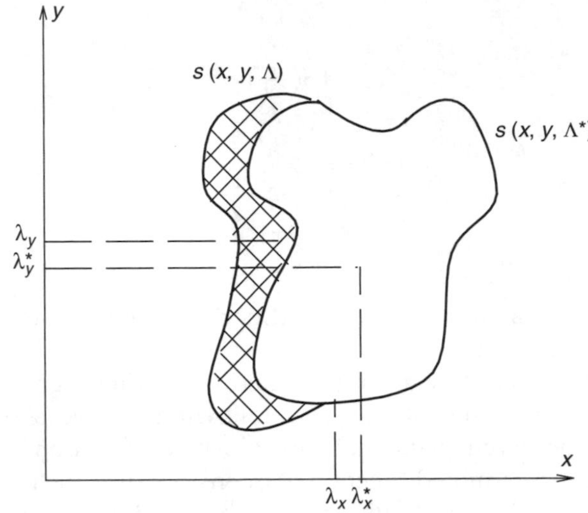

FIGURE 12.15
The displacement of the model image in the coordinate system *XOY*.

for each channel of the generalized detector, where

$$X = X(x,t) = a_i(x - \lambda_i, t) + n(x,t), \tag{12.101}$$

$$< n(x + \Delta x, t + \Delta t) \cdot n(x,t) > = C_0 \delta(\Delta x, \Delta t), \tag{12.102}$$

$f(\lambda_i)$ is the *a priori* probability distribution density of the parameter λ_i. Equation (12.102) is true only at the inputs of the preliminary and additional filters because, after passing the preliminary and additional filters of the generalized detector, the Gaussian noise will be limited by the spectral bandwidth.

The statistics at the output of each *i*-th channel are compared with the purpose of defining the greatest value and to make a decision regarding the hypothesis H_1 for the *i*-th channel. Thus, the pattern recognition problem is solved by the definition of the conditional probability distribution density $f_{X|H_i}[(X|H_i)|\lambda_i]$ at the output of the generalized detector, which can also be called the likelihood functional. In the case of the space–time additive noise, we can write

$$f_{X|H_i}[(X|H_i)|\lambda_i] = c$$
$$\times \exp\left\{-\frac{1}{8\sigma_n^4}\int_X\int_T [2X(x,t)a^*(x-\lambda_i^*,t) - X(x,t)X(x,t) + \eta(x,t)\eta(x,t)]\,dx\,dt\right\}. \tag{12.103}$$

If the dimensions of the observed region X are much more than a correlation length of the signal $a_i(x, t)$, we can write

$$f_{X|H_i}[(X \mid H_i) \mid \lambda_i] = c \cdot \exp\{-\tfrac{1}{8\sigma_n^4}[2R_{ii^*} - R_{ii} + 2R_{i^*n} - 2R_{in} + R_n]\}, \tag{12.104}$$

where

$$R_{ii^*} \cong \frac{1}{XT} \int_X \int_T a_i(x - \lambda_i, t) \cdot a_i^*(x - \lambda_i^*, t)\, dx\, dt; \tag{12.105}$$

$$R_{ii} \cong \frac{1}{XT} \int_X \int_T a_i^2(x - \lambda_i, t)\, dx\, dt; \tag{12.106}$$

$$2R_{i^*n} \cong \frac{1}{XT} \int_X \int_T 2a_i^*(x - \lambda_i^*, t) \cdot \xi_i(x, t)\, dx\, dt; \tag{12.107}$$

$$2R_{in} \cong \frac{1}{XT} \int_X \int_T 2a_i(x - \lambda_i, t) \cdot \xi_i(x, t)\, dx\, dt; \tag{12.108}$$

$$R_n \cong \frac{1}{XT} \int_X \int_T [\eta_i^2(x, t) - \xi_i^2(x, t)]\, dx\, dt; \tag{12.109}$$

R_{ii^*} is the correlation function between the information signal $a_i(x - \lambda_i, t)$ in the moving image $X_i(x, t)$ and the model image $a_i^*(x - \lambda_i^*, t)$ at the output of the i-th channel of the generalized detector; R_{ii} is the correlation function of the information signal $a_i(x - \lambda_i, t)$ in the moving image $X_i(x, t)$; R_{i^*n} is the correlation function of the correlation channel noise component caused by the interaction between the model image $a_i^*(x - \lambda_i^*, t)$ and the noise $\xi_i(x, t)$ at the output of the i-th channel; R_{in} is the random component correlation function of the autocorrelation channel caused by the interaction between the information signal $a_i(x - \lambda_i, t)$ in the moving image $X_i(x, t)$ and the noise $\xi_i(x, t)$ at the output of the i-th channel; R_n is the correlation function of the background noise; $\xi_i(x, t)$ is the noise at the output of the preliminary filter; and $\eta_i(x, t)$ is the noise at the output of the additional filter; the noise $\xi_i(x, t)$ and $\eta_i(x, t)$ are uncorrelated (see Chapter 11).

Because the signal and noise are uncorrelated at the input of the generalized detector and the difference between the correlation functions R_{i^*n} and R_{in} tends to approach zero in the statistical sense if the coefficient of correlation between the model image $a_i^*(x - \lambda_i^*, t)$ and the information signal $a_i(x - \lambda_i, t)$ in the moving image $X_i(x, t)$ is equal to unity, and the background noise at the output of the generalized detector tends to approach zero in the statistical sense, the probability distribution density $f_{X|H_i}[(X|H_i)|\lambda_i]$ is defined by the accuracy of the position definition λ_i of the information signal in the moving image $X_i(x, t)$. Actually, under these conditions, we can write

$$f_{X|H_i}[(X|H_i)|\lambda_i] = c \cdot \exp\left\{-\tfrac{1}{8\sigma_n^4}[R_{ii^*} - R_{ii}]\right\}. \qquad (12.110)$$

The exponent in Equation (12.110) varies monotonically depending on values of R_{ii^*} and R_{ii}. Therefore, the correlation functions R_{ii^*} and R_{ii} define completely the character and behavior of the probability distribution density $f_{X|H_i}[(X|H_i)|\lambda_i]$. In the condition $i = i^*$, i.e., when the moving image and the model image belong to the same group and an error $\Delta\lambda = \lambda_i - \lambda_i^*$ of the information signal position measurement in the moving image $X_i(x, t)$ is close to zero, the probability of true recognition is maximal. Actually, due to low errors during the model image tuning and at the condition $i = i^*$, we can assume that

$$R_{ii^*} - R_{ii} \cong R'_{ii^*}\Delta\lambda, \qquad (12.111)$$

where R'_{ii^*} is the derivative of the correlation function at zero; $\Delta\lambda = \lambda_i - \lambda_i^*$; and λ_i^* is the estimation of the moving image shift. Thus, $f_{X|H_i}[(X|H_i)|\lambda_i]$ is the *a posteriori* probability distribution density of shift estimation of the observed i-th moving image $X_i(x, t)$. The generalized receiver has p channels. Each channel uses its own *i-th* model image with the initial tuning defined by the expected position of the i-th moving image on the observation plane. If the i-th information signal is in the moving image, then the i-th channel of the generalized receiver defines the optimal by the maximum *a posteriori* probability distribution density estimation of its position λ_i^*. The multichannel generalized receiver is analyzed in more detail in Tuzlukov.[8]

Thus, the composition of the p-channel generalized receiver allows us to solve the pattern recognition problem in the following manner. The moving image $X(x,t)$ comes in at each input of the p-channel generalized receiver. The moving image $X(x,t)$ is the discrete stochastic process. There is a discrete model image for each channel of the p-channel generalized receiver. The position of the model image at the initial instant of time is defined by the *a priori* knowledge and the estimation λ_i^* obtained earlier. If for the i-th channel of the generalized detector we obtain the less value of $\sigma^2_{ii^*}$, where $\sigma^2_{ii^*} = <(\lambda_i - \lambda_i^*)^2>$, we can determine the "yes" information signal in the moving image, and the information signal belongs to the same group as the model image.

In the sequential pattern recognition technique, we can use the one-channel generalized detector with a set of model images. In this case, we should have p model images. This technique is used in a small sample of moving images. If the sample is large, the computing cost is very high.

12.5 Peculiarities of Optical Signal Formation

Let us consider the peculiarities of optical signal formation in navigational systems in the definite order caused by specific conditions. The input stochastic process is formed in the space called the object space in the form of characteristics of the landmark totality. The Earth and sea surfaces, the cosmic space, etc., can be considered the object space. The information signals generated by the object space come in at the input of the navigational system. The information signals are transferred by the channel that can be the propagation medium of the optical signals. The input information signal is processed by the optical receiver containing the preliminary filter for the ordinary frequency range and the spatial frequencies. The optical receiver forms the object images within the limits of vision field, i.e., the moving image. The moving image is reproduced by the receiver, the form of which defines the procedure of signal processing with the use of the generalized approach in the presence of noise. Using this sequence of the procedure, let us consider the process of optical signal formation. Depending on the spectrum range in which the transmitter or source generates signals, we use the energy and photoelectric characteristics to define the peculiarities of the optical signal source. The transition from energy characteristics to photoelectrical characteristics is not difficult.[38]

The procedure of optical bearing is based on the use of image contrast of the observed objects (landmarks) in the background of noise. The image contrast exists because they have features that radiate or rereflect the energy of electromagnetic waves into the optical wave range to a greater or lesser degree in comparison with sources of interference and noise that are around these objects. The summation of the radiation of the object is caused by two components: internal heat radiation and reflected radiation from natural or manmade sources (sun, moon, stars, the Earth, the atmosphere, manmade lighting, etc.). The type of these sources defines the character of the natural or manmade optical field. However, it should be noted that in the majority of cases, the summing field is formed simultaneously under the stimulus of natural and manmade sources. This applies restrictions on signal processing in a wide range of optical wavelengths. Therefore, we should use additional techniques such as spectral selection.

The range 0.2 ... 3 μm is traditionally recommended for employment in reflected and scattered solar radiation. The range 10 ... 12 μm is recommended for employment in radiation of the Earth's surface. Evidently, a redistribution of energy between own and reflected radiation can be carried out within the limits of short time intervals and sufficiently wide limits.[39,40]

All these components of the summing object radiation depend essentially on the shape, construction, material, orientation, position on terrain, regime of functioning, etc., of the object. In addition, a state of the environment, location of external sources, and other factors play a large role. These factors generate a high deviation of radiation characteristics that can be defined using the theory of statistical decision making. Thus, the problem of definition of manmade objects or landmarks can be simplified due to their definite uniformity of shape, dimensions, and material with which these objects are produced. Unlike natural noise and interferences, which can be considered as a background, manmade object-landmarks possess a low deviation of various radiation components. The dependence on direction and power of the wind and seasonal conditions is less for manmade object-landmarks. Incidentally, the direction and power of the wind and seasonal conditions can significantly change the background, for example, orientation and color of tree leaves, the slope of plant stems, sea and lake surfaces, etc.

The internal radiation of objects is a function of the surface temperature and physical features. This radiation can be both coherent and noncoherent. Coherent radiation is a peculiarity of electromagnetic radiation that keeps the difference in phases within the limits of the time interval that is required to detect and measure this difference. Evidently, in the majority of cases, any navigational system processes noncoherent signals. The most informative characteristic of internal noncoherent radiation is the spectral energy brightness of the heated object. This characteristic can be determined in the following form:

$$\mathcal{L}_\lambda = \varepsilon_\lambda \cdot \frac{1.19 \cdot 10^4}{\lambda^5 (e^{\frac{14388}{\lambda T}} - 1)}, \tag{12.112}$$

where T is the temperature of the radiation source, λ is the wavelength, and ε_λ is the spectral power of the blackness of the radiating surface at the given temperature and definite bearing. To define the coherent radiation of the object, it is necessary to take into consideration its interference and diffracting characteristics.

Characterizing the efficiency of internal radiation of the heat sources, we can define three kinds of emitters: blackbody, gray emitter, and selective emitter. The parameter ε_λ defines the efficiency of radiation for the given wavelength. Sometimes, the parameter ε_λ is called the coefficient of radiation. The absolute blackbody has a black power $\varepsilon_\lambda = 1$ for the whole range of the wavelength. In the case of the gray emitter, we obtain $\varepsilon_\lambda = const < 1$ within the limits of the definite range of the wavelength. In the case of the selective emitter, we have $0 \leq \varepsilon_\lambda < 1$, and the parameter ε_λ is an unambiguous function of λ of any kind. Speaking rigorously, the degree of blackness of real objects $\varepsilon_{\lambda T}$ is always a function of the wavelength and temperature. For this reason, the constant value of ε_λ can be defined within the definite limits of change of λ and T. This function depends also on the angle of the vision field. However, these angles are very low in value in navigational systems of any kind, as a rule. Therefore, we can consider the parameter ε_λ as independent of the angle of vision field.

The working principles of the majority of objects are based on the use of energy equipment. A large amount of heat energy is released as a result of the functioning of energy equipment. A part of this heat energy is discharged into the environment. For this reason, all objects have one more peculiarity: the internal radiation of the vapor phase that is often called a flare (the vapor stream of reactive engines, ship pipes, explosion stacks, etc.) is added to the internal surface radiation of the object. The dependence of the spectral degree of blackness on the wavelength of those vapor sources takes the form of oscillations unlike with the smooth character of solid metallic objects.

The reflection features of objects depend on the relationships between the dimensions of structure heterogeneities of the object surface and the wavelength of the incident radiation. If these dimensions are much less than the wavelength, we can consider that there is a mirror reflection. If these dimensions are commensurable with the wavelength, there is a scattered reflection. The main characteristic of reflection features is the reflection coefficient μ_0 characterized by the ratio between the radiation stream reflected by the object surface and the incident radiation stream on the object surface. In mirror reflection from the flat surfaces of objects, the coefficient of reflection can be defined as the ratio between the surface brightness after reflection and the initial surface brightness under the condition that a space angle, within the limits of which the incident radiation is propagated, is kept constant after radiation. This condition is not satisfied with scattered reflection when the space angle, within the limits of which the reflected radiation is propagated, is greater than that within the limits of which the incident radiation is propagated.[41]

The limit of the space reflection angle for a flat surface is equal to 2π. In this case, the surface is called a diffuse reflection surface and the reflection is called the diffuse or Lambert reflection. The brightness of this surface is the same in value for all directions of radiation propagation, i.e., it is independent of both the angle θ in the meridian plane and the azimuth angle φ of the viewfinder plane. In addition, the brightness is independent of both the antiaircraft angle θ' and the azimuth angle φ' defining the position of the radiation source. The brightness of the ideal scattered surface is functionally related to its illumination by an extraneous radiation source $E(\theta', \varphi')$ and can be determined in the following form:

$$\mathcal{L}_0(\theta', \varphi') = \frac{E(\theta', \varphi')}{\pi}. \tag{12.113}$$

Thus, a scattered radiation by its character can be uniform and nonuniform if the brightness distribution in space depends on the direction of the viewfinder and location of the radiation source. The coefficient of the brightness characterizes this feature of scatterers.

The coefficient of brightness $\mu_b(\theta, \varphi, \theta', \varphi')$ is the ratio between the brightness characteristic of the object surface region along the given direction and the brightness $\mathcal{L}_0(\theta', \varphi')$ of the ideal scattering diffuse surface having the coefficient of reflection equal to unity, i.e., the absolutely "white" surface, which is in the same conditions of brightness and observation,

$$\mu_b(\theta,\varphi,\theta',\varphi') = \frac{\mathcal{L}(\theta,\varphi,\theta',\varphi')}{\mathcal{L}_0(\theta',\varphi')}. \tag{12.114}$$

Therefore, the brightness of the real object surface along the direction of observation or the viewfinder is given by

$$\mathcal{L}(\theta, \varphi, \theta', \varphi') = \frac{E(\theta',\varphi')}{\pi} \cdot \mu_b(\theta,\varphi,\theta',\varphi'). \tag{12.115}$$

The specific conditions of functioning of the navigational system and the feature of objects in absorbing and reflecting the radiation selectively allow us to define the spectral reflection ability by a set of coefficients of the spectral brightness $\mu_{b\lambda}(\theta, \varphi)$, which are ratios of the spectral brightness of the objects and the spectral brightness of the ideal scattering surface under the same conditions of illumination and observation, i.e.,

$$\mu_{b\lambda}(\theta,\varphi) = \frac{\mathcal{L}_\lambda(\theta,\varphi,\theta',\varphi')}{\mathcal{L}_{0\lambda}(\theta',\varphi')}. \tag{12.116}$$

Consequently, the coefficients of brightness are the function of the zenith and azimuth angles in the definition of the radiation source position, as a rule of the sun. This function cannot be defined in analytical form and is presented, as a rule, in the diagrammatic or graph form using polar diagrams or graphs, the radius-vector lengths of which are proportional to the values of the coefficients of brightness in corresponding directions. These diagrams or graphs are usually called the indicatrices of reflection or scattering. As a rule, the indicatrices of brightness are plotted in the form of cross sections of the indicatrices of reflection using polar or Cartesian coordinate systems.

The brightness and, consequently, the coefficient of brightness are the functions of the coordinates x, y, z of the radiation point in the object space, the time t, and the wavelength λ. This fact causes some difficulties in describing radiation models. However, the conditions of optical-electronic equipment employment in various navigational systems carrying out specific functions allow us to simplify this problem. For example, in aircraft navigational systems, the observation of the Earth's surface relief and generation of the corresponding image are carried out by scanning the local transmission and receiving diagrams. Many of the other optical-electronic equipment in navigational systems have the same local transmission and receiving diagrams, for example, devices for observation. This allows us to change the object space when the depth of this space is not so high in value and the object space is far from the optical receiver by plane, the brightness of which is defined by the function of two coordinates x and y along the direction of the viewfinder and the time and wavelength $\mathscr{L}(x, y, t, \lambda)$. In addition, the interaction of such optical-electronic equipment with the majority of the surface and sea objects is carried out in the course of short time periods. This allows us to consider that the brightness is not varied in time except in the case of fast-moving objects or objects possessing the fluctuating brightness.[42]

The distribution function of brightness within the area limits of the observed object can have various forms. Thus, we can simplify the problem of definition of the distribution function of brightness due to specific conditions of employment of the pattern recognition optical-electronic systems. One of the peculiarities is that the procedure of object detection starts for distances that are larger in comparison with the dimensions of the object. For this reason, the optical system must be photosensitive. This implies that aberrations of the optical equipment cannot be reduced to negligible values taking into consideration the finite dimension of the vision field and the bandwidth of the spectral range. Similar optical systems cannot distinguish the components of the shape of the highly remote objects and considers these objects in the form of the point sources. Therefore, the total dimensions of the object's shape play a secondary role. In this case, the energy characteristic J_{en} is characterized as a power P_{en} radiated by the point source or by the source, the dimensions of which are negligible in comparison with the distance between the radiation source and the receiver or detector of the navigational system, in the solid angle Ω. Based on these statements, we can write

$$J_{en} = \frac{dP_{en}}{d\Omega}. \qquad (12.117)$$

When the receiver and the radiation source are approaching, the condition that the angle dimensions of the object be negligible is not satisfied and the representation of the object in the form of the point radiation source leads to high-level errors. For this reason, we use the statement of the optical power.

Let us consider briefly the main characteristics and peculiarities of the background radiation sources. In the consideration of the object detection problem, we can assume that the natural radiation sources limited by the radar range and noise immunity of the optical-electronic navigational systems are the background noise. For this reason, the background noise can be the atmosphere, the clouds, the Earth's surface, the sea surface, the stars in sky, etc. During navigation operations, some of the background radiation sources can serve as landmarks. In that case, it is necessary to solve the object detection problem using other background noise radiation sources.[43,44]

Summing the background radiation noise, as well as object radiation, is defined by two components: (1) the internal radiation and (2) radiation scattered by the sun and other external sources. The characteristics of the background noise sources, as noted previously, have more variety compared to the characteristics of the detecting objects. This phenomenon can be explained by a set of features of the background noise source formation. First, there is a variety of configurations and their omponents, which have various physical natures. The dimensions of the background noise sources are various but, as a rule, are greater than the dimensions of the object. Therefore, it is necessary to consider the components of the background noise sources in the three-dimensional space, for example, in the analysis of cloud forms. The essential differences arise in the definition of the characteristics of the background noise sources because of the unpredicted appearance and interchange of various gradations of the background noise sources and their elements under scanning by optical–electronic navigational systems. It should be noted that the radiation characteristics of the background noise sources depend highly on external conditions, i.e., the weather, the season, rain, snow, wind, etc. All these factors allow us to consider the signal as a stochastic process.

The total definition of the three-dimensional stochastic field structure of the background noise sources leads to the multidimensional distribution laws of brightness, which is a very complex problem. In practice, the problem of defining the statistical characteristics of the stochastic field of brightness is reduced to more simple interpretations. For example, we often use the one-dimensional probability distribution densities, correlation functions, and power spectral densities to analyze the stochastic fields in the form of underlying surfaces or the Earth's surface relief. We can use these statistical characteristics in the definition both of the one-dimensional and of the two-dimensional stochastic fields at their stationary state and isotropy conditions. For this purpose, it is necessary to estimate these statistical characteristics for the given Earth's surface relief. This estimation allows us to define the degree of error and the applicability of the estimation in solving the specific problem. Moreover, we can consider the probability distribution density of brightness as a stationary process inside the individual characteristic zones, in which the total surface of the object can be divided into parts. The character of each zone is defined, for example, by the homogeneity of the underlying surface. Sometimes it is possible to obtain the theoretical models of the background noise sources both with nonrandom and random parameters.[45] However, a great amount of data can be obtained by experimental investigations of the main radiation characteristics, such as the coefficient of reflection, coefficient of brightness, and radiant emittance.

As the main characteristic of radiation of the background noise sources we use the parameters ε_λ and μ_λ. The values of these coefficients can be considered as the basis for the diagram or graph of the background noise source regions in the coordinate system of primary features of the pattern recognition problem or space of states, and, in the first approximation, the coordinates. In the last case, two variants caused by the totality of the representations of the coefficients ε_λ and μ_λ can arise. As an example of the first variant, we can consider the case when statistical characteristics such as the mean and variance of the coefficients ε_λ and μ_λ are given. Then the problem of the graph representation of the background noise sources is solved in the simplest way if the *a priori* probability distribution density of the coefficients ε_λ and μ_λ is known.

The problem of the graph representation of the background noise sources for the second case is more complex, especially if there is a deficit of statistical data. At the present time, this case is more typical because, in spite of the large amount of references we are not able to obtain a completely satisfactory statistical picture of radiation and scattering for the background noise sources. This circumstance forces us to use a set of techniques and tools that will allow us to reach sequentially a satisfactory approximation of the complete definition.

As one of these simplest procedures, we can use the technique of sequential or systematic approximation with the attraction of additional information about variations of chosen features at each step. The geometric representation of this technique using the mean of the spectral coefficients of radiation and reflection as initial data is very evident. The mean of the coefficients of radiation and reflection are used as the coordinate axes. We can define the regions corresponding to various forms of the background noise source images using these coordinates. The first step is completed using this procedure. During the second step, an extension of the definition of these regions caused by the attraction of additional information data is carried out. These additional information data would be the results of experimental investigations obtained by the angle characteristics of radiation discussed in many references. The third step is devoted to a more precise definition of the region of the given form of the background noise sources obtained by the knowledge of brightness conditions, etc. We repeat this operation until we will have sufficient information about all parameters influencing the variations of the coefficients ε_λ and μ_λ. Obviously, the regions obtained this way cannot be considered as the model of features of the investigated object image. However, there is a possibility to correct them using further procedures steps obtained from receiving new and more precise information. In view of the fact that there is a relationship between the statistical characteristics and physico-geographical essence of the natural Earth's surface relief, it is worthwhile to consider the seasonal background noise source regions for the various regions of the Earth's surface relief.

The environment in which optical–electronic navigational systems operate greatly stimulates the statistical characteristics of the optical signals. During propagation, optical signals have a great stimulus of the environment in comparison with the propagation of radio signals. In the general case, three main phenomena define the laws of propagation of optical signals:[46,47] absorption, scattering, and turbulence. The first and second phenomena define the average fading of the electromagnetic field under fixed atmospheric conditions and comparatively slow variations of the electromagnetic field characteristics under changed meteorological conditions. The third phenomenon is the turbulence that causes rapid variations of the electromagnetic field characteristics observed under any meteorological conditions. Moreover, the structure of the received optical signals can be significantly changed compared to the original signal, which because of turbulence, generates a multibeam effect. The ratio between the radiation past the atmospheric layer with the thickness equal to l and the incident radiation is characterized by the coefficient of attenuation $\tilde{\beta}$ depending on the wavelength, in the general case.

Thus, based on the preceding statements, we can consider that

$$\tilde{\beta} = \beta_{ab} + \beta_{sc} + \beta_{tub}, \tag{12.118}$$

where β_{ab} is the coefficient of molecular absorption, β_{sc} is the coefficient of scattering by particles, and β_{tub} is the coefficient of scattering by nonhomogeneities caused by turbulence. The value $\varpi_0 = \tilde{\beta}l$ is called the optical thickness. The value $\varpi = e^{-\tilde{\beta}l} = e^{-\varpi_0}$ is called the coefficient of transparency. There is also the effect of back scattering, which makes the operating characteristics of navigational systems poor. If the source radiation is concentrated within the limits of the definite solid angle, the receiver processes both the radiation reflected from the surface and the source radiation scattered by particles being within the limits of the solid angle. This scattered radiation with the high concentration of aerosol particles having various dimensions within the limits of the viewfinder of the navigational system receiver can create a very high level of background noise and the detection of landmarks will be very difficult.[48,49] The mathematical model of radiation attenuation caused by the phenomena discussed in the preceding text is very complex and is based on quantum mechanics mathematics. Therefore, we use approximated methods or results of experimental investigations during the analysis. Similar techniques are discussed in References 45 and 50.[45,50]

Optical receivers or detectors employed in aircraft navigational systems, in their principal schemes and constructions, are of a great variety. The general features of all optical receivers and detectors are image-forming, amplification of brightness at the input eye of the optical receiver or detector, and radiation filtering by energy spectrum. Moreover, optical receivers and detectors can carry out other functions, for example, scanning, separation or summation of the radiation stream, filtering by polarization power, ensurance of variables increasing with various view fields, etc.

The view field of optical receivers and detectors, as a rule, is not so large so that the angle coordinates of individual elements of the image are proportional to the linear dimensions in planes that are orthogonal to the optical axis of the receiver or detector. Therefore, the measurement of the linear position of the observed object is equivalent to bearing. As was noted in the preceding text, object radiation can be coherent or noncoherent. Therefore, we often call the signal processing in the optical receiver or detector as coherent or noncoherent signal processing. In noncoherent radiation, the distribution of lightness $E(x, \lambda_x, y, \lambda_y)$ is formed in the image space in which the lightness is proportional to the brightness $\mathcal{L}(x, y)$ of associated points in the object space. If the object radiation is coherent, the optical receiver or detector is able to fulfill the transformation operation of the complex amplitudes $E(x, y)$ in the two-dimensional space-frequency Fourier spectrum. These transformations are accompanied by losses in the energy of the signal and distortions of the signal due to aberrations of the optical receiver or detector. If these distortions of the signal are absent, i.e., the optical receiver or detector is the ideal system, then, in the large distances between the optical receiver or detector and the observed object, the plane of the optical receiver or detector forming the moving image coincides with its back focal plane. Thus, each point of the observed object with the coordinates λ_x and λ_y is represented in the focal plane by the point with the coordinates λ_x' and λ_y'.

The lightness corresponding to the information signal forming in the moving image in this way is given by

$$E(x-\lambda_x, y-\lambda_y) = \varpi_{op}\pi\,\mathcal{L}(x-\lambda_x, y-\lambda_y)\sin^2\vartheta, \qquad (12.119)$$

where $\mathcal{L}(x, y)$ is the brightness contrast of the information signal in the moving image, which is defined in the plane of the input eye of the optical receiver or detector; ϖ_{op} is the coefficient of transparency of the optical receiver or detector; and ϑ is the back aperture angle.

The no-ideal optical receiver or detector possesses such phenomena as diffracting scattering and abberation. These are the reasons for a fuzzified image forming in the plane of the object observation. The fuzzified image is characterized by the function of scattering $\mathcal{H}(x, \lambda_x, y, \lambda_y)$, the physical form of which is that this function is the radiation at the point (λ_x', λ_y') when the current equal to unity is directed to the point (x', y'). This definition explains the normalization of the function of scattering

$$\int_{-\infty}^{\infty}\mathcal{H}(x, \lambda_x, y, \lambda_y)\,dx\,dy = 1. \qquad (12.120)$$

The condition of normalization is the following. The fuzzified stream at the point (λ_x, λ_y) must be equal to the initial incident radiation stream. After normalization, the function of scattering is often called the weight function.

The function of scattering is related to the couple of the Fourier transform with the aperture function defining the view field of the optical receiver or detector. In the infinite diameter of the optical lenses of the receiver or detector, the aperture function takes the square waveform shape, the base of which tends to approach infinity. Therefore, for this optical lens, the function of scattering is defined by the delta function that corresponds to the ideal optical receiver or detector. For the no ideal optical receiver or detector, the total lightness at the point (x', y') at the object image plane will be equal to sum of lightnesses in terms of the stream scattering along the direction to all elements (dx', dy'). If the fuzzified images are the same at all points of the view field of the optical receiver or detector, we can write

$$E(x,y) = \int_{-\infty}^{\infty}\int_{-\infty}^{\infty} E(\lambda_x - x, \lambda_y - y)\cdot \mathcal{H}(\lambda_x, \lambda_y)\, dx\, dy. \tag{12.121}$$

For all cases of optical observations, navigational systems use optical receivers or detectors having responses by the energy of the incoming signals, i.e., optical receivers or detectors are able to fix the radiant contrast of the observed object. This statement is caused by the wideband signals used and by the uncertainty in phase of their frequency components.[51]

The optical receivers or detectors operating by the radiant contrast or brightness contrast can be divided into the following classes: integral optical receivers or detectors, optical receivers or detectors with sequential searching, and multielement optical receivers or detectors. In the first, the total energy of the incoming signals changes the parameters of the optical receiver or detector. An example of such receivers or detectors is the photo resistor, the resistance of which changes proportionally to the incident light stream because of the internal photo effect. Optical receivers or detectors of the second class are used for sequential in-time construction of the object image in the observation plane. Transformation of the two-dimensional object image into the one-dimensional electric signal is carried out by the sweep in time on the observation plane. Examples of optical receivers or detectors of this kind are TV transmitters with multiframe sweep and heat vision receiver that allow us to construct the observed object image in the infrared optical range.

The multielement optical receiver or detector functioning is based on simultaneous representation of all radiating elements of the observed object in the plane in which the object image is formed. For these optical receivers or detectors, we use the matrix structure. The matrix receivers or detectors are the totality of elements spaced into the same plane that are sensitive to the incident radiation stream. In other words, the matrix receivers or detectors are constructed based on a set of the integral optical receivers or detectors with individual outputs. The output signals of the optical receivers or detectors correspond by level to that part of the total radiation energy which covers an area occupied by the optical receivers or detectors. An example of these optical receivers or detectors is the matrix receiver or detector consisting of photo elements.

In the majority of cases, we use the integral and spectral sensitivities by the current and by the voltage $\varepsilon(x', y')$. Thus, the main characteristic of the optical receiver or detector is the ratio between the energy characteristic of the signal at the receiver or detector output and the energy characteristic of the radiation stream causing this output signal. The output signal of the optical receiver or detector can be written in the following form:

$$\mathcal{U} = \int_{-\infty}^{\infty}\int_{-\infty}^{\infty} \varepsilon(x', y') \cdot E(x' - \lambda'_x, y' - \lambda'_y)\, dx'dy'. \tag{12.122}$$

The formula in Equation (12.122) is true under the following conditions: the optical receiver or detector is inertialess and the variable $\mathcal{U}$ is the total response of the individual element responses. Otherwise, the formula in Equation (12.122) is cumbersome.

12.6 Peculiarities of the Formation of the Earth's Surface Radar Image

To estimate the operating ability of the radar navigational system, taking into consideration features of the landmarks and model images, the preliminary generation of radar maps is widely used. However, the preliminary radar image obtained with the use of radar field models is considered a cheaper operational process.

The following initial premises operate in the construction of radar images of the Earth's surface based on experimental study: the amplitude of the target return signals from the Earth's surface relief, for example, the steppe, the forest, the arable land, and so on, obeys the Rayleigh probability distribution density, and the phase of these target return signals is distributed uniformly within the limits of the interval $[0, 2\pi]$; the mean and variance of the target return signal amplitude are defined by the specific effective scattering area S° of the background noise source; S° is defined by the Earth's surface relief form, such as the steppe, the forest, the arable land, the fields, the city districts, etc. The background noise source distribution $S^\circ(x, y)$ is defined by the specific relief observed by the radar navigational system. There are powerful individual pulsed target return signals reflected from the bridges, the electric power transmission lines, etc., the amplitudes of which obey the Rayleigh probability distribution density.

The amplitude of the target return signal from the background noise sources can be expressed as two cofactors:

$$S_1(x, y) = A(x, y) \cdot \sqrt{S^\circ}\ , \tag{12.123}$$

where $\sqrt{S^\circ}$ corresponds to the Earth's surface relief background (see Table 12.3). The dimensions of the Earth's surface relief background regions with

TABLE 12.3

Values of the Specific Effective Scattering Area

S°, dB	Background
–40…+45	Water
–30	Concrete
–20	Steppe
–15	Forest
–10…+10	City

the same value of $S°$ are greater in comparison with the dimensions of the resolution elements of the output radar image. The spectral bandwidth of spatial frequencies of the background noise sources and the correlation function are defined by the Earth's surface relief types. The cofactor $A(x, y)$ is a stationary stochastic process with the variance σ_A^2 and obeys the Rayleigh probability distribution density[52]

$$f(a) = \frac{2A}{\sigma_A^2} \cdot e^{-\frac{A^2}{\sigma_A}}. \tag{12.124}$$

The amplitudes of the target return signals from the point contrast landmarks can be presented, as a rule, by the delta function

$$S_2(x, y) = A(x, y) \sum_{k=1}^{N} \sqrt{S_k} \cdot \delta(x - x_k, y - y_k), \tag{12.125}$$

where $\sqrt{S_k}$ is the specific effective scattering area of the k-th landmark with the coordinates x_k and y_k. The value of S_k is defined by the landmark type (see Table 12.4). If the point landmarks are very close to each other, they form a lengthy image with definite configuration, for example, bridge, road, etc. Thus, the total electromagnetic field strength distribution, without taking into account the directional diagram of the navigational system, can be written in the following form:

TABLE 12.4

Kinds of Landmarks

S_k, m²	Object
1…10	Car
3…5	Small aircraft
15…20	Large aircraft
150	Small ship
$14 \cdot 10^3$	Tanker

$$S(x,y) = S_1(x,y) + S_2(x,y) = A(x,y)\Big[\sqrt{S^{\circ}(x,y)} + \sum_{k=1}^{N} \sqrt{S_k} \cdot \delta(x - x_k, y - y_k)\Big]. \tag{12.126}$$

The generalized detector for the radio signal with the envelope has the same structure as in the case of the generalized receiver detecting the same envelope in the background additive wideband noise. Therefore, the power spectral density is doubled in comparison with the power spectral density of the additive high-frequency noise. Based on this we can consider Equation (12.126) as the basic formula in the generation of the radar image. The model of the radar navigational system forming the Earth's surface relief image can be presented in the form of the connected sequential receiver–transmitter channel, the signal processing system, and the decision-making device (see Figure 12.16). The receiver–transmitter channel can be presented in the form of the linear system with the pulse response $g(x, y)$ with respect to the amplitude envelope of the signals. In scanning the Earth's surface relief region, the amplitude of the generated signal can be determined in the following form:

$$\mathcal{U}(\lambda_x, \lambda_y) = \hat{c} \sum_{k=1}^{N} A(x_k, y_k) \cdot \sqrt{S_k} \cdot g(x - \lambda_x - x_k, y - \lambda_y - y_k) + n(x,y), \tag{12.127}$$

where λ_x and λ_y are the coordinates of the image on the observation plane XOY, and $\hat{c}$ is the transmission coefficient of the radar channel.

The function $g(x, y)$ can be expressed as the product of the two functions $g(x)$ and $g(y)$. The first function $g(x)$ is defined by the width θ_0 of the directional diagram and sloped plain. The second function $g(y)$ is defined by the duration τ_p of the pulsed searching signal. In side scanning (see Figure 12.17), the axis x is the course distance, the axis y is the side distance, and the directional diagram can be approximated by the Gaussian law

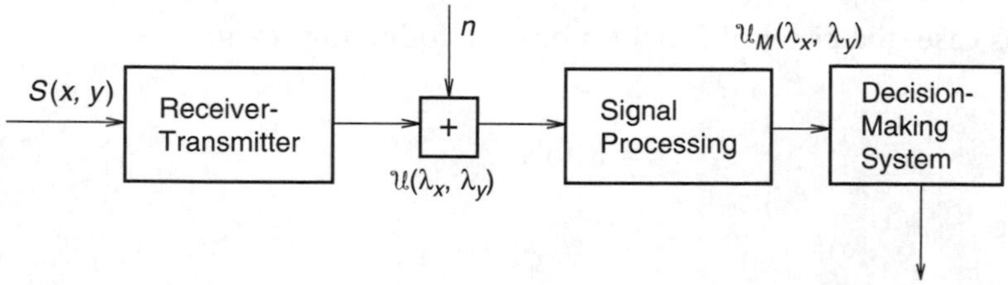

FIGURE 12.16
Radar channel model.

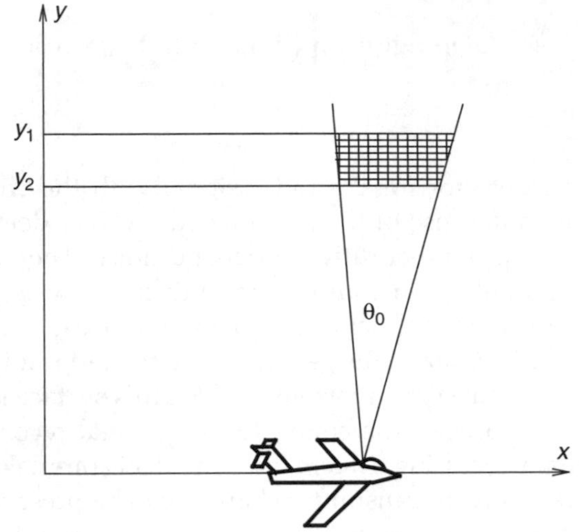

FIGURE 12.17
Side scanning.

$$g(x,y)=e^{-a_x x^2-a_y y^2},\quad a_x=\frac{c_1}{y^2\theta_0},\quad \text{and}\quad a_y=\frac{c_2}{\tau_p^2},\tag{12.128}$$

where c_1 and c_2 are the constant coefficients.

Thus, the resulting image is the totality of bright flashes on the indicator observed in the background noise. The dimensions of each flash are defined by the coefficients a_x and a_y, i.e., by the width θ_0 of the directional diagram and the duration τ_p of the pulsed searching signal. The information signal in the moving image during M periods of scanning in accordance with Equation (12.127) takes the following form:

$$\mathcal{U}_M(\lambda_x,\lambda_y)=\hat{c}M\sum_{k=1}^{M}A(x_k,y_k)\cdot S_k\cdot e^{-2a_x(x-\lambda_x-x_k)^2-2a_y(y-\lambda_y-y_k)^2}.\tag{12.129}$$

In this case, the criterial function has the following form:

$$I=R(0)\cdot e^{-a_x\Delta x_k^2-a_y\Delta y_k^2},\tag{12.130}$$

where

$$R(0)=\hat{c}^2M^2\sum_{k=1}^{N}\sum_{l=1}^{N}\sqrt{S_kS_l}\cdot A(x_k,x_l),\ \Delta x_k=x_k-x_k^*,\ \text{and}\ \Delta y_k=y_k-y_k^*.\tag{12.131}$$

The discrimination characteristics in measuring the parameters λ_x and λ_y are determined by

$$\frac{\partial I}{\partial \lambda_x^*} = 2R(0)a_x \Delta x_k \cdot e^{-a_x \Delta x_k^2 - a_y \Delta y_k^2}, \qquad (12.132)$$

$$\frac{\partial I}{\partial \lambda_y^*} = 2R(0)a_y \Delta x_y \cdot e^{-a_x \Delta x_k^2 - a_y \Delta y_k^2}. \qquad (12.133)$$

The formulae in Equation (12.131)–Equation (12.133) are written without taking into account the background noise.

Thus, in the observation of the point landmarks, the shape of the directional diagram and the duration of the pulsed searching signal play the main role in the process of forming the discrimination characteristic. In practice, this dependence is very complex. The phenomenon is explained by the following factors: the high-level brightness contrast landmark cannot be defined by the delta function; the linear model of the radar channel has an approximated character; the background noise source leads to the fuzzified image of the observed landmark totality; and the characteristics of the images of the same Earth's surface relief hardly depend on weather conditions, seasonal conditions, or observation direction.

There are some other techniques of forming the Earth's surface image using a topographical relief map. The topographical map allows us to define the area occupied by cities, forests, steppes, etc. The total energy reflected by the area is limited by the values of $0.5c\tau_p$, where c is the velocity of electromagnetic radiation and of the width of the horizontal-coverage directional diagram. This total energy is the sum of the energies caused by area elements. In forming the images, we use the coordinate lattice with the distance between the horizontal lines equal to $\frac{c\tau_p}{2\sin\varphi}$, where φ is the angle of scanning. A distance between the vertical lines is defined by the shape of the horizontal-coverage directional diagram.

12.7 Foundations of Digital Image Processing

The wide use of matrix receivers or detectors generates the application of digital signal processing. Digital signal processing methods are widely used to determine the criterial functions in navigational systems for the continuous signals also. Two procedures form the basis of digital signal processing: (1) the observed stochastic space–time process sampling and (2) the quantization of the observed stochastic space–time process.

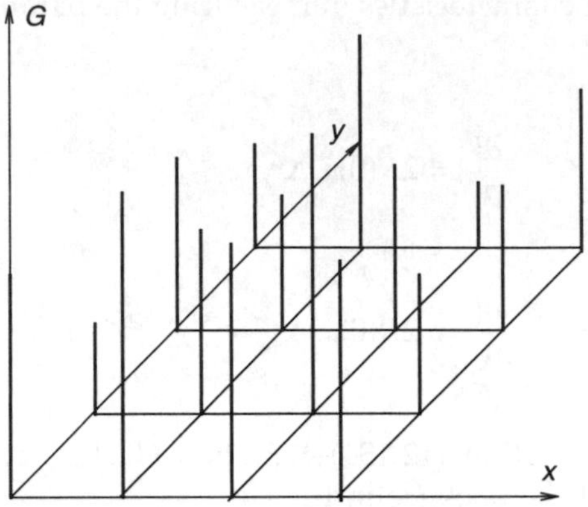

FIGURE 12.18
The lattice function.

Sampling is a representation of the stochastic space–time process in the form of the totality of readings corresponding to the chosen discrete values of the arguments ($x_1, x_2, \ldots, x_k; y_1, y_2, \ldots, y_l$). The values of the arguments (x_k, y_l) are the chosen multiples to intervals Δx and Δy, respectively, which are called sampling intervals. If we consider the space–time signal $a(x, y, t)$, the discrete readings with respect to the variables x and y are carried out for the fixed instant of time. In the discrete representation of a stochastic process, it is very convenient to use the lattice function (see Figure 12.18):

$$\sum_{i=-\infty}^{\infty}\sum_{j=-\infty}^{\infty} a_{lat}(i\Delta x, j\Delta y) = \sum_{i=-\infty}^{\infty}\sum_{j=-\infty}^{\infty}\int_{-\infty}^{\infty} a(x,y)\cdot\delta(i\Delta x - x)\cdot\delta(j\Delta y - y)\, dx\, dy. \tag{12.134}$$

As follows from the interpolation theory of continuous spatial signals, the only realization $a(x, y)$ can be constructed for the use of the given lattice function. The correlation lengths Δx_0 and Δy_0 of the realization $a(x, y)$ are satisfied by the following condition:

$$\Delta x \le \Delta x_0 \quad \text{and} \quad \Delta y \le \Delta y_0 . \tag{12.135}$$

In other words, the continuous stochastic image can be presented in a discrete form always, i.e., in the form of the lattice function with discrete intervals for the condition given by Equation (12.135). Therefore, the continuous image $a(x, y)$ can be reconstructed using the readings of the discrete image by the following interpolation:[18]

$$\tilde{a}(x,y) = \sum_{i=-\infty}^{\infty}\sum_{j=-\infty}^{\infty} a_{lat}(i\Delta x, j\Delta y)\cdot f(x - i\Delta x, y - j\Delta y), \tag{12.136}$$

where $f(x, y)$ is the deterministic interpolation function.

We can assume that the higher interpolation certainty, i.e., the reconstructed field and the initial image are equivalent in the statistical sense, can be obtained in the case if the interpolation function $f(x, y)$ coincides with the coefficient of the space correlation function of the original signal. If the discrete sequence $a_{lat}(i\Delta x, j\Delta y)$ comes in at the input of the spatial filter with the pulse characteristic $f(x, y)$, the filter response corresponds to the initial image $a(x, y)$. The correspondence is more precise, the closer the function $f(x, y)$ is to the coefficient of spatial correlation of this image.

The most convenient and widely used discrete representation of the continuous image is the presentation in the form of the step envelope lattice function[53] (see Figure 12.19):

$$a(i,j) = a(x,y)\cdot f_1(x,y), \tag{12.137}$$

where

$$f_1(x,y) = \sum_{i=1}^{K}\sum_{j=1}^{M} p(x - i\Delta x, y - j\Delta y) \tag{12.138}$$

is a totality of the $K \times L$ identical square waveform normalized pulses

$$\int_{-\infty}^{\infty}\int_{-\infty}^{\infty} p(x,y)\, dx\, dy = 1\,. \tag{12.139}$$

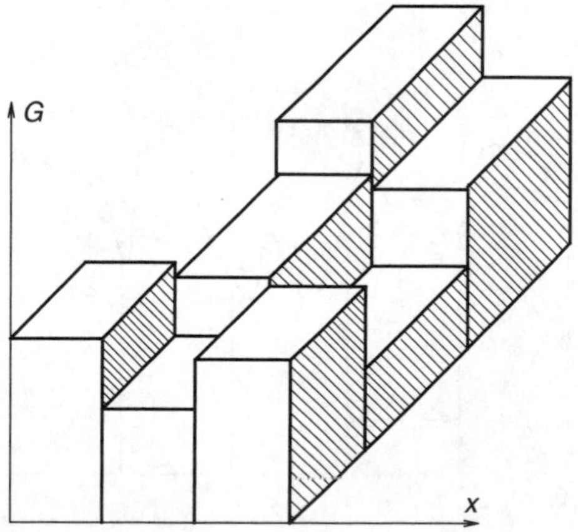

FIGURE 12.19
The step envelope of the lattice function.

Up till now, we discussed the quantization of the image $a(x, y)$ without taking into consideration the spatial noise $n(x, y)$. Let us consider the moving image in the following form:

$$X(x,y) = a(x,y) + n(x,y). \tag{12.140}$$

We assume that the correlation length of the noise is much less than that of the signal. In this case, the use of the condition given by Equation (12.135) can give rise to additional distortions in the reconstruction of the image. The essence of these distortions is the following. Let the information signal $a(x)$ with the correlation length Δx_0 be observed in the presence of additive noise with the correlation length $\Delta x_n << \Delta x_0$ (see Figure 12.20). If the sampling interval is equal to $\Delta x_n << \Delta x_0$, we can lose some information (see Figure 12.21) in the reconstruction of the initial moving image. To avoid these losses we should carry out filtering for the moving image before sampling. Here, the correlation length of the noise is close to the correlation length of the information signal, and additional distortions of the information signal $a(x)$ are absent. In practice, the space–time signal $a(x, y, t)$ at the fixed instant of time defined by the lattice function $a_{lat}(i\Delta x, j\Delta y)$, $i = 1, 2, \ldots, K$, $j = 1, 2, \ldots, L$ can be determined in the form of the step envelope $A(i, j)$ or in the form of the rectangular matrix of readings with the number of elements equal to $K \times L$. The criterial function can be determined in the following form:

$$I(m,n) = \hat{c}\sum_{i=1}^{K}\sum_{j=1}^{L} A(i,j)\cdot X(i-m, j-n), \tag{12.141}$$

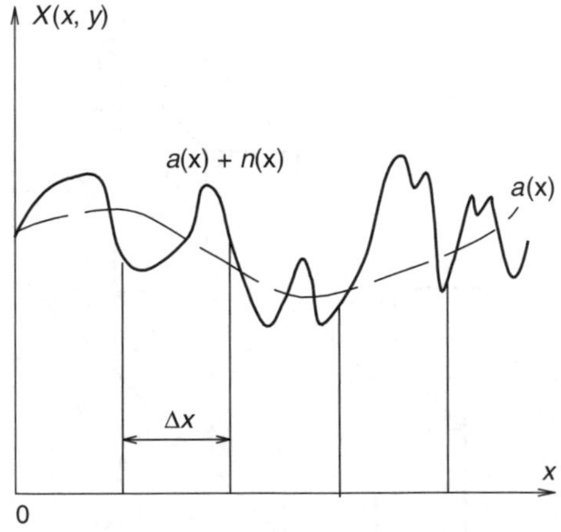

FIGURE 12.20
Additive mixture of signal and noise.

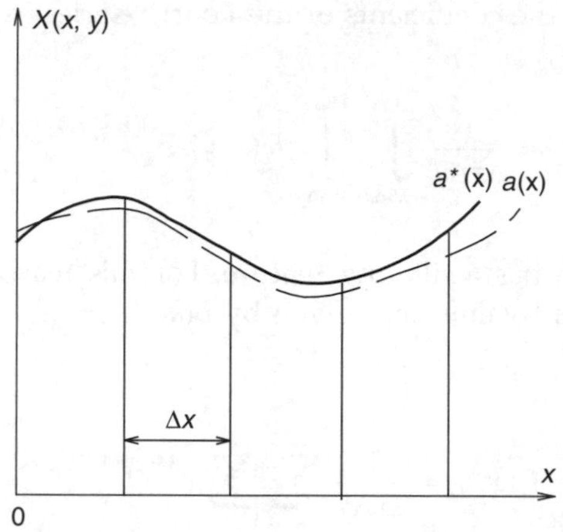

FIGURE 12.21
Quantization of signal and noise.

where $\hat{c} = \frac{1}{KL}$. The criterial function $I(m, n)$ given by Equation (12.141) is the fixed moment of the second order for the digital image. The moments of higher order can be determined in the following form:

$$\mu_{pq} = \hat{c} \sum_{i=1}^{K} \sum_{j=1}^{L} A^p(i,j) \cdot X^q(i,j). \tag{12.142}$$

In the case when the means $<A(i, j)>$ and $<X(i, j)>$ are not equal to zero, the central moments are given by

$$\mu_{pq}^0 = \hat{c} \sum_{i=1}^{K} \sum_{j=1}^{L} [A(i,j) - < A(i,j) >]^p \cdot [X(i,j) - < X(i,j) >]^q, \tag{12.143}$$

where $\hat{c} = \frac{1}{KL}, K = \frac{X}{\Delta x}, L = \frac{Y}{\Delta y}$. The discrete power spectral density of the stationary two-dimensional stochastic field $A(i, j)$ takes the following form:

$$S(\omega_x, \omega_y) = \frac{1}{MN} \sum_{m=1}^{M} \sum_{n=1}^{N} I(m,n) \cdot e^{-j\left(\frac{m\omega_x}{M} + \frac{n\omega_y}{N}\right)}. \tag{12.144}$$

We can represent the continuous signal in the discrete form in the spectral range. For this purpose, the totality of the delta functions used in forming the lattice functions can be determined by the Fourier-series expansion

$$f_2(x,y) = \sum_{k_1=-\infty}^{\infty} \sum_{k_2=-\infty}^{\infty} \delta(x - k_1\Delta x, y - k_2\Delta l) = \sum_{k_1=-\infty}^{\infty} \sum_{k_2=-\infty}^{\infty} B(k_1,k_2) \cdot e^{j\left(k_1 \frac{2\pi}{\Delta x} x + k_2 \frac{2\pi}{\Delta y} y\right)}, \tag{12.145}$$

where $B(k_1, k_2)$ are the coefficients of the Fourier-series expansion:

$$B(k_1, k_2) = \frac{1}{\Delta x \Delta y} \int_{-0.5\Delta x}^{0.5\Delta x} \int_{-0.5\Delta y}^{0.5\Delta y} f_2(x, y) \cdot e^{-j\left(k_1 \frac{2\pi}{\Delta x} x + k_2 \frac{2\pi}{\Delta y} y\right)} dx\, dy. \tag{12.146}$$

The delta function has a filtering feature. For this reason, the coefficients $B(k_1, k_2)$ are constant values and given by $B(k_1, k_2) = \frac{1}{\Delta x \Delta y}$. Consequently, we can write

$$f_2(x, y) = \frac{1}{\Delta x \Delta y} \sum_{k_1=-\infty}^{\infty} \sum_{k_2=-\infty}^{\infty} e^{j\left(k_1 \frac{2\pi}{\Delta x} x + k_2 \frac{2\pi}{\Delta y} y\right)}. \tag{12.147}$$

The power spectral density for this function takes the following form:

$$S(\omega_x, \omega_y) = \frac{1}{\Delta x \Delta y} \sum_{k_1=-\infty}^{\infty} \sum_{k_2=-\infty}^{\infty} \int_{-\infty}^{\infty} \int_{-\infty}^{\infty} e^{j\left(k_1 \frac{2\pi}{\Delta x} x + k_2 \frac{2\pi}{\Delta y} y\right)}$$

$$\times e^{-j(\omega_x x + \omega_y y)} dx\, dy = \frac{4\pi^2}{\Delta x \Delta y} \sum_{k_1=-\infty}^{\infty} \sum_{k_2=-\infty}^{\infty} \delta\left(\omega_x - k_1 \tfrac{2\pi}{\Delta x}, \omega_y - k_2 \tfrac{2\pi}{\Delta y}\right). \tag{12.148}$$

The power spectral density of the digital image can be defined using the power spectral density of the continuous image:

$$S(\omega_x, \omega_y) = \frac{4\pi^2}{\Delta x \Delta y}$$

$$\times \sum_{k_1=-\infty}^{\infty} \sum_{k_2=-\infty}^{\infty} \int_{-\infty}^{\infty} \int_{-\infty}^{\infty} S(\xi, \eta) \cdot \delta\left(\xi - \omega_x - k_1 \tfrac{2\pi}{\Delta x}, \eta - \omega_y - k_2 \tfrac{2\pi}{\Delta y}\right) d\xi\, d\eta$$

$$= \frac{4\pi^2}{\Delta x \Delta y} \sum_{k_1=-\infty}^{\infty} \sum_{k_2=-\infty}^{\infty} S\left(\omega_x - k_1 \tfrac{2\pi}{\Delta x}, \omega_y - k_2 \tfrac{2\pi}{\Delta y}\right). \tag{12.149}$$

Let us assume that the space–time signal $a(x, y, t)$ obeys the probability distribution density $f(a)$ in the range of the possible values Ω. Quantization assumes the partition of the range Ω for the p quantization intervals Ω_p. In doing so, all values $a(x, y)$ within the limits of the interval Ω_p have the same estimation value a_p. The parameter p is called the quantization volume. The parameter a_p is called the quantization level. The error of quantization of the signal $a(x, y)$ can be determined in the following form:

$$\varepsilon^2 = \sum_{p=1}^{P} \int_{\Omega_p} (a - a_p)^2 \cdot f(a)\, da \,. \tag{12.150}$$

If the number of quantization levels is high, the probability distribution density of the quantized signal can be thought as constant. In this case, we can write

$$\varepsilon^2 = \sum_{p=1}^{P} f(a_p) \int_{\Omega_p} (a - a_p)^2 da \approx 0.33 \sum_{p=1}^{P} f(a_p)\left[(d_{p+1} - a_p)^3 - (d_p - a_p)^3\right], \tag{12.151}$$

where d_p and d_{p+1} define the p-th quantization interval Ω_p. It is well known that the optimal level of quantization is determined by $a_p = 0.5(d_{p+1} + d_p)$. In other words, if the quantization level a_p is at the middle of the quantization interval, the value ε is minimum.

12.8 Conclusions

The totality of landmarks used in solving navigational problems such as the Earth's surface during navigation of the aircraft flight track, the image of starry sky, etc., is observed, as a rule, in the background of interferences and noise generated by various sources. For this reason, the effectiveness of solving the navigational problems is defined by the quality of the components used in obtaining the moving image, i.e., by the quality of signal and image processing of information components of the moving image containing useful information regarding the totality of landmarks. It is obvious that the methods and techniques of space–time signal and image processing in navigational systems are defined by the nature of the used signals.

The moving image of the Earth's surface relief is formed due to periodic changes in the aircraft flight altitude. If we solve the navigational problem of the moving ship, the sea depth is measured. In both these cases the moving image is a function of time due to the moving radar navigational system. Here, the signal and image processing is similar to that in time just as in the observation of the geophysical field. Thus, methods and techniques of spatial target return signal can be classified into space signal and image processing, space–time signal and image processing, and signal and image processing in time.

The solution of the navigational problems, i.e., the definition of location of the navigational object, is carried out under ambiguous conditions. This uncertainty is formed by the following main factors: external interference, sensor noise, interference that can arise with the moving radar of navigational system, indefinite information regarding landmark coordinates, etc. Incorrect knowledge of location of landmarks, i.e., the relative position of the navigational system in the coordinate system, is referred to as *a priori* uncertainty in the definition of the coordinates of the navigational object. For reasons already noted, we can discuss only the estimation of some coordinates of the navigational object that can be obtained anyway. In the case when the estimation of the coordinates of the navigational system is definitely the best, we can consider that this estimation is optimal. Thus, the main problem in navigational systems is the definition of the optimal estimation of the navigational object coordinates, for example, the definition of the true aircraft flight track, using information in the signals of various physical sources.

The results discussed in this chapter indicate that we have a good chance to use the generalized detector in navigational systems in the signal processing of the target return signal with stochastic parameters, and that the estimation of the definition of the navigational object coordinates has high accuracy. We proceed as follows: construct the amplitude and phase tracking systems that control the appropriate parameters of the model signal generated at the output of the model signal generator of the generalized detector. In principle, the construction of the amplitude and phase tracking systems is made possible by using the condition $T >> \tau_c$.

The space–time signal and image processing by the use of the generalized approach to signal processing in the presence of noise is to define the maximum of the integral given by Equation (12.68). If the dimensions of the observed moving image are much more than the correlation length of the signal and the signal is the stationary stochastic process, the integral in Equation (12.68) matches the maximum of the likelihood functional. The definition of this maximum is carried out by the use of the generalized detector tracking systems discussed in Section 12.2. Thus, the discrimination characteristics are defined along the corresponding axes l_x and l_y. Navigational systems that realize the generalized approach to signal processing in noise are called correlation-extremal systems because Equation (12.68) allows us to define the maximum of the likelihood functional only under the main conditions of the generalized detector functioning. The functions given by Equation (12.68) and Equation (12.69) are often called criterial functions. The criterial functions given by Equation (12.69) are proportional to the measurement errors because the background noise in the generalized approach to signal processing tends to approach zero in the statistical sense and the noise component $2\int_X\int_Y a^*(x,y,\mathbf{\Lambda}^*,t)n(x,y,t)dxdy$ of the correlation channel of the generalized detector, caused by the interaction between the model image and noise, and the random component $2\int_X\int_Y a(x,y,\mathbf{\Lambda},t)n(x,y,t)dxdy$ of the autocorrelation channel of the generalized detector, caused by interaction between the moving image and noise, are compensated by each other in the statistical sense (see Chapter 11).

Correlation extremal receivers or detectors constructed according to the generalized approach to signal processing in the presence of noise may be used in navigational systems without tracking systems. The use of the generalized receivers or detectors of this form implies the employment of the signals with known amplitude-phase-frequency structure. The comparison of the incoming moving image at the input of the navigational system with the model image defined by the form of the signal and the *a priori* knowledge regarding the location of the signal on the observation plane allows us to estimate the true value of the parameter vector of $\mathbf{\Lambda}$. Thus, an automatic control of this value of the parameter vector $\mathbf{\Lambda}$ is carried out. The main condition of the navigational system functioning without tracking devices is the correlation between the moving image $\mathbf{X}(x, y, t)$ and the model image $\mathbf{a}^*(x, y, \mathbf{\Lambda}^*, t)$. In other words, we cannot recognize the information signal or image $\mathbf{a}(x, y, \mathbf{\Lambda}, t)$ in the input moving image $\mathbf{X}(x, y, t)$. The use of the generalized detector with tracking systems (see Section 12.2) allows us to solve this problem, to recognize the type of the signal $\mathbf{a}(x, y, \mathbf{\Lambda}, t)$, and to define and measure the parameter vector $\mathbf{\Lambda}$.

According to the character of the additive noise $n(x, y)$ at the output of the navigational system, we can use the preliminary filter of the generalized detector of various forms. For simplicity, we assume that the noise $n_1(x, y)$ is the Gaussian narrowband process. Then, the criterial function for each estimation channel λ_x and λ_y is defined by the derivative of the space–time correlation function of the signal derivatives with the weight coefficients that are inversely proportional to the spectral bandwidth of the noise $n_1(x, y)$ in the region of spatial and ordinary frequencies. To realize this space–time signal in practice and the image processing algorithm based on the generalized approach to signal processing in the presence of noise to solve navigational problems, we should have a special memory block to store both the map of the field (the model image) and the map of the first partial derivatives of the map of the field (the first partial derivatives of the model image).

In the observation of the spatial signals $a(x, y)$ and space–time signals $a(x, y, t)$, the pattern recognition principles are the following. We should find a region on the observed plane XOY in which the function of intensity is similar to the predetermined function of intensity called the model function. In this case, the criterion of similarity is the quadratic measure given by Equation (12.91), which can be reduced to the correlation function between the moving and model images given by Equation (12.99) in the case of the Gaussian probability distribution density of the noise, where the parameters λ_x and λ_y define the shift between the model image relative to the moving image belonging to the same group of images. The maximum of the correlation function given by Equation (12.99) corresponds to the complete match between the model image and the moving image. The position of the model image in the coordinate system *XOY* is known *a priori* (see Figure 12.15). Thus, the definition of the maximum of the correlation function given by Equation (12.99) allows us both to solve the pattern recognition problem and to define and measure the position of the moving image in the coordinate system *XOY*.

The composition of the *p*-channel generalized receiver allows us to solve the pattern recognition problem in the following manner. The moving image $X(x, t)$ comes in at each input of the *p*-channel generalized receiver. The moving image $X(x, t)$ is the discrete stochastic process. There is the discrete model image for each channel of the *p*-channel generalized receiver. The position of the model image at the initial instant of time is defined by *a priori* knowledge and the estimation λ_i^* obtained before. If for the *i*-th channel of the generalized detector we obtain the reduced value of $\sigma^2_{ii^*}$, where $\sigma^2_{ii^*} = <(\lambda_i - \lambda_i^*)^2>$, we can make the decision a "yes" information signal in the moving image, and the information signal belongs to the same group as the model image. With the sequential pattern recognition technique, we can use the one-channel generalized detector with a set of model images. In this case, we should have *p* model images. This technique is used in a small sample of the moving image. If the sample of the moving image is large, the computing cost is very high.

The environment in which the optical–electronic navigational systems operate acts as a great stimulus on the statistical characteristics of optical signals. During propagation, optical signals are greatly stimulated by the environment in comparison with the propagation of the radio signals. In the general case, three main phenomena define the laws of propagation of optical signals: absorption, scattering, and turbulence. The first and second phenomena define the average fading of the electromagnetic field in fixed atmospheric conditions and comparatively slow variations of the electromagnetic field characteristics under changed meteorological conditions. The third phenomenon is the turbulence that causes rapid variations of the electromagnetic field characteristics observed under any meteorological conditions. Moreover, the structure of the received optical signals can be significantly changed in comparison to the initial signal that, due to the turbulence, generates a multibeam effect. The ratio between the radiation past the atmospheric layer with the thickness equal to l and incident radiation is characterized by the coefficient of attenuation β depending on the wavelength, in the general case.

Optical receivers or detectors employed in aircraft navigational systems are of a great variety in terms of their principal schemes and constructions. The general features of all optical receivers and detectors are the forming of the image, amplification of lightness at the input eye of the optical receiver or detector, and radiation filtering by the energy spectrum. Moreover, optical receivers and detectors can perform other functions, for example, scanning, separation or summation of the radiation stream, filtering by polarization power, and ensuring the increase of the variable with various view fields, etc.

Optical receivers or detectors operating by radiant contrast or brightness contrast can be divided into the following classes: integral optical receivers or detectors, optical receivers or detectors with sequential searching, and multielement optical receivers or detectors. In the first class, the total energy of the incoming signals changes the parameters of the optical receiver or detector. An example of such receivers or detectors is the photo resistor, whose resistance changes proportionally to the incident light stream because of the internal photo effect. Optical receivers or detectors of the second class are used for the time sequential construction of the object image in the observation plane. The transformation of the two-dimensional object image into the one-dimensional electric signal is performed by the sweep in time on the observation plane. Examples of optical receivers and detectors in this case are the TV transmitter with multiframe sweep and the heat vision receiver, allowing us to construct the observed object image in the infrared optical range.

The following are the initial premises in the construction of radar images of the Earth's surface based on experimental study: the amplitude of the target return signals from the Earth's surface relief, for example, the steppe, the forest, the arable land, etc., obeys the Rayleigh probability distribution density, and the phase of these target return signals is distributed uniformly within the limits of the interval $[0, 2\pi]$; the mean and variance of the target return signal amplitude are defined by the specific effective scattering area S° of the background noise source; the specific effective scattering area S^{0} is defined by the Earth's surface relief form, such as the steppe, the forest, the arable land, the fields, the city districts, etc. The background noise source distribution $S^{\circ}(x, y)$ is defined by the specific relief observed by the radar navigational system. There are powerful individual pulsed target return signals reflected from bridges, electric power transmission lines, etc., the amplitudes of which obey the Rayleigh probability distribution density.

References

1. Papoulis, A., *Systems and Transforms with Application in Optics,* Krieger, Malabar, FL, 1981.
2. Hansen, T. and Yaghjian, A., *Plane-Wave Theory of Time-Domain Fields: Near-Field Scanning Applications,* IEEE Press, Piscataway, NJ, 1999.
3. Felsen, L. and Marcuvitz, N., *Radiation and Scattering of Waves,* Prentice Hall, Englewood Cliffs, NJ, 1973.
4. Franceschetti, G., Iodice, A., Migliaccio, M., and Riccio, D., Scattering from natural rough surfaces modeled by fractional Brownian motion two-dimensional processes, *IEEE Trans.,* Vol. AP-47, No. 9, 1999, pp. 1405–1415.

5. Tuzlukov, V., A new approach to signal detection theory, *Digital Signal Process. Rev. J.*, Vol. 8, No. 3, 1998, pp. 166–184.
6. Tuzlukov, V., *Signal Processing in Noise: A New Methodology,* IEC, Minsk, 1998.
7. Tuzlukov, V., *Signal Detection Theory,* Springer-Verlag, New York, 2001.
8. Tuzlukov, V., *Signal Processing Noise,* CRC Press, Boca Raton, FL, 2002.
9. Bochkarev, A., Optimal correlation navigational systems, *A Broad Radio Electronics,* No. 9, 1981, pp. 37–45 (in Russian).
10. Vorob'ev, V., *Optical Location for Radio Engineers,* Radio and Svyaz, Moscow, 1983 (in Russian).
11. Pavlov, Yu, Selevnev, A., and Tolstousov, G., *Geoinformatic Systems,* Mashinostroenie, Moscow, 1978 (in Russian).
12. Krasovsky, A., Beloglazov, I., and Chigin, G., *Theory of Optimal Correlation Navigational Systems,* Nauka, Moscow, 1979 (in Russian).
13. DeSanto, J. and Brown, G., Analytical techniques for multiple scattering from rough surfaces, in *Progress in Optics XXIII,* E. Wolf, Ed., North-Holland, Amsterdam, 1986.
14. Bossavit, A. and Mayergoyz, I., Edge elements for scattering problems, *IEEE Trans.,* Vol. MG-25, No. 7, 1989, pp. 2816–2821.
15. Born, M. and Wolf, E., *Principles of Optics,* Pergamon, Oxford, 1980.
16. Thompson, A., Moran, J., and Swenson, G., *Interferometry and Synthesis in Radio Astronomy,* 2nd ed., John Wiley & Sons, New York, 2001.
17. Baldauf, J., Lee, S., Ling, H., and Chou, R., On physical optics for calculating scattering from coated bodies, *J. Electromagnet. Waves Appl.,* Vol. 3, No. 8, 1989, pp. 725–746.
18. Baklitzky, V. and Yur'ev, A., *Correlation Extremal Methods of Navigation,* Radio and Svyaz, Moscow, 1982 (in Russian).
19. Goldsmith, P., *Quasioptical Systems: Gaussian Beam Quasioptical Propagation and Applications,* IEEE Press, New York, 1998.
20. Dogaru, T. and Carin, L., Time-domain sensing of targets buried under a rough air-ground interface, *IEEE Trans.,* Vol. AP-46, No. 3, 1998, pp. 360–372.
21. Jiao, D. and Jin, J., Three-dimensional orthogonal vector basis functions for time-domain finite element solution of vector wave equations, *IEEE Trans.,* Vol. AP-51, No. 1, 2003, pp. 59–66.
22. Baklitzky, V., The use of Kalman filtering under synthesis of correlation-extremal systems, *News of the USSR Universities. Radio Electronics,* Vol. 25, No. 3, 1982, pp. 65–73 (in Russian).
23. Deans, S., *The Random Transform and Some of Its Applications,* Krieger, Malabar, FL, 1993.
24. Krasovsky, A., Optimal estimation in distributed systems defined by the Green function, *Automatics and Telemechanics,* No. 10, 1981, pp. 25–32 (in Russian).
25. Capolino, F. and Felsen, L., Time-domain Green's function for an infinite sequentially excited periodic planar array of dipoles, *IEEE Trans.,* Vol. AP-51, No. 2, 2003, pp. 160–170.
26. Felsen, L. and Capolino, F., Time-domain Green's function for an infinite sequentially excited periodic line array of dipoles, *IEEE Trans.,* Vol. AP-48, No. 6, 2000, pp. 921–931.
27. Capolino, F. and Felsen, L., Frequency and time-domain Green's function for a phased semi-infinite periodic line array of dipoles, *IEEE Trans.,* Vol. AP-50, No. 1, 2002, pp. 31–41.

28. Krasovsky, A., Field estimation under vector fuzzy measurements, *Reports of Academy of the USSR. Cybernetics and Control Theory,* Vol. 256, No. 5, 1981, pp. 387–393 (in Russian).
29. Ozaktas, H., Zalevsky, Z., and Kutay, M., *The Functional Fourier Transform With Applications in Optics and Signal Processing*, John Wiley & Sons, Chichester, U.K. 2001.
30. Zverev, V., *Radio Optics,* Soviet Radio, Moscow, 1975 (in Russian).
31. Zhao, L. and Cangellaris, A., GT-PML: Generalized theory of perfectly matched layers and its application to the reflectionless truncation of finite-difference time-domain grids, *IEEE Trans.,* Vol. MTT-44, No. 12, 1996, pp. 2555–2563.
32. Jin, J., *The Finite Element Method in Electromagnetics,* John Wiley & Sons, New York, 1993.
33. Vishnevsky, V., Method of forming the standards that are invariant relative to shifts and rotations of objects, in *Proceedings of the 4th Samara's University Conference,* Samara, Russia, 1980, pp. 127–131 (in Russian).
34. Galdi, V., Felsen, L., and Castanon, D., Quasi-ray Gaussian beam algorithm for short-pulse two-dimensional scattering by moderately rough dielectric interfaces, *IEEE Trans.,* Vol. AP-51, No. 2, 2003, pp. 171–183.
35. Galdi, V., Felsen, L., and Castanon, D., Quasi-ray Gaussian beam algorithm for time-harmonic two-dimensional scattering by moderately rough interfaces, *IEEE Trans.,* Vol. AP-49, No. 9, 2001, pp. 1305–1314.
36. Hansen, T. and Johansen, P., Inversion scheme for ground penetrating radar that takes into account the planar air-soil interface, *IEEE Trans.,* Vol. GRS-38, No. 1, 2000, pp. 496–506.
37. Chew, W. and Jin, M., Perfectly matched layers in the discretized space: An analysis and optimization, *Electromagnetics,* Vol. 16, 1996, pp. 325–340.
38. Galdi, V., Felsen, L., and Castanon, D., Time-domain radiation from large two-dimensional apertures via narrow-wasted Gaussian beams, *IEEE Trans.,* Vol. AP-51, No. 1, 2003, pp. 78–88.
39. Lamb, J., Low-noise, high-efficiency optics design for ALMA receivers, *IEEE Trans.,* Vol. AP-51, No. 8, 2003, pp. 2035–2047.
40. Wootten, A., Ed., *Science With the Atacama Large Millimeter Array,* ASP, San Francisco, CA, 2001, Vol. 253.
41. Chew, W. and Weedon, W., A 3-D perfectly matched medium from modified Maxwell's equations with stretched coordinates, *Microwav. Opt. Tech.,* Vol. 7, No. 13, 1994, pp. 599–604.
42. Jiao, D. and Jin, J., An effective algorithm for implementing perfectly matched layers in time-domain finite-element simulation of open-region EM problems, *IEEE Trans.,* Vol. AP-50, No. 11, pp. 1615–1623.
43. Jiao, D., Jin, J., Michielssen, E., and Riley, D., Time-domain finite-element simulation of three-dimensional scattering and radiation problems using perfectly matching layers, *IEEE Trans.,* Vol. AP-51, No. 2, 2003, pp. 296–305.
44. Jiao, D. and Jin, J., Time-domain finite element modeling of dispersive media, *IEEE Microwav. Wireless Components Lett.,* Vol. 11, No. 5, 2001, pp. 220–222.
45. Levshin, V., *Spatial Filtering in Optical Bearing Systems,* Soviet Radio, Moscow, 1971 (in Russian).
46. Berenger, J., A perfectly matched layer for the absorption of electromagnetic waves, *J. Comput. Phys.,* Vol. 144, No. 2, 1994, pp. 185–200.

47. Felsen, L. and Carin, L., Diffraction theory of frequency- and time-domain scattering by weakly aperiodic truncated thin-wire gratings, *J. Opt. Soc. Amer. A,* Vol. 11, No. 4, 1994, pp. 1291–1306.
48. Gedney, S., An anisotropic perfectly matched layer-absorbing medium for the truncation of FDTD lattices, *IEEE Trans.,* Vol. AP-44, No. 12, 1996, pp. 1630–1639.
49. Sacks, Z., Kingsland, D., Lee, R., and Lee, J., A perfectly matched anisotropic absorber for use as an absorbing boundary condition, *IEEE Trans.,* Vol. AP-43, No. 12, 1995, pp. 1460–1463.
50. Iznar, A., Pavlov, A., and Fedorov, B., *Optical Electronic Devices for Cosmic Apparatus,* Mashinostroenie, Moscow, 1972 (in Russian).
51. Shestov, N., *Detection of Optical Signal in Noise,* Soviet Radio, Moscow, 1967 (in Russian).
52. Hubral, P. and Tygel, M., Analysis of the Rayleigh pulse, *Geophysics,* Vol. 54, No. 5, 1989, pp. 654–658.
53. Boland, J., et al., Design of a correlator for real-time video comparisons, *IEEE Trans.,* Vol. AES-15, No. 1, 1980, pp. 63–75.

13

Implementation Methods of the Generalized Approach to Space–Time Signal and Image Processing in Navigational Systems

13.1 Synthesis of Quasioptimal Space–Time Signal and Image Processing Algorithms Based on the Generalized Approach to Signal Processing

As shown in Chapter 12, the optimal computer calculator constructed on the basis of the generalized approach to space–time signal and image processing in the presence of noise must determine the spatial correlation function between the moving and model images in definite conditions. Digital implementation of the generalized approach requires a high computational cost that, in some cases, limits the possibilities of practical use of this approach. The use of more simple algorithms based on the generalized approach, which allows us to decrease computational costs with predetermined accuracy, is a very real problem.

Many quasioptimal algorithms based on the generalized approach were constructed in heuristic ways by different approaches to techniques of signal and image processing and the presentation of results. Such a circumstance makes comparative analysis of the algorithms more difficult. Let us consider the following procedure in which we are able to synthesize such algorithms and to carry out a comparative analysis. This procedure is based on the theory of signal processing and spectral analysis.

The synthesis of the quasioptimal generalized space–time signal and image processing algorithm based on the generalized approach to signal processing in noise consists of the following steps: definition of the preliminary signal and image processing form, definition of the criterial correlation function, and definition of the procedure to define an extremum of the criterial correlation function. The first and second steps have a great influence on the characteristics of the constructed algorithms. The third step is not specific to algorithms and, for this reason, we do not consider this step in the subsequent discussion.

The first step of synthesis is that the moving image contains excessive information. If we can remove this excessive information in the absence of noise, the probability of signal and image detection and accuracy of object coordinate definition are not decreased. Reducing by decreasing the initial information image informativeness is one possible way to make computational costs less in employing the generalized space–time signal and image processing algorithms based on the generalized approach to signal processing in the presence of noise. The possible level of decrease in image informativeness is defined by the predetermined noise immunity of navigational systems. The procedure of image informativeness decrease can be considered as a linear filtering. In this case, the problem of synthesis of the generalized space–time signal and image processing algorithm is reduced to the problem of the definition of the filter transfer function, which ensures the required transformation of the moving image.

There are some approaches to decreasing the informative excess of space–time signals and images. The most widely used technique is to reduce image resolution. Let us consider briefly this technique. The image forming with lower resolution at the k-th point includes the two-dimensional image filtering at the $(k-1)$-th point using the low-frequency preliminary filter of the generalized receiver linear tract with further quantization of images at a half frequency of quantization at the $(k-1)$-th point. The transfer function of the low-frequency preliminary filter takes the following form:[1]

$$C(\omega_x, \omega_y) = \left[\cos(0.5\omega_x) \cdot \cos(0.5\omega_y)\right]^n \tag{13.1}$$

where ω_x and ω_y are spatial frequencies and $n = 1, 2, \ldots$. Sequential decrease in image resolution lies at the root of the hierarchical signal and image processing algorithms constructed based on the generalized approach to signal processing in the presence of noise. The disadvantage of this procedure is a requirement to increase the memory block because it is necessary to store a set of object images with different resolutions.

In the quantization of images, the rectangular raster is often used. For this, the values of each coordinate are chosen at equidistance points with Kotelnikov's frequency. Digitization based on the rectangular raster is a particular case of the general quantization procedure, in accordance with which the nonrectangular raster is used to choose the values of the continuous object image. In some cases, it is worthwhile to use a hexagonal raster that allows us to decrease the required number of readings to 13. 4% in comparison with the rectangular digitization. There are several complex procedures of image preprocessing that allow us to keep the geometrical images of definite forms such as the line, the circle, etc., in the preliminary filtered object image. These procedures are used in the image processing of contour lines.

The second step in the synthesis of the quasioptimal generalized space–time signal and image processing algorithm is the choice of the criterial function type or characteristic allowing us to estimate the degree or measure based on which we can make a decision about the similarity between the compared object images [see Equation (12.68) and Equation (12.69)]. A set of probability relationship characteristics is known. This set should satisfy the following requirements. The functional measure R_{XY} exists for some pairs of the random variables X and Y always and the probability that the random variables X and Y are not constant is equal to unity. R_{XY} is symmetrical, i.e., $R_{XY} = R_{YX}$. The functional measure is within the limits of the interval [0,1], i.e., $0 \leq R_{XY} \leq 1$; if the random variables X and Y are independent the equality $R_{XY} = 0$ is true. The equality $R_{XY} = 1$ corresponds to the functional dependence between the random variables X and Y, i.e., $X = f(Y)$ or $Y = g(X)$, where $f(Y)$ and $g(X)$ are the measured Boreal functions. If the measured Boreal functions $f(Y)$ and $g(X)$ coincide with the real coordinate system axes, the following equality

$$R\left[f(Y), g(X)\right] = r_{XY} \tag{13.2}$$

is true. If the random variables X and Y obey the Gaussian probability distribution density, the equality $R_{XY} = |r_{XY}|$ is true, where r_{XY} is the coefficient of correlation between the random variables X and Y. It should be noted that at the present time, there are no universally adopted and rigorously proven mathematical requirements for the criterial correlation functions. Therefore, we consider briefly the main classes of the criterial correlation functions.

13.1.1 Criterial Correlation Functions

Criterial correlation functions are based on the determination of the mutual correlation function of a stochastic process. The mutual correlation function

$$R_{1.1}(\alpha_x) = M\{[a^*(x) - M[a^*(x)]] \cdot [X(x+\alpha_x) - M[X(x+\alpha_x)]]\} \tag{13.3}$$

is the mixed central moment of the second order, where $a^*(x)$ is the model image, $X(x)$ is the moving image, and M denotes the mean. The normalized mutual correlation function can be determined in the following form:

$$R_{1.2}(\alpha_x) = \frac{R_{1.1}(\alpha_x)}{\sigma[a^*(x)] \cdot \sigma[X(x)]}, \tag{13.4}$$

where σ is the operator of the mean square deviation. The functions $R_{1.1}(\alpha_x)$ and $R_{1.2}(\alpha_x)$ are functionally related in the following form:

$$R_{1.1}(\alpha_x) = R_{1.2}(\alpha_x) \cdot \sigma[a^*(x)] \cdot \sigma[X(x)]. \tag{13.5}$$

The correlation function of the mixed initial moment of the second order is determined by

$$R_{1.3}(\alpha_x) = M\{a^*(x) \cdot X(x + \alpha_x)\}. \tag{13.6}$$

The functions $R_{1.1}(\alpha_x)$ and $R_{1.3}(\alpha_x)$ are related by

$$R_{1.1}(\alpha_x) = R_{1.3}(\alpha_x) - M[a^*(x)] \cdot M[X(x)]. \tag{13.7}$$

The weighted criterial correlation function of the second order moment is denoted by $R_{1.4}(\alpha_x)$. In spite of the similar structure of the criterial functions $R_{1.1}(\alpha_x)$ and $R_{1.3}(\alpha_x)$, there are big differences in their peculiarities. Let us consider these peculiarities using the example of binary (0 and 1) image processing. In the criterial correlation function $R_{1.3}(\alpha_x)$, only elements of the moving image $X(x)$ having the brightness characteristics corresponding to 1 and coinciding with elements of the model image $a^*(x)$, which also have the brightness characteristics corresponding to 1, can influence the computer-calculated results. Other elements of the moving image $X(x)$ are multiplied by the corresponding elements of the model image $a^*(x)$ with the brightness characteristics corresponding to 0. Thus, these elements do not contribute to the criterial correlation function magnitude. The great advantage of the criterial correlation function form is that the moving images observed under various conditions of lightness give the same maximal magnitudes of the criterial correlation function $R_{1.3}(\alpha_x)$ under the condition that the contour lines of the moving and model images are the same. At that time, the use of the criterial correlation function $R_{1.3}(\alpha_x)$ is limited in navigational systems that are critical with respect to the probability of a false alarm. This is so because, even in the absence of the investigated object image, the sufficient number of elements of the moving image $X(x)$ with the brightness characteristics corresponding to 1 caused by noise and underlying background noise sources can coincide with the elements of the model image $a^*(x)$, the brightness characteristics of which also correspond to 1. Therefore, the probability of a false alarm increases. The use of $R_{1.1}(\alpha_x)$ allows us to decrease the probability of a false alarm, but the probability of object image detection also decreases.[2] The use of the weighted criterial correlation function of the second-order moment ensures a predetermined ratio between the probability of object image detection and the probability of a false alarm

$$R_{1.4}(\alpha_x) = R_{1.3}(\alpha_x) - \beta M[a^*(x)] \cdot M[X(x)], \tag{13.8}$$

where $0 \leq \beta \leq 1$.

13.1.2 Difference Criterial Functions

In the general case, the difference criterial correlation function can be represented in the following form:

$$R_{2.0}(\alpha_x) = M\left[\,\left| a^*(x) - X(x+\alpha_x)\right|^p\right], \tag{13.9}$$

where $p = 1, 2, \ldots$. There are also the following difference criterial correlation functions. The average square difference criterial correlation function can be determined in the following form:

$$R_{2.1}(\alpha_x) = M\left[a^*(x) - X(x+\alpha_x)\right]^2. \tag{13.10}$$

Reference to Equation (13.10) shows that, after squaring the first and third terms that are the variances, the second term is the double criterial correlation function $R_{1.3}(\alpha_x)$. Thus, we can write

$$R_{2.1}(\alpha_x) = -2R_{1.3}(\alpha_x) + D\left[a^*(x)\right] + D\left[X(x)\right]. \tag{13.11}$$

The average absolute difference criterial correlation function takes the following form:

$$R_{2.2}(\alpha_x) = M\left[\,\left| a^*(x) - X(x+\alpha_x)\right|\,\right]. \tag{13.12}$$

Note that for binary image processing, the difference criterial correlation functions $R_{2.1}(\alpha_x)$ and $R_{2.2}(\alpha_x)$ are the same, i.e., $R_{2.1}(\alpha_x) = R_{2.2}(\alpha_x)$. $R_{2.2}(\alpha_x)$ is connected with the normalized mutual correlation function $R_{1.2}(\alpha_x)$ by the relationship

$$R_{1.2}(\alpha_x) = 1 - \frac{R_{2.2}^2(\alpha_x)}{2\mu^2\sigma\left[a^*(x)\right]\cdot\sigma\left[X(x)\right]}, \tag{13.13}$$

where μ is the coefficient depending on the probability distribution density of image brightness characteristics. In the case of the Gaussian probability distribution density of image brightness characteristics, we can assume that $\mu = \sqrt{\frac{2}{\pi}}$. The peculiarity of $R_{2.2}(\alpha_x)$ is that there is no need to carry out the operation of centering, and obtained estimations are invariant with respect to some form of nonstationary state.

The Minkovsky function[3] takes the following form:

$$R_{2.3}(\alpha_x) = \left[M\sqrt{\left| a^*(x) - X(x+\alpha_x)\right|}\,\right]^2. \tag{13.14}$$

The Camber function has the following form:

$$R_{2.4}(\alpha_x) = M\left\{\frac{\left|a^*(x) - X(x+\alpha_x)\right|}{\left|a^*(x) + X(x+\alpha_x)\right|}\right\}. \tag{13.15}$$

The Chebyshev function takes the following form:

$$R_{2.5}(\alpha_x) = \max_x \left| a^*(x) - X(x+\alpha_x) \right|. \tag{13.16}$$

The remarkable superiority of the difference criterial correlation functions compared to the other correlation functions is the absence of the product. Thus in using the difference criterial correlation functions, the computational cost is 4 to 10 times less compared with the use of the other correlation functions. At the same time, in using the difference criterial correlation functions, the detection performance is worse in comparison with when we use the correlation functions at low values of the signal-to-noise ratio.

The difference criterial correlation functions are equal to zero in the case of exact matching between the same object images and are high in value in the case of mismatching between object images. When we use the difference criterial correlation function $R_{2.0}(\alpha_x)$, with an increase in the value of the parameter p, the detection performances are improved, but if the condition $p \geq 3$ is satisfied, this improvement is negligible. Note that the difference can be replaced by summing for the criterial correlation functions $R_{2.0}(\alpha_x)$, …, $R_{2.5}(\alpha_x)$.

13.1.3 Spectral Criterial Functions

The criterial correlation function in the spectral range takes the following form:

$$R_{3.1}(\alpha_x) = \mathcal{F}^{-1}\{S_{a^*}(\omega_x) \cdot S_{\dot{X}}(\omega_x)\} = \mathcal{F}^{-1}\{S_{a^*X}(\omega_x)\}, \tag{13.17}$$

where $S_{a^*}(\omega_x)$ is the power spectral density of the model image, $S_X(\omega_x)$ is the power spectral density of the moving image, $S_{a^*X}(\omega_x)$ is the mutual power spectral density, $\mathcal{F}^{-1}$ is the inverse Fourier transform, and the sign $(\cdot)$ denotes the complex conjugate value. The Rot function[4] takes the following form:

$$R_{3.2}(\alpha_x) = \mathcal{F}^{-1}\left(\frac{S_{a^*X}(\omega_x)}{S_X(\omega_x)}\right). \tag{13.18}$$

The function of coherency is determined by[4]

$$R_{3.3}^2(\alpha_x) = \mathscr{F}^{-1}\left\{\frac{S_{a^*X}^2(\omega_x)}{\sqrt{S_{a^*}(\omega_x)\cdot S_X(\omega_x)}}\right\} = \mathscr{F}^{-1}\{S_{a^*X}(\omega_x)\cdot\gamma(\omega_x)\}\,. \tag{13.19}$$

The Knapp function[4] takes the following form:

$$R_{3.4}(\alpha_x) = \mathscr{F}^{-1}\left\{\frac{S_{a^*X}(\omega_x)\cdot|\gamma(\omega_x)|^2}{|S_{a^*X}(\omega_x)|\left[1-|\gamma(\omega_x)|^2\right]}\right\}. \tag{13.20}$$

The spectral criterial functions are based on the relationship between the mutual correlation function and the mutual power spectral density using the Fourier transform [see Equation (12.56) and Equation (12.57)]. These functions allow us to decrease the computational cost due to the use of the fast Fourier transform algorithms. Moreover, signal processing in the spectral range allows us to amplify the components of those spatial frequencies at which the signal-to-noise ratio is maximal and to compensate for components distorted by noise.

13.1.4 Bipartite Criterial Functions

The bipartite criterial correlation functions suppose digital image processing if the number of quantization is equal to two or more. If each element of the first object image with the relative object image shift α_x has the number of quantization equal to i, and each element of the second object image has the number of quantization equal to j, then the bipartite criterial correlation function $F_{ij}(\alpha_x)$, $0 \le i, j \le 2^n - 1$, is increased by unity. Here 2^n is the number of quantization. Consequently, under the condition $i = j$, the bipartite criterial correlation function $F_{ij}(\alpha_x)$ is equal to the number of elements having the same brightness characteristics. Under the condition $i \ne j$, the bipartite criterial correlation function $F_{ij}(\alpha_x)$ is equal to the number of elements having different brightness characteristics. If the object images with dimensions $N \times N$ are identical, we can write

$$\sum_{i=0}^{2^n-1}\sum_{j=0}^{2^n-1} F_{ij}(\alpha_x) = \begin{cases} N^2 & \text{at} \quad i = j, \\ 0 & \text{at} \quad i \ne j. \end{cases} \tag{13.21}$$

As an example, consider two criterial correlation functions obtained by the use of bipartite criterial correlation functions in the course of multilevel quantization:[5]

$$R_{4.1}(\alpha_x) = \frac{1}{N}\sum_{i=0}^{n-1} F_{ii}(\alpha_x) , \tag{13.22}$$

$$R_{4.2}(\alpha_x) = \prod_{i=0}^{2^n-1}\left\{\frac{F_{ii}(\alpha_x)}{\sum_{j=0}^{2^n-1} F_{ij}(\alpha_x)}\right\}. \tag{13.23}$$

The bipartite criterial correlation functions are widely used in binary image processing. For this case, let us introduce the following notation:

$$F_{11}(\alpha_x) = \alpha;\; F_{01}(\alpha_x) = \beta;\; F_{10}(\alpha_x) = \gamma;\; F_{00}(\alpha_x) = \zeta; \tag{13.24}$$

d_* is the number of elements with brightness characteristics equal to 1 in the model image; d is the number of elements with brightness characteristics equal to 1 in the moving image; f_* is the number of elements with brightness characteristics equal to 0 in the model image; f is the number of elements with brightness characteristics equal to 0 in the moving image. The well-known bipartite criterial correlation functions[3] fall into the following types:

the Rao function

$$R_{4.3} = 2(\alpha + \beta + \gamma + \zeta)^{-1}; \tag{13.25}$$

the Jakard function

$$R_{4.4} = \alpha(\alpha + \beta + \gamma)^{-1}; \tag{13.26}$$

the Deak function

$$R_{4.5} = 2\alpha(2\alpha + \beta + \gamma)^{-1}; \tag{13.27}$$

the Soucal and Snit function

$$R_{4.6} = \alpha(\alpha + 2\beta + 2\gamma)^{-1}; \tag{13.28}$$

the Kulzinsky function

$$R_{4.7} = \alpha(\beta + \gamma)^{-1}; \tag{13.29}$$

the first Rodgers and Tanimoto function

$$R_{4.8} = \alpha(d_* + d - \alpha)^{-1}; \tag{13.30}$$

the second Rodgers and Tanimoto function

$$R_{4.9} = (\alpha + \zeta)\cdot(\alpha + 2\beta + 2\gamma + \zeta)^{-1}; \tag{13.31}$$

the Soucal and Mishner function

$$R_{4.10} = (\alpha + \zeta)\cdot(\alpha + \beta + \gamma + \zeta)^{-1}; \tag{13.32}$$

the Yule function

$$R_{4.11} = (2\zeta - \beta\gamma)\cdot(2\zeta + \beta\gamma)^{-1}; \tag{13.33}$$

the Chamman function

$$R_{4.12} = (\alpha + \zeta - \beta - \gamma)\cdot(\alpha + \beta + \gamma + \zeta)^{-1}. \tag{13.34}$$

The choice of the criterial correlation functions is defined by the relative importance of elements with brightness characteristics equal to 1 or 0 and by the relative importance of events that are coincident or not coincident with the brightness characteristics of elements for the considered problem. For example, the bipartite criterial correlation function $R_{4.5}$ has a double weight in the case of coincidence of elements with the brightness characteristics equal to 1. The bipartite criterial correlation function $R_{4.6}$ has a double weight in the case of noncoincident elements, i.e., the brightness characteristics of elements are equal to 1 and 0. If we have equivalent elements with the brightness characteristics equal to 1 and 0, it is worthwhile to use the bipartite criterial correlation function $R_{4.3}$.

The sign functions can be also considered as the bipartite criterial correlation functions. The normalized mutual criterial correlation function $R_{1.2}$ (α_x) for the stationary ergodic images obeying the Gaussian probability distribution density of brightness can be defined using the probabilities of coinciding signs in the following form:[6]

$$R_{1.2}(\alpha_x) = -\cos 2\pi \cdot P^{(++)}_{a^*X}(\alpha_x) = -\cos 2\pi \cdot P^{(--)}_{a^*X}(\alpha_x), \tag{13.35}$$

where $P^{(++)}_{a^*X}(\alpha_x)$ is the probability of coincidence of positive signs only, and $P^{(--)}_{a^*X}(\alpha_x)$ is the probability of coincidence of negative signs only.

13.1.5 Rank Criterial Functions

The rank criterial correlation functions define the ranking of object image elements by the brightness characteristics. In other words, the rank criterial correlation function defines the position of the object image elements according to the order of the decreasing or increasing brightness characteristics. Each object image element is denoted by the corresponding number — the rank. The Spearman function[6] is most widely used:

$$R_{5.1}(\alpha_{xj}) = 1 - \frac{6\sum_{i=1}^{N}(r_{a^*i+\alpha_{xj}} - r_{Xi})}{N(N^2-1)}, \tag{13.36}$$

where r_{a^*i} is the rank of the i-th element of the model image and r_{Xi} is the rank of the i-th element of the moving image. At the present time, a formalized method of choosing the type of preliminary image processing and the criterial correlation function is absent. Therefore, the synthesis of quasioptimal generalized signal and image processing algorithms based on the generalized approach to signal processing in the presence of noise is carried out only in a heuristic way.

13.2 The Quasioptimal Generalized Image Processing Algorithm

Let us consider the example of the use of the procedure discussed in the previous section. In the synthesis of the quasioptimal generalized image processing algorithm, we are guided by the following main statements.

First, the algorithm should operate both with object image identification (the condition of searching) and identified object image tracking (the condition of nonsearching). In object image identification, the main characteristics of the algorithm are the probabilities of object omission and false alarm. The accuracy of identification is not so important and plays a secondary role. In identified object image tracking (nonsearching), first, the quasioptimal generalized image processing algorithm should guarantee high accuracy of object image tracking because an incorrect mismatching of object images is improbable. Therefore, in object image identification, the space–time criterial correlation function between the moving and model images must ensure a maximal ratio between the main and side lobes. In other words, this function should tend to approach the delta function in the limiting case. In object image identification, the determination of the moving object image coordinates is carried out using a derivative of the criterial correlation function. Therefore, the main lobe must be extended a little. Second, the algorithm should ensure image processing in the spectral range, which allows us to reduce computational costs due to the use of the fast Fourier transform. Third, the algorithm has to be functionally related to the universally adopted generalized approach to signal processing in the presence of noise.

In accordance with the preceding statements, the sequence of the quasioptimal generalized image processing algorithm based on the generalized approach to signal processing in the presence of noise[7–10] can be represented in the following form. During the first step, the preprocessing of the moving and model images is carried out. Its main purpose is to decrease the object image informativeness. Hereinafter, it is necessary to carry out additional processing of the moving image to obtain the criterial correlation function with the minimal width of the main lobe. We can present these two steps in the form of linear filtering (see Figure 13.1).

Let us introduce the following notation: $h_1(x, y)$ is the weight function and $C_1(\omega_x, \omega_y)$ is the transfer function of the filter providing a reduction in the object image informativeness; $\hat{a}^*(x, y)$ and $\hat{S}_{a^*}(\omega_x, \omega_y)$ are the model image and the power spectral density of the model image, respectively, with decreased informativeness; $\hat{X}(x,y)$ and $\hat{S}_X(\omega_x, \omega_y)$ are the moving image and the power spectral density of the moving image, respectively, with decreased informativeness; $h_2(x, y)$ is the weight function and $C_2(\omega_x, \omega_y)$ is the transfer function, respectively, of the filter during the second step of image preprocessing (additional processing of the moving image).

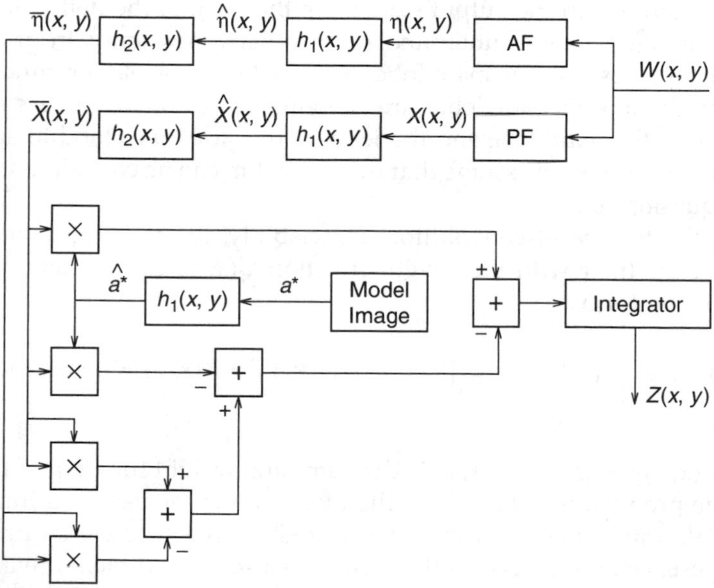

FIGURE 13.1
The generalized receiver.

Taking into consideration this notation, the algorithm takes the following form:

$$Z(x,y) \cong \int_X \int_Y \{2\hat{X}(x,y) * h_2(x,y) \cdot \hat{a}^*(x-\lambda_x, y-\lambda_y) - \hat{X}(x,y) * h_2(x,y) \cdot \hat{X}^*(x,y) * h_2(x,y) + \hat{\eta}^2(x,y) * h_2(x,y) \cdot \hat{\eta}^{*2}(x,y) * h_2(x,y)\}\, dx\, dy = \delta(x-\lambda_x, y-\lambda_y) \quad (13.37)$$

where

$$\hat{a}^*(x-\lambda_x, y-\lambda_y) = a(x-\lambda_x, y-\lambda_y) * h_1(x,y); \quad (13.38)$$

$$\hat{X}(x,y) = X(x,y) * h_1(x,y); \quad (13.39)$$

$$\hat{\eta}(x,y) = \eta(x,y) * h_1(x,y); \quad (13.40)$$

the symbol (∗) denotes convolution.

The formula in Equation (13.37) is true and conditional in the statistical sense. We can obtain the output statistic in the form of the delta function in processing object images unlimited in informativeness only. In practice, in real image processing, the main lobe of the criterial correlation function has a finite width and the side lobes are present. However, we can assume that the width of the main lobe and the level of the side lobes tend to approach zero. Therefore, we can assume that the algorithm can be considered as ideal, even if quasioptimal.

Using the feature of convolution associability, the moving image at the output of the filter with the weight function $h_2(x, y)$ can be determined in the following form:

$$[X(x,y) * h_1(x,y)] * h_2(x,y) = [h_2(x,y) * h_1(x,y)] * X(x,y) = h_\Sigma(x,y) * X(x,y), \quad (13.41)$$

where $h_\Sigma(x, y) = h_2(x, y) * h_1(x, y)$ is the summing weight function of the filter. Thus, the problem of synthesis of the quasioptimal generalized image processing algorithm based on the generalized approach to signal processing in the presence of noise can be thought of as a solved problem if we are able to define the weight functions $h_1(x, y)$ and $h_\Sigma(x, y)$ or $h_2(x, y)$ of filters and corresponding transfer functions.

Let us define the weight function $h_1(x, y)$ of the filter forming the object image with decreased informativeness. In navigational systems, it is worthwhile to have information at the center of the image scene, where the object is located, in more detail, and information at the periphery of the image scene in less detail with the purpose of reducing the effect of background noise. It is possible to solve this problem in the course of image processing using the weighted functions having the maximal value equal to unity at the center of the image scene and decreasing to zero at the periphery of image scene.[11] However, the introduction of the weight functions leads to an increase in the computational costs for image processing.

The filter, the output signal of which can be presented by the totality of readings reckoned on radial lines emanating from the coordinate system origin and uniformly distributed within the limits of the interval $[0, 2\pi]$, can be used as an alternative method of reducing image informativeness (see Figure 13.2). Thus, the requirement to reduce the object image informativeness with predetermined nonuniformity is carried out without increase in computational costs. The transfer function of the filter ensuring image processing by the use of filtering features of the delta function can be determined in the following form:

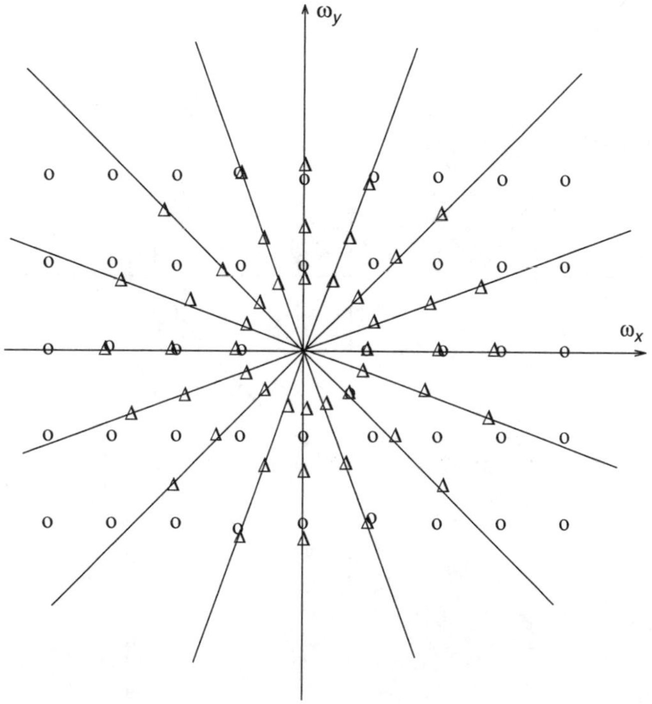

FIGURE 13.2
Radial lines: o — readings in the Cartesian coordinate system; Δ — readings on radial lines.

$$C_1(\omega_x,\omega_y)=\sum_{k=1}^{n}\sum_{m=1}^{p}\delta\{q_k(\omega_x,\omega_y)\}\cdot\delta\{u_m(\omega_x,\omega_y)\}, \tag{13.42}$$

where $q_k(\omega_x, \omega_y)$ is a function of the line bundle and $u_m(\omega_x, \omega_y)$ is a function of concentric circles with the center at the coordinate system origin. Therefore, we assume that the well-known features of the delta function are true for the two-dimensional case: $\delta^2(x) = c\delta(x)$, where c is an arbitrary constant and $\delta(x - a)(x - b) = 0$ if the condition $a \neq b$ is satisfied.[12] Thus, the transfer function of the filter has the form (see Figure 13.3)

$$C_1(\omega_x,\omega_y)=\sum_{k=1}^{n}\sum_{m=1}^{p}\delta\left\{\omega_y-\omega_x\cdot\mathrm{tg}\left(\theta_\mathrm{n}k\right)\right\}\cdot\delta\left\{\omega-m\omega_0\right\}, \tag{13.43}$$

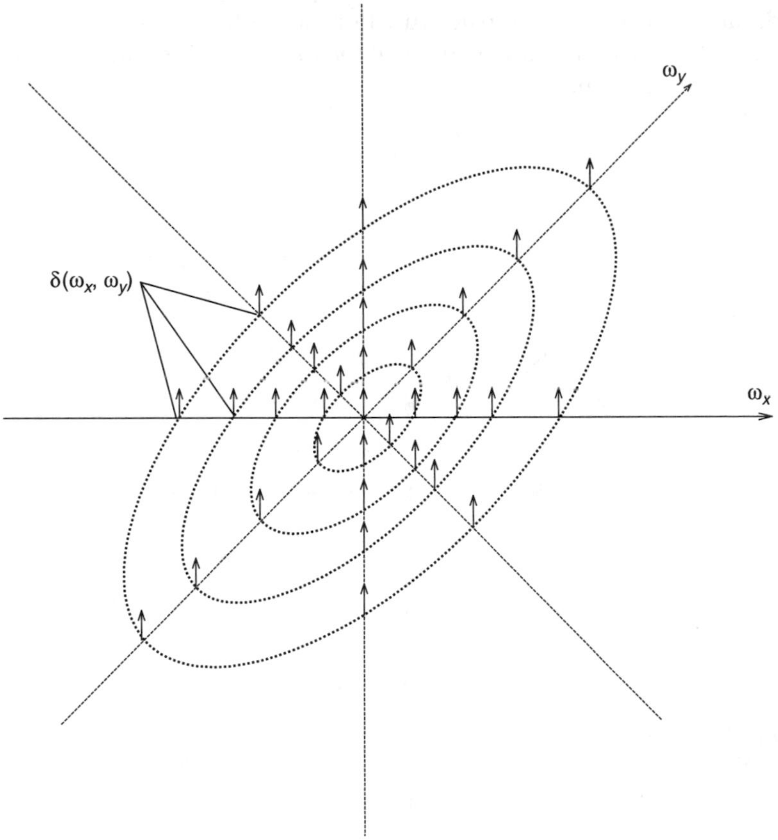

FIGURE 13.3
The transfer function.

where n is the number of lines in the bundle, $\theta_n = \frac{\pi}{n}; \omega = \sqrt{\omega_x^2 + \omega_y^2}; \omega_0$ is Kotelnikov's frequency, $p = \frac{\Omega}{\omega_0}; k \neq 0.5n$, and Ω is the highest frequency of the power spectral density $S_a(\omega_x, \omega_y)$ of the signal. The power spectral density of the signal at the output of the filter with the weight transfer function $C_1(\omega_x, \omega_y)$ takes the following form:

$$\hat{S}_a(\omega_x, \omega_y) = S_a(\omega_x, \omega_y) \sum_{k=1}^{n} \sum_{m=1}^{p} \delta\{\omega_y - \omega_x \cdot \mathrm{tg}\,(\theta_n k)\} \cdot \delta\{\omega - m\omega_0\} \tag{13.44}$$

and tends to approach the power spectral density of the initial signal as $n \to \infty$.

Now, let us consider the synthesis of the filter with the transfer function $C_2(\omega_x, \omega_y)$. Let us consider the filter functioning for the searching condition. In this case, the filter would ensure the processing of the moving image, under which the criterial correlation function is transformed into the delta function for the limiting case. The moving image at the output of the filter with the weight function $h_2(x, y)$ can be written in the following form:

$$\bar{X}(x,y) = \hat{X}(x,y) * h_2(x,y)\,. \tag{13.45}$$

In the spectral range, Equation (13.45) can be written in the following form:

$$\bar{S}_X(\omega_x, \omega_y) = \hat{S}_X(\omega_x, \omega_y) \cdot C_2(\omega_x, \omega_y)\,. \tag{13.46}$$

The criterial correlation function of statistic at the generalized detector output can be written in the following form:

$$R_Z(x,y) = \int_X \int_Y \{2\bar{X}(x,y) \cdot \hat{a}^*(x,y) - \bar{X}(x,y) \cdot \bar{X}^*(x,y) + \bar{\eta}(x,y) \cdot \bar{\eta}^*(x,y)\}\, dx\, dy, \tag{13.47}$$

where $\bar{\eta}(x,y) = \hat{\eta}(x,y) * h_2(x,y)$. In the spectral range, Equation (13.47) can be written in the following form:

$$S_z(\omega_x,\omega_y)=\int_{\Omega_X}\int_{\Omega_Y}\{2\bar{S}_X(\omega_x,\omega_y)\cdot\hat{S}_{a^*}(\omega_x,\omega_y)-\bar{S}_X(\omega_x,\omega_y)\cdot\bar{S}_x^*(\omega_x,\omega_y)$$
$$+\bar{S}_\eta(\omega_x,\omega_y)\cdot S_\eta^*(\omega_x,\omega_y)\}dxdy, \tag{13.48}$$

where

$$\begin{aligned}\bar{S}_\eta(\omega_x,\omega_y)&=\hat{S}_\eta(\omega_x,\omega_y)\cdot C_2(\omega_x,\omega_y)\\&=[S_\eta(\omega_x,\omega_y)\cdot C_1(\omega_x,\omega_y)]\cdot C_2(\omega_x,\omega_y),\end{aligned} \tag{13.49}$$

$S_\eta(\omega_x, \omega_y)$ is the power spectral density of the reference sample noise $\eta(x, y)$ at the additional filter output of the generalized detector. The noise $\xi(x, y)$ at the preliminary filter output of the generalized detector included in the moving image $X(x, y)$ can be presented by analogous formulae.

Because we require that the criterial correlation function of statistic at the generalized detector output would tend to approach the delta function $\delta(x, y)$, then, representing the power spectral densities of the moving and model images in the complex form and substituting Equation (13.46) in Equation (13.48), we can write

$$\delta(x,y)=\mathcal{F}^{-1}\{\{2|\hat{S}_X(\omega_x,\omega_y)|e^{j(\varphi_{X_x}+\varphi_{X_y})}\cdot|\hat{S}_{a^*}^*(\omega_x,\omega_y)|e^{-j(\varphi_{a_x^*}+\varphi_{a_y^*})}$$
$$-|\hat{S}_X(\omega_x,\omega_y)|e^{j(\varphi_{X_x}+\varphi_{X_y})}\cdot|\hat{S}_X^*(\omega_x,\omega_y)|e^{-j(\varphi_{X_x}+\varphi_{X_y})}$$
$$+|\hat{S}_\eta(\omega_x,\omega_y)|e^{j(\varphi_{X_x}+\varphi_{X_y})}\cdot|\hat{S}_\eta^*(\omega_x,\omega_y)|e^{-j(\varphi_{X_x}+\varphi_{X_y})}\}\cdot C_2(\omega_x,\omega_y)\}, \tag{13.50}$$

where $\mathcal{F}^{-1}(x)$ is the inverse Fourier transform. Based on Equation (13.50), it is not difficult to ensure that the transfer function $C_2(\omega_x, \omega_y)$ should have the following form:

$$\begin{aligned}C_2(\omega_x,\omega_y)=&\Big|2\hat{S}_X(\omega_x,\omega_y)\cdot\hat{S}_{a^*}^*(\omega_x,\omega_y)-\hat{S}_X(\omega_x,\omega_y)\\&\times\hat{S}_X^*(\omega_x,\omega_y)+\hat{S}_\eta(\omega_x,\omega_y)\cdot\hat{S}_\eta^*(\omega_x,\omega_y)\Big|^{-1}.\end{aligned} \tag{13.51}$$

Substituting Equation (13.51) in Equation (13.50), we obtain

$$\mathcal{F}^{-1}\{e^{j(\varphi_{X_x}-\varphi_{a_x^*}+\varphi_{X_y}-\varphi_{a_y^*})}\}=\delta(x,y)\,. \tag{13.52}$$

The result obtained can be interpreted as follows. The device with the transfer function $C_2(\omega_x, \omega_y)$ is the phase filter, i.e., the filter at the output of which the phase component of the complex power spectral density of the object image does not vary and the power spectral density of the amplitude component of the object image is normalized. The fact that the phase component plays the main role in solving the pattern recognition problem is discussed in References 11 and 13.[11,13] The phase component is important because the total information regarding the shift of the object image is included in it. For instance, let us consider that the object image $a(x, y)$ with the power spectral density $S_a(\omega_x, \omega_y)$ of $a(x, y)$ shifted by the value λ_x along the axis x, and by the value λ_y along the axis y, can be written in the following form:

$$\mathcal{F}\{a(x-\lambda_x, y-\lambda_y)\} = \int_{-\infty}^{\infty}\int_{-\infty}^{\infty} a(x-\lambda_x, y-\lambda_y)\cdot e^{-j(\omega_x x+\omega_y y)}dx\,dy\,, \quad (13.53)$$

where F denotes the Fourier transform. Let us introduce the notation $u = x - \lambda_x$ and $\upsilon = y - \lambda_y$. Then, Equation (13.53) can be written in the following form:

$$\begin{aligned}\mathcal{F}\{a(x-\lambda_x, y-\lambda_y)\} &= \int_{-\infty}^{\infty}\int_{-\infty}^{\infty} a(u,\upsilon)\cdot e^{-j[\omega_x(u+\lambda_x)+\omega_y(\upsilon+\lambda_y)]}du\,d\upsilon \\ &= S(\omega_x,\omega_y)\cdot e^{-j(\omega_x\lambda_x+\omega_y\lambda_y)}.\end{aligned} \quad (13.54)$$

Reference to Equation (13.54) shows that the phase component of the complex power spectral density of the object image varies. The possibility of using only the phase component of the object image complex power spectral density is proved by the experimental investigations discussed in References 11 and 13.[11,13]

In the use of the nonsearching condition, the filter transfer function can be generalized by introducing the power degree ℓ in the exponent of Equation (13.51):

$$\begin{aligned}C_2(\omega_x,\omega_y) = \big|\, & 2\hat{S}_X(\omega_x,\omega_y)\cdot\hat{S}^*_{a^*}(\omega_x,\omega_y) - \hat{S}_X(\omega_x,\omega_y) \\ & \times \hat{S}^*_X(\omega_x,\omega_y) + \hat{S}_\eta(\omega_x,\omega_y)\cdot\hat{S}^*_\eta(\omega_r,\omega_y)\big|^{-\ell}.\end{aligned} \quad (13.55)$$

If $\ell = 1$, the criterial correlation function takes the form of the delta function and the quasioptimal generalized image processing algorithm based on the generalized approach to signal processing in the presence of noise ensures the searching condition. In the condition $0 \le \ell < 1$, the delta function becomes fuzzy so that it ensures the nonsearching condition for the functioning of the tracking device in the navigational system, the discrimination characteristic of which is defined by the derivative with decreased informativeness. In the nonsearching condition, the value of ℓ is defined by the statistical characteristics of the object image and background noise.[14]

Thus, the quasioptimal generalized image processing algorithm can be written in the following form:

$$Z(\lambda_x, \lambda_y) = \max\left\{\mathcal{F}^{-1}\left[2\bar{S}_X(\omega_x, \omega_y)\cdot \bar{S}^*_{a^*}(\omega_x, \omega_y) - \bar{S}_X(\omega_x, \omega_y) \times \bar{S}^*_X(\omega_x, \omega_y) + \bar{S}_\eta(\omega_x, \omega_y)\cdot \bar{S}^*_\eta(\omega_x, \omega_y)\right]\right\}, \tag{13.56}$$

where

$$\bar{S}_X(\omega_x, \omega_y) = S_X(\omega_x, \omega_y)\cdot C_\Sigma(\omega_x, \omega_y); \tag{13.57}$$

$$\hat{S}^*_{a^*}(\omega_x, \omega_y) = S^*_{a^*}(\omega_x, \omega_y)\cdot C_1(\omega_x, \omega_y); \tag{13.58}$$

$$\bar{S}_\eta(\omega_x, \omega_y) = S_\eta(\omega_x, \omega_y)\cdot C_\Sigma(\omega_x, \omega_y); \tag{13.59}$$

$$C_1(\omega_x, \omega_y) = \sum_{k=1}^{n}\sum_{m=1}^{p}\delta\left\{w_y - \omega_x \cdot tg(\theta_n k)\right\}\cdot\delta\left\{\omega - m\omega_0\right\}; \tag{13.60}$$

$$C_\Sigma(\omega_x, \omega_y) = \frac{\sum_{k=1}^{n}\sum_{m=1}^{p}\delta\{\omega_y - \omega_x \cdot \mathrm{tg}\,(\theta_n k)\}\cdot\delta\{\omega - m\omega_0\}}{\left|2\hat{S}_X(\omega_x, \omega_y)\cdot \hat{S}^*_{a^*}(\omega_x, \omega_y) - \hat{S}_X(\omega_x, \omega_y)\cdot \hat{S}^*_X(\omega_x, \omega_y) + \hat{S}_\eta(\omega_x, \omega_y)\cdot \hat{S}^*_\eta(\omega_x, \omega_y)\right|^{-\ell}}; \tag{13.61}$$

$$0 \le \ell \le 1;\ k \ne 0.5n;\ p = \frac{\Omega}{\omega_0};\ \text{and}\ \ \theta_n = \frac{\pi}{n}. \tag{13.62}$$

The formula in Equation (13.56) defines the quasioptimal generalized image processing algorithm for the searching condition. In the nonsearching condition, the difference is that the shift parameters λ_x and λ_y are defined by the zero value of the first-order partial derivatives with respect to λ_x and λ_y and by the total derivatives with respect to time. As $n \to \infty$ and at $\ell = 0$, the algorithm corresponds to the universally adopted generalized signal processing algorithm based on the generalized approach to signal processing in the presence of noise. This fact verifies the quasioptimality of the generalized image processing algorithm given by Equation (13.56).

Let us consider the practical realization of the quasioptimal generalized image processing algorithm based on the generalized approach to signal processing in the presence of noise. In practice, it is very difficult to realize the algorithm given by Equation (13.56). The difficulties are caused by the fact that in object image processing by a filter with the transfer function $C_1(\omega_x, \omega_y)$ given in the spectral range by concentric circles, we assume that the power spectral density of the object image is defined in the polar coordinate system u and θ, not in the Cartesian coordinate system ω_x and ω_y.

In this case, the power spectral density of the object image can be determined in the following form:[14]

$$S_a(u,\theta) = 2\pi \sum_{n=-\infty}^{\infty} \bar{C}_{nn}(u) \cdot e^{-jn\theta} \tag{13.63}$$

where

$$\bar{C}_{nn}(u) = \int_{-\infty}^{\infty} \rho C_n(\rho) J_n(pu) d\rho \,, \tag{13.64}$$

$$C_n(\rho) = \frac{1}{2\pi} \int_{-\pi}^{\pi} a(\rho,\phi) \cdot e^{-jn\phi} d\phi. \tag{13.65}$$

Equation (13.63) represents the power spectral density of the object image $a(\rho, \phi)$ in using the Fourier-series expansion. The coefficients of the Fourier-series expansion are the Chancel transform of the n-th order for the sequence $C_n(\rho) = 0$, where $n = 0, \pm 1, \pm 2\ldots.$

The theorem of readings has a more complex form using polar coordinates. The object image $a(\rho, \phi)$ having the power spectral density $S_a(u, \theta) = 0$ at $\rho \geq \lambda$ can be reconstructed by its readings in accordance with Stark and Woods:[15]

$$\begin{aligned} a(\rho,\phi) = \sum_{i=1}^{\infty}\sum_{k=0}^{2K} a(\alpha_{0_i}, \tfrac{2\pi k}{2K+1}) \cdot \theta_{0_i}(\rho)\,(2K+1)^{-1} + 2(2K+1)^{-1} \\ \times \sum_{i=1}^{\infty}\sum_{k=0}^{2K}\sum_{n=1}^{K} a(\alpha_{n_l}, \tfrac{2\pi k}{2K+1}) \cdot \theta_{n_i}(\rho) \cos\left[n(\phi - \tfrac{2\pi k}{2K+1})\right], \end{aligned} \tag{13.66}$$

where

$$\theta_{n_i}(\rho) = \frac{2\alpha_{n_i} J_n(\rho\lambda)}{\lambda J_{n+1}(\alpha_{n_i}\lambda)\cdot(\alpha_{n_i}^2 - \rho^2)}, \tag{13.67}$$

$\alpha_{n_i} = \frac{z_{n_i}}{\lambda}$, z_{n_i} is the i-th zero of the Bessel function $J_n(x)$.

Reference to Equation (13.63) and Equation (13.66) shows that various transforms of the object image $a(\rho, \phi)$ given in polar coordinates include operations with the Bessel functions and, in particular, the Chancel transform. The computational effectiveness of the Chancel transform is less than that of the fast Fourier one.[16] Therefore, we consider the possibility of excluding the Chancel transform from Equation (13.67). For this purpose, we use the sequential determination of the correlation functions of rows and columns of the two-dimensional rectangular database. Thus, the transition from processing of the two-dimensional functions to one-dimensional ones guarantees a decrease in computational costs.

Let us use this technique for the considered case. The totality of values of spectral components lying along the line of the bundle $q(\omega_x, \omega_y)$ and given by Equation (13.45) and Equation (13.46) is chosen as a partial one-dimensional definition of the object image. Note that these spectral components are the central cross section of the power spectral density of the object image, which are uniformly distributed within the limits of the interval $\theta \in [0, 2\pi]$

$$\mathcal{S}_a(\omega_u, \theta) = S_a(\omega_x, \omega_y)\Big|_{\substack{\omega_x = \omega_u \cos\theta \\ \omega_y = \omega_u \sin\theta}}. \tag{13.68}$$

The power spectral density of the object image is shown in Figure 13.4. Its cross sections at $\theta = 0°$ and at $\theta = 90°$ are shown in Figure 13.5 and Figure 13.6, respectively.

Obviously, with an increase in the number n of central cross sections of the power spectral density, we can define the total power spectral density of the object image with predetermined accuracy and, consequently, we can define the space–time correlation function of the moving and model images. Therefore, Equation (13.56) defining the synthesized quasioptimal generalized image processing algorithm based on the generalized approach to signal processing in the presence of noise is divided into n analogous formulae, each of which is derived from the correlation function determined under the same angles of the central cross sections of the moving and model image power spectral densities. The zero value of the derivative of the space–time correlation function defines the shift α_θ of cross sections and with it the shift of compared images along the direction θ. To define values λ_x and λ_y of the object image shift in the Cartesian coordinate system, it is necessary to define the correlation function of two cross sections of the power spectral densities

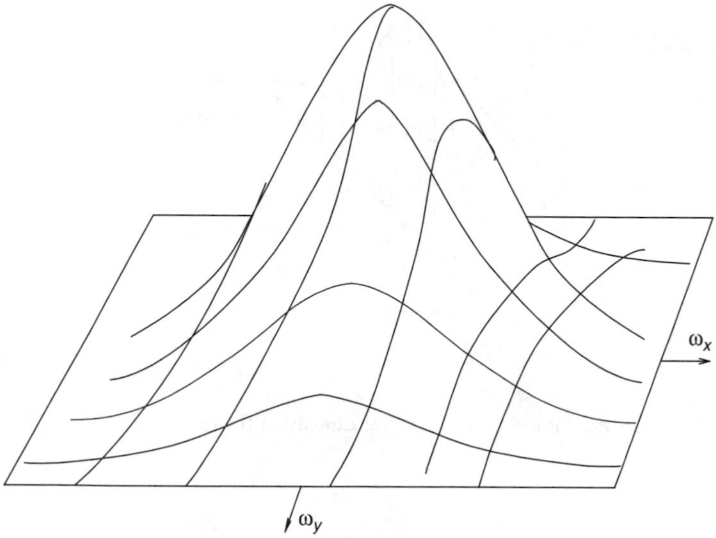

FIGURE 13.4
The power spectral density of the object image.

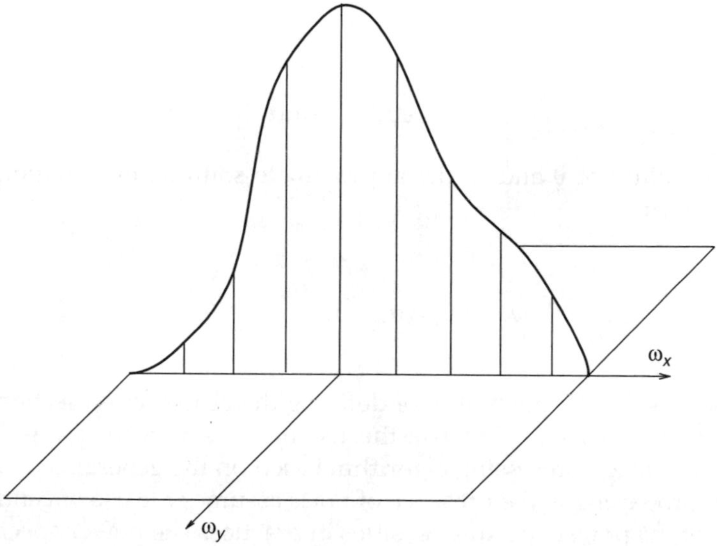

FIGURE 13.5
The central cross section of the power spectral density of the object image, $\theta = 0°$.

under arbitrary angles θ_k and θ_ℓ. The shift is defined by formulae in the condition $\theta_k < \theta_\ell$

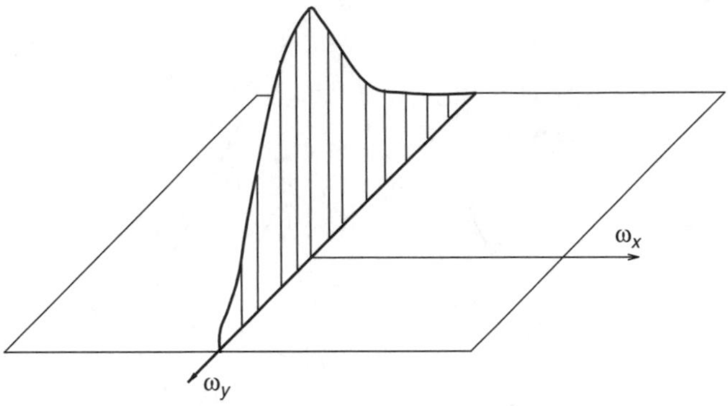

FIGURE 13.6
The central cross section of the power spectral density of the object image, θ = 90°.

$$\begin{cases} \alpha = \pm \arccos \dfrac{\alpha_{\theta_k}}{v} + \theta_k + 2\pi n, \\ v = \dfrac{\alpha_{\theta_\ell}}{\cos(\alpha - \theta_\ell)}, \end{cases} \tag{13.69}$$

where

$$v = \frac{\lambda_x}{\cos\alpha} = \frac{\lambda_y}{\sin\alpha}. \tag{13.70}$$

With low values of θ and α, the approximate solution of Equation (13.69) takes the form

$$\lambda_x = \lambda_y = \alpha_{\theta_k} - \theta_k \frac{\alpha_{\theta_k} - \alpha_{\theta_\ell}}{\theta_k - \theta_\ell}. \tag{13.71}$$

Let us consider the possibility of defining the central cross sections of the power spectral densities because the use of the synthesized quasioptimal generalized image processing algorithm based on the generalized approach to signal processing in the presence of noise assumes the use of central cross sections of the power spectral densities in practice. The power spectral density of the object image $a(x, y)$ can be written in the following form:

$$S_a(\omega_x, \omega_y) = \int_{-\infty}^{\infty}\int_{-\infty}^{\infty} a(x, y) \cdot e^{-j(\omega_x x + \omega_y y)} dx\, dy. \tag{13.72}$$

Let us assume that $\theta = 0°$, i.e., let us consider the cross section $S_a(\omega_x)$ of the power spectral density $S_a(\omega_x, \omega_y)$, which can be determined in the following form:

$$\mathcal{S}_a(\omega_x) = S_a(\omega_x, \omega_y)\Big|_{\omega_y=0} . \tag{13.73}$$

Substituting Equation (13.72) in Equation (13.73), we obtain

$$\mathcal{S}_a(\omega_x) = \int_{-\infty}^{\infty}\int_{-\infty}^{\infty} a(x, y) \cdot e^{-j\omega_x x} dx\, dy . \tag{13.74}$$

Changing the order of integration in Equation (13.74), we can write

$$\mathcal{S}_a(\omega_x) = \int_{-\infty}^{\infty}\left\{\int_{-\infty}^{\infty} a(x, y)\, dy\right\} \cdot e^{-j\omega_x x} dx . \tag{13.75}$$

Let us denote the integral in braces in Equation (13.75) as $s_a(x)$. It is easy to see that the central cross section of the power spectral density $\mathcal{S}_a(\omega_x)$ is the Fourier transform

$$s_a(x) = \int_{-\infty}^{\infty} a(x, y)\, dy. \tag{13.76}$$

The function $s_a(x)$, being the result of the transform given by Equation (13.76), is called the projection. We can rewrite Equation (13.76) in the following form:

$$s_a(i) = \sum_{j=1}^{N} a(i, j), \tag{13.77}$$

where N is the number of elements in the expansion of the object image $a(i, j)$ along the axis j. Reference to Equation (13.77) shows that the projection $s_a(i)$ is the result of summing the brightness characteristics of elements of the object image $a(i, j)$ over the coordinate j. Thus, if the object image $a(i, j)$ is given in the right matrix form, we can obtain the projection of this object image at the angle $\theta = 0°$ by summing the brightness characteristics of the elements of the matrix by columns. The effectiveness of the computer-calculated projection is evident. In accordance with Equation (13.75), it is necessary to determine the Fourier transform of the projection to obtain the central cross section of the power spectral density of the object image. The relationship between the cross sections and projections of the power spectal density of the object image is defined by the theorem about projections and cross sections of multidimensional signals, the meaning of which is the following: $(N-1)$-dimensional Fourier transform with respect to the projection $s_{ax1}(x_2, \ldots, x_N)$ is the central cross section of the N-dimensional Fourier transform with respect to the function $f(x_1, \ldots, x_N)$ or, in other words, the power spectral density of the projection is the cross section of the power spectral density of the object image.[17] Equations (13.72)–(13.75) are variants of this theorem about projections and cross sections in the case of two-dimensional signals. A function of the cross section of the power spectral density and projections (the Fourier transform) is used in tomography and molecular biology to reconstruct multidimensional signals using their projections.[17] Projections are also used to solve pattern recognition problems.[4]

The formula in Equation (13.76) defines the projection of the object image at $\theta = 0°$. Let us define the projection $s_a(x, \theta)$ for the arbitrary value of the angle θ within the limits of the interval $[0, 2\pi]$.Note that the angle θ is a parameter, not a variable. For this purpose, we can introduce new variables or rotate the object image $a(x, y)$ using the previous coordinate system. Rotating the object image, we can write

$$s(x,\theta) = \int_{-\infty}^{\infty} a\big[(x,y)\mathbf{A}\big]\, dy \tag{13.78}$$

where

$$\mathbf{A} = \begin{Vmatrix} \cos\theta & \sin\theta \\ -\sin\theta & \cos\theta \end{Vmatrix}. \tag{13.79}$$

It is well known that if $a(x, y)$ and $S_a(\omega_x, \omega_y)$ form a pair of Fourier transforms, then $a[(x, y)\mathbf{A}]$ and $S_a[(\omega_x, \omega_y)\mathbf{A}]$ are also a pair of Fourier transforms, if the matrix $\mathbf{A}$ is orthogonal, i.e., $\mathbf{A}^T = \mathbf{A}^{-1}$.[18] The rotation matrix $\mathbf{A}$ is orthogonal. Because of this, the rotation of the cross section of the power spectral density of the object image by the angle θ corresponds to the rotation of the projection $s(x, \theta)$ by the same angle. Figure 13.7 shows us the determination of the projection $s(x, \theta)$ for an arbitrary angle θ. The elements of the projection $s(k, \theta)$ are equal to the sum of the brightness characteristics of the object image elements $a(i, j)$ within the limits of the region dy or of the beam width $b(k, \theta)$, which is limited by lines y_k and y_{k-1} or x_k and x_{k-1}. At $\theta = 0°$ and $\theta = 90°$, the set of lines takes the following form:

$$x_k = k + 0.5, \quad k = 0, 1, 2, \ldots, N. \tag{13.80}$$

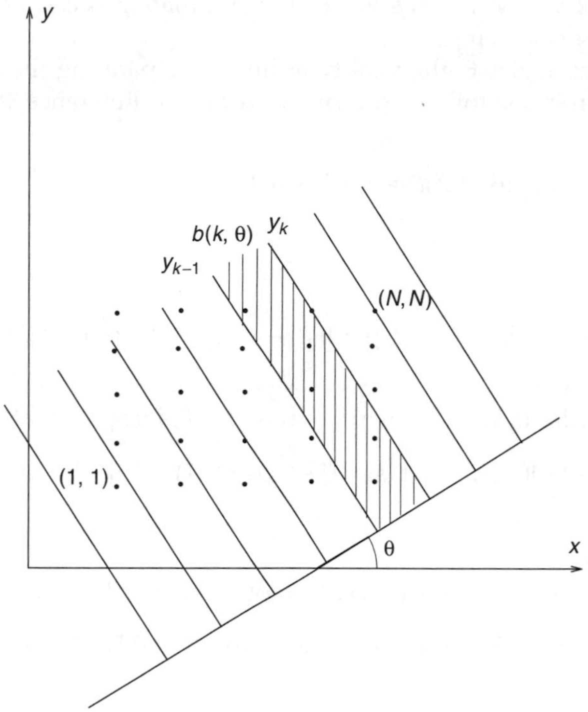

FIGURE 13.7
Projection determination.

The elements of the projection are determined by

$$s(k,\theta) = \sum_{j=1}^{N} a(i,j), \tag{13.81}$$

where $k = 1, 2, \ldots, N$; $x_{k-1} \le i \le x_k$, i.e., the summation is carried out by columns of object image elements. At $\theta = 90°$ and $\theta = 270°$ the totality of lines takes the form

$$y_k = k + 0.5, \quad k = 1, 2, \ldots, N. \tag{13.82}$$

The elements of the projection are determined by

$$s(k,\theta) = \sum_{i=1}^{N} a(i,j), \tag{13.83}$$

where $k = 1, 2, \ldots, N$; $y_{k-1} \le j \le y_k$, i.e., the summation is carried out by rows of object image elements.

For all other angles θ, the totality of lines y_k separating the beams $b(k, \theta)$ can be written in the following form according to Reference 19:[19]

$$y_k = x\ \mathrm{tg}\ (\theta + 0.5\pi) + y_0 + k\,|\sin\theta|^{-1}, \quad k = 0, 1, 2, \ldots, C_\theta, \tag{13.84}$$

where

$$y_0 = 0.5(N+1) - L\,|\sin\theta|^{-1} - 0.5(N+1)\ \mathrm{tg}\ (\theta + 0.5\pi); \tag{13.85}$$

$$L = \begin{cases} 0.5N + Int[0.5(N-1)\cdot(|\sin\theta| + |\cos\theta| - 1) + 0.5], & \text{at} \quad [.] > Int[.], \\ 0.5N + Int[0.5(N-1)\cdot(|\sin\theta| + |\cos\theta| - 1) + 0.5] - 1, & \text{at} \quad [.] = Int[.], \end{cases} \tag{13.86}$$

$$C_\theta = \begin{cases} N + 2Int[0.5(N-1)\cdot(|\sin\theta| + |\cos\theta| - 1) + 0.5], & \text{at} \quad [.] > Int[.], \\ N + 2Int[0.5(N-1)\cdot(|\sin\theta| + |\cos\theta| - 1) + 0.5] - 1, & \text{at} \quad [.] = Int[.], \end{cases} \tag{13.87}$$

where $Int[.]$ is the integer part of [.]. The projection elements for an arbitrary angle θ are determined by

$$s(k,\theta) = \sum_{(i,j)\in b(k,\theta)} a(i,j)\ , \quad k = 1, 2, \ldots, C_\theta. \tag{13.88}$$

As is well known from the theory of multidimensional signals, there is a finite set of rational angles $\{\theta_1, \ldots, \theta_n\}$ among infinite sets of projection angles, the use of which ensures an unambiguous reconstruction of signals with minimal numbers of projections. Therefore, we define these projection angles. The rational angles can be determined by[20]

$$\theta_q = \mathrm{arctg}\ \frac{T_q}{Q_q}, \tag{13.89}$$

where $q = 1, \ldots, n$; T and Q are integer coprime numbers satisfying the conditions $|T| \le N$ and $|Q| \le N$, and n is the number of projection angles.

The rational angles are within the limits of the interval $[-0.5\pi, 0.5\pi]$. Because the projections are periodic functions of the angle θ with the period π, i.e.,

$$s(u,\theta) = s\left[(-1)^k u, \theta + k\pi\right], \tag{13.90}$$

we can use the interval [0, π] to determine the projections. Note that the rational angles θ_{2q-1} and $\theta_{2q}(\theta = 1, \ldots, n)$ are shifted with respect to each other by 0.5π. This allows us to simplify Equation (13.69) and Equation (13.71) in the definition of the estimation λ_x and λ_y in the course of determining the criterial correlation function of projections at the angles θ_{2q-1} and θ_{2q}. In this case, the estimations of the object image shift along the q-th pair of orthogonal projections in the Cartesian coordinate system and take the form

$$\lambda_x(q) = \alpha_{\theta_{2q-1}} \cos\theta_{2q-1} - \alpha_{\theta_{2q}} \sin\theta_{2q-1}, \tag{13.91}$$

$$\lambda_y(q) = \alpha_{\theta_{2q-1}} \sin\theta_{2q-1} + \alpha_{\theta_{2q}} \cos\theta_{2q-1}. \tag{13.92}$$

Equation (13.91) and Equation (13.92) are shown in Figure 13.8, in which an example of the processing of the object image given by the point $a(x, y) = \delta(x, y)$ is illustrated. Final estimations of the shifts λ_x^* and λ_y^* of the object image are formed due to statistical averaging of the estimations $\lambda_x(q)$ and $\lambda_y(q)$ for all $q = 1, \ldots, n$.

The performed analysis of possibilities of using the quasioptimal generalized image processing algorithm based on the generalized approach to signal processing in the presence of noise shows the appropriateness of the

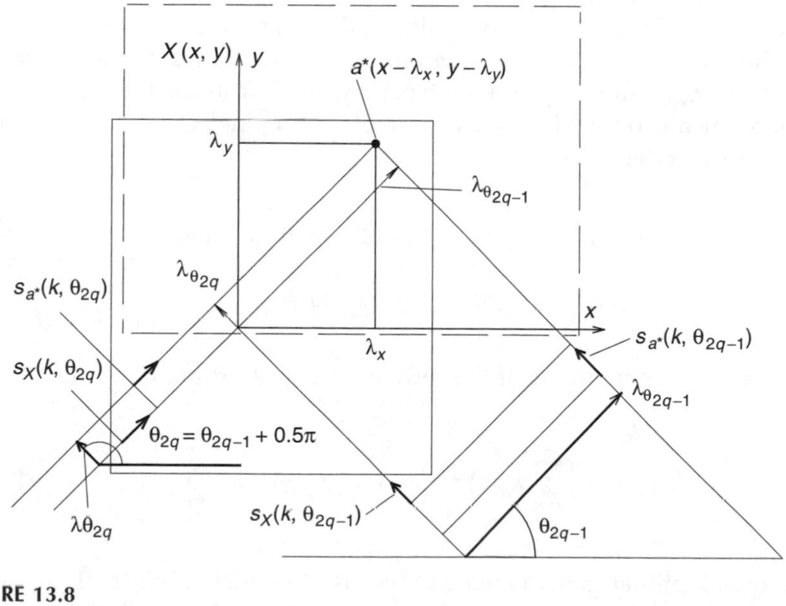

FIGURE 13.8
An example of image processing.

sequence of steps discussed in the following text. The first is forming an ensemble from the n projections $s_a(k, \theta_q)$ of the model image and $s_X(k, \theta_q)$ of the moving image:

$$s(k,\theta_q) = \sum_{(i,j)\in b(k,\theta_q)} a(i,j). \tag{13.93}$$

The elements $(i, j) \in b(k, \theta_q)$ are given by Equation (13.84) — Equation (13.87). The projection angles θ_q are defined in accordance with Equation (13.89). The second is the determination of the central cross sections of the object image power spectral densities

$$\mathcal{S}(\omega_u,\theta_q) = \sum_{k=1}^{N} s(k,\theta_q)\cdot e^{-j\frac{2\pi}{N}k\omega_u} = \left|\mathcal{S}(\omega_u,\theta_q)\right|\cdot e^{j\phi(\omega_u,\theta_q)}. \tag{13.94}$$

The third is the determination of the correlation functions of the moving and model image projections carried out according to

$$R(k,\theta_q) = \mathcal{F}^{-1}\{2\left|\mathcal{S}_X(\omega_u,\theta_q)\cdot\mathcal{S}^*_{a^*}(\omega_u,\theta_q)\right|^{(1-\ell)}\cdot e^{j[\varphi_X(\omega_u,\theta_q)-\varphi_{a^*}(\omega_u,\theta_q)]}\}, \tag{13.95}$$

where $\mathcal{F}^{-1}$ is the inverse Fourier transform, $0 \le \ell < 1$. The fourth is that in the searching condition, we have $\ell = 1$; in the nonsearching condition, we have $0 \le \ell < 1$. Using the maximal value of $R(k, \theta_q)$ for the searching condition, the estimations α_{θ_q} of the shift along the direction θ_q are defined. Fifth, the estimations $\lambda_x(q)$ and $\lambda_y(q)$ of the object image shift in the Cartesian coordinate system are defined by each pair of orthogonal cross sections of the power spectral densities

$$\begin{aligned}\lambda_x(q) &= \alpha_{\theta_{2q-1}}\cos\theta_{2q-1} - \alpha_{\theta_{2q}}\sin\theta_{2q-1} \quad \text{and}\\ \lambda_y(q) &= \alpha_{\theta_{2q-1}}\sin\theta_{2q-1} - \alpha_{\theta_{2q}}\cos\theta_{2q-1}.\end{aligned} \tag{13.96}$$

Sixth, the final estimations of the moving image shift are determined by

$$\lambda^*_x(q) = n^{-1}\sum_{q=1}^{n}\lambda_x(q) \quad \text{and} \quad \lambda^*_y(q) = n^{-1}\sum_{q=1}^{n}\lambda_y(q). \tag{13.97}$$

The quasioptimal generalized receiver is a multichannel tracking device constructed based on the generalized approach to signal processing in the presence of noise. Each channel of the receiver is processed by the central cross sections of the power spectral densities of moving and model images. The discussed quasioptimal generalized image processing algorithm allows us to reduce computational costs significantly. This effect is caused by reducing the initial information $\frac{N}{n}$ times due to formation of one-dimensional projections and use of the fast Fourier transform. The coefficient of the computational cost reduction in comparison with the universally adopted generalized signal processing algorithm can be determined in the following form:

$$\chi = \frac{N^3}{n[12\log_2 N + 0.1(N-1)]} . \qquad (13.98)$$

To determine the coefficient χ, it is worthwhile to use the procedure discussed in Clary and Russell.[21]

13.3 The Classical Generalized Image Processing Algorithm

In the use of the classical generalized image processing algorithm based on the generalized approach to signal processing in the presence of noise, we use the criterial correlation function. Image preprocessing is carried out by the filter with the transfer function $C(\omega_x, \omega_y) = 1$. In this case, we should define the correlation function maximum of the generalized receiver output statistic (see Section 12.3). The disadvantage of the use of the classical generalized image processing algorithm is the high computational cost because the determination of the correlation function of the generalized receiver output statistic is carried out for all possible relative shifts of observed object images. The total number of product operations in processing object images with dimensions $M \times M$ and $N \times N$ can be determined in the following form:[21]

$$\mathcal{N}_1 = N(N+1) \cdot M(M+1) . \qquad (13.99)$$

This is a very large number.

In the use of the Fourier transform with respect to the classical generalized image processing algorithm, we should use the criterial correlation function given by Equation (13.17). In some cases, we can obtain definite advantages, for example, if a navigational system uses optimal analog devices. In the use of the fast Fourier transform, the number of product operations can be determined by[21]

$$\mathcal{N}_2 = 12MN\log_2 N + 4M, \quad M \geq N, \qquad (13.100)$$

TABLE 13.1

Number of Product Operations

Object Image Dimensions	Analog Generalized Image Processing Algorithm	Generalized Image Processing Algorithm With Fast Fourier Transform
4×4	400	392
8×8	5184	2320
16×16	73,984	12,320
32×32	1,115,136	61,504
64×64	17,305,600	295,042
128×128	272,646,144	1,376,518
256×256	4,328,587,264	6,291,960

and we are able to decrease computational costs significantly.

The number of product operations in the use of the classical generalized image processing algorithm based on the generalized approach to signal processing in the presence of noise both for signal processing in time and in the spectral range using the fast Fourier transform is shown in Table 13.1. Table 13.1 shows that the use of the fast Fourier transform allows us to decrease the computational cost significantly during image processing.

Let us estimate the probability of true and false location of object images with reference to a control point. Let us consider the case of the matched one-dimensional object images

$$f(i)\,, \quad i = 1, 2, \ldots, N \quad \text{and} \quad g(i)\,, \quad i = 1, 2, \ldots, M\,, \tag{13.101}$$

where $M > N$ and $g(i + \bar{r}) = f(i)$, $i = 1, 2, \ldots, N$, the parameter $\bar{r}$ corresponds to complete matching between the model image and the moving image. Let us introduce the following notation

$$\sigma_f^2 = \frac{1}{N}\sum_{i=1}^{N}\left\{f(i) - \frac{1}{N}\sum_{j=1}^{N} f(j)\right\}^2; \tag{13.102}$$

σ_n^2 is the variance of the Gaussian background noise with zero mean;

$$\sigma_g^2(\bar{r}) = \frac{1}{N}\sum_{i=1}^{N}\left\{g(i+\bar{r}) - \frac{1}{N}\sum_{j=1}^{N} g(j+\bar{r})\right\}^2; \tag{13.103}$$

$$\rho_{fg}(\bar{r}) = \frac{\frac{1}{N}\sum_{j=1}^{N}\left\{f(j) - \frac{1}{N}\sum_{i=1}^{N} f(i)\right\}\cdot\left\{g(j+\bar{r}) - \frac{1}{N}\sum_{i=1}^{N} g(i+\bar{r})\right\}}{\sigma_f \sigma_g(\bar{r})} \tag{13.104}$$

is the correlation coefficient in the course of matching the model image and the moving image, where $\overline{r} = 0, 1, \ldots, M - N$.

The low boundary of the true location probability of the object image with reference to the control point in the use of the classical generalized image processing algorithm based on the generalized approach to signal processing in the presence of noise can be defined using the procedure discussed in Stubberud[22]

$$P_{tr} = 1 - \sum_{i=1}^{N} \left[1 - P\left(z < \tfrac{M\{Y_i\}}{\sigma\{Y_i\}}\right)\right], \tag{13.105}$$

where z is the Gaussian random variable with zero mean and the variance equal to unity;

$$M\{Y_i\} = \left[1 - \rho_{fg}(\overline{r} + i)\right]\left(1 - \tfrac{\sigma_n^2}{\sigma_f^2}\right); \tag{13.106}$$

$$\sigma\{Y_i\} = \sqrt{\tfrac{1}{N}\left\{0.5[1 - \rho_{fg}(\overline{r} + i)]^2 \tfrac{\sigma_n^4}{\sigma_f^4} + [1 - \rho_{fg}(\overline{r} + i)]\tfrac{\sigma_n^2}{\sigma_f^2}\right\}}, \tag{13.107}$$

$i = -\overline{r}, -\overline{r} + 1, \ldots, M - N - \overline{r}, i \neq 0$. The probability of false location of the object image with reference to the control point can be defined based on the procedure discussed in Rockmore[23] in the following form. It can be considered as the totality of the false location probabilities of the object image with reference to the control point $P'_{fr}(\Delta x, \Delta y)$ for all possible shifts of the object image

$$P_{fr} = 1 - \prod_{\Delta x, \Delta y}^{M_i^2} \left[1 - P'_{fr}(\Delta x, \Delta y)\right], \tag{13.108}$$

where M_i^2 is the number of possible relative shifts.

The determination of the probability of the event that the correlation function $R(\Delta x, \Delta y)$ formed at the generalized receiver output does not exceed the threshold K_g can be carried out using the Edgeworth-series expansion:

$$\begin{aligned}
\Pr\{R(\Delta x, \Delta y) \leq K_g\} &= 1 - T(x) + \varphi(x)\left\{N_i^{-1}[0.667\Gamma_3(x^2 - 1)]\right. \\
&+ N_i^{-2}[0.041\Gamma_4(x^3 - 3x) + 0.0014\Gamma_3^2(x^5 - 10x^3 + 15x)] \\
&+ N_i^{3}[0.008\Gamma_5(x^4 - 6x^2 + 3) + 0.007\Gamma_3\Gamma_4(x^6 - 15x^4 + 45x^2 - 15) \\
&\left. + 0.00013\Gamma_3^3(x^8 - 28x^6 + 210x^4 - 420x^2 + 105]\right\},
\end{aligned} \tag{13.109}$$

where

$$T(x) = \frac{1}{\sqrt{2\pi}} \int_{-\infty}^{\infty} \varphi(t)dt \int_{-\infty}^{\infty} e^{-0.5t^2} dt; \tag{13.110}$$

$$x = \frac{K_g - N_i^2 \mu_1}{N_i \sqrt{\mu_2}}; \tag{13.111}$$

$$\Gamma_3 = \frac{\mu_3}{\sqrt{\mu_2^3}}; \quad \Gamma_4 = \frac{\mu_4 - 3\mu_2^2}{\mu_2^2}; \quad \text{and} \quad \Gamma_5 = \frac{\mu_5 - 10\mu_2\mu_3}{\sqrt{\mu_2^5}\, N_i^2} \tag{13.112}$$

are the average number of series-expansion elements of the object image, μ_1 is the average value of $R(\Delta x, \Delta y)$, and μ_n is the central moment of the n-th order of the correlation function $R(\Delta x, \Delta y)$.

The main peculiarity of this technique is the following. First, there is no limitation on the shape of the correlation function in the case of the Gaussian probability distribution density. Second, because the false location probability of the object image with reference to a control point is determined for each relative shift of the object image, we can take into consideration variations of the statistical characteristics of the object image as a function of the shift.

Let us consider the classical generalized image processing algorithm based on the generalized approach to signal processing in the presence of noise. We can represent the moving and model images in the following form:

$$X(x) = P_d(x) + n(x) \quad \text{and} \quad a(x) = P(x), \quad x \in M, \tag{13.113}$$

where $x = (x_1, x_2)$ is a two-dimensional variable. The differences between the moving image and the model image are caused by the additive noise $n(x)$ and geometrical distortions, which can be presented by the affine transform of object image coordinates[24–26]

$$P_d(x) = P(\mathbf{A}x + t_0), \tag{13.114}$$

where

$$\mathbf{A} = \alpha \begin{Vmatrix} \cos\theta & \sin\theta \\ -\sin\theta & \cos\theta \end{Vmatrix} \tag{13.115}$$

is the matrix of the relative shift of the object image by the angle θ, and α is the coefficient of changes in scale.

The average maximum of the correlation function of the generalized receiver output statistic, which is normalized to the average maximum of this correlation function if noise is absent, depends only on the brightness characteristics of distortions and at low values of θ and $|1-\alpha|$ takes the following form:

$$d = \sqrt{(1-\alpha)^2 + \theta^2}. \qquad (13.116)$$

The normalized average maximum of the correlation function of the generalized receiver output statistic as a function of the normalized width η of the object image is shown in Figure 13.9 for various values of the distortions d. The quality of the classical generalized image processing algorithm is characterized by the ratio between the average maximum and the mean square deviation of the correlation function in the range of side lobes

$$\kappa = \frac{M\{R(t_0)\}}{\sqrt{\sigma^2\{R(t)\}}}. \qquad (13.117)$$

The dependence κ on η is shown in Figure 13.11. Reference to Figure 13.10 shows that there is an optimal dimension of the model image for various values of distortions. The dependence of the false location probability of the

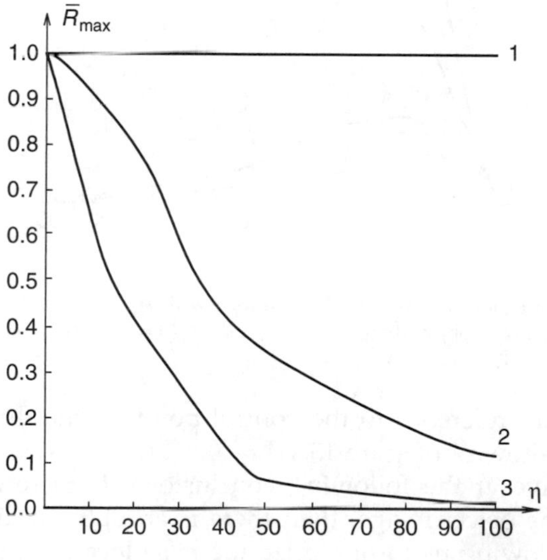

FIGURE 13.9
The normalized average peak of the correlation function as a function of the normalized object image width: (1) $d = 0$ ($\theta = 0°$); (2) $d = 0.087$ ($\theta = 5°$); (3) $d = 0.174$ ($\theta = 10°$).

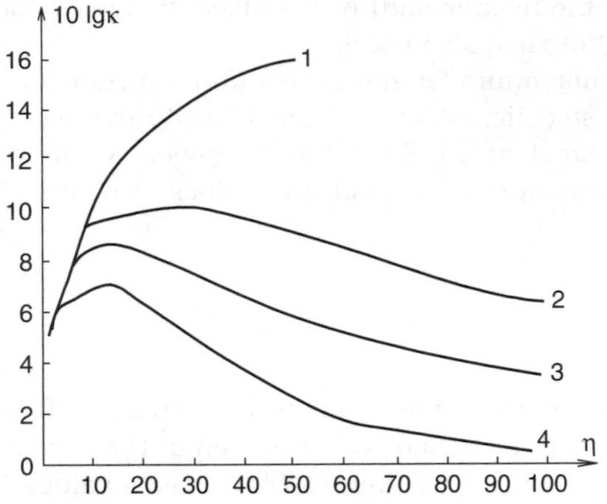

FIGURE 13.10
The dependence r on the normalized object image width: (1) $d = 0$ ($\theta = 0°$); (2) $d = 0.087$ ($\theta = 5°$); (3) $d = 0.122$ ($\theta = 7°$); (3) $d = 0.174$ ($\theta = 10°$).

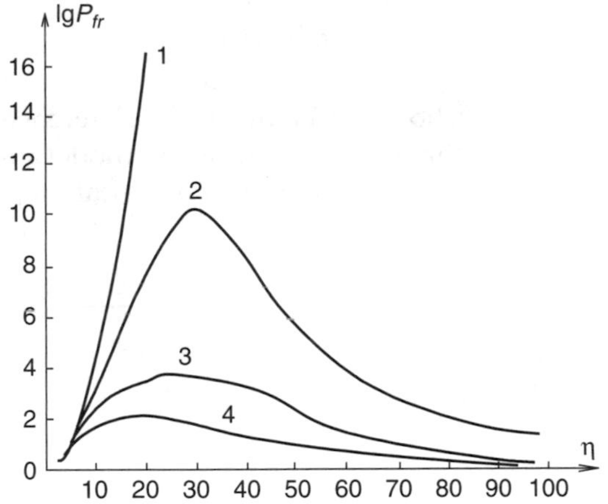

FIGURE 13.11
The probability of false location of the object image with reference to the control point as a function of the normalized object image width: (1) $d = 0$ ($\theta = 0°$); (2) $d = 0.087$ ($\theta = 5°$); (3) $d = 0.122$ ($\theta = 7°$); (3) $d = 0.174$ ($\theta = 10°$).

object image with reference to the control point on the dimensions of the object image is shown in Figure 13.11.

Thus, we arrive at the following conclusions. If there are geometrical distortions of the object image, then there is an optimal dimension of the model image allowing us to minimize the false location probability of the object image with reference to the control point. This optimal dimension is proportional to the correlation function width and decreases with an increase in geometrical distortions. In the general case, the minimization of the false location probability of the object image with reference to the control point must be carried out by the choice of both the dimensions and the shape and orientation of the model image. For the discussed case, we assume that the correlation function of the moving image takes a circular symmetry, because of which the optimal model image must be of square form. There are analogous dependencies in the minimization of the mean square deviation of matching between the moving and model images. However, with the given level of distortions, the dimensions of the object image in which the error of matching is minimal, is less than the dimensions of the object image that are required to minimize the false location probability of the object image with reference to the control point.

13.4 The Difference Generalized Image Processing Algorithm

The difference generalized image processing algorithm based on the generalized approach to signal processing in the presence of noise uses the criterial function in the form $R_{2.0} \ldots R_{2.5}$ (see Section 13.12). Image preprocessing is carried out by the filter with the transfer function $C(\omega_x, \omega_y) = 1$, i.e., the difference generalized image processing algorithm is based on the element-by-element determination of differences of the object image brightness characteristics. Let us consider the algorithm given by Equation (13.10) in more detail. Note that using the analogous formula, we can define the distance in Euclidean space. The formula given by Equation (13.10) can be rewritten in the following form:

$$R_{2.1}(m,n) = \frac{1}{N^2}\sum_{i=1}^{N}\sum_{j=1}^{N}\left[a^{*2}(i,j) - 2a^{*}(i,j)X(i-m,j-n) + X^2(i-m,j-n)\right], \tag{13.118}$$

where summation is carried out over all i and j, in which the arguments of the moving image are in the domain of definition. The first term in Equation (13.118) is constant for all m and n and can be included into the threshold value in the decision making. In terms of the reference noise $\eta(i, j)$ forming at the output of the additional filter of the generalized receiver, the criterial function can be determined in the following form:

$$-R_{2.1}(m,n) = \frac{1}{N^2}\sum_{i=1}^{N}\sum_{j=1}^{N}\left[2a^{*}(i,j)X(i-m,j-n) - X^2(i-m,j-n) + \eta^2(i,j)\right]. \tag{13.119}$$

The formula in Equation (13.119) is the criterial correlation function of the classical generalized image processing algorithm based on the generalized approach to signal processing in the presence of noise (see Chapter 12). Consequently, the main characteristics of the difference generalized image processing algorithm are not significantly different from the main characteristics of the classical generalized image processing algorithm.

13.5 The Generalized Phase Image Processing Algorithm

In the use of the generalized phase image processing algorithm, the criterial correlation function can be determined in the following form:

$$
\begin{aligned}
R(k) &= \frac{1}{N}\sum_{m=0}^{N-1}\{\{2|S_{a^*}(m)|\cdot|S_X(m)|\}^{(1-L)}\cdot e^{j[2\pi mkN^{-1}+\varphi_{a^*}(m)-\varphi_X(m)]} \\
&= \{|S_X(m)|\cdot|S_X^*(m)|\}^{(1-L)} - \{|S_\eta(m)|\cdot|S_\eta^*(m)|\}^{(1-L)} \\
&\times e^{j[2\pi mkN^{-1}+\varphi_X(m)-\varphi_{a^*}(m)]}\} \\
&\cong \frac{1}{N}\sum_{m=0}^{N-1}\{2|S_{a^*}(m)|\cdot|S_X(m)|\}^{(1-L)}\cdot e^{j[2\pi mkN^{-1}+\varphi_{a^*}(m)-\varphi_X(m)]},
\end{aligned}
\tag{13.120}
$$

where

$$|S(m)| = \sqrt{\mathrm{Re}\{S(m)\}^2 + \mathrm{Im}\{S(m)\}^2}; \tag{13.121}$$

$$\varphi(m) = \mathrm{arctg}\,\frac{\mathrm{Im}\{S(m)\}}{\mathrm{Re}\{S(m)\}}; \tag{13.122}$$

$$\mathrm{Re}\{S(m)\} = \sum_{k=0}^{N-1} S(k)\cos\frac{2\pi km}{N}; \tag{13.123}$$

$$\mathrm{Im}\{S(m)\} = -\sum_{k=0}^{N-1} S(k)\sin\frac{2\pi km}{N}; \tag{13.124}$$

N is the number of elements of image expansion, and $0 \le L \le 1$. At $L = 0$, we obtain the classical generalized image processing algorithm:

$$R(\lambda_x) = \mathcal{F}^{-1}\{2|S_{a^*}(\omega_x)|\cdot|S_X^*(\omega_x)|\cdot e^{j[\varphi_{a^*}(\omega_x)-\varphi_X(\omega_x)]}$$
$$-|S_X(\omega_x)|\cdot|S_X^*(\omega_x)|\cdot e^{j[\varphi_X(\omega_x)-\varphi_X(\omega_x)]}$$
$$+|S_\eta(\omega_x)|\cdot|S_\eta^*(\omega_x)|\cdot e^{j[\varphi_X(\omega_x)-\varphi_X(\omega_x)]}\} = \mathcal{F}^{-1}\{2S_{a^*}(\omega_x)\cdot S_X^*(\omega_x)\}. \tag{13.125}$$

At $L = 1$, Equation (13.120) corresponds to the generalized phase image processing algorithm consisting of the following operations: the definition of the discrete Fourier transform with respect to the moving and model images, forming of the phase difference matrix at each spatial frequency by the definition of the mutual energy spectrum divided by the absolute value, the definition of the inverse Fourier transform with respect to the normalized energy spectrum, and the definition of the criterial correlation function maximum. At $0 < L < 1$, Equation (13.120) corresponds to the modified generalized phase image processing algorithm.

The total use of the information included in the phase component of the complex power spectral density of the signal and the partial use or modification of the amplitude component information of the signal are characteristic of the generalized phase image processing algorithm. The phase component of the complex power spectral density of the signal is caused, in particular, by the fact that all information with respect to the relative shift of the object image is concentrated in the phase component. For example, let us consider the object image $a(x, y)$ having the power spectral density $S_a(\omega_x, \omega_y)$. The power spectral density of the object image $a(x, y)$ shifted by the value λ_x along the x axis and by the value λ_y along the y axis can be written in the following form:

$$\mathcal{F}[a(x-\lambda_x, y-\lambda_y)] = \int_{-\infty}^{\infty}\int_{-\infty}^{\infty} a(x-\lambda_x, y-\lambda_y)\cdot e^{-j(\omega_x x+\omega_y y)}dx\,dy\ , \tag{13.126}$$

where $\mathcal{F}$ is the symbol of the Fourier transform. Introduce the variables $u = x - \lambda_x$ and $\upsilon = y - \lambda_y$. Then, Equation (13.126) takes the form

$$\mathcal{F}[a(x-\lambda_x, y-\lambda_y)] = \int_{-\infty}^{\infty}\int_{-\infty}^{\infty} a(u,\upsilon)\cdot e^{-j[\omega_x(u+\lambda_x)+\omega_y(\upsilon+\lambda_y)]}\,du\,d\upsilon$$
$$= S_a(\omega_x,\omega_y)\cdot e^{-j(\omega_x\lambda_x+\omega_y\lambda_y)}. \tag{13.127}$$

Reference to Equation (13.127) shows us that only the phase component of the power spectral density of the object image has been changed.

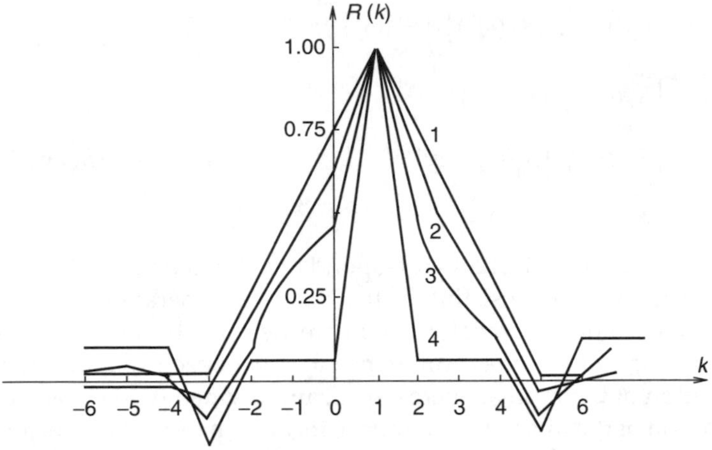

FIGURE 13.12
The normalized criterial correlation functions: (1) $L = 0$; (2) $L = 0.2$; (3) $L = 0.6$; (4) $L = 1.0$.

The continuous narrowing of the criterial correlation function corresponds to the variations of the parameter L from zero to unity in Equation (13.120). If the object image has infinite dimensions, the criterial correlation function is transformed into the delta function at $L = 1$:

$$R(\Delta\lambda) = \mathcal{F}^{-1}\{e^{j\Delta\varphi}\} = \delta(\Delta\lambda). \tag{13.128}$$

The normalized criterial correlation functions for two square waveform signals shifted relative to each other by one discrete are shown in Figure 13.12. Reference to Figure 13.12 shows that at $L = 1$ the criterial correlation function has a sharp peak at the matching point of the object images.

In narrowing the spatial correlation function, the steepness of the discrimination characteristic defined by the derivative of the spatial correlation function of the moving $X(x - \lambda_x)$ and model $a^*(x - \lambda_x^*)$ images is increased:

$$D(\Delta\lambda) = -\frac{dR_{a^*}(\Delta\lambda)}{d(\Delta\lambda)}, \qquad \Delta\lambda = \lambda_x - \lambda_x^*. \tag{13.129}$$

It is well known that estimations of the mutual positions of the moving and model images and the steepness of the discrimination characteristic are inversely proportional to each other. Therefore, the dependence of the steepness of the discrimination characteristic on the parameter L is of great interest.

Let us assume that the mutual power spectral density of object images $|S_a(\omega_x)|^2$ takes a bell-shaped form:[27]

$$|S_a(\omega_x)|^2 = e^{-0.5k^2\omega_x^2}. \tag{13.130}$$

In this case, the criterial correlation function has the form

$$R(\Delta\lambda) = \mathcal{F}\{e^{-0.5k^2(1-L)\omega_x^2}\} = \frac{e^{-\frac{\Delta\lambda^2}{2k^2(1-L)}}}{k\sqrt{2\pi(1-L)}}. \tag{13.131}$$

Substituting Equation (13.131) in Equation (13.129), we obtain

$$D(\Delta\lambda) = -\frac{d}{d(\Delta\lambda)}\left\{\frac{e^{-\frac{\Delta\lambda^2}{2k^2(1-L)}}}{k\sqrt{2\pi(1-L)}}\right\} = \frac{\Delta\lambda \cdot e^{-\frac{\Delta\lambda^2}{2k^2(1-L)}}}{k^3\sqrt{2\pi(1-L)^3}}. \tag{13.132}$$

The ratio between the steepness of the discrimination characteristics at the condition $0 \leq L \leq 1$ (the generalized image processing algorithm) and the condition $L = 0$ (the classical generalized processing algorithm) takes the following form:

$$Q = \frac{1}{\sqrt{(1-L)^3}}. \tag{13.133}$$

As shown in Vasilenko,[28] if the parameter L is varied from 0.2 to 0.4, the variance of the estimation error of mutual position between the moving and model images can be decreased by 1.5 to 2 times.

13.6 The Generalized Image Processing Algorithm: Invariant Moments

The detection of the invariable peculiarities of object images is a very important landmark in pattern recognition theory. The invariable features of object images allow us to identify the object image independently of its location, dimensions, and orientation. The theory of algebraic invariants is the basis to detect the invariable or invariant peculiarities of object images.

The theory of algebraic invariants studies a class of algebraic functions that do not vary under the transformations of coordinates. The first attempt to use invariant moments with the pattern recognition of two-dimensional object images is discussed in Hu.[29] The discussed technique is widely used in the recognition of object images on photos, in the recognition of aircraft in optical and radar range, and in searching moving clouds by satellite. The use of invariant moments in generalized image processing algorithm is preferred for the case when the image is the totality of the object and background, and therefore the transition to the background should be sufficiently sharp.

The central moments of the third order take the following form according to Equation (12.143):

$$\begin{cases} \mu_{00} = m_{00};\ \mu_{11} = m_{11} - \overline{X}m_{10}; \\ \mu_{10} = 0;\ \mu_{30} = m_{30} - 3\bar{a}m_{20} + 2m_{10}\overline{a^2}; \\ \mu_{01} = 0;\ \mu_{12} = m_{12} - 2\overline{X}m_{11} - \bar{a}m_{02} + 2\overline{X^2}m_{01}; \\ \mu_{20} = m_{20} - \bar{a}m_{10};\ \mu_{21} = m_{21} - 2\bar{a}m_{11} - \overline{X}m_{20} = 2\overline{a^2}m_{01}; \\ \mu_{02} = m_{02} - \overline{X}m_{01};\ \mu_{03} = m_{03} - 3\overline{X}m_{02} + 2\overline{X^2}m_{01}. \end{cases} \tag{13.134}$$

The central moments are invariant with respect to the shift of object images. The normalized central moments can be determined in the following form:

$$\nu_{pq} = \frac{\mu_{pq}}{\mu_{00}^{y}}, \quad \text{where} \quad y = 0.5(p+q) \quad \text{and} \quad p+q = 2, 3, \ldots,. \tag{13.135}$$

Using the central moments of the second and third order, we can generate a system consisting of seven invariant moments:[30]

$$\begin{cases} \varphi_1 = \nu_{20} + \nu_{02}; \\ \varphi_2 = (\nu_{20} + \nu_{02})^2 + 4\nu_{11}^2; \\ \varphi_3 = (\nu_{30} - 3\nu_{12})^2 + (3\nu_{21} + \nu_{03})^2; \\ \varphi_4 = (\nu_{30} + \nu_{12})^2 + (\nu_{21} + \nu_{03})^2; \\ \varphi_5 = (\nu_{30} - 3\nu_{12})(\nu_{30} + \nu_{12})[(\nu_{30} + \nu_{12})^2 - 3(\nu_{21} + \nu_{03})^2] \\ \quad + (3\nu_{21} - \nu_{03})(\nu_{21} + \nu_{03})[3(\nu_{30} + \nu_{21})^2 - (\nu_{21} + \nu_{03})^2]; \\ \varphi_6 = (\nu_{20} - \nu_{02})[(\nu_{30} + \nu_{12})^2 - (\nu_{21} + \nu_{03})^2] + 4\nu_{11}(\nu_{30} + \nu_{12})(\nu_{21} + \nu_{03}); \\ \varphi_7 = (3\nu_{12} - \nu_{30})(\nu_{30} + \nu_{12})[(\nu_{30} + \nu_{12})^2 - 3(\nu_{21} + \nu_{03})^2] \\ \quad + (3\nu_{21} - \nu_{03})(\nu_{21} + \nu_{03})[3(\nu_{30} + \nu_{12})^2 - (\nu_{21} + \nu_{03})^2]. \end{cases} \tag{13.136}$$

This system of moments is invariant with respect to the shift, rotation up to 45°, and double change of image scale.[29,30]

The criterial correlation function can be determined in the following form:

$$R(\Delta x, \Delta y) = \frac{\sum_{i=1}^{7} M_i N_i(\Delta x, \Delta y)}{\sqrt{\sum_{i=1}^{7} M_i^2 \sum_{i=1}^{7} N_i^2(x, y)}}, \tag{13.137}$$

where M_i is the i-th moment of the first object image, and N_i is the i-th moment of the second object image shifted by the value $(\Delta x, \Delta y)$. The probability of true location of the object image with reference to a control point can be determined by

$$P_{tr} = \frac{1}{\sqrt{2\pi}\,\sigma} \int_{K_g}^{\infty} e^{-\frac{[R(x,y)-R(x^*,y^*)]^2}{2\sigma^2}}\, dR(x, y), \tag{13.138}$$

where $R(x^*, y^*)$ is the maximum of the criterial correlation function for all points, K_g is the threshold, and σ^2 is the variance of the probability distribution density of the criterial correlation function $R(x, y)$. The probability of the false location of the object image with reference to the control point takes the following form:

$$P_{fr} = \frac{1}{\sqrt{2\pi}\,\bar{R}} \int_{0}^{K_g} e^{-\frac{[R(x,y)-\bar{R}]^2}{2\bar{R}^2}}\, dR(x, y), \tag{13.139}$$

where $\bar{R}$ is the average of the criterial correlation function for all points.

13.7 The Generalized Image Processing Algorithm: Amplitude Ranking

The criterial correlation function of rank form and hierarchical preprocessing are used in the implementation of the generalized image processing algorithm with amplitude ranking based on the generalized approach to signal processing in the presence of noise. The algorithm with amplitude ranking is used when there is a very large region for searching and there are high requirements for calculation accuracy and efficiency. The algorithm is a totality of algorithms, the complexity and calculation accuracy and efficiency of which can be optimized for definite parameters of the object image, for example, the dimensions of the object image, searching regions, etc.

Let us consider the binary generalized image processing algorithm with amplitude ranking based on the generalized approach to signal processing in the presence of noise, which was constructed using the procedure discussed in Ormsby.[31] This algorithm consists of two steps. The preliminary image processing is carried out at the first step, in the course of which the smallest images are coded to binary correlation matrices. Therefore, we assume that either the moving image or the model image is much less in dimensions. These binary matrices are correlated sequentially with the decreasing set of the matrix having the larger dimension during the second step.[32–34]

The purpose of image preprocessing is to distribute the image elements in the order of the decreasing brightness characteristic and assigning the binary rank for each image element. This ranking procedure is shown in Figure 13.13 and Table 13.2. If the object image has 2^D resolution elements, then the object image element with the lowest brightness characteristic takes the rank consisting of D zero and the object image element with the highest brightness characteristic takes the rank consisting of D units. Other object

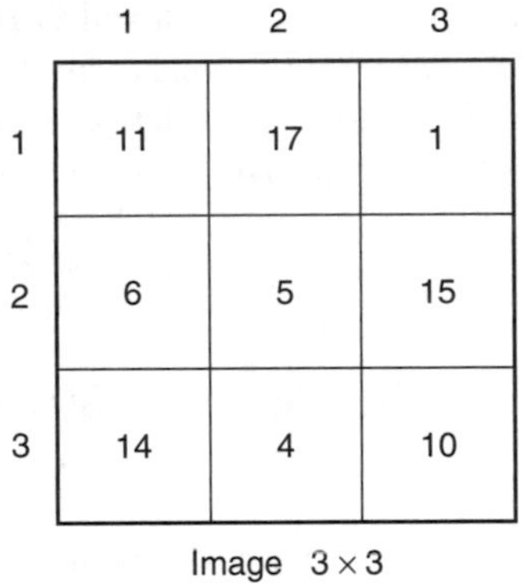

FIGURE 13.13
Ranking procedure.

TABLE 13.2
Ranking Procedure

i, j	Brightness Characteristics	Binary Rank
1,2	17	111
2,3	15	110
3,1	14	101
1,1	11	100
3,3	10	XXX
2,1	6	011
2,2	5	010
3,2	4	001
1,3	1	000

image elements take a rank that is an element binary number in the ordered list. After ordering of the brightness characteristics, the object-image element list is divided into two different classes. Those from the first class (object image elements with high brightness characteristics) take 1 and those from the second class take 0. If the number of the object image elements in a class is odd, then the object image element having a middle brightness characteristic (between 1 and 0) takes the index X. This procedure is iterated for each class of object image elements. Those with the index X keep this index until the procedure stops. The procedure stops when there is only one object image element in each class.

After this procedure, the correlation matrices are formed. The first correlation matrix C_1 is formed by mapping the first signs of the binary rank onto the initial object image. The D correlation matrices are formed in an analogous way (see Figure 13.14). The generalized image processing algorithm with amplitude ranking based on the generalized approach to signal processing in the presence of noise begins with the processing of the first correlation matrix C_1 and the largest object image L using the following technique:

$$B_1(x,y) = \sum_{i,j=1|_{C_1(i,j)=1}}^{N,M} L_{x+i,y+j} - \sum_{i,j=1|_{C_1(i,j)=0}}^{N,M} L_{x+i,y+j}\,, \tag{13.140}$$

where $N \times M$ is the dimension of the object image. Thus, for each position of the matrix C_1 on the largest object image, the sum of the object image elements at points corresponding to 0 in the matrix C_1 is subtracted from the sum of the object image elements at points corresponding to 1 in the matrix C_1. The object image elements at points corresponding to X are not taken into consideration. The function $B_1(x, y)$ is the primary correlation surface and characterizes the process of matching between the largest object image L and approximate structural information about the smallest object image.

During this step, the points for which the value of $B_1(x, y)$ is low are rejected. For this purpose, a definite threshold is established. The secondary correlation surface $B_2(x, y)$, which is a more precise definition of the first correlation surface $B_1(x, y)$, is determined by

$C_1 =$

1	1	0
0	0	1
1	0	X

$C_2 =$

0	1	0
1	1	1
0	0	X

$C_3 =$

0	1	0
1	0	0
1	1	X

FIGURE 13.14
Formation of N correlation matrices.

$$B_2(x,y) = B_1(x,y) + 0.5\left\{ \sum_{i,j=1|_{C_1(i,j)=1}}^{N,M} L_{x+i,y+j} - \sum_{i,j=1|_{C_1(i,j)=0}}^{N,M} L_{x+i,y+j} \right\}. \quad (13.141)$$

Calculations are carried out only for the object image elements, the brightness characteristics of which exceed the predetermined threshold. The described procedure is carried out for all correlation matrices and, in the general case, can be determined in the following form:

$$B_n(x,y) = B_{n-1}(x,y) + 2^{-(n-1)}\left\{ \sum_{i,j=1|_{C_n(i,j)=1}}^{N,M} L_{x+i,y+j} - \sum_{i,j=1|_{C_n(i,j)=0}}^{N,M} L_{x+i,y+j} \right\}. \quad (13.142)$$

After determination of the last correlation surface, the point with the maximal magnitude of $B_1(x, y)$ is determined. This point is the point of matching between the moving image and the model image.

The computational costs of the generalized image processing algorithm based on the generalized approach to signal processing in the presence of noise are characterized by the number of summation operations and determined by

$$Q = \mathfrak{K}(N - n) \cdot (M - m) \cdot nm, \quad (13.143)$$

where $\mathfrak{K}$ is the coefficient $1 < \mathfrak{K} < 2$, $N \times M$ is the dimension of the largest object image, and $n \times m$ is the dimension of the smallest object image. The generalized image processing algorithm is an analog of the classical generalized signal processing algorithm. The performance and the probabilities of true and false location of the object image with reference to the control point are the same in practice.

13.8 The Generalized Image Processing Algorithm: Gradient Vector Sums

The generalized image processing algorithm with gradient vector sums based on the generalized approach to signal processing in the presence of noise is invariant to rotation of the object image for the wide-angle range. During the first step, the grey color gradient of the moving image and the model image are determined. During the second step, the histogram of the gradient vector sums as a function of the angle is formed for each object image if the gradient vectors are within the limits of discrete intervals of angles. Thereafter, these functions are processed either by the classical correlation function or by the criterial correlation function.

The generalized image processing algorithm with gradient vector sums is based on the following principle. If the moving or model image is rotated by a definite angle, then the gradient vector determined for corresponding points of the moving and model images is rotated by the same angle. The brightness characteristics of corresponding points of the moving and model images are determined in the spherical coordinate system in the following form:

$$G(\rho,\theta) = H(\rho',\theta'). \tag{13.144}$$

If, for instance, the model image is rotated with respect to the moving image by the angle φ relative to the coordinate system, then $\rho = \rho'$ and $\theta' = \theta + \varphi$. In this case, the gradients at the corresponding points have the same value of ρ and are shifted with respect to each other by the angle determined by

$$H(\rho,\theta) = G(\rho,\theta') \cdot e^{j\varphi}. \tag{13.145}$$

Because this condition is true for each pair of points, it is true for average values. For instance, if the average values of the object image gradients are determined in the following form

$$\nabla G_{av} = \frac{1}{\pi\rho_0^2} \int_0^{2\pi} \int_0^{\rho_0} \nabla G(\rho,\theta)\, d\rho\, d\theta; \tag{13.146}$$

$$\nabla H_{av} = \frac{1}{\pi\rho_0^2} \int_0^{2\pi} \int_0^{\rho_0} \nabla H(\rho,\theta)\, d\rho\, d\theta, \tag{13.147}$$

where object images are defined on the distance ρ_0 from the origin of the coordinate system, then we can consider that

$$H_{av} = G_{av} \cdot e^{j\varphi}, \tag{13.148}$$

because

$$H(\rho,\theta) = G(\rho,\theta - \varphi) \cdot e^{j\varphi}. \tag{13.149}$$

Consequently, we can define without any difficulty the angle of relative shift of these object images $\varphi = \alpha_1 - \alpha_2$, determining the average magnitude of gradients for two object images

$$G_{av} = \mathcal{U}_1 \cdot e^{i\alpha_1} \quad \text{and} \quad H_{av} = \mathcal{U}_2 \cdot e^{i\alpha_2}. \tag{13.150}$$

The main disadvantage of the generalized image processing algorithm with gradient vector sums is that the values of the average gradient are so close to zero that it makes this algorithm very critical with respect to errors caused by variations of low spatial frequencies. Therefore, the vector function determined by

$$\mathbf{S}(\beta) = \int_0^{2\pi} \int_{S_{ar}} \nabla G \delta(\theta - \beta)\, dS_{ar} d\theta \tag{13.151}$$

can be an alternative, where S_{ar} is the area of object image. The vector function $\mathbf{S}(\beta)$ is defined within the limits of the interval and can be considered as the sum of all gradient vectors of the object image having the angle β.

In digital image processing, the gradient can be approximated by finite differences

$$\nabla G(x_m, y_m) \equiv \frac{\partial G}{\partial x_m}\hat{i} + \frac{\partial G}{\partial y_m}\hat{j}, \tag{13.152}$$

where

$$\frac{\partial G}{\partial x_m} = G_{m+2,n} + 2G_{m+2,n+1} + G_{m+2,n+2} - G_{m,n} - 2G_{m,n+1} - G_{m,n+2}; \tag{13.153}$$

$$\frac{\partial G}{\partial y_m} = G_{m,n+2} + 2G_{m+1,n+2} + G_{m+2,n+2} - G_{m,n} - 2G_{m+1,n} - G_{m+2,n}. \tag{13.154}$$

Experimental results confirming the possibility of identifying two object images rotated relative to each other by angles of up to 30° are discussed in Davies and Bouldin.[35]

13.9 The Generalized Image Processing Algorithm: Bipartite Functions

Let us consider the following criterial bipartite correlation functions:[36,37]

$$R_1(u,\upsilon) = \prod_{i=0}^{2^n-1} T_i; \tag{13.155}$$

$$R_2(u,\upsilon) = \frac{\sum_{i=0}^{2^n-1}\sum_{j=0}^{2^n-1}(i,j)F_{ij}(u,\upsilon)}{\sqrt{\sum_{i=0}^{2^n-1} i^2\left[\sum_{j=0}^{2^n-1}F_{ij}(u,\upsilon)\right]\sum_{j=0}^{2^n-1} j^2\left[\sum_{i=0}^{2^n-1}F_{ij}(u,\upsilon)\right]}}. \tag{13.156}$$

The bipartite correlation function $R_1(u, \upsilon)$ given by Equation (13.155) is a weighted product and depends on the number of quantization levels and weighted estimation of coincidences and noncoincidences of the brightness characteristic levels. For example, for two and four quantization levels we can write

$$T_0 = \frac{F_{00}}{R_0 + F_{01}} \quad \text{and} \quad T_1 = \frac{F_{11}}{R_1 + F_{10}}; \tag{13.157}$$

$$T_0 = \frac{2F_{00} + F_{01}}{2R_0 + F_{02} + 2F_{03}}; \tag{13.158}$$

$$T_1 = \frac{2F_{11} + F_{10} + F_{12}}{2R_1 + F_{13}}; \tag{13.159}$$

$$T_2 = \frac{2F_{22} + F_{21} + F_{23}}{2R_2 + F_{20}}; \tag{13.160}$$

$$T_3 = \frac{2F_{33} + F_{32}}{2R_3 + F_{31} + 2F_{30}}, \tag{13.161}$$

where $R_i = \sum_{j=0}^{2^n-1} F_{ij}$ is the number i for the model image. Thus, noncoincidences for one quantization level can be considered as coincidences for another. The bipartite criterial function $R_2(u, \upsilon)$ is very close to the classical generalized signal processing algorithm based on the generalized approach to signal processing in the presence of noise.

The bipartite criterial correlation functions given by Equation (13.22) and Equation (13.23) are the products of ratios of the number of object image elements with the same brightness characteristics and the number of possible coincidences of all forms of the brightness characteristics. Note that the number of bipartite functions increases as the square of the number of quantization levels. Consequently, we should try to reduce the number of quantization levels.

In binary image processing, there are four forms of the bipartite function united into a matrix

$$\begin{Vmatrix} F_{00} & F_{01} \\ F_{10} & F_{11} \end{Vmatrix}. \tag{13.162}$$

If the object images are identical, then only the functions on the main diagonal of the matrix given by Equation (13.162) are nonzero functions. The binary object images can be defined by four bipartite functions. In this case, based on Equation (13.22) and Equation (13.23), we can write

$$R(u,\upsilon) = \frac{F_{00}}{F_{00} + F_{01}} \cdot \frac{F_{11}}{F_{11} + F_{10}}. \tag{13.163}$$

The criterial function given by Equation (13.163) has a narrow main lobe and the level of the side lobes is also more narrow.

13.10 The Hierarchical Generalized Image Processing Algorithm

The basis of the hierarchical generalized image processing algorithm is a procedure in image preprocessing in which a set of object images with sequentially decreasing resolutions is formed from the initial object images. Various criterial functions are used to compare the obtained sets of object images. The hierarchical generalized image processing algorithm begins with the comparison of object images having the lowest resolution level. In image processing, a set of most probable control points is chosen at each level. At the next levels, computer calculations are carried out only for those points that were chosen at the previous levels. Therefore, the number of processed points is reduced sharply, which ensures the high effectiveness of the algorithm.

In the hierarchy generalized image processing algorithm, the number of processed points at the k-th level is determined by

$$n_k = \left[\frac{N-M+1}{2^L}+1\right]^2. \tag{13.164}$$

The value of n_k is 2^{2L} times less than the number of processed points of the object image with an initial high resolution. The decrease in the number of processed points obeys the logarithmic law $B \log (N - M + 1)^2$, where B is a constant. As a rule, the object image $X_{k-1}(x, y)$ at the $(k - 1)$-th level is filtered by the linear low-frequency filter $h(x, y)$ with the transfer function $C(\omega_x, \omega_y)$:

$$X_0(x,y) = X_{k-1}(x,y) * h(x,y). \tag{13.165}$$

The digitization function takes the following form:

$$d(x,y) = \sum_{j_2=-\infty}^{\infty}\sum_{j_1=-\infty}^{\infty} \delta(x-2j_1\Delta x, y-2j_2\Delta y), \tag{13.166}$$

where Δx and Δy are the intervals of digitization at the $(k - 1)$-th level. The Fourier transform at the k-th level after digitization is determined by

$$S_k(\omega_x,\omega_y) = \frac{\pi^2}{\Delta x \Delta y}\sum_{j_1=-\infty}^{\infty}\sum_{j_2=-\infty}^{\infty} S_{k-1}\left(\omega_x - \tfrac{j_1\pi}{\Delta x}, \omega_y - \tfrac{j_2\pi}{\Delta y}\right)\cdot C\left(\omega_x - \tfrac{j_1\pi}{\Delta x}, \omega_y - \tfrac{j_2\pi}{\Delta y}\right). \tag{13.167}$$

The transfer function of the low-frequency filter can be written in the following form:

$$C(\omega_x,\omega_y) = [\cos(0.5\omega_x)\cdot\cos(0.5\omega_y)]^n, \tag{13.168}$$

where n characterizes the degree of signal damping by the high-frequency filter. According to Hall,[18] sufficient signal damping is ensured at $n = 8$.

In digital signal processing, we assume that $X_{k-1}(n_1, n_2)$ is the moving image at the filter input, and $C(l_1, l_2)$ is the transfer function of the filter. The object image at the k-th level at the filter output can be determined in the following form:

$$X_k(m_1, m_2) = \sum_{n_1=m_1}^{m_1-l_1+1} \sum_{n_2=m_2}^{m_2-l_2+1} X_{k-1}(n_1, n_2) \cdot C(n_1 - m_1 + 1, n_2 - m_2 + 1). \tag{13.169}$$

At $n = 8$, we obtain

$$C(l_1, l_2) = \frac{1}{10} \cdot \begin{Vmatrix} 1 & 1 & 1 \\ 1 & 2 & 1 \\ 1 & 1 & 1 \end{Vmatrix}. \tag{13.170}$$

Let us consider the two-level hierarchy generalized image processing algorithm. In the initial step, the processing of the moving image and the part with the most informative features of the model image is carried out with all possible relative shifts of images. During the second step, the total model image is used, but a comparison is carried out only at the points with maximal correlation that were chosen during the first step. Computational costs J in using the hierarchy generalized image processing algorithm can be estimated in the following form:

$$J = A + P_{thr} V, \tag{13.171}$$

where A is a computational cost during the first step; P_{thr} is the probability of exceeding the threshold during the first step, that is, the probability of definition of the point that will be processed during the second step and V is the computational cost in using the total model image during the second step of the hierarchy generalized image processing algorithm. In the general case, the greatest part of the model image is used during the first step and the computational cost is high, but, as a result, the probability P_{thr} is decreased.

Let $\{t_1, \ldots, t_n\}$ be the values of the grey level of n points of the model image and $\{g_1, \ldots, g_n\}$ the values of the grey level of the moving image part arbitrarily chosen. The criterial correlation function can be written in the following form:

$$R = \sum_{i=1}^{n} | g_i - t_i |. \tag{13.172}$$

In this case, the computational cost is proportional to the dimensions of the model image if the grey level of background obeys the Gaussian probability distribution density and the absolute values of the brightness characteristics of the moving and model images are independent.

13.11 The Generalized Image Processing Algorithm: The Use of the Most Informative Area

In this case, the peculiarity of the generalized image processing algorithm is the image preprocessing in the course of which the most informative areas are detected.[38] This technique ensures a decrease in computational cost. Object image elements, the brightness characteristics of which are high compared to the average value of the brightness characteristics, can be used as the most informative areas. To choose the most informative areas, it is necessary to obtain the probability distribution density of the object image elements as a function of the brightness characteristic $\mathscr{L}(x)$ (see Figure 13.15), where m and σ are the mean and the variance of the brightness characteristic $\mathscr{L}(x)$, respectively, and $\rho \geq 0$ is a constant. The object image $g'(i, j)$ is formed as a result of preprocessing the object image $g(i, j)$. The object image $g'(i, j)$ contains the elements of the initial object image in accordance with the formula

$$g'(i,j) \in \{\mathscr{L}(m - \rho\sigma) \geq g(i,j) \geq \mathscr{L}(m + \rho\sigma)\} . \tag{13.173}$$

Thus, the volume of reduced information depends on the behavior of the probability distribution density of the brightness characteristics $\mathscr{L}(x)$ and the parameter ρ.

For example, if the brightness characteristic $\mathscr{L}(x)$ obeys the Gaussian probability distribution density

$$f_{\mathscr{L}}(x) = \left(\sqrt{2\pi}\ \sigma\right)^{-1} \cdot e^{-\frac{(x-m)^2}{2\sigma^2}} , \tag{13.174}$$

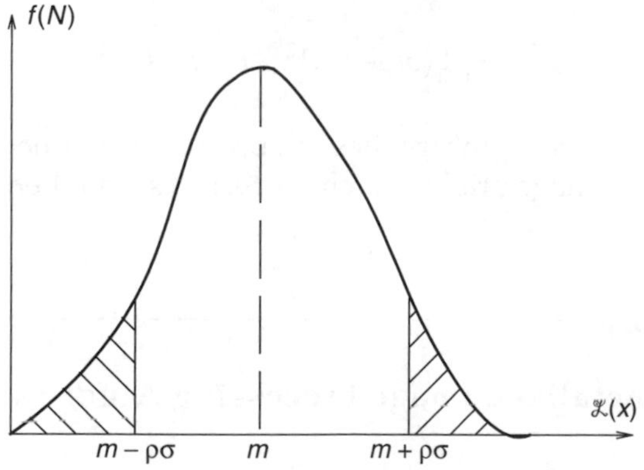

FIGURE 13.15
Numbers of object image elements as a function of the brightness characteristics.

then the ratio of the element numbers of the object image $g'(i, j)$ to the element numbers of the object image $g(i, j)$ can be determined in the following form:

$$\beta = 1 - \int_{m-\rho\sigma}^{m+\rho\sigma} f_{\mathcal{L}}(x)\, dx = 1 - \frac{\sqrt{2}}{\pi} \int_{0}^{\rho} e^{-0.5x^2} dx\ . \tag{13.175}$$

Thus, the coefficient of the computational cost reduction in comparison with the classical generalized image processing algorithm can be determined by

$$\gamma = \frac{200}{\sqrt{2\pi}} \int_{0}^{\rho} e^{-0.5x^2} dx. \tag{13.176}$$

The dependence of the coefficient γ on the parameter ρ, if the condition $0 < \rho < 1$ is satisfied, has a linear character. For example, at $\rho = 1$, we obtain $\gamma = 65\%$.

If the probability distribution density $f_L(x)$ can be presented as the totality of two Gaussian laws with the parameters m_1, σ_1, m_2, and σ_2, then the coefficient γ can be determined by

$$\begin{aligned} \gamma = P_1\Big[\operatorname{erf}\Big(\frac{P_2(m_1 - m_2) + \rho\sigma}{\sigma_1}\Big) - \operatorname{erf}\Big(\frac{P_1(m_2 - m_1) + \rho\sigma}{\sigma_1}\Big)\Big] \\ + P_2\Big[\operatorname{erf}\Big(\frac{P_1(m_1 - m_2) + \rho\sigma}{\sigma_2}\Big) - \operatorname{erf}\Big(\frac{P_1(m_1 - m_2) + \rho\sigma}{\sigma_2}\Big)\Big] \end{aligned}, \tag{13.177}$$

where

$$\operatorname{erf}(x) = \left(\sqrt{2\pi}\right)^{-1} \int_{0}^{x} e^{-0.5\lambda^2} d\lambda; \tag{13.178}$$

$$\sigma^2 = P_1 P_2 (m_1 - m_2)^2 + P_1\sigma_1^2 + P_2\sigma_2^2, \tag{13.179}$$

P_1 and P_2 are the *a priori* probabilities of appearance of one of the two main levels of the object image brightness characteristics, and therefore, $P_1 + P_2 = 1$.

13.12 The Generalized Image Processing Algorithm: Coding of Images

The main peculiarity of the generalized image processing algorithm in the coding of images based on the generalized approach to signal processing in the presence of noise is the preliminary image processing based on the use of various techniques of economic coding and removal of the excessive informativeness of images. One of the methods that is used in the generalized image processing algorithm is cut block coding. This technique realizes principles of adaptive block coding applied to the adaptation of local information structures, not the average brightness characteristics of the object image.

The essence of cut block coding is the following. It is necessary to divide the initial image $X(i, j)$ into blocks consisting of $n \times m$ elements with the sequential binary quantization of each block at the levels a and b. It is recommended that the linear dimensions of blocks be 3 to 5 times less than the correlation length of the object image. There are some image preprocessing procedures that use the technique of choosing the levels and quantization threshold K_{g_d} in the generalized image processing algorithm with cut block coding. The determination of the levels and quantization threshold can be carried out in the following manner:

$$a = \bar{x} - \sigma\sqrt{\frac{q}{p-q}}, \qquad b = \bar{x} + \sigma\sqrt{\frac{p-q}{q}}, \qquad \text{and} \qquad K_{g_d} = \overline{x_1^2}, \tag{13.180}$$

where

$$\bar{x} = \frac{1}{mn}\sum_{i=1}^{n}\sum_{j=1}^{m} X(i,j); \tag{13.181}$$

$$\sigma^2 = \overline{x_1^2} - \bar{x}; \tag{13.182}$$

$$\overline{x_1^2} = \frac{1}{mn}\sum_{i=1}^{n}\sum_{j=1}^{m} X^2(i,j); \tag{13.183}$$

$\bar{x}$ and σ^2 are the moment and variance defined for each block, $p = mn$, and q is the number of elements of the object image $X(i, j)$ exceeding the threshold K_g.

There is another technique, based on which the parameters of the quantizer are chosen to maximize the criterial correlation function. For this purpose, the elements of the object image block $X_k(i, j)$ are ranked in increasing order. Based on the obtained sequence $z_{k_1}, \ldots, z_{k_p}$, we determine the quantization levels

$$a_k = \frac{1}{p - q_k}\sum_{i=1}^{p-q_k} z_{k_i} \qquad \text{and} \qquad b_k = \frac{1}{q_k}\sum_{i=p-q_k+1}^{p} z_{k_i}, \tag{13.184}$$

where q_k is the number of elements z_{k_i} exceeding the threshold K_{gd}. The threshold value is defined by the maximization of the criterial correlation function compared to corresponding blocks of the initial object image and the coding object image:

$$R = a_k \sum_{i=1}^{p-q_k} z_{k_i} + b_k \sum_{i=p-q_k+1}^{p} z_{k_i}. \tag{13.185}$$

It is also possible to define simultaneously the quantization levels in accordance with Equation (13.180) and the maximum of the criterial correlation function given by Equation (13.185).

13.13 The Multichannel Generalized Image Processing Algorithm

The multichannel generalized image processing algorithm is used to increase the accuracy of shift estimation between the moving and model images that could be made using channels added in parallel to the main channel of the generalized receiver. These additional channels of the generalized receiver process components of the moving and model images. Preliminary image processing is carried out by nonlinear filters with Π-form characteristics to obtain these components of the moving and model images.

As is well known, the accuracy of shift estimation between the moving and model images is proportional to the steepness of the discrimination characteristic which, taking into consideration additional channels of the generalized receiver, can be determined in the following form:

$$D_N(\alpha) = -\left\{\frac{d}{da} R_{a^*X}(a) + \frac{d}{da} \sum_{i=1}^{N} r_i R^*_{a^*X_i}(a)\right\}, \tag{13.186}$$

where $R_{a^*X}(a)$ is the correlation function between the moving and model images, r_i are the coefficients, N is the number of additional channels of the generalized receiver, and $R^*_{a^*X_i}$ (a) is the correlation function between components of the moving and model images. If the brightness characteristics of these components obey the Gaussian probability distribution density, we can write the correlation function $R^*_{a^*X_i}$ (a) in the following form:

$$R^*_{a^*X_i}(a) = \sum_{k=0}^{\infty} (k!)^{-1} h^2_{k_i} R^k_{a^*}(a); \tag{13.187}$$

$$h_{k_i} = \frac{a_i}{\sqrt{2\pi}\ \sigma^k}\left\{H_{k-1}(\tfrac{b_i}{\sigma})\cdot e^{-\frac{b_i^2}{2\sigma^2}} - H_{k-1}(\tfrac{b_i+c_i}{\sigma})\cdot e^{-\frac{(b_i+c_i)^2}{2\sigma^2}}\right\}, \tag{13.188}$$

where $H_n(.)$ is the Hermitian polynomial. In the only additional channel of the generalized receiver, the accuracy of shift estimation is increased 2.7 times. Thus, the use of the multichannel generalized image processing algorithm allows us to increase significantly the accuracy of matching between the moving image and the model image.

13.14 Conclusions

In this chapter, we consider briefly some quasioptimal generalized image processing algorithms in the use of the various types of criterial correlation functions that are constructed based on of the generalized approach to signal processing in the presence of noise. Each quasioptimal generalized image processing algorithm has both advantages and disadvantages. We will discuss these advantages and disadvantages briefly.

The classical generalized image processing algorithm is highly efficient with low values of signal-to-noise ratio but has a very high computational cost. The difference generalized image processing algorithm allows us to reduce computational costs significantly and has better detection performance with high values of signal-to-noise ratio in comparison with the classical generalized image processing algorithm. With low values of the signal-to-noise ratio, the detection performance of the difference generalized image processing algorithm is less compared to the classical generalized image processing algorithm. The generalized image processing algorithm with bipartite functions displays a very high efficiency in the computational cost. However, the use of the sign functions in the implementation of the generalized image processing algorithm with the bipartite functions is possible only in the case of the Gaussian probability distribution density. The phase generalized image processing algorithm ensures very spiked peaks of the criterial correlation function, displays a high efficiency in the computational cost, and is nonsensitive to narrow-band noise. However, it is very highly sensitive to high-frequency distortions. The use of the hierarchy generalized image processing algorithm ensures a low computational cost, but it requires an increase in the volume of computer memory. Its significant disadvantage is the very high complexity of detection of informative features. The generalized image processing algorithm with amplitude ranking has approximately the same detection performance as the classical generalized image processing algorithm. Low computational cost is realized only in the case when the object image has large dimensions. The generalized image processing algorithm with invariant moments is invariant to rotation of object images, but its computational efficiency in image processing is low.

References

1. Wang, R. and Hall, E., Sequential hierarchical scene matching, *IEEE Trans.*, Vol. C-27, No. 4, 1978, pp. 359–366.
2. Arsenault, H. et al., Incoherent method for rotation-invariant recognition, *Appl. Opt.*, Vol. 21, No. 4, 1982, pp. 610–615.
3. K. Fu, Ed., *Digital Pattern Recognition*, Springer-Verlag, Berlin, 1976.
4. Knapp, C. and Carter, G., The generalized correlation method for estimation of time delay, *IEEE Trans.*, Vol. ASSp-24, No. 4, 1976, pp. 320–327.
5. Novak, L., Correlation algorithms for radar map matching, *IEEE Trans.*, Vol. AES-14, No. 4, 1978, pp. 641–648.
6. Mirskiy, G., *Characteristics of Stochastic Correlation and Its Measurement*, Energoizdat, Moscow, 1982 (in Russian).
7. Tuzlukov. V., A new approach to signal detection theory, *Digital Signal Process. Rev. J.*, Vol. 8, No. 3, 1998, pp. 166–184.
8. Tuzlukov, V., *Signal Processing in Noise: A New Methodology*, IEC, Minsk, 1998.
9. Tuzlukov, V., *Signal Detection Theory*, Springer-Verlag, New York, 2001.
10. Tuzlukov, V., *Signal Processing Noise*, CRC Press, Boca Raton, FL, 2002.
11. Kuglin, C. et al. Map-matching techniques for terminal guidance using Fourier phase information, in *Proceedings of the SPIE*, Vol. 186, 1979, pp. 21–29.
12. Vladimirov, V., *Equations in Mathematical Physics*, Nauka, Moscow, 1971. (In Russian.)
13. Kuglin, C. and Hines, D., The phase correlation image alignment method, in *Proceedings of the IEEE International Conference on Cybernetics and Society*, 1975, pp. 163–165.
14. Ausherman, D., Kozma, A., Walker, J., Jones, H., and Poggio, E., Developments in radar imaging, *IEEE Trans.*, Vol. AES-20, No. 3, 1984, pp. 363–279.
15. Stark, H. and Woods, J., Polar sampling theorems and their applications to computer-aided tomography, in *Proceedings of the SPIE*, Vol. 231, 1980, pp. 230–241.
16. Candel, S., Dual algorithms for fast calculation of the Fourier-Bessel transformations, *IEEE Trans.*, Vol. ASSP-29, No. 5, 1981, pp. 963–972.
17. Herman, G., *Image Reconstruction from Projections*, Springer-Verlag, New York, 1979.
18. Hall, E., *Computer Image Processing and Recognition*, Academic Press, New York, 1979.
19. Budinger, T. and Gullbery, G., Three-dimensional reconstruction in nuclear medicine emission imaging, *IEEE Trans.*, Vol. NS-21, No. 6, 1974, pp. 2–20.
20. Cook, G. et al., Optimal reconstruction angles, in *Proceedings of the IEEE Conference on Computers in Radiology*, 1979, pp. 291–304.
21. Clary, J. and Russell, R., All-digital correlation for missile guidance, in *Proceedings of the SPIE*, Vol. 119, 1977, pp. 36–46.
22. Stubberud, A., Acquisition probability for a correlation algorithm, in *Proceedings of the IEEE Circuits and Systems Conference*, 1981, pp. 263–265.

23. Rockmore, A., The probability of false acquisition for image registration, *Image Science Mathematics Symposium*, 1976, pp. 252–255.
24. Xu, X. and Narayanan, R., Three-dimensional interferometric ISAR imaging for target scattering diagnosis and modeling, *IEEE Trans.*, Vol. IP-10, No. 7, 2001, pp. 1094–1102.
25. Zebker, H. and Lu, Y., Phase unwrapping algorithms for radar interferometry: reside-cut, least-squares, and synthesis algorithms, *J. Opt. Soc. Am.*, Vol. 15, No. 3, 1998, pp. 586–598.
26. Herrero, J., Portas, J., and Corredera, J., Use of map information for tracking targets on airport surface, *IEEE Trans.*, Vol. AES-39, No. 2, 2003, pp. 675–693.
27. Johnson, M., Analytical development and test results of acquisition probability for terrain correlation devices used in navigation systems, *AIAA*, Paper N72–122.
28. Vasilenko, G., *Holography Pattern Recognition*, Soviet Radio, Moscow, 1977. (In Russian.)
29. Hu, M.-K., Visual pattern recognition by moment invariants, *IRE Trans.*, Vol. IT-8, 1962, No. 1, pp. 179–187.
30. Wong, R. and Hall, E., Scene matching with invariant moments, *Comput. Vision Grap.*, Vol. 8, No. 1, 1978, pp. 16–24.
31. Ormsby, C., Advanced scene matching techniques, in *Proceedings of IEEE NAECON*, 1979, pp. 68–78.
32. Bryant, M., Gostin, L., and Soumekh, M., 3-D E-CSAR imagimg of a T-72 tank and synthesis of its SAR reconstructions, *IEEE Trans.*, Vol. AES-39, No. 1, 2003, pp. 211–227.
33. Collins, N. and Baird, C., Terrain aided passive estimation, in *Proceedings of the IEEE National Aerospace and Electronics Conference*, 1989, Vol. 3, pp. 909–916.
34. Nabaa, N. and Bishop, R., Validation and comparison of coordinated turn aircraft maneuver models, *IEEE Trans.*, Vol. AES-36, No 1, 2000, pp. 250–259.
35. Davies, D. and Bouldin, D., Correlation of rotated images by the method of gradient vector sums, in *Proceedings of IEEE SOUTHEASTCON*, 1979, pp. 367–372.
36. Garett, G. et al., Detection threshold estimation for digital area correlation, *IEEE Trans.*, Vol. SMC-6, No. 1, 1976, pp. 65–70.
37. Wong, R., Sequential scene matching using edge features, *IEEE Trans.*, Vol. AES-14, No. 1, 1978, pp. 128–140.
38. Ranganath, H. et al., Feature extraction technique for fast digital image registration, in *Proceedings of IEEE SOUTHEASTCON*, 1980, pp. 225–228.

14

Object Image Preprocessing

14.1 Object Image Distortions

The quasioptimal generalized signal and image processing algorithms based on the generalized approach to signal processing in the presence of noise discussed in Chapter 13 are intended for the processing of identical object images having a definite relative shift. It is assumed that distortions are absent, which is not always true. In practice, images of the same object obtained at different times or by various sensors can be fundamentally different from each other. So, object image preprocessing is necessary to bring the images into one-to-one correspondence. The quality of the preprocessing will have an impact on the general parameters of any navigational system. Let us consider some distortions generated by electromagnetic fields.

All object image distortions in form can be divided into geometrical distortions and distortions in brightness characteristics. Both types of distortions are caused by various reasons. For example, geometrical distortions of object images are possible due to inaccurate information about the object's position in space. In this case, distortions for various spatial coordinates are shown in Figure 14.1. Geometrical distortions arise during transmission of object images by various sensors when an implementation of these sensors assumes various tracking angles. Distortions in brightness characteristics of object images arise, as a rule, due to changes in meteorological or seasonal conditions.

14.2 Geometrical Transformations

Object images obtained by various sensors, for example, by radar side-scanning and optical scanning, can be brought into one-to-one correspondence with a geometrical transformation or correction. We consider two techniques that can be applied with the use of the generalized image

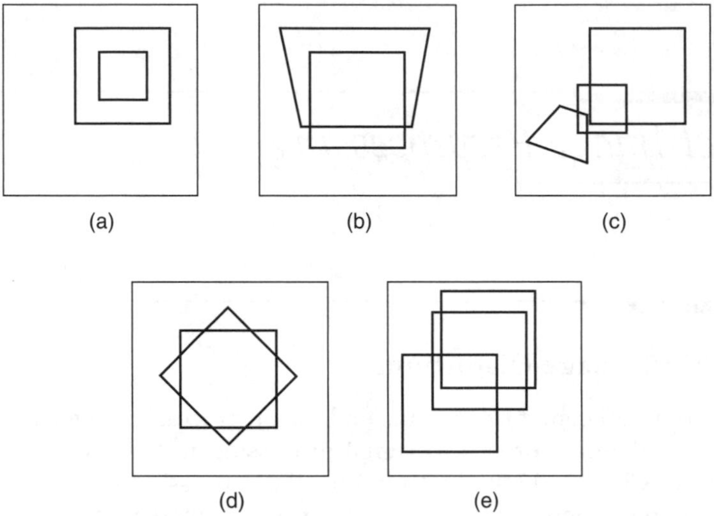

FIGURE 14.1
Distortions for various spatial coordinates: (a) altitude; (b) pitch; (c) complex error; (d) heading; (e) displacement.

processing algorithm based on the generalized approach to signal processing in the presence of noise: the perspective transformation and polynomial estimation.[1]

14.2.1 The Perspective Transformation

The principal scheme of object image generation, which is also of general applicability, uses two sensors, with geometrical features of optical systems and radar side-scanning included in the navigational system, and is shown in Figure 14.2. The image plane of an optical system and the object plane (the Earth's surface) are parallel (see Figure 14.3). The optical axis is orthogonal to both planes. Because the distance between the optical system and the object is much greater than the focus f, the image plane and the focal plane coincide and $x_3 = f$. Each point of the object image $u' = (u'_1, u'_2, u'_3)$ has a corresponding point on the focal plane $x = (x_1, x_2, x_3)$. The components of the point x can be determined as follows

$$x_1 = f \cdot \frac{u'_1}{u'_3}, \qquad x_2 = f \cdot \frac{u'_2}{u'_3}, \qquad \text{and} \qquad x_3 = f. \tag{14.1}$$

The object image obtained by radar side-scanning (see Figure 14.4) depends on the angles of pitch θ, roll γ, and hunting ψ, which define the position of the navigational system. The object point $u = (u_1, u_2, u_3)$ on the radar side-

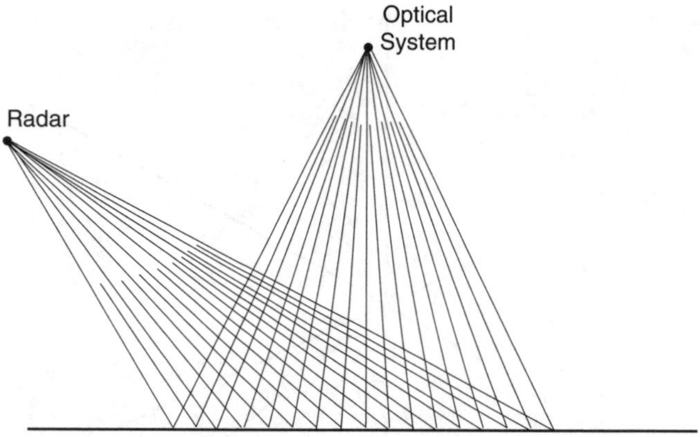

FIGURE 14.2
Optical system and radar side-scanning.

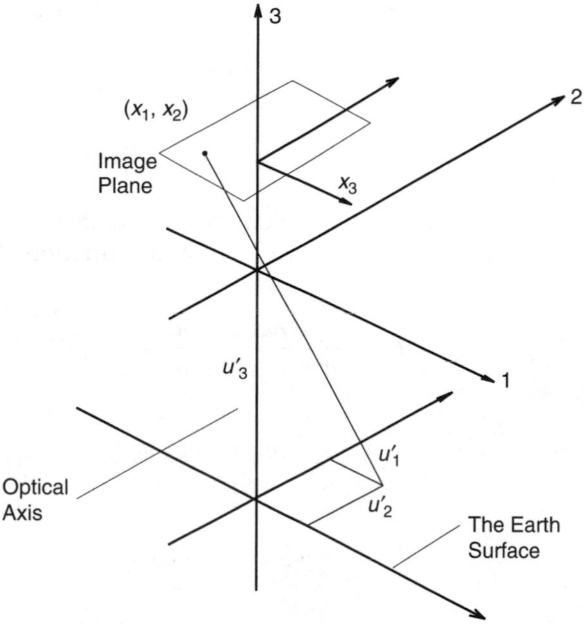

FIGURE 14.3
Optical system and the Earth's surface.

scanning plane can be defined as the point $u' = (u_1', u_2', u_3')$ in the coordinate system of the optical sensor by a formula in matrix form

$$\mathbf{u}' = \mathbf{M} \cdot \mathbf{u} \tag{14.2}$$

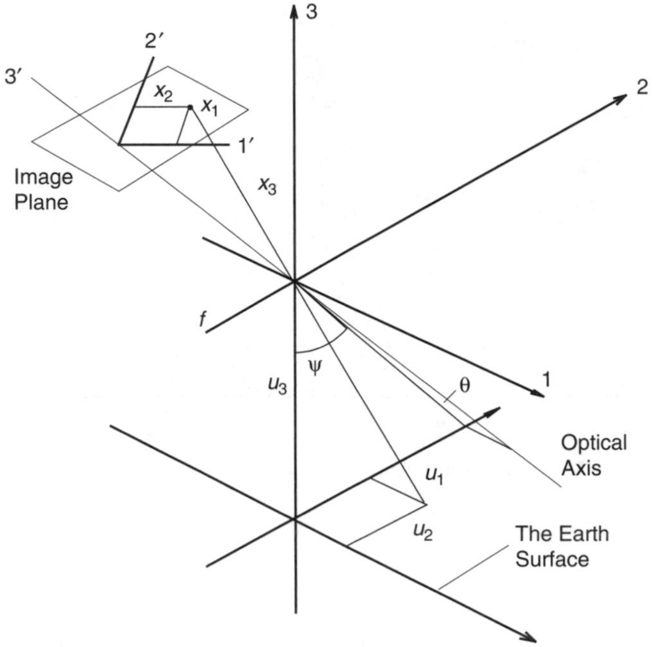

FIGURE 14.4
Radar side-scanning object image.

where **M** is the matrix of rotation, the elements of which are functions of the angles θ, γ, and ψ.[1] Substituting Equation (14.2) in Equation (14.1), we get

$$x_1 = f \cdot \frac{m_{11}u_1 + m_{12}u_2 + m_{13}u_3}{m_{31}u_1 + m_{32}u_2 + m_{33}u_3}; \tag{14.3}$$

$$x_2 = f \cdot \frac{m_{21}u_1 + m_{22}u_2 + m_{23}u_3}{m_{31}u_1 + m_{32}u_2 + m_{33}u_3}; \tag{14.4}$$

$$x_3 = f, \tag{14.5}$$

where m_{ij} are the elements of the rotation matrix **M**. Using Equation (14.3) and Equation (14.4), the object image point with coordinates (u_1, u_2) can be transformed into the coordinates (x_1, x_2) of the optical system.

Note that the transforms given by Equation (14.3)–Equation (14.5) are characteristic of object images obtained by optical and heat vision apparatus. Radar side-scanning represents the object image in the following coordinates: sloped distance and azimuth. So, it is necessary to first transform the object image in the radar-side-scanning plane to the coordinate system of the angles θ, γ, and ψ to use this technique.

14.2.2 Polynomial Estimation

Let $f(u_1, u_2)$ be the radar image and $f(x_1, x_2)$ be the optical image. The polynomial approach is to use the following formulae:[2]

$$x_1 = \sum_{i=0}^{N}\sum_{j=0}^{N-i} a_{ij} u_1^i u_2^j; \tag{14.6}$$

$$x_2 = \sum_{i=0}^{N}\sum_{j=0}^{N-i} b_{ij} u_1^i u_2^j, \tag{14.7}$$

where a_{ij} and b_{ij} are the constant polynomial coefficients. In practice, we use polynomials of the second order, i.e., $N = 2$, as a rule. To obtain coefficients it is necessary to choose the most significant features of the radar and optical object images, such as the finite points of long lines, of intersections, and so on. With the use of a polynomial of the second order, it is necessary to choose no less than six conjugate points. The values of these points are as follows:

$$\left\| \begin{matrix} x_1^1 \\ x_1^2 \\ x_1^3 \\ \vdots \\ x_1^n \end{matrix} \right\| = \left\| \begin{matrix} 1 \; u_1 \; u_2 \; u_1^2 \; u_2^2 \; u_1 \; u_2 \\ \vdots \\ \vdots \\ \vdots \\ \vdots \end{matrix} \right\| \cdot \left\| \begin{matrix} a_{00} \\ a_{10} \\ a_{01} \\ a_{20} \\ a_{02} \\ a_{11} \end{matrix} \right\|, \tag{14.8}$$

where $n \geq 6$. The analogous formulae can be defined for x_2 and b_{ij} and can be expressed in matrix form: $\mathbf{x}_1 = \mathbf{u} \cdot \mathbf{A}$ and $\mathbf{x}_2 = \mathbf{u} \cdot \mathbf{B}$. Estimates **A** and **B** are defined by the following equations:

$$\mathbf{A} = (\mathbf{u}^{\mathrm{T}}\mathbf{u})^{-1}\mathbf{u}^{\mathrm{T}}\mathbf{x}_1 \quad \text{and} \quad \mathbf{B} = (\mathbf{u}^{\mathrm{T}}\mathbf{u})^{-1}\mathbf{u}^{\mathrm{T}}\mathbf{x}_2 . \tag{14.9}$$

The polynomial estimation technique is preferred because additional information is not required; information included in the obtained object images is sufficient.

Consider the method of preprocessing active sensor signals by measurement of radar range with a radar or a scanning laser system.[3] Sensor data can be presented in spherical coordinates: radar range and two angles. Because the generalized signal and image processing algorithms are in the

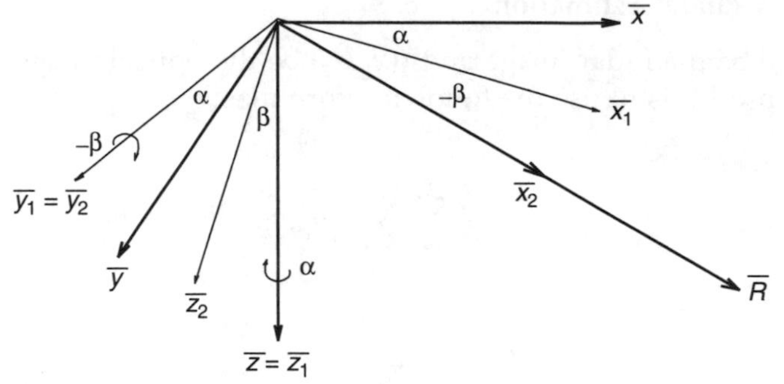

FIGURE 14.5
Coordinate systems.

Cartesian coordinate system, it is necessary to transform the object image to the Cartesian coordinate system and carry out additional processing to reduce geometrical distortions during the preprocessing step. These coordinate systems are shown in Figure 14.5. Here, $\overline{\mathbf{R}} = \mathbf{R} \cdot \overline{\mathbf{x}}_2$, where $\overline{\mathbf{R}}$ is the radar-range vector

$$\begin{Vmatrix} \overline{x}_1 \\ \overline{y} \\ \overline{z} \end{Vmatrix} = \mathbf{A} \cdot \begin{Vmatrix} \overline{x} \\ \overline{y} \\ \overline{z} \end{Vmatrix}; \tag{14.10}$$

$$\mathbf{A} = \begin{Vmatrix} \cos\alpha & \sin\alpha & 0 \\ -\sin\alpha & \cos\alpha & 0 \\ 0 & 0 & 1 \end{Vmatrix}; \tag{14.11}$$

$$\begin{Vmatrix} \overline{x}_2 \\ \overline{y}_1 \\ \overline{z}_2 \end{Vmatrix} = \mathbf{B} \cdot \begin{Vmatrix} \overline{x}_1 \\ \overline{y}_1 \\ \overline{z}_1 \end{Vmatrix} = \mathbf{B} \cdot \mathbf{A} \cdot \begin{Vmatrix} \overline{x} \\ \overline{y} \\ \overline{z} \end{Vmatrix}; \tag{14.12}$$

$$\mathbf{B} = \begin{Vmatrix} \cos\beta & 0 & \sin\beta \\ 0 & 1 & 0 \\ -\sin\beta & 0 & \cos\beta \end{Vmatrix}. \tag{14.13}$$

The coordinate transformation takes the following form:

$$\begin{Vmatrix} x \\ y \\ z \end{Vmatrix} = \mathbf{A}^{-1}\mathbf{B}^{-1} \begin{Vmatrix} R \\ 0 \\ 0 \end{Vmatrix}, \tag{14.14}$$

where $\mathbf{A}^{-1}$ and $\mathbf{B}^{-1}$ are the matrices obtained by replacing the variables α and β with $-\alpha$ and $-\beta$ in the matrices **A** and **B**, respectively. The matrices $\mathbf{A}^{-1}$ and $\mathbf{B}^{-1}$ are the reciprocal matrices.

Coordinate transformation does not influence the brightness characteristic I for any value of the matrix **R**. The vector of parameters I, **R**, α, and β is transformed into the vector of parameters I, x, y, and z, respectively. The second step of image preprocessing is to obtain a projection of the object image on the plane that is orthogonal to the tracking line. This transformation is carried out by two sequential rotations of the Cartesian coordinates (x, y, z): the first rotation by the angle α_p with respect to the axis z and the second rotation by the angle β with respect to the axis y. The projection plane coordinates are x_p, y_p, and z_p. Because of this, we can write

$$\begin{Vmatrix} x_p \\ y_p \\ z_p \end{Vmatrix} = \mathbf{B}_\mathbf{p}\mathbf{A}_\mathbf{p} \begin{Vmatrix} x \\ y \\ z \end{Vmatrix}, \tag{14.15}$$

where $\mathbf{A}_\mathbf{p}$ and $\mathbf{B}_\mathbf{p}$ are the matrices obtained by replacing the variables α and β with α_p and β_p in the matrices **A** and **B**, respectively. Taking into account Equation (14.13), we can write

$$\begin{Vmatrix} x_p \\ y_p \\ z_p \end{Vmatrix} = \mathbf{B}_\mathbf{p}\mathbf{A}_\mathbf{p}\mathbf{A}^{-1}\mathbf{B}^{-1} \begin{Vmatrix} R \\ 0 \\ 0 \end{Vmatrix}. \tag{14.16}$$

The values of **R**, $\mathbf{A}^{-1}$, and $\mathbf{B}^{-1}$ vary from point to point of the sensor searching area, whereas the values of $\mathbf{A}_\mathbf{p}$ and $\mathbf{B}_\mathbf{p}$ are constant. As the radar range vector **R** lies along the axis x_2 (see Figure 14.5), it is necessary to determine the elements of the first column of the matrix $\mathbf{B}_\mathbf{p}\mathbf{A}_\mathbf{p}\ \mathbf{A}^{-1}\mathbf{B}^{-1}$. The second step of the transformation is illustrated by Figure 14.6, in which the point (x_p, y_p, z_p) in the projection plane corresponds to the point (R, α, β) in the spherical coordinate system.

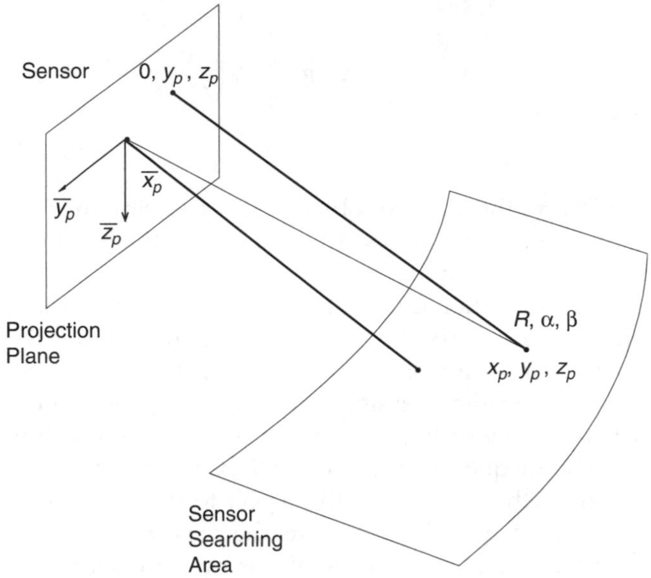

FIGURE 14.6
The second step of the coordinate transformation.

14.2.3 Transformations of Brightness Characteristics

Images of the same objects obtained by various sensors have different distributions of brightness characteristics. If, for example, the optical object image is positive, then the radar image of the same object is negative. An example of the brightness characteristics distribution of optical and radar images of the same object is shown in Figure 14.7 and Figure 14.8. The generalized signal and image processing algorithm for these images cannot bring positive results. The object images must be brought into one-to-one correspondence. Let us consider two transformation techniques of object image brightness characteristics that are discussed in References 2 and 4.[2,4] The main feature of the first technique is the use of statistical characteristics of object images. The second technique does not require this information, because the radar object image is transformed so that the brightness characteristics correspond to the exponential distribution law.

The first transformation step of brightness characteristics is a transition from the negative radar image to the positive radar image. This transformation is carried out by replacing the brightness characteristic of each image element by the brightness level e_{ij}. Then,

$$\overline{e}_{ij} = 2^n - e_{ij} - 1 , \tag{14.17}$$

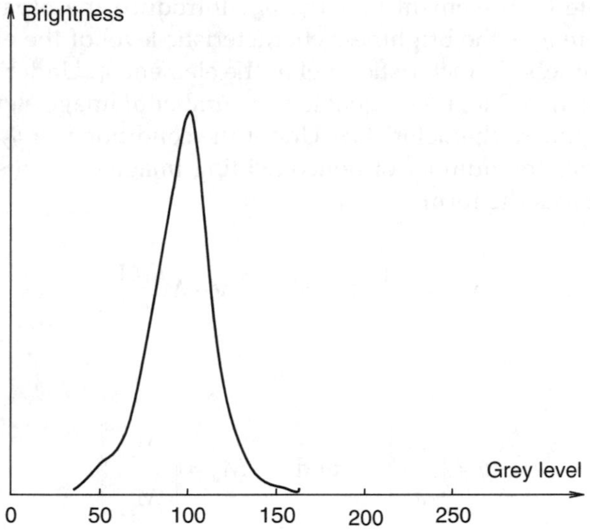

FIGURE 14.7
The brightness characteristic distribution of an optical object image.

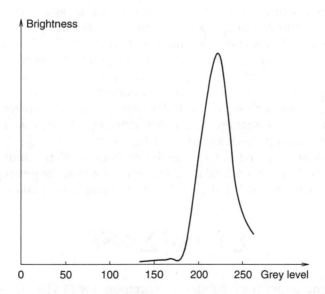

FIGURE 14.8
The brightness characteristic distribution of a radar object image.

where 2^n is the number of quantization levels. Assume that the radar image **R** and the optical image **P** have the dimensions $N \times N$:

$$\mathbf{R} = (r_1, \dots, r_{N\times N}) \qquad \text{and} \qquad \mathbf{P} = (p_1, \dots, p_{N\times N})\,. \tag{14.18}$$

We investigate each element pair (r_k, p_k). Introduce the bipartite function $F(y_1, y_2)$, where y_1 is the brightness characteristic level of the element r_k and y_2 is the brightness characteristic level of the element p_k. Under the condition $y_1 = y_2$ the function $F(y_1, y_2)$ is equal to the number of image elements having the same brightness characteristics. Under the condition $y_1 \neq y_2$, the function $F(y_1, y_2)$ defines the number of noncoinciding image elements. The covariance matrix takes the form

$$\mathbf{C}_y = M\left\{\| \, y - M_y \, \| \cdot \| \, y - M_y \|^T \right\}, \tag{14.19}$$

where

$$y = \left\| \begin{matrix} y_1 \\ y_2 \end{matrix} \right\| \quad \text{and} \quad M_y = \left\| \begin{matrix} M_{y_1} \\ M_{y_2} \end{matrix} \right\|, \tag{14.20}$$

M_{y_1} and M_{y_2} are the expected levels y_1 and y_2 of brightness characteristics, respectively; M is the symbol of the mean.

The radar image is transformed in the following manner. The number of image elements of each brightness characteristic level coincides with the corresponding number of optical image elements. For this purpose, we introduce a new coordinate system g_1 and g_2 so that $\mathbf{C}_\lambda = \mathbf{G}\mathbf{C}_y\mathbf{G}^{-\mathrm{T}}$, where $\mathbf{G}$ is the matrix whose columns are the eigenvectors $\mathbf{C}_y$ with the eigenvalues λ_1 and λ_2. The required correction of brightness characteristics is reached when the angle between g_1 and y_1 becomes approximately equal to 45°.

With the second technique, the radar object image is transformed into a positive form [see Equation (14.6)]. A further transformation replaces the brightness characteristics $\overline{e}'_{ij}$ so that $N_0 \leq \overline{e}_{ij} \leq N_k$ and $M_0 \leq \overline{e}'_{ij} \leq M_L$. The transformation is carried out so that the probability distribution law at the output $P[\overline{e}'_{ij} = M_\ell]$, $0 \leq \ell \leq L$ takes an expected shape at the given probability distribution law $P[\,\overline{e}_{ij} = N_k]$, $0 \leq k \leq K$. The transformation takes the form:

$$\sum_{q=0}^{k} P[e_{ij} = N_q] = \sum_{x=0}^{\ell} P[\overline{e}'_{ij} = M_k]\,. \tag{14.21}$$

To obtain the uniform probability distribution law $P[\overline{e}'_{ij}] = (M_L - M_0)^{-1}$, the transformation should take the following form

$$\overline{e}'_{ij} = [M_L - M_0] \cdot P\,[\overline{e}_{ij}] + M_0\,. \tag{14.22}$$

To obtain the Rayleigh law with the parameter $\beta > 0$, the following equality must be satisfied:

$$P\,[\bar{e}'_{ij}] = \frac{\bar{e}'_{ij}}{\beta^2} \cdot e^{-\frac{(\bar{e}'_{ij})^2}{2\beta^2}}, \tag{14.23}$$

and the transformation function is given by

$$\bar{e}'_{ij} = \sqrt{2\beta^2 \lg \frac{1}{(1-\bar{e}_{ij})}}\,. \tag{14.24}$$

14.3 Detection of Boundary Edges

The most stable features of compared object images are the boundary lines.[5] This is characteristic of images of the same object across various ranges of the electromagnetic spectrum. There are many boundary edge detection algorithms. Methods of boundary edge detection are based, as a rule, on estimation of the gradient for each resolution element of the object image. Elements whose gradients exceed a certain brightness level are united as boundary lines with the same brightness characteristics, e.g., 0 or 1. Consider two boundary edge detection algorithms that can be applied in navigational systems constructed on the basis of the generalized approach to signal processing in the presence of noise.

The first algorithm is known as the Roberts operator or detection operator of boundary lines by four object image elements. This algorithm is based on the estimation and choice of image fragments with a high gradient level. If the digital object image is represented by the two-dimensional function $g(i, j)$, then a gradient level at the point (i, j) is given by

$$\|\nabla g(i,j)\| \approx R(i,j) = \sqrt{[g(i,j)-g(i+1,j+1)]^2 + [g(i,j+1)-g(i+1,j)]^2}\,. \tag{14.25}$$

Reference to Equation (14.24) shows that for image fragments with a constant brightness characteristic, the function $R(i, j)$ is equal to zero. The function $R(i, j)$ increases when the brightness characteristics of image elements change. The difference algorithm of absolute sum determination of diagonal gradients is more efficient from a computational viewpoint:[5]

$$F(i,j) = |g(i,j)-g(i+1,j+1)| + |g(i,j+1)-g(i+1,j)|\,. \tag{14.26}$$

After gradient determination, we can carry out quantization according to

$$F_q(i,j) = \begin{cases} 1, & F(i,j) \geq K_g, \\ 0, & F(i,j) < K_g, \end{cases} \tag{14.27}$$

where K_g is the threshold. The use of Equation (14.26) and Equation (14.27) allows us to form a boundary image in addition to the initial object image.

For the second boundary edge detection algorithm, we take into account eight neighboring elements of the image element (i, j) for estimation of gradient.[5] The gradient is determined by

$$(G,W)_{ij} = \sum_{K=1}^{3}\sum_{L=1}^{3} g(i+K-2, j+L-2) \cdot W(K,L)\ , \tag{14.28}$$

where $g(i, j)$ are the digital image elements; and $\mathbf{W}(K, L)$ is the weight function in matrix form (the matrix dimension is 3×3). The weight function has many forms:[6]

- The weight functions of a smooth gradient

$$\mathbf{W}_1 = \begin{Vmatrix} 1 & 1 & 1 \\ 0 & 0 & 0 \\ -1 & -1 & -1 \end{Vmatrix} \quad \text{and} \quad \mathbf{W}_2 = \begin{Vmatrix} 1 & 0 & 1 \\ 1 & 0 & -1 \\ 1 & 0 & 1 \end{Vmatrix}\ . \tag{14.29}$$

- The Soibel weight functions

$$\mathbf{W}_1 = \begin{Vmatrix} 1 & 2 & 1 \\ 0 & 0 & 0 \\ -1 & -2 & -1 \end{Vmatrix} \quad \text{and} \quad \mathbf{W}_2 = \begin{Vmatrix} 1 & 0 & -1 \\ 2 & 0 & -2 \\ 1 & 0 & -1 \end{Vmatrix}\ . \tag{14.30}$$

- The isotropic weight functions

$$\mathbf{W}_1 = \begin{Vmatrix} 1 & \sqrt{2} & 1 \\ 0 & 0 & 0 \\ -1 & -\sqrt{2} & -1 \end{Vmatrix} \quad \text{and} \quad \mathbf{W}_2 = \begin{Vmatrix} 1 & 0 & -1 \\ \sqrt{2} & 0 & -\sqrt{2} \\ 1 & 0 & -1 \end{Vmatrix}\ . \tag{14.31}$$

The gradient level at the point (i, j) is equal to

$$\| \nabla g(i,j) \| = \sqrt{S_x^2(i,j) + S_y^2(i,j)}\ , \tag{14.32}$$

where

$$S_x(i,j) = (G, \mathbf{W}_1) \quad \text{and} \quad S_y(i,j) = (G, \mathbf{W}_2)\ . \tag{14.33}$$

By analogy with Equation (14.25) we can use the following formula:

$$F(i,j) = | S_x(i,j) | + | S_y(i,j) |\ . \tag{14.34}$$

Quantization is carried out in accordance with Equation (14.26).

The choice of the threshold K_g in Equation (14.27) has a great effect on the efficiency of the generalized signal and image processing algorithm. The threshold value depends on the object image characteristics: number of boundary lines, clearness of boundary lines, etc. There are three methods to automatically define the threshold.[7]

The first method is called the method of the main boundaries. With this method, the threshold is chosen so that only the main boundary lines remain after digital image processing. The numbers of 1 and 0 are very high for an image; so the probability of false peaks of the correlation function is high, too. To circumvent these phenomena we use only the image elements with level 1, a technique that gives good results. The second method is called the method of averaging by image fragment. In this case, the gradient images obtained with Soibel weight functions are quantized on two levels based on the average value for nine neighboring image elements with the central point (i, j):

$$A = 0.11 \sum_{m=i-1}^{i+1} \sum_{n=j-1}^{j+1} F(m,n)\ . \tag{14.35}$$

The quantization equation takes the following form

$$F_q(i,j) = \begin{cases} 1, & \text{if} \quad F(i,j) \geq kA, \\ 0, & \text{otherwise,} \end{cases} \tag{14.36}$$

where k is the scale coefficient. With a low value of k, the boundary lines of the image become thick and with high values of k, they become thin. An analogous approach is used in Reference 8.[8] The third method is called the method of the mean and mean square deviation. In this method, there are two thresholds:

$$L = m - K\sigma \quad \text{and} \quad H = m + K\sigma, \tag{14.37}$$

where k is the scale coefficient ($k \geq 0$). The quantization equation takes the following form:

$$F_q(i,j) = \begin{cases} 1, & F(i,j) \geq H, \\ 0, & F(i,j) \leq L. \end{cases} \tag{14.38}$$

Thus, the image elements with gradient levels belonging to the interval $[H, L]$ are not taken into consideration with the use of the generalized image processing algorithm. Rejection of image elements with gradient levels that are close to the mean reduces noise.

14.4 Conclusions

The discussion in this chapter allows us to draw the following conclusions. There are two types of object image distortions in form: the geometrical distortions and the distortions in brightness characteristics. Sources of these distortions are dissimilar. A source of the geometrical distortions is inaccurate information regarding the object's position in space. As a rule, the geometrical distortions can appear during transmission of object images by various sensors when an implementation of these sensors assumes various tracking angles. Sources of the distortions in brightness characteristics of object images are changes in meteorological or seasonal conditions.

If there are various types of sensors, for example, the radar side-scanning sensors and the optical scanning sensors, it is necessary to carry out one-to-one correspondence with a geometrical transformation or correction. Two techniques can be applied with the use of the generalized image processing algorithm based on the generalized approach to signal processing in the presence of noise. These are the perspective transform and the polynomial estimation. With the use of the first technique, it is necessary to first transform the object image in the radar-side-scanning plane to the coordinate system of angles θ, γ, and ψ. The second technique is widely used because additional information is not required and the information included in the obtained object images is sufficient.

If the same objects are obtained by various sensors, the distributions of brightness characteristics are different. It is necessary to carry out one-to-one correspondence with the object images. For this purpose, two transformation techniques can be used. The first technique is based on knowledge of statistical characteristics of object images. The second technique does not require that information because the radar object image is transformed so that the brightness characteristics correspond to the exponential distribution law.

The boundary lines are the most stable features of compared object images. The boundary lines can be considered as a characteristic of images of electromagnetic spectrum. Many boundary edge detection algorithms are constructed. The main principle of boundary edge detection is based on estimation of the gradient for each resolution element of the object image.

References

1. Wong, R., Sensor transformations, *IEEE Trans.*, Vol. SMC-7, No. 12, 1977, pp. 836–841.
2. Lipkin, B. and Rosenfeld, A., Eds., *Picture Processing and Psychopictories*, New York, 1970.
3. Berry, J. and Yoo, J., Geometric preprocessing of sensor data used for image matching, in *Proceedings of the SPIE*, Vol. 186, 1979, pp. 2–11.
4. Wong, R. and Hall, E., Image transformations, in *Proceedings of the IEEE 4th International Joint Conference on Pattern Recognition*, 1978, pp. 939–942.
5. Boland, J., Peters, E. et al., Automatic correlation of non-compatible imaging systems, in *Proceedings of IEEE SOUTHEASTCON*, 1979, pp. 230–233.
6. Frei, W. and Chen, C., Fast boundary detection: a generalization and a new algorithm, *IEEE Trans.*, Vol. C-26, No. 10, 1977, pp. 988–998.
7. Boland, J. et al., A pattern recognition technique for scene matching of dissimilar imagery, in *Proceedings of the IEEE 18th Conference on Decision and Control*, 1979, pp. 806–811.
8. Robinson, G., Detection and coding of edges using directional masks, *Opt. Eng.*, Vol. 16, No. 6, 1977, pp. 653–667.

Appendix I

Classification of Stochastic Processes

In analysis of the considered problems, we investigated many types of stationary and nonstationary stochastic processes (the target return signal). So it is necessary to classify the analyzed stochastic processes. The basis of this classification is the behavior of the two-dimensional correlation function $R(t, \tau)$ of fluctuations of the target return signal, where $\tau = t_2 - t_1$ and $t = 0.5(t_1 + t_2)$. We take into consideration only those general features of the behavior of $R(t, \tau)$ that significantly influence the structure of the two-dimensional power spectral density $S(\omega, \Omega)$ of target return signal fluctuations, namely, the periodic or nonperiodic dependence of $R(t, \tau)$ on the variable τ and on the variable t in the case of the nonstationary target return signal. The function of τ defines stochastic (random) features of the target return signal and the function of t defines regular (deterministic) features of the target return signal. From the viewpoint of regular features of the target return signal, we can define stationary processes (see Figure I.1–Figure I.3) and two forms of nonstationary processes: nonperiodic (Figure I.4–Figure I.6) and periodic (Figure I.7–Figure I.9) target return signals, depending on the time t. From the viewpoint of stochastic processes, we can define nonperiodic target return signals (Figure I.1, Figure I.4, and Figure I.7), periodic target return signals (Figure I.2, Figure I.5, and Figure I.8), and the quasiperiodic target return signals (Figure I.3, Figure I.6, and Figure I.9), in which the correlation function $R(t, \tau)$ can be both periodic and nonperiodic with respect to the variable τ. The simplest example of this correlation function is the product discussed in Chapter 3, Chapter 6, and Chapter 9: $R(\tau) = R_1(\tau) \times R_2(\tau - nT_p)$. In addition, we discussed more complex correlation function (see Chapter 4 and Chapter 10).

We now list some problems where we used the correlation functions and power spectral densities of target return signal fluctuations shown in Figure I.1–Figure I.9:

1. Doppler fluctuations of the target return signal with the simple harmonic searching signal; chaotic motion of scatterers with the simple harmonic searching signal; fluctuations of the transformed target return signal with the linear frequency-modulated searching signal in the case of stationary radar (see Figure I.1); line or conical

scanning with the simple harmonic searching signal if the radar is stationary (Figure I.2); spiral scanning with the simple harmonic searching signal if the radar is stationary; line scanning with chaotic motion of scatterers and the simple harmonic searching signal if the radar is stationary; rotation of the radar antenna polarization plane and chaotic motion of scatterers with the simple harmonic searching signal if the radar is stationary (Figure I.3).

2. Doppler fluctuations of the target return signal under the continuous searching signal with nonperiodic variations in power or the radar moves with acceleration (Figure I.4); line or conical scanning under the continuous searching signal with nonperiodic variations in power if the radar is stationary (Figure I.5); spiral scanning under the simple harmonic searching signal with nonperiodic variations in power if the radar is stationary; line scanning under chaotic motion of scatterers and the simple harmonic searching signal with nonperiodic variations in power if the radar is stationary; rotation of the radar antenna polarization plane and chaotic motion of scatterers with nonperiodic variations in power under the simple harmonic searching signal if the radar is stationary (Figure I.6).
3. Doppler fluctuations of the target return signal under the continuous searching signal with periodic variations in power or in the velocity of moving radar (Figure I.7); target return signal fluctuations under scanning of the three-dimensional (space) target with the pulsed searching signal if the radar is stationary; segment scanning with the simple harmonic searching signal if the radar is stationary (Figure I.8); target return signal fluctuations under scanning of the three-dimensional (space) target by the pulsed searching signal if the radar is moving; line scanning with the simple harmonic searching signal if the radar is moving (Figure I.9).

The main reason for the nonstationary state of the target return signal in these problems is the dependence of the amplitude variance (or power) of the target return signal on time (the constant component in all considered cases is equal to zero because we consider the radio signal). Using the main results discussed in Chapter 8–Chapter 10, we can obtain the correlation functions and power spectral densities of target return signal fluctuations, in which the frequency of the searching signal is a function of time. The proposed classification of stochastic processes given here (the target return signal) is true, naturally, only within the limits of correlation theory and does not exclude other forms of classification.[1,2]

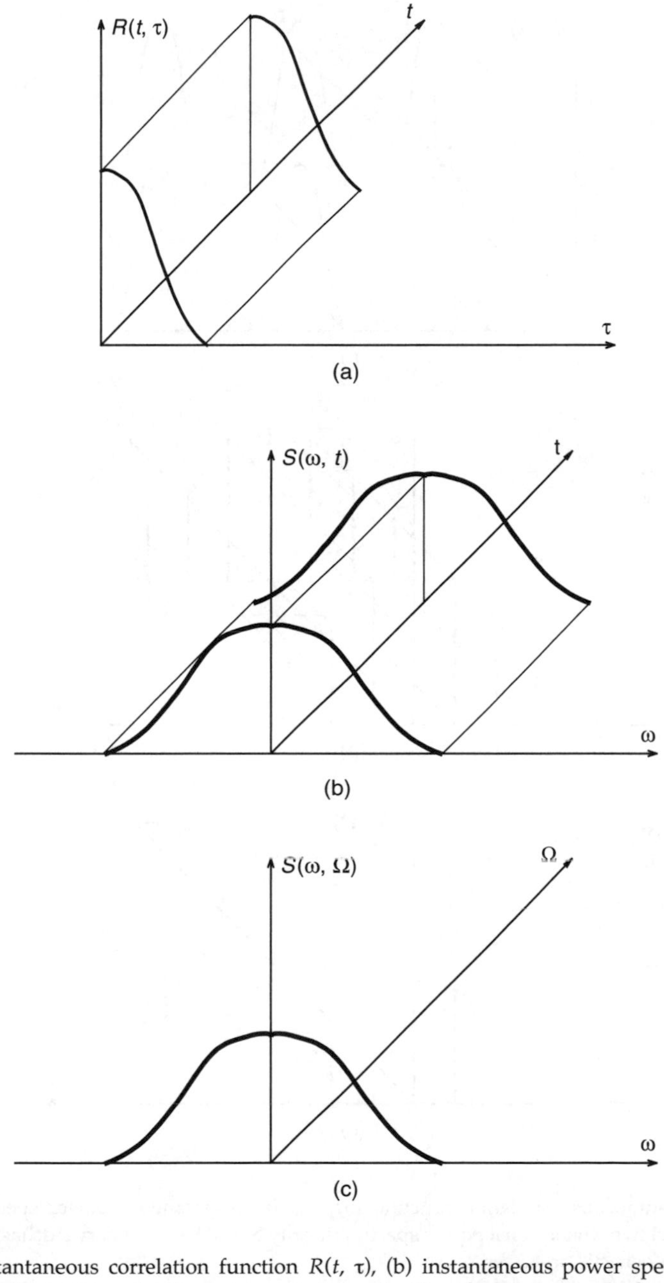

FIGURE I.1
(a) The instantaneous correlation function $R(t, \tau)$, (b) instantaneous power spectral density $S(\omega, t)$, and (c) two-dimensional power spectral density $S(\omega, \Omega)$ of stationary stochastic processes — nonperiodic stochastic processes.

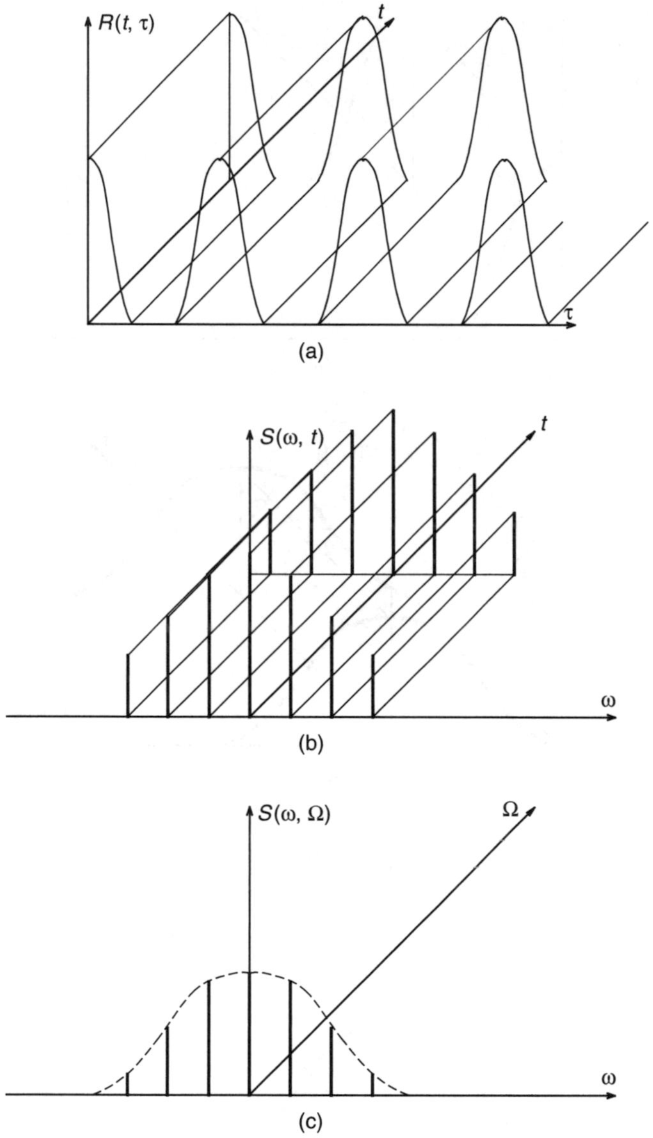

FIGURE I.2
(a) The instantaneous correlation function $R(t, \tau)$, (b) instantaneous power spectral density $S(\omega, t)$, and (c) two-dimensional power spectral density $S(\omega, \Omega)$ of stationary stochastic processes — periodic stochastic processes.

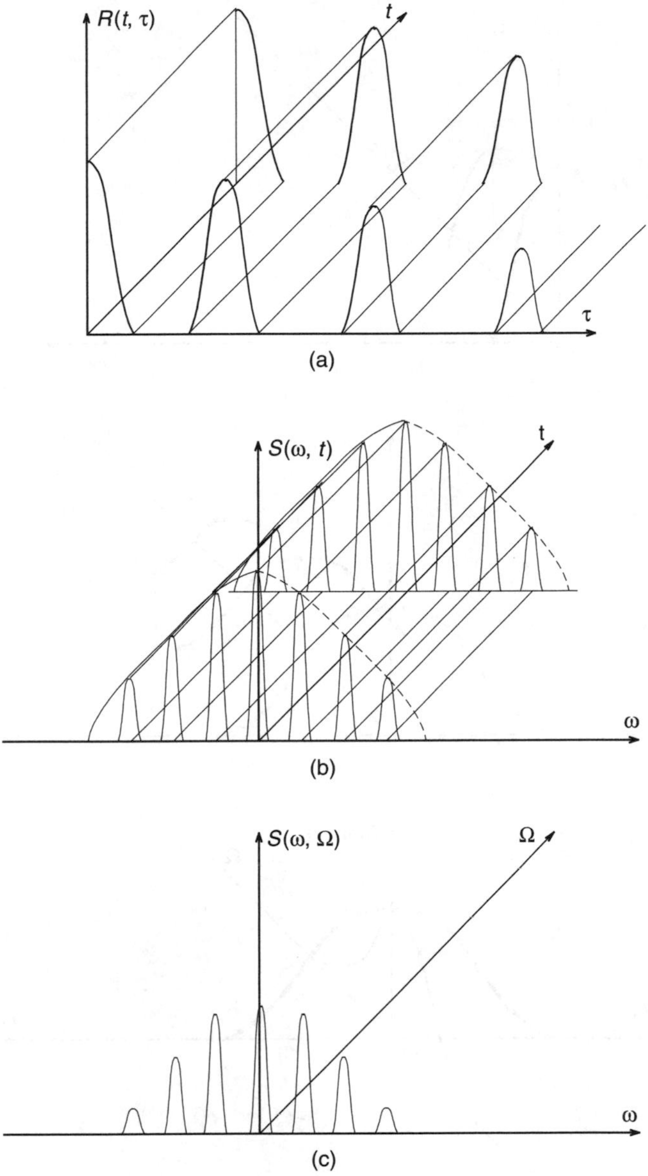

FIGURE I.3
(a) The instantaneous correlation function $R(t, \tau)$, (b) instantaneous power spectral density $S(\omega, t)$, and (c) two-dimensional power spectral density $S(\omega, \Omega)$ of stationary stochastic processes — quasiperiodic stochastic processes.

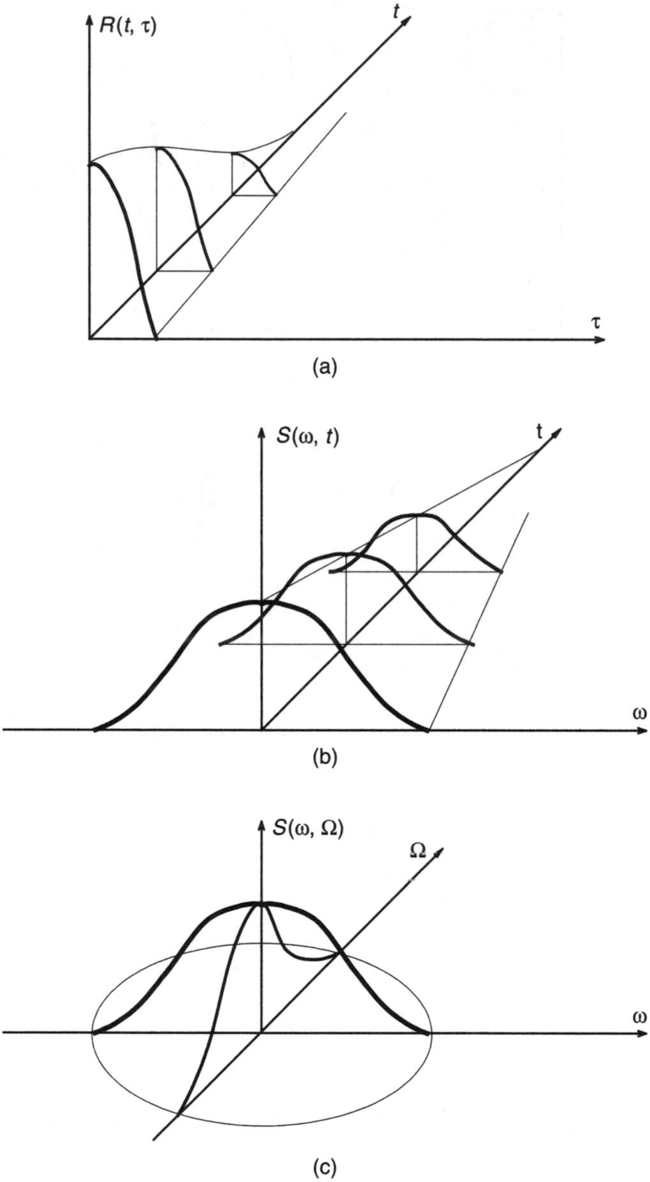

FIGURE I.4
(a) The instantaneous correlation function $R(t, \tau)$, (b) instantaneous power spectral density $S(\omega, t)$, and (c) two-dimensional power spectral density $S(\omega, \Omega)$ of nonstationary nonperiodic processes — nonperiodic stochastic processes.

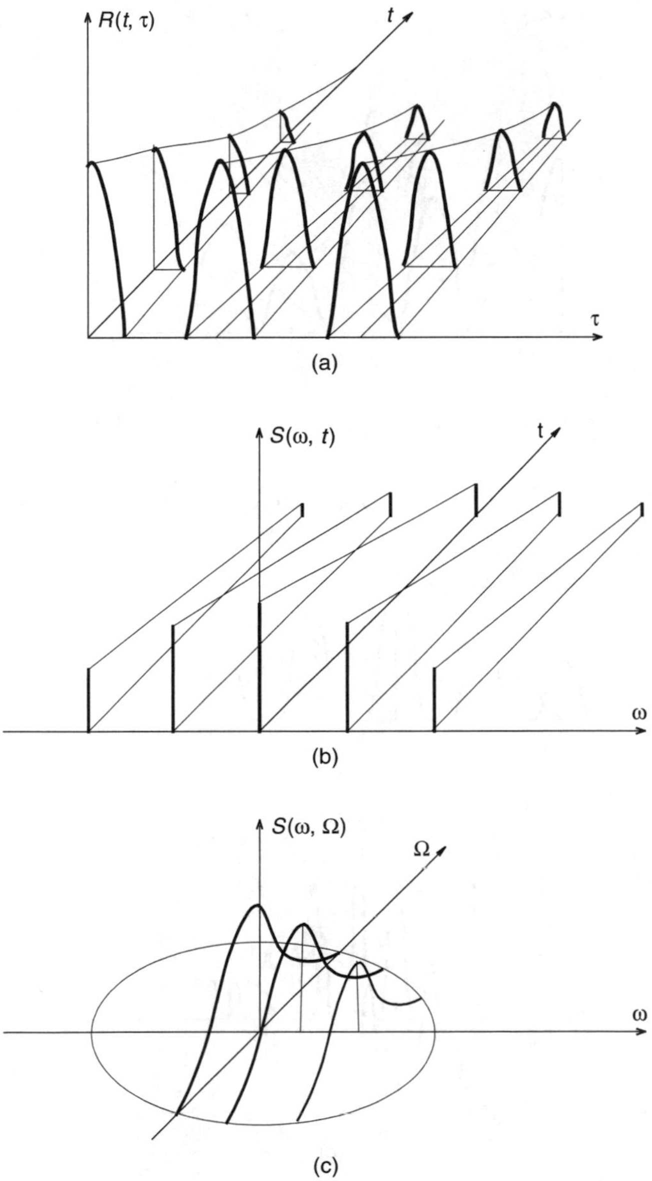

FIGURE I.5
(a) The instantaneous correlation function $R(t, \tau)$, (b) instantaneous power spectral density $S(\omega, t)$, and (c) two-dimensional power spectral density $S(\omega, \Omega)$ of nonstationary nonperiodic processes — periodic stochastic processes.

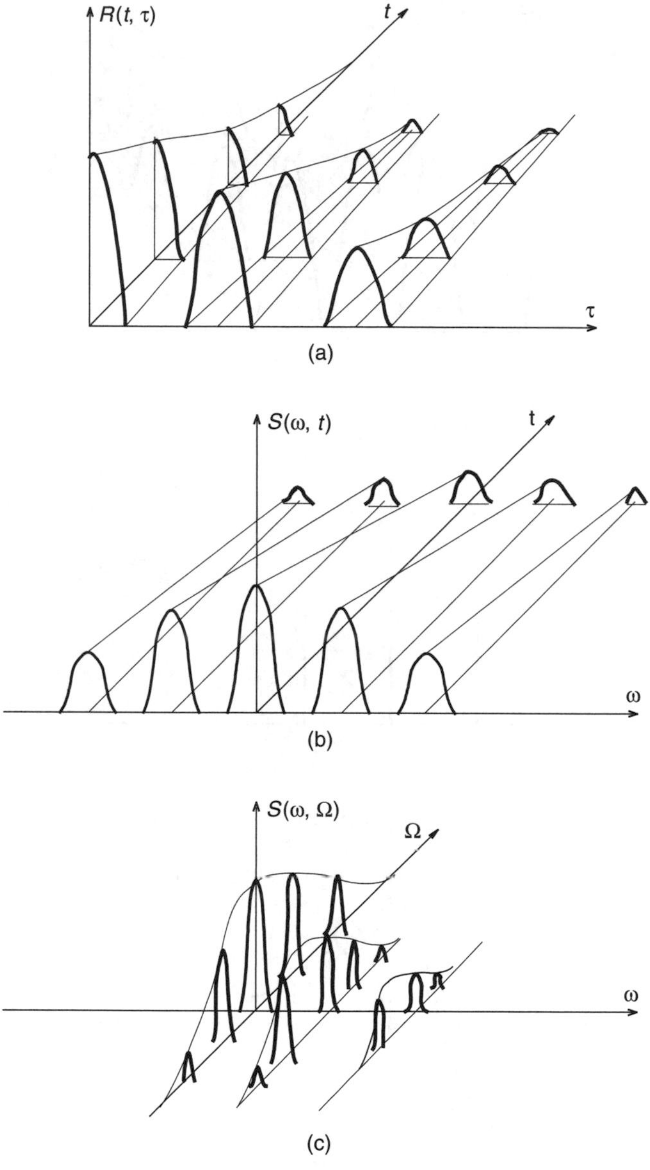

FIGURE I.6
(a) The instantaneous correlation function $R(t, \tau)$, (b) instantaneous power spectral density $S(\omega, t)$, and (c) two-dimensional power spectral density $S(\omega, \Omega)$ of nonstationary nonperiodic processes — quasiperiodic stochastic processes.

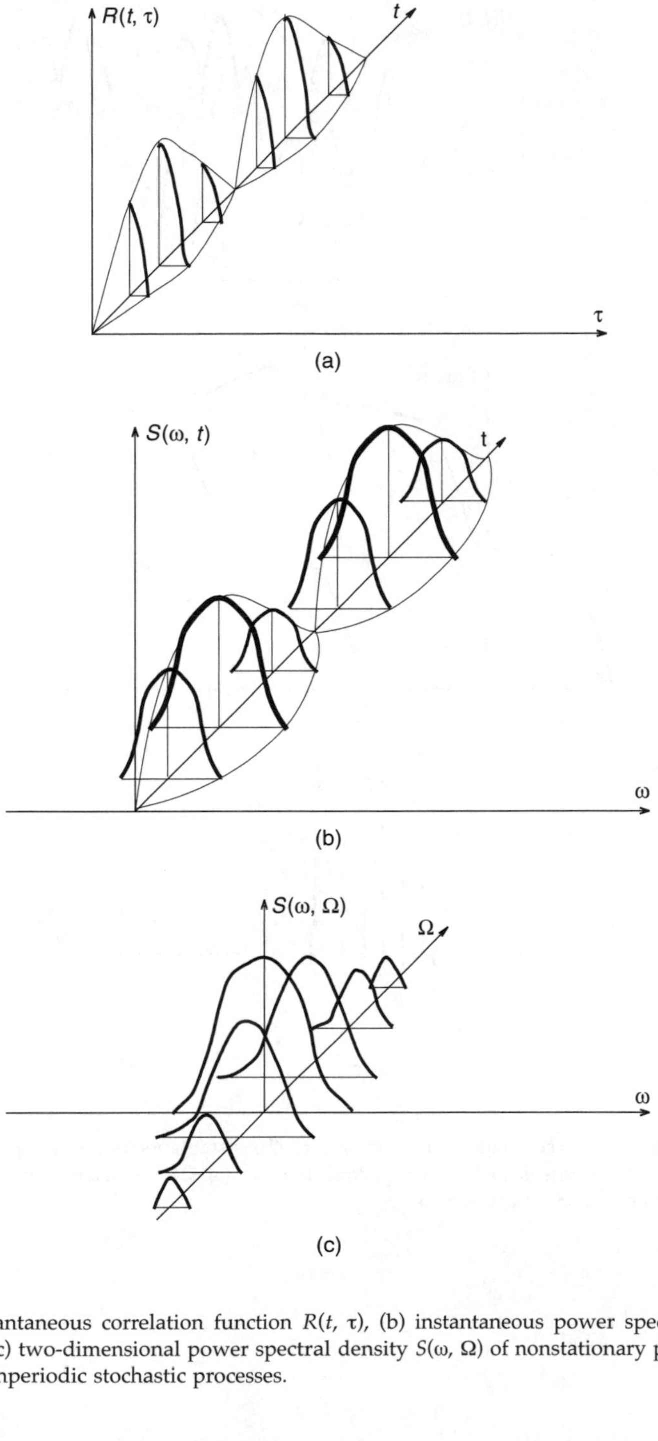

FIGURE I.7
(a) The instantaneous correlation function $R(t, \tau)$, (b) instantaneous power spectral density $S(\omega,t)$, and (c) two-dimensional power spectral density $S(\omega, \Omega)$ of nonstationary periodic processes — nonperiodic stochastic processes.

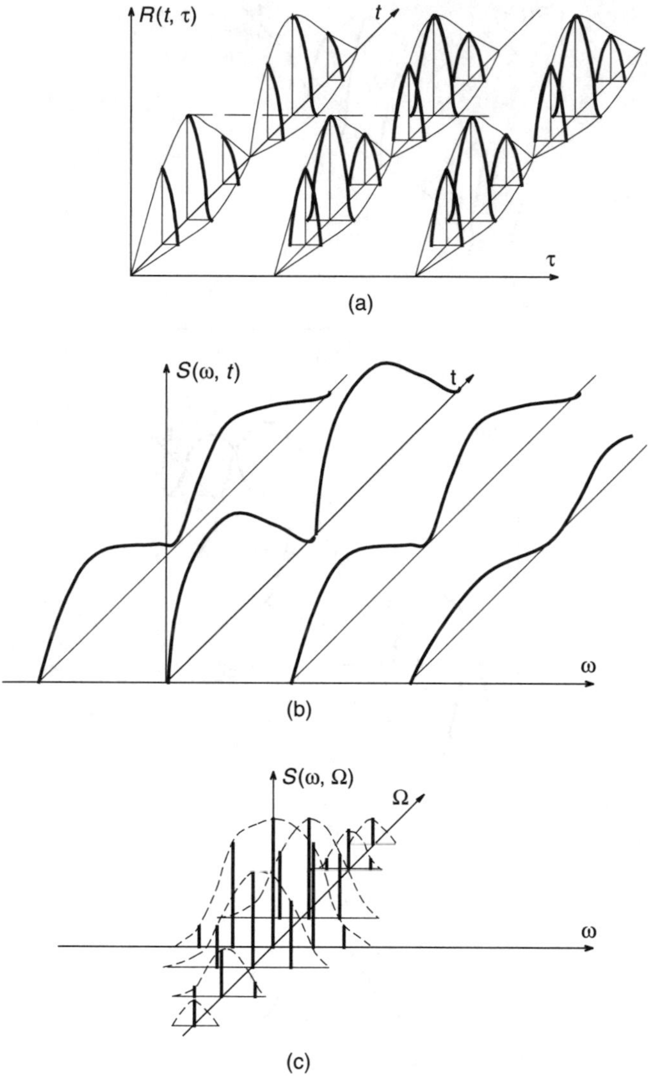

FIGURE I.8
(a) The instantaneous correlation function $R(t, \tau)$, (b) instantaneous power spectral density $S(\omega, t)$, and (c) two-dimensional power spectral density $S(\omega, \Omega)$ of nonstationary periodic processes — periodic stochastic processes.

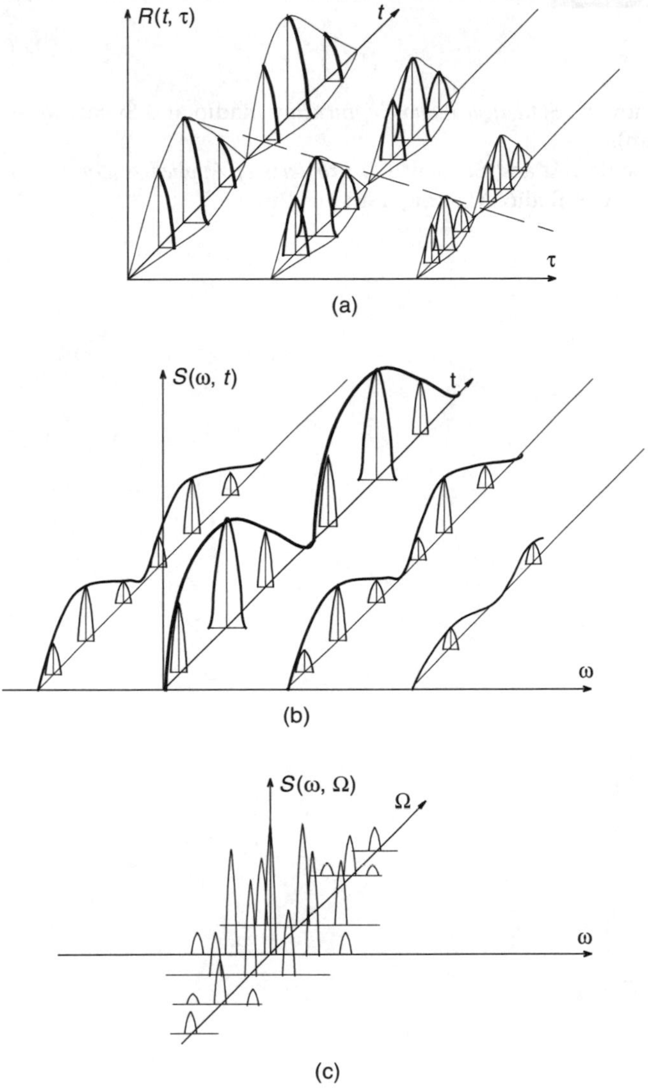

FIGURE I.9
(a) The instantaneous correlation function $R(t, \tau)$, (b) instantaneous power spectral density $S(\omega, t)$, and (c) two-dimensional power spectral density $S(\omega, \Omega)$ of nonstationary periodic processes — quasiperiodic stochastic processes.

References

1. Tikhonov, V., *Statistical Radio Engineering,* Radio and Svyaz, Moscow, 1982 (in Russian).
2. Romanenko, A. and Sergeev, G., *Problems of Applied Analysis of Stochastic Processes,* Soviet Radio, Moscow, 1968 (in Russian).

Appendix II

The Power Spectral Density of the Target Return Signal with Arbitrary Velocity Vector Direction of Moving Radar in Space and with the Presence of Roll and Pitch Angles

The main parameters of the target return signal from the three-dimensional (space) target — the power, the instantaneous and average frequency, and the power spectral density bandwidth of target return signal fluctuations — depend on the radar velocity and position of the radar antenna directional diagram with respect to the velocity vector of moving radar and the reflecting surface. Before, we believed that the position of the directional diagram was defined by angles reckoned from the velocity vector of moving radar. In fact, the direction of the velocity vector of moving radar is defined relative to the axes of the immovable coordinate system of the Earth, and the position of the directional diagram is defined with respect to the axes of aircraft. Due to vibrations of aircraft, the directional diagram can change its position with respect to the velocity vector of moving radar and the underlying reflecting surface that leads to changes in all characteristics and parameters of the power spectral density of target return signal fluctuations.

Introduce three Cartesian coordinate systems: the immovable horizontal coordinate system $OXYZ$ (see Figure II.1), whose axes are parallel to the axes of the immovable coordinate system of the Earth and whose center O is at the aircraft's center of gravity; the coordinate system $OX_1Y_1Z_1$ localized by the aircraft, whose center coincides with the center of the horizontal coordinate system $OXYZ$ (see Figure II.1), and whose axes, in the general case, are rotated by the angles of heading ψ_c, pitch θ_c, and roll γ_c with respect to the horizontal coordinate system $OXYZ$. Because we assume that the radar antenna is hardly connected with the aircraft, $\psi_c = 0$, and under the condition $\psi_c = \theta_c = \gamma_c = 0°$, the coordinate system $OXYZ$ coincides with the coordinate system of the aircraft $OX_1Y_1Z_1$; the intermediate coordinate system $OX'Y'Z'$ is shown in Figure II.1 to illustrate a jump from the coordinate system $OXYZ$ to the coordinate system $OX_1Y_1Z_1$. In the general case, a vector defining the Doppler shift in the target return signal frequency is the sum of three vectors

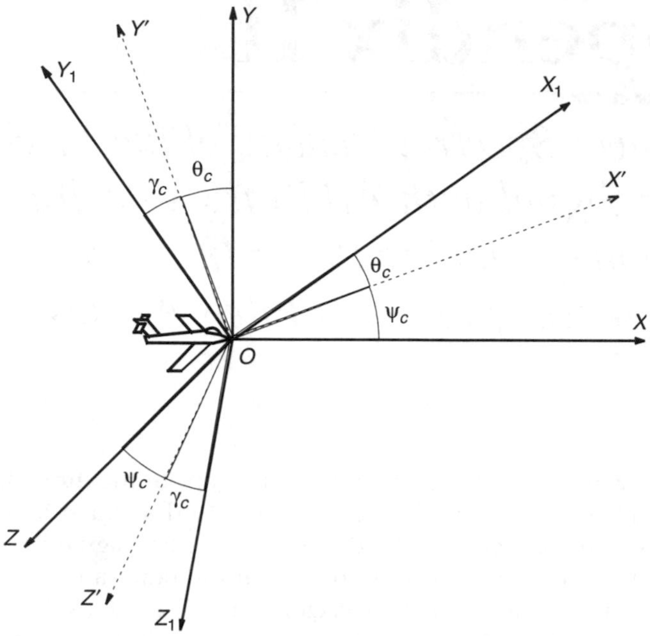

FIGURE II.1
The coordinate systems $OXYZ$ and $OX_1Y_1Z_1$.

$$\mathbf{V}_\Sigma = \mathbf{V} - \mathbf{U} + [\mathbf{r}, \boldsymbol{\omega}] \,, \tag{II.1}$$

where **V** is the velocity vector of the moving aircraft's center of gravity relative to the immovable reflecting surface; **U** is the velocity vector of moving scatterers of the underlying surface, for example, a sea surface; $[\mathbf{r}, \boldsymbol{\omega}]$ is the vector of linear velocity of the moving radar antenna caused by angle displacement of the aircraft relative to its center of gravity when the radar antenna is spaced at a distance **r** from the center of gravity of the aircraft, and $\boldsymbol{\omega}$ is the vector of angular velocity of the aircraft. Based on Equation (II.1) we can express the vector **V** in the horizontal coordinate system in the following form (see Figure II.2):

$$\mathbf{V} = \{\, V_x, V_y, V_z \} = \{V \cos\varepsilon_0 \cos\alpha, \;\; V \sin\varepsilon_0, \;\; -V\cos\varepsilon_0 \sin\alpha\} \,. \tag{II.2}$$

The direction "radar–locality" defined by the basis vector $\mathbf{I}_1$ can be represented in the coordinate system $OX_1Y_1Z_1$ in the following form (see Figure II.3):

$$\boldsymbol{I}_1 = \{\, l_x, l_y, l_z \} = \{\cos\gamma_1 \cos\beta_1, \;\; -\sin\gamma_1, \;\; -\cos\gamma_1 \sin\beta_1\} \,, \tag{II.3}$$

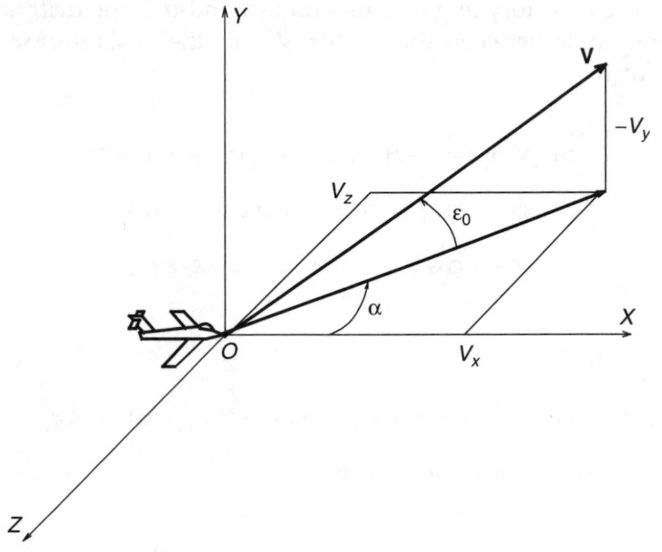

FIGURE II.2
The velocity vector **V** of moving radar in the coordinate system *OXYZ*.

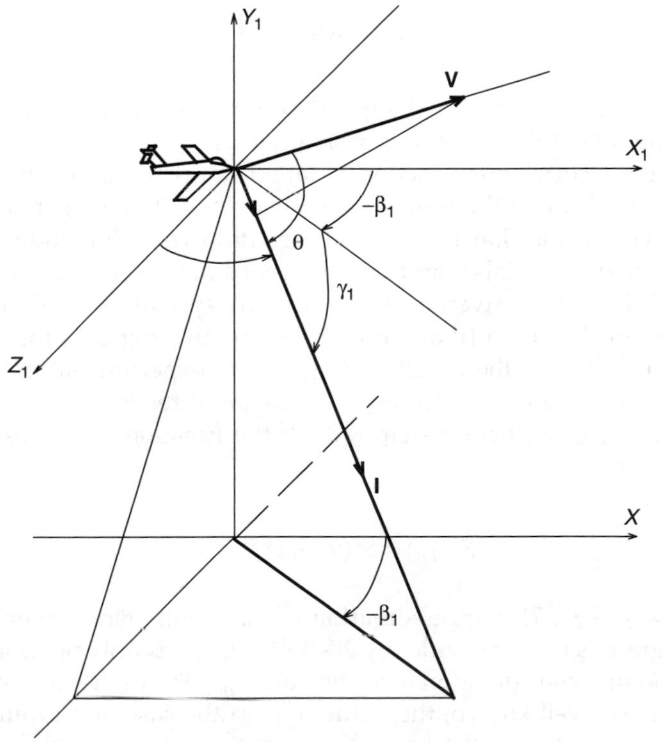

FIGURE II.3
The basis vector **I** in the coordinate system $OX_1Y_1Z_1$.

where ε_0 is the trajectory angle of the aircraft and α is the drift angle of the aircraft. The angle between the vector $\mathbf{V}$ and the basis vector $\mathbf{I}_h$ has the following form:

$$\begin{aligned}\cos(\mathbf{V}\cdot\mathbf{I}_h) &= \cos\theta = \cos\varepsilon_0\cos\alpha\cos\gamma_h\cos\beta_h \\ &-\sin\varepsilon_0\sin\gamma_h - \cos\varepsilon_0\sin\alpha\cos\gamma_h\sin\beta_h \quad , \\ &= \cos\varepsilon_0\cos\gamma_h\cos(\alpha+\beta_h) - \sin\varepsilon_0\sin\gamma_h\end{aligned} \tag{II.4}$$

where

$$\begin{aligned}\cos\gamma_h\cos\beta_h &= \cos\gamma_1\cos\beta_1\cos\theta_c + \sin\gamma_1\sin\theta_c\cos\gamma_c \\ &- \cos\gamma_1\sin\beta_1\sin\gamma_c\sin\theta_c\end{aligned} \quad ; \tag{II.5}$$

$$\begin{aligned}\sin\gamma_h &= \sin\gamma_1\cos\theta_c\cos\gamma_c - \cos\gamma_1\cos\beta_1\sin\theta_c \\ &- \cos\gamma_1\sin\beta_1\sin\gamma_c\cos\theta_c\end{aligned} \quad ; \tag{II.6}$$

$$\cos\gamma_h\sin\beta_h = \cos\gamma_1\sin\beta_1\cos\gamma_c + \sin\gamma_1\sin\gamma_c \ . \tag{II.7}$$

The angles γ_h and β_h define the position of the basis vector $\mathbf{I}_h$ in the coordinate system $OXYZ$, as with the basis vector $\mathbf{I}_1$.

To define the correlation function of target return signal fluctuations, it is necessary to substitute the formula $\Delta\rho = -V\cdot\tau\cos\theta$ in the general formula using the $\cos\theta$-expansion in terms of Equation (II.5)–Equation (II.7) with respect to the differentials φ and ψ in the coordinate system $OX_1Y_1Z_1$. As the directional diagram is given in the coordinate system $OX_1Y_1Z_1$, it does not vary. Based on Equation (II.6) we must define the angle γ_h, for which it is necessary to estimate the function $S^\circ(\gamma_h)$ using experimental or theoretical data. Because the angle γ_{h_0} corresponds to the direction of the directional diagram in the coordinate system $OXYZ$, the function can be estimated in the following form:

$$S^\circ(\gamma_h) = S^\circ(\gamma_{h_0})\cdot e^{k_1\psi_h + k_2\psi_h^2} , \tag{II.8}$$

where $\psi_h = \gamma_h - \gamma_{h_0}$. This approximation is sufficiently close when the directional diagram is not very wide (by 20–30°). Also, it is only necessary to take the factor k_2 into consideration near the value $\gamma_h \approx 90°$ in the case of a smooth sea surface. As well known, the value of k_1 in the case of a ground surface is not very high, but in the case of a smooth sea surface, the value of k_1 reaches 20–30 at $\gamma_h = 60$–$80°$ and the value of k_2 reaches approximately 200.[1]

In the case of slope scanning, the parameter $k_1(\gamma_h)$ can be defined by experimental data

$$k_1(\gamma_{h_0}) = 13.2 \cdot \frac{10\lg S^\circ\left(\gamma_{h_0} - \frac{\Delta_v}{2\sqrt{2}}\right) - 10\lg S^\circ\left(\gamma_{h_0} + \frac{\Delta_v}{2\sqrt{2}}\right)}{\frac{\Delta_v}{\sqrt{2}}}, \tag{II.9}$$

where the value of Δ_v is given in degrees. We can define other parameters depending on the angle γ_h, for example,

$$T_d = \frac{2h}{c\sin\gamma_h}. \tag{II.10}$$

For this purpose, we have to express the function of the angle γ_h in the form of a series expansion with respect to the variables φ and ψ to define a function relating the angle ψ_h to the variables φ and ψ using Equation (II.6).

For example, define the power spectral density of target return signal fluctuations with the nonmodulated searching signal in the case of slope scanning. We assume that the directional diagram $g(\varphi, \psi)$ can be approximated by the Gaussian law and that the function $S^\circ(\gamma_h)$ can be approximated by the exponent under the condition $k_2 = 0$. We can show that the power spectral density is Gaussian:[2]

$$S(\omega) = \frac{p}{\Delta\Omega} \cdot e^{-\pi\frac{(\omega-\Omega)^2}{\Delta\Omega^2}}; \tag{II.11}$$

$$\Omega = \omega_0 + \Omega_0 + \Delta\Omega_k + \Delta\Omega_\rho; \tag{II.12}$$

$$p = \frac{P_S\lambda^2 G_0^2 S^\circ(\gamma_{h_0})\Delta_h\Delta_v \sin\gamma_{h_0}}{128\pi^3 h^2} \cdot e^{0.125\pi(k_{h_1}^2\Delta_h^{(2)} + k_{v_1}^2\Delta_v^{(2)})}; \tag{II.13}$$

$$\Delta\Omega = 2\sqrt{2}\,\pi\lambda^{-1}\sqrt{\Delta_v^{(2)}V_1^2 + \Delta_h^{(2)}V_2^2}; \tag{II.14}$$

$$\begin{aligned} V_1 &= -(V_{x_1}\sin\gamma_0\cos\beta_0 + V_{y_1}\cos\gamma_0 + V_{z_1}\sin\gamma_0\sin\beta_0) \quad \text{and} \\ V_2 &= -(V_{x_1}\sin\beta_0 - V_{z_1}\cos\beta_0); \end{aligned} \tag{II.15}$$

$$\begin{cases} V_{x_1} = V_x\cos\theta_c; \\ V_{y_1} = -V_x\sin\theta_c\cos\gamma_c + V_y\cos\gamma_c\cos\theta_c + V_z\sin\gamma_c; \\ V_{z_1} = V_x\sin\gamma_c\sin\theta_c - V_y\sin\gamma_c\cos\theta_c + V_z\cos\gamma_c; \end{cases} \tag{II.16}$$

Here, the angles γ_0 and β_0 define the direction of the directional diagram axis in the coordinate system $OX_1Y_1Z_1$; V_{x_1}, V_{y_1}, and V_{z_1} are the components of the vector $\mathbf{V}$ in the coordinate system $OX_1Y_1Z_1$;

$$\begin{aligned}
&\Omega_0 = 4\pi V\lambda^{-1}\cos\theta_0 = \Omega_{D_{x1}} + \Omega_{D_{y1}} + \Omega_{D_{z1}};\\
&\Omega_{D_{x1}} = 4\pi V_{x_1}\lambda^{-1}\cos\gamma_0\cos\beta_0;\\
&\Omega_{D_{y1}} = -4\pi V_{y_1}\lambda^{-1}\sin\gamma_0; \quad \text{and}\\
&\Omega_{D_{z1}} = -4\pi V_{z_1}\lambda^{-1}\cos\gamma_0\sin\beta_0;
\end{aligned} \tag{II.17}$$

$$\begin{aligned}
&\Delta\Omega_k = \delta_{x_1}\Omega_{D_{x1}} + \delta_{y_1}\Delta\Omega_{D_{y1}} + \delta_{z_1}\Delta\Omega_{D_{z1}} \quad \text{and}\\
&\Delta\Omega_\rho = \delta_{\rho_x}\Omega_{D_{x1}} + \delta_{\rho_y}\Omega_{D_{y1}} + \delta_{\rho_z}\Omega_{D_{z1}};
\end{aligned} \tag{II.18}$$

$$\begin{aligned}
&\delta_{x_1} = -\frac{k_v\Delta_v^{(2)}\sin\gamma_0\cos\beta_0 + k_h\Delta_h^{(2)}\sin\beta_0}{4\pi\cos\gamma_0\cos\beta_0};\\
&\delta_{y_1} = \frac{k_1\Delta_v^{(2)}\operatorname{ctg}\gamma_0}{4\pi}; \quad \text{and}\\
&\delta_{z_1} = -\frac{k_v\Delta_v^{(2)}\sin\gamma_0\sin\beta_0 + k_h\Delta_h^{(2)}\cos\beta_0}{4\pi\cos\gamma_0\sin\beta_0};
\end{aligned} \tag{II.19}$$

$$\delta_{\rho_x} = \frac{\delta_{x_1}\operatorname{ctg}\gamma_{h_0}}{k_1}; \quad \delta_{\rho_y} = \frac{\delta_{y_1}\operatorname{ctg}\gamma_{h_0}}{k_1}; \quad \delta_{\rho_z} = \frac{\delta_{z_1}\operatorname{ctg}\gamma_{h_0}}{k_1}; \tag{II.20}$$

$$\begin{cases}
\alpha_1 = \dfrac{\cos\gamma_0\cos\theta_c\cos\gamma_c + \sin\gamma_0\cos\beta_0\sin\theta_c + \sin\gamma_0\sin\beta_0\sin\gamma_c\cos\theta_c}{\sin\gamma_{h_0}};\\[2ex]
\alpha_2 = \dfrac{\sin\beta_0\sin\theta_c - \cos\beta_0\sin\gamma_c\cos\theta_c}{\sin\gamma_{h_0}};
\end{cases} \tag{II.21}$$

$$\begin{aligned}
&k_v = \alpha_1 k_1\operatorname{tg}\gamma_{h_0};\\
&k_{v_1} = \alpha_1(k_1 + \operatorname{ctg}\gamma_{h_0})\operatorname{tg}\gamma_{h_0};\\
&k_h = \alpha_2 k_1\operatorname{tg}\gamma_{h_0}; \quad \text{and}\\
&k_{h_1} = \alpha_2(k_1 + \operatorname{ctg}\gamma_{h_0})\operatorname{tg}\gamma_{h_0},
\end{aligned} \tag{II.22}$$

where p is the target return signal power; and P_S is the power of the searching signal [see Equation (II.13)]. It should be pointed out that in the majority of cases, with low values of the parameters Δ_h and Δ_v, we can limit ourselves to the first factor and there is no need to take the second factor into consideration to define the target return signal power. For example, assuming that $\Delta_h = \Delta_v = 6°$, $\gamma_c = \theta_c = 0°$, $\gamma_0 = 65°$, $k_1 = 20$, we can write

$$e^{\frac{(k_1 + \text{ctg}\,\gamma_0)^2 \Delta_v^{(2)}}{8\pi}} = 1.172 \quad \text{or} \quad 0.7\ \text{dB}\ . \tag{II.23}$$

The effective bandwidth $\Delta\Omega$ given by Equation (II.14) depends on the directional diagram width in both the planes Δ_h and Δ_v. The parameter $\Delta\Omega_k$ given by Equation (II.18)–Equation (II.22) is the shift in the average frequency of the power spectral density of target return signal fluctuations caused by the effect of the function $S°(\gamma)$. The parameter $\Delta\Omega_\rho$ given by Equation (II.18) — Equation (II.22) is the shift in the average frequency of the power spectral density of target return signal fluctuations caused by distance variations within the directional diagram. The parameter $\Delta\Omega_\rho$ is k_1 times less than the parameter $\Delta\Omega_k$ [see Equation (II.18) — Equation (II.22)].

Assume that roll and pitch are absent, i.e., that the condition $\gamma_c = \theta_c = 0°$ is satisfied. In this case, the components of the velocity vector of moving radar for both coordinate systems $OXYZ$ and $OX_1Y_1Z_1$ are the same:

$$\begin{aligned}
&\gamma_{h_0} = \gamma_0,\\
&\alpha_1 = \text{ctg}\,\gamma_0,\\
&\alpha_2 = 0, k_h = 0, k_v = k_1;\\
&\delta_x = \delta_z = -\frac{k_1 \Delta_v^{(2)}\ \text{tg}\,\gamma_0}{4\pi};\\
&\delta_y = \frac{k_1 \Delta_v^{(2)}\ \text{ctg}\,\gamma_0}{4\pi};\\
&\delta_{\rho_x} = \delta_{\rho_z} = -\frac{\Delta_v^{(2)}}{4\pi}\ ;\ \text{and}\\
&\delta_{\rho_y} = \frac{\Delta_v^{(2)}\ \text{ctg}^2\gamma_0}{4\pi}\ ;
\end{aligned} \tag{II.24}$$

$$\Delta\Omega = 2\sqrt{2}\ \pi\lambda^{-1}\sqrt{\Delta_v^{(2)} V_1^2 + \Delta_h^{(2)} V_2^2}\ ; \tag{II.25}$$

$$\begin{aligned}
&V_1 = -V[\cos\varepsilon_0 \sin\gamma_0 \cos(\alpha + \beta_0) + \sin\varepsilon_0 \cos\gamma_0] = Va_1 \quad \text{and}\\
&V_2 = -V\cos\varepsilon_0 \sin(\alpha + \beta_0) = Vb_1\ .
\end{aligned} \tag{II.26}$$

It should be noted that the relative shifts δ_x, δ_z, δ_{ρ_x}, and δ_{ρ_z} of the horizontal components Ω_{D_x} and Ω_{D_z} are the same and increase with an increase in the angle γ_0 [see Equation (II.24)]. The shifts δ_y and δ_{ρ_y} have a different sign because the effects of the function $S^\circ(\gamma)$ and distance ρ increase the effective angle γ_0, and the shifts δ_y and δ_{ρ_y} increase with a decrease in the angle γ_0. An increase in the angle γ_0 leads to a decrease in the view of the vector **V** on the *X* and *Z* axes and to an increase in the view of the vector **V** on the *Y* axis. The values of the shifts δ_{ρ_x} and δ_{ρ_z} are constant for a given directional diagram and independent of the angle γ_0, but the shift δ_{ρ_y} depends mainly on the angle γ_0. We define the effect of each component of the velocity vector of moving radar on the effective power spectral density bandwidth of target return signal fluctuations. If only one component of the velocity vector of moving radar is different from zero, we can write

$$\begin{aligned}
&\Delta\Omega_x = 2\sqrt{2}\,\pi V_x \lambda^{-1}\sqrt{\Delta_v^{(2)}\sin^2\gamma_0\cos^2\beta_0 + \Delta_h^{(2)}\sin^2\beta_0}\ ;\\
&\Delta\Omega_y = 2\sqrt{2}\,\pi V_y \lambda^{-1}\Delta_v\cos\gamma_0\ ;\ \text{and} \qquad\qquad \text{(II.27)}\\
&\Delta\Omega_z = 2\sqrt{2}\,\pi V_z \lambda^{-1}\sqrt{\Delta_v^{(2)}\sin^2\gamma_0\sin^2\beta_0 + \Delta_h^{(2)}\cos^2\beta_0}\ .
\end{aligned}$$

It should be pointed out that in the case of the directional diagram with axial symmetry, there is no need to carry out complex mathematics for aircraft vibrations. It is necessary to define only the angles β_{h_0} and γ_{h_0} using Equation (II.5)–Equation (II.7) in the coordinate system $OXYZ$ that corresponds to the direction of the directional diagram axis and to define the parameter $k_1(\gamma_{h_0})$ using the function $S^\circ(\gamma)$ for the determined angle γ_0. All characteristics of the power spectral density of target return signal fluctuations are defined for these values of the angles under the condition $\gamma_c = \theta_c = 0°$. This technique of defining the power spectral density $S(\omega)$ of target return signal fluctuations (with a nonhorizontally moving radar and with the angles of roll γ_c and pitch θ_c) can be used for all types of searching signals.

References

1. Kolchinsky, V., Mandurovsky, I., and Konstantinovsky, M., *Doppler Devices and Navigational Systems*, Soviet Radio, Moscow, 1975 (in Russian).
2. Winitzky, A., *Basis of Radar under Continuous Generation of Radio Waves*, Soviet Radio, Moscow, 1961 (in Russian).

Notation Index

$A(t)$	signal amplitude factor
$\mathbf{a}^*(x, y, t)$	model image
$a(t)$	signal
$a(x, y, \mathbf{\Lambda}, t)$	information space–time signal
$a^*(t)$	model signal
$a_{lat}(i\Delta x, j\Delta y)$	lattice function
$\mathbf{C}(x)$, $\mathbf{S}(x)$	Fresnel integrals
$C(\omega_x, \omega_y)$	transfer function
c	velocity of light
d_a	diameter of the radar antenna
$E(x, y)$	lightness corresponding to the information signal in the moving image
$\mathscr{F}(\omega)$	Fourier transform
$\mathscr{F}^{-1}(\omega)$	inverse Fourier transform
$f(S, \phi)$	two-dimensional probability distribution density of the amplitude and phase of the target return signal
$f_{s,\vartheta}(s, \vartheta)$	two-dimensional probability distribution density of the amplitude and phase of elementary signals
$f(S)$	probability distribution density of normalized amplitude of the target return signal
$f(x, y)$	two-dimensional probability distribution density
$f(\phi)$	probability distribution density of the phase of the target return signal
G_0	amplifier coefficient of the radar antenna
$G(\bar{x}, \bar{x}_z)$	Green function
$\lvert\vec{g}\rvert$	gradient of velocity of the wind
$g(\varphi, \psi)$	normalized two-dimensional directional diagram of the radar antenna
$\tilde{g}(\varphi, \psi)$	generalized two-dimensional directional diagram of the radar antenna

$g_h(\varphi)$	normalized radar antenna directional diagram by power in the horizontal plane
$g_t(\varphi, \psi)$	normalized radar antenna directional diagram under the condition of transmission
$g_r(\varphi, \psi)$	normalized radar antenna directional diagram under the condition of receiving
$g_v(\varphi)$	normalized radar antenna directional diagram by power in the vertical plane
h	altitude
$h(x, y)$	weight function
$\mathbf{H}(x)$	Hermite polynomial
$\mathcal{H}(x, \lambda_x, y, \lambda_y)$	function of scattering
I, I_x, I_y	criterial functions
$I_k(x)$	modified Bessel function
J	Jacobian
$J_0(x)$	Bessel function of the first order
k_a	coefficient of the shape of the radar antenna directional diagram
k_a^h	coefficient of the shape of horizontal-coverage directional diagram of the radar antenna
k_a^v	coefficient of the shape of vertical-coverage directional diagram of the radar antenna
k_p	coefficient of the pulsed searching signal shape
k_ω	velocity of frequency variation
$L(\mathbf{\Lambda})$	likelihood functional
$\mathcal{L}_\lambda$	spectral energy brightness of the heated object
$\mathcal{L}(\mathbf{\Lambda})$	likelihood function
m_0	mean with respect to an ensemble of numbers of scatterers
$\mathbf{n}$	ℓ-dimensional noise vector
P	probability
P_D	probability of true detection
P_F	probability of false alarm
P_S	power of the searching signal
P_{tr}	probability of true location of the object image with reference to the control point
P_{fr}	probability of false location of the object image with reference to the control point

p	power of the target return signal
$Q(j\omega)$	frequency response of the linear system
$q(\xi, \zeta)$	function representing the dependence of the amplitude of the target return signal on orientation of scatterer in space
$q^2(\xi, \zeta)$	effective scattering area with fixed values of the angles ξ and ζ
$R(t, \tau)$	correlation function of fluctuations of the nonstationary target return signal
$R(\tau)$	correlation function of fluctuations of the stationary target return signal
$R_{\Delta x}(t_1, t_2)$	correlation function of space and time fluctuations of the nonstationary target return signal
$R_{\Delta\rho,\Delta x}(t, \tau, \Delta\omega)$	high-frequency correlation function of space and time fluctuations of the nonstationary target return signal
$R^{en}_{\Delta\rho,\Delta \mathbf{x}}(t, \tau, \Delta\omega)$	envelope of high-frequency correlation function of space and time fluctuations of the nonstationary target return signal
$R_{\Delta\rho,\Delta\varphi,\Delta\psi,\Delta\xi,\Delta\zeta}(t, \tau)$	total correlation function of fluctuations of the target return signal caused by the moving radar, displacements and rotation of scatterers, and antenna scanning
$R^{en}_{\Delta\rho,\Delta\varphi,\Delta\psi,\Delta\xi,\Delta\zeta}(t, \tau)$	envelope of the total correlation function of the target return signal fluctuations caused by the moving radar, displacements and rotation of scatterers, and antenna scanning
$R(\Delta x, \Delta y, \Delta t)$	space–time correlation function
$R_{i,j}(x)$	criterial correlation function
$R(t, \tau)$	normalized correlation function of fluctuations of the nonstationary target return signal
$R(\tau)$	normalized correlation function of fluctuations of the stationary target return signal
$R_g(\Delta\ell, \Delta\beta_0, \Delta\gamma_0)$	normalized correlation function of Doppler fluctuations of the target return signal
$R_p(\tau, \Delta\ell)$	normalized correlation function of rapid fluctuations of the target return signal
$R_q(\Delta\xi)$	normalized correlation function of space fluctuations of the target return signal caused by the rotation of the radar antenna polarization plane

$R_{mov,sc}(\Delta\ell, \Delta\beta_0, \Delta\gamma_0)$	total normalized correlation function of fluctuations of the target return signal caused by the moving radar with simultaneous radar antenna scanning
$R_{mov}(\Delta\ell)$	normalized correlation function of fluctuations of the target return signal caused only by the moving radar
$R_{sc}(\Delta\beta_0, \Delta\gamma_0)$	normalized correlation function of space fluctuations of the target return signal caused only by radar antenna scanning
$R_\beta(\Delta\ell, \Delta\beta_0)$	azimuth normalized correlation function of fluctuations of the target return signal
$R_\gamma(\Delta\ell, \Delta\gamma_0, t, \tau)$	aspect-angle normalized correlation function of fluctuations of the target return signal
$R_{\Delta x}(t_1, t_2)$	correlation function of space and time fluctuations of the nonstationary target return signal
$R_{\Delta\rho,\Delta\varphi,\Delta\psi,\Delta\xi,\Delta\zeta}(t, \tau)$	total normalized correlation function of fluctuations of the target return signal caused by the moving radar, displacements and rotation of scatterers, and antenna scanning
$R^{en}_{\Delta\rho,\Delta\varphi,\Delta\psi,\Delta\xi,\Delta\zeta}(t,\tau)$	the envelope of the total normalized correlation function of fluctuations of the target return signal caused by the moving radar, displacements and rotation of scatterers, and antenna scanning
$R_{\Delta\rho,\Delta\varphi,\Delta\psi}(t,\tau)$	particular normalized correlation function of fluctuations of the target return signal caused by the moving radar, displacement of scatterers, and antenna scanning
$R_{\Delta\xi,\Delta\zeta}(\Delta\xi, \Delta\zeta)$	particular normalized correlation function of fluctuations of the target return signal caused by the rotation of scatterers and polarization plane of the radar antenna
$S(t)$	amplitude of the searching signal
$S(\rho, \mathbf{x})$	amplitude of the received target return signal
$S(\omega)$	power spectral density of fluctuations of the target return signal
$S^{en}(\omega)$	envelope of the regulated power spectral density of fluctuations of the target return signal
$S(\omega, t)$	instantaneous power spectral density of fluctuations of the nonstationary target return signal
$S_{a^*}(\omega_x)$	power spectral density of the model image
$S_X(\omega_x)$	power spectral density of the moving image

$\overline{S(\omega, t)}$	average power spectral density of the nonstationary target return signal
$S^{en}(\omega, t)$	envelope of the corrugated power spectral density of fluctuations of the target return signal
$S_{mov}(\omega, t)$	continuous power spectral density of fluctuations of the target return signal caused only by the moving radar
$S_{mov,sc}(\omega, t)$	power spectral density of fluctuations of the target return signal caused by the moving radar with simultaneous radar antenna scanning
$S_{sc}(\omega, t)$	regulated power spectral density of fluctuations of the target return signal caused only by radar antenna scanning
$S(\omega, \Omega)$	two-dimensional power spectral density of fluctuations of the nonstationary target return signal
$S(\omega_1, \omega_2)$	two-dimensional power spectral density of fluctuations of the target return signal
$S(\omega_x, \omega_y)$	power spectral density of spatial frequency
$S_g(\omega)$	power spectral density of the Doppler fluctuations of the target return signal caused by variations in radial velocity within the limits of whole width of the radar antenna directional diagram
$S_h(\omega)$	power spectral density of the Doppler fluctuations of the target return signal caused by variations in radial velocity within the limits of the horizontal width of the radar antenna directional diagram
$S_v(\omega)$	power spectral density of the Doppler fluctuations of the target return signal caused by variations in radial velocity within the limits of the vertical width of the radar antenna directional diagram
$S_\gamma(\omega)$	power spectral density of the Doppler fluctuations of the target return signal caused by variations in radial velocity within the limits of the aspect-angle plane of the radar antenna directional diagram
$S_\beta(\omega)$	power spectral density of the Doppler fluctuations of the target return signal caused by variations in radial velocity within the limits of the azimuth plane of the radar antenna directional diagram
$\mathcal{S}(\omega)$	power spectral density of fluctuations of the target return signal shifted in frequency
$\mathcal{S}(\omega_x, \omega_y)$	cross section of the power spectral density $S(\omega_x, \omega_y)$
$s_i(t)$	amplitude of the i-th elementary signal

$\boldsymbol{S}_t$	effective scattering area
$\boldsymbol{S}^\circ$	specific effective scattering area
$\boldsymbol{S}_N^\circ$	specific effective scattering area under vertical scanning
$S(t)$	normalized amplitude of the target return signal
T_a	period of radar antenna hunting
T_d	delay of the target return signal
T_p	period of the pulsed signal
T_r	effective duration of the target return signal
T_{sc}	period of radar antenna scanning
t	observed instant of time
$\boldsymbol{T}(x)$	Toronto function
$U(t, \omega)$	searching signal
V_a	velocity of the moving radar relative to the Earth's surface
V_r	radial component of velocity of the moving radar
V_{r_0}	projection of velocity of the moving radar on the axis of radar antenna directional diagram
V_w	velocity of the wind
V_{w_0}	velocity of the wind on the altitude h
V_{w_r}	radial component of the velocity of scatterers
$W(t)$	target return signal
$W(\mathbf{x}, t)$	resulting target return signal
$W_h(t)$	heterodyne signal
$W_{tr_i}(t)$	transformed target return signal from the i-th scatterer
$w_i(t)$	i-th elementary signal
$w(\mathbf{x}, t)$	elementary signal reflected by individual scatterer
$X(t)$, $Y(t)$	quadrature components of the target return signal
$\mathbf{X}$	m-dimensional vector of the observed signal at the input of the navigational system receiver
$\mathbf{X}(x, y, t)$	moving image
α, θ	angles defining the position of scatterers in the polar coordinates
α_0, θ_0	angles defining the position of the axis of the radar antenna directional diagram in the polar coordinates
β, γ	azimuth and aspect angle of scatterer in the spherical coordinates

β_0, γ_0	azimuth and aspect angle of the axis of the radar antenna directional diagram in the spherical coordinates
β_{es}, γ_{es}	azimuth and aspect angle of equisignal direction in the spherical coordinates
$\tilde{\beta}$	coefficient of attenuation
β_{ab}	coefficient of molecular absorption
β_{sc}	coefficient of scattering by particles
β_{tub}	coefficient of scattering by nonhomogeneities caused by turbulence
γ_*	aspect angle of the center of observed radar range element
Δ_a	effective width of radar antenna directional diagram
Δ_h	effective width of horizontal-coverage directional diagram of the radar antenna
$\Delta_h^{(2)}$	effective width of square of horizontal-coverage directional diagram of the radar antenna
Δ_v	effective width of vertical-coverage directional diagram of the radar antenna
$\Delta_v^{(2)}$	effective width of square of vertical-coverage directional diagram of the radar antenna
$\Delta F, \Delta\Omega$	effective bandwidth of the power spectral density of fluctuations of the target return signal
ΔF_d	effective bandwidth of the power spectral density of Doppler fluctuations of the target return signal
ΔF_{mov}	effective bandwidth of the power spectral density of fluctuations of the target return signal caused only by the moving radar
ΔF_{sc}	effective bandwidth of the power spectral density of fluctuations of the target return signal caused only by radar antenna scanning
$\Delta \ell$	displacement of radar
$\Delta\beta_0, \Delta\gamma_0$	shifts of the axis of the radar antenna directional diagram
$\Delta\zeta, \Delta\xi$	angles of rotation of scatterers
$\Delta\rho$	displacement of scatterers
$\Delta\rho_0$	radial shift along the axis of the radar antenna directional diagram
$\Delta\rho_{\varphi,\psi}$	deviation of radial shifts for various scatterers within the limits of the radar antenna directional diagram
$\Delta\varphi, \Delta\psi$	angles of displacement of scatterers

$\Delta\varphi_{sc}$, $\Delta\psi_{sc}$	angles of displacement of scatterers caused by radar antenna scanning
$\Delta\varphi_{rm}$, $\Delta\psi_{rm}$	angles of displacement of scatterers caused by moving radar
$\Delta\omega_M$	deviation in frequency
$\Delta\mathbf{\Lambda}$	error vector
$\delta(x)$	delta function
$\delta\mathbf{x}$	elementary volume
ε_0	trajectory angle or angle between the velocity vector of the moving radar and the horizon
ε_λ	coefficient of radiation
$\boldsymbol{\eta}(x, y, t)$	additional reference noise vector
ζ	angle defining the position of the scatterer in space relative to the polarization plane of the radar antenna and the direction of the beam
θ	angle between the velocity vector of the moving radar and the direction of the scatterer
θ_0	angle between the velocity vector of the moving radar and the axis of the radar antenna directional diagram
λ	wavelength
λ_x, λ_y	views of the parameter vector $\mathbf{\Lambda}$
$\mathbf{\Lambda}$	n-dimensional vector of parameters of the navigational object coordinates
$\mathbf{\Lambda}^*$	vector estimation of the vector of parameters of the navigational object coordinates
$\mu, \bar{\mu}$	scale factor
μ_0	coefficient of reflection
μ_b	coefficient of brightness
ξ	angle defining the position of the scatterer in space relative to the polarization plane of the radar antenna and direction of the beam
$\xi(t)$	noise at the preliminary filter output of the generalized detector
$\Pi(t)$	envelope of the high-frequency pulsed searching signal (video signal)
ρ	radar range or distance between the radar antenna and scatterer
ρ_*	distance between the center of pulse volume and radar

ρ_{sr}	conditional boundary of the short range
σ	root mean square deviation
σ^2	variance
$\sigma^2(t)$	variance of the stochastic process at the instant of time t
ϕ	phase of the target return signal
ϑ	phase of elementary signal
τ_p	duration of the pulsed searching signal
$\Phi(x)$	error integral
φ, ψ	angles defining the position of scatterer relative to the axis of the radar antenna directional diagram
φ_0	random initial phase of the signal
χ	coefficient of asymmetry
$\Psi(t)$	phase modulation law
Ω_0	Doppler frequency corresponding to the center of pulse volume
Ω_M	modulation frequency
$\Omega_{\max}$	maximum Doppler frequency
Ω_p	instantaneous frequency of the periodic pulsed searching signal
Ω_{sc}	angular velocity of radar antenna scanning
$\Omega(t)$	instantaneous frequency
$\Omega(\varphi, \psi)$	Doppler frequency
Ω_ρ	range finder frequency
ω	frequency of the target return signal
ω_0	carrier frequency of the signal
ω_{av}	averaged within the limits of the modulation period high frequency
ω_h	frequency of the heterodyne signal
ω_{im}	intermediate frequency
$<\ldots>$	mean
sinc (x)	sinc-function

Index

A

Aberration, 507
Absolutely "white" surface, 502
Absorption, 505
Affine transform, 558
Altitude, 63, 474
Amplitude ranking, 567
Amplitude tracking system, 471
Angular velocity, 135, 260
Apex angle, 96, 110
Aspect angle, 54, 63
Asymmetric frequency modulation, 324
Averaged Doppler frequency, 260
Azimuth, 54

B

Back aperture angle, 507
Background noise, 432
Bayesian criterion, 424
Bessel function, 173, 546
Bipartite criterial function
Blackbody, 500
Boreal function, 529
Boundary edge, 595
Brightness contrast, 507

C

Camber function, 532
Cartesian coordinate system, 473
Chamman function, 535
Chancel transform, 545
Chaotic rotation of scatterers, 296
Chebyshev function, 532
Coding of image, 578
Coefficient of attenuation, 506
Coefficient of brightness, 501
Coefficient of computational cost reduction, 555
Coefficient of correlation, 529, 556
Coefficient of directional diagram shape, 175
Coefficient of molecular absorption, 506
Coefficient of radiation, 500
Coefficient of reflection, 501
Coefficient of scale, 256
Coefficient of scattering, 506
Coefficient of the Fourier-series expansion, 518
Coefficient of transmission, 511
Coefficient of transparency, 506
Comb function, 224
Conditional boundary, 192
Conical scanning, 219, 243
Correlation-extremal systems, 483
Correlation interval, 54, 113, 164
    effective —, 196
    space —, 107
Correlation function
    aspect-angle-normalized —, 67, 142, 160
    azimuth-normalized —, 67, 142, 170
    conditional normalized —, 303
    high-frequency —, 47
    low-frequency —, 47
    nonstationary —, 38, 41
    normalized —, 35, 41
    space normalized —, 67
    space–time —, 36, 479
    total —, 178
Correlation length, 516
Correlation radius, 460
Correlation signal processing algorithm, 424
Correlation surface, 569
Covariance matrix, 465
Criterial function, 483
    bipartite —, 533
    difference —, 531
    average square —, 531
    average absolute —, 531
    rank —, 536
    spectral —, 532
Criterial correlation function, 529

average of —, 567
weighted —, 530
Criterion of similarity, 495
Block coding
adaptive —, 579
cut —, 579

D

Deak function, 534
Decision function, 434
Delta function, 22, 433, 473
Detection parameter, 425
Diffracting scattering, 507
Digitization function, 575
Direct interferences, 463
Directional diagram, 51
horizontal-coverage , 69, 241
Gaussian —, 59, 179, 184
generalized —, 143
receiving —, 247
sinc —, 59
transmitting —, 247
vertical-coverage —, 69
— width, 107
Dirichlet conditions, 477
Displacement of scatterers, 281
Distortions in brightness characteristics, 585
Distortions of object image, 585
Doppler beats, 30
Doppler effect, 30
secondary —, 30
shift in frequency, 171
Doppler frequency, 90
Duhamel integral, 485

E

Edgeworth-series expansion, 557
Eigenfunction, 422
Eigenvalue, 422
Effect of back scattering, 506
Effective duration, 60, 155
Effective scattering area, 51
— of scatterer, 58
specific —, 58, 194
Effective signal bandwidth, 434
"Effect of dazzley," 63
Efficiency of internal radiation, 500
Elementary signal, 4
Error integral, 12, 96
Error vector, 464
Estimation of the minimum variance, 464

F

Field potential, 474
Filter of spatial frequencies, 485
First-kind model of the field, 473
Fluctuations, 29
interperiod —, 31, 198, 202, 405
intraperiod —, 31, 193
Doppler —, 30, 101
frequency —, 30
polarization —, 30
rapid —, 31, 446
scanning —, 30
slow —, 33, 443
space —, 35, 47, 220, 249
time —, 35
Fourier transform, 477
Fresnel integral, 98
Function of coherency, 532
Function of concentric circles, 540
Function of line bundle, 540
Function of scattering, 507

G

Generalized approach to signal processing, 421
Generalized detector, 436
Generalized image processing algorithm
classical —, 555
difference —, 561
hierarchical —, 574
multichannel —, 580
phase —, 562
quasioptimal , 536
Generalized signal processing algorithm, 433
Geometrical transformations, 585
Gradient of the wind velocity, 288
Gradient vector sum, 570
Gray emitter, 500
Green function, 474

H

Harmonic frequency modulation, 358, 366
Hermitian polynomial, 155, 581
High-deflected radar antenna, 174
Hydrometeors, 3

I

Ideal optical receiver, 508
Indicatrices of brightness, 502
Indicatrices of reflection, 502
Informative area, 577
Integral detector, 459
Internal heat radiation, 499
Interpolation function, 515
"Interrevolution" correlation, 234
Invariant moments, 565
Inverse Fourier transform, 477
Isodope, 208

J

Jakard function, 534
Jointly sufficient statistics, 428

K

Knapp function, 533
Kotelnikov's frequency, 541
Kulzinsky function, 534

L

Lambert reflection, 501
Landmark, 457
Latitude, 474
Lattice function, 514
Layered wind, 286
Likelihood function, 430, 464
Likelihood ratio, 428
Line circular scanning, 241
Line scanning, 219
Line segment scanning, 235
Linear frequency modulation, 98
Longitude, 474
Long-range area, 190
Low-deflected radar antenna, 176

M

Manmade fields, 458
Matrix detector, 508
Mean of projections, 6
Mean square of projections, 6
Minkovsky function, 531
Mini-max criterion, 424
Minimum of the mean square error, 464
Mirror reflection, 501
Modified Bessel function, 345
Modulation frequency, 326
Multielement receiver, 460
Multiple-line circular scanning, 233
Mutual correlation function, 529

N

Natural field, 458
Navigational parameter ratio, 321
Neyman–Pearson criterion, 424
Noise component, 242
Nonmodulated signal, 143
Nonperiodic variations in frequency, 401
Nontransformed target return signal, 378
Nonzero phase region, 478
Normalized amplitude, 7
Normalized mutual correlation function, 529

O

Object image preprocessing, 585
Observation plane, 457
One-line circular scanning, 226
Optical signal formation, 499
Optical thickness, 506

P

Parseval theorem, 423
Pattern recognition, 492
Period of modulation, 314
Periodic frequency modulation, 410

Periodicity of function, 326
Perspective transformation, 586
Phase tracking system, 469
Poisson stochastic process, 30–35
Polarity coordinates, 7
Polynomial estimation, 589
Polarization plane, 249
Power spectral density, 35
    average —, 407
    generalized —, 42
    instantaneous —, 43, 270
    nonstationary —, 42
    normalized —, 202
    two-dimensional —, 41
    of spatial frequency, 480
Probability distribution density
    a posteriori —, 464
    a priori —, 464
    Beckman —, 8
    conditional —, 464
    first moment of the —, 6
    Gaussian —, 5, 25, 465
    Hoyt —, 9
    — of the amplitude, 8–11
    — of the phase, 11–14
    multidimensional —, 481
    Rayleigh —, 9
    Rice —, 9
    second moment of the —, 6
    "triangular" —, 24
    two-dimensional —, 3–7
    uniform —, 22
Probability of detection, 434
Probability of false alarm, 434
Projection of the steady vector, 7
p-channel generalized detector, 498

Q

Quantization level, 519
Quantization volume, 518
Quasi-stationary radar antenna scanning, 263

R

Radar antenna diameter, 107
Radar range, 51, 90
    fixed —, 129
    glancing —, 129
Radar resolution, 333
Radial displacement, 281
Radiant emittance, 504
Radiation attenuation, 506
Radius of the Earth, 474
Range-finder frequency, 319
Rao function, 534
Raster system, 461
Receiver, 3
Reciprocal matrix, 591
Reference region, 428
Reference sample, 428
Reflected radiation, 499
Resolution volume, 61
Resulting vector, 5
Roberts operator, 595
Rodgers and Tanimoto function
    first —, 535
    second —, 535
Rot function, 532
Rotation of polarization plane, 304
Rotation of scatters, 288

S

Saw-tooth frequency modulation
    asymmetric —, 324
    symmetric —, 350
Scattering, 505
Scattered reflection, 501
Second-kind model of the field, 474
Selective emitter, 500
Short-range area, 189, 265
Sign function, 535
Signal energy, 423
Signal noise, 463
Signal-to-noise ratio, 425
Sinc-function, 22, 23
Sloping scanning, 329
Solid angle, 503
Soucal and Mishner function, 535
Soucal and Snit function, 534
Space–time signal, 472
Space signal and image processing, 460
Space–time signal and image processing, 458
    parallel —, 458
    sequential —, 458
Spatial frequency, 477
Spatial image, 459
Spatial signal, 472
Spearman function, 536
Specific effective scattering area, 510
Spectral energy brightness, 500
Spectral power of blackness, 500

Spectral technique, 484
Spherical coordinates, 474
Spiral scanning, 233
Stationary radar, 148
System *radar–scatterer*, 32, 91
Steady vector, 8–25
Sufficient statistic, 425
Surface potential, 474

T

Time section, 32
Theorem of shift, 485
Three-dimensional (space) target, 4
Threshold, 423
Toronto function, 81
Transfer function, 537
Transformation of brightness characteristics, 592
Transformed target return signal, 392
Transmitter, 3
Two-dimensional (surface) target, 4
Two-dimensional theory of Brownian motion, 5
Turbulence, 505

U

Unconditional estimation, 464

V

Velocity of the wind, 282
Velocity of the moving radar, 283
Vertical scanning, 340, 345
Vision field, 499

W

Weight function, 537
  isotropic —, 596
  Soibel —, 596
  summing —, 538
  — of a smooth gradient, 596
Weight transfer function, 541
"White" Gaussian noise, 434
Wiener–Heanchen theorem, 480

Y

Yule function, 535

Z

Zenith angle, 502
Zero-phase region, 477